GENETICS

A Modern Approach

GENETICS

A Modern Approach

Second Revised & Enlarged Edition

Author

Suraksha Agrawal

2025

Daya Publishing House®

A Division of

Astral International Pvt. Ltd.

New Delhi – 110 002

ISBN: 9789359196961

Publisher's Note:

Every possible effort has been made to ensure that the information contained in this book is accurate at the time of going to press, and the publisher and author cannot accept responsibility for any errors or omissions, however caused. No responsibility for loss or damage occasioned to any person acting, or refraining from action, as a result of the material in this publication can be accepted by the editor, the publisher or the author. The Publisher is not associated with any product or vendor mentioned in the book. The contents of this work are intended to further general scientific research, understanding and discussion only. Readers should consult with a specialist where appropriate.

Every effort has been made to trace the owners of copyright material used in this book, if any. The author and the publisher will be grateful for any omission brought to their notice for acknowledgement in the future editions of the book.

Published by : **Daya Publishing House®**
A Division of
Astral International Pvt. Ltd.
– ISO 9001:2015 Certified Company –
4736/23, Ansari Road, Darya Ganj
New Delhi-110 002
Ph. 011-43549197, 23278134
E-mail: info@astralint.com
Website: www.astralint.com

Preface

Gene is the basic unit of inheritance. Since the discovery of molecular structure of DNA too many facts have come into the light and a new branch in the biological sciences have started *i.e.* study of inheritance or genetics. This is one of the most rapidly growing areas of biology today. The goal of present book is for the undergraduate and post graduate courses working in the fields of human genetics. The text is targeted to examines how human genes are discovered and transmitted from one generation to the next, further how the normal genes behave and, once a gene is known, how the mutated versions(s) causes a particular disorder. The initiative started with my own fundamental knowledge and then to go more and more into the details of it. To the best of my capacity I have kept the text very simple so that it could be understood by all the students opting for this field of biology

The progress in this field was phenomenal hence many unanswered questions are still there. Further to keep pace with the developing knowledge is also very difficult. Hardly a week passes without a report in a major journal stating about the discovery of the new gene, therefore we are in the midst of the genomic era in the field of human genetics.

Present book has been structured to introduce the students to human genetics, with an emphasis on its applications to medicine, genetic counseling *etc.* The focus of the book will be on new developments in the field afforded by present day techniques in molecular biology and the completion of sequencing the human genome. Our efforts will permit students to learn basic principles used in the analysis of human genes, the structure of the human genome and the molecular basis of many human diseases.

Other topics included are Fundamentals of transmission genetics, pedigree analysis of inheritance of dominant and recessive genes. Linkage analysis (linkage disequilibrium and LOD score analysis). Sex linked inheritance; sex influenced and sex limited inheritance. Determination of sex in humans. Fundamentals of human genome structure. Gene and chromosome structures. Organization and expression of the human genes. Human multi gene families. Human repetitive DNA, VNTR analysis, mini and microsatellites. Mutation and instability of human DNA. Chromosomal abnormalities. Mapping of the human genome. Physical mapping, genetic mapping. Human Genome Project. SNP mapping *etc.* a special afford has been made to include some of the Human genetic diseases. Molecular pathology of most common genetic diseases. Complex diseases. Genetic testing. Gene therapy and novel therapeutic approaches to human genetic abnormalities. Genetics of cancer. Oncogenes and tumor suppressor genes.

It is extremely difficult to write a book without taking the ideas from the other books, journals or internet sources. I fully acknowledge various authors whose books and ideas I have taken for the completion of this book. I am deeply thankful to Mr. Sanjay Kumar Johari for helping me at all stages of the book to see it in the published version. In case there is similarity in the text at certain places I acknowledge it and agree that it is completely daunting.

The books and internet sites which I have used for the completion of this hard task are mentioned below:

Books

1. An Introduction to Human Molecular Genetics: Mechanisms of Inherited Diseases. 2nd Edition 2005 by Jack J. Pasternak, Publisher: John Wiley and Sons.
2. Problems and Approaches- Human Genetics. 4th Edition 2010, by Michael Speicher, Stylianos E. Antonarakis, Publisher: Arno G. Motulsky.
3. Genetics: From Genes to Genomes, 5th Edition, 2015 by Leland H. Hartwell, Michael L. Goldberg, Janice A. Fischer, Leroy Hood, Charles F. Aquadro, Publisher: McGraw-Hill Education
4. Fundamentals of Genetics, 2nd Edition 2000, by Peter J. Russell, Publisher: Addison Wesley Longman.
5. Concepts of Genetics 10th Edition 2011 by William S. Klug, Michael R. Cummings, Charlotte A. Spencer, Michael A. Palladino. Publisher: Benjamin Cummings.
6. Cell and Molecular Biology: Concepts and Experiments 6th Edition by Gerald Karp, Publisher: Wiley.
7. Essential Cell Biology 3rd Edition March 27, 2009 by Bruce Alberts, Dennis Bray, Karen Hopkin, Publisher: Garland Science;
8. Genetics: A Conceptual Approach, 5th Edition December 27, 2013 by Benjamin A. Pierce, W. H. Freeman
9. Human Molecular Genetics, 3rd Edition by Peter Sudbery Publisher: Pearson Education Ltd,
10. Principles of Genetics, 3rd Edition 2003, by Snustad, D Peter - Simmons, Michael J.
11. Essential Genetics, 2nd Edition November 1998 by Daniel L. Hartl, Elizabeth W. Jones, Publisher: Jones and Bartlett
12. Molecular Biology of the Gene 7th Edition, by James D. Watson, Tania A. Baker, Stephen P. Bell

Internet sites

http://bio3400.nicerweb.com/

www.coursehero.com.

detectingdesign.com

www.webmedcentral.com

www.pearsonhighered.com

cmbi.bjmu.edu.cn

genetica.fcien.edu.uy

ajrcmb.atsjournals.org

Prof. Suraksha Agrawal

Contents

1

Introduction to Human Genetics

1.1 Introduction

Human genetics is an exciting field in the biological sciences as it involves both basic and functional approaches. Fundamental genetics deals with organisms' lives from conception until death. The transmission of characters from one generation to the next can be understood using the basic principles of inheritance, *i.e.*, the similarity of offspring to their parents. The principles of genetics apply even to single-cell organisms like bacteria and protozoa and to multicellular plants and animals. In an organism, functional DNA fragments are units of inheritance that regulate all characteristic features of the organism, *i.e.*, behavior, reproduction, survival, *etc.*, in each environment. These functional units of DNA are known as genes, and the study of genes is called genetics. As this science deals with different life features, it occupies a crucial position in the biological sciences. Genetics is a comparatively young science from a historical perspective. Engravings in Chaldea in Babylonia date back to at least more than 6000 years and show pedigrees of horses. These pedigrees indicate that some characters are inherited. Other old carvings show cross-pollination of date palm trees. Most of the mechanisms of heredity, however, remained a mystery until the 19th century. However, knowledge about genetics started with the dawn of civilization. Long before genetics became a documented science, our ancestors were improving crops through the process of selective breeding, or the selection of desirable traits. Not only was plant and animal breeding practiced, but they were also concerned about why children resemble their parents, why siblings resemble each other, and why certain diseases run in families. At the beginning of the 20th century, certain discoveries began to clarify our understanding of the physical basis of living organisms and their relationship to others. Several related ideas were gaining acceptance. These are: (i) matter is composed of small atoms; (ii) cells are like bricks, which are the fundamental unit of living organisms. (iii) nuclei somehow serve as the living force of cells, and (iv) chromosomes play an important role in heredity.

1.2 Prehistoric Times

There was an era when people were unaware of the scientific details of how babies were conceived and how different characters were inherited. On the other hand, they discovered a similarity between parents and children, but there was no knowledge of the mechanisms. The Greek philosopher Theophrastus gave diverse ideas from biology and physics to ethics and metaphysics. **Theophrastus** was born around 371 BC in Lesbos, Greece. Theophrastus was a successor to Aristotle. Most of what is known about his life comes from Diogenes Lartius and his *Lives and Opinions of Eminent Philosophers*, written at an unknown time, most likely around 300 AD. Hippocrates was an ancient Greek physician who lived during Greece's classical period and is traditionally regarded as the father of medicine. He speculated that "seeds" were produced by various body parts and transmitted to offspring at the time of conception, and Aristotle thought that male and female semen are mixed. Aeschylus, in 458 BC, proposed that the male act as the parent and the female nurture the young life sown within her. Anton van Leeuwenhoek was a Dutch microscopist from 1632–1723. He discovered "animalcules" in the sperm of humans and other animal species. At that time, scientists thought they had seen a little man (homunculus) inside each sperm. They further believed that sperm was responsible for life on this earth, hence being known as "spermists". They believed that the females carried the little man who grew inside them. An opposing school of thought was that the future human was inside the egg and sperm acted as a stimulatory factor for the growth of the egg; these were called ovists. Ovists thought women carried eggs containing boy and girl children; the sex of the children was determined well before conception.

The other theory was the blending theory of inheritance, supported by the spermists and ovists of the 19th century. The sperm and egg intermingle; hence, the characteristics of both parents are mixed. Sperm and eggs are called gametes. This word is taken from the Greek word gamos, meaning marriage. Scientists who believed in the blending theory thought that the children resembled their parents because of the blending hypothesis. However, blending theory ignores characteristics that are skipped sometimes. Charles Darwin had to deal with the implications of blending theories to explain the process of evolution. He overruled the blending theory, however; the correct answer was not available till mid-1800. The final answer came from Gregor Mendel; unfortunately, Darwin was not aware of Mendel's work. Mendel proposed that the characters are inherited from parents to offspring. During the last hundred years, three milestones have become remarkable. These are the discoveries of rules governing the inheritance of traits in organisms, the discovery of the material responsible for inheritance, and the elucidation of its structure; however, modern genetics lies in many more discoveries. Some of the important discoveries are shown below:

1.2.1 Landmark Discoveries in Genetics

- ☆ In Britain, Chaim Weizemann (1874–1952) developed bacterial fermentation processes for producing organic chemicals such as acetone and cordite propellants. During WWII, he worked on synthetic rubber and high-octane gas.
- ☆ In 1950s: The first synthetic antibiotic is created.
- ☆ 1951: Artificial insemination of livestock is accomplished using frozen semen.
- ☆ In 1953, JD Watson and FHC Crick for the first time cleared the mysteries around the DNA as genetic material, by giving a structural model of DNA, popularly known as, 'Double Helix Model of DNA.'
- ☆ 1954: Dr. Joseph Murray performs the first kidney transplants between identical twins.
- ☆ 1955: An enzyme, DNA polymerase, involved in the synthesis of nucleic acid, is isolated for the first time.
- ☆ 1955: Dr. Jonas Salk develops the first polio vaccine. The development marks the first use of mammalian cells (monkey kidney cells) and the first application of cell culture technology to generate a vaccine.
- ☆ 1957: Scientists prove that sickle-cell anemia occurs due to a change in a single amino acid in hemoglobin cells
- ☆ 1958: Dr. Arthur Kornberg of Washington University in St. Louis makes DNA in a test tube for the first time.
- ☆ Edward Tatum (1909–1975) and Joshua Lederberg (1925–2008) shared the 1958 Nobel Prize for showing that genes regulate the metabolism by producing specific enzymes.
- ☆ 1960: French scientists discover messenger RNA (mRNA).
- ☆ 1961: Scientists understand genetic code for the first time.
- ☆ 1962: Dr. Osamu Shimomura discovers the green fluorescent protein in the jellyfish *Aequorea victoria*. He later develops it into a tool for observing previously invisible cellular processes.
- ☆ 1963: Dr. Samuel Katz and Dr. John F. Enders develop the first vaccine for measles.
- ☆ 1964: The existence of reverse transcriptase is predicted.
- ☆ At a conference in 1964, Tatum laid out his vision of "new" biotechnology: "Biological engineering seems to fall naturally into three primary categories of means to modify organisms. These are 1. The recombination of existing genes, or eugenics. 2. The production of new genes by a process of directed mutation, or genetic engineering. 3. Modification or control of gene expression, or to adopt Lederberg's suggested terminology, euphenic engineering."
- ☆ 1967: The first automatic protein sequencer is perfected.

- ☆ 1967: Dr. Maurice Hilleman develops the first American vaccine for mumps.
- ☆ 1969: An enzyme is synthesized in vitro for the first time.
- ☆ 1969: The first vaccine for rubella is developed.
- ☆ 1970: Restriction enzymes are discovered.
- ☆ 1971: The measles/mumps/rubella combo-vaccine was formed.
- ☆ 1972: DNA ligase, which links DNA fragments together, is used for the first time.
- ☆ 1973: Cohen and Boyer perform the first successful recombinant DNA experiment, using bacterial genes.
- ☆ In 1974, Stanley Cohen and Herbert Boyer developed a technique for splicing together strands of DNA from more than one organism. The product of this transformation is called recombinant DNA (rDNA).
- ☆ Kohler and Milestein in 1975 came up with the concept of cytoplasmic hybridization and produced the first-ever monoclonal antibodies, which have revolutionized diagnostics.
- ☆ Techniques for producing monoclonal antibodies were developed in 1975.
- ☆ 1975: Colony hybridization and Southern blotting are developed for detecting specific DNA sequences.
- ☆ 1976: Molecular hybridization is used for the prenatal diagnosis of alpha thalassemia.
- ☆ 1978: Recombinant human insulin is produced for the first time.
- ☆ 1978: with the development of synthetic human insulin the biotechnology industry grew rapidly.
- ☆ 1979: Human growth hormone is synthesized for the first time.
- ☆ In the 1970s-80s, the path of biotechnology became intertwined with that of genetics.
- ☆ By the 1980s, biotechnology grew into a promising real industry.
- ☆ 1980: Smallpox is globally eradicated following a 20-year mass vaccination effort.
- ☆ In 1980, The U.S. Supreme Court (SCOTUS), in Diamond v. Chakrabarty approved the principle of patenting genetically engineered life forms.
- ☆ 1981: Scientists at Ohio University produce the first transgenic animals by transferring genes from other animals into mice.
- ☆ 1981: The first gene-synthesizing machines are developed.
- ☆ 1981: The first genetically engineered plant is reported.
- ☆ 1982: The first recombinant DNA vaccine for livestock is developed.
- ☆ 1982: The first biotech drug, human insulin produced in genetically modified bacteria, is approved by the FDA. Genentech and Eli Lilly developed the product. This is followed by many new drugs based on biotechnologies.
- ☆ 1983: The discovery of HIV/AIDS as a deadly disease has helped tremendously to improve various tools employed by life-scientists for discoveries and applications in various aspects of day-to-day life.
- ☆ In 1983, Kary Mullis developed the polymerase chain reaction (PCR), which allows a piece of DNA to be replicated repeatedly. PCR, which uses heat and enzymes to make unlimited copies of genes and gene fragments, later becomes a major tool in biotech research and product development worldwide.
- ☆ 1983: The first artificial chromosome is synthesized.
- ☆ In 1983, the first genetic markers for specific inherited diseases were found.
- ☆ 1983: The first genetic transformation of plant cells by TI plasmids is performed.
- ☆ In 1984, the DNA fingerprinting technique was developed.
- ☆ 1985: Genetic markers are found for kidney disease and cystic fibrosis.
- ☆ 1986: The first recombinant vaccine for humans, a vaccine for hepatitis B is approved.
- ☆ 1986: Interferon becomes the first anticancer drug produced through biotech.
- ☆ 1986: University of California, Berkeley, chemist Dr. Peter Schultz describes how to combine antibodies and enzymes (abzymes) to create therapeutics.
- ☆ 1988: The first pest-resistant corn, Bt corn, is produced
- ☆ 1988: Congress funds the Human Genome Project, a massive effort to map and sequence the human genetic code as well as the genomes of other species.
- ☆ In 1988, chymosin (known as Rennin) was the first an enzyme produced from a genetically modified source-yeast-to be approved for use in food.
- ☆ In 1988, only five proteins from genetically engineered cells had been approved as drugs by the United States Food and Drug

Administration (FDA), but this number would skyrocket to over 125 by the end of the 1990s.

- ☆ In 1989, microorganisms were used to clean up the Exxon Valdez oil spill.
- ☆ 1990: The first successful gene therapy is performed on a 4-year-old girl suffering from an immune disorder.
- ☆ In 1993, The U.S. Food and Drug Administration (FDA) declared that genetically modified (GM) foods are "not inherently dangerous" and do not require special regulation.
- ☆ 1993: Chiron's Betaseron is approved as the first treatment for multiple sclerosis in 20 years.
- ☆ 1994: The first breast cancer gene is discovered.
- ☆ 1995: Gene therapy, immune-system modulation, and recombinantly produced antibodies enter the clinic in the war against cancer.
- ☆ 1995: The first baboon-to-human bone marrow transplant is performed on an AIDS patient.
- ☆ 1995: The first vaccine for Hepatitis A is developed.
- ☆ 1996: A gene associated with Parkinson's disease is discovered.
- ☆ 1996: The first genetically engineered crop is commercialized.
- ☆ 1997: Ian Wilmut, an Irish scientist, was successful in cloning an adult animal, using sheep as a model and naming the cloned sheep 'Dolly.'
- ☆ 1997: The first human artificial chromosome is created.
- ☆ 1998: A rough draft of the human genome map is produced, showing the locations of more than 30,000 genes.
- ☆ 1998: Human skin is produced for the first time in the lab.
- ☆ 1999: A diagnostic test allows quick identification of Bovine Spongiform Encephalopathy (BSE, also known as "mad cow," disease) and Creutzfeldt-Jakob Disease (CJD).
- ☆ 1999: The complete genetic code of the human chromosome is deciphered.
- ☆ 2000: Kenya field-tests its first biotech crop, virus-resistant sweet potato.
- ☆ Craig Venter, in 2000, was able to sequence the human genome.
- ☆ 2001: The sequence of the human genome is published in Science and Nature, making it possible for researchers all over the world to begin developing treatments.
- ☆ 2001: FDA approves Gleevec® (imatinib), a gene-targeted drug for patients with chronic myeloid leukemia. Gleevec is the first gene-targeted drug to receive FDA approval.
- ☆ 2002: EPA approves the first transgenic rootworm-resistant corn.
- ☆ 2002: The banteng, an endangered species, is cloned for the first time.
- ☆ 2003: China grants the world's first regulatory approval of a gene therapy product, Gendicine (Shenzhen SiBiono GenTech), which delivers the p53 gene as a therapy for squamous cell head and neck cancer.
- ☆ In 2003, TK-1 (GloFish) went on sale in Taiwan, as the first genetically modified pet.
- ☆ 2003: The Human Genome Project completes the sequencing of the human genome.
- ☆ 2004: UN Food and Agriculture Organization endorses biotech crops, stating biotechnology is a complementary tool to traditional farming methods that can help poor farmers and consumers in developing nations.
- ☆ 2004: FDA approves the first antiangiogenic drug for cancer, Avastin®.
- ☆ 2005: The Energy Policy Act is passed and signed into law, authorizing numerous incentives for bioethanol development.
- ☆ 2006: FDA approves the recombinant vaccine Gardasil®, the first vaccine developed against human papillomavirus (HPV), an infection implicated in cervical and throat cancers, and the first preventative cancer vaccine.
- ☆ 2006: USDA grantsDow AgroSciences the first regulatory approval for a plant-made vaccine.
- ☆ 2006: The National Institutes of Health begins a 10-year, 10,000-patient study using a genetic test that predicts breast-cancer recurrence and guides treatment.
- ☆ In 2006, the artist Stelarc had an ear grown in a vat and grafted onto his arm.
- ☆ 2007: FDA approves the H5N1 vaccine, the first vaccine approved for avian flu.
- ☆ 2007: Scientists discover how to use human skin cells to create embryonic stem cells.
- ☆ 2008: Chemists in Japan create the first DNA molecule made almost entirely of artificial parts.
- ☆ 2009: Global biotech crop acreage reaches 330 million acres.

- ☆ In 2009, Sasaki and Okana produced transgenic marmosets that glow green in ultraviolet light (and pass the trait to their offspring).
- ☆ 2009: FDA approves the first genetically engineered animal to produce a recombinant form of human antithrombin.
- ☆ In 2010, Craig Venter was successful in demonstrating that a synthetic genome could replicate autonomously.
- ☆ 2010: Dr. J. Craig Venter announces the completion of "synthetic life" by transplanting synthetic genome capable of self-replication into a recipient bacterial cell.
- ☆ 2010: Harvard researchers report building "lung on a chip" – technology.
- ☆ In 2010, scientists created malaria-resistant mosquitoes.
- ☆ 2011: Trachea derived from stem cells transplanted into human recipients.
- ☆ 2011: Advances in 3-D printing technology led to "skin-printing."
- ☆ 2012: For the last three billion years, life on Earth has relied on two information-storing molecules, DNA and RNA. Now there's a third: XNA, a polymer synthesized by molecular biologists Vitor Pinheiro and Philipp Holliger of the Medical Research Council
- ☆ The United Kingdom. Just like DNA, XNA can store genetic information and then evolving through natural selection. Unlike DNA, it can be carefully manipulated.
- ☆ 2012: Researchers at the University of Washington in Seattle announced the successful sequencing of a complete fetal genome using nothing more than snippets of DNA floating in its mother's blood.
- ☆ 2013: Two research teams announced a fast and precise new method for editing snippets of the genetic code. The so-called CRISPR system takes advantage of a defense strategy used by bacteria.
- ☆ 2013: Researchers in Japan developed functional human liver tissue from reprogrammed skin cells.
- ☆ 2013: Researchers published the results of the first successful human-to-human brain interface.
- ☆ 2013: Doctors announced that a baby born with HIV had been cured of the disease.
- ☆ 2014: Researchers showed that blood from a young mouse can rejuvenate an old mouse's muscles and brain.
- ☆ 2014: Researchers figured out how to turn human stem cells into functional pancreatic β cells–the same cells that are destroyed by the body's immune system in type 1 diabetes patients.
- ☆ 2014: All life on Earth as we know it encodes genetic information using four DNA letters: A, T, G, and C. Not anymore! In 2014, researchers created new DNA bases in the lab, expanding life's genetic code and opening the door to creating new kinds of microbes.
- ☆ 2014: For the first time, a woman gave birth to a baby after receiving a womb transplant.
- ☆ 2014: An international team of scientists reconstructed a synthetic and fully functional yeast chromosome. A breakthrough seven years in the making, the remarkable advance could eventually lead to custom-built organisms (human organisms included).
- ☆ 2014 and Ebola: Until this year, ebola was merely an interesting footnote for anyone studying tropical diseases. Now it's a global health disaster. But the epidemic started at a single point with one human-animal interaction – an interaction which has now been pinpointed using genetic research. A total of 50 authors contributed to the paper announcing the discovery, including five who died of the disease before it could be published.
- ☆ 2014: Doctors discovered a vaccine that blocks infection altogether in the monkey equivalent of the disease – a breakthrough that is now being studied to see if it works in humans.
- ☆ 2015: Scientists from Singapore's Institute of Bioengineering and Nanotechnology designed short strings of peptides that self-assemble into a fibrous gel when water is added for use as a healing nanogel.
- ☆ 2015 and CRISPR: scientists hit several breakthroughs using the gene-editing technology CRISPR. Researchers in China reported modifying the DNA of a nonviable human embryo, a controversial move. Researchers at Harvard University inserted genes from a long-extinct woolly mammoth into the living cells – in a petri dish – of a modern elephant. Elsewhere, scientists reported using CRISPR to potentially modify pig organs for

human transplants and modify mosquitoes to eradicate malaria.

- ✰ 2015: Researchers in Sweden developed a blood test that can detect cancer at an early stage from a single drop of blood.
- ✰ 2015: Scientists discovered a new antibiotic, the first in nearly 30 years, that may pave the way for a new generation of antibiotics and fight growing drug-resistance. The antibiotic, teixobactin, can treat many common bacterial infections, such as tuberculosis, septicemia, and C. diff.
- ✰ 2015: A team of geneticists finished building the most comprehensive map of the human epigenome, a culmination of almost a decade of research. The team was able to map more than 100 types of human cells, which will help researchers better understand the complex links between DNA and diseases.
- ✰ 2015: Stanford University scientists revealed a method that may be able to force malicious leukemia cells to change into harmless immune cells, called macrophages.
- ✰ 2015: Using cells from human donors, doctors, for the first time, built a set of vocal cords from scratch. The cells were urged to form a tissue that mimics vocal fold mucosa – vibrating flaps in the larynx that create the sounds of the human voice.
- ✰ 2016: A little-known virus first identified in Uganda in 1947–Zika–exploded onto the international stage when the mosquito-borne illness began spreading rapidly throughout Latin America. Researchers successfully isolated a human antibody that "markedly reduces" infection from the Zika virus.
- ✰ 2016: CRISPR, the revolutionary gene-editing tool that promises to cure illnesses and solve environmental calamities, took a major step forward this year when a team of Chinese scientists used it to treat a human patient for the very first time.
- ✰ 2016: Researchers found that an ancient molecule, GK-PID is the reason single-celled organisms started to evolve into multicellular organisms approximately 800 million years ago.
- ✰ 2016: Stem Cells Injected into Stroke Patients Re-Enable Patient To Walk.
- ✰ 2016: Cloning does not cause long-term health issues, study finds
- ✰ 2016: For the first time, bioengineers created a completely 3D-printed 'heart on a chip.
- ✰ 2017: Researchers at the National Institute of Health discovered a new molecular mechanism that might be the cause of severe premenstrual syndrome known as PMDD.2017: Scientists at the Salk Institute in La Jolla, CA, said they're one step closer to being able to grow human organs inside pigs. In their latest research, they were able to grow human cells inside pig embryos, a small but promising step toward organ growth.
- ✰ 2017: The first step is taken toward epigenetically modified cotton.
- ✰ 2017: Research reveals different aspects of DNA demethylation involved in the tomato ripening process.
- ✰ 2017: Sequencing of green alga genome provides a blueprint to advance clean energy, bioproducts.
- ✰ 2017: Fine-tuning 'dosage' of mutant genes unleashes long-trapped yield potential in tomato plants.
- ✰ 2017: Scientists engineer disease-resistant rice without sacrificing yield.
- ✰ 2017: Blood stem cells were grown in the lab for the first time.
- ✰ 2017: Researchers at Sahlgrenska Academy – part of the University of Gothenburg, Sweden – generated cartilage tissue by printing stem cells using a 3D-bioprinter.
- ✰ 2017: Two-way communication in the brain-machine interface achieved for the first time.
- ✰ 2018: Zebrafish: Development of a Vertebrate Model Organism
- ✰ 2019: To Sleep Is to Heal: How the Immune System Regulates Sleep When Sickness Strikes
- ✰ 2019: role of genetics in shaping sexual behavior, no one "gay gene" – instead, there are many, many genes that influence a person's likelihood of having had same-sex partners.
- ✰ 2019: Use CRISPR To Correct Mutation in Duchenne Muscular Dystrophy Model
- ✰ 2020 A useful RNA that caps the chromosome – a new protective mechanism against chromosome shortening discovered [Switzerland, October 2020].
- ✰ 2020 The previously unknown risks of spaceflight brought to light – flying in space means damaging one's chromosomes.
- ✰ 2020 Our genes remember the suffering of ancestors: genetic analysis complements and improves historical records of the slave trade.
- ✰ 2020 A new genetic cause of hearing loss

was discovered due to damage in a gene called *RIPOR2*. [Netherlands, July 2020].

- ✰ 2020 Reversing aging through epigenetics – using a complex of factors involved in epigenetic regulation helps restore vision [USA, October 2020]. injection of OSK into mice with glaucoma led to the slow restoration of nerve endings in the eyes and improving their condition.
- ✰ 2021 study indicates that the genetics of eye color is not straightforward (UK, Jan 2021).
- ✰ 2021Researchers identify 14 genes that contribute to obesity (USA, Oct 2021).
- ✰ 2021 In patients with severe schizophrenia, researchers discover genetic variations linked to rare diseases (USA, Dec 2021)
- ✰ 2022 A Neanderthal family portrait. Deep inside a cave in southern Siberia, a team of researchers discovered the fossilized bone fragments of the first-known Neanderthal family.

1.3 Basic Concepts of Genetics

1.3.1 Classical Genetics

Classical genetics is concerned with the inheritance of an organism's traits. Dominant traits refer to those that appear in the next generation, while recessive traits remain hidden. When traits appear partially, they are intermediate. Sometimes, multiple genes are responsible for expressing a trait, which is called polygenic inheritance. These traits may be either sex-linked, resulting from a gene on the sex chromosome, or autosomal, resulting from a gene on a non-sex chromosome. Classical genetics originated from Mendel's study of inheritance in garden peas and has since expanded to include many different plants and animals. Today, classical genetics is primarily used for gene discovery, which involves identifying and assembling a set of genes that influence a particular biological property. Classical genetics deals with how an organism's traits are passed down through inheritance. Dominant traits refer to those that appear in the next generation, while recessive traits remain hidden. When traits appear partially, they are intermediate. Sometimes, multiple genes are responsible for expressing a trait, which is called polygenic inheritance. These characteristics can be determined by either sex-linked genes located on the sex chromosome or autosomal genes located on non-sex chromosomes. Classical genetics originated from Mendel's study of inheritance in garden peas and has since expanded to include many different plants and animals. Currently, classical genetics is mainly utilized for gene discovery. This process involves identifying and compiling a group of genes that affect a specific biological characteristic. Classical genetics is the primary tool used to discover genes. It involves the meticulous identification and compilation of a set of genes that impact a particular biological characteristic.

1.3.2 Cytogenetics

Cytogenetics is a highly skilled microscopic study that expertly combines the knowledge of cytologists, who specialize in researching cell structure and activities, with the expertise of geneticists, who focus on studying genes. The discovery of chromosomes, and the way they replicate and separate during cell division, was a groundbreaking discovery by cytologists that coincided with geneticists' deep understanding of gene behavior at the cellular level. The strong correlation between these two disciplines led to their successful merger, making cytogenetics an essential field in modern science. The study of plant cytogenetics is vital to the field of cytogenetics as plant chromosomes are generally larger compared to those found in animals. The squash technique, which involves flattening entire cells onto a glass slide for observation under a microscope, has also been used in animal cytogenetics. This method was especially important in identifying and numbering human chromosomes.

Various techniques are accessible to affix molecular labels to specific genes, chromosomes, RNAs, and proteins. The utilization of these labels facilitates the differentiation of these molecules from other cellular components, providing significant advantages in cytogenetic research.

1.3.3 Microbial Genetics

Early geneticists ignored microorganisms due to their small size and perceived lack of variable traits and sexual reproduction. However, it was later discovered that microorganisms possess various physical and physiological characteristics that can be studied. As such, they became of great interest to geneticists due to their size and rapid reproduction. Bacteria became essential model organisms for genetic analysis, leading to numerous general genetic discoveries. Bacterial genetics is integral in cloning technology, and viral genetics also plays a critical role in microbial genetics. The genetics of viruses that attack bacteria were the first to be understood, and subsequent findings have been applied to viruses pathogenic in plants and animals, including humans. Viruses are also utilized as vectors in DNA technology to introduce modified genetic material into organisms. Microorganisms were once disregarded by early geneticists due to their small size and lack of perceived variability in traits and sexual reproduction. However, it was later discovered that these tiny organisms possess a range of physical and physiological characteristics that can be studied. As a result, geneticists became intrigued by these organisms as they offer unique advantages like their rapid reproduction and size. Bacteria, in particular, became pivotal model organisms for genetic analysis, leading to numerous breakthroughs in general genetics. Bacterial genetics is now a vital component of cloning technology, while viral genetics plays an essential role in microbial genetics. The genetics of viruses that attack bacteria were the first to be understood, and subsequent findings have been applied to viruses that are pathogenic in plants, animals, and even humans. In addition, viruses are now commonly utilized as vectors in DNA technology to introduce modified genetic material into organisms. Early geneticists once dismissed microorganisms due to their size and perceived lack of variability in traits and sexual reproduction. However, it was later discovered that these tiny organisms possess numerous physical and physiological characteristics that can be studied. Geneticists became fascinated by these organisms due to their unique advantages, such as their rapid reproduction and size. Bacteria became a crucial model organism for genetic analysis, leading to many breakthroughs in general genetics. Bacterial genetics is now a vital component of cloning technology, while viral genetics plays an essential role in microbial genetics. The genetics of viruses attacking bacteria were the first to be understood, and subsequent findings have been applied to viruses that are pathogenic in plants, animals, and even humans. Additionally, viruses are now commonly used as vectors in DNA technology to introduce modified genetic material into organisms.

1.3.4 Molecular Genetics

The field of molecular genetics delves into the molecular structure of DNA and its various cellular activities, including replication, to determine an organism's overall characteristics. Genetic engineering, also known as recombinant DNA technology, is used to manipulate organisms by adding foreign DNA to create transgenic organisms. This technique has been extensively used in biological research since the 1980s and is integral to the biotechnology industry, which focuses on producing agricultural and medical products. Transgenic organisms are also the basis for gene therapy, which aims to treat genetic diseases by adding normally functioning genes from external sources.

1.3.5 Genomics

The field of genomics has become dominant in genetics research due to the advancements in technology that have enabled routine sequencing of whole genomes. It involves the study of entire genomes, including their structure, function, and evolutionary comparison, leading to a broader understanding of gene function. Researchers can now identify sets of genes that work together to affect various biological properties of interest. To analyze large sets of biological information, particularly genomic data, bioinformatics, a computer-based discipline, is used. This has led to significant progress in the field of genetics research.

1.3.6 Population Genetics

Studying genes within populations of animals, plants, and microbes can provide valuable insights into past migrations, evolutionary relationships, and how different varieties and species adapt and interact in their environment. Statistical methods can examine gene distributions and chromosomal variations in populations.

Population genetics uses mathematical analysis to understand the frequencies of alleles and the genetic makeup of populations. The Hardy-Weinberg formula, p2 + 2pq + q2 = 1, can predict the occurrence of individuals with homozygous dominant (AA), heterozygous (Aa), and homozygous recessive (aa) genotypes in a randomly mating population. Mathematical models can also incorporate selection, mutation, and other changes to explain and predict evolutionary changes at the population level. These methods can analyze alleles with known phenotypic effects, such as the recessive allele for albinism, or DNA segments with any known or unknown function.

Human population geneticists have used gene tracing to trace the origins, migration, and invasion routes of modern humans, Homo sapiens. DNA comparisons among present-day populations have shown that Homo sapiens originated in Africa. Geneticists have used gene tracing to infer the likely migration routes from Africa to the areas that are now inhabited by humans. Similar studies also show the extent to which present populations have intermixed through recent travel patterns.

1.3.7 Behavior Genetics

Analyzing human behavior can be challenging due to the significant impact of environmental factors. However, genomics studies offer an effective approach to investigating the genetic aspects of behavioral genetics.

1.3.8 Human Genetics

Medical genetics is a specialized field of study that focuses on understanding and treating genetic disease and related health issues in humans. Expert geneticists in this field work to investigate the processes of human gene function, as well as malfunctions, while also exploring pharmaceutical and other potential treatments. To gain important insights, researchers often study model organisms like bacteria, fungi, and fruit flies, which offer a high degree of evolutionary conservation between organisms.

While some diseases are caused by a single gene mutation, such as Tay-Sachs disease and phenylketonuria (PKU), more complex heredity components involving multiple genes can contribute to other diseases like depression, schizophrenia, and heart disease. Scientists are working tirelessly to better understand these more complicated diseases.

Clinical genetics is another important area of activity in this field, which involves providing parents with advice on the likelihood of their children being affected by genetic disease caused by mutant genes and abnormal chromosome structure and number. In genetic counseling, individuals and families' medical records undergo careful examination, along with diagnostic procedures that can detect unexpressed, abnormal forms of genes. Physicians or specially trained non-physicians with expertise in human genetics can carry out counseling.

1.4 Chromosomal Theory of Inheritance

Through the study of chromosomes during mitosis and meiosis, scientists speculated that understanding heredity may be found within them. Theodor Boveri's observation in 1902 that chromosomes were required for proper embryonic development in sea urchins, and Walter Sutton's discovery that chromosomes separated into daughter cells during meiosis, led to the establishment of the Chromosomal Theory of Inheritance. This theory stated that genes are located on chromosomes and are responsible for Mendelian inheritance. Although Charles Darwin proposed the idea of hereditary units in 1868, it was later discarded with the advent of more advanced tools for studying genetics. However, Hugo de Vries' 1900 rediscovery of Gregor Mendel's principles of heredity and his theory of biological mutation played a significant role in the discovery of the Chromosomal Theory of Inheritance. This resolution of previously ambiguous concepts concerning the nature of species variation allowed for the universal acceptance and investigation of Charles Darwin's system of organic evolution Hugo de Vries, a scientist who taught at the University of Amsterdam between 1878 and 1918, made a significant contribution to our understanding of evolution. In 1886, while studying the evening primrose, he discovered new varieties that appeared randomly, which he called mutations. Unlike Darwin's theory of natural selection, de Vries believed that these sudden changes were a result of evolution that could be experimentally studied. However, other scientists also made noteworthy contributions to genetics. William Bateson, a British biologist, founded and named the science of genetics. In 1894, he published a theory stating that species' evolution could not occur through continuous variation, but rather through distinct and sudden changes in traits. Bateson championed Mendel's principles experimentally, after discovering an article by Gregor Mendel and published a series of breeding experiments that extended Mendel's principles to animals. He demonstrated that certain features were consistently inherited together, which was later understood to be the result of genes located nearby on the same chromosome. Bateson's work transformed the John Innes Horticultural Institution at Merton, South London, into a center for genetic research. The terms "gene" and "genetics" were adopted from the Latin word "gen," meaning origin. The union of cytology with Mendelian breeding analysis became known as the "chromosome theory of heredity."Mendel didn't know about genes or discover genes, but he did speculate that there were 2 factors for each basic trait and that 1 factor was inherited from each parent. He described them as stable units that seemed to disappear in a hybrid (a plant grown from a cross between two parent plants that show differing traits) but would reappear among some of the progeny of such hybrids. Mendel gave two laws the law of segregation and the law of independent assortment which together governed the movement of factors from parents to offspring's which showed the inheritance of these factors were discrete physical objects. Just before Mendel's work was rediscovered the microscope was also invented that allowed scientists to make careful observations of cell division. This led to the discovery of colored bodies (which were named as chromosomes) in the cell nucleus that appeared to double and divide just before each division.

In 1907, Thomas Hunt Morgan and his students conducted laboratory studies on the fruit fly, Drosophila melanogaster. Their research built upon Sutton's chromosome theory of heredity and resulted in new discoveries. For instance, they revealed X-linked inheritance in 1920, which showed that females have two X chromosomes, while males have only one. Morgan's team also found that some homologous chromosomes don't separate during meiosis, leading to deviations from the expected 3:1 ratio for white-eyed flies. These findings further supported the chromosomal theory of inheritance.

Morgan's discovery of crossing over allowed for the mapping of genes along the length of the X chromosome. Alfred Sturtevant, one of Morgan's students, created the first linkage map using Morgan's data. Later, Barbara McClintock independently demonstrated the same phenomenon while working on maize.

However, most traits are governed by multiple genes, making it difficult to formulate a hypothesis. Hermann Joseph Muller showed that fruit fly wing shapes and lengths are controlled by many factors, including modifier genes that could intensify or diminish their expression. Muller predicted the percentages of wing shape and length among the progeny by combining different modifier genes and the chief gene in two parents. He also attempted to explain the Darwinian model of natural selection of traits whose variations owe their origins to the highly heterozygous state of natural populations and to new mutations that arise in each generation.

1.5 Transmission of Genetic Information

Living organisms inherit their traits from their parents through genetic material. Gregor Mendel, an Austrian monk, proposed the laws of heredity and conducted experiments with garden peas to demonstrate how traits are passed down from generation to generation. From his experiments, Mendel concluded that genes or factors determine inherited traits, with each organism carrying two copies of each gene inherited from their parents. Additionally, Mendel discovered that alternative versions of genes, known as alleles, exist.

Organisms with a pair of identical alleles for a trait are considered homozygous, while those with two different alleles are called heterozygous. The complete genetic makeup of an organism is referred to as its genotype, while its observable traits are its phenotype. However, the genotype alone does not determine the phenotype, as it interacts with both the internal and external environment to produce observable traits.

Mendel identified factors that control the phenotype and realized that each individual carries two paired factors, one from each parent. He also discovered that these factors segregate randomly into gametes (Mendel's first law, the principle of segregation) and that factors controlling one trait independently assort from those controlling another trait during gamete formation (Mendel's second law, the principle of independent assortment), as illustrated in Figure 1.1.

P (parental) generation	YY	X	yy
Parental phenotype	yellow		Green
	seeds		Seeds
Haploid gametes	Y		Y

F_1 generation — Yy (Fusion of the gametes)

F_1 phenotype — yellow seeds because Y allele is dominant over y allele; or y is recessive to Y.

Parents	Tall, yellow T/T Y/Y x Dwarf green t/t y/y				
F1	All tall yellow T/t Y/y x T/t Y/y				
F2		TY	Ty	tY	ty
	TY	T/T Y/Y	T/T Y/Y	T/t Y/Y	T/t Y/y
	Ty	T/T Y/y	T/T y/y	T/t Y/y	T/t y/y
	tY	T/t Y/Y	T/t Y/y	t/t Y/Y	t/t Y/y
	ty	T/t Y/y	T/t y/y	t/t Y/y	t/t y/y
	Ratio: 9 tall yellow; 3 tall green; 3 dwarf yellow; 1 dwarf green				

Figure 1.1: Principle of Independent Assortment.

(Fusion of the gametes)

When F_1 plants are crossed:

$F_1 \times F_1$ Yy x Yy

F_1 gametes Y y Y y

F_2 generation

	Y	y
Y	YY	Yy
y	Yy	Yy

The genotypic ratio is 1 YY: 2 Yy: 1 yy, where Y dominates over y. This leads to a phenotypic ratio of 3 yellow to 1 green. It took nearly two decades after Mendel's demise in 1884 for the fundamental basis of gene segregation to be uncovered. In 1902, Walter Sutton and Theodor Boveri put forth the chromosome theory of heredity, which established that genes reside on chromosomes and that gene segregation occurs during meiosis, when chromosomes are transmitted from one generation to the next.

Mendel's experiments showed that genes segregate from one generation to the next. It wasn't until 1902 that the chromosome theory of heredity was presented, which demonstrated that genes are located on chromosomes and segregate during meiosis.

1.6 Genetic Variation

The existence of genetic variation is critical for evolution, and it can arise through either mutations or recombination. Organisms possess repair mechanisms capable of correcting mutations caused by internal or external factors. Mutations that remain uncorrected become part of the genetic code. Recombination, on the other hand, shuffles genes from both parents and is influenced by their proximity. Linkage occurs when genes are inherited together due to their position on the same chromosome. Linkage disequilibrium is used by scientists to map genes related to diseases.

The selection process, first proposed by Charles Darwin, plays a significant role in genetic variation by changing gene frequencies. Beneficial genes are selected, and harmful ones are eliminated. Genetic variation occurs through three mechanisms: mutation, recombination, and selection, which work together to create new genetic variations that is crucial for evolution.

Phenotypic variation can result from genetic variation in quantitative or discrete traits. Polymorphic genes have more than one allele at each locus, and enzyme variations can also identify genetic variation. Advances in sequencing technology have enabled scientists to detect even more genetic variation. Differences in nucleotides can alter the amino acid sequence, which can affect enzyme function and cause phenotypic variation.

1.7 Chemical Nature of Genes

In 1944, is an evolutionary the moment for nucleic acid research, Oswald Avery, Maclyn McCarty, and Colin MacLeod, at Rockefeller Institute (now University) Hospital, New York, proved that DNA is the genetic material and not the proteins as initially, it was thought.

After proving that DNA constitutes genetic material it was important to know about the structure of the DNA molecule and the mechanisms through which all the activities of the living organism are governed.

One of the great discoveries of the 19th century was made by Watson and Crick who showed that two strands of DNA run anti-parallel to one another and are also complementary to each other. The molecular complementarity of double-stranded DNA is its most important property and the key to understanding how DNA replicates transmit mutates, and also store genetic information. The sugar-phosphate backbone is outside the helix, and the bases are towards the inside. The backbone can be taken as the sides of a ladder, whereas the bases in the middle form the stairs of the ladder. The biological information is stored in units known as nucleotides. The DNA alphabets are G (Guanine), C (Cytosine), A (Adenine), T (Thymine). These nucleotides are very important as their sequence determines which protein an organism will make as well as when and where protein synthesis will occur.

Each rung is composed of two base pairs. Either an Adenine-Thymine pair that form a two-hydrogen bond together or a Cytosine-Guanine pair form three-hydrogen bonds. Base pairing is hence restricted.

This restriction is essential when the DNA is copied: the DNA-helix is first "unzipped" in two long stretches of the sugar-phosphate backbone with a line of free bases sticking up from it, resembling the teeth of a comb. Each half forms the template for a new, complementary strand. There are biological mechanisms that can correct if the mistake occurs. After replication, there are two daughter copies of the original DNA. In the DNA strand, there are certain sequences that encode proteins these are called exons while those which flank the coding region are called introns. Initially, it was thought that introns are junk DNA but now we know that introns are also involved in the regulation of gene expression.

RNA is another nucleic acid that is chemically like DNA. Structurally it is single-stranded and contains ribose instead of deoxyribose. In place of thymine, it has uracil. Like DNA, RNA has the capacity to store, replicate, mutate, and express information like proteins; RNA can fold in three dimensions to produce molecules capable of catalyzing the chemistry of life. RNA is more unstable as compared to DNA. The proteins have greater diversity, three dimensional foldings of RNA holds the intermediary position between DNA and proteins.

1.8 DNA to Phenotype

Proteins are created through two critical stages in gene expression: transcription and translation. RNA polymerase plays an essential role in transcription by generating an mRNA strand from a specific DNA region. This mRNA molecule then models the process of translation, where amino acids are synthesized to produce proteins. The genetic code for amino acids is expressed through DNA or RNA codons, which remain consistent across all living organisms.

Genetic mutations can cause changes in protein function and result in altered phenotypes. For instance, sickle-cell anemia is a human genetic disorder caused by a mutation in hemoglobin, which is responsible for transporting oxygen. To function correctly, hemoglobin requires a specific structure that is determined by the precise sequence of As, Cs, Ts, and Gs that encode the protein. The hemoglobin molecule consists of four polypeptide chains that contain alpha and beta-globin chains.

The heme group is a flat ring molecule that holds an iron ion through four nitrogen ligands from the porphyrin ring. In sickle-cell anemia, mutations in the beta-globin gene cause a change in one of the DNA nucleotides, resulting in a change in the amino acid sequence of beta-globin from glutamic acid to valine. This alteration in the protein structure ultimately leads to sickle-cell anemia (**Figure 1.2**).

Sickle-cell anemia is a medical condition caused by a genetic mutation that leads to the formation of abnormal hemoglobin molecules in red blood cells. These abnormal molecules can cause the cells to take on a sickle-like shape, resulting in severe pain, tissue damage, and potential heart attacks and strokes if left untreated. The mutation affects only one amino acid in the beta-globin molecule, highlighting the strong connection between genotype and phenotype.

Although the beta-globin gene is inactive for a few days after birth, prenatal testing using molecular techniques can detect the mutation. This testing can help individuals and families determine if they carry a mutant copy of the gene and are at risk of passing it on to their children.

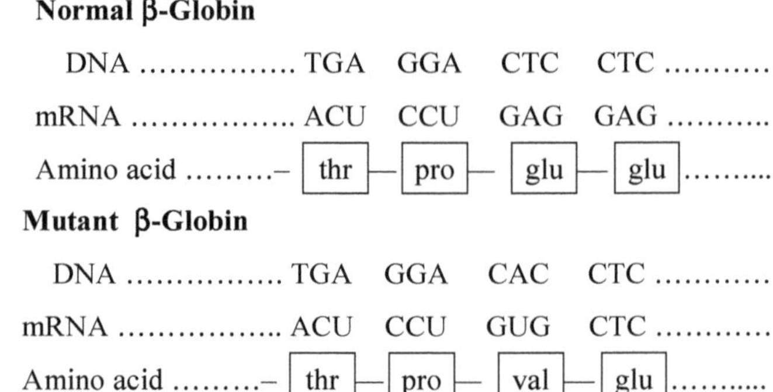

Figure 1.2: When the DNA Encoding the Beta-globin Gene Experiences a Single Nucleotide that changes from CTC to CAC, it causes a Change in the mRNA Codon from GAG to GUG. This alteration results in the insertion of a different amino acid, Val, instead of Glu, which leads to the production of an altered form of the beta-globin protein, ultimately causing sickle cell anemia.

1.9 Function of Proteins

Proteins are incredibly diverse, with the potential for a 100-amino-acid chain to contain any of the 20 different amino acids. There are numerous unique sequences of 100-amino-acid proteins, totaling 20100, demonstrating the vast possibilities. Proteins are classified by function, with enzymes being the most extensive group. Inherited defects in enzymes that result in disease are known as inborn errors of metabolism. One example of this is alkaptonuria, a rare disorder that affects 1 in 200000 individuals. Garrod discovered that an abnormal substance in urine could lead to this disease. The PKU diet is now used to treat phenylketonuria, a condition caused by a lack of phenylalanine hydroxylase (**Figure 1.3**).

Proteins play essential roles in hormone production, including insulin and growth hormone. They also aid in the immune response and neurotransmitters. Fibrinogen helps form clots, while actin and myosin are crucial components of muscles. Proteins also assist in the storage of nutrients, such as iron in the protein ferritin. In plants, proteins are vital for storing nutrients necessary for new plant growth are a major component of seeds, and storage of proteins is carried out.

There are several ways that proteins contribute to the structure of living things. Glycoproteins play a major role in the structures of cell membranes. While lipids are more in number in the membrane, many of the key functions take place at the membrane level, such as transport of materials, which are carried out by transmembrane proteins. Receptors, including those in nerve cells, are proteins that help cells interact with their external environment. Some tissues like

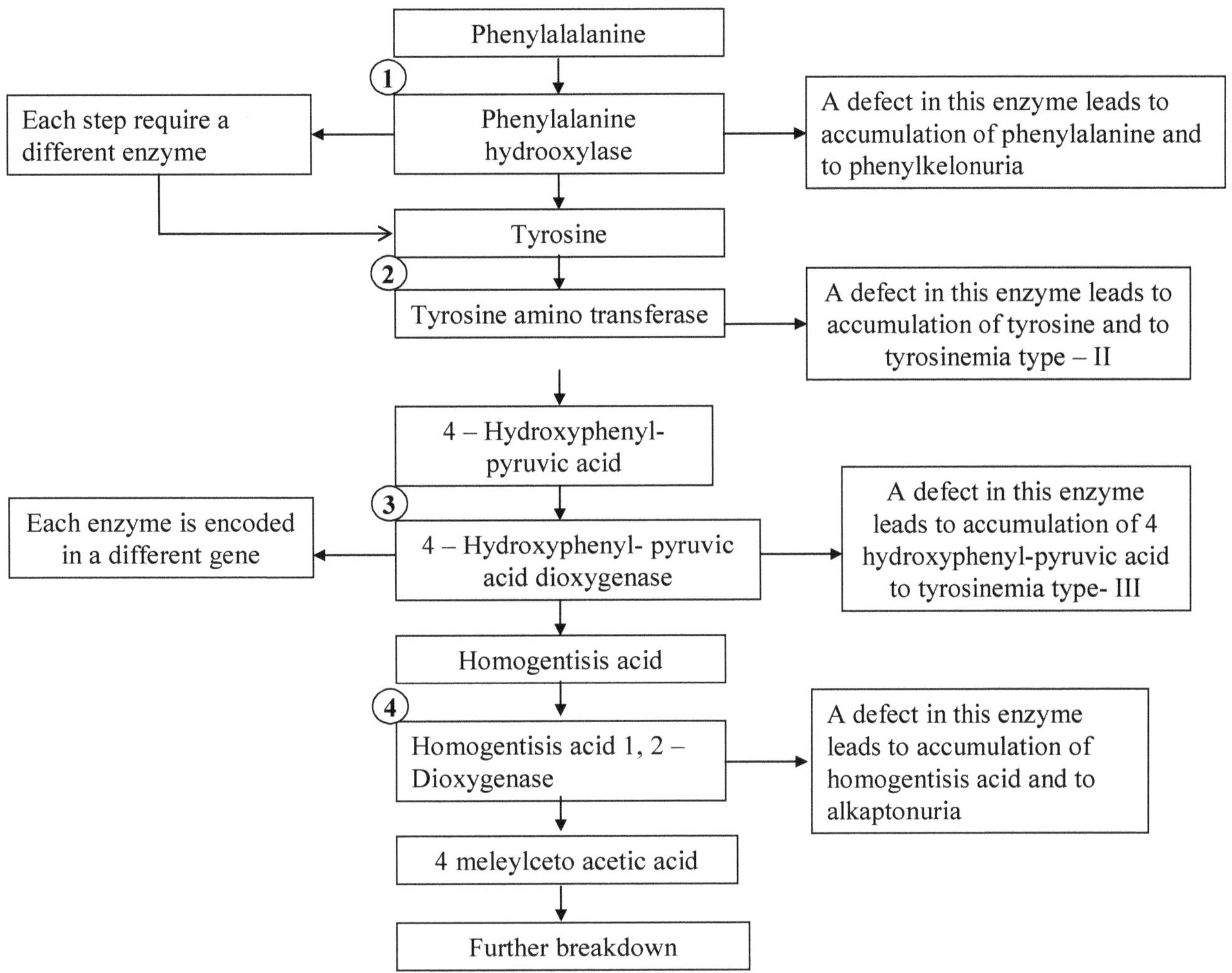

Figure 1.3: Breakdown of Phenylalanine and Tyrosine.

cartilage have proteins that help to provide structure on a larger scale. Hemoglobin is also a protein that binds to CO_2 and carries it to the lungs to be released.

1.10 Genetics and Medicine

Garrod, in 1902 concluded that alkaptonuria was inherited according to Mendelian principles. Garrod discovered a defective or mutant gene that gets blocked in a metabolic pathway causing an inherited disorder. In recognition of his contributions, Garrod is often referred to as "the father of biochemical genetics" as he was first to discover the metabolic pathway. Since Garrod's time, enormous progress has been made in establishing links between defective genes and disease. The genes in normal individuals are called *wild –type* genes; those with defects that result in the synthesis of nonfunctional products are called mutant genes.

Hemophilia, or bleeding disease was one of the first human disorders found to be linked to a mutant gene and its nonfunctional product. In the past, individuals with this inherited defect in blood clotting usually died during childhood. Today, they often live full and normal lives. This is due to the effective treatment with clotting factors produced by genetically engineered mammalian cells growing in culture. A fatal neurological disease, *i.e.*, Huntington's disease, is caused by a mutation in a gene. This was isolated in 1993 after intense 10-year research. The mutant gene which causes cystic fibrosis has been isolated and its gene product has been identified. Because of these discoveries it is now possible to treat the disease.

Gene therapy involves the introduction of normal genes into the cells of patients with defective genes-offers effective treatment of inherited diseases. Unfortunately, to date, gene therapy has not lived up to its promise. Gene therapy has been used in combination with other treatments on children with a devastating immune system disorder called severe combined immunodeficiency disease, but most of the results have been discouraging. In most of the experiments, the introduced genes were expressed for only a short period of time. Recently, however, one type of severe combined immunodeficiency disease has been treated successfully by somatic-cell gene therapy, and hence there is a hope that gene therapy will be effective in treating other inherited disorders, in the future.

Some forms of cancer are familial (*i.e.*, they occur frequently within families); others occur sporadically among all members of a population. However, all cancers are genetic diseases in the sense that they are caused by the change in the genetic information that is transmitted to progeny cells. The available evidence indicates that all cancers result from the accumulation of damage to genes that control or influence cell division, cell differentiation, and related processes. Although there are hundreds of different types of cancers, they all have one thing in common loss of the normal control of cell division.

Mutant genes that cause an increased risk of cancer are being identified and studied intensely. One such gene is designated as *BRCA1* for breast cancer gene *I, which* results in the increased susceptibility to breast and ovarian cancer in certain families. The *BRCA1* gene was isolated, characterized, and shown to be located on chromosome 17. A second breast cancer gene, *BRCA2,* was subsequently identified. There is a long list of genes that cause familial cancers, including Wilms' tumor, cancer of kidneys in children, and Gardner syndrome, a type of colon cancer. Genetic mutations in cancer lead us to understand the gene function and this further leads to effective therapeutic treatments.

Even with information about the function of a gene, designing a therapy is another problem of major concern. For example, the gene for Huntington's disease was isolated in 1993, and its protein product, *huntingtin* was identified shortly after that. The function of *huntingtin* is still unknown. The normal function of *huntingtin* is poorly understood, but it has been proposed to play a role in neurogenesis, apoptotic cell death, and vesicle trafficking. However, once the function of the Huntington's disease gene is fully understood, gene therapy may be possible. In contrast to Huntington's disease, gene product, called *CFTR* (cystic fibrosis transmembrane conductance regulator) function is better characterized. It is possible to treat cystic fibrosis by gene therapy, that is, by introducing functional copies of the *CFTR* gene into cells of individuals with defective versions of the gene. In one strategy, a normal form of the *CFTR* gene was inserted into a common cold virus that had been genetically altered so that it couldn't complete its normal life cycle. This engineered virus was then applied to the nasal passages of CF patients. The envelope that encapsulates the cold virus functioned as the vehicle for transferring the normal *CFTR* gene into nasal epithelial cells. Once in the cells, it was hoped that the *CFTR* gene would direct the synthesis of a normal gene product, and the cells-previously crippled due to the lack of *CFTR*-would regain some of their normal function. However, this was not successful.

The isolation and characterization of the mutant *BRCAl* and *BRCA2* genes that increase the risk of breast cancer are creating important social and ethical issues. *BRCA1* is having approximately 100 mutations as it is a large gene. It is an expensive and labor-intensive job to screen women for all possible variants of this gene on a routine basis. To complicate matters further, some variants of *BRCAI* do not predispose women to breast cancer. Therefore, it appears that the testing will be done primarily on women with a family history of breast cancer. Laboratory expertise is another issue that adds to the complications. In cases where a diagnostic test indicates a woman is carrying a mutant *BRCA 1* or *BRCA2* gene that may predispose her to develop breast cancer, what treatment options can be offered? It is always in the interest of the woman to be told that she is carrying a mutation that predisposes her to develop breast cancer. Prophylactic mastectomy is a recommended treatment for carriers of mutant *BRCA* genes; however, this may or may not be always successful. *BRCA* mutations also may cause ovarian cancer hence testing for these genes raises many other issues.

1.11 Other Factors Affecting the Phenotype

The easiest way of studying genetics is to investigate the physical appearance of the organism. Physical appearance is called phenotype which is the result of the genetic makeup of the individual. The effect of a single gene is dependent on many other genes in the genome and many environmental factors such as nutrition, climate, or exposure to infectious agents. Although the complete DNA sequence analysis of a genome has provided more insight into the understanding of many genes responsible for the phenotypes the identification of environmental factors is slower because the understanding of these factors requires careful and controlled experimentation and observation. Such conditions are difficult to study, particularly in humans.

The effect of the environment is evident in many experimental organisms, such as fungi, were "conditional" mutants are artificially constructed and utilized in genetic studies. For instance, temperature-sensitive mutants grow well at certain temperatures but die at a slightly elevated temperature. Even in humans, some environmental factors contribute to the manifestation of the disease. One of the best-known examples is phenylketonuria (PKU). The mental retardation associated with the disease can be prevented by giving phenylalanine restricted diet.

1.12 Epistasis

Genes do not work in isolation. One gene has been found to influence or mask the expression of the other gene; the former gene is epistatic to the latter. In epistasis, the interaction between genes is antagonistic, such that one gene masks or interferes with the expression of another. "Epistasis" is a word composed of Greek roots that mean "standing upon." The alleles that are being masked or silenced are said to be hypostatic to the epistatic alleles that are doing the masking. Often the biochemical basis of epistasis is a gene pathway in which the expression of one gene is dependent on the function of a gene that precedes or follows it in the pathway. Examples of epistasis abound in nonhuman organisms. In mice, as in humans, the gene for albinism has two variants: the allele for non albino and the allele for albino. The latter allele is unable to synthesize the pigment melanin. Mice, however, have another pair of alleles involved in melanin placement. These are the agouti allele, which produces dark melanization of the hair except for a yellow band at the tip, and the black allele, which produces melanization of the whole hair. If melanin cannot be formed (the situation in the mouse homozygous for the albino gene), neither agouti nor black can be expressed. Hence, homozygosity for the albinism gene is epistatic to the agouti/black alleles and prevents their expression.

1.13 Complementation

The phenomenon of complementation is another form of interaction between nonallelic genes. For example, there are mutant genes that in the homozygous state produce profound deafness in humans. One would expect that the children of two persons with such hereditary deafness would be deaf. This is frequently not the case, because the parents' deafness is often caused by different genes. Since the mutant genes are not alleles, the child becomes heterozygous for the two genes and hears normally. In other words, the two mutant genes complement each other in the child. Complementation thus becomes a test for allelism. In the case of congenital deafness cited above, if all the children had been deaf, one could assume that the deafness in each of the parents was owing to mutant genes that were alleles. This would be more likely to occur if the parents were genetically related (consanguineous).

1.14 Polygenic Inheritance

The greatest difficulties of analysis and interpretation are presented by the inheritance of many quantitative or continuously varying traits. Inheritance of this kind produces variations in degree rather than in kind, in contrast to the inheritance of discontinuous traits resulting from single genes of major effect. The yield of milk in different breeds of cattle; the egg-laying capacity in poultry; and the stature, shape of the head, blood pressure, and intelligence in humans range in continuous series from one extreme to the other and are significantly dependent on environmental conditions. Crosses of two varieties differing in such characters usually give F_1 hybrids intermediate between the parents. At first sight, this situation suggests a blending inheritance through "blood" rather than Mendelian inheritance; in fact, it was probably observations of this kind of inheritance that suggested the folk idea of "blood theory."

It has, however, been shown that these characters are polygenic–*i.e.*, determined by several or many genes, each taken separately producing only a slight effect on the phenotype, as small as or smaller than that caused by environmental influences on the same characters. That Mendelian segregation does take place with polygenes, as with the genes having major effects (sometimes called oligogenic), is shown by the variation among F_2 and further-generation hybrids being usually much greater than that in the F_1 generation. By selecting among the segregating progenies, the desired variants–for example, individuals with the greatest yield, the best size, or a desirable behavior–it is possible to produce new breeds or varieties sometimes exceeding the parental forms. Hybridization and selection are consequently potent methods that have been used for the improvement of agricultural plants and animals.

Polygenic inheritance also applies to many of the birth defects (congenital malformations) seen in humans. Although expression of the defect itself may be discontinuous (as in clubfoot, for example), susceptibility to the trait is continuously variable and follows the rules of polygenic inheritance. When a developmental threshold produced by a polygenically inherited susceptibility, and a variety of environmental factors are exceeded the birth defect results.

1.15 Penetrance and Expressivity

Penetrance refers to the probability of a gene or trait being expressed. In some cases, despite the presence of a dominant allele, a phenotype may not be present. One example of this is polydactyly in humans (extra fingers and/or toes). A dominant allele produces polydactyly in humans but not all humans with the allele display extra digits. "Complete" penetrance means the gene or genes for a trait are expressed in all the population who have the genes. "Incomplete" or 'reduced' penetrance means the genetic trait is expressed in only part of the population. The penetrance of expression may also change in different age groups of a population. Reduced penetrance probably results from a combination of genetic, environmental, and lifestyle factors, many of which are unknown. This phenomenon can make it challenging for genetics professionals to interpret a person's family medical history and predict the risk of passing a genetic condition to future generations. Expressivity on the other hand refers to variation in phenotypic expression when an allele is a penetrant. Back to the polydactyly example, an extra digit may occur on one or more appendages. The digit can be full size or just a stub. Hence, this allele has reduced penetrance as well as variable expressivity. Variable expressivity refers to the range of signs and symptoms that can occur in different people with the same genetic condition. As with reduced penetrance, variable expressivity is probably caused by a combination of genetic, environmental, and lifestyle factors, most of which have not been identified. If a genetic condition has highly variable signs and symptoms, it may be challenging to diagnose.

1.16 Pleiotropy

Sickle cell disease is a very common genetically inherited disorder that is often fatal in childhood. This is characterized by anemia, extreme pain and fatigue, heart failure, spleen enlargement, severe microbial infection, and impaired mental abilities. Early genetic studies have shown that the disease is inherited in an autosomal, recessive manner.

A responsible allele of sickle cell disease expresses a mutated version of a protein subunit of hemoglobin. Normal hemoglobin is found in red blood cells and is essential for the transport of oxygen to all tissues of the body. In sickle cell disease, the conformation of the mutant hemoglobin sub-unit is altered, by virtue of a single amino acid change from the wild-type subunit, causing the proteins to associate with each other into abnormal fibers. This association inhibits the binding of oxygen and the red blood cells appear stretched out in a characteristic "sickle" shape. The sickle cells clog capillaries, seriously compounding the effect by reducing all blood flow to various tissues.

This disease provides a useful example of how a highly complex phenotype can result from a single genotype (in this case a single nucleotide mutation). This is called pleiotropy, which arises due to a cascade of consequences from the single mutated protein. In this case, the abnormal hemoglobin causes poor oxygen transport and sickling of red blood cells, which in turn causes anemia, poor blood circulation, and accumulation of sickle cells in the spleen. The poor blood circulation in turn causes heart failure, lung damage (often followed by pneumonia), and brain damage. Overloading the spleen with sickle cells causes extreme pain and an increased frequency of infectious disease.

1.17 Recombinant DNA Technology

Use of recombinant deoxyribonucleic acid (DNA) technology can manipulate DNA sequences. The development of recombinant DNA technology in the 1970s has revealed an exciting invention. This technique has profound effects on society, with better health through better disease diagnosis, much-improved understanding of human genetic variation, improved drug and pharmaceutical production, forensics, and manufacture of genetically modified organisms that meaningfully improve yields and nutritional value of different crops while declining reliance on pesticides and artificial fertilizers. Recombinant DNA and the transgenic technology that has been created through recombinant DNA technology have already been used in day-to-day life.

It has made a major economic impact on the pharmaceutical industry, single human proteins; specific antibodies have been produced for beneficial use. Harvesting human insulin created in bacterial cells is far easier than isolating it from a pig or human cadaver pituitary glands. The financial basis for recombinant DNA technology should continue to improve as genetically modified organisms are becoming widely used in agriculture; the soybean crop now consists of strains that are genetically modified to reduce the number of herbicides necessary to have a good yield. Some of the FDA approved drugs used in the Indian market are shown in **Table 1.1**.

1.18 Future of Recombinant DNA Technology

Perhaps the most profound impact of molecular biology on the discovery of new drugs is that it enables a target-driven approach, with many advantages over the historical approach of identifying active compounds in a therapeutically relevant biological assay, but with no real knowledge of its target or mechanism. The fundamental question that comes in the mind about any compound during its discovery and development is related to its feasibility of working as well as safety of the recombinant technology. Building a drug-discovery programme

Table 1.1: Recombinant Products for Human Therapeutic Use Indian Scenario

Therapeutic Category	*Product*	*Expression Host*	*Abbreviated Indication*
Hormone of therapeutic interest	Human insulin	*E. coli/S. cerevisiae*	Treatment of diabetes
	Insulin as part	*S. cerevisiae*	Treatment of diabetes
	Insulin glargine	*E. coli*	Treatment of diabetes
	Insulin lispro	*E. coli*	Treatment of diabetes
	Insulin glulisine	*E. coli*	Treatment of diabetes
	Human choriogonadotropin	CHO cells	Treatment of women undergoing superovulation prior to assisted reproductive techniques such as *in vitro* fertilization
	Follicle-stimulating hormone	CHO cells	Treatment of infertility
	Luteinizing hormone	CHO cells	Induction of ovulation
	Somatotrophin	*E. coli*	Treatment of deficiency of growth failure
Haemopoietic growth factors	Erythropoietin alpha	CHO cells	Treatment of anemia associated with renal failure, HIV infection, cancer
	Erythropoietin beta	CHO cells	Treatment of anemia associated with renal failure
	Erythropoietin omega	BHK cells	Treatment of anemia associated with renal failure and cancer
	Darbepoetin	CHO cells	Treatment of anemia associated with renal failure and cancer
	Filgrastim	*E. coli*	Reduction in duration of neutropenia and incidence offebrile neutropenia in patients treated with cytotoxicchemotherapy for malignancy
	Sargramostim.	*Saccharomyces cerevisiae*	Treatment of neutrophil recovery
Human blood coagulation factors	Factor VIII	CHO cells	Treatment of haemophilia A
	Factor IX	CHO cells	Treatment of haemophilia B
	Factor VII A	BHK cells	Treatment of haemophilia A and B
Thrombolytic agents	Tissue plasminogen activator Alteplase	CHO cells	Treatment of acute myocardial infarction
	Reteplase	*E. coli*	Treatment of acute myocardial infarction
	Tenecteplase	CHO cells	Treatment of acute myocardial infarction
	Saruplase	*E. coli*	Thrombolytic therapy for acute myocardial infarction
Anticoagulants	Lepirudin	*S. cerevisiae*	Anticoagulation therapy for heparin associated thrombocytopenia
	Desirudin	*S. cerevisiae*	Prevention of venous thrombosis
Human interferons	Interferon alpha-2b	*E. coli*	Treatment of hairy cell leukaemia, chronic hepatitis B and C, AIDS, cancer
	Interferon beta-1b	*E. coli*	Treatment of multiple sclerosis
	Interferon gamma	*E. coli*	Chronic granulomatous disease
Human interleukins	Interleukin-2	*E. coli*	Renal cell carcinoma
	Interleukin-11	*E. coli*	Treatment of thrombocytopenia
Therapeutic enzymes	Dorsase alpha	CHO cells	Cystic fibrosis
	Glucocerebrosidase	CHO cells	Replacement therapy in patients with Gaucher disease

around a known target enables us with greater precision and accuracy to identify new drugs that prove to be highly potent and efficacious in humans while being reassured of a low risk of toxicity (**Figure 1.4**).

1.19 Use of Model Organisms in Genetics

Most insights into human disease are a result of experiments on animals that would be unethical or not feasible to perform on humans. Instead, we can use animal models to look at the functions of

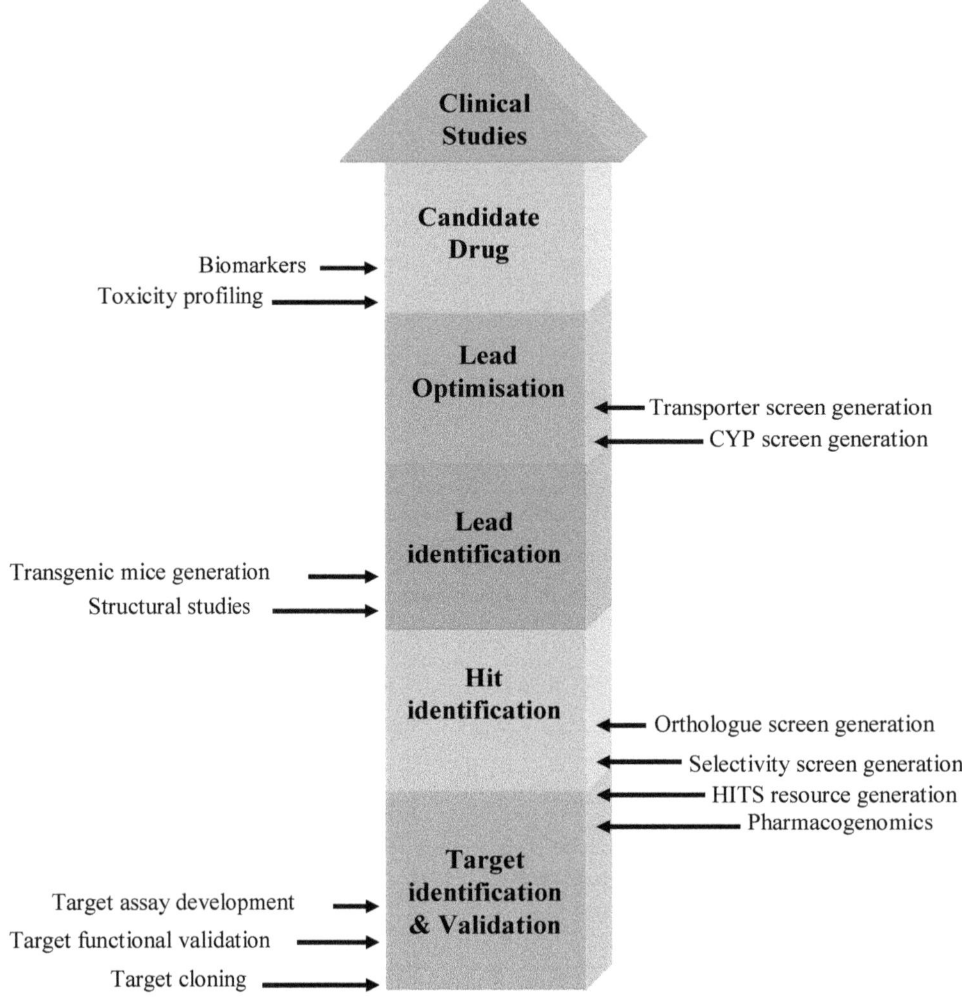

Figure 1.4: Drug Discovery and Role of Recombinant DNA Technology.

genes involved in maintaining healthy organisms to obtain vital clues about the causes and progression of human diseases. These animals are known as model organisms and are used in biology (**Table 1.2**). As a result, there is a wide range of characteristics common to model organisms, including i) rapid development with short life cycles, ii) small adult size, iii) the ready availability, and iv) controllability. Being small, growing rapidly, and being readily available is crucial in terms of housing them in each budget and space limitations of the research and in teaching laboratories. Tractability relates to the ease with which they can be employed. For example, *C. elegans* is a popular research organism as it possesses all the characteristics mentioned above. It is widely used for many experiments as it shares many essential properties with humans. Scientists are studying *C. elegans* for apoptosis studies that provide hope of finding treatments for certain types of human cancers, such as leukemia, *etc.* the results obtained by these studies may provide researchers with a tool for treating the rapid proliferation of cancer cells.

Table 1.2: Model Organisms Used to Study Human Diseases

Organism	*Human Diseases*
E. coli	DNA repair, colon cancer, and other cancers.
Yeast	Cell cycle, cancer, Werner Syndrome
Drosophila	Cell signaling cancer
C. elegans	Cell signaling diabetes
Zebra fish	Developmental pathways; cardiovascular disease
Mouse	Gene expression; Lesch-Nyhan Syndrome, Cystic fibrosis, fragile – X syndrome and many other diseases

1.20 *Drosophila*

The fruit fly, *Drosophila melanogaster*, is an important eukaryote. It is used to understand human genetics and developmental processes. *Drosophila* is a very popular and successful model organism because it has a short life cycle of only (14 days) hence numerous generations can be studied in one year.

It is easy to culture and inexpensive to house large numbers. Its size is amenable, neither too small nor too large hence can be used in various experiments. It can be seen with the naked eye or under low-power magnification. Moreover, it has a very long history in biological research and there are many useful tools to facilitate genetic studies. For example, the use of antibodies makes the scoring of specific cells or cell types possible in *Drosophila*. *Drosophila* has been used in research in molecular genetics. It has been extensively used to study developmental genetics.

A complex human disease of the retina called retinitis pigmentosa is identical to Drosophila retinal degeneration genes such as rdg B and rdg C. Hence studies in Drosophila are helpful in dissecting this complex disease and identifying the function of genes involved. Drosophila is also used to study the diseases of the human nervous system by transferring a human disease gene into flies. This technique is in use to study human diseases like Huntington disease, Machado Joseph disease, Myotonic Dystrophy, and Alzheimer's disease.

The zebrafish (*Brachydanio rerio*) with its short reproductive cycle and a high number of eggs per clutch is one of the most popular vertebrate systems used in genetic analysis. Its rapid development has also made it popular in the developmental biology and embryology communities. Its eggs are externally fertilized and translucent, making it possible to follow its developmental progress easily using a dissecting microscope. The tractability of this fish to gene knockout methods and the existence of techniques for producing large numbers of clones have also helped to establish it as a favorite system in developmental genetics.

1.21 Frog (*Xenopus laevis*)

The African clawed frog has a long history of use in the study of vertebrate embryology and development. As aquatic animals, they undergo external development, making it relatively easy to manipulate the fertilized egg as compared to animals in which the embryo grows within the uterus. Their eggs are also quite large, about 1 mm in diameter, which is about ten times the size of a mouse oocyte. Ovulation can be induced relatively simply using the luteinizing hormone, which eliminates some of the problems associated with extended reproductive cycles. They follow rapid progress starting from fertilization to neurulation. It is completed in less than 24 hours. This provides scientists with a dynamic picture of the first stages of life.

1.22 Chicken (*Gallus gallus*)

The chick embryo has been in use as a model organism since the time of Aristotle, making it one of the most exhaustively studied systems in biology. The egg's extra-uterine development made it attractive in the days before surgical techniques. The chicken's regular laying cycle and the fact that fertilized eggs can be cold stored for several weeks, temporarily freezing their development, ensure constant access to embryos. Chick embryonic development shows some similarity to that of mammals, and the flat blastodisc of the chick is morphologically like the human embryo at the same stage of development. The introduction of techniques for embryonic sectional explanation and the production of chick-quail chimeras have only served to add to the popularity of the chick embryo as a model system.

1.23 Mouse (*Mus musculus*)

Approximately one hundred years ago the mouse was the animal of choice for modeling human genetics and disease. The whole genomes have been sequenced and show quite a large homology with the human genome. Humans and mice also share many physiological traits and susceptibilities, making the mouse an attractive model in biomedical studies. Its rapid breeding cycle and small size further gives it an added advantage hence it is chosen to be used as transgenic and for gene knockout experiments.

1.24 *Escherichia coli*

Escherichia coli, a prokaryotic organism without a nuclear membrane, is living material often used in laboratories. It reproduces rapidly (under optimal conditions 0.5 hr/generation). This rapid growth provides a lot of material for many experiments. Some of the *Escherichia coli undergo mutations, the* mutant forms are not able to express certain proteins at the time of saturation growth, and, therefore, die. *E. coli was* also the organism used to elucidate the regulation of the *lac* operon in genetics. *E. coli's* ability to take up exogenous genetic material under the procedure known as DNA-mediated cell transformation has also made it a popular model for studies involving recombinant DNA. The value of *E. coli in* recombinant DNA makes it a good model organism for students to study the genetic material.

1.25 Eugenic

The inadequate knowledge about genetics led to the development of preferred marriages for various traits. This is known as Eugenics. In the United States, the Eugenics movements quickly became popular and show a significant effect on the formulation of federal immigration laws were made. The government was so strict that it passed laws for sterilization of people who were not fit for one reason or the other to have children. One of the examples is Nazi attempts to wipe out groups of people in the 1940's. By 1950 human genetic studies had made considerable contributions in identifying inherited diseases and complex conditions with genetic composition but still the relationship between a genetic disease and defective gene remained a complete enigma. It was with the advances in molecularization of genetics which lead to the recent advances in genetics which abolished the term eugenics.

1.26 Genetic Databases and Maps

Bioinformatics play a very major role in the understanding of molecular genetics. Presently there are various important databases available. Some of these are given in the **Table 1.3**.

1.27 Atlas of Genetics

The Atlas of Genetics and Cytogenetics in Oncology and Haematology is a peer-reviewed on-line journal and database exist in free access on the internet devoted to genes, cytogenetics, and clinical entities in cancer, and cancer-prone diseases. Some of these are given web site in the **Table 1.4**.

Table 1.3 Genetic Databases

		Hotlinks	*Gen Bank*	*EBI*	*NCBI*	*Medline*
Image databases	Genomics	Health Care	Laboratory	Lipidomics	Medicine	Microbiology
Oncology	Patents	Pharmacology	Physics/ Chemistry	Proteomics	Toxicology	Veterinary Medicine
cDNA	DB Indexes	DNA	Educational	Epigenetics	Ethnic Variation	Exons/Introns
Functional elements	Gene DB	Gene Expression	Gene Therapy	Genetic Maps	Genomic DB	Metabolomics
Mitochondrion	Mutations	Nucleotide DB	Pathology	Pathways	Polymorphism	Probes
Retrotransposons	Ribosome	RNA	Splicing	Structure DB	Taxonomy	Tools
Transcription	Vector	—	—	—	—	—

Table 1.4 Atlas of Genetics

CBS	*Genome Atlas Database"*
	Database of sequenced bacterial genomes.http://www.cbs.dtu.dk/services/GenomeAtlas/
Ensembl	Ensembl genome browser
	ENSEMBL is a joint project between EMBL-EBI and the Sanger Center to develop a system which automatically tracks all the sequenced pieces of the human genome, attempts to assemble them into large single stretches and then analyze the assembled DNA to find genes and other features of interest to biologists and medical researchers. http://www.ensembl.org/
Entrez Gene	**Entrez Gene**
	Gene provides a unified query environment for genes defined by sequence and/or in NCBI's Map Viewer. You can query on names, symbols, accessions, publications, GO terms, chromosome numbers, E.C. numbers, and many other attributes associated with genes and the products they encode.>br> Entrez integrates the scientific literature, DNA and protein sequence databases, 3D protein structure and protein domain data, population study datasets, expression data, assemblies of complete genomes, and taxonomic information into a tightly interlinked system. http://www.ncbi.nlm.nih.gov/entrez/query.fcgi?db=gene
GeneLoc	**Gene location**
	GeneLoc presents an integrated map for each human chromosome, based on data integrated by the GeneLoc algorithm. GeneLoc includes further links to GeneCards, NCBI's Human Genome Sequencing, UniGene, Genome Database, and mapping resources. GeneLoc is presented by the Weizmann Institute of Science, Rehovot, Israelhttp://genecards.weizmann.ac.il/geneloc/
Genome reviews	**Genome reviews**
	The goal of the Genome Reviews project is to provide an up-to-date, standardized and comprehensively annotated view of the genomic sequence of organisms with completely deciphered genomes. http://www.ebi.ac.uk/GenomeReviews/

CBS	*Genome Atlas Database”*
HapMap	**International HapMap project**
	The goal of the International HapMap Project is to develop a haplotype map of the human genome, the HapMap, which will describe the common patterns of human DNA sequence variation. The HapMap is expected to be a key resource for researchers to use to find genes affecting health, disease, and responses to drugs and environmental factors. http://www.hapmap.org/index.html.en
Human genome resources	**Human genome resources (NCBI)**
	NCBI's Genome Browser.http://www.ncbi.nlm.nih.gov/genome/guide/human/
Intronless genes in eukaryotes	**Intronless genes in eukaryotes**
	Database about intronless genes.http://sege.ntu.edu.sg/wester/intronless
NCBI Map View	**NCBI map view**
	Graphical view of chromosomes of different organisms, including homo sapiens. http://www.ncbi.nlm.nih.gov/mapview/
NCBI map viewer	**NCBI map viewer (Homo Sapiens Genome)**
	The NCBI Map Viewer displays genome assemblies using sets of aligned chromosomal maps. A new Map Viewer home page organizes the available organisms by taxonomic group and provides links to both Map Viewer and Genomic BLAST pages. http://www.ncbi.nlm.nih.gov/mapview/map search.cgi?chr=hum chr.inf and query#availdoc
OMIM Morbid Map	**OMIM Morbid Map**
	The OMIM Morbid Map presents the cytogenetic map location of disease genes described in OMIM.http://www.ncbi.nlm.nih.gov/Omim/getmorbid.cgi?
TCGA	**The cancer genome atlas**
	The Cancer Genome Atlas (TCGA) is a comprehensive and coordinated effort to accelerate our understanding of the molecular basis of cancer through the application of genome analysis technologies, including large-scale genome sequencing. TCGA is a joint effort of the National Cancer Institute (NCI) and the National Human Genome Research Institute (NHGRI), which are both parts of the National Institutes of Health, U.S. Department of Health and Human Services.http://cancergenome.nih.gov/
TraitMap	**TraitMap: Genetic-Map database**
	TraitMap contains curator-checked data records based on published papers of mapping results in Homosapiens, Mus musculus, and Arabidopsis thaliana.TraitMap is the first database of mapping charts in genetics, and it is integrated into a web-based retrieval framework: termed Genome <—> Phenome Superhighway (GPS) system, where it is possible to combine and visualize multiple mapping records in a two-dimensional display. http://omicspace.riken.jp/gps/

http://atlasgeneticsoncology.org/

1.28 Genetic Maps

Representation of the genes on a chromosome arranged in a linear order with distances between loci expressed as per cent recombination (map units, centimorgans) is known as the genetic map this is also called the linkage map. Genes on the same chromosome are described as linked/syntenic. Humans have 24 linkage groups, corresponding to the 22 autosomes, X, and Y chromosomes. Groups of genes that are widely separated on a chromosome may show independent assortment; however, all such groups can eventually be tied together by mapping additional loci between them. The genes on a chromosome can be represented as a single linear structure that goes from one end of the chromosome to the other. Genetic distance is measured by the frequency of crossing-over between loci on the same chromosome. One map unit = one centimorgan (cM) = 1 per cent recombination between loci. If two loci are located far apart, more likely is that a crossover will occur; however, if two loci are lying close together, a crossover is less likely to occur between them. Recombination can only be detected between two loci, both of which are heterozygous. The dominant/recessive relationships must allow the detection of recombinants. The most useful systems involve co-dominant alleles. Efficient mapping requires polymorphic loci, *i.e.*, loci with two or more common alleles. Loci that have a single common allele are described as monomorphic. Any variations in DNA, whether in coding regions of genes or genes in the noncoding regions, can be used as genetic markers, *i.e.*, as a label for a particular point on a chromosome. Odd numbers of crossovers create recombinant allelic combinations and are counted as one crossover. If there is a 50 per cent recombination, then it shows independent assortment. Therefore, only distances less than 50 map units can be measured depending upon recombination. However, greater distances can be constructed by adding up distances between closer loci.

Closely linked genes show the association of alleles within families but may not be within populations. Random haplotypes are generated by crossing over within populations. If the loci are very closely placed, equilibrium among the possible combinations may take many generations. If two loci are linked, the alleles that are on the same chromosome are described as coupled; alleles on opposite homologous chromosomes are in the repulsion phase. A physical map describes the physical location of genes on chromosomes. The genes on the physical and genetic maps should be in the same order, as crossing over may occur more often in some regions than in others. Following techniques may be used to construct the physical maps:

1. Hybrids of human/mouse cell tend to lose human chromosomes at random, causing hybrid cell lines which have one or a few human chromosomes in case a gene is always present or absent when a particular chromosome is present or absent, it may be concluded that the gene is on that chromosome.
2. There are fluorescent or radioactive probes which bind to a particular gene can be observed microscopically and are used for localization purposes in the metaphase spread.
3. During metaphase the chromosomes can be sorted with high-speed electronic sorters.

1.29 Genetics in the Genomics and Proteomics Era

Genetics, genomics, and proteomics are opening new perspectives in diagnosis and therapy. A new era has been started, which may be termed as the era of genomics and proteomics, an era in which scientists can study all the genes (the complete genome) and all the proteins of an individual, an organ, a tissue, or a cell. Almost every disease is affected by both genetic and environmental factors. Environmental factors are toxins, radiation, infections, diet, age, stress which may cause certain diseases to occur, however, all the individuals may not get affected by the environmental factors this is because of the genetic predisposition for various diseases to occur. Small changes in our genes can trigger, prevent, promote or alleviate diseases. It is precisely the interplay of environment, genes, and proteins which are responsible for an individual's health.

Presently many drugs are being used for different treatments here the knowledge of the structure, *i.e.*, the three-dimensional form, of a target molecule makes it possible to decide in advance whether a given substance has any potential use as a drug. Computer-based drug design can take various factors into consideration and can reduce toxicity. If the genetic preconditions for a disease are known, a patient's individual risk can be determined, and appropriate preventive action can be taken. Sickle-cell anemia is an example of this. In this condition, an inherited modification of certain components of the gene for hemoglobin, the red blood pigment, results in the production of an altered protein that changes shape when the oxygen supply is inadequate. Under these conditions, the red blood cells assume the form of a sickle, clump together and block the blood vessels. Carriers of this trait, therefore, need to avoid traveling to great heights as this may change in air pressure.

Many diseases require intervention at the genetic level. If we know the genetic basis of the disease, we can formulate gene therapy if possible. Genetics, genomics, and proteomics thus provide medicine with a variety of new ways of intervening in the development and progression of diseases. Nevertheless, the intervention has not become easier. In the past methods to overcome complex diseases were based largely on trial and error, precisely because such disorders are not caused by simple infections or gene mutation, but rather arise because of a combination of external and internal, predisposing, and protective, and variable or unchangeable influences. This is true for most of the major diseases that afflict people in industrialized countries, *e.g.*, cancer, Alzheimer's disease, diabetes, and cardiovascular disease. Every ray of light that genetics, genomics, and proteomics cast on the factors that contribute to these diseases helps to fight against them. This is because disease-inducing environmental influences can generally be modified – whereas our genetic makeup generally cannot. Among the risk factors that contribute to the development of the disease, our genetic predisposition is one of the constant factors. In the 1980s scientists succeeded in identifying the genetic basis of a number of severe hereditary diseases like Huntington's disease (Huntington's chorea), cystic fibrosis (mucoviscidosis), and hemophilia. Scientists can investigate the genetic causes of more complex diseases in which various genes can exert positive and negative influences. Monogenic diseases such as Huntington's disease, cystic fibrosis, and hemophilia follow the classical Mendelian laws of inheritance. The pattern of occurrence and non-occurrence of such diseases within affected families is determined where only one or both copies of the gene in question need to be altered. Identification of such genes has become relatively easy by comparing the genetic material of affected and unaffected members of a family. Genetic tests are available and can be applied to find out if parents are at risk of having a baby that will be affected by a heritable disease, such as cystic fibrosis, *etc.* No such tests are available for polygenic diseases as many genes are involved, however, it has been seen that such diseases tend to cluster in certain families, but not in such a way that the precise distribution of affected and unaffected individuals can be predicted. With the knowledge of the human genome in hand, scientists have started using single nucleotide substitutions in important sections of a gene which may have profound effects on the function of the corresponding gene product. The finding on the increased frequency of certain SNPs in association with a disease indicates that the genes concerned play an important role in the disease in question. However, it is not always easy to separate environmental influences from genetic influences, especially since the environment can influence the behavior of our genes. To study the effect of genes and environment family studies and twin studies play a very important role in dissecting the role of environment and heredity. Where genes that play a role in the development of a disease (or, for example, in intolerance of a drug or failure of a drug to exert its expected effects) are known, an individual's risk of developing that disease can be determined by using appropriate genetic tests. This knowledge will help the individual to take appropriate precautions and to modify his/her lifestyle and if required to take preventive medicines. Early prevention is therefore one of the potential applications of molecular medicine. Nowadays, however, scientists are searching for the sites where the drugs can act. Modern molecular medicine is the amalgamation of therapy and diagnosis. Now it is possible to determine the causes of a disease based on a patient's genetic composition/predisposition to predict the effect of drugs on the disease based on the molecular characteristics of the drugs concerned, and finally to choose the therapy that is optimal for the individual patient. Knowledge of the molecular level of the disease process thus opens completely new approaches to the treatment: new targets, new strategies, early prevention, and a better understanding of the effects and side effects of drugs. Genetics, genomics, and proteomics will help doctors to avoid ineffective therapies.

Information about genetics has increased. The genome sequences of many organisms are added to DNA databases every year, and new details about gene structure and function are continually expanding. Recent research on COVID-19 shows how important is to know about the link between medicine, genetics, and evolution to search for the best strategy and find the therapeutic options for Corona. CORONA-19 biology provides us with a better understanding of numerous biological processes and evolutionary relationships which will further enhance our knowledge.

2

The Chromosome Structure and Variation

Wheat farming has a rich history, dating back to the Middle East around 10,000 years ago. This versatile crop can thrive in varying environmental conditions, with an estimated 17,000 different types of wheat. Triticum aestivum, the wheat we know today, is a hybrid of three distinct wheat species. Its ancestral forms were low-yielding grasses that grew in regions such as Syria, Iran, Iraq, and Turkey, and were cultivated by ancient peoples in these areas. It's unclear when interbreeding began to improve the crop, but when it did, people recognized the importance of hybridization and discovered that hybrid varieties are more resistant. Consequently, double and triple hybrid varieties have been developed. Modern wheat is believed to result from a triple hybrid plant, with each wheat species contributing its chromosomes. Chromosomal makeup can be used to create genetic hybrids and various cytological techniques are used to study chromosomes.

2.1 History of Cytogenetics

In the mid-1800s, researchers discovered that cells came from other cells and that genetic information was in the nucleus. However, because basic microscopes had limited resolution, they had to use stains to improve the contrast of cell contents. Walther Flemming, a German anatomist, meticulously studied the fibrous network within the nucleus, which he named chromatin. Flemming's research led to the discovery of chromosomes, and he accurately determined the sequence of chromosome movements during mitosis. He also observed that chromosomes divided lengthwise during mitosis, realizing that chromosomal movements played a crucial role in distributing nuclear material during cell division. Flemming's work helped pave the way for the discovery of hereditary mechanisms.

In the late 1800s, advancements in cytological techniques and microscopy allowed Theodor Boveri, a German embryologist, to provide evidence that chromosomes of germ cell lineage provide continuity between generations. Boveri's observations were based on the early development of Ascaris megalocephala, now known as Parascaris equorum, which was an ideal experimental model due to its two pairs of chromosomes and distinct somatic cell and germ cell lineages. Boveri traced the fate of chromosomes in individual cell lineages with great precision and discovered that the full complement of chromosomes was retained only in the Ascaris germ cell lineage. He also recognized that chromosome numbers were reduced in gametes. Boveri's work provided one of the first descriptions of meiosis, and he noted that Ascaris eggs retained only two chromosomes after the polar body formed, with the normal number of four chromosomes being restored after the fusion of the sperm and egg pronuclei. Working independently near the dawn of the twentieth century, American graduate student Walter Sutton confirmed and expanded upon Boveri's observations using a superior cytological model. Sutton, a Kansas farm boy, had made the serendipitous discovery that it was possible to distinguish individual chromosomes in cells undergoing meiosis in the testes of the lubber grasshopper, *Brachystola magna*. Thus, in his classic paper published in 1902, Sutton described the configurations of individual chromosomes in cells at various stages of meiosis. Sutton was able to identify 11 pairs of chromosomes that could be distinguished by their sizes, as well as an accessory singleton that he correctly presumed to be a sex chromosome. Due to its apparent lack of a partner, this chromosome sorted into only half of the sperm cells, while the remaining half never received a copy. Though the accessory chromosome was unpaired, it still replicated and entered stages of mitosis in the same manner as all other chromosomes, which prompted Sutton to declare it a "true chromosome" and not simply an accessory. Sutton postulated that all chromosomes have a stable structure, or "individuality," that is maintained between generations, and he used this property to follow the behavior of individual chromosomes through the various stages of meiosis, including synapsis. Most notably, Sutton recognized that his observations were consistent with the, whose findings had only recently been rediscovered. In fact, Sutton closed his 1902 paper with the statement, "I may finally call attention to the probability that the association of paternal and maternal chromosomes in pairs and their subsequent separation during the reducing division as indicated above may constitute the physical basis of the Mendelian law of heredity." With these words, Sutton articulated the chromosomal theory of inheritance.

Sutton explains that hereditary characteristics differ because chromosomes are randomly positioned during metaphase, without any consistent maternal or paternal influence. Each chromosome is self-contained and can have a unique mix of parental traits when they divide into gametes. The number of possible combinations of chromosomes in each gamete can be calculated based on the number of chromosomes in the organism, resulting in 2n potential combinations. When two gametes are paired, this results in (2n)2 variations in zygotes, leading to many possibilities. Sutton provided hypothetical examples of organisms with gamete chromosome numbers ranging from 1 to 19, demonstrating the potential for variation. Notably, each chromosome carries multiple traits, increasing the total variation beyond what is shown in Table 2.1.

Sutton's discovery of potential variation strengthened his assertion that he had found the physical basis for Mendel's principle of independent assortment. However, scientists still lacked concrete proof and required an experimental system that could connect genetic inheritance with chromosome behavior. This opportunity came when a distinct mutation was discovered in the fruit fly, Drosophila melanogaster.

Table 2.1: Number of Possible Combinations in Organisms with 1 to 19 Chromosome Pairs in Pre-synaptic Cells

Chromosomes		*Combinations in Gametes*	*Combinations in Zygotes*
Somatic series	*Reduced series*		
2	1	2	4
4	2	4	16
6	3	8	64
8	4	16	256
10	5	32	1,024
12	6	64	4,096
14	7	128	16,384
16	8	256	65,536
18	9	512	262,144
20	10	1024	1,048,576
22	11	2048	4,194,304
24	12	4096	16,777,216
26	13	8192	67,108,864
28	14	16384	268,435,456
30	15	32768	1,073,741,824
32	16	65536	4,294,967,296
34	17	131072	17,179,869,184
36	18	262144	68,719,476,736

In the early 1900s, Thomas Hunt Morgan and his colleagues at Columbia University identified hundreds of Drosophila genes and made significant discoveries about genetic transmission. Through cytological examination, they found that Drosophila has four pairs of chromosomes, including a pair of sex chromosomes. Unlike humans, a fruit fly's sex is determined by the number of X chromosomes, not the presence of a Y chromosome.

One day, Morgan's colleagues discovered a male fly with unusual white eyes in their cultures. Breeding experiments revealed that the white eye color was caused by a recessive mutation in the gene responsible for normal red eye color, located on the X chromosome. The chromosome theory predicted that male flies would always display the eye color encoded on their single X chromosome, while female flies would develop white eye color only when they inherited two mutant versions of the eye color gene. This prediction was confirmed by thousands of mattings, but the group discovered "exceptional" white-eyed females among the progeny of a heterozygous female fly and a normal male, contradicting the chromosome theory.

Lilian Vaughn Morgan had the insight to recognize that these "exceptional" females might have an unusual chromosome composition. Indeed, these females had two X chromosomes and a Y chromosome, resulting from defects in meiosis that caused a high frequency of nondisjunction. When eggs containing two non-disjoined X chromosomes, each carrying the mutant white gene, were fertilized by a Y-bearing sperm, the product was an XXY female with white eyes. This discovery provided strong experimental support that genes were located on chromosomes, rather than disproving the chromosome theory.

Cytogenetics is the science of studying chromosomes in terms of number and structure, and European biologists discovered chromosomes in the 19th century. Cytogenetics deals with both direct and indirect chromosomal analysis, and it is used in medicine to determine whether genetic disease conditions are associated with chromosomal abnormalities. It is also used in cancer genetics to explore chromosomal rearrangements, deletions, and duplications that occur during tumor formation, resulting in an unstable genome. Chromosomal alterations can be explored using cytogenetic techniques.

2.2 Analysis of Chromosomes

Chromosomes occur in pairs, except for sex chromosomes, and a pair of chromosomes is called a homologous chromosome. The haploid number (n) is the number of chromosome pairs. Chromosomes can be examined during the middle of mitosis using fast-growing tissues such as animal embryos or plant root tips. Advances in cell culture techniques have allowed the study of chromosomes in other types of cells, such as white blood cells collected from humans. Chromosome bands can be revealed by some treatments, allowing not only the identification of normal and abnormal chromosomes but also providing insight into fundamental aspects of chromatin structure and genome compartmentalization. Various steps involved in the karyotying of human chromosomes is shown in **Figure 2.1**.

The growth of the dividing cells can be arrested by treating these cells with hypotonic solutions. Water is taken up through osmosis. This technique greatly simplifies successive analysis, especially if the chromosome number is large. In humans, the correct number of chromosomes is 46. Feulgen's reagent which is a purple dye was used until the late 1960s and early 1970s, this stained the chromosome spreads. Feulgen's reagent reacts with the sugar molecules in DNA, or with acetocarmine, a deep red dye. However, presently the stains used are the ones that can go into the DNA. When the stain can insert into the base pairs of DNA it is called an intercalating agent.

Another reagent used is quinacrine, which is the cousin of the antimalarial rial drug quinine. If chromosomes are stained with quinacrine they reveal a distinguishing pattern of bands. The Quinacrine is a fluorescent compound, hence the chromosome has to be visible or in other words exposed to ultraviolet (UV) light. Ultraviolet irradiation results in some of the quinacrine molecules that have been inserted into the chromosome to emit energy. Interestingly some parts of the chromosome shine luminously, whereas other parts are dark. This differential fluorescence pattern of bright and dark bands is highly reproducible. Using these bands, cytogeneticists can differentiate chromosomes and investigate structural abnormalities. This procedure is known as Q banding, and the bands are called Q bands.

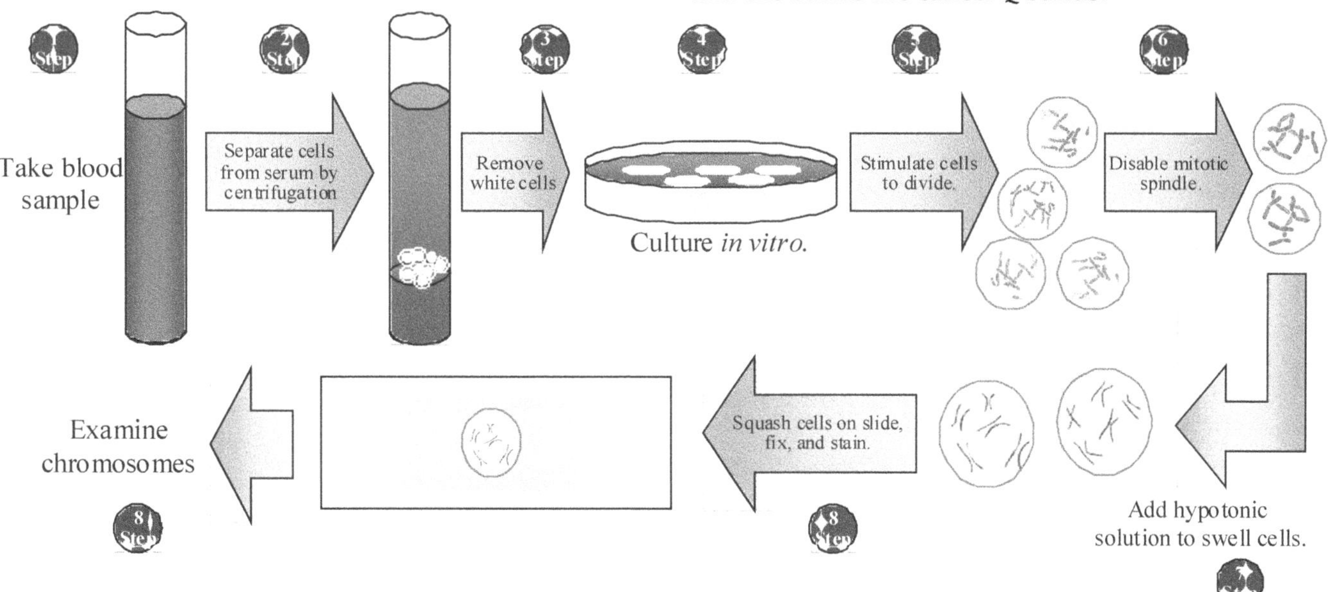

Figure 2.1: Various Steps Involved in the Karyotyping of Human Chromosomes.

Chromosome banding methods are either based on staining chromosomes with a dye or on assaying for a particular function. The most common methods of dye-based chromosome banding are G-(Giemsa), R-(reverse), C-(centromere), and Q-(quinacrine) banding. Bands that show strong staining are referred to as positive bands; weakly stained bands are negative bands. However, the staining patterns are not black and white, different bands stain to different intensities. G-positive bands are usually just called G-bands and likewise for R-positive (R-) bands. Positive C-bands contain constitutive heterochromatin. Q-bands are considered equivalent to G bands. The most widely used function-based banding method is replication banding which is because different bands replicate their DNA at different times during the S phase of the cell cycle. Generally, during R-banding DNA is replicated earlier than G-bands. The most widely used function-based banding method is replication banding which is based on the fact that different bands replicate their DNA at different times during the S phase of the cell cycle. Generally, R-band DNA is replicated earlier than G-bands. G-bands also correspond to the condensed staining of meiotic chromosomes and R-bands to inter-chromomeric regions.

G-and R-banding is the most used technique for chromosome identification (karyotyping) and for identifying abnormalities of chromosome number, translocations of material from one chromosome to another, and deletions, inversions, or amplification of chromosome segments. This has an invaluable impact on human genetics and medicine and the power of this approach has been augmented by combining cytogenetics with fluorescence in situ hybridization (FISH).

The detection of chromosome deletions is associated with disorders, very often contiguous gene syndromes provided some of the first disease gene localizations in humans. Similarly, translocations have been important in pinpointing the location of disease-associated genes, and the characteristic translocation associated with some leukemia's are important, not only for understanding the molecular basis of these cancers but also for their diagnosis and prognosis. One of the best examples of this is the translocation between human chromosomes 9 and 22 – t (9; 22) (q34: q11 – or the Philadelphia chromosome diagnostic of chronic myelogenous leukemia (CML).

Studying the banding patterns of chromosomes is an important way to identify evolutionary relationships between different species and detect any changes in karyotypes that may have contributed to the development of new species. It's interesting to note that the banding patterns of human, gorilla, and chimpanzee chromosomes are strikingly similar. It's also worth noting that human chromosome 2 results from two ape chromosomes fusing together, and human chromosome bands have many similarities with those of lower primates. Although Q-, G-, and R-banding patterns are limited to some eukaryotes, replication banding is almost always observed in organisms with large chromosomes that can be viewed under a microscope.

The study of chromosomes in mammals, birds, and reptiles is made easier by using G and R banding techniques. These techniques provide clear patterns in lower vertebrates. The evolution of chromosome banding patterns suggests that the first discernible compartmentalization in the genomes of eukaryotes was the temporal control of replication, differences in chromatin packaging, and the segregation of some chromosomal domains into heterochromatin. C-banding is a useful tool for visualizing heterochromatin in all eukaryotic groups with visible chromosomes, while T-bands consist of gene-rich and gene-free constitutive heterochromatin found at opposite extremes of a chromatin flavor continuum.

In humans, the primary C-bands are found on the long arm of the Y-chromosome and near the centromeres of chromosomes 1, 9, and 16. The size of these C-bands varies between individuals, with smaller C-bands located at the centromere of each chromosome and on the p arms of the five acrocentric chromosomes. In mice, primary blocks of visible heterochromatin are located near the centromere of each chromosome, and the amount of heterochromatin at these sites varies between different strains of the house mouse (Mus musculus). In Drosophila, the main sites of constitutive heterochromatin are found at the chromocenter, telomeres, and on the fourth/Y-chromosome.

Heterochromatin is generally associated with repression of transcription and recombination, along with late replication. However, no endogenous genes or CpG islands have yet been mapped to regions of C-banded constitutive heterochromatin in the human or mouse genomes. Transgenes that integrate near constitutive heterochromatin are frequently silenced, and a few endogenous genes in Drosophila are found in a heterochromatic location and cease to function when moved to euchromatic locations.

C-band-positive constitutive heterochromatin has a particular chromatin structure. It is the site of most methylation in human and mouse chromosomes, and the 5MeCpG-binding protein MECP2 is highly concentrated over constitutive heterochromatin in vertebrates in most cell types. Constitutive heterochromatin is also associated with very low levels of histone acetylation, except for the chromocenter of Drosophila, which is enriched in histones acetylated at Lys12, and histone H4 within the pericentromeric heterochromatin is acetylated in mammalian embryonic stem cells prior to differentiation. Distinctive chromosomal proteins can also be found concentrated in heterochromatin, many of which have motifs, such as the chromo box, that are characteristic of proteins involved in the formation of multi-protein heterochromatin complexes. Antibodies against M31, a chromo box containing protein that is a homolog of Drosophila heterochromatin protein HP1, highlight the pericentric heterochromatin of mouse and human chromosomes. After establishing cytogenetic techniques, it was possible to know the somatic and gamete numbers of different species. The somatic and gamete chromosomal numbers of some common plants and animals are shown in **Table 2.2**.

Table 2.2: Somatic and Gamete Chromosome Numbers of some Common Plants and Animals

Sl. No.	*Common Name*	*Scientific Name*	*Chromosome Number*	
			Somatic	*Gamete*
		Plant		
1	Onion	*Allium cepa*	16	8
2	Mold	*Aspergillus nidulans*	16	8
3	Pigeonpea	*Cajanus cajan*	22	11
4	Green alga	*chlamydomonas reinhardtii*	32	16
5	Chickpea	*Cicer arietinum*	16	8
6	Chickpea	*Crepis capillaries*	6	3
7	Carrot	*Daucus carota*	20	10
8	Strawberry	*Fragaria virginiana*	56	28
		Haplopappus gracilis	4	2
		Haplopappus ravenii	8	4
9	Barley	*Hordeum vulgare*	14	7
10	Tomato	*Lycopersicon esculentum*	24	12
11	Bread mold	*Neurospora crassa*	14	7
12	Tobacco	*Nicotiana tabacum*	48	24
13	Rice	*Oryza sativa*	24	12
14	Pearl millet	*Pennisetum americanum*	14	7
15	Rajma, kidney bean	*Phaseolus vulgaris*	22	11
16	Pepper	*Piper nigrum*	128	64
17	Pea	*Pisum sativum*	14	7
18	Rye	*Secale cereale*	14	7
19	Potato	*Solanum tuberozum*	48	24
20	Jowar, sorghum	*Sorghum bicolor*	20	10
21	Spinach	*Spinacea oleracea*	12	6
22	Wake –robin	*Trillium grandiflorum*	10	5
23	Diploid wheat	*Triticum monococcum*	14	7
24	Durum wheat	*Triticum durum*	28	14
25	Bread wheat	*Triticum aestivum*	42	21
26	Faba bean	*Vicia faba*	12	6
27	Maize	*Zea mays*	20	10
		Animals		
1	Honey bee	*Apis mellifera*	32	16
2	Round worm (horse nematode)	*Ascaris megalocephala univalens* (renamed as *parascaris equorum*)	2	1
3	Silk worm	*Bombyx mori*	56	28
4	Toad	*Bufo viridis*	22	11
5	Cattle (cow)	*Bos Taurus*	60	30
6	Dog	*Canis familaris*	78	39
7	Pigeon	*Columbia livia*	80	40
8	Fruit fly	*Drosophila melanogaster*	8	4
9	Donkey	*Equus asinus*	62	31
10	Horse	*Equus caballus*	64	32
11	Cat	*Felis domesticus*	38	19
12	Chicken	*Gallus gallus domesticus*	78	39
13	Man	*Homo sapiens*	46	23
14	Rhesus monkey	*Macaca mulatta*	48	24
15	House fly	*Musca domestica*	12	6
16	Rat	*Rattus norvegicus*	42	21

2.3 Size of the Chromosome

The chromosome size can be determined by knowing the stage of cell division. During interphase chromosomes are thin and long hence not visible under a light microscope. As the prophase starts, there is a progressive decrease in the length and an increase in thickness (diameter); chromosomes are the smallest at the time of anaphase. Chromosomes can be easily studied at mitotic metaphase at this stage they are thick, quite short, and well spread in the cell. Therefore, the measurements can be taken during mitotic metaphase, but chromosome size varies during the metaphase stage (early and late metaphase stages). In addition, chromosome size is also affected by the type of pretreatment given to cells if the cells are pretreated with certain chemicals, *e.g.*, colchicines = 8-hydroxyquinoline, *etc.*, or low temperature (4–50C) show relatively shorter chromosomes at metaphase than untreated cells.

The size of mitotic metaphase chromosomes of different animal and plant species generally varies between 0.5μ and 32μ in length and between 0.2μ and 3.0μ in diameter. However, the smallest known chromosome is only about 1/80,000 in the length of the longest reported chromosome; such macrochromosomes are known both in plants and animals. The longest metaphase chromosomes are found in Trillium; its longest chromosome is 32μ long. The giant chromosomes found in Diptera are permanently in the prometaphase stage and may be as long as 300μ and up to 10μ in diameter. Therefore, giant chromosomes are easily visible in the interphase nucleus. Chromosomes differ in size in different human disease conditions (**Table 2.3**).

2.4 Human Chromosome

There are 22 pairs of autosomes and sex chromosomes (2 pairs) in human cells, sex chromosomes are XX in females and XY in males. All the chromosomes can be seen at the time of the mitotic phase. The categorization of the chromosomes is based on size, shape, and banding patterns. For cytological analysis, well-stained metaphase spreads are required which are photographed, and then each of the chromosome images is either cut manually and matched with its partner and arranged according to the size and placed on a chart or one can use a computer to arrange the chromosome from largest to smallest and see them on the computer screen and then take out the prints. The mitotic spread is shown and in karyotype is depicted the largest chromosome is chromosome number 1, and the smallest is chromosome number 21 the size of the X chromosome is intermediate, however; the Y chromosome size is like chromosome 22. Once the chromosomes are arranged as per their size this arrangement is called a karyotype.

The largest chromosome group is called A followed by B and so on. The banding technique has made it easier to recognize the chromosome and fit them into groups. Based on banding the short and long arms of the chromosomes can be identified. The two arms of the chromosomes are attached in the middle of this portion and is called the centromere. The centromere position determines the length of the chromosome. The short arm of the chromosome is called the p arm while the long arm is known as the q arm. Chromosomes are further classified according to the position of the centromere. When the centromere occurs near the middle of the chromosome, the chromosome is said to be metacentric. An acrocentric chromosome has its centromere near the tip, and the submetacentric chromosome has a centromere somewhere between the middle and the tip. The tip of each chromosome is called the telomere. The arms are roughly equal in length in metacentric chromosomes (**Figure 2.2**).

A normal female karyotype is 46, XX; a normal male karyotype is 46, XY. The size of the normal human chromosome varies as shown in Table 2.3. Once the chromosomal abnormalities occur the nomenclature is changed as shown in **Table 2.4**.

2.5 Chromosome Banding

Initially, karyotypes were thought to be counting the number of chromosomes this missed number of abnormalities. The main problem was how to

Table 2.3: Five Autosomes

Genes of Interest	*Chromosome*	*Percentage of Genome*	*Size in Megabases (millions of bases)*
Atherosclerosis, Type I diabetes mellitus, DNA repair	19	2	60.00
Alzheimer disease, amyotrophic lateral sclerosis, bipolar disorder susceptibility, Homocystinuria, Usher syndrome	21	1	33.55
Cat eye syndrome, Chronic myelogenous leukemia, DiGeorge syndrome, Schizophrenia susceptibility	22	1	33.46
Adult polycystic kidney disease Breast cancer, Crohn disease, Prostate cancer	16	3	98.00
Acute myelogenous leukemia basal cell carcinoma, Colorectal cancer, Dwarfism, Salt resistant hypertension	5	6	194.00

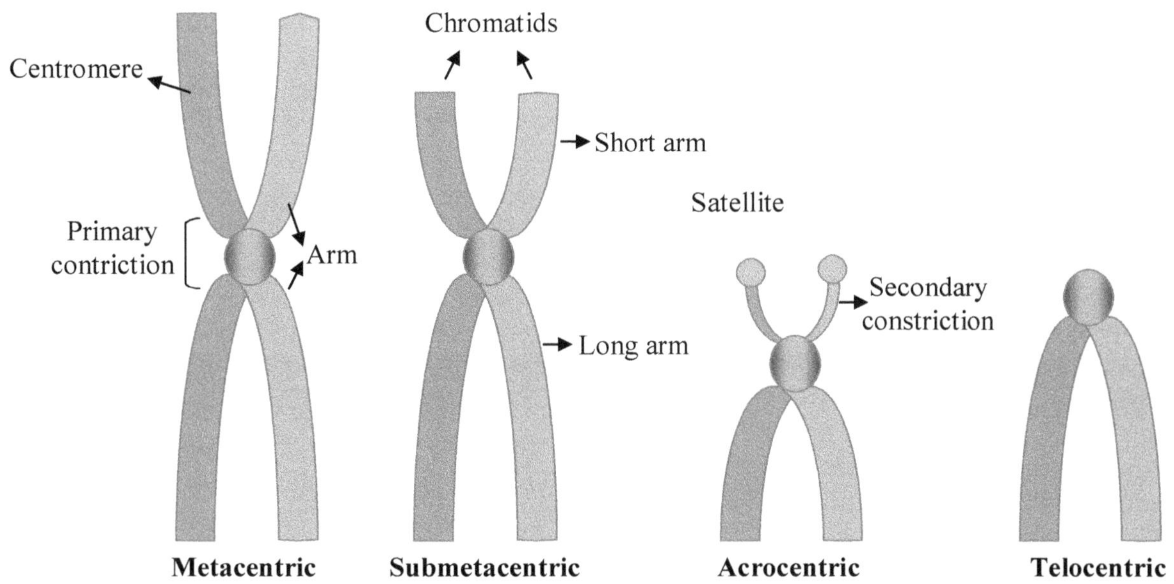

Figure 2.2: Position of the Centromere in the Chromosomes.

Table 2.4: Standard Nomenclature for Chromosome Karyotypes

Karyotype	*Description*
46, XY	Normal male chromosome
46 XX	Normal female chromosome
47, XX, +21	Female with trisomy 21, (Down syndrome)
47, XY, 21/46,XY	Male with a mosaic of trisomy 21 cells and also normal cells
46,XY,del (4) (p14)	Male with distal deletion of the short arm of chromosome 4 band designated 14
46, XY,dup(5p)	Male with a duplication of the short arm of chromosome 5
45,XY,-13, -14,t (13q;14q)	A male with a balanced robertsonian translocation of chromosomes 13 and 14; karyotype shows that one normal 13 and one normal 14 are missing
46,XY,t(11;22) (q23;q22)	A male with a balanced reciprocal translocation between chromosomes 11 and 22; the break points are at 11q23 and 22q22
46,Xx,inv(3) (p21;q13)	An inversion on chromosome 3 that extends from p21 to q13; because it includes the centromere, this is a pericentric inversion
46,X,r(X)	A female with one normal X chromosome and one ring X chromosome
46, X,i(Xq)	A female with one normal X chromosome and an isochromosome of the long arm of the X chromosome

+: Addition of chromosome; del: Deletion of chromosome; –: Deletion, t: Translocation; inv: Inversion; r: Ring; i: Isochromosome.

find out the structural abnormalities and this was solved by developing better staining techniques in the 1970s. This technique helped to carry out chromosomal banding which is representative of up-to-date karyotypes. Chromosome banding assists in the detection of deletions, duplications, and other structural abnormalities. This helped scientists in this field to detect the correct identification of individual chromosomes. The major bands on each chromosome are numbered. Thus, 14q32 refers to the second band in the third region of the long arm of chromosome 14. Sub-bands are characterized by decimal points following the band (*e.g.*, 14q32.3 is the third sub-band of band 2).

Several chromosome–banding techniques are used in cytogenetic laboratories. Quinacrine banding (Q banding) was the first staining method used to produce specific banding patterns. This is no longer widely used as Giemsa banding (G-banding). To produce G bands, Giemsa stain is applied after the chromosomal proteins are partially digested with the help of trypsin. Along with Q banding, G banding, and C banding techniques, there is another technique called reverse banding (R-banding) which requires heat treatment that reverses the usual white and black pattern that is seen in G-bands and Q –bands. This method is particularly helpful in staining the distal ends of chromosomes. Other staining techniques include C-banding and NOR (nucleolar organizing region) stains. These latter methods specifically stain only certain portions of the chromosome. C banding stains the constitutive heterochromatin, which usually lies near centromeres, and NOR staining highlights the satellites and stalks of acrocentric chromosomes (**Figure 2.2**).

High-resolution banding involves the staining of chromosomes at the time of maximal condensation which is seen at the time of prophase or early metaphase (prometaphase). Prophase and prometaphase chromosomes are more extended than metaphase chromosomes, the number of bands observed for all chromosomes increased from about 300 to 450 to as many as 800. This allows the detection of even less understandable abnormalities that are otherwise not familiar with conventional banding. High-resolution banding is difficult and takes a lot of time, so it is usually used when the clinician is looking for a specific, fine chromosomal abnormality. The deletion of the 15q11-13 bands is noticed in the Prader–Willi syndrome.

2.6 Fluorescence *in situ* Hybridization and Comparative Genomic Hybridization

Fluorescence in situ Hybridization (FISH) is a recent technique. It involves the preparation of two main components: the DNA probe and the target DNA to which the probe can be hybridized. The DNA probe typically comes from cloned sources such as plasmids, Cosmides, PACs, YACs, or BACs; where the insert may contain a specific gene or originate from a specific chromosomal locus. Whole-chromosome paints may also be used but are usually applicable to metaphase preparations. The purified DNA can then be labeled and detected indirectly using haptens or labeled directly using fluorochrome or dye-conjugated nucleotides. Labeling strategies are also variable, employing standard nick translation or PCR labeling methods. The target DNA can take the form of chromosome spreads or interphase nuclei. The sources of interphase targets may come from cytogenetic preparations or from paraffin-embedded tissues. Both the labeled DNA probe and DNA target are denatured to a single-stranded state and permitted to hybridize with each other. Post-hybridization washes and fluorescently labeled antibody incubations follow the 24-hour hybridization, and the specimen is ready for visualization by fluorescent microscopy. Successful interpretation of FISH experiments is dependent on the quality of the starting materials, hybridization efficiencies, and stringency of post-hybridization washes and antibody detections. Some -time the probe may hybridize at three places indicating that the chromosome is having extra material. A combination of FISH probes can also be used to spot chromosome rearrangements such as translocations *etc.*

A child who is missing a small piece of the short arm of chromosome 17 the centromere probe can be used which will hybridize to both copies of chromosome 17; however, a probe corresponding to a specific region of 17p will hybridize to only one chromosome. This demonstrates that there is a deletion that causes Smith-Magenis syndrome.

Multiple probes can be used in FISH each labeled with a different color; this enables us to find out the common numerical abnormalities simultaneously in the same cell (*e.g.*, abnormal numbers of chromosomes 13, 18, 21, X, and Y). In spectral karyotyping varying combinations of five different fluorescent probes can be used, however, this needs special cameras and image-processing software so that each chromosome is painted for ready identification. The rearrangement of chromosomes can be studied by using spectral karyotyping.

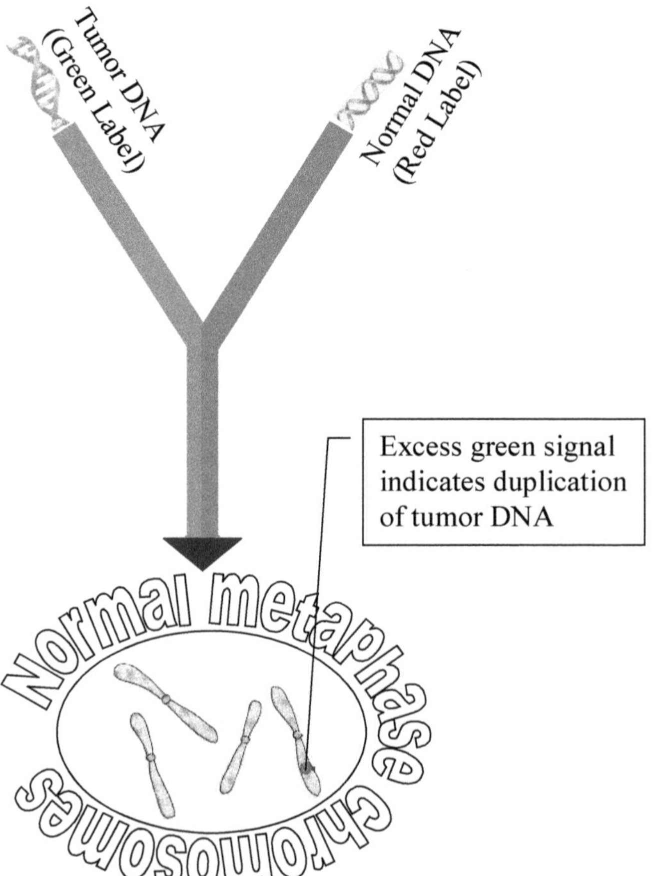

Figure 2.3: Diagrammatic Representation of Comparative Genomic Gybridization (CGH).

Comparative Genomic Hybridization can be used to study duplications (CGH) (**Figure 2.3**). DNA is taken from a test source (*e.g.*, tumor cells) it is labeled with a fluorescence dye that shows only one color (*e.g.*, green) under a fluorescence microscope; DNA taken from normal control cells is labeled with a second color (*e.g.*, red). Both sets of DNAS are then hybridized to normal metaphase chromosomes. If any chromosome region is duplicated in the tumor cell, then this region of the metaphase chromosome will hybridize with excess quantities of the green-labeled DNA. This region will exhibit a green color under the microscope. Conversely, if a region is deleted in the tumor cell, then the corresponding region of the metaphase chromosome will hybridize with excess, red-labeled DNA, and the region will exhibit a red color under the microscope. CGH is useful in cancer genetics as in cancer most of the time there occur deletion or duplication of the chromosomes.

If chromosomal numbers are changed it results in the altered phenotype in an organism these numerical changes are usually described as variations in the ploidy of the organism (from the Greek word meaning "fold" as in "twofold"). Organisms with complete, or normal, sets of chromosomes are said to be euploid (from the Greek words meaning "good" and "fold"). Organisms that carry extra sets of chromosomes are said to be polyploid (from the Greek words meaning "many" and "fold"), and the level of polyploidy is described by referring to a basic chromosome number, usually denoted as n. A cell is diploid having 2n chromosomes; triploids, with 3n; tetraploids, with 4n; and so forth.

Organisms in which a specific chromosome, or chromosome section, is under or overrepresented are said to be aneuploid (from the Greek words meaning "not," "good," and "fold"). This results in a genetic imbalance that is specific to the chromosome and hence is under- or over-represented. The distinction between aneuploidy and polyploidy is related to numerical numbers. There occurs a numerical change in part of the genome, usually, just a single chromosome, whereas polyploidy refers to a numerical change in a whole set of chromosomes. Aneuploidy causes a genetic imbalance, however, polyploidy does not.

The general effect of polyploidy is that cell size is greater than before, apparently because there is an extra number of chromosomes in the nucleus. Often this intensifies the size is correlated with an overall increase in the size of the organism hence stronger and healthier organisms. However, some time polyploidy leads to infertility or sterility. Let us suppose there is a 3n number instead of 2n at the time of meiosis the two homologous chromosomes will pair. These two homologs may pair completely along their length, leaving the third without a partner; this single chromosome is univalent in nature. Another option is that all three homologs may synapse, forming a trivalent in which each member is partly paired with each other. In either case, it is not easy to predict how the chromosomes will move during anaphase at the time of the first meiotic division. The maximum probability is that two homologs will move to one pole, and one homolog will move to the other, resulting in the gametes with one or two copies of the chromosome. All three homologs might move to one pole, producing gametes with three or zero copies of the chromosome. Because this segregation uncertainty applies to each trio of the chromosomes in the cell, the total number of chromosomes in a gamete can vary from zero to 3n. Zygotes formed by fertilization with such gametes are mostly lethal and certain to die; thus, most triploids are completely sterile. This can be avoided in agricultural science by allowing species to reproduce asexually. Many methods of asexual propagation include cultivation from cuttings (bananas), grafts (Winesap, Gravenstein, and Baldwin apples), and bulbs (tulips). In nature, polyploid plants can also reproduce asexually. One mechanism is apomixes which involve modified meiosis that produces unreduced eggs; these eggs then form seeds that germinate into new plants. A plant that reproduces in this way is the dandelion, Taraxacum officinal, a highly successful polyploid weed.

Polyploidy is much less in humans as extra genetic material creates harmful effects Triploidy and tetraploidy are lethal. Triploidy is rare. It occurs in 1/10,000 and causes approximately 15 per cent of the chromosomal abnormalities occurring at the time of conception. The most common cause of triploidy is the fertilization of an egg by two sperm (dispermy). Triploidy can also be caused by the fusion of an ovum and a polar body, containing 23 chromosomes, and subsequent fertilization by a sperm cell. Meiotic failure can also produce a triploid zygote. Tetraploidy is rare and is caused by a mitotic failure in the early embryo, in which all the duplicated chromosomes migrate to one of the two daughter cells. Two diploids zygote fusions may also cause this.

When polyploids unite at the time of fertilization they result in a zygote with an abnormal number of chromosomes hence this zygote dies. However, most polyploids are sterile.

2.7 Fertile Polyploids

Tetraploids with four identical chromosome sets are sterile. However, some tetraploids can produce viable progeny. The probable mechanism is that fertile tetraploids seem to have arisen by chromosome duplication in a hybrid that was produced by a cross

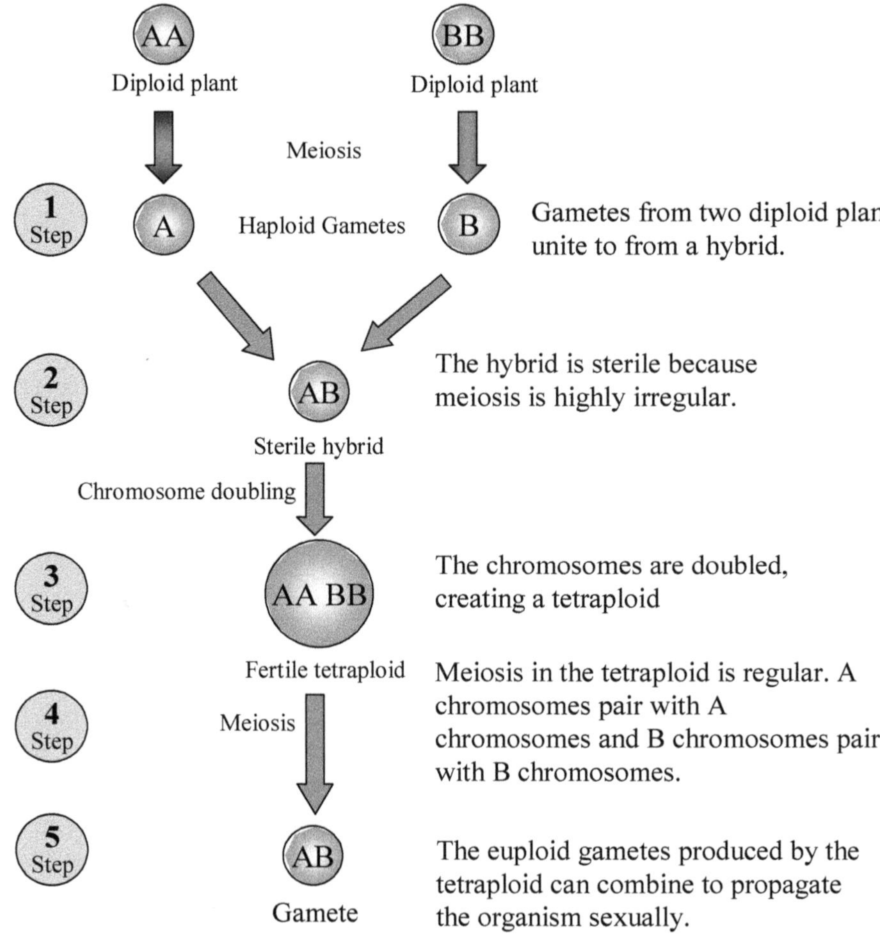

Figure 2.4: Tetraploidy is caused when Two Diploids' Gametes Hybridize.

of two different, but related, diploid species; most often these species have the same or very similar chromosome numbers hence fertile two diploids, denoted A and B, (**Figure 2.4**) are crossed to produce a hybrid that receives one set of chromosomes from one of the parental species. Such a hybrid will probably be sterile because the A and B chromosomes cannot pair with each other.

However, if the chromosomes in this hybrid are duplicated the resultant meiosis will proceed. Each of the A and B chromosomes will be able to pair with a perfectly homologous partner. Meiotic segregation can, therefore, produce gametes with a complete set of A and B chromosomes.

The state of hybridization between unlike but interrelated species followed by chromosome doubling occurs during plant evolution. In some cases, this results in complex polyploids with distinct chromosome sets. In Triticum aestivum (wheat) it is found quite often (**Figure 2.5**) so much so that wheat is found in hexaploidy condition hexaploid containing three different chromosome sets, each of which has been duplicated. There are seven chromosomes in each set, for a total of 21 in the gametes and 42 in the somatic cells.

Existing wheat has been developed by combining two diploid species that combined to form a tetraploid, and the second involved a combination between this tetraploid and another diploid, to produce a hexaploidy this has taken place as a part of the evolution in agricultural progress this process is called autopolyploid. Another example is of American cranberry (Vaccinium macrocarpon A) which is a diploid (2n=24). There are potential advantages of higher ploidy cranberries, particularly enhanced flower bud set, and increased fruit size; but these advantages are compromised by reduced fertility. People are interested in developing fertile polyploid cranberries for the potential to increase yield, as well as a method of transgene containment to assist genetic engineering programs. Three promising colchicine-induced polyploid individuals were selected for breeding to improve fertility in subsequent generations: polyploids of 'Pilgrim', a high fruit color 'Stevens' x 'Ben Lear' selection, and an embryo of mixed heritage. Selfed progeny of these selections and hybrids between the 'Pilgrim'

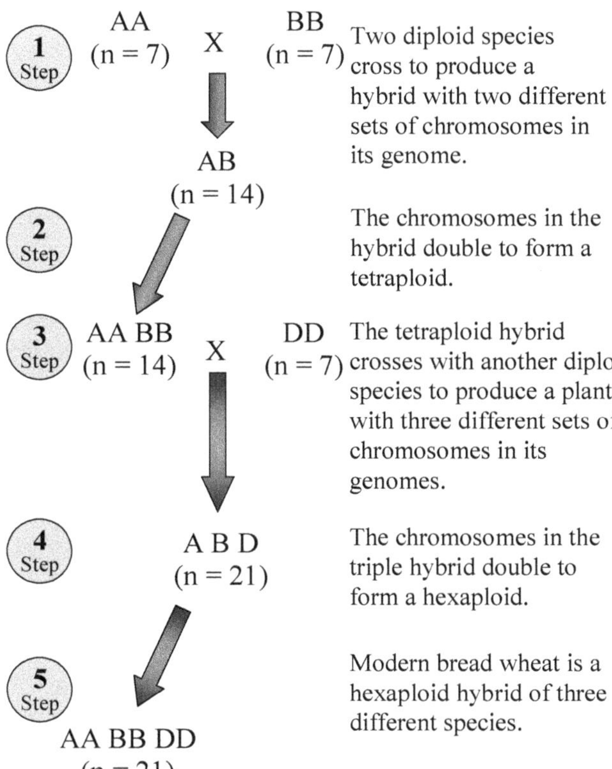

Figure 2.5: Origin of Hexaploid Wheat by Sequential Hybridization of different Species.

colchiploid and the other two parents were generated. Pollen germination, fruit set, fruit size, seed set, and seed germination were evaluated on a subset of the progeny that flowered in the greenhouse in 2000.

The best pollen germination obtained by individuals within each progeny group exceeded that of the respective parents, up to 90 per cent. Fruit set in the progeny ranged from 0 per cent to 94 per cent on cross-pollinated flowers. The seed set ranged from zero up to an average of 12 seeds per berry. Fruit size cannot be reliably measured in the greenhouse, but some of the polyploid plants had fruit sizes up to double that typically seen for diploid plants in the greenhouse (3.3 vs. 1.6 g). In general, selfed fruit had a smaller size and lower seed set than cross-pollinated fruit. Seed germination was variable in the resulting seed populations. Some populations of seed had germination rates exceeding 90 per cent, while seeds from other lines failed to germinate. These preliminary results suggest that there is good potential to enhance fertility through breeding in polyploid cranberry.

During development, certain tissues may become polyploid in some organisms. This is called endomitosis and involves chromosome duplication, followed by separation of the resulting sister chromatids. In human beings, one round of endomitosis produces tetraploid cells in the liver and kidney.

Sometimes polyploidization occurs without the separation of sister chromatids. In these cases, the duplicated chromosomes pile up next to each other, forming a bundle of strands that are aligned in parallel. The resulting chromosomes are said to be polytene, from the Greek word meaning "many threads". The most spectacular examples of polytene chromosomes are found in the salivary glands of Drosophila larvae. Each chromosome undergoes about nine rounds of replication, producing a total of about 500 copies in each cell. All the copies pair tightly, forming a thick bundle of chromatin fibers. This bundle is so large that it can be seen under low magnification under a dissecting microscope. The denser chromatin of these chromosomes' stains more deeply, having a combination of dark and light bands.

The polytene chromosomes of Drosophila show two additional features:

1. Homologous polytene chromosomes pair. In many insect species, the somatic chromosomes also pair-probably as a way of organizing the chromosomes within the nucleus. When Drosophila polytene chromosomes pair, the large chromatin bundles become even larger.
2. All the centromeres of Drosophila polytene chromosomes combine into a body called the chromocenter. Material flanking the centromeres is also drawn into this mass.

The result is that the chromosome arms seem to emanate out of the chromocenter. These arms, which are banded, consist of euchromatin, that portion of the chromosome that contains most of the genes; the chromocenter, which is uniformly stained, consists of heterochromatin, a gene-poor material that surrounds the centromere. The distinction between these different materials is that heterochromatin always stains darkly with dyes such as Feulgen and acetocarmine.

In the 1930s C. B. Bridges published detailed drawings of the polytene chromosomes. Bridges arbitrarily divided each of the chromosomes into sections, which he numbered; each section was then divided into subsections, which were designated by the letters A to F.

The polytene chromosomes of Drosophila are trapped in the interphase of the cell cycle. Thus, although most cytological analyses are performed on mitotic chromosomes, the most thorough and detailed analyses are performed on polytenized interphase chromosomes. Such chromosomes are found in many species within the insect order Diptera, including flies and mosquitoes.

2.8 Aneuploidy

Aneuploidy is caused due to alterations in the chromosomal number as a result dosage of a single chromosome is changed. Individuals who have an extra chromosome, or missing a chromosome, or have a combination of these anomalies are aneuploid in some cases chromosome arm may be deleted and is also considered to be aneuploid. Aneuploidies disturb the delicate balance of gene products in cells. Aneuploid cells have an abnormal number of chromosomes. Because each chromosome contains hundreds of genes, the addition or loss of even a single chromosome disrupts the existing equilibrium in cells, and in most cases, is not compatible with life.

Aneuploidy was originally studied in plants, where it was shown that a chromosome imbalance usually has a phenotypic effect. The classic study was one by Albert Blakeslee and John Belling they analyzed chromosome anomalies in Jimson weed, Datura stramonium. This diploid species has 12 pairs of chromosomes. Blakeslee collected plants with altered phenotypes and discovered that in some cases the phenotypes were inherited in an irregular way. These peculiar mutants were apparently caused by dominant factors that were transmitted primarily through the female. By examining the chromosomes of the mutant plants, Belling found that in every case an extra chromosome was present.

Detailed analysis established that the extra chromosome was different in each mutant strain. Altogether there were 12 different mutants, each corresponding to a triplication of one of the Datura chromosomes. Such triplications are called trisomy's. The transmission irregularities of these mutants were due to anomalous chromosome behavior during meiosis.

Belling also revealed the cause for the preferential transmission of the trisomic phenotypes through the female. During pollen tube growth, aneuploid pollen-in particular, pollen with n + 1 chromosomes-does does not compete well with euploid pollen. These results in trisomic plants almost always inherit their extra chromosome from the female parent. Belling's work with Datura demonstrated that each chromosome must be present in the proper dosage for normal growth and development.

Since Belling's work, aneuploids have been identified in many species, including our own. An organism in which a chromosome, or a piece of a chromosome, is underrepresented is referred to as a hypoploid (from the Greek prefix for "under"). When a chromosome or chromosome segment is overrepresented, the organism is said to be hyperploid (from the Greek prefix for "over"). Both haploid and hyperploid conditions result in a wide range of chromosomal abnormalities.

2.9 Sex Chromosomes

The females may have one extra chromosome as compared to males in some animal species (**Figure 2.6a**). It was named X chromosome and was observed initially in some insects. There are two X chromosomes in these species and males have only one; thus, females are XX and males are XO, where the "O" means the absence of a chromosome. The two X chromosome pair at the time of meiosis latter on separate at the time of gamete formation hence each egg has only a single chromosome. In males the single X chromosome moves independently of all other chromosomes and forms one-half of the sperms; the other half receive no X chromosome. Hence, when sperms and eggs unite, two kinds of zygotes are produced: XX, which develop into females, and XO, which develop into males. This event has an equal chance. Nature has kept this mechanism to preserve a 1:1 ratio of males to females in all the species where this kind of reproduction takes place.

Human beings both males and females have the same number of chromosomes (**Figure 2.6b**). This holds true for many other animal species also.

Every female constitutes XX, and every male is XY the Y chromosome is much shorter than the X, and its centromere is closer to one of the ends (**Figure 2.7**). The region of similarity between X and Y chromosomes is very less the short segments near the two ends of the chromosomes have some similarities. During meiosis in the male, the X and Y chromosomes separate from each other, producing two kinds of sperms, X –bearing and Y- bearing; the frequencies of the two types are almost equal. XX females produce only one egg, which is X – bearing. If fertilization occurs randomly, approximately half the zygotes would be XX, and the other half would be XY, leading to a 1:1 sex ratio at conception. However, in humans, Y-bearing sperm show the fertilization advantage, and the zygotic sex ratio, is 1.3:1. During development, the excess of males, is reduced due to the differential viability of XX and XY embryos, at birth, males are only slightly more in number than females (sex ratio 1.07:1). The excess of males gets eliminated to keep the sex ratio very close to 1:1.

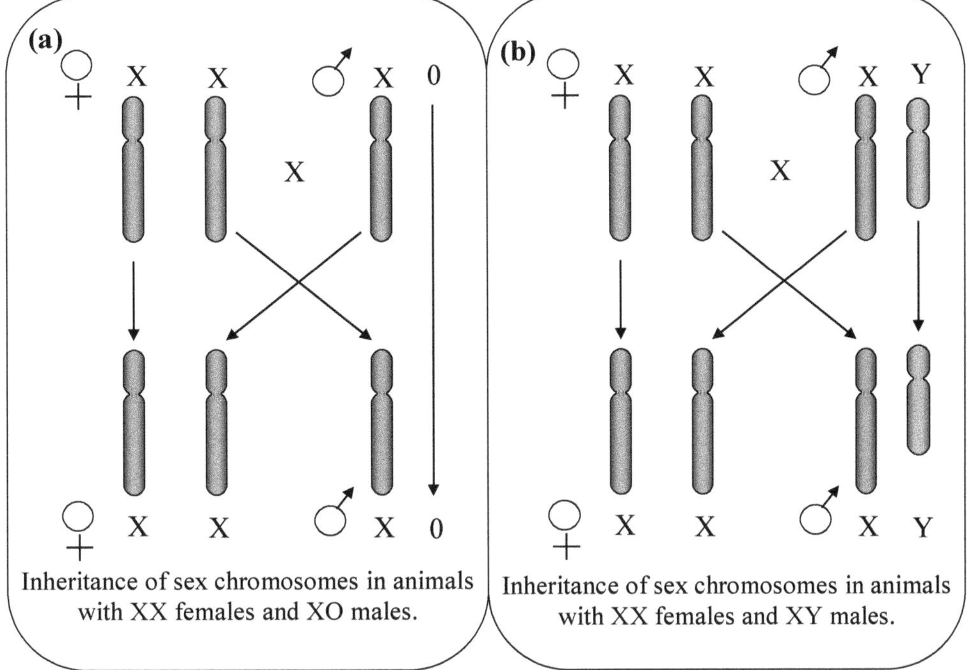

Figure 2.6: Inheritance of Sex Chromosomes.

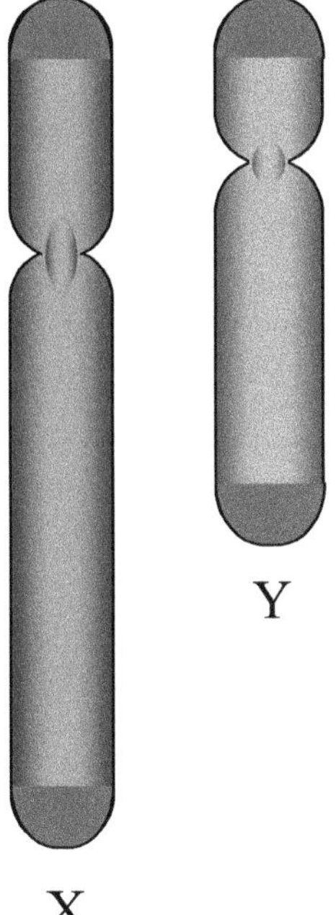

Figure 2.7: Human X and Y Chromosomes.

2.10 The Chromosome Theory of Inheritance

To explain the chromosomal theory of inheritance, Morgan carried out experiments where males with white eyes instead of the red eyes of wild –type flies. Further, these males got crossed with wild –type females, and the resultant progeny showed red eyes, revealing that the white eye was recessive over red-eye. Upon intercrossing with each other, a typical segregation pattern observed. All the daughters, but only half of the sons, showed red eyes; the other half of the sons had white eyes. This pattern of segregation suggested that the inheritance of eye color linked to the sex chromosome. Morgan proposed that a gene for eye color was present on the X chromosome, but not on the Y, chromosome. The white and red phenotypes were due to two different alleles, a mutant allele as w and wild –a type allele as w+.

The wild-type females in the first cross are homozygous for the w+ allele. Their mate is assumed to carry the mutant w allele on its X chromosome and none of the alleles on its Y chromosome. If an organism has only one copy of the genes is hemizygous, one copy of a gene is called a hemizygote. Sons inherit an X chromosome from their mother and a Y chromosome from their father; because the maternal X carries the w+ allele, the sons have red eyes. The w+ is dominant to w, so the heterozygous F1 females will also have red eyes. (**Figure 2.8**)

All first-generation offspring of a mutant, white-eyed male and a normal red-eyed female would have red eyes because every chromosome pair would contain at least one copy of the X chromosome with the dominant trait. But half the females from this union would now possess a copy of the white-eyed male's recessive X chromosome. This chromosome would be transmitted, on average, to one-half of second-generation offspring–one-half of which would

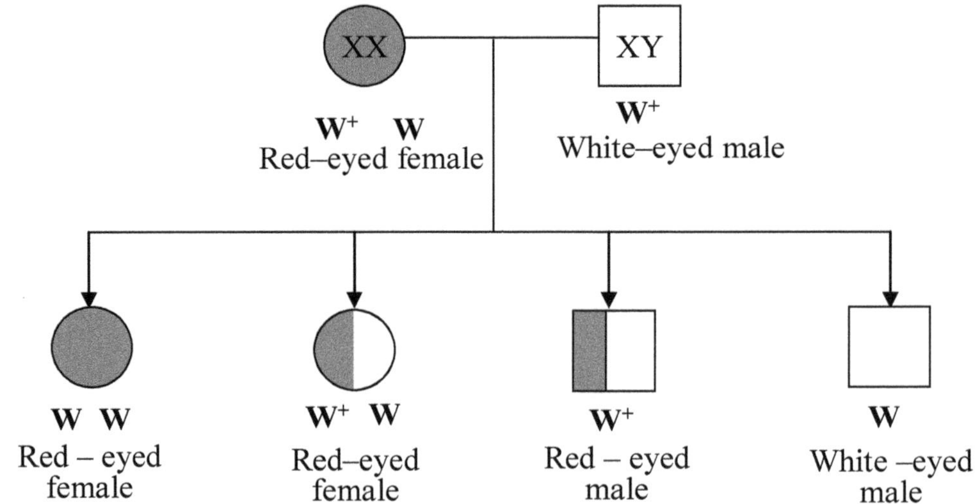

Figure 2.8: Morgan's Experiment: Eye Color in *Drosophila* Expressed a Sex-linked Trait.

be male. Thus, second-generation would include one-quarter with white eyes–and all of these would be male.

When the F1 males and females get interbred, four genotypic classes produced, each representing a different combination of sex chromosomes. The XX flies, which are female, have red eyes because at least one w+ allele is present. The XY flies, which are male, have either red or white eyes, depending on which X chromosome gets inherited from the heterozygous F1 females. Segregation of the w and w+ alleles in these females is the reason for half of the F2 males have white eyes.

Morgan carried out additional experiments to confirm his hypothesis, in one of the experiments (**Figure 2.9a**), he crossed F1 females assumed to be heterozygous for the eye color gene to mutant white males. *As* he expected, half the progeny of each sex had white eyes, and the other half had red eyes. In another experiment (**Figure 2.9b**), he crossed white-eyed females to red-eyed males. This time, all the females had red eyes, and all the males had white eyes. When he intercrossed these progenies, Morgan observed the expected segregation: half the progeny of each sex had white eyes, and the other half had red eyes. Morgan formulated the hypothesis that eye color is is linked to the X chromosome.

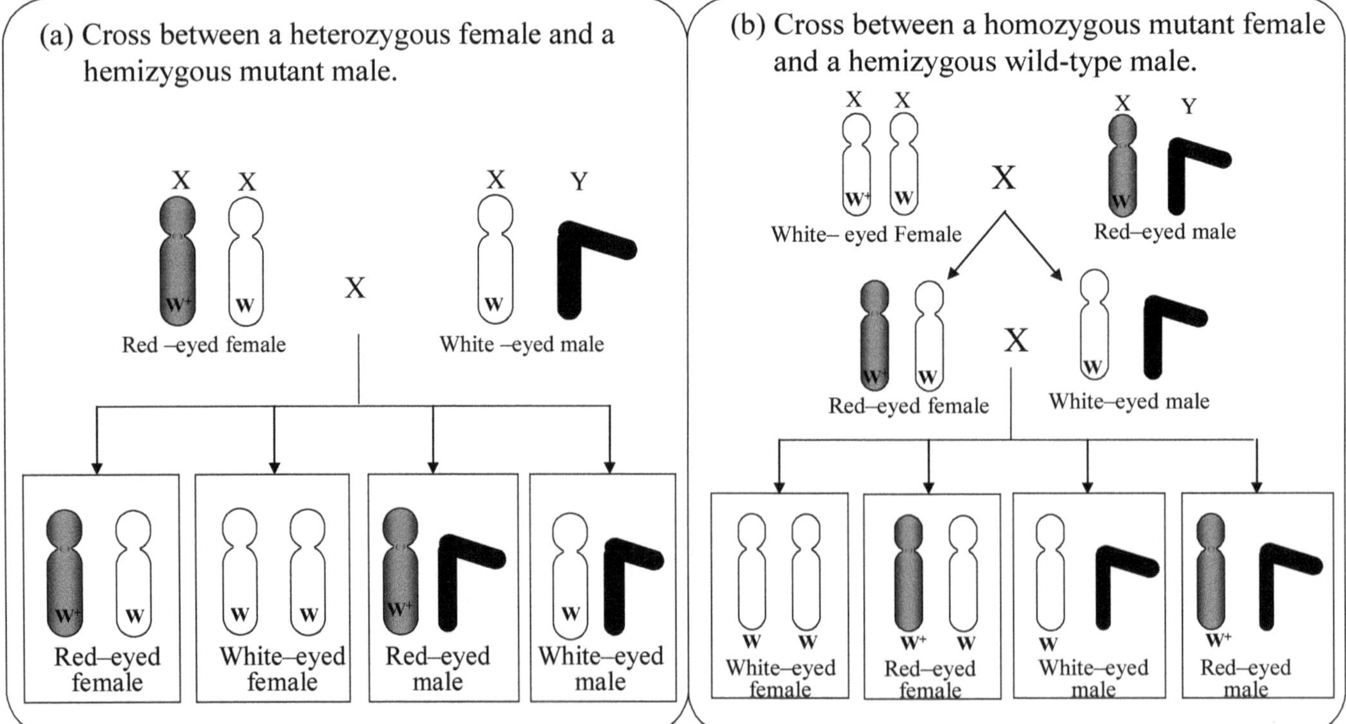

Figure 2.9: Morgan's Proposed that the Gene for Eye Color in *Drosophila* is X-linked.

2.11 Chromosomes as Arrays of Genes

Simple breeding experiments by Morgan and his students revealed that some recessive mutations transmitted along with the X chromosome. More evidence suggest that many more genes are found on the X chromosome, while other genes are not on the X chromosome. These genes also followed the law of segregation; however, they were not segregating with sex as an eye color gene as Morgan showed that these genes are on three autosomes in the *Drosophila* genome. Thus, each *Drosophila* chromosome appeared to contain a different set of genes or arrays of genes.

Morgans students were able to show that genes situated at different sites or loci (from the Latin word for "place"; singular: locus), on a linear structure. Morgans experiments pioneered the methods for genetic map-making and laid the foundation for subsequent research on the physical structure of chromosomes. The linearity of the chromosome is connected with the linear structure of DNA (**Figure 2.10**).

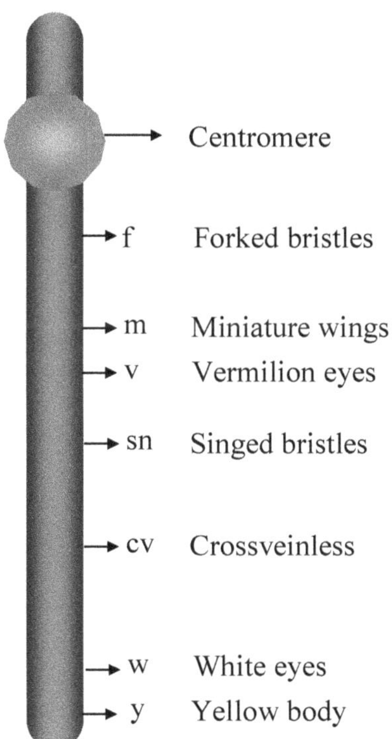

Figure 2.10: X Chromosome Map of Drosophila.

2.12 Nondisjunction as the Proof of the Chromosome Theory

Bridges repeated one of Morgan's experiments on a larger scale. He crossed white-eyed female Drosophila to red-eyed males and examined the number of F1 progeny. As per expectation, all the F1 flies were either red-eyed females or white-eyed males, few exceptional flies that are white-eyed females, and red-eyed males. He further crossed these exceptions to know the mechanisms. Interestingly, the resultant males were sterile; however, the exceptional females were fertile, and when he crossed these with the normal, red-eyed males, they produced many progenies, including large numbers of white-eyed daughters and red-eyed sons. Thus, the exceptional F1 females, though rare in their own right, were prone to produce many progeny that are exceptional.

Bridges explained these results by proposing that the exceptional F1 flies were the result of abnormal X chromosome behavior during meiosis in the females of the P generation. Ordinarily, the X chromosomes in these females should disjoin, or get segregated, during meiosis. Occasionally, however, they might fail to separate, producing an egg with two X chromosomes or an egg with no X chromosome at all. **Figure 2.11** illustrates the different possibilities.

When an egg with two X chromosomes fertilized with Y-bearing sperm, the zygote will be XwXwY. Since each of the X chromosomes in this zygote carries a mutant w allele, the resulting fly will have white eyes. If an egg without an X chromosome (usually called a null -X egg) is fertilized by an X – bearing sperm (X+), the zygote will be X+ O. (Once again, "O" denotes the absence of a chromosome.) Because the single X in this zygote carries a w+ allele, the zygote will develop into a red-eyed fly. Bridges inferred that XXY flies were female and that XO flies were male.

The exceptional, white-eyed females that he observed were, therefore, XwXwY, and the red-eyed males were X+O. Bridges confirmed the chromosome constitutions of the exceptional flies by direct cytological observation. Because the XO flies were male, bridges concluded that in *Drosophila* the Y chromosome has nothing to do with the determination of the sexual phenotype. However, because the XO males were always sterile, he realized that this chromosome must be important for male sexual function.

Bridges recognized that the fertilization of abnormal eggs by normal sperm could produce two additional kinds of zygotes: XwXwX+, arising from the union of a Diplo-X egg, and an X-bearing sperm, and YO, arising from the union of a null-X egg and a Y-bearing sperm. The XwXwX+ zygotes develop into females that are red-eyed, but weak and sickly. These "meta females" can be distinguished from XX females by a syndrome of anatomical abnormalities, including ragged wings and *etc.*hed abdomens. Generations of geneticists have inappropriately called them "super females" – a term confined by Bridges – even though there is nothing super about them. The YO zygotes

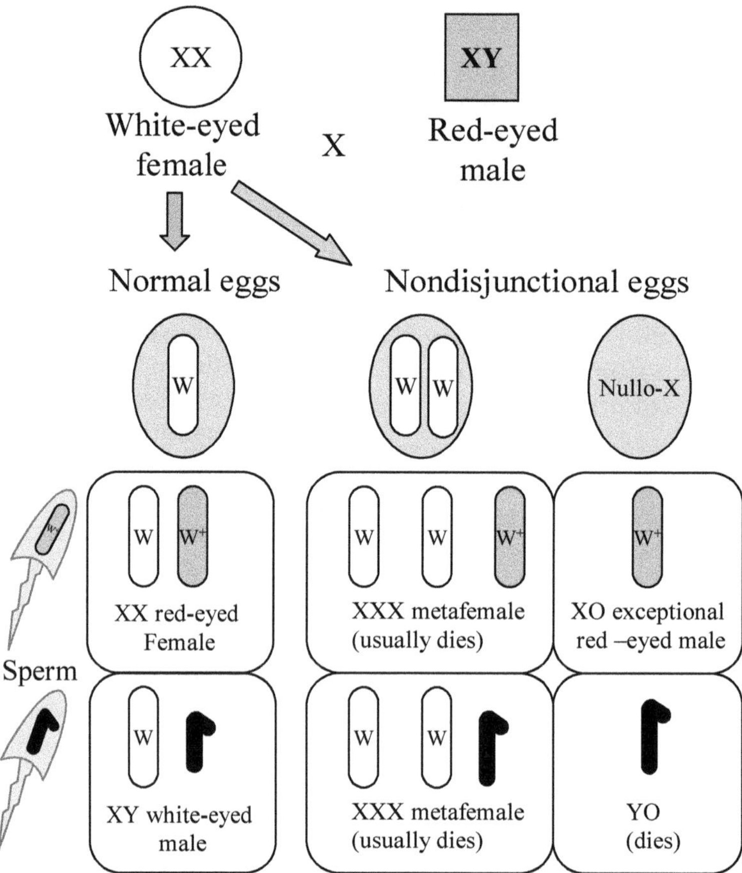

Figure 2.11: X Chromosome Non-disjunction is Responsible for the Exceptional Progeny that Appeared in Bridges' Experiment. The XXY zygotes develop into white-eyed females, the XO zygotes develop into red-eyed, sterile males and the XXX and YO zygotes die.

turn out to be completely viable; that is, they die. In Drosophila, as in most other organisms with sex chromosomes, at least one X chromosome is required for viability.

Bridges' ability to explain the exceptional progeny that came from these crosses showed the power of chromosome theory. Each of these exceptions was due to aberrant chromosome behavior during meiosis. Bridges called the anomaly non-disjunction because it involved a failure of the chromosomes to disjoin during one of the meiotic divisions. This failure could result from faulty chromosome movement, imprecise or incomplete pairing, or centromere malfunction. From Bridges' data, it is difficult to specify the exact cause. However, Bridges found that the exceptional progeny, presumably because their sex chromosomes can disjoin in different ways: the X chromosomes can disjoin from each other, or either X can disjoin from the Y. In the latter case, a Diplo- or null- X egg is produced because the X that does not disjoin from the Y is free to move to either pole during the first meiotic division. When fertilized by normal sperm, these abnormal eggs will produce exceptional zygotes.

2.13 Application of Chromosomal Basis of Inheritance to Mendel's Laws

According to Mendel genetic inheritance takes place under two laws (i) the alleles of a gene segregate from each other, and (ii) the alleles of different genes assort independently. These laws are called the law of segregation and the law of independent assortment. Both these laws are true for autosomal inheritance.

2.14 Sex-Linked Genes in Human Beings

In humans also, recessive X–linked traits are more easily identified than the autosomal recessive traits. A male need only to inherit one recessive allele to show an X–linked trait; however, a female needs to inherit two–one from each of her parents. Thus, the preponderance of individuals having –linked trait is male.

2.14.1 Hemophilia, an X-linked Blood – Clotting Disorder

A factor is essential for blood clotting and, the people who lack it suffer from the clotting defect.

The clotting factor is present on the X chromosome, therefore, called X linked. In patients who do not have clotting factors and have a cut/wound/bruises patient continues to bleed and, no treatment results in death. Nearly all the individuals affected with X–linked hemophilia are male. Other blood–clotting disorders are found in both males and females because they are due to mutations in autosomal genes. If the father suffering from hemophilia and the mother is a heterozygous carrier female the daughters may have this disease, they will be homozygous for hemophilia.

The most famous case of X- linked hemophilia was seen in a Russian imperial family at the beginning of the twentieth century (**Figure 2.12**). Czar Nicholas and Czarina Alexandra had four daughters and one son. The son, Alexis, suffered from hemophilia. The X– linked mutation responsible for Alexis' disease was transmitted to him by his mother, who was a heterozygous carrier. Czarina Alexandra was a granddaughter of Queen Victoria of Great Britain, who was also a carrier. Pedigree records show that Victoria transmitted the mutant allele to three of her nine children: Alice, who was Alexandra's mother, Beatrice, who had two sons with the disease, and Leopold, who had the disease himself. The allele that Victoria carried arose as a new mutation in her germ cells, or came through her mother, father, or a more distant maternal ancestor.

2.14.2 Color Blindness, an X-linked Vision Disorder

n human beings, color perception is mediated by light-absorbing proteins in the specialized cone cells of the retina in the eye. Three such proteins have been identified – one to absorb blue light, one to absorb green light, and one to absorb red light. Color blindness may be caused by an abnormality in any of these receptor proteins. The classic type of color blindness, involving faculty perception of red and green light follows an X-linked pattern of inheritance. About 5 to 10 per cent of human males are red-green colorblind while females are also found to be color blind but only less than 1 percent, has this disability, showing that the mutant alleles are recessive. Molecular studies have shown that there are two distinct genes for color perception on the X chromosome; one encodes the receptor for green light, and the other encodes the receptor for red light. Detailed analyses have demonstrated that these two receptors are structurally very similar, perhaps the genes encoding them evolved from an ancestral color-receptor gene. A third gene for color perception, the one encoding the receptor for blue light, is situated on an autosome.

In **Figure 2.13** color blindness demonstrates the procedures for finding out the risk of inheriting a

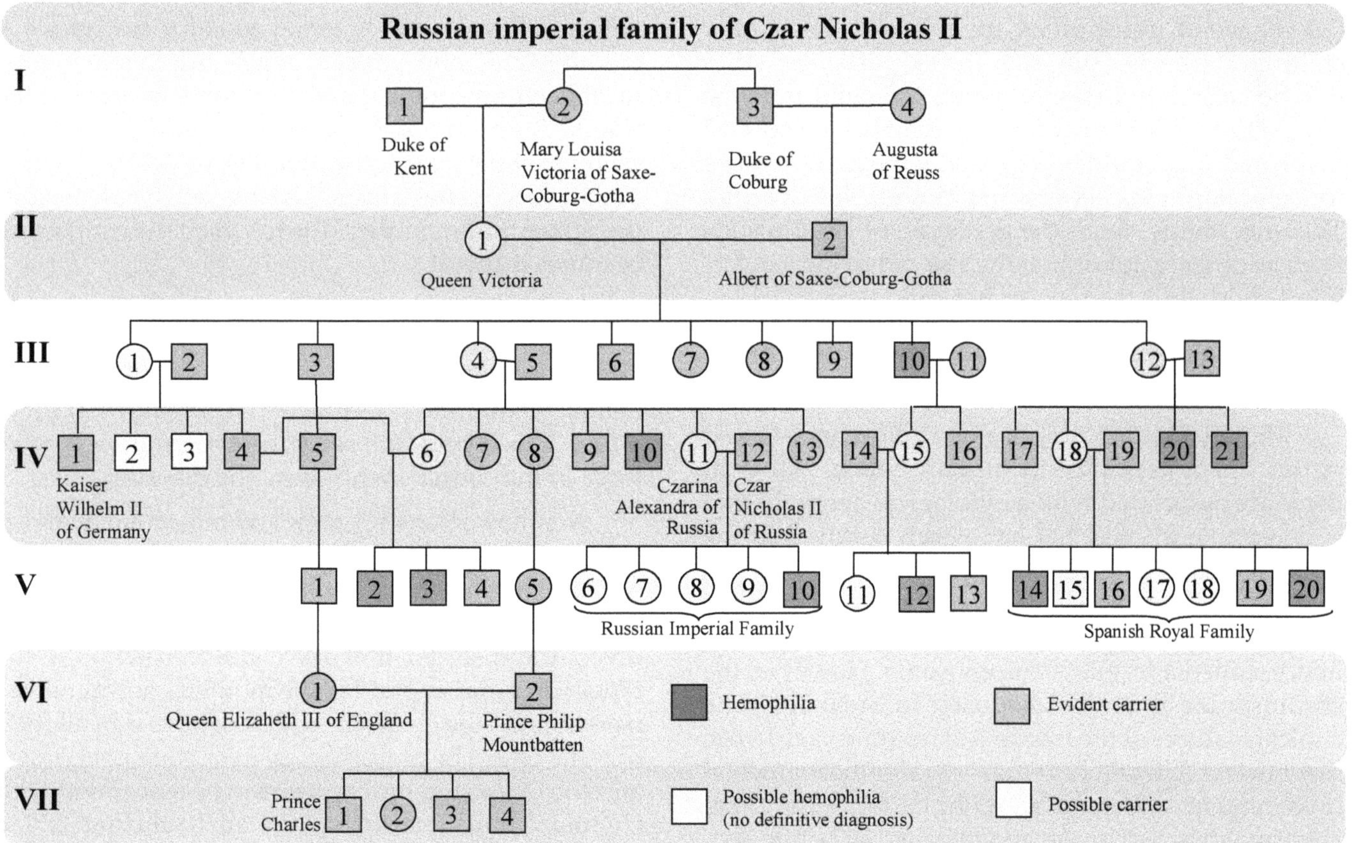

Figure 2.12: X-linked Hemophilia in Royal Families of Europe.

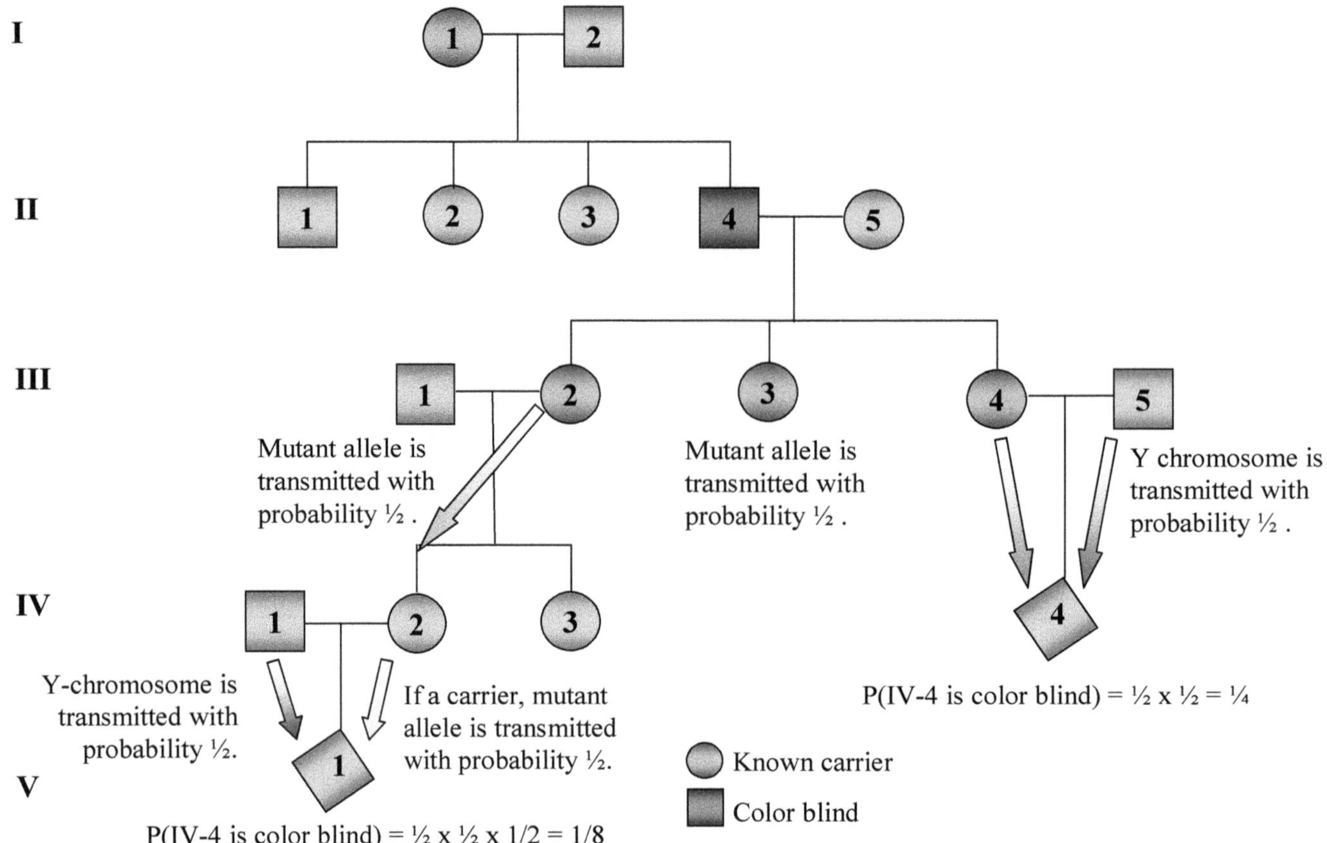

Figure 2.13: Segregation X-linked Colour Blindness.

recessive X- linked condition. A heterozygous carrier, such as in 3rd generation 4th child has a 50 per cent (½) chance of transmitting the mutant allele to her children. However, the risk that a particular child will be color blind is only since the child must be a male to manifest the trait. The female labeled IV-2 in the pedigree could be a carrier of the mutant allele for color blindness because her mother was a carrier. The uncertainty about the genotype of IV -2 results from another factor of ½ at the risk of having a color-blind child; thus the risk for her child is x ½ = 1/8.

2.14.3 The Fragile X Syndrome and Mental Retardation

Mental retardation appears to follow an X-linked pattern of inheritance in human beings. Most of these are associated with a cytological anomaly that is detectable in cells that have been cultured in the absence of nucleotides. During this anomaly, the tip of the X chromosome looks as if it will get detached from the rest of the chromosome (**Figure 2.14a**); hence, called a fragile X chromosome. However, the chromosome is not so disposed to breakage. The clinical features of the fragile X syndrome vary hence diagnosis is difficult. Patients show significant mental involvement, and some show facial and behavioral abnormalities; both males and females can be affected. Its incidence is 1/2000.

The fragile X syndrome is defined as an X linked dominant disorder with incomplete penetrance. Affected females are heterozygous for the fragile X chromosome, and affected males are heterozygous for this chromosome. However, some carriers of the fragile X chromosome, both male and female, are asymptomatic for the disorder (**Figure 2.14b**). There is a lack of full penetrance which makes the analysis of pedigrees quite complex due to which the counseling becomes difficult.

Molecular techniques have been used to isolate and analyze the fragile X site. It consists of a short repeat unit in the DNA adjacent to FMRI (Fragile X mental retardation 1) its function is not known (**Figure 2.14c**). This repeat unit is variable in length it is quite large in individuals who show the disorder.

A combination of genetic and molecular analyses has shown that the repeating unit tends to increase in size, due to faulty replication of chromosomes. This instability explains why individuals who do not reveal the disorder may have children who do; the repeat unit may expand in the mother's germline and passed on to the children, who will have it in all their somatic cells. The physiological details are not known, and the expanded unit appears to be associated with chemical modification of the surrounding DNA. This modification shows an adverse effect on the

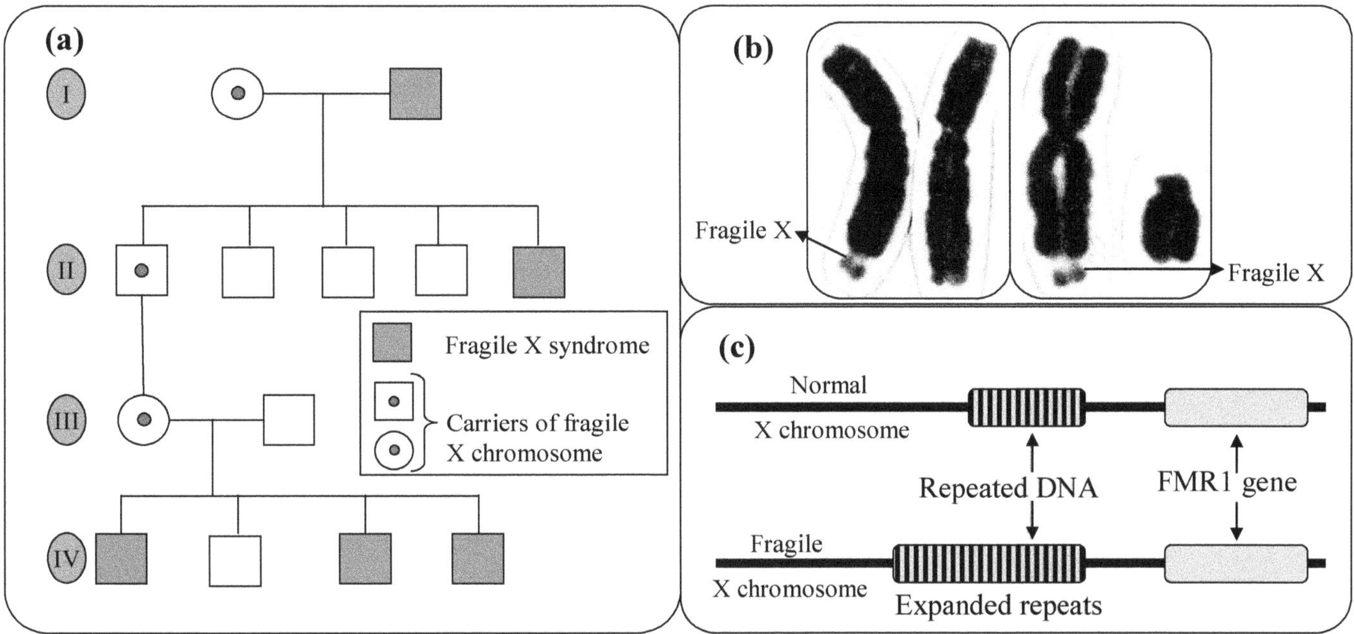

Figure 2.14: The Fragile X Chromosome.
(a) A pedigree with fragile X syndrome. II -1 male is a carrier. (b) The fragile X and a normal X chromosome from a female (left), and the fragile X and a normal Y chromosome from a male (right). (c) Molecular basis of the fragile X syndrome. The mutation in the fragile X chromosome is due to repeat expansion in the DNA flanking the FMR1 gene.

expression of the FMR1 gene and is probably the actual cause of the fragile X syndrome.

2.15 Genes on the Human Y Chromosome

There are very few genes on the Y chromosome. If the mutations take place the resultant genes will be passed on to all the son's however, one of the daughters would be carrying these genes. The Y-linked genes can be identified through pedigree analysis. To date, only a few Y-linked genes have been found. One is responsible for the synthesis of a male-specific substance called the H-Y antigen, which is seen on the cell surface, the other is involved in the production of a factor that is critical for the differentiation of the testes and the subsequent acquisition of male sexual characteristics.

2.16 X and Y Chromosomes Genes

Some genes are present on both X and Y chromosomes, mostly near the ends of the short arms. Alleles of these genes do not follow a distinct X- or Y-linked pattern of inheritance. Instead, they are transmitted from mothers and fathers to sons and daughters alike, mimicking the pattern of inheritance of an autosomal gene. Such genes are therefore called pseudo autosomal genes. In males, the regions that contain these genes seem to mediate pairing between the X and Y chromosomes.

2.17 Sex Chromosomes and Sex Determination

Among animal kingdom, sex is perhaps the most conspicuous phenotype. Animals that are identified as male and female sex are sexually dimorphic. Sometimes this dimorphism is established by environmental factors. In one species of turtles, for example, sex is determined by temperature. Eggs that have been incubated above 30ºC hatch into females, while eggs incubated at a lower temperature hatch into males. In many other species, sexual dimorphism is established by genetic factors, often involving a pair of sex chromosomes.

2.18 Sex Determination in Humans

Y chromosome is the dominant sex-determining chromosome in humans. It is evident from the fact that individuals with an abnormal number of sex chromosomes XO animals are females, and XXY animals are males (**Figure 2.15**). The study of Y chromosomes governs the primordial gonads which develop into testes. Once the testes formed, they produce testosterone, a hormone that stimulates the development of male secondary sexual characteristics.

TDF formed when SRY (for sex-determining region Y) gene is present just outside the pseudo autosomal region in the short arm of the Y chromosome. The discovery of SRY was made possible by the

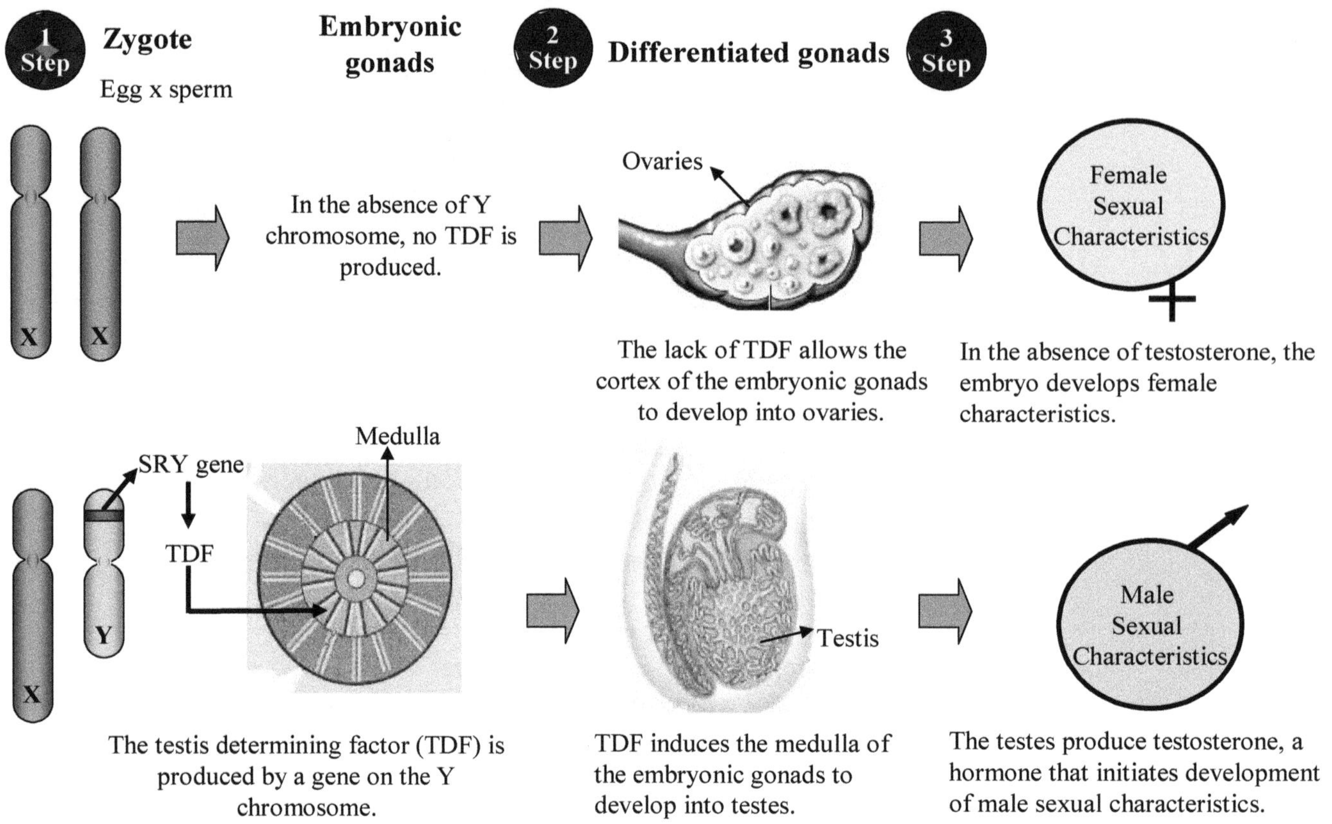

Figure 2.15: Sex Determinations among Human Beings.
Sexual development depends upon the production of the testis-determining factor (TDF).

identification of unusual individuals whose sex was inconsistent with their chromosome constitution-XX males and XY females (**Figure 2.16**). Some of the XX males found to carry a small piece of the Y chromosome inserted into one of the X chromosomes. This piece carried a gene responsible for maleness. Some of the XY females carry an incomplete Y chromosome. The part of the Y chromosome that was missing corresponded to the piece that was present in the XX males; its absence in the XY females apparently prevented them from developing testes. These complementary lines of evidence showed that a particular segment of the Y chromosome required for male development. Molecular analyses subsequently identified the SRY gene in this male-determining segment. Additional research has revealed that an SRY gene is present on the Y chromosome of the

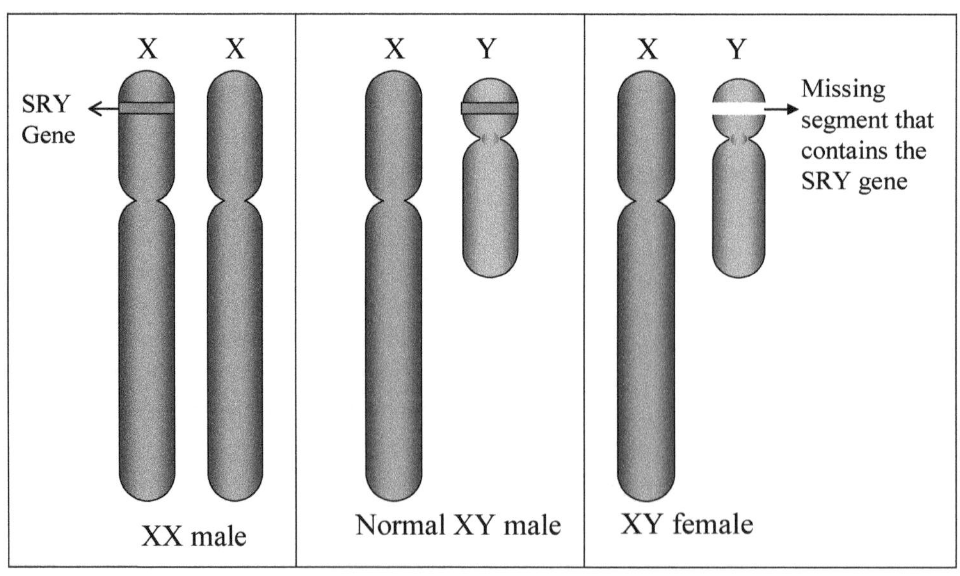

Figure 2.16: SRY Ggenes on Y Chromosome.

mouse, and that-like the human SRY gene-it specifies male development.

Testes produce testosterone that initiates the development of male sexual characteristics. It is a hormone that binds to the receptors present on many types of cells. Once binding takes place, the hormone-receptor complex transmits a signal to the nucleus, educating, the cell to differentiate. The differentiation of different types of cells results in the development of male characteristics such as heavy musculature, beard, and deep voice. If the testosterone signaling system fails, these characteristics do not appear, and the individual develops as a female. One reason for the failure is an inability to make the testosterone receptor (**Figure 2.17**).

XY individuals with the inability to make the testosterone receptor primarily develop as male-testes and testosterone are formed. However, the testosterone has no influence because it cannot transmit the developmental signal inside the target cells. Individuals missing the testosterone receptor therefore obtain female sexual features. They do not, have ovaries and are therefore germ-free. This is called testicular feminization and results from a mutation in an X-linked gene, Tfm, which codes the testosterone receptor. The tfm mutation is transmitted from mothers to their hemizygous XY descendants (who are phenotypically female) in a typical X-linked pattern.

2.19 Sex Determination in *Drosophila*

Sex determination in *Drosophila* is very interesting as the Y chromosome has no role to play it is the ratio of X chromosomes to autosomes. This mechanism was first demonstrated by Bridges in 1921 through an analysis of flies with uncommon chromosome compositions.

(a) Normal male with the wild-type Tfm gene.

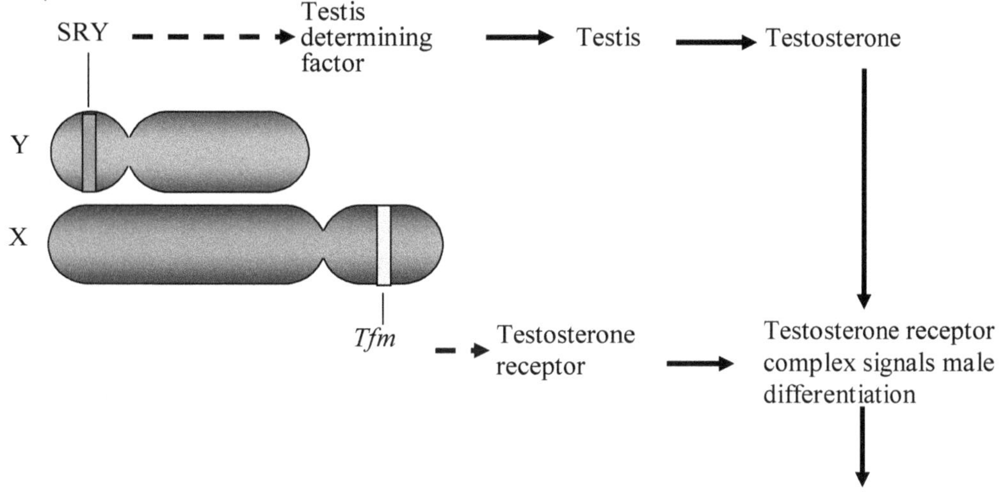

(b) Male with the tfm mutation and testicular feminization.

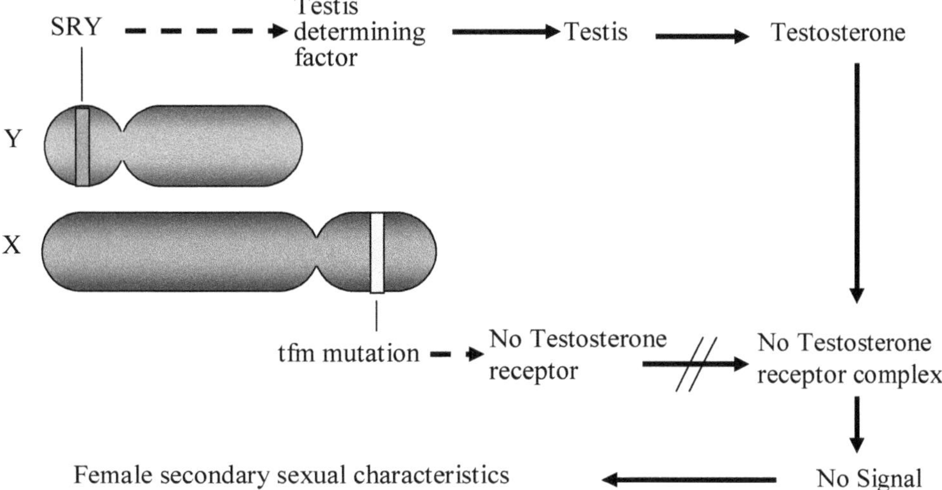

Figure 2.17: Testicular Feminization, due to X-linked Mutation in the Tfm Gene.

Normal diploid flies have a pair of sex chromosomes, either XX or XY, and three pairs of autosomes, usually denoted AA; here, each A represents one haploid set of autosomes. In complex experiments, Bridges took flies with abnormal numbers of chromosomes. He detected that whenever the ratio of X's to it was 1.0 or greater, the fly was female, and whenever it was 0.5 or less, the fly was male. Flies with an X: A ratio between 0.5 and 1.0 developed characteristics of both sexes; thus, Bridges called them intersexes. In none of these flies did the Y chromosome have any consequence on the sexual phenotype. It was, however, required for male fertility. If there are 1X and two autosomes phenotype is male and X: A ratio is 0.5 2X and 2A will be female and X: A ratio would be 1. 3X and 2A give rise to meta females, 4X 4A to tetraploid females 3X 3A to triploid females. 3X 4A to intersex, 2X 3A again to intersex 2X 4A to tetraploid male and 1x 3A to Meta male.

2.20 Sex Determination in other Animals

When two kinds of gametes *i.e.*, X-bearing and Y-bearing are produced the species is referred to as heterogametic sex; in these species, females are the homogametic sex. In birds, butterflies, and some reptiles, this situation is reversed (**Figure 2.18**). Males are homogametic (usually denoted ZZ) and females are heterogametic (ZW). The mechanism of sex determination in the Z- W sex chromosome system is still not very clear.

In honeybees, sex is determined by whether the animal is haploid or diploid. Diploid embryos, are the result of fertilized eggs, these develop into females; the embryos which are not fertilized remain haploid and they develop into males the reproductive female is called queen she governs the larva. In this system, a queen can control the ratio of males to females by regulating the proportion of unfertilized eggs that she lays. Because this number is small, most of the progeny are female, albeit sterile, and serve as workers for the hive. In the sex determination, eggs are produced through meiosis by the queen, and sperms are produced through mitosis by the male. This system ensures that fertilized eggs will have the diploid chromosome number and that unfertilized eggs will have the haploid number.

2.20.1 Dosage Compensation of X-linked Genes

If chromosomal or gene imbalance occurs, it will result in an abnormal phenotype. Sometimes this can be lethal also. Hence it is a little puzzling why in the process of evolution females have two X chromosomes, and males have only one X chromosome. To keep the balance of the genes there may be two mechanisms that may compensate for this difference: (i) each X-linked gene could work twice as hard in males as it does in females, or (ii) one copy of each X-linked gene get inactivated in females. In drosophila X linked genes in males work two times harder than females. This phenomenon is called hyperactivation which involves a complex of different proteins that bind to many sites on the X chromosome in males and trigger a doubling of gene

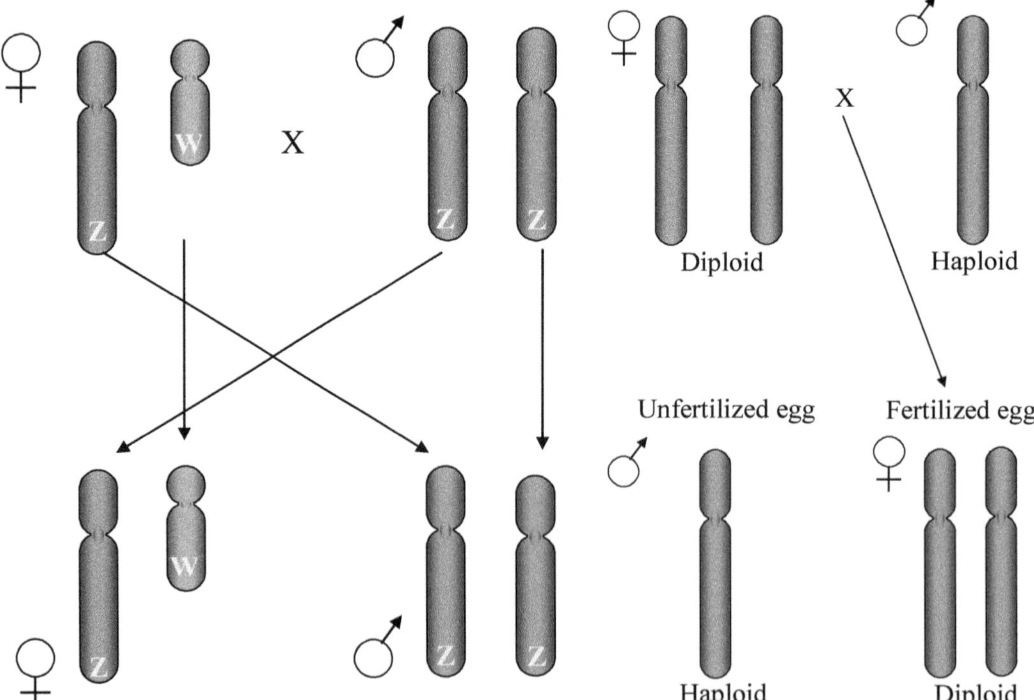

Figure 2.18: Sex Determinations in Birds. Females are ZW and male are ZZ. The sex of the offspring is determined by which of the sex chromosomes, Z or W, is transmitted by the female.

activity. When this protein complex does not bind, as is the case in females, hyperactivation of X-linked genes does not occur. Among mammals' genes on one chromosome get inactivated.

2.20.2 Inactivation of X-linked Genes in Female Mammals

In placental mammals, dosage compensation of X-linked genes may be achieved by the inactivation of one of the female's X chromosomes (**Figure 2.19**). This mechanism was first proposed in 1961 by the British geneticist Mary Lyon, by carrying out studies on mice. The inactivation starts in the mouse embryo only when it has thousands of cells. Each cell takes an independent decision to silence one of the X chromosomes. It is a random process.

Once chosen, it remains inactivated in all the descendants of that cell. Thus, female mammals are the genetic mosaics, containing two types of cell lineages; the maternally inherited X chromosome is inactivated in roughly 50 per cent of these cells, and the paternally inherited X is inactivated in the other half. A female that is heterozygous for an X-linked gene is, therefore, able to show two different phenotypes.

One of the best examples of this phenotypic mosaicism comes from the study of fur coloration in cats and mice. In both of these species, the X chromosome carries a gene for pigmentation of the fur. Females heterozygous for different alleles of this gene show patches of light and dark fur representing light patches with one allele, and dark patches with the other. This happens in cats, where one allele produces black pigment and the other produces orange pigment. Different patches of fur define a clone of pigment producing cells, or melanocytes, which have been derived by mitosis from a precursor cell present at the time of X chromosome inactivation.

An X chromosome gets inactivated hence does not look or acts like other chromosomes. Methylation of DNA takes place by the addition of various methyl groups. DNA condenses into a darkly staining

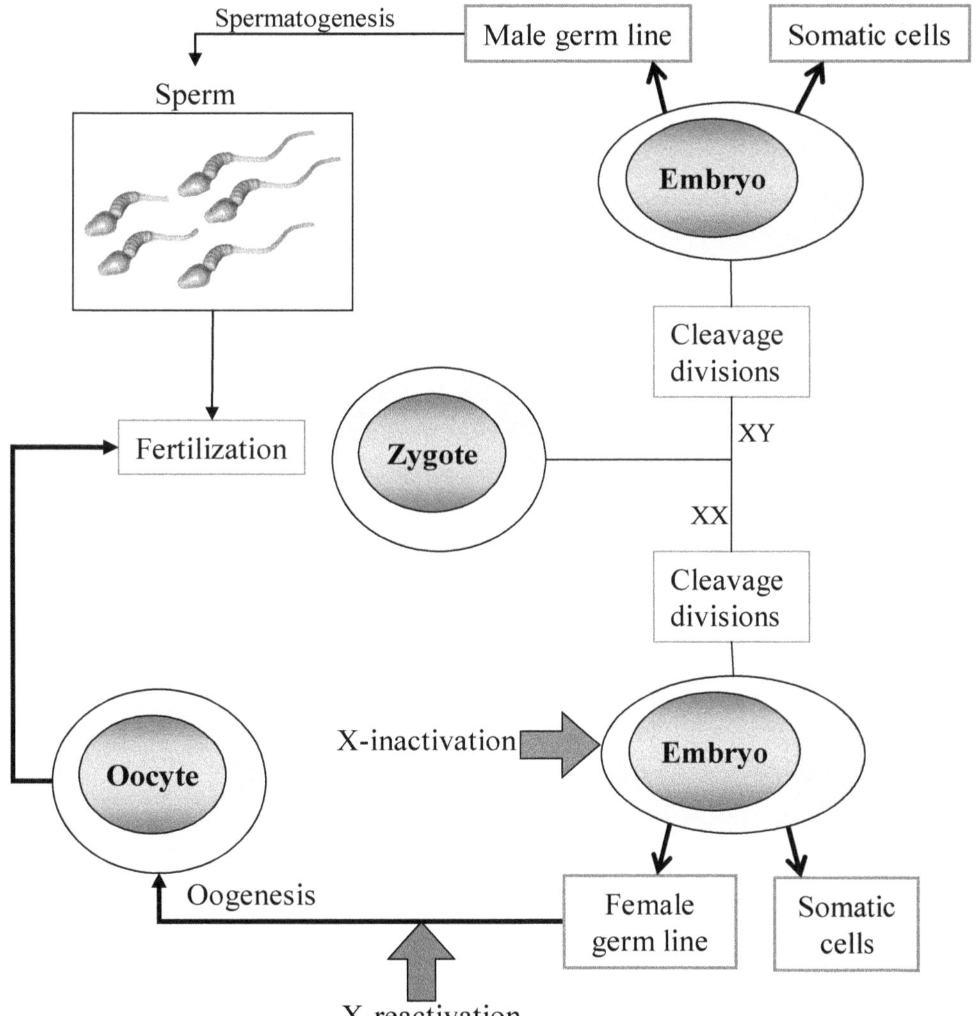

Figure 2.19: X Inactivation in Mammals.

structure called a Barr Body, after the Canadian geneticist Murray Barr, who first time observed it. The inactivated X chromosome remains in the altered state in somatic tissues. However, in the germ tissues, it is reactivated, perhaps because two copies of some X-linked genes are required for the successful completion of oogenesis.

Cytological studies have identified human beings with more than two X chromosomes. Mostly they are phenotypically normal females, are because one of their X chromosomes is inactivated. Often all the inactivated X's result into a single Barr body.

Some cells contain multiples of 23 chromosomes these are aneuploid cells. Aneuploidies of the autosomes are clinically important as they may result in some chromosome abnormalities. They contain primarily monosomy and trisomy (three copies of a chromosome). When monosomies are seen in autosomes, these may be incompatible, therefore, difficult survival, the trisomy's produce less severe concerns than monosomies. The principle is that the body can tolerate excess genetic material more readily than it can tolerate less genetic material.

The most common cause of aneuploidy is non-disjunction when the chromosomes are not able to disjoin normally during meiosis (**Figure 2.20**). Aneuploidy can occur during meiosis I/meiosis II. The resulting gamete either lacks a chromosome or has two copies of it, producing a monosomic or trisomic zygote.

2.20.3 Trisomy 21

The best-known and most common chromosome abnormality in humans is down syndrome, a condition associated with an extra chromosome 21. This syndrome was first described in 1866 by a British physician, Langdon Down, but its chromosomal basis not clearly understood until 1959. People with Down syndrome are short stature with loose-joints, particularly in the ankles; they have broad skulls, wide nostrils, large tongue with a distinctive furrowing, and stubby hands with a crease on the palm. Impaired mental abilities require training and care.

The extra chromosome 21 in Down syndrome is an example of trisomy. There are 47 chromosomes, including two X chromosomes as the extra chromosome 21. The karyotype of this individual is therefore written 47, XX, +21.

Trisomy 21 can be caused by chromosome non-disjunction in one of the meiotic cell divisions (**Figure 2.21**). The non-disjunction event can occur in either parent, but it seems to be more likely in females. In addition, the frequency of non-disjunction increases with maternal age. Thus, among mothers younger than 25 years old, the risk of having a child with Down syndrome is about 1 in 1500, whereas, among mothers with or above 40 years, it is 1 in 100. This increased risk is due to factors that adversely affect meiotic chromosome behavior as a woman's age increases. In human females, meiosis begins in the fetus, but

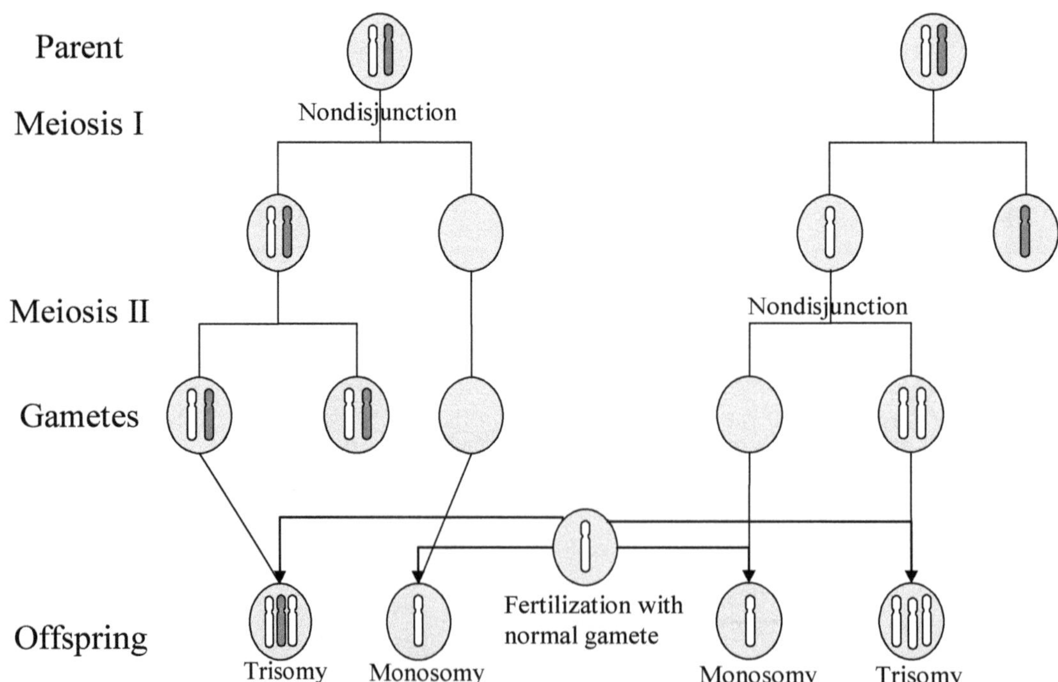

Figure 2.20: Meiotic Non-disjunction.

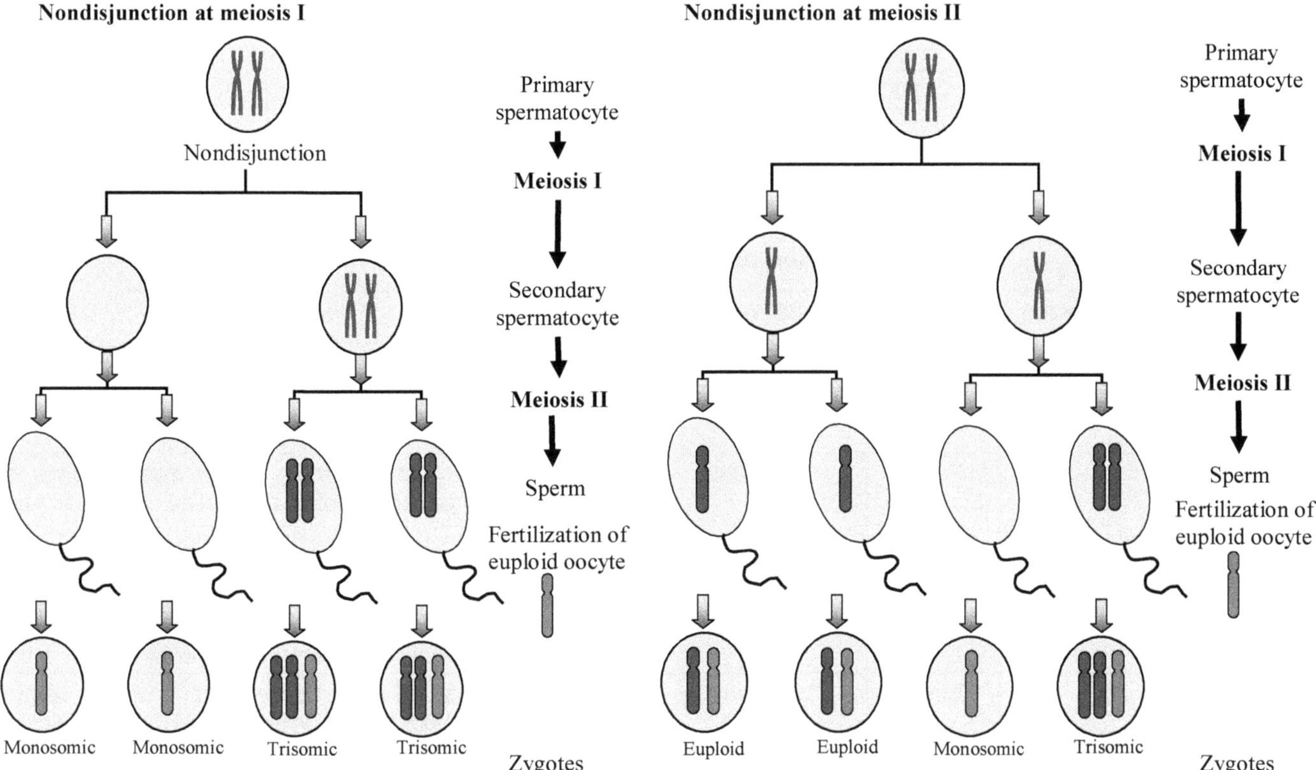

Figure 2.21: Meiotic Non-disjunction and the Origin of Down Syndrome.

it is not completed until after the egg is fertilized. During a long time prior to fertilization, the meiotic cells are arrested in the prophase of the first division. In this suspended state, the chromosomes may not remain in pairs. The longer the time in prophase, the greater is the chance for un-pairing and subsequent chromosome non-disjunction. Older females are therefore more likely than younger females to produce aneuploid eggs.

2.20.4 Trisomy 18 (47, XY, +18)

This trisomy is known as Edwards's syndrome, an autosomal trisomy; with approximately a prevalence of 1/6,000 live births. It is the most common chromosome abnormality among stillborns with congenital malformations. Trisomy 18 patients have prenatal growth deficiency characteristics like facial features and a distinctive hand abnormality that helps clinicians to make the first diagnosis. Minor anomalies of diagnostic importance include small ears with unraveled helices, a small mouth that is often hard to open, a short sternum, and rather short, big toes. Most babies with trisomy 18 have major malformations. Congenital heart defects, particularly ventricular septal defects, are the most common. Other medically significant congenital malformations include omphalocele (protrusion of the bowel into the umbilical cord), diaphragmatic hernia, and occasionally, spina bifida.

Table 2.5: How Non-disjunction Leads to Sex Chromosome Aneuploids

Situation	*Sperm*	*Oocyte*	*Consequence*
Normal	Y	X	46, XY normal male
	X	X	46, XX normal female
Female non-disjunction	Y	XX	47, XXY Klinefelter syndrome
	X	XX	47, XXX triplo-X
	Y		45, Y nonviable
	X		45, X Turner syndrome
Male non-disjunction		X	45, X Turner syndrome
(Meiosis I)	XY	X	47, XXY Klinefelter syndrome
Male non-disjunction	XX	X	47, XXX triple-X
	XX	X	47, XYY Jacobs syndrome
		X	45, X Turner syndrome

Recent studies have indicated that about 50 per cent of children with trisomy 18 die in the first month, and only about 10 per cent are still alive at 12 months of age A combination of factors, including aspiration pneumonia, predisposition to infections and apnea, and congenital heart defects, account for the high mortality rate.

Table 2.6: Comparing and Contrasting Trisomic 13, 18 and 21

Type of Trisomy	*Incidence of Birth*	*Percent of Conceptions that Survive 1 Year after Birth*
21 (Down)	1/800 – 1/826	85 per cent
18 (Edward)	1/6,000 – 1/10,000	<5 per cent
13 (Patau)	1/12,500 – 1/21,700	<5 per cent

Table 2.7: Non-disjuction Causes Aneuploidy

Karyotype	*Chromosome Formula*	*Clinical Syndrome*	*Estimated Frequency of Birth*	*Phenotype*
47, +21	2n + 1	Down	1/700	Short, broad hands with palmar crease, short stature, hyper flexibility of joints, mental retardation, broad head with round face, open mouth with large tongue, epicanthal fold.
47, +13	2n + 1	Patau	1/20,000	Mental deficiency and deafness, minor muscle seizures, cleft lip and/or palate, cardiac anomalies, posterior heel prominence.
47, + 18	2n + 1	Edward	1/8000	Congenital malformation of many organs, low – set, malformed ears, receding mandible, small mouth and nose, mental deficiency, horseshoe or double kidney, short sternum; 90 per cent die within first six months after birth.
45, X	2n – 1	Turner	1/2500 females' births	Female with retarded sexual development, usually sterile, short stature, webbing of skin in neck region, cardiovascular abnormalities, hearing impairment.
47, XXY 48, XXXY 48, XXYY 49, XXXXY 50, XXXXXY	2n + 12n + 22n +22n + 32n + 4	Klinefelter	1/500 male births	Male, sub fertile with small testes, developed breasts, feminine – pitched voice, knock –knees, long limbs.
47, XXX	2n + 1	Triple –X	1/700	Female with usually normal I genitalia and limited fertility, slight mental retardation.

Those with trisomy 18 who have survived infancy have a marked developmental disability. The degree of delay is much more significant than in Down syndrome, and most children are not able to walk. However, children with trisomy 18 achieve their milestones, and older infants learn some communication skills. In **Table 2.6** the trisomies 13, 18 and 21 are compared. The nm disjunction caused aneuploidy is illustrated in **Table 2.7**.

More than 95 per cent of individuals with Edwards's syndrome have complete trisomy 18. A small percentage has mosaicism. As in trisomy 21, there is a significant maternal age effect. In trisomy 18 cases are the result of a maternally contributed extra chromosome.

2.20.5 Trisomy 13 (47, XY, +13)

This condition causes Patau syndrome the incidence of which is 1/10,000 births. The malformation pattern is quite distinctive and usually permits clinical recognition. It consists primarily of oral-facial clefts, micro-ophthalmia (small, abnormally formed eyes), and postaxial polydactyly. Malformations of the central nervous system are seen frequently, and cutis aplasia (a scalp defect on the posterior occiput) may also be seen.

The survival rate is very similar to that of trisomy 18, with 90 per cent of live-born cases dying during the first year of life. As in trisomy 18, the children who survive infancy have significant developmental retardation, with skills seldom progressing beyond those of a child of 2 years. However, as in trisomy 18, children with trisomy 13 do progress somewhat in their development and are often able to communicate with their families.

About 80 per cent of cases of Patau syndrome have full trisomy 13. Most of the remaining cases have trisomy of the long arm of chromosome 13 due to a translocation. As in trisomies 18 and 21, the risk of bearing a child with this condition increases with advanced maternal age. It is estimated that 95 per cent or more of trisomy 13 and 18 conceptions are spontaneously lost during pregnancy due to their lethal nature.

2.20.6 Trisomies, Non-disjunction, and Maternal Age

Among mothers under the age of 30, the risk is less than 1/1,000. It increases to about 1/400 among women at age 35, to approximately 1/100 at age 40, and to 1/50 or more after age 45. Most other trisomies, with those in which the fetus does not continue to term, also rise in prevalence as maternal age increases. This threat is one of the primary indications for prenatal diagnosis among women over age 35.

There could be several reasons for this like frequent non-disjunction among older mothers. Female oocytes are formed during embryonic development. They remain suspended in prophase -I until they are shed at the time of ovulation. Thus, an ovum produced by a 45-year–old woman is itself about 45 years old. This long period of suspension in prophase- I may cause normal disjunction. The exact mechanism is not known. The consequences of sex chromosome non-disjunction led to aneuploids these are shown in **Table 2.8.**

Table 2.8 How non-disjunction leads to sex chromosome aneuploids

Situation	*Sperm*	*Oocyte*	*Consequence*
Normal	Y	X	*46, XY normal male*
	X	X	*46, XX normal female*
Female non-disjunction	Y	XX	*47, XXY Klinefelter syndrome*
	X	XX	*47, XXX triplo-X*
	Y		*45, Y nonviable*
	X		*45, X Turner syndrome*
Male non-disjunction		X	*45, X Turner syndrome*
(meiosis I)	XY	X	*47, XXY Klinefelter syndrome*
Male non-disjunction	XX	X	*47, XXX triple-X*
	XX	X	*47, XYY Jacobs syndrome*
		X	*45, X Turner syndrome*

The other factors could be hormone levels, cigarette smoking, autoimmune thyroid disease, alcohol consumption, and radiation (the latter does increase nondisjunction when administered in very large doses to experimental animals). None of these has shown consistent correlations with nondisjunction in humans, however, maternal age remains the only known correlated factor. Although maternal age is strongly correlated with the risk of Down syndrome, it should be pointed out that approximately three-fourths of Down syndrome children are born to mothers under the age of 35. This is because the great majority of children (>90 per cent) are borne by women in this age group. Male effects are insignificant. The reason could be spermatogenesis takes place throughout the male life. The comparison of various trisomies is shown in **Table 2.9**.

Table 2.9: Comparing and Contrasting Trisomic 13, 18 and 21

Type of Trisomy	*Incidence of Birth*	*Percent of Conceptions that Survive 1 Year after Birth*
21 (Down)	1/800 – 1/826	85 per cent
18 (Edward)	1/6,000 – 1/10,000	<5 per cent
13 (Patau)	1/12,500 – 1/21,700	<5 per cent

2.21 Sex Chromosome Aneuploidy

Among live births, about 1/400 males and 1/650 females have a certain form of sex chromosome aneuploidy. This may be because of X inactivation, these aneuploidies are less severe than those caused by autosomal aneuploidy. With the exception of a complete absence of an X chromosome, all of the sex chromosome aneuploidies are compatible with survival in at least some cases.

2.21.1 Monosomy of the X Chromosome (Turner syndrome)

Henry Turner in 1938 first time discovered 45X females with a characteristic phenotype. They are short stature, sexual infantilism, ovarian digenesis, and a pattern of major and minor defects. The face is triangle shaped face; posteriorly rotated external ears; and abroad, "webbed" neck.

In addition, the chest is broad and shield-like in shape. At birth lymphedema of the hands and feet are seen. Patients with Turner syndrome have congenital heart defects, most common obstructive lesions of the left side of the heart (bicuspid aortic valve in 50 per cent of patients and narrowing of the aorta in 15 per cent to 20 per cent). Severe obstructions can be repaired surgically. About 50 per cent of individuals with Turner syndrome have structural kidney defects, but with no medical problems. Girls with Turner syndrome exhibit proportionate short stature and do not undergo an adolescent growth spurt. Growth hormone administration produces increased height in these girls. Gonadal dysgenesis (lack of ovaries) is commonly seen in Turner syndrome. Instead of ovaries, most females with turner syndrome have streaks of connective tissue. Lacking normal ovaries, they do not usually develop secondary sexual characteristics, and most women with this condition are infertile (about 5 per cent to 10 per cent have sufficient ovarian development to undergo menarche, and a small number give birth to children). Teenagers with turner syndrome if treated with estrogen may develop secondary sexual characteristics. The dose is then continued at a reduced level to maintain these characteristics.

Frequently the diagnosis is made in the newborn infant, especially if there is a striking neck webbing coupled with a heart defect. The facial features are more subtle than in the autosomal abnormalities described previously, but the experienced clinician will often diagnose turner syndrome because of one or more of the listed clues. If turner syndrome is not diagnosed in infancy or childhood, it will often be diagnosed later because of short stature and/or amenorrhea.

About 50 per cent of these patients have a 45, X karyotype in their peripheral lymphocytes. About 30 per cent to 40 per cent are mosaics, most commonly 45, X/46, XX, and less commonly 45, X/46, XY. Mosaics that have Y chromosomes in some cells are predisposed to developing malignancies (gonadoblastomas) in their gonadal streaks. About 10 per cent to 20 per cent of patients with Turner syndrome have structural X chromosome abnormalities involving a deletion of some or the entire short arm. This variation in chromosome abnormality helps to explain the considerable phenotypic variation seen in this syndrome.

Molecular studies have shown that approximately 80 per cent of monosomy X cases are caused by a meiotic error in the father. The prevalence of the 45, X karyotype is low compared to those of other sex chromosome abnormalities, with only about 1/2,500 to 1/5,000 live-born females having the disorder. The 45, X karyotype is very common among conceptions, however, accounting for 15 per cent to 20 per cent of the chromosomal abnormalities seen among spontaneous abortions. Thus, the great majority (>99 per cent) of 45, X conceptions do not survive prenatally. Those who do not survive till term are chromosomal mosaics with mosaicism of the placenta alone (confined placental mosaicism) being especially common and survive. The presence of some normal cells in mosaic fetuses contributes to survival.

Mutations in the SHOX gene, which encodes a transcription factor, produce short stature. The gene is located on the distal tip of the X and Y short arms (in the pseudoautosomal region). Thus, it is normally present in two copies in both males and females. In females with Turner syndrome, however, this gene would be present in only one active copy, and the resulting haploinsufficiency is likely to contribute to short stature.

It has been recently reported that females with Turner syndrome who receive the X chromosome from their father have higher verbal IQ scores and better social cognition than those who receive the X chromosome from their mother. This difference implies the presence of a genomic imprinting effect on a specific region of the X chromosome. Interestingly partial deletions of the X chromosome show that the imprinted region escapes X inactivation. Because normal females inherit active copies of this region from both parents, while males inherit a copy only from their mothers (the existence of a Y homolog is not yet known), this finding could have important implications for general male-female differences in social cognition and verbal intelligence.

2.21.2 Klinefelter Syndrome

In 1942 Harry Klinefelter described this syndrome. It is seen in approximately 1/1,000 male births. Although Klinefelter syndrome is a common cause of primary hypogonadism in the male, the phenotype is less striking than those of the syndromes described thus far. Males with Klinefelter syndrome tend to be taller than average, with disproportionately long arms and legs. Clinical examination of post-pubertal patients reveals small testes (<10 ml), and most males with Klinefelter syndrome are sterile as a result of atrophy of the seminiferous tubules. Gynecomastia (breast development) is seen in approximately one-third of affected males and leads to an increased risk of breast cancer. In addition, there is a predisposition for learning disabilities and sub-average intelligence. Although males with Klinefelter syndrome are not usually mentally retarded, the IQ is on average 10 to 15 points below that of the affected individual's siblings.

The extra X chromosome is derived maternally in about 50 per cent of Klinefelter cases, and the syndrome increases in incidence with advanced maternal age. Mosaicism is seen in about 15 per cent of patients. Despite the relatively mild phenotypic features of this disorder, it is estimated that at least half of 47, XXY conceptions are spontaneously aborted. It is detected most of the time until after puberty.

Individuals with the 48, XXXY and 49, XXXXY karyotypes have also been reported. Because they have a Y chromosome, they still have a male phenotype, but the degree of mental deficiency and physical abnormality gets enhanced with each additional X chromosome addition.

Secondary sex characteristics in post pubertal males with Klinefelter syndrome may get improved by testosterone therapy. Mastectomy for gynecomastia is sometimes recommended Klinefelter syndrome (47, XXY) males are taller than average, may have a reduction in IQ, they are usually sterile.

2.21.3 Trisomy X

In approximately 1/1,000 females 47, XXX karyotype occurs which is usually because of the benign consequences. Rarely overt physical abnormalities are seen. The females with trisomy X sometimes suffer from sterility, menstrual irregularity, or mild mental retardation. As in Klinefelter syndrome, females with the 47, XXX karyotype are often first confirmed infertility clinics. The majority of 47, XXX females are the result of nondisjunction in the mother, and, the incidence increases among older mothers as seen in other trisomies. Females have also been seen with four, five, or even more X chromosomes. Each additional X chromosome causes increased mental retardation and physical abnormality.

2.21.4 47, XYY Syndrome

There are 47, XYY karyotype due to the presence of Y chromosomes they are male. They tend to be taller than average, and they have a 10 to 15 per cent reduction in average IQ. Interestingly, various studies show that such type of karyotype was high (1/30) in male prison populations as compared to the normal male population 1/1000. This revealed that this type of karyotype may confer a predisposition to violent, criminal behavior. These individuals suffer from minor behavioral disorders, such as hyperactivity, attention deficit disorder, and learning disabilities.

The 47, XYY karyotype is another viable trisomy in human beings. These individuals are male, and except for a tendency to be taller than 46, XY men, they do not show a consistent syndrome of characteristics. A possible connection between the XYY karyotype and antisocial behavior has not been studied thoroughly enough to draw a firm conclusion. All the other trisomy's in human beings are embryonic lethal, demonstrating the importance of correct gene dosage. In plants like datura trisomies are viable but in human beings most of the time these are lethal some of the chromosomal imbalances are shown in **Table 2.10**.

2.21.5 Monosomy

Monosomy means one chromosome is missing in a diploid individual. In human beings, the 45, X karyotype is a monosomic condition but viable. Henry H. Turner first described the condition in 1938; thus, it is now known as Turner syndrome after his name. These individuals diploid complement of autosomes but single X. They are female phenotypically they are almost always sterile as ovaries are rudimentary. 45, X individuals are usually short stature; webbed necks, hearing deficiencies, and significant cardiovascular abnormalities.

The mechanism of 45, X is that egg or sperm may lack a sex chromosome as a sex chromosome is lost after fertilization (**Figure 2.22**). People with the 45, X

Table 2.10: Non-disjuction causes Aneuploidy

Karyotype	*Chromosome Formula*	*Clinical Syndrome*	*Estimated Frequency of Birth*	*Phenotype*
47, +21	2n + 1	Down	1/700	Short, broad hands with palmar crease, short stature, hyper flexibility of joints, mental retardation, broad head with round face, open mouth with large tongue, epicanthal fold.
47, +13	2n + 1	Patau	1/20,000	Mental deficiency and deafness, minor muscle seizures, cleft lip and/or palate, cardiac anomalies, posterior heel prominence.
47, + 18	2n + 1	Edward	1/8000	Congenital malformation of many organs, low – set, malformed ears, receding mandible, small mouth and nose, mental deficiency, horseshoe or double kidney, short sternum; 90 per cent die within first six months after birth.
45, X	2n – 1	Turner	1/2500 females' births	Female with retarded sexual development, usually sterile, short stature, webbing of skin in neck region, cardiovascular abnormalities, hearing impairment.
47, XXY 48, XXXY 48, XXYY 49, XXXXY 50, XXXXXY	2n + 12n + 22n +22n + 32n + 4	Klinefelter	1/500 male births	Male, sub fertile with small testes, developed breasts, feminine – pitched voice, knock –knees, long limbs.
47, XXX	2n + 1	Triple –X	1/700	Female with usually normal I genitalia and limited fertility, slight mental retardation.

(a) Origin of monosomy at fertilization.

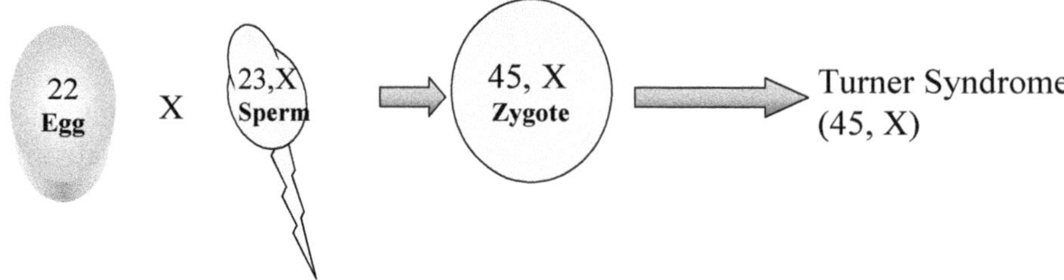

(b) Origin of monosomy in the cleavage division following fertilization.

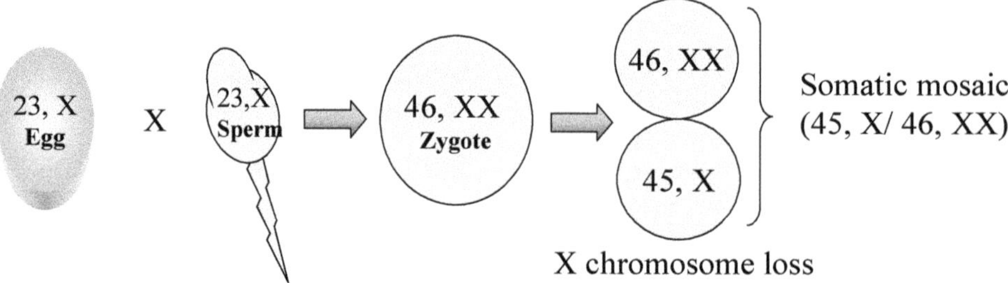

Figure 2.22: Origin of Turner Syndrome Karyotype.

karyotype have no Barr bodies in their cells, indicating that the single X chromosome that is present is not inactivated. Why, then, should turner patients, who have the same number of active X chromosomes as normal XX females, show any phenotypic abnormalities at all? The answer probably involves a small number of genes that remain active on both of the X chromosomes in normal 46, XX females.

These non-inactivated genes are apparently needed in a double dose for proper growth and development. The finding that at least some of these special X –linked genes are also present on the Y chromosome would explain why XY males grow and develop normally. In addition, the X chromosome that has been inactivated in 46, XX females is reactivated during oogenesis, presumably because two copies of some X-linked genes are required for normal ovarian function. 45, X individuals, who have only one copy of these genes, cannot meet this quantitative requirement and therefore are sterile. This latter possibility is supported by the finding that many turner individuals are somatic mosaics. These people have two types of cells in their bodies; some are 45, X, and others are 46, XX. This karyotypic mosaicism arises when an X chromosome is lost during the development of a 46, XX zygote. All the descendants of the cell in which there is a loss are 45 X. If the loss occurs early in development, most of the body's cells will be aneuploid and the individual will show the features of Turner syndrome. If the loss occurs later, the aneuploid cell population will be less hence the severity of the syndrome is likely to be less.

XX/XO chromosome mosaics are also seen in fruit flies (Drosophila). As discussed before sex in these species is determined by the ratio of X chromosomes to autosomes, such flies are part female and part male. XX cells become male. Flies with both male and female structures are a mixture of male and female and called gynandromorphy (from Greek words meaning "woman" "man" and "form").

Curiously, the cognate of the XO turner karyotype in the mouse exhibits no anatomical abnormalities. This finding implies that the mouse homologs of the human genes that are involved in Turner syndrome need to be present in one copy for normal growth and development.

2.21.6 Chromosome Segments may get Deleted or Duplicated

Deletion means the loss of a segment of a chromosome. The deletion can occur at the end of the chromosome either at short arm or long arm of the chromosome. This type of deletion is called terminal deletion. Deletions can occur within the chromosome. These are called interstitial deletions. Deletion can take place in any of the chromosomes. There are some deletions that have typical phenotype these are Wolf-Hirschhorn syndrome (4p-), and cri-du-chat syndrome (5p-). They both involve the loss of the distal end of the short arm. The deletions of 18p, 18q, 22q, 21q, 15q, 11p, 17p and 4q are clinically recognizable. Some of the microdeletion syndromes are listed in **Table 2.11**. Duplication means the addition of a segment from the homologous chromosome.

Table 2.11: Micro Deletion Syndromes

Syndrome	Chromosomal Deletion	Clinical Features
Aniridia/ Wilms tumor	11p13	Mental retardation, aniridia, predisposition to Wilms tumor, genital defects
Williams	7q1	Developmental disability, characteristic facies, supravalvular aortic stenosis
DiGeorge anomaly/velo-cardio-facial	22q11	DiGeorge anomaly/ characteristic facies, cleft palate, heart defect
Smith – Magenis	17p11.2	Mental retardation, hyperactivity, dysmorphic features, self –destructive behavior
Langer – Giedion	8q24	Characteristic facies, sparse hair, exocytosis, variable mental retardation
Miller-Dieker	17q13.3	Lissencephaly, characteristic facies
Prader-Willi	15q11-13	Mental retardation, short stature, obesity, hypotonia, characteristic facies, small feet

A missing chromosome segment is referred to either as a deletion or as a deficiency cytologically large deletions can be detected but not the small ones. In a diploid organism, the deletion of a chromosome segment makes part of the genome hypoploid. This hypoploidy may be associated with a phenotypic effect, especially if the deletion is large. A classic example is the *cri-du-chat* syndrome (from the French words for "cry of the cat") in human beings (**Figure 2.23**). This condition is caused by a conspicuous deletion in the short arm of chromosome 5. Approximately half the chromosomal arm is missing.

A missing chromosome segment is referred to either as a deletion or as a deficiency cytologically large deletions can be detected but not the small ones. In a diploid organism, the deletion of a chromosome segment makes part of the genome hypoploid. This hypoploidy may be associated with a phenotypic effect, especially if the deletion is large. A classic example is the *cri-du-chat* syndrome (from the French words for "cry of the cat") in human beings (**Figure 2.23**). This condition is caused by a conspicuous deletion in the short arm of chromosome 5. Approximately half the chromosomal arm is missing.

An extra segment can come from the other chromosome also. The effect remains the same. These aberrations may be seen by examining the mitotic chromosomes through staining with banding agents such as quinacrine or Giemsa. However, small aberrations are difficult to detect in this way and are usually identified by other genetic and molecular techniques. The best organism for studying deletions and duplications is Drosophila, where the polytene chromosomes show an unparalleled opportunity for detailed cytological analysis.

In the 1930s C. B. Bridges analyzed X chromosomes carrying the Bar mutation and found that the 16A region, which contains a gene for eye shape, had been tandemly duplicated. Tandem triplications

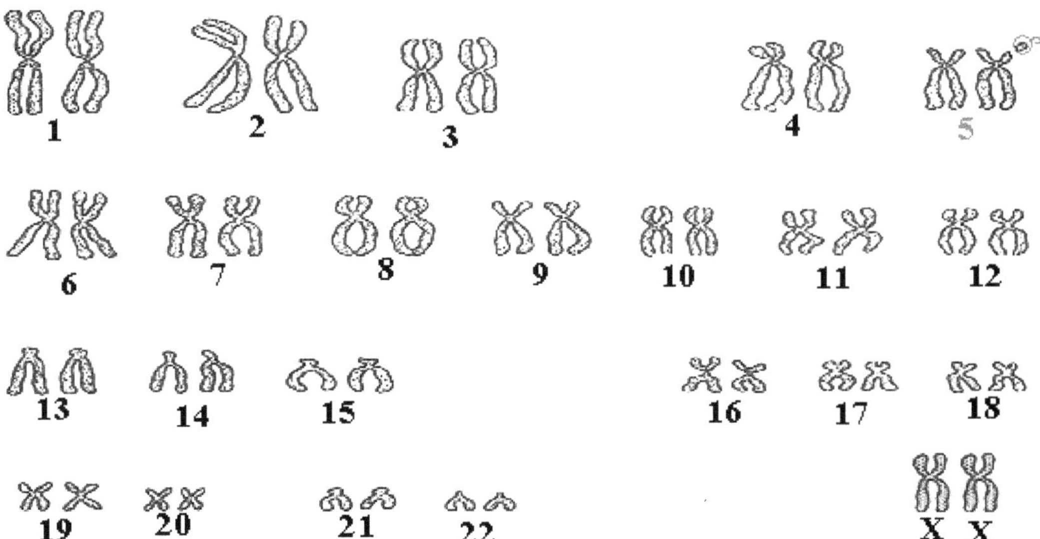

Figure 2.23: Cri-du-chat Syndrome. Patient with cri-du-chat syndrome. Karyotype of infant with cri-du-chat syndrome show there is a deletion in the short arm of chromosome 5 indicated by arrow.

of 16A were also observed, and in these cases, the compound eye was extremely small-a phenotype referred to as double-bar. Many other tandem duplications have been found in *Drosophila,* where polytene-chromosome analysis makes their detection relatively easy.

Today, molecular techniques have made it possible to detect very small tandem duplications in a wide variety of organisms. For example, the genes that encode the hemoglobin proteins have been tandemly duplicated in mammals. Gene duplications appear to be relatively common and provide a significant source of variation for evolution.

2.22 Rearrangements of Chromosome Structure

Many kinds of chromosomal rearrangements have been identified. The description of which is shown below.

2.23 Inversions

Inversions involve only one chromosome in which breaks occur at two points and, in the process of repair, the intervening segment is rejoined in an inverted or opposite fashion. Since there is no loss or a gain of chromosomal material, inversion carriers are normal. It is paracentric if the inverted segment is on the long arm or the short arm and does not include the centromere. The inversion is pericentric if breaks occur on both the short arm and the long arm and the inverted segment contains the centromere. In meiosis, the normal chromosome and the inverted chromosome will form a loop to allow the pairing of specific DNA sequences. The exchange of chromatids (a normal event in meiosis between homologous chromosomes), occurs within the inversion loop result in gametes having both deletions and duplications that are often not compatible with life and result in a high incidence of abortions (**Figure 2.24**).

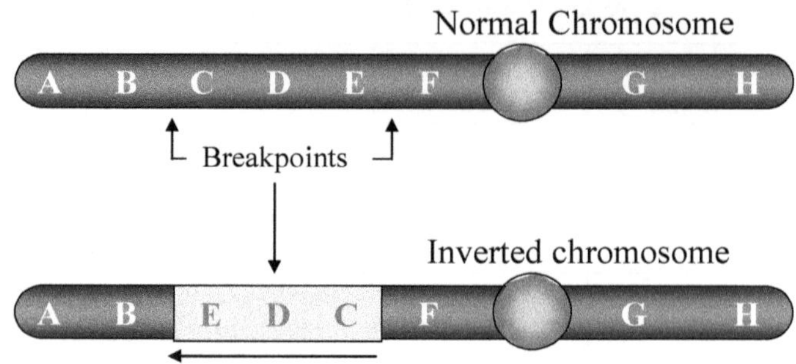

Figure 2.24: Example of Inversion.
The chromosome has been broken at two points, and the segment between them (containing regions C, D, and E) has been inverted.

If an acrocentric chromosome acquires an inversion with a breakpoint in one of the chromosome's arms (that is, a pericentric inversion) it can be transformed into a metacentric chromosome. However, if an acrocentric chromosome acquires an inversion in which both the breaks are in the chromosome's long arm (that is, a paracentric inversion) the morphology of the chromosome will not be changed (**Figure 2.25**).

2.24 Translocations

Translocation involves two non-homologous chromosomes (*e.g.*, chromosome 2 and chromosome 6). First, the break occurs and then broken chromosomes reunite, and a segment of chromosome 2 can become attached to chromosome 6 and vice versa. It is the most clinically significant chromosomal rearrangement (**Figure 2.26**).

In most cases, there is no loss or gain of chromosomal material during the exchange process. Translocation carriers are estimated to be normal phenomena in 1/500 individuals.

On occasion, balanced translocation carriers (on karyotype studies) are clinically abnormal. The reason could be that the break may have occurred in the middle of a gene resulting in the formation of an abnormally short, nonfunctional protein.

(a) Pericentric inversion – includes centromere

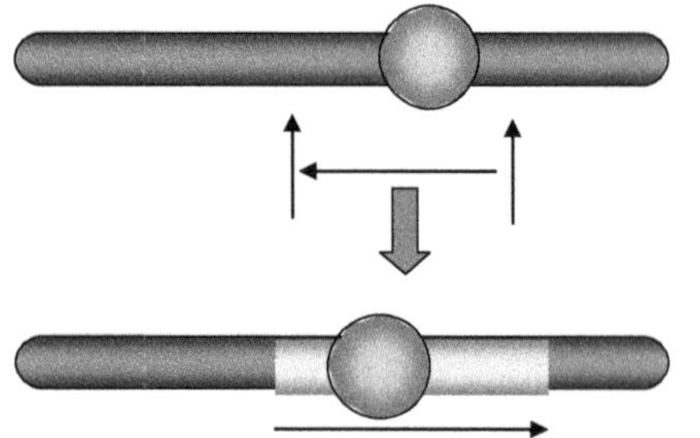

(b) Pericentric inversion – excludes centromere

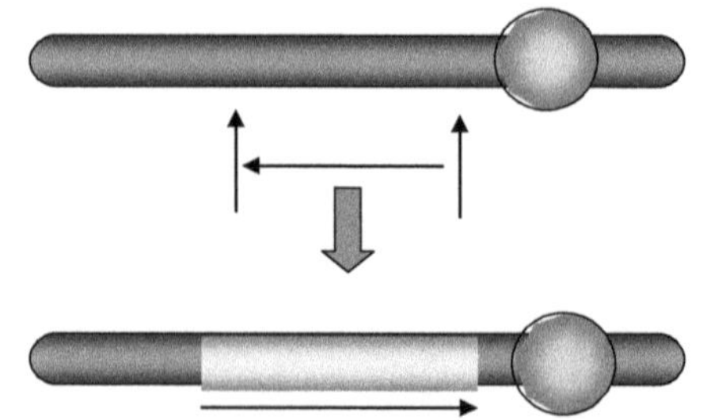

Figure 2.25: Pericentric and Paracentric Inversions.
A pericentric inversion changes the size of the chromosome because the centromere is included within the inversion.

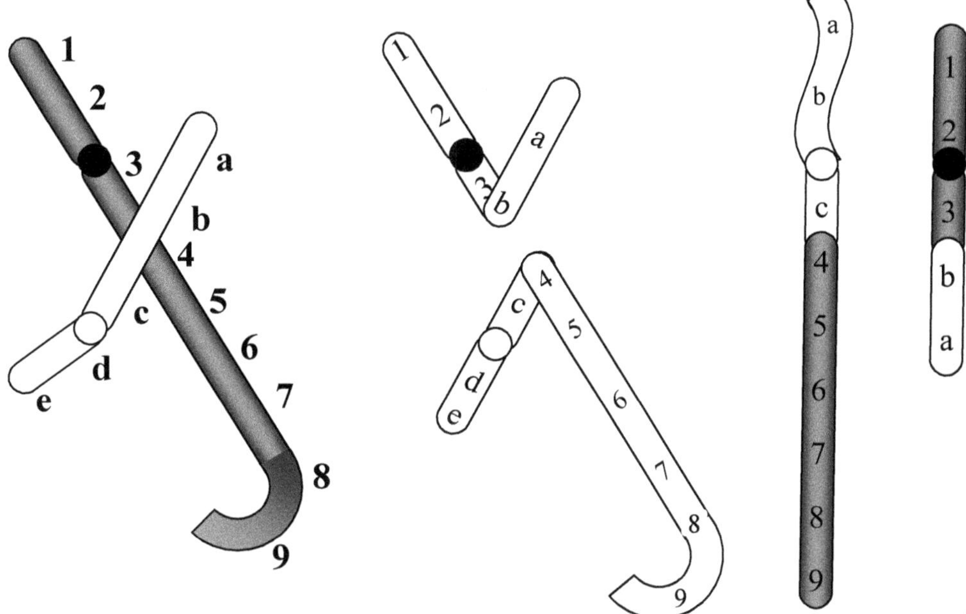

Figure 2.26: Translocation.

Individuals and families have described with a translocation chromosome abnormality and a concurrent genetic condition; the genetic condition occurring because of the chromosome breakpoint is amid a gene. Various studies of the informative families have led to the localization of specific genes to specific chromosome segments (*e.g.*, Duchene muscular dystrophy on Xp21, neurofibromatosis on 17q, retinoblastoma on 13q14, *etc.*).

When the balanced translocation is found at the time amniocentesis, both the parents are investigated for their chromosomes. If the translocation is familial (one of the parents is a normal translocation carrier), then it is safe to assume that the fetus carrying a similar translocation is going to be normal. If the translocation is *de novo* (the parents are not translocation carriers), then the fetus has a 10 per cent empiric risk of having a possible genetic abnormality. It may seem surprising that the risk is not much greater. This is because only 10 per cent of the genome carries coding sequences for genes. The other 90 per cent of the genome consists of non-coding sequences.

As previously noted, people who carry balanced translocations have a high risk of miscarriages and abnormal offspring because they are more likely to create unbalanced gametes. In meiosis, the normal chromosomes and the translocated chromosomes pair up by creating a cross figure (quadrivalent) (**Figure 2.27**). The chromosomes are distributed to the daughter cells by the centromeres which are attached to spindle fibers. The spindle fibers contract and draw the attached chromosome to each of the poles.

Reciprocal translocation is defined as exchange of genetic material between two different chromosomes.

2.25 Use in Clinical Context

Reciprocal translocations occur when part of one chromosome is exchanged with another. Translocations can disrupt functional parts of the genome and have implications for protein production with phenotypic consequences.

Reciprocal translocations are usually balanced and so may not have apparent functional implications. However, they can become unbalanced in the parents' offspring if the pair of chromosomes where genetic material has been exchanged is not inherited together. Unbalanced translocations can result in missing genes or extra copies of genes which can have clinical impact. It is possible to see large translocations using a karyotyping. Smaller translocations can be detected using microarrays and sequencing. Reciprocal translocation is shown in **Figure 2.28**.

Noted above is the usual 2:2 segregation in meiosis where two chromosomes of the original four chromosomes, are distributed to each of the two daughter cells. On occasion, a 3:1 segregation occurs in meiosis where three chromosomes go to one daughter cell and one chromosome goes to the other daughter cell; in effect an aneuploidy nondisjunction affecting translocation chromosomes.

The risk of abnormal offspring in translocation families depends to some extent on how the family was assessed. If the family is being examined because they

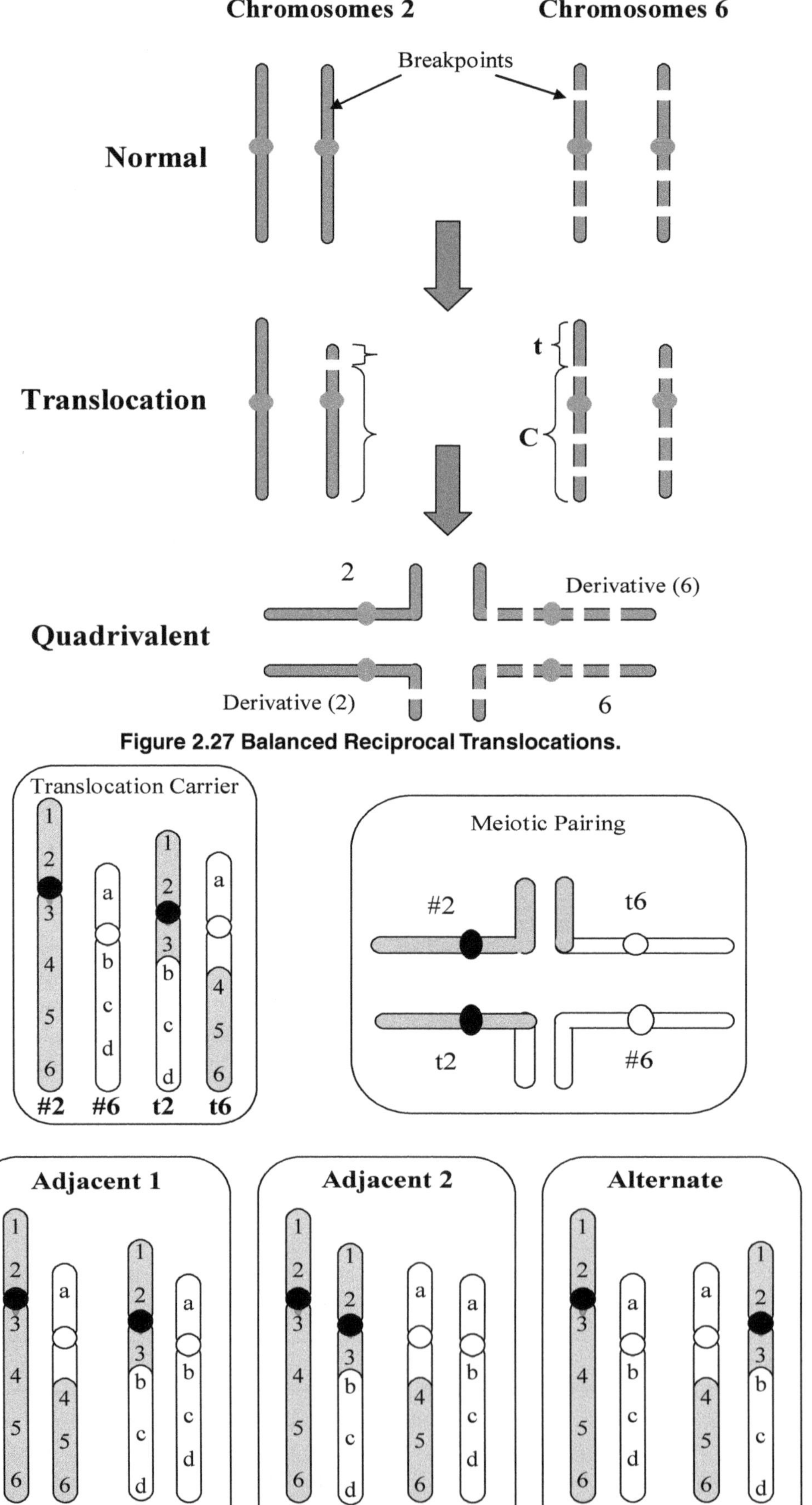

Figure 2.27 Balanced Reciprocal Translocations.

Figure 2.28: Reciprocal Translocation its Segregation and Six Possible Gamete Formation.

have a child with a chromosome abnormality, then obviously the unbalanced translocation karyotype is compatible with life. In such family's empiric data suggests that the risk of recurrence is approximately 15 per cent. If, on the other hand, the family comes to clinical attention because of multiple miscarriages, or as part of an infertility workup, then the unbalanced translocation is probably not compatible with life and the risk of abnormal offspring is approximately 1.5 per cent.

A Robertsonian translocation is a particular type of translocation involving the reciprocal transfer of the long arms of two of the acrocentric chromosomes: 13, 14, 15, 21, or 22. On rare occasions, other non-acrocentric chromosomes undergo Robertsonian translocation, a reciprocal transfer of the whole long or short arms close to the centromere. A relatively common Robertsonian translocation is between chromosome 14 and chromosome 21. During meiosis, a trivalent will be formed.

If there occurs adjacent 1 and adjacent 2 segregation, the resultant gametes cause trisomy 14, monosomy 14, trisomy 21, or monosomy 21 after fertilization. With alternate segregation, the gametes produced have a balanced translocation 14; 21 or the normal chromosome complement following fertilization. There is occurrence of following gametes one normal, one balanced translocation, and four unbalanced translocation complements.

Of the latter four, only trisomy 21 reaches term.

Theoretically, a person who carries a 14; 21 translocations has a 1/3 chance of having a normal child, a 1/3 chance of having a child who carries a balanced translocation, and a 1/3 chance of having a child with Down syndrome. However, the actual risk for Down syndrome is much smaller because many of the trisomy 21 fetuses are aborted spontaneously. The empirical risk of having a child with Down syndrome is around 3 to 5 per cent if the father is the carrier and 10 to 15 per cent if the mother is the carrier. The male translocation Down syndrome patient will have a karyotype of 46, XY,t (14;21) inferring that there are two normal chromosome 21 plus a third 21 that is attached to chromosome 14. The carrier mother of such a patient will have a karyotype of 45, XX,t (14;21) inferring a single chromosome 21, and a second 21 attached to chromosome 14.

Coordinately distributed to opposite poles in the first meiotic division, chromosome disjunction in translocation heterozygotes is somewhat uncertain process, prone to produce aneuploid gametes. There are three possible disjunction events, as shown in **Figure 2.29**. If centromeres 2 and 4 move to the same pole forcing 1 and 3 to the opposite pole, all the resulting gametes will be aneuploid –because some chromosome segments will be deficient for genes, and others will be duplicated (**Figure 2.29**). Similarly, if centromeres 1 and 2 move to one pole and 3 and 4 to the other, only aneuploid gametes will be produced (**Figure 2.29**). Each of these cases is referred to as adjacent disjunction because centromeres that were next to each other in the cruciform pattern moved to the same pole. When the centromeres that move to the same pole are from different chromosomes (homologous), the disjunction is referred to as adjacent I (**Figure 2.29**); when the centromeres that move to the same pole are from the same chromosome (that

Adjacent disjunction I

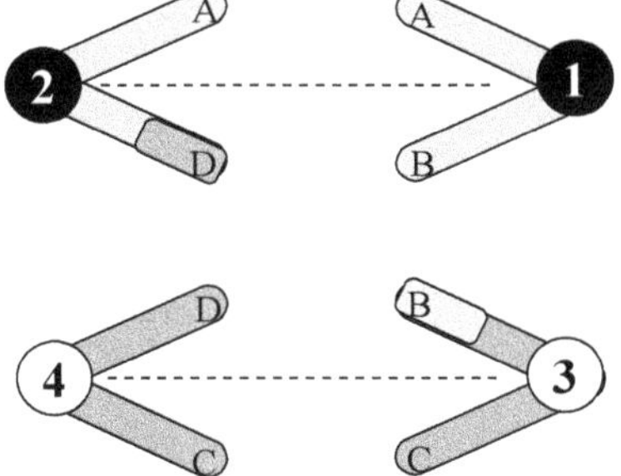

Centromeres 1 and 3 go to one pole and centromeres 2 and 4 go to the other pole, producing aneuploid gametes.

Adjacent disjunction II

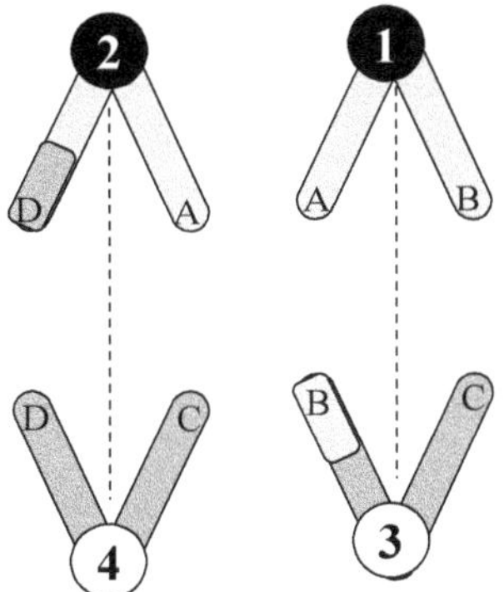

Centromeres 1 and 2 go to one pole and centromeres 3 and 4 go to the other pole, producing aneuploid gametes.

Alternate disjunction

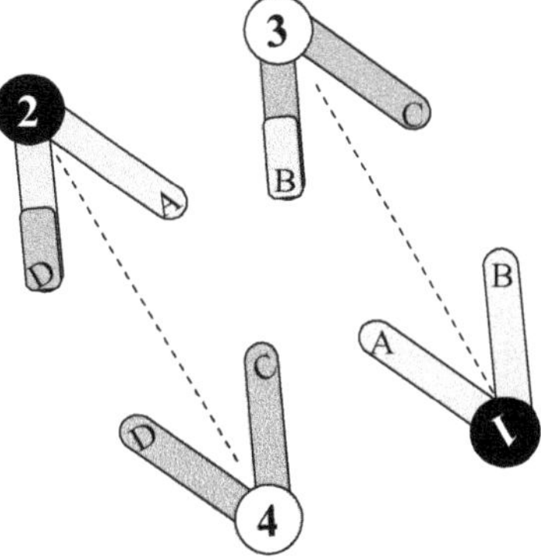

Centromeres 2 and 3 go to one pole and centromeres 1 and 4 go to the other pole, producing euploid gametes.

Figure 2.29: Types of Disjunction in a Translocation Heterozygote.

is, they are homologous), the disjunction is referred to as adjacent II (**Figure 2.29**). Another possibility is that centromeres 1 and 4 move to the same pole, forcing 2 and 3 to the opposite pole. This is called alternate disjunction, resulting into euploid gametes, although half of them will carry only translocated chromosomes (**Figure 2.29**).

The production of aneuploid gametes by adjacent disjunction explains why translocation heterozygotes have reduced fertility. When such gametes fertilize a euploid gamete, the resulting zygote will be genetically unbalanced and therefore will be unlikely to survive. In plants, aneuploid gametes are themselves often in viable, especially on the male side, and fewer zygotes are produced. Translocation heterozygotes are therefore characterized by low fertility.

2.26 Compound Chromosomes and Robertsonian Translocations

The compound chromosome was discovered in 1922 for the first time by Lillian Morgan, she was wife of T.H. Morgan. This compound chromosome was apparently formed by the fusion of the two X chromosomes in Drosophila, creating double–X or attached–X chromosomes. The discovery was made through genetic experimentation rather than cytological analysis. Lillian Morgan crossed females homozygous for a recessive X–linked mutation to wild –type males. From such a cross, it is expected that all the daughters to be of wild –type and all the sons to be mutant. However, Morgan observed just the opposite: all the daughters were mutant, and all the sons were wild–type. She worked further and found that the X chromosomes in the mutant females were attached to each other. **Figure 2.30** illustrates the genetic significance of this attachment. The attached – X females produce two kinds of eggs, Diplo-X and nullo-X, and their mates produce two kinds of sperm, X –bearing and Y-bearing. The union of these gametes in all possible ways produces two kinds of viable progeny: mutant XXY females, which inherit the attached–X chromosomes from their mothers and Y chromosomes from their fathers; and phenotypically wild–type XO males, which inherit an X chromosome from their fathers and no sex chromosomes from their mothers. Because the Y chromosome is needed for fertility, these XO males are sterile. Lillian Morgan was able to propagate the attached–X chromosomes by backcrossing XXY females to wild–type XY males from another stock. Because the males of this cross inherited a Y chromosome from their mothers, they were fertile and could be crossed to their XXY sisters to establish a stock in which the attached –X

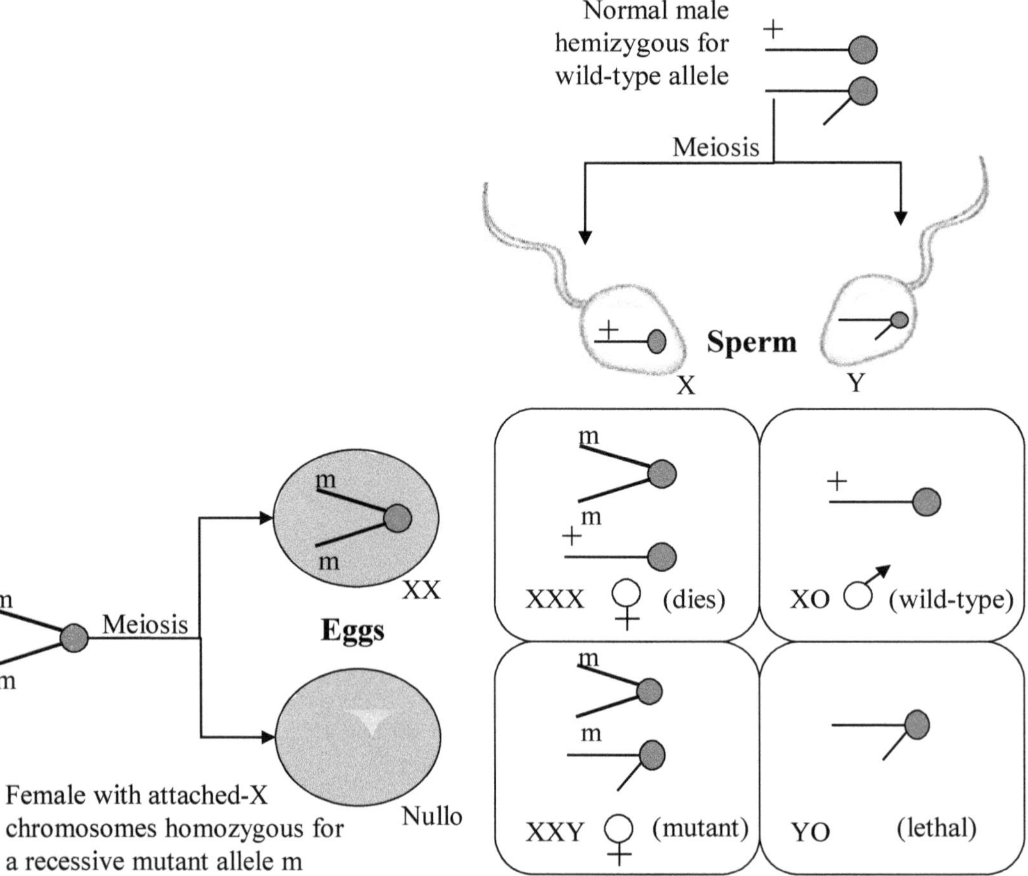

Figure 2.30: Attached X Chromosome.

chromosomes were permanently maintained in the female line.

Non homologous chromosomes can also fuse at their centromeres, forming a structure called a Robertsonian translocation (**Figure 2.31**), named after F.W. Robertson. Fusion of two acrocentric chromosomes results in a metacentric chromosome; the tiny short arms of the participating chromosomes are simply lost in this process.

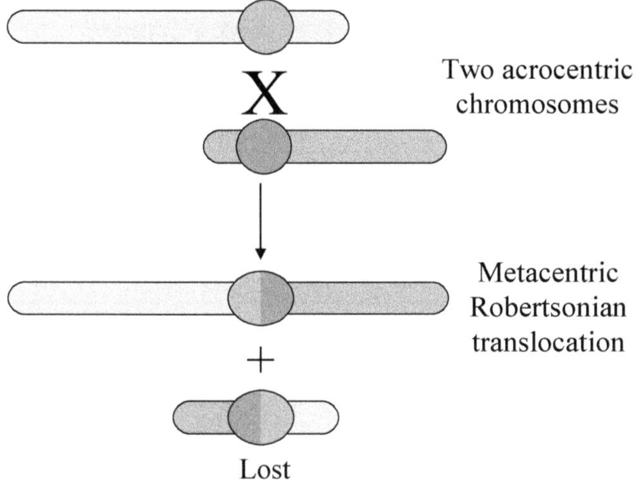

Figure 2.31: Formation of a Metacentric Robertsonian Translocation by Exchange between Two Non-homologous Acrocentric Chromosomes.

Chromosomes can also fuse end–to–end and have two centromeres. If one of the centromeres is inactivated, the chromosome fusion will be stable. Such type of fusion may cause evolution of our own species. Human chromosome 2, which is a metacentric, has arms that correspond to two different acrocentric chromosomes in the genomes of the great apes. Detailed cytological analysis has shown that the ends of the short arms of these two chromosomes apparently fused to create human chromosome 2.

2.27 Isochromosome

Isochromosome formation is a relatively frequent chromosomal aberration, mainly in X chromosomes. The chromosomes are not divided along their length, but they divide transversely. The resulting isochromosomes (karyogram) either have two short or two long arms. Persons with this X chromosome anomaly have the same phenotype as patients suffering from Turner's syndrome (45, X0). This is because one of the X chromosome arms is missing.

2.28 Pregnancy Loss due to Chromosomal Abnormalities

Chromosome abnormalities are the most important known cause of pregnancy loss. It is found in approximately 10 per cent to 15 per cent of conceptions have a chromosome abnormality. Approximately 95 per cent of these chromosomally abnormal conceptions are lost before term. Karyotype studies of abortus material show that about 50 per cent of the chromosome abnormalities are trisomies, 20 per cent are monosomies, 15 per cent are triploids, and the remainder consists of tetraploids and structural abnormalities. Some chromosome abnormalities which are common at conception very rarely or never survive to term. For example, trisomy 16 is thought to be the most common trisomy at conception, but it is never seen in live births.

Cytogenetic studies are important for the couple undergoing recurrent fetal losses. This can be done either by conventional karyotyping or FISH.

2.29 Chromosome Abnormalities and Clinical Phenotypes

1. Most chromosome abnormalities (especially those involving autosomes) are associated with developmental delay in children and mental retardation in older individuals. This reflects the fact that many our genes, perhaps 25,000 or more, participate in the development of the central nervous system. Thus, a chromosome abnormality, which typically may affect hundreds or thousands of genes, is very likely to involve genes affecting nervous system development.
2. Most chromosome syndromes involve alterations of facial morphogenesis that produce characteristic facial features. For this reason, the patient often resembles other individuals with the same disorder more than members of his or her own family. It is usually accompanied with facial features and minor anomalies of the head and neck are the best aids to diagnosis.
3. Growth delay (short stature and/or poor weight gain in infancy) is commonly seen in autosomal syndromes.
4. Congenital malformations, especially congenital heart anomalies, occur with increased frequency in most autosomal chromosome disorders. These defects occur in specific patterns. For example, while AV canals and ventricular septal defects are common in children with Down syndrome, other congenital heart defects, such as aortic coarctation or hypoplastic (underdeveloped) left ventricle, are seldom seen in these children but may be seen in those with Turner syndrome.

The most common clinical indications for a chromosome analysis are the newborn with multiple congenital malformations or a child with developmental delay. Chromosome abnormalities typically result in developmental delay, mental retardation, characteristic facial features, and various types of congenital malformations. Despite some overlap of phenotypic features, many chromosome abnormalities can be recognized by clinical examination.

2.30 Cancer and Cytogenetics

Most of the chromosomal rearrangements take place at the time of meiosis *i.e.* reduction division at the time of gamete formation. However, at the time of mitosis chromosomal rearrangements are also seen, and these are responsible for large number of malignancies. The first such cancer reported was chronic myelogenous leukemia. There is a chromosomal translocation between chromosome 22 onto the long arm of chromosome 9 in fact a small distal portion of 9q is translocated to chromosome 22 (**Figure 2.32**). Now the genes near the translocation position have been investigated. A proto-oncogene abl1 is translocated from 9q to 22q. Translocin an increase into increase in the tyrosine kinase activity. Increase in the tyrosine activity affects the lymphocytes therefore resulting in CML.

Burkitt lymphoma is caused due to a reciprocal translocation involving chromosomes 8 and 14 where the *myc* proto oncogene moves from 8q24 to 14q32, which is located near the immunoglobulin heavy chain loci. This activates *myc* gene resulting in Burkit lymphoma.

More than 100 different rearrangements involving nearly every chromosome have been observed in over 40 different types of cancer. Some of these are summarized in **Table 2.12**.

Table 2.12: Some known Chromosomal Aberrations in Cancers

Condition	*Most Common Chromosome Aberration*
Solid tumors	
Meningioma	Monosomy 22
Retinoblastoma	del(13) (q14)
Wilms tumor	del(11)(p13)
Burkitt lymphoma	t(8;14)(q24;q32)
Ewing sarcoma	t(11;22)(q21;q12)
Leukemias	
Acute promyelocytic leukemia	t(15;17)(q22;q11-12)
Acute non-lymphocytic leukemia	+8, -7, -5, del(5q),del(20q)
Chronic myelogenous leukemia	t(9;22)(q34;q11)
Acutre myeloblastic leukemia	t(8;21)(q22;q22)

Balanced translocations in somatic cells can sometimes cause malignancies by interrupting or altering genes or their regulatory sequences.

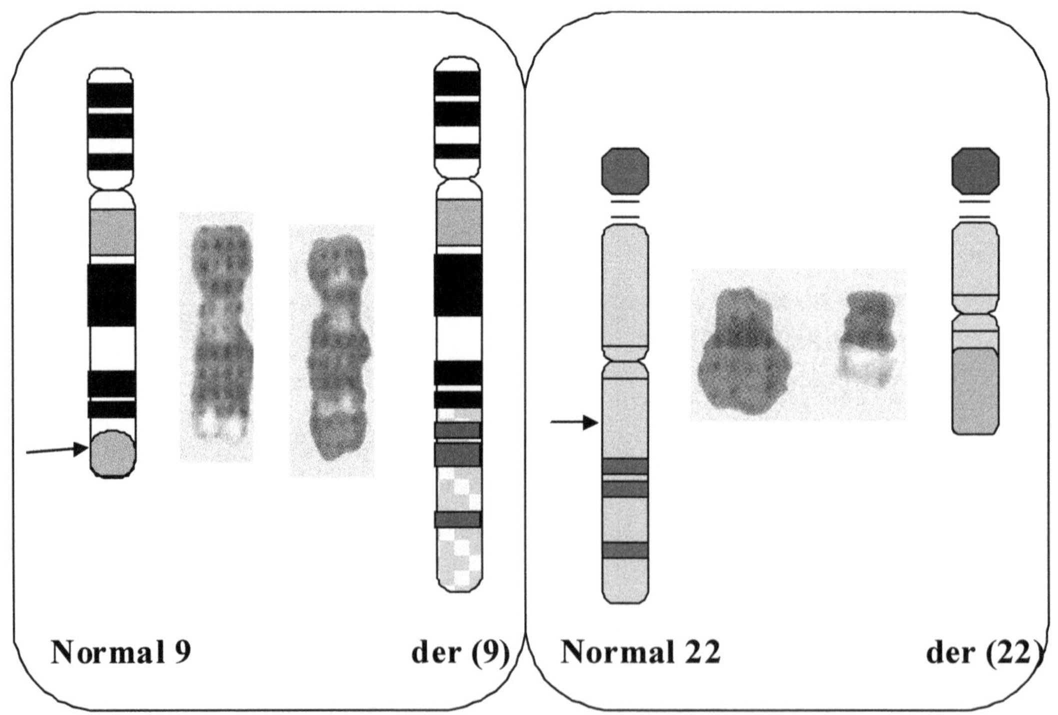

Figure 2.32: Philadelphia Chromosome is a Reciprocal Translocation between Chromosome 22 and the Long Arm of Chromosome 9.

2.31 Syndromes due to Chromosome Instability

Chromosomal breaks are seen more often in many autosomal recessive diseases. These are called chromosomal instability syndromes, for example ataxia-telangiectasia B loom syndrome (high incidence of somatic cell sister chromatid exchange) Fanconi anemia (exposure to certain alkylating agents) and Xeroderma Pigmentosa.

Chromosomal breaks disturb the DNA repair mechanisms hence these patients are at higher risk to develop cancer. Recently over 60 000 probes are spotted on a standard microscope slide. These are known as microarrays The DNA samples are hybridized to these arrays. Probes are cloned genomic DNA inserts, PCR amplicons, or oligonucleotides targeted towards the sequences of interest. Microarray is such a robust technique that inserting clones of over 100 kb in size can be used for higher resolution analysis of DNA–protein interaction or chromatin modifications, over three million sequences of 1 kb in size. So far it is not yet achieved hence higher-resolution arrays are mostly restricted to smaller areas of the genome or specific chromosomes.

In a typical experiment using a spotted clone array, a test DNA and a reference DNA are labeled with different fluorochromes and co-hybridized (usually in the presence of unlabeled DNA to block shared repeat sequences) to the array after hybridization and washing, the microarray is scanned and the fluorescence intensity of each fluorochrome for each feature spot is quantified about the reference DNA. The relative fluorescence intensity of the test to reference sample is determined and a difference between the abundance of complementary sequences in the two hybridized DNAs is quantified. This is like the comparative genomic hybridization (CGH) were differentially labeled test and reference genomes are hybridized to metaphase spreads and the relative genomic copy number is determined along each chromosome, thus finding out the aberration in chromosome structure. In CGH, the detection of gains and losses in the genome is limited to a resolution of 5 Mb by the highly condensed target DNA in the metaphase chromosomes. Microarrays are now replacing metaphase spreads as the target for hybridization in these studies so that in array-CGH, resolution is now limited only by the size and density of the target sequences. Some studies utilize microarrays that constitute large insert clones spaced at approximately one clone per Mb, but higher resolution arrays comprising overlapping clone sets for chromosomes or chromosome regions and more recently for the whole genome are being employed for more detailed and sensitive studies.

Micro-array or Array-CGH has been used to analyze patients with chromosomal rearrangements. Microarray or CGH is useful to identify small microdeletions and microduplications in patients with learning disability and dysmorphology who otherwise show balanced karyotype. Most of the microdeletions detected are *de nova* while microduplications are inherited. There is another technique called array painting. The two derivative chromosomes of imbalanced translocation are separated by flow cytometry and are labeled and hybridized with a genomic clone array. The ratio change is estimated, and data is plotted according to chromosomal location.

If higher-resolution arrays are used the specific breakpoints and the mechanisms involved in chromosome rearrangement can be studied. 15-20 kb resolution can be obtained. If we use single coding exons of genes and define copy number changes a resolution is up to 150 bp.

Another powerful technique used is chromatin immunoprecipitation (ChIP) with the help of this technique modification in chromatin material or protein complexes can be studied, quantified and mapped using microarrays. The combined use of immunoprecipitation with arrays is called ChIP-on-chip or ChIP-chip.

ChIP-on-chip is performed using the following steps. The first important step is the cross-linking of protein to the DNA with formaldehyde. The cross-linked DNA is then broken and further immunoprecipitated with a specific antibody against the protein of interest. This precipitates the protein along with the DNA cross-linked to it. The cross-linking is then reversed, and the associated DNA is extracted. In many studies, the yield of extracted DNA is very low. Multiple extractions are required, and these are then pooled or an amplification step is employed before hybridization of the extracted DNA to the microarray. The enriched extracted DNA is compared on the array with un-enriched DNA so that the pattern and extent of enrichment of the regions represented on the array can be analyzed.

Microarrays have been used to study acetylation, methylation, phosphorylation, *etc.* by using specific antibodies with chromatin immunoprecipitation. With the help of this technique, it was possible to study the gene silencing in cancer.

Microarray analysis using ChIP-on-chip is particularly suited to the study of direct interaction of proteins and protein complexes with DNA. The coordinated regulation of genes by both E2F1 and E2F4 using ChIP in combination with an array containing the promoters' of 1500 genes has been investigated. Transcription-factor–DNA interactions have also been assayed using high-resolution tile paths of human chromosomal regions and entire chromosome tile paths, thus providing detailed information on the spatial organization of transcription-factor binding sites for genes and other genome features.

When chromosomal rearrangement takes place, the centromere gets disturbed a neo centromere has created the function of the neo centromere is to provide stability during the process of mitosis CHIP-on-Chip analysis can be used to identify this neo centromere and also to study their function, it has been reported that neo centromere formation is a sequence-independent epigenetic mechanism.

While high-resolution arrays are most suited for the analysis of protein–DNA interactions due to the range over which binding occurs, recent technology limits the scope of these arrays such that the whole genome cannot be scanned. Where this is compulsory, it is currently essential to use large insert clone arrays spanning throughout the whole genome. People have both 1 Mb resolution and chromosome tiling arrays to map the dispersal of proteins linked with cellular senescence. Chromatin gets immunoprecipitated with an antibody against phosphorylated histone H2AX. A marker of DNA damage which is associated with DNA damage and DNA checkpoint proteins revealed that H2AX was specifically enriched at human chromosome telomeres. It is proposed that telomere-initiated senescence shows a DNA damage checkpoint response that is activated by direct influence from dysfunctional telomeres.

The methylation of DNA in most CpG dinucleotides is important in gene regulation in that methylation is an inhibitor of gene expression. In promoter regions associated with CpG islands, the methylated DNA is bound by proteins such as MeCP2 which in turn recruit's histone deacetylase. The combination of methylated DNA and deacetylated histones generates a transcriptionally inactive chromatin state and gene silencing. The methylation and histone deacetylation changes during development and differentiation as different genes are switched on and off. These processes are also involved in genomic imprinting. Errors in DNA methylation or the enzymes associated with methylation lead to diseases such as Craniofacial and Rett Syndromes, Beckwith–Weidemann, and Prader Willi/Angelman syndromes due to errors in imprinting. Methylation and deacetylation plays an important role in cancer where the silencing of tumor suppressor genes is thought to be causative in many sporadic tumors. DNA methylation is intimately involved in gene regulation it is associated with many diseases. DNA microarrays are used to assess the methylation status of many genes and several approaches are employed. These approaches are Methylation-sensitive restriction enzymes and DNA size selection or specific PCR amplification to differentially amplify methylated and unmethylated loci. These products are then hybridized to an array where stronger hybridization means hypermethylation. Using this approach, it has been possible to correlate hypermethylation with hormone status in breast tumors and progression-free survival in ovarian cancer. In an alternative approach, Use of bisulfite modification of DNA to convert unmethylated cytosines to uracil which after PCR is converted to thymine. Unmethylated sites can be identified from methylated sites by knowing the conversion of cytosine to thymine after the bisulfite treatment. The conversion was detected by hybridization of the bisulfite-treated DNA to oligonucleotide arrays designed to be sensitive to the base change.

The application of microarrays to the study of chromosome structure and function is still not fully matured. However, with the help of microarray, we can quantify the gene expression at higher resolution which helps to understand the function of genomic DNA.

3

Mendelian Genetics

Mendelian genetics can be best understood by taking examples from human diseases. One such disease is alkaptonuria (AKU; MIM 203500), an iconic disease that holds an important position in the history of genetic disease. It was the first human disorder that was recognized to conform to the principles of Mendelian autosomal recessive inheritance by Sir Archibald Garrod over 100 years ago. It was in his Croonian Lecture at the Royal College of Physicians in 1908 that Garrod introduced the concept of "inborn errors of metabolism" to describe AKU and three other inherited disorders: albinism, cystinuria, and pentosuria. Fifty years later, Bert La Du and colleagues identified the enzymatic defect as a deficiency of homogentisate 1,2 dioxygenase (HGD; E.C.1.13.11.5), an enzyme in the metabolism of tyrosine and phenylalanine (**Figure 3.1**).

AKU is an ultra-rare disease with a low prevalence of 1:200,000–250,000 in most ethnic groups. However, several hot spots have been identified, including the Dominican Republic and northwest Slovakia, where

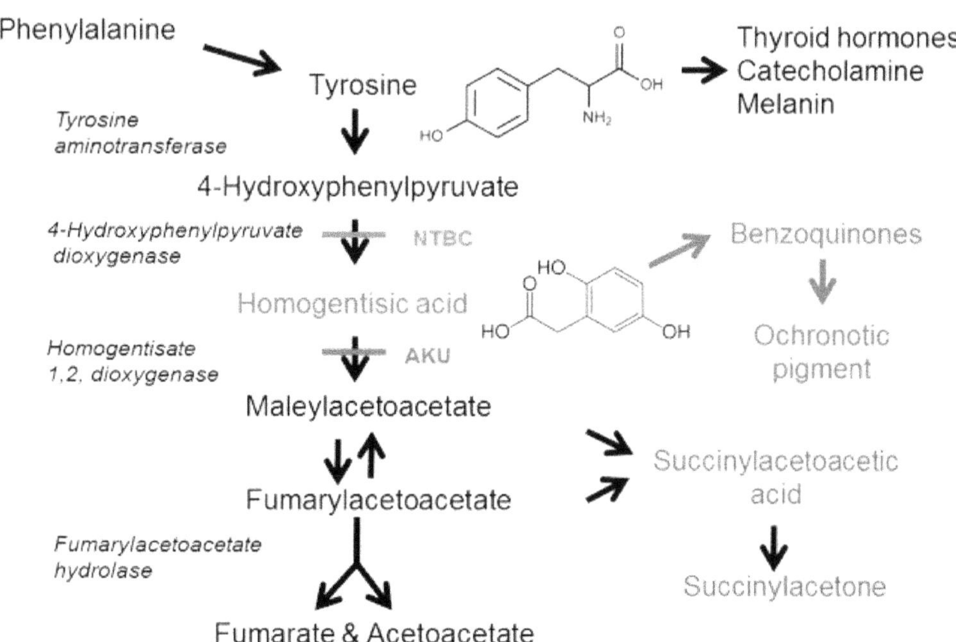

Figure 3.1: The Metabolism of Tyrosine shows the Location of the Metabolic Block in AKU and the Site of Action of NTBC.

the incidence is greater than 1:20,000. Recently, a high incidence of AKU was also discovered in specific regions of Jordan and India, and there is likely a vast reservoir of undiagnosed AKU worldwide.

The presenting symptom of alkaptonuria is urine that turns dark brown or black on prolonged exposure to oxygen or rapidly after the addition of an alkali. Alkaptonuria is a genetic cause of ochronosis, in which a slate blue or gray pigmentation of the sclera of the eyes and the cartilage of the ears becomes evident by the second or third decade of life and the cartilage becomes progressively black. Although the disorder is often diagnosed in childhood, it may not be detected until adulthood, when arthritis appears or dark pigmentation of the cartilage is noted. In addition to arthritis, with progressing age, individuals with alkaptonuria frequently develop degenerative disk disease and back pain, aortic and mitral valve stenosis, coronary artery calcifications, nephrolithiasis, and tendons that tear easily.

Oxidation of homogentisic acid leads to the formation of a pigmented polymer, the structure of which has not been elucidated. The pigment is preferentially deposited in the cartilage, leading to arthritis of the hip and knee, which occurs earlier in men. There is degeneration and calcification of the intervertebral discs of the lumbar spine, followed by fusion. The radiographic appearance is nearly pathognomonic. The mechanism of damage to the cartilage is not known. Ochronosis also occurs in the cardiac valves, and there is a high percentage of alkaptonuria patients older than 50 years with aortic valve thickening, stenosis, and insufficiency. There is currently no cure for alkaptonuria. Some have advocated the consumption of up to 1g of ascorbate per day to diminish pigment deposition in the cartilage by increasing renal excretion of homogentisic acid or its oxidized metabolite. However, this approach seems to have a little clinical effect. NTBC (nitisinone), an inhibitor of pHPPA dioxygenase, decreases the production of homogentisic acid and has been proposed as a potential therapy, but would probably require a reduction of tyrosine intake in the diet similar to tyrosinemia type I.

Nobel Prize winner George Wells Beadle, together with Edward Tatum, declared the famous theory one gene, one enzyme. He gave this theory based on the work they carried out on Neurospora. Beadle and Tatum appreciated that genes are physical realities, the structure of which embodies hereditary information. However, the basic principles of inheritance were laid down by Gregor Mendel, who is now considered the father of genetics.

Gregor Mendel (1822–1844) was an Augustinian monk born in a small village in Heinzendorf (now Hynice, Czech Republic). His parents, Anton and Rosina Mendel, were farmers by profession and named him Johann. He was an excellent student. Mendel decided to join an Augustinian monastery in 1843 in the town of Brunn to continue his studies. There, he was required to choose a new name; he changed his name to Gregor Johann Mendel and became a priest in 1847. At the monastery, the priest in charge sent Mendel to the University of Vienna so that he could study physics, chemistry, zoology, and mathematics. After his return to Brunn, Mendel's interest in evolution and gardening led him to take up the hobby of hybridization. During this period, he carried out numerous experiments on pea plants in the beautiful garden that surrounded the monastery. Mostly, he conducted breeding experiments (1856–1863) and presented his results publicly at meetings of the Brno Natural Science Society. In 1863, Mendel's work was published; however, its impact on research in the scientific community at that time was limited. At that time, scientists never realized that he had developed the laws of inheritance. However, he was aware of the importance of his work and quoted, "*My scientific studies have afforded me great gratification, and I am convinced that it will not be long before the whole world acknowledges the results of my work.*"

In 1868, Mendel was elected abbot of his monastery, and due to increasing administrative duties and teaching, the genetic experiments were stopped by him. He died at the age of 61 on January 6th, 1884, without getting due recognition for his work. In 1900, Dutch botanist and geneticist Hugo de Vries, German botanist and geneticist Carl Erich Correns, and Austrian botanist Erich Tschermak von Seysenegg independently reported their work on hybridization experiments that were like Mendel's; according to them, they were not aware of Mendel's work.

3.1 Mendel's Success

The success of Mendel was due to his scientific and systematic approach to the experiments, which used simple organisms like the garden pea. Hence, he was able to control pollination in the experimental plants, and most importantly, he used true-breeding plants with observable characteristics (*e.g.*, flower color and height). The most important part of his experimentation was keeping meticulous records, which resulted in consistent ratios involving thousands of plant-breeding experiments for more than eleven years. Interestingly, Mendel, while a student at the University of Vienna, studied physics and math under Christian Doppler (1803–1853), who later discovered the Doppler Effect. Mendel's basic training in statistics helped him determine the ratios and laws for which he is considered the father of genetics. Another important observation was that he studied the inheritance pattern of these plants for several generations. As the pea plant has a short span of life, Mendel was able to successfully carry out all his experiments. Moreso, the plants used by Mendel were diverse, hence differed for dissimilar traits, and were genetically pure. The traits of the pea plant chosen by him to observe had extremely simplified inheritance patterns. In his own words, Mendel stated, "*The value and utility of any experiment are determined by the fitness of the material to the purpose for which it is used.* They were unlinked and exhibited simple dominant or recessive inheritance; he concentrated on those traits that are present in two easily distinguished forms, such as white versus gray seed coats, round versus wrinkled seeds, and inflated versus constricted pods. Being good at statistics, he was able to formulate the hypothesis based on his initial observations and then conduct additional tests to test his hypotheses. He kept excellent records of the numbers of progeny having each type of trait and calculated the ratios of the different types. He noted all the details and conducted his experiments for almost ten years before attempting to analyze and write up his results. Mendel was not even aware of the term gene.

Some of the terms that are commonly used in genetics are shown below. It was coined in 1909 by a Danish geneticist, Wilhelm Johannsen. The definition of a gene varies with the context of its use. In the context of genetic crosses, a gene is defined as an inherited unit that determines a characteristic.

The genotype is defined as the set of alleles that an individual organism possesses. If there are two identical loci for a character, it is known as the homozygous locus. If there are two different alleles, the organism is heterozygous for the locus.

Some of the terms commonly used in genetics are shown in **Table 3.1**. The term *gene* was a word that Mendel never knew. It was not coined until 1909, when the Danish geneticist Wilhelm Johannsen first used it. The definition of a gene varies with the context of its use, and so its definition will change as we explore different aspects of heredity. For our present use in the context of genetic crosses, a gene may be defined as an inherited factor that determines a characteristic. The genetic makeup of an organism is called the genotype. A given phenotype is the result of the genotype that is manifested in a particular environment. However, the phenotype depends on the differential effects of other genes and environmental factors, and the balance between these influences varies from character to character. For some characters, the differences between phenotypes depend on differences in the genotype. Seed shape in Mendel's peas is a good example of a characteristic for which the phenotypic differences are largely dependent upon genetic composition. For other characters, environmental differences are more important. The height of the plants is a phenotype that is strongly influenced by environmental factors like the availability of water, sunlight, and nutrients. However, genetic factors are still playing their role in determining the height of an oak tree that will never grow to be 300 m tall no matter how much sunlight, water, and fertilizer are provided. For many characteristics, both genes and the environment are important in determining phenotypic differences. Hence, the distinction between genotype and phenotype is one of the most important principles of modern genetics (**Figure 3.2**).

Table 3.1: Important Genetic Terms

Term	*Definition*
Gene	A hereditary unit consisting of a sequence of DNA that occupies a specific location on a chromosome and determines a particular characteristic in organism.
Allele	One of two or more alternate forms of a gene
Locus	Specific place on a chromosome occupied by an allele
Genotype	Set of alleles that an individual possesses
Heterozygote	An individual possessing two different alleles at a locus
Homozygote	An individual possessing two of the same alleles at a locus
Phenotype or trait	The appearance or manifestation of a character
Character or characteristic	An attribute or feature

3.2 Mendel's Experiments

Mendel first tried to understand how a single character is inherited. Mendel selected a factor like the height of the plant. The plants can be tall or dwarf. The tall plants are on average 2 meters tall, and the dwarf plants are 1.5 meters tall. He studied several generations of tall plants and concluded that the offspring of tall plants are tall; hence, some determinant is there in tall plants that determine tallness. Mendel identified this "determinant" as a factor in tallness. Similar observations were also made

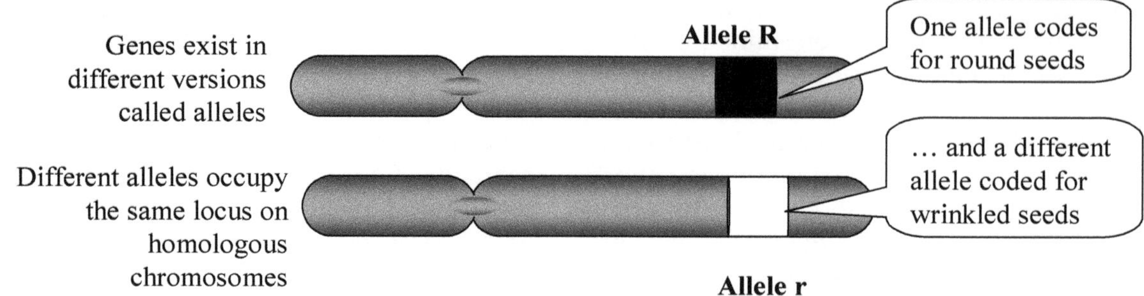

Figure 3.2: The Diploid Organism has Two Alleles Located on different Homologous Chromosomes.

for dwarf plants. Based on these experiments, Mendel concluded that the factor for dwarfness is present in the dwarf plant. Hence, both of these plants were considered pure for the expression. This means that tall plants have factors only for tallness and not for dwarfness. A dwarf plant has a factor for dwarfness but not for tallness. For all his experiments, he selected pure plants and carried out cross-fertilization among them. Monohybridization is the cross-fertilization of plants that differ in the expression of one character and are called monohybrids. Parents P are the plants with which the experiments were initiated (**Figure 3.3**).

First filial generation F1 is the offspring following the crossing between these parent P plants. The generations immediately after the first filial generation (F1) are F2, F3, *etc.* The original pure tall and pure dwarf pea plants selected by Mendel were used as parent plants, and the crossing was carried out among them. The tall or dwarf plants were selected as male or female because pea plants are bisexual. The pure tall and pure dwarf pea plants that crossed all F1 generation plants that developed were tall, with an average height of 2 meters. Mendel carried out the self-fertilization of F1 plants. The offspring of the F1 plants are the F2 generation. In the F2 generation, 34 plants were tall, while the other 1/4 were dwarfs.

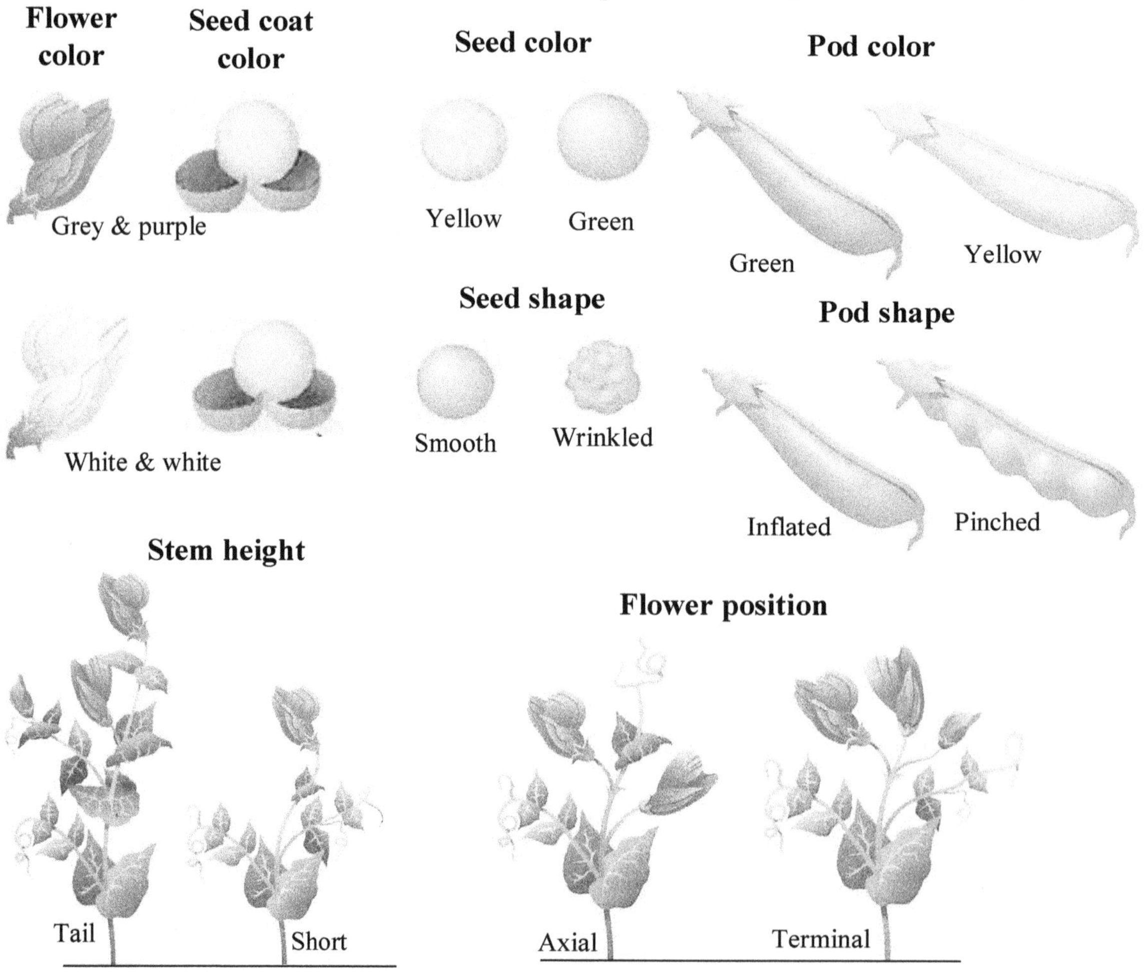

Figure 3.3: Different Features of Pea Plant Selected by Mendel for his Experiments.

Thus, the ratio of tall people to dwarfs was 3:1. This ratio is called the monohybrid ratio. Mendel postulated certain laws based on the above results. The inheritance of height was controlled by some factors. Mendel presumed that these factors were in pairs. Out of the paired factors in the heterozygous state, one appears dominant, *i.e.*, dominant, while others do not appear to be recessive. The expression of the recessive factor is prevented by the dominant factor. In the above example, F1 plants had both factors for tallness and dwarfness, but only the factor for tallness was expressed. Thus, the factor for tallness is the dominant factor, and the factor for dwarfness, which is not expressed, is the recessive factor. These two factors are distributed to different gametes during gametogenesis. The gamete is either a factor in tallness or dwarfism. What Mendel called a factor is now known as a gene; these are arranged on chromosomes.

A capital letter is used to show the dominant expression of the gene, *e.g.*, tallness is denoted by 'T' for recessive expression. Normally, a small letter is used, *e.g.*, dwarfness is denoted by t. Pure tall plants of the parent generation can be denoted by the sign 'TT, as chromosomes occur in pairs. Members of a pair are homologous chromosomes. Pure tall plants with similar chromosomes are homozygous plants and are pure tall plants. Similarly, a pure dwarf plant of the P' generation does not have any genes for tallness and is denoted by the sign 'tt'. This plant is pure and homozygous for dwarfism.

All characters are determined by genes. These are found in pairs in diploid organisms. Thus, T and t are allelic genes. When parentally pure tall and dwarf plants were crossed, all F1-generation plants were found to be tall. The gene composition of the parental tall plant is TT. The gene composition of the parental dwarf plant is tt. Reduction division takes place during gamete formation. Hence, chromosomal numbers are reduced to half the original number, and they become haploid. During gamete formation, paired alleles are segregated. Mendel's record-keeping and meticulousness helped him carry out the rigorous analysis of the data. Characters selected by him were very simple, like flower color, pod color, seed color, seed shape, pod shape, stem height, and flower position of the plant.

Mendel's observations from these experiments can be summarized in two basic principles: (i) the principle of segregation; and (ii) the principle of independent assortment, as shown in **Table 3.2.**

Mendel used the test-cross method for crossing different characters. In the F1 generation, the ratio of tall people to dwarfs was 3:1. If tall plants (TT) were crossed with dwarf plants (TT). All plants would be tall, but with a genotype (Tt). Now if tall heterozygous plants, *i.e.*, Tt, are crossed to dwarf plants with genotype (tt), the resultant progeny will be 50 per cent tall plants and 50 per cent dwarf plants. The ratio of the tall to the dwarf would be 1:1. This method of testcrossing showed that there are some factors that control tall and dwarf traits (**Figure 3.4**).

In the next stage, he took two characters independently and crossed them and called them di-hybrids. He crossed yellow round seeds with green wrinkled and yellow wrinkled and green, round in F1 generation. The results were all yellow and round. He crossed the resultant product of F1 generation *i.e.*,

Table 3.2: Results of Mendel's Observation

Character	*Contrasting Trait*	*F_1 Results*	*F_2 Results*	*F_2 Ratio*
Seeds	Round/wrinkled	All round	5474 rounds	2.96:1
	Yellow/Green	All yellow	1850 wrinkled	
			6022 yellow	3.01:1
			2001 green	
Pods	Full/constricted green/yellow	All full	882 full	2.95:1
		All green	299 constricted	2.82:1
			428 green	
			152 yellow	
Flower colour	Violet/white	All violet	705 violets	3.15:1
			224 white	
Flower position	Axial/terminal	All axial	651 axials	3.14:1
			207 terminals	
Stem height	Tall/dwarf	All tall	787 tall	2.84:1
			277 dwarfs	

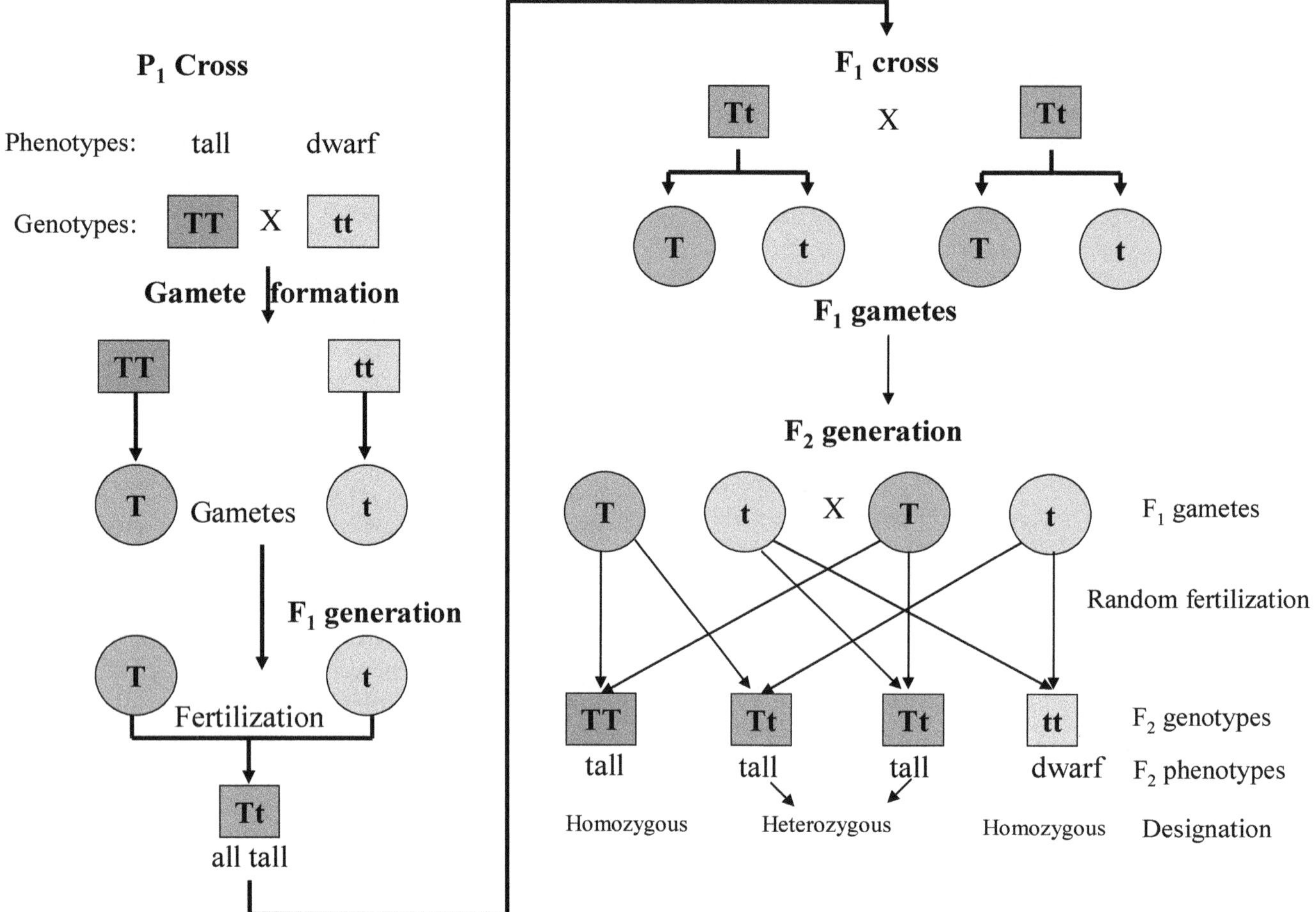

Figure 3.4: Tall and Dwarf Monohybrid Pea Plants after Crossing Over.

yellow round and yellow wrinkled. In F2 generation he got 9/16 yellow round 3/16 yellow wrinkled and 1/16 green wrinkled (**Figure 3.5**).

These results are quite simple to follow. Suppose that two sets of traits are inherited independently of one another. As yellow is dominant over green so in the F1 generation all are yellow. Predicted F2 of the first cross are yellow, green. In the second cross rounds and wrinkled. The dihybrid cross 12/16 of F2 plants are yellow while 4/16 are green exhibiting the 3:1 (3/4:) ratio. Similarly, 12/16 of all F2 plants have round seeds, while 4/16 have wrinkled seeds, again revealing a 3:1 (3/4:) ratio.

If we use the product law for independent events, the combined probability of two outcomes is the product of their individual chances of occurring, as per the results. During the formation of gametes, pairs of unit factors separate independently, resulting in all possible gamete combinations being formed with equal frequency. **Figure 3.6** displays four different traits, and the resulting product is a ratio of 9:3:3:1, which means 9/16 is green and round, while 1/16 is green and wrinkled. This ratio is known as Mendel's 9:3:3:1 dihybrid ratio.

In the next stage, trihybrid crosses were created, and Mendel observed that both the law of segregation and independent assortment applied to the three pairs of contrasting characters also. The law of segregation suggests that when gametes (sperm, egg, *etc.*) are formed each gamete will receive one allele or the other. The Law of the independent assortment may be defined as that two or more alleles will separate independently of each other when gametes are formed.

Let us consider a cross with three characters *i.e.*, Aa, Bb, and Cc. The cross would be AABBCC x aabbcc individuals, the resultant offspring would be heterozygous, and their genotypes would be AaBbCc phenotypically they will be A, B, and C. When these individuals produce gametes, there would be different gametes in equal frequencies. A punnet square with 64 combinations can be drawn. However, it is difficult hence another approach is followed known by the name of the forked line method. In a monohybrid cross *i.e.*, AA x aa the phenotype in the next generation would be A in the F2 generation it would either be A or with a ratio of 3:1. The 3 hybrids cross *i.e.* A/a, B/b, C/c in F2 generation will express A, B, and C in and will express a, b and c.

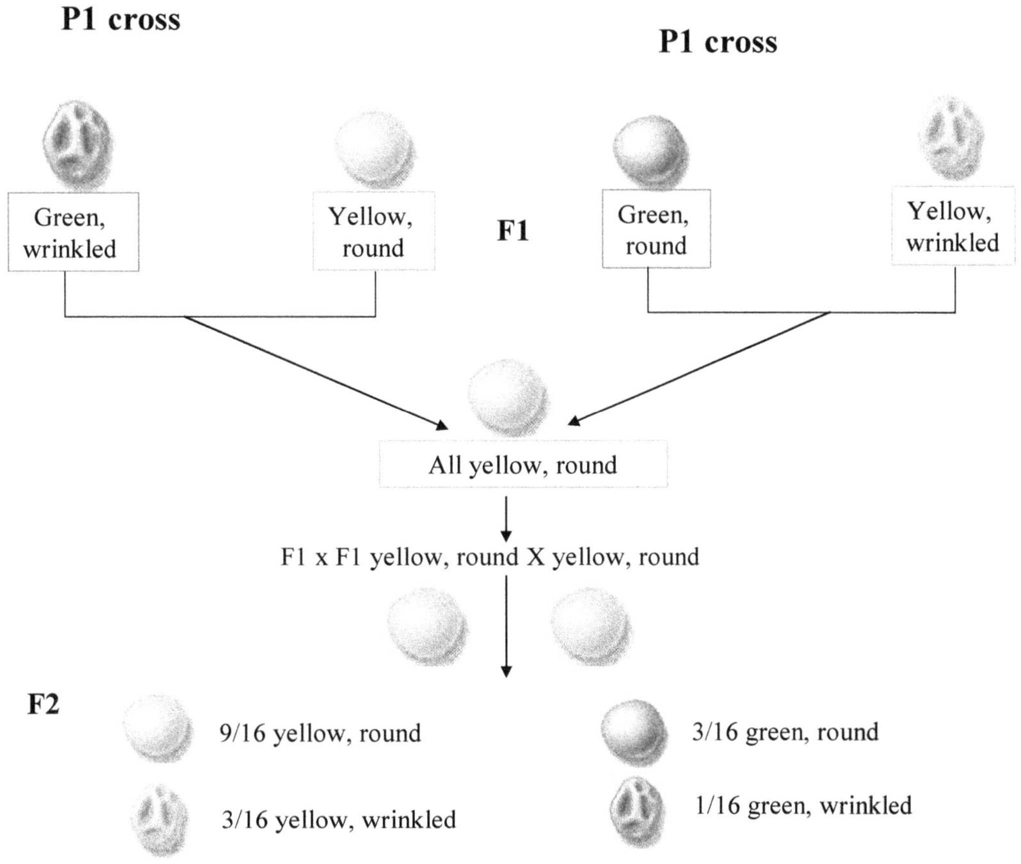

Figure 3.5: F_1 and F_2 Generation of di-hybrid Crosses.

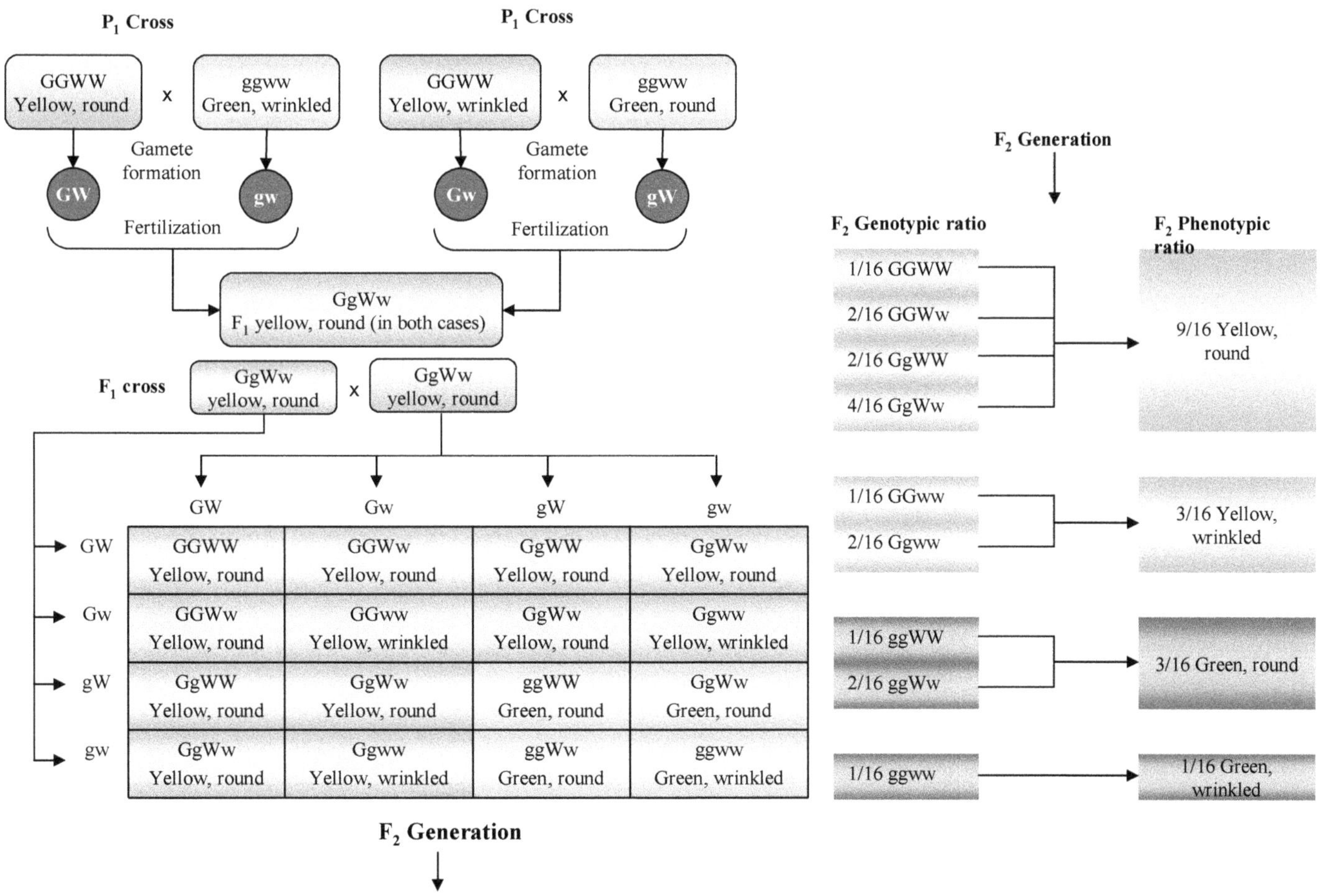

P_1 Cross

GGWW Yellow, round x ggww Green, wrinkled

Gamete formation: GW, gw

Fertilization

P_1 Cross

GGWW Yellow, wrinkled x ggww Green, round

Gamete formation: Gw, gW

Fertilization

GgWw

F_1 yellow, round (in both cases)

F_1 cross: GgWw yellow, round x GgWw yellow, round

	GW	Gw	gW	gw
GW	GGWW Yellow, round	GGWw Yellow, round	GgWW Yellow, round	GgWw Yellow, round
Gw	GGWw Yellow, round	GGww Yellow, wrinkled	GgWw Yellow, round	Ggww Yellow, wrinkled
gW	GgWW Yellow, round	GgWw Yellow, round	ggWW Green, round	GgWw Green, round
gw	GgWw Yellow, round	Ggww Yellow, wrinkled	ggWw Green, round	ggww Green, wrinkled

F_2 Generation

F_2 Generation

F_2 Genotypic ratio	F_2 Phenotypic ratio
1/16 GGWW, 2/16 GGWw, 2/16 GgWW, 4/16 GgWw	9/16 Yellow, round
1/16 GGww, 2/16 Ggww	3/16 Yellow, wrinkled
1/16 ggWW, 2/16 ggWw	3/16 Green, round
1/16 ggww	1/16 Green, wrinkled

Figure 3.6: Combination of Four Characters of di-hybrid Cross.

The ratio would be 27/64 ABC, 9/64 ABc, 9/64 AbC, 3/64 Abc, 9/64 aBc, 3/64 aBc, 3/64 abC, 1/64 abc *i.e.* 27:9:9:3:9:3:3:1. This is applicable only if genes are assorted independently of one another (**Figure 3.7**). In using a forked line method, each gene pair must be considered separately.

In the following stage, trihybrid crosses were generated, and Mendel noticed that both the law of segregation and independent assortment were applicable to all three sets of contrasting traits. The law of segregation states that during gamete formation, each gamete will receive one allele or the other. The law of independent assortment states that two or more alleles will separate independently of each other during gamete formation.

Let's take a cross with three traits, namely Aa, Bb, and Cc. The cross would be between AABBCC and aabbcc individuals, resulting in heterozygous offspring with genotypes AaBbCc and phenotypic traits A, B, and C. When these individuals produce gametes, there would be different gametes in equal frequencies. A 64-combination Punnett square can be drawn, but a more practical method is the forked line method.

In a monohybrid cross, AA x aa, the phenotype in the next generation would be A, with a ratio of 3:1 in the F2 generation. The phenotype of the 3 hybrids cross, A/a, B/b, C/c, in the F2 generation will be A, B, and C in of the offspring, and a, b, and c in of the offspring. The ratio would be 27/64 ABC, 9/64 ABc, 9/64 AbC, 3/64 Abc, 9/64 aBc, 3/64 aBc, 3/64 abC, and 1/64 abc, respectively. This applies only when genes are independently assorted (Figure 3.7). In using the forked line method, each gene pair must be considered separately. In the following stage, trihybrid crosses were generated, and Mendel noticed that both the law of segregation and independent assortment were applicable to all three sets of contrasting traits. The law of segregation states that during gamete formation, each gamete will receive one allele or the other. The law of independent assortment states that two or more alleles will separate independently of each other during gamete formation.

Let's take a cross with three traits, namely Aa, Bb, and Cc. The cross would be between AABBCC and aabbcc individuals, resulting in heterozygous offspring with genotypes AaBbCc and phenotypic traits A, B, and C. When these individuals produce gametes, there would be different gametes in equal frequencies. A 64-combination Punnett square can be drawn, but a more practical method is the forked line method.

In a monohybrid cross, AA x aa, the phenotype in the next generation would be A, with a ratio of 3:1 in the F2 generation. The phenotype of the 3 hybrids cross, A/a, B/b, C/c, in the F2 generation will be A, B, and C in of the offspring, and a, b, and c in of the offspring. The ratio would be 27/64 ABC, 9/64 ABc, 9/64 AbC, 3/64 Abc, 9/64 aBc, 3/64 aBc, 3/64 abC, and 1/64 abc, respectively. This applies only when genes are independently assorted (Figure 3.7). In using the forked line method, each gene pair must be considered separately.

3.3 The Law of Probability

Laws of probability help to explain genetic events. Probability means the occurrence of an event taking all possible outcomes of the process. The collection of all events is called sample space. For a coin toss, there are two events *i.e.*, either head or a tail will come upon tossing. If a card is drawn from a deck the probability is 52 times for a single card.

Let us take a mating between GG Gg and gg. The probability associated with each progeny would be (for GG) ½ (for Gg) and (for gg).

There are two issues that one must keep in mind (i) what is the probability of two events occurring

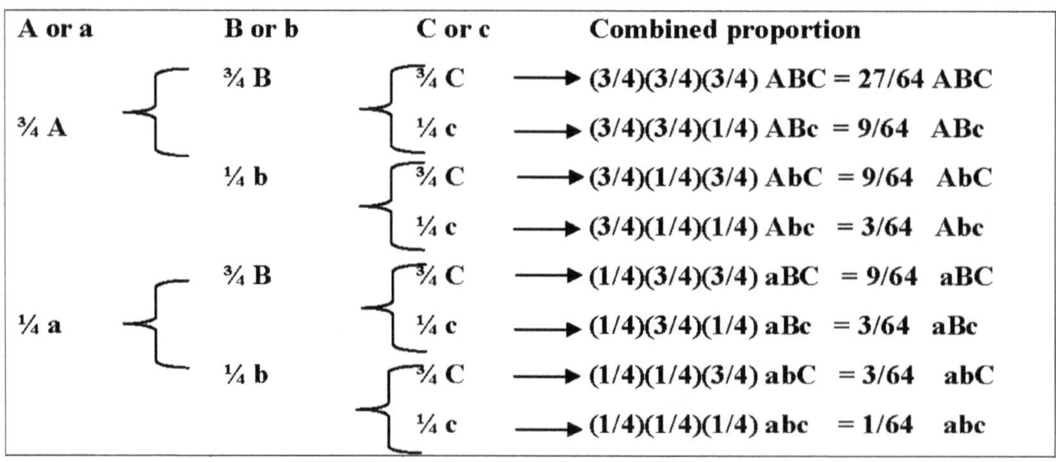

Figure 3.7: F2 Trihybrid Ratios using Forked Line Method.

together? (ii) What is the probability that at least one of the events is occurring? This is very easy to answer *i.e.*, 50 per cent probability for two events to occur together.

However, there are two rules of probability estimation *i.e.* multiplicative rule and additive rule to calculate the probabilities.

Multiplicative rule: If the event A and B are independent the probability with which they will occur together can be P (A and B) is P(A) x P(B). In this case, P(A) and P(B) are the probabilities of the individual events. The independent events do not mean that they are not overlapping in the sample space. In fact, non-overlapping, or disjoint, events are not independent, for if one occurs, then the other cannot. In probability theory, independent means that one event provides no information about the other, B for example, if a card drawn from a deck turns out to be an ace, and we have no clue about the Card's sent. Thus, drawing the ace of hearts represents the joint occurrence of two independent events that the card is an ace (A) and it is heart (H). According to multiplicative rule P (A and H) = P(A) x P(H) and because P(A) = 4/52 and P(H) = (P and H) = (4/52) x (1/4) = 1/52.

The additive rule: If the events A and B are independent the probability that at least one of them occurs is denoted P(A or B) is P(A) + P(B) – [P(A) x P(B)].

Here the term P(A) x P(B), which is the probability that A and B occur together, is subtracted from the sum of the probabilities, P(A) + P(B), because the straight sum includes this term twice. As an example, suppose we seek the probability that a card drawn from a deck is either an ace or a heart. According to the Additive Rule, P (A or H) = P(A) + P(H) – [P(A) x P(H)] = (4/52) + (1/4) – [(4/52) x (1/4)] = 16/52.

If the two events do not overlap in the sample space, the Additive Rule reduces to a simpler expression: P (A or B) = P(A) + P(B). For example, suppose we seek the probability that a card drawn from a deck is either an ace or a king (K). These two events do not overlap in the sample space; they are said to be mutually exclusive. Thus, P (A or K) = P(A) + P(K) = (4/52) + (4/52) = 8/52.

The Punnett square and forked –line methods – are also gametes, half contain based upon the principle of Probability–**Figure 3.7**. Mendelian segregation is like a coin toss; when a heterozygote produces one allele, and the other half contains the other allele. If two segregating heterozygotes are crossed, their gametes are combined randomly, producing the zygotic genotypes. Let us suppose the cross is Aa x Aa. The chance that a zygote will be AA is simply the probability that each of the uniting gametes contains A, or (1/2) x (1/2) = ( ), since the two gametes are produced independently. The chance for an aa homozygote is also 1/4. However, the chance for an Aa heterozygote is ½ because there are two ways of creating a heterozygote – A may come from the egg and a from the sperm, or vice versa. Because each of these events has a one-quarter chance of occurring, the total probability that an offspring is heterozygous is (1/4) + (1/4) = (1/2). We, therefore, obtain the

Cross: Aa Bb X Aa Bb

		Segregation of A gene	
		A – (3/4)	aa (1/4)
Segregation of B gene	B– (3/4)	A– B– (3/4) x (3/4) = 9/16	aa B– (1/4) x (3/4) = 3/16
	bb (1/4)	A – bb (3/4) x (1/4) = 3/16	aa bb (1/4) x (1/4) = 1/16

Progency:

Genotype	Frequency	Phenotype	Frequency
A– B–	9/16	Dominant for both genes	9/16
aa B– A – bb aa bb	3/16 3/16 1/16	Recessive for at least one gene	7/16

Figure 3.8: Probability of Dominant and Recessive Traits when Intercross take place between Two Genes.

following probability distribution of the genotypes from the mating Aa x Aa:

AA	
Aa	½
Aa	¼

Hence (1/4) + (1/2) = (3/4) of the progeny will have the dominant phenotype and will have the recessive **(Figure 3.8).**

What fraction of the progeny will be homozygous for all four recessive alleles? To answer this question, we consider genes one at a time. For the first gene, the fraction of offspring that will be recessive homozygote is 1/4, as it will be for the second, third, and fourth genes. Therefore, according to the principle of independent assortment of Mendel, the fraction of offspring that will be quadruple recessive homozygotes is (1/4) x (1/4) x (1/4) x (1/4) = (1/256). The probability method is a better approach than drawing Punnett square with 256 blocks.

Let us think what fraction of the offspring will be homozygous for all the four genes in this case we must first decide what genotypes can satisfy the question. There are dominant and recessive homozygotes, and they constitute half of the progeny. The fraction of progeny that will be homozygous for all four genes will be (1/2) x (1/2) x (1/2) x (1/2) = (1/16).

Suppose the cross is AaBb X AaBb and we want to know what fraction of the progeny will show the recessive phenotype for at least one gene. Three kinds of genotypes would satisfy this condition: (1) A-bb (the dash stands for either A or a), (2) aa B-, and (3) aa bb. The answer to the question must therefore be the sum of the probabilities corresponding to each of these genotypes. The probability for A-bb is (3/4) x (1/4) = (3/16), that for aa B- is (1/4) x (3/4) = (3/16), and that for aa bb is (1/4) x (1/4) = (1/16). Adding these together, we find that the answer is 7/16.

3.4 Conditional Probability

Sometimes we may wish to calculate the probability of an outcome that is dependent on a specific condition related to that outcome. For example, in the F2 of Mendel's monohybrid cross involving tall and dwarf plants, what is the probability that a tall and dwarf plant, what is the probability that a tall plant is heterozygous (and not homozygous)? The condition we have set is to consider only tall F2 offspring since we know that all dwarf plants are homozygous. Because the outcome and specific conditions are not independent, we cannot apply the product law of probability. This is referred to as a conditional probability. In the simplest terms, what is the probability that one outcome will occur, given the specific condition upon which this outcome is dependent? Let us call this probability as pc.

To find out the solution for pc, we must consider both the probability of the outcome of interest and that of the specific condition that includes the outcome. These are (a) the probability of an F2 plant being heterozygous as a result of receiving both a dominant and a recessive allele (pa) and (b) the probability of the condition under which the event is being assessed, that is, being tall (pb).pa = plant inheriting one dominant and one recessive allele (*i.e.*, being a heterozygote)

= 1/2

pb = probability of an F_2 plant of a monohybrid cross being tall

=

To calculate the conditional probability (pc), we divide pa by pb:

Pc = pa/pb

= (1/2)/(3/4) = (1/2) (4/3)

= 4/6

pc = 2/3

The conditional probability of any tall plant being heterozygous is two–thirds (2/3). On average, two-thirds of the F_2 tall plants will be heterozygous **Figure 3.9**.

Conditional probability has many applications in genetics. During genetic counseling, for example, it is possible to calculate the probability (pc) that an unaffected sibling of a brother or sister expressing a recessive disorder is a carrier of the disease-causing allele (*i.e.* a heterozygote). Assuming that both parents are unaffected (and are therefore carriers), the calculation of pc is identical to the preceding example. The value of pc = 2/3.

3.5 Deviation from Mendelian Genetics

Mendelian genetics may not be followed always. After Mendel gave his laws many more new discoveries have been made which are discussed below.

To appreciate the various modes of inheritance, the probable function of an allele should be known. In some organisms where wide populations have been studied, the allele that occurs most commonly and behaves normally is called the wild-type allele. The common allele is most of the time, dominant. Wild type allele gives rise to wild type phenotype. For all practical purposes if a mutation occurs, we must compare the mutated type with the wild type allele.

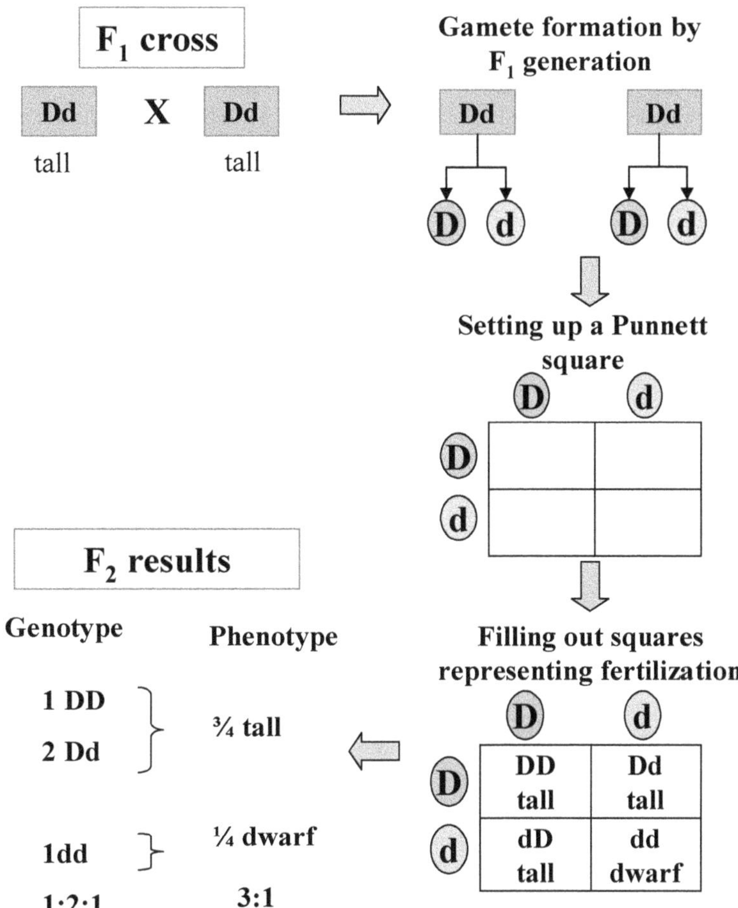

Figure 3.9: A Punnett Square can be Generated in the F_2 Generation when F_1 x F_1 Cross Occurs.

A mutant allele contains modified genetic information hence have the altered gene product. In human populations, there are many known alleles β-chain of human hemoglobin. All such alleles store information necessary for the synthesis of the, β chain polypeptide, but each allele specifies a slightly different form of the same molecule. Once manufactured, the product of an allele may or may not have its function altered.

The process of mutation is the source of alleles. For a new allele to be recognized when observing an organism, it must cause a change in the phenotype. A new phenotype results from a change in the functional activity of the cellular product specified by that gene. Often, the mutation causes the loss of function. This is designated as a loss of function mutation. If the loss is complete, the mutation has resulted in what is called a null allele.

Some mutations may enhance the function of the wild type of product. Most often when this occurs, it is the result of increasing the quantity of the gene product. In such cases, the mutation affects the regulation of transcription of the gene. Such cases are designated gain of function mutations, which generally result in dominant alleles since one copy in a diploid organism is sufficient to alter the normal phenotype. An example of the gain of function mutation is that of oncogenes. Some mutations do not cause any change of function therefore, there is no apparent effect. Many times, more than one gene is required to carry out the function well-known examples are of complex metabolic pathways.

Various symbols are used to denote the various forms of alleles Capital letter for the dominant trait, small for a recessive trait, wild type is written as W. The normal wild-type body color of drosophila is gray; ebony is denoted by the symbol e, while gray is denoted by e+. The responsible locus may be occupied by either the wild-type allele (e+) or the mutant allele (e). A diploid fly may thus exhibit one of the three possible genotypes.

e+/e+ Gray homozygote (wild type)

e+/e Gray heterozygote (wild type)

E/e Ebony homozygote (mutant)

The slash between the letters indicates that the two allele designations represent the same locus on two homologous chromosomes. Let us suppose a mutant allele that is dominant to the normal wild–type allele, such as wrinkled wing in drosophila, the initial uppercase letter is used to designate dominance (Wr.). Thus, the three possible genotypes would be.

Wr/Wr Wrinkled wings

Wr/Wr+ Wrinkled wings

Wr +/Wr+ Normal, straight wings

It is because of dominance that the initial two genotypes express the mutant wrinkled –wing phenotype. One advantage of the above system is that further abbreviation can be used when convenient: the wild–type allele may simply be denoted by the + symbol. With ebony as an example, the designations of the three possible genotypes become.

+/+ Gray homozygote (wild type)

+/e Gray homozygote (wild type)

E/e Ebony homozygote (mutant)

Two other points are important. First, although we have adopted a standard convention for assigning genetic symbols, there are many diverse systems of genetic nomenclature used to identify genes in various organisms. Usually, the symbol selected reflects the function of the gene or occasionally, a disorder caused by a mutant gene. For example, the CDK gene from yeast refers to encoding cyclin-dependent kinases. In bacteria, leu- refers to a mutation that interrupts the biosynthesis of the amino acid leucine, where the wild-type gene is designated leu+. The symbol dnaA represents a bacterial gene involved in DNA replication. In humans, italicized capital letters are used to name genes: BRCA1 represents one of the genes associated with susceptibility to breast cancer. Although these different systems may sometimes be confusing.

3.6 Types of Dominance

There exist three forms of dominance: complete, incomplete, and co-dominance. Gregor Johann Mendel's Principle of Dominance explains how traits are passed down. When parents with different versions of a trait mate, one version disappears in the hybrid offspring, while the other reappears unchanged in the third generation, according to this principle. The visible version of the trait in hybrids is the dominant, and the invisible one is the recessive.

It is important to understand that complete dominance is just one type of dominant relationship identified by Mendelian genetics. Dominance is a result of interactions between gene products of different alleles of the same gene. Dominance is typically defined in terms of the phenotype of the heterozygote.

Phenylketonuria (PKU) is a type of autosomal recessive condition caused by a mutant form of the enzyme PAH. The AA or Aa genotype guarantees normal breakdown of phenylalanine (Phe) to tyrosine. Nonetheless, the aa genotype indicates that there is no functional PAH, leading to nervous system damage. Although the mechanism of normal gene action in PKU is incomplete dominance, the disease is recessive in nature as two recessive alleles (aa) are required to show a disease phenotype.

In complete dominance, the heterozygote is like the dominant, while in incomplete dominance, the heterozygote is intermediate between the two homozygotes. In pea plants, red is dominant, and white is recessive. The d+d+ genotype is red, the d+d genotype is also red, and the dd genotype is white.

In incomplete dominance for flower color, the c+c+ genotype is red, the c+c genotype is pink, and the cc genotype is white. The heterozygote's intermediate nature is due to the protein produced by the A allele of a gene showing incomplete dominance being less functional than the protein produced by the A allele of a gene showing complete dominance.

Complete dominance is called haploinsufficiency as only one copy of the A allele produces enough enzyme to negate the effects of the mutant allele. In incomplete dominance, having only half the number of enzymes is insufficient. It is important to understand that the genotypes are identical in both complete and incomplete dominance, and the phenotypic difference results from the effectiveness of the molecule that causes the phenotype.

3.7 Co Dominance

We normally write alleles as if all genes had only two (*e.g.*, we use capital and lower case letters or plus and minus signs), but all alleles can mutate at one or more of the nucleotide spots. Most nucleotide changes will change the function of the protein. Thus, there are many ways that an allele (A) could be mutated, and nearly all those changes will make a nonfunctional allele (a). Thus, there will always be many different alleles of a given gene.

The human ABO blood system group reveals co-dominant alleles because both express a separate immunological protein on the surface of the blood cell. Both the alleles are equally expressed. Thus, they have different co-dominant phenotypes. The O allele does not express the protein on the surface, so it is not phenol-typically noticeable, and is recessive to both A and B. Dominance is a feature of the protein made by the gene, not of the gene itself. The "completeness" of dominance depends only on how effective the protein works. Incomplete dominance simply means that half as many molecules (in the heterozygote) are not as effective as the number produced by the homozygote. Co-dominance means that both proteins are equally effective in the cell.

3.8 Multiple Alleles

The control of a trait by genetic loci can involve multiple highly polymorphic alleles. For instance, the immune response genes of the major histocompatibility complex have numerous different alleles at each locus. An individual's susceptibility or resistance to a particular gene depends on the combination of their alleles. In humans, the ABO blood groups effectively demonstrate multiple alleles by having four possible blood types: A, B, AB, and O. These blood types are determined by two specific carbohydrate molecules found on the surface of red blood cells, which distinguish them from the MN blood groups. Blood type A is caused by the A carbohydrate, blood type B by the B carbohydrate, blood type AB by both carbohydrates, and blood type O by neither carbohydrate.

The ABO blood groups arise from three different allele combinations: IA, which codes for carbohydrate A; IB Cross, which codes for carbohydrate B; and i, which codes for the absence of any carbohydrate. As shown in the figure, an individual with one or two IA alleles will have blood type A, one or two IB alleles will result in blood type B, both IA and IB alleles will lead to blood type AB, and the genotype ii will yield blood type O (Figure 3.10).

The Punnett square can be used to predict the genotype frequencies resulting from multiple allele crosses. However, one cannot be sure of an individual's genotype if they are blood type A or B because there are two possible genotypes for each of these blood types. Therefore, many cross problems that examine blood types are like test crosses; that is, the parental genotype is uncertain. A few examples will aid in your understanding.

The genes that determine ABO blood type encode enzymes that add particular sugar groups to proteins in blood cells. A person's specific blood type is due to the presence or absence of A and B sugar-protein complexes on the surface of red blood cells.

There are three alleles involved A, B, and O, and six possible genotypes: AA, BB, OO, AB, AO, and BO. The various genotypes result in the four different phenotypes or blood types: A, B, O, and AB. Individuals have blood type A if their genotypes are AA or AO. Individuals have blood type B if their genotypes are BB or BO. Individuals have blood type O if their genotype is OO, and they have blood type AB if their genotype is AB as illustrated below:

Genotype	*Antigen*	*Phenotype*
$I^A I^A$	A	A
$I^A I^O$	A	
$I^B I^B$	B	t
$I^B I^O$	B	
$I^A I^B$	A,B	AB
$I^O I^O$	None of the above	O

In these assignments, the I^A and I^B alleles behave dominantly to the I^O allele, but not co-dominantly to each other. The co-dominant pattern can be checked by studying the whole family for a particular trait **Table 3.3** and **Figure 3.10**.

Table 3.3: Potential Phenotypes in the Offspring of Parents with all Possible ABO Blood Group Combinations

Parents		*Potential Offspring*			
Phenotypes	*Genotypes*	*A*	*B*	*AB*	*O*
A X A	I^AI^O X I^AI^O	¾	-	-	¼
B X B	I^BI^O X I^BI^O	-	¾	-	¼
O X O	I^OI^O X I^OI^O	-	-	-	ALL
A X B	I^AI^O X I^BI^O	¼	¼	¼	¼
A X AB	I^AI^O X I^AI^B	½	¼	¼	-
A X O	I^AI^O X I^OI^O	½	-	-	½
B X AB	I^BI^O X I^AI^B	¼	½	¼	-
B X O	I^BI^O X I^OI^O	-	½	-	½
AB X O	I^AI^B X I^OI^O	½	½	-	-
AB X AB	I^AI^B X I^AI^B	¼	½	½	-

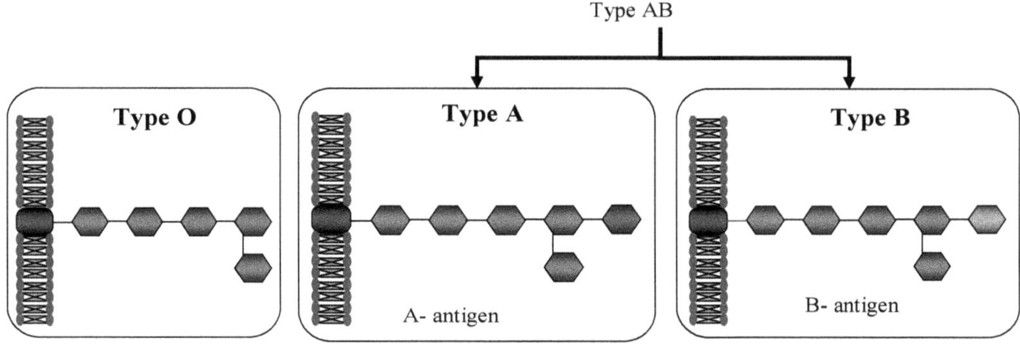

Phenotypes (blood group)	Genotypes
O	ii
A	$_IA_IA$ or $_IA_i$
B	$_IB_IB$ or $_IB_i$
AB	$_IA_IB$

Figure 3.10: ABO Blood Groups and Multiple Alleles.

3.9 Pleiotropy

Pleiotropy is a well-known genetic phenomenon in the fields of medical and physiological genetics, as well as evolutionary biology. It occurs when a single genetic locus affects two or more distinct traits. Ludwig Plate, a German geneticist, introduced this term over a century ago in 1910. Although researchers have differing opinions on pleiotropy, modern molecular data has changed how we understand this phenomenon since 1910.

It is common for genes to influence multiple physical characteristics rather than just one. It is referred to as pleiotropy and is observed in many genetic disorders. For example, albino individuals often experience pigment loss in their hair and skin and are more prone to crossed eyes. This is because the gene responsible for albinism can also cause problems in the nerve connections between the brain and the eyes. These traits are not always related, highlighting the complex interplay between genetics and physical characteristics. The phenomenon of pleiotropy occurs when a single genetic locus impacts two or more distinct traits. This term was first introduced by Ludwig Plate, a geneticist from Germany, over a century ago in 1910. Pleiotropy has played a significant role in the fields of medical and physiological genetics, as well as evolutionary biology. However, different researchers have varying perceptions of pleiotropy, and various approaches to its study have resulted in disparities in its understanding. Besides, modern molecular data has changed how we comprehend pleiotropy since 1910.

It is widely known that a single gene can affect multiple traits, which is called pleiotropy. This means that a gene can impact many different characteristics instead of just one. Albinism is a good example of this phenomenon since individuals with this condition experience pigment loss in their skin and hair, as well as a higher risk of crossed eyes. The gene responsible for albinism can also cause defects in the nerve connections between the brain and the eyes, leading to these additional symptoms. This highlights the complex relationship between genetics and observable traits. Mendel, the famous pea plant researcher, also recognized pleiotropy. He observed that pea plants with red flowers had different coloration in their leaf-stem juncture and seed coats than plants with white flowers. He concluded that these combinations were likely controlled by the same hereditary unit (gene).

Marfan syndrome is another example of pleiotropy. It is caused by a mutation in the gene that encodes connective tissue protein fibrillin. Fibrillin is present in many tissues in the body, leading to multiple effects of such a defect. The syndrome is characterized by lens dislocation, an increased risk of aortic aneurysm, and lengthened long bones in limbs. Abraham Lincoln is even believed to have had this disorder.

Porphyria variegata is another autosomal dominant disorder that shows pleiotropy. Individuals with this condition cannot properly metabolize the porphyrin component of hemoglobin, leading to excess porphyrins accumulating in the body, particularly in the brain. This can cause a range of symptoms such as abdominal pain, muscular weakness, fever, a racing pulse, insomnia, headaches, vision problems, delirium, and ultimately convulsions. King George III is believed to have suffered from all these symptoms. In summary, most mutations have multiple effects when expressed, and there are many other examples of pleiotropy. It can be difficult to determine which trait best characterizes a disorder.

3.10 Epistasis

The concept of epistasis, which refers to the interaction of different genes, has garnered significant attention in complex disease genetics. However, the lack of consensus in the literature regarding its definitions and interpretations has posed major challenges in detecting and comprehending these interactions. Through historical research, epistatic interaction effects have been discovered, revealing the discrepancies between commonly used definitions. This phenomenon is especially concerning for complex disorders like diabetes, asthma, hypertension, and multiple sclerosis, which have complicating factors such as an increased number of contributing loci and susceptibility alleles, incomplete penetrance, and environmental effects. These factors can alter or mask the effect of one locus, reducing the ability to detect it and impeding the elucidation of the joint effects at the two loci through interaction. The situation becomes even more complex with the involvement of more than two loci, leading to complex multiway interactions among some or all the contributing loci.

3.11 Lethal Alleles Represent Essential Genes

Lethal alleles are genes that are essential for the survival of an organism, and many genes are required to sustain life. Although a loss of function mutation resulting in a non-functional gene product may sometimes be tolerated in the heterozygous state, such mutations are recessive lethal. Homozygous recessive individuals cannot survive, and the time of death depends on when the product is essential, which could be during development, early childhood, or even adulthood in mammals. In some cases, the allele responsible for a lethal effect in the homozygous state may also result in a distinct mutant phenotype when present in the heterozygous state. For instance, a mutation that causes yellow coat color in mice was discovered in the early 1900s and resulted in a yellow coat phenotype that differed from the normal agouti coat phenotype, as shown in Figure 3.10. Crosses between various combinations of the two strains yielded unexpected outcomes.

Crosses

(A) Agouti x agouti ® All agouti

(B) Yellow x yellow ® 2/3 yellow: 1/3 agouti

(C) Agouti x yellow ® ½ yellow: ½ agouti

These results are explained based on a single pair of alleles concerning coat color, mutant yellow allele is *AY*. It is dominant to the wild-type agouti allele *A*, hence heterozygous mice will have a yellow coat color. However, the yellow allele also behaves as a homozygous recessive lethal. Mice of the genotype *AY AY* die before birth, so no homozygous yellow mice are ever recovered. The genetic basis for these three crosses is shown in **Figure 3.11**.

Molecular analysis of the *A* gene in both normal agouti and mutant yellow mice has provided insight into how mutation can be both dominant for one phenotypic effect (hair color) and recessive for another (embryonic development). The *AY* allele is a classic example of a "gain of function" mutation. Animals homozygous for the wild-type A allele have yellow pigment deposited as a band on the otherwise black hair shaft, resulting in the agouti phenotype (**Figure 3.11**). Heterozygotes deposit yellow pigment along the entire length of hair shafts because of the deletion of the regulatory region preceding the DNA coding region of the *AY* allele. Without any means to regulate gene expression, one copy of the *AY* allele is always turned on in heterozygotes, resulting in the gain of function leading to the dominant effect.

The homozygous lethal effect has also been explained as a result of molecular analysis of the mutant gene. The extensive deletion of genetic material characterizing the *AY* allele extends into the coding region of an adjacent gene (*Merir*, rendering it nonfunctional). It is this gene that is critical for embryonic development. It is the "loss of function" in *AY*/*AY* homozygotes that causes lethality.

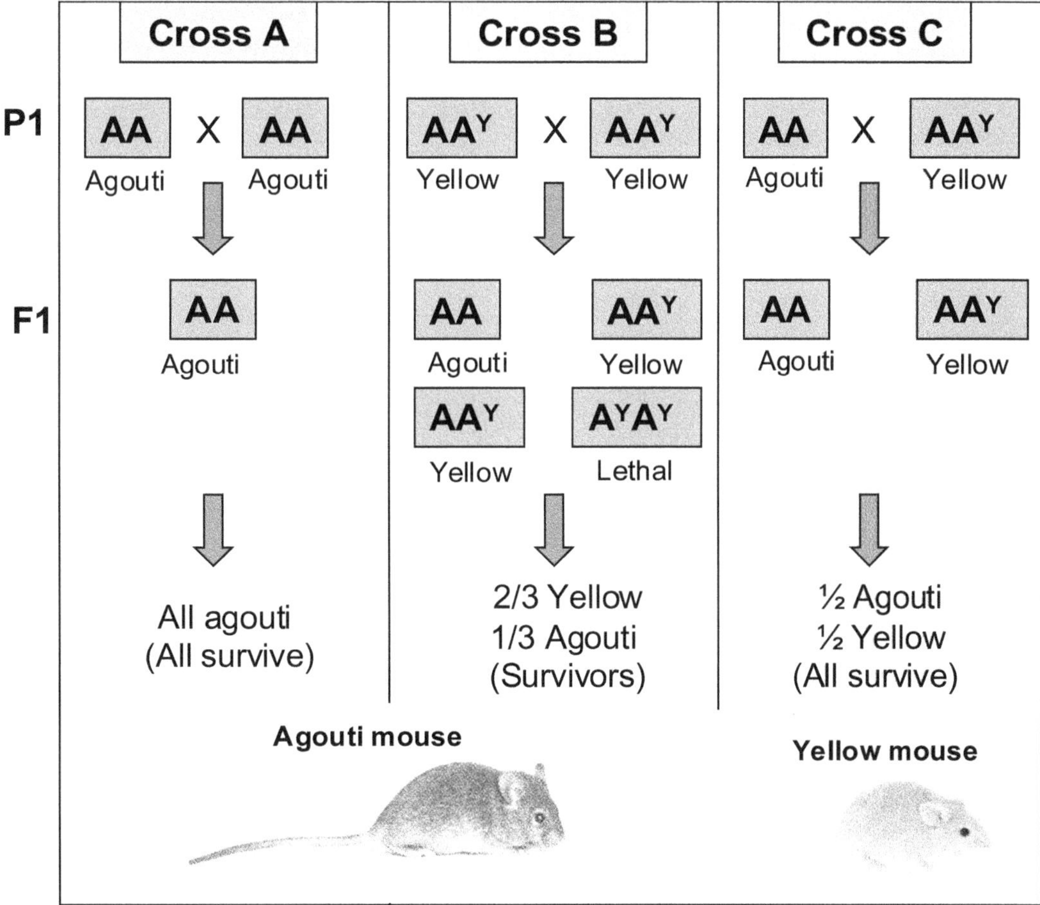

Figure 3.11: Three Crosses involving the Normal Wild Type Agouti Allele (A) and the Mutant Yellow Allele (AY) and Respective Phenotype in Mouse.

Heterozygotes exceed the threshold level required for the wild type *Merir* gene product.

Many genes are known to exhibit similar properties in other organisms. In *Drosophila, Curly* wing (*Cy*), *Plum* eye (*Pm*), *Diehaete* wing (*D*), *Stubble* bristle (*Sb*), and *Lyra* wing (*Ly*) behave as homozygous lethal but are dominant for the expression of the mutant phenotype when heterozygous.

3.12 Dominant Lethal Mutations

In some cases, one copy of the wild-type gene is insufficient for normal development; hence the mutation is called a dominant lethal mutation. Sometimes the presence of the mutant gene product may somehow override the normal function of the wild-type product. In Huntington's disease (HD), the gene for the protein huntingtin mutates by expanding the CAG repeats number which results in the protein having an expanding glutamine tract with an expanded glutamine tract. The length of the expanded CAG repeats correlates inversely with the age of onset of disease. Individuals with CAG repeats below a certain threshold (about 35 repeats) may live longer till old age with no symptoms, while those above 40 repeats manifest symptoms at some point. Between 35 and 39 repeats, there is variable penetrance, a phenomenon once thought not to occur in HD families. More than 60 repeat results in juvenile-onset HD.

However, when the variation in repeat length is controlled for, variation in onset is still highly heritable, suggesting that other genetic factors modify the disease process. Moreover, genetic factors may also modify the phenotypic expression of the disease gene, resulting in variable symptoms. The lethal disorder is particularly tragic because an affected individual may be married and have children before discovering the condition as it is late on the set disease. Affected individuals are heterozygous (Hh), having received the mutant allele from one of their parents. Thus, each of their offspring has a 50 per cent probability of inheriting the lethal allele and developing the disease.

Another interesting situation is when the combinations of two gene pairs involving two modes of inheritance modify the 9:3:3:1 ratio (**Figure 3.12**).

Monohybrid ratio of 3:1 modified due to incomplete dominance, co-dominance multiple alleles, *etc.* Let us take two different modes of inheritance in humans. Albinism is caused by homozygous autosomal recessive alleles. While the AB blood group is due to the co-dominance of A and B alleles of ABO blood groups. Let us suppose two individuals of AB blood group and heterozygous for albinism marry. The conventional Mendelian ratio of 9:3:3:1 will vary. This is demonstrated in **Figure 3.12.**

It is puzzling to note that there are only two genes that interact in such a manner that they produce green, tan, gray, and brown seed colors in lentils in the F2 generation. The ratio is 9:3:3:1 *i.e.,* 9 brown, 3 tan, 3 grey, and 1 green. The two genes assort independently, showing that the seed color coat is determined by two independently assorting genes in a dihybrid cross (**Figure 3.13**)

Two gene hypotheses can explain the above observations. In this case, the green genotype is pure breeding *i.e.* aa, bb, two types of tans: pure -breeding AA bb as well as tan–and green producing Aa bb, two types of grays: pure –breeding aa BB and gray– and green– producing aa Bb, yet, four types of browns– and tan-producing AA Bb, brown– and gray– producing Aa BB, and Aa Bb dihybrids that give rise to plants producing lentils of all four colors.

In short, for the two genes that determine seed coat color, both dominant alleles must be present to yield brown (A- B-); the dominant allele of one gene produces tan (A- bb); the dominant allele of the other specifies gray (aa B-); and the complete absence of dominant alleles (that is, the double recessive) yields green (aa bb). Thus, the four-color phenotypes arise from four genotypic classes, with each class defined in terms of the presence or absence of the dominant alleles of two genes: (1) both present (A- B-), (2) one present (A – bb), (3) the other present (aa B-), and (4) neither present (aa bb). Note that the A – notation means that the second allele of this gene can be either A or a while B- denotes the second allele of either B or b. Note also that only with a two-gene system in which the dominance and recessiveness of alleles at both genes are complete can the nine different genotypes of the F2 generation be categorized into the four genotypic classes described. With incomplete dominance or codominance, the F2 genotypes could not be grouped in this simple way, as they would give rise to more than four phenotypes.

Further crosses between plants carrying lentils of different colors confirmed the two –gene hypothesis (**Figure 3.13c**). Thus, the 9:3:3:1 phenotypic ratio of brown to tan to gray to green in an F2 descended from pure-breeding tan and pure breeding gray lentils tell us not only that two genes assorting independently interact to produce the seed coat color, but also that each genotypic class (A-, B-, A- bb, aa B- and aa bb) determines a particular phenotype.

This is not always the case. In some two –gene interactions, the four F2 genotypic classes have been defined by Mendel and they produce fewer than four observable phenotypes because some of the

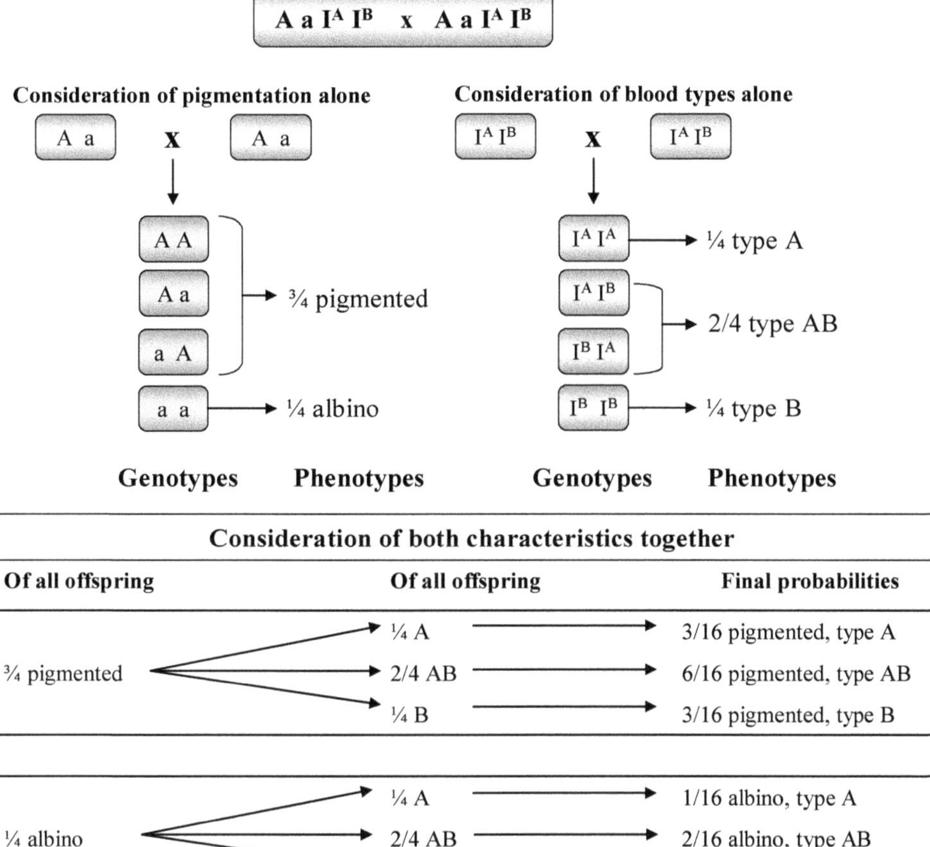

Final phenotypic ratio = 3/16 : 6/16 : 3/16:1/16 : 2/16 : 1/16

Figure 3.12: Fork Lines Demonstrating the Probabilities in a Mating involving the ABO Blood Type and Albinism in Humans.

(a) A dihybrid cross with lentil coat colors

P AA bb X aa BB

Gametes A b a B

F_1 (all identical) ? Aa Bb ? Aa Bb

F_2

9	**A – B – (brown)**
3	**A – bb (tan)**
3	**aa B – (gray)**
1	aa bb (green)

	A B	A b	a B	a b
A B	AA BB	AA Bb	Aa BB	Aa Bb
A b	AA Bb	AA bb	Aa Bb	Aa bb
a B	Aa BB	Aa Bb	aa BB	aa Bb
a b	Aa Bb	Aa bb	aa Bb	aa bb

(b) Self pollination of the F2 to produce an F3

Phenotypes of F_2 individual	Observed F_3 phenotypes	Expected proportion of F_2 population
Green	Green	1/16
Tan	Tan	1/16
Tan	Tan, green	2/16
Gray	Gray, green	2/16
Gray	Gray	1/16
Brown	Brown	1/16
Brown	Brown, tan	2/16
Brown	Brown, gray	2/16
Brown	Brown, gray, tan, green	4/16

This 1:1:2:2:1:1:2:2:4 F2 genotypic ratio corresponds to a 9 bro wn: 3 tan: 3 gray: 1green F2 phenotypic ratio.

(C) Sorting out the dominance relation by select crosses

Seed coat color of parents	F_2 phenotypes and frequencies	Ratio
Tan X green	231 tan, 85 green	3:1
Gray X green	2586 gray, 867 green	3:1
Brown X gray	964 brown, 312 gray	3:1
Brown X tan	255 brown, 76 tan	3:1
Brown X green	57 brown, 18 gray, 13 tan, 4 green	9:3:3:1

Figure 3.13: Seed Colour in Lentils due to the Interaction of Gene A and B.

phenotypes include two or more genotypic classes. For example, in the first decade of the twentieth century, William Bateson conducted a cross between two lines of pure-breeding white-flowered sweet peas (**Figure 3.13**). Quite unexpectedly, all of the F1 progeny were purple. Self–pollination of these novel hybrids produced a ratio of 9 purple: 7 white in the F2 generation. The explanation is that two genes work in tandem to produce purple sweet–pea flowers and a dominant allele of both genes must be present to produce that color. A simple biochemical explanation for this type of complementary gene action is shown in **Figure 3.14**.

Because it takes two enzymes catalyzing two separate biochemical reactions to change a colorless precursor into a colorful pigment, only the A- B- genotypes, which produce active forms of both required enzymes, can generate colored flowers. The other three genotypes (A- bb, aa B-, and aa bb) become grouped for phenotype since they do not specify functional forms of one or the other requisite enzyme and thus give rise to no color, which is the same as white. It is easy to see how the "7" part of the 9:7 ratio encompasses the 3:3:1 of the 9:3:3:1 ratio of two genes in action. The 9:7 ratio is the phenotypic signature of this type of complementary gene interaction in which the dominant alleles of two genes acting together (A-B-) produce color or some other trait, while the other three genotypic classes (A- bb, aa B-, and aa bb) do not.

3.13 Mendelian Principles in Human Genetics

The application of Mendelian principles to human genetics began soon after the rediscovery of Mendel's paper in 1900. However, its growth was limited because of incomplete family records, a smaller number of offspring's, no controlled environment, mistaken paternity and time of manifestation of genetic conditions as many diseases manifest latter in life. However, in spite of all these limitations to date we have learned about thousands of human genes. To apply Mendelian genetics to humans it is important to know some of the terms. A pedigree is a representation of our family tree. It shows how individuals within a family are related to each other. We can also indicate which individuals have a particular trait or genetic

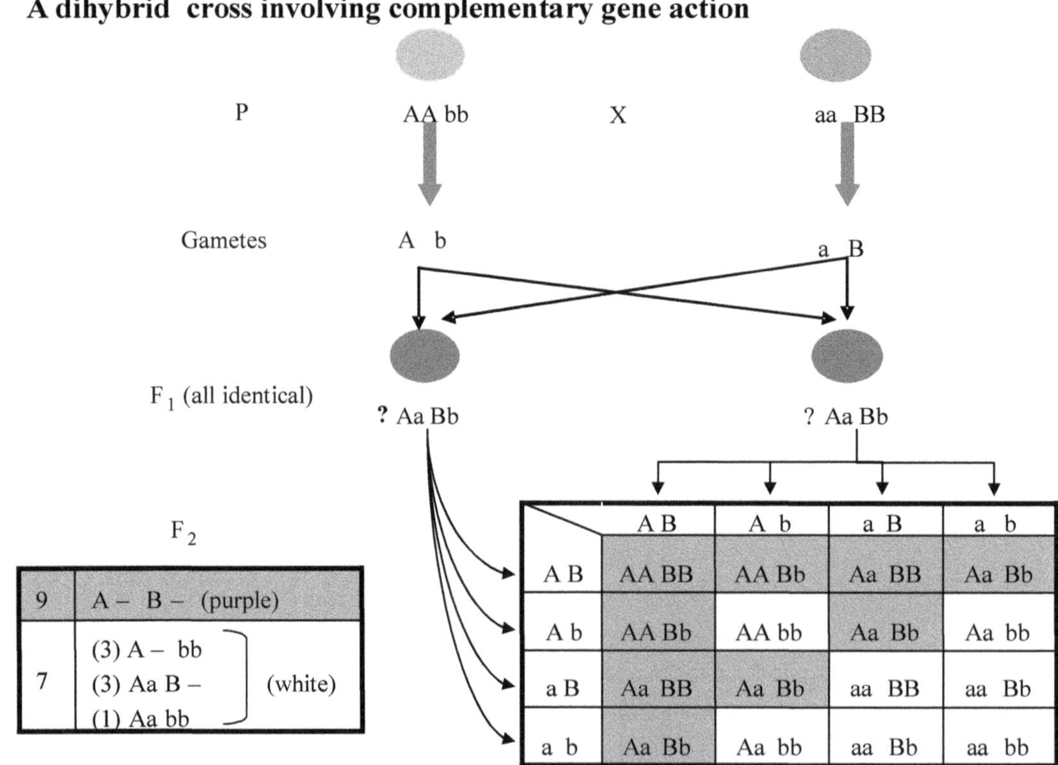

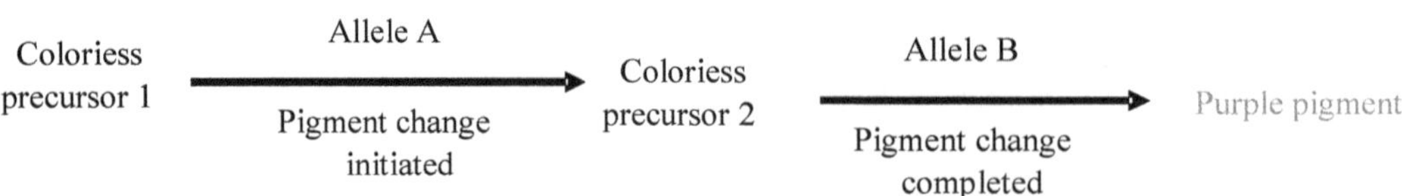

Figure 3.14: Complementary Gene Action and Purple Color in Sweet Pea Flower.

condition. If we take a pedigree, which we usually try to include at least three generations, we might be able to determine how a particular trait is inherited. Using that information, we might be able to tell the chance that a given individual will have the trait themselves or could pass it on to their children. There are standard ways to draw pedigrees so that we can all look at a pedigree and understand it. We use squares to represent males and circles to represent females. We then can number our generations with roman numerals, so the top generation would be generation one, or Roman numeral I. Along this line, we'd indicate males and females. We would indicate marriages between individuals with a horizontal line connecting the two individuals. If an individual has a genetic trait, we will blacken those individuals in or shade them so that it would be understood that they had a particular trait. We would then draw a line, a vertical line, off the horizontal line where we would indicate any of their children that they had, and we would then indicate if any of their children were infected. And we can do this for as many generations as we have. It's important when we draw pedigree that we try to put in as much information as possible. So for example, if there have been children that died in early infancy or were stillborn, we also want to include those individuals. And those are typically shown as very small blackened-in symbols to indicate there was a loss of a child, either in pregnancy or early in life. Pedigree is the diagrammatic representation where relationship among the members of a family is shown. Male is represented as square and female as circles he or she is the consult and (the person who is referred to the clinic). A horizontal line connecting circle and square represents mating. The offspring is shown below the males the firstborn is on the left and proceeds through the birth order to the right. The affected individual is shown as a filled square or circle. The generations are depicted as in Roman numerals and particular individuals are depicted in Arabic numerals. Various symbols used in the pedigree analysis are shown in **Figure 3.15**.

Pedigree analysis is an art, one has to be careful in asking systematically about each person, note their name, date of birth, if dead the data and cause of death, and any relevant clinical information. Ask about miscarriages or reproductive problems and if possible, place of origin of each person. Even if the consultants are convinced that they know which side of the family is the source of the problem, it is wise to collect full details of both sides of family myths can be very misleading.

3.13.1 Autosomal Dominant Inheritance

There are approximately 200 human genetic conditions known to be caused by dominant genes with loci disbursed among all the autosomes. These autosomal dominant (ad, AD) disorders can affect any organ system and occur with different frequencies (**Table 3.4**).

A typical pedigree of an autosomal dominant condition has distinguishing features, a vertical pedigree pattern; with multiple generations affected

Table 3.4: Human Autosomal Dominant Disorders

Disorder	*Prevalence*	*Description*
Huntington disease	1/10,000	Late onset; degeneration in the cerebral cortex and basal ganglia; involuntary movement (chorea); dementia
Neurofibromatosis type 1	1/5,000	Multiple neurofibromas on the nerves of the head, neck and body; pigmented (café-au-lait) spots
Tuberous sclerosis	1/5,800	Multisystem; growths (haematomas) within the brain, eyes, skin, kidneys, heart, lungs, and skeleton
Myotonic dystrophy	1/8,000	Multisystem; prolonged muscle contraction (myotonia); variable muscle weakening and wasting (atrophy; cataracts; defective impulse conduction by the heart; inadequate gonadal function (hypogonadism)
Polycystic kidney disease	1/1,000	Genetically heterogeneous; variable age of onset; kidney cysts; decreased kidney concentrating ability; enlarged kidneys; hypertension
Retinitis pigmentosa	1/4,000	Genetically heterogeneous; progressive loss of night vision and visual acuity
Marfan syndrome	1/10,000	Multisystem; long fingers and toes (arachnodactyly); skeletal deformities; loose joints; ocular lens dislocation; impaired vision; cardiovascular disorders; lateral curvature of the spine (scoliosis); rupture of the aorta
Waardendurg syndrome	1/1,00,000	White forelock; premature graying; different colored eyes; deafness
Hypercholesterolemia	1/500	High serum cholesterol levels; early –onset coronary artery disease
Osteogenesis imperfecta	1/10,000	Genetically heterogeneous; clinically heterogeneous; bone deformity; brittle bones; deafness; blue sclera

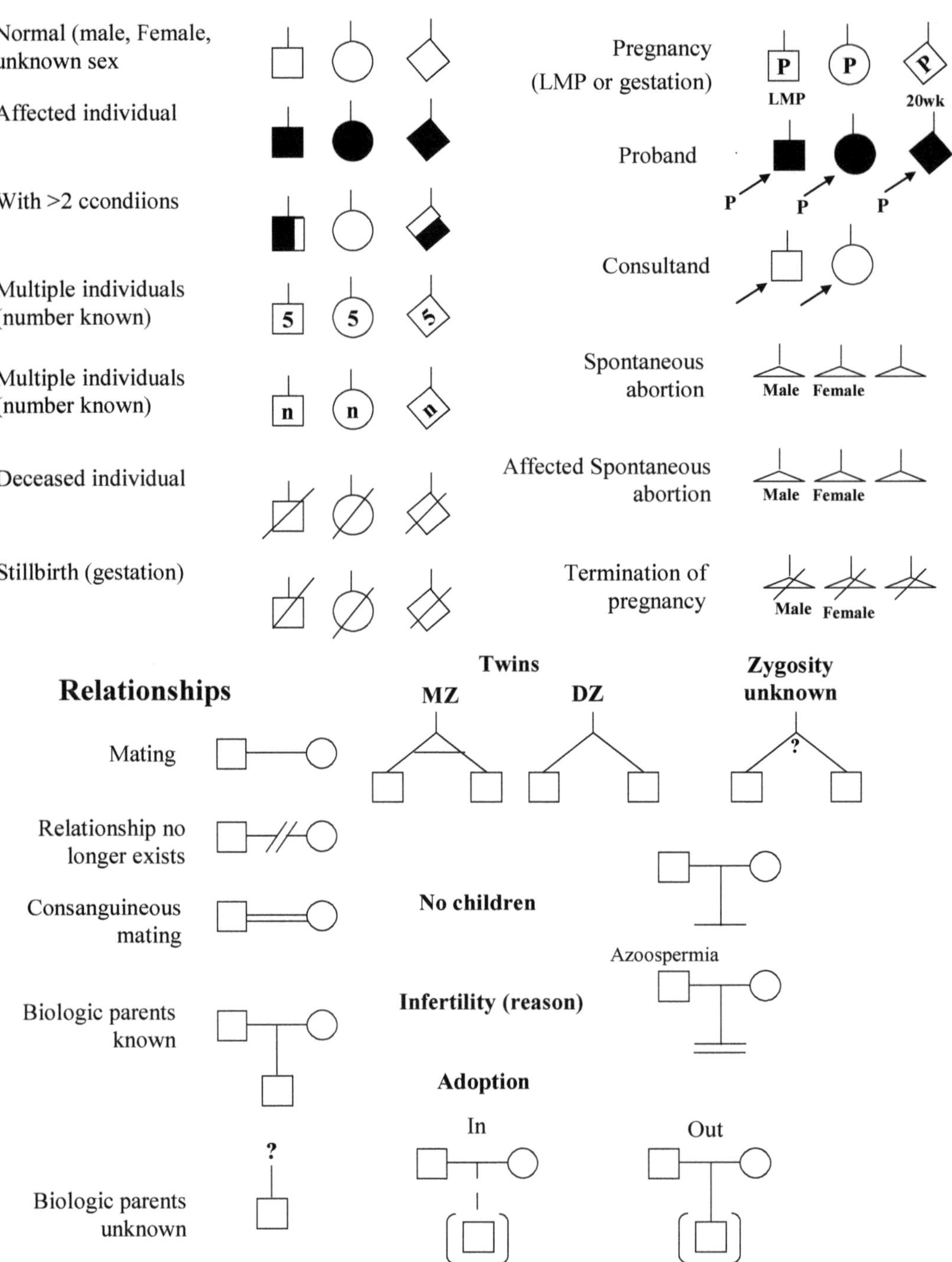

Figure 3.15: Symbols Commonly Used in Pedigree Charts.

each affected person normally has one affected parent equal numbers of affected males and affected females, (**Figure 3.16**). Each child of the affected person has a 1 in 2 chance of being affected.

3.13.2 Autosomal Recessive Inheritance

Approximately 900 different autosomal recessive (AR) disorders affect a range of organ systems and the genes responsible for them are scattered among all the autosomes. Two defective alleles at the same locus cause these disorders. Some human autosomal recessive disorders are shown in **Table 3.5**.

The distinctive features of a pedigree with an autosomal recessive disorder include unaffected parents who have affected offspring, equal numbers of affected males and females, all offspring affected when both parents are affected, and in rare disorders, affected children who are the offspring of marriages between first cousins or other closely related family members (**Figure 3.16-3.17**).

The interpretation of the mode of inheritance from a single pedigree is not always reliable. If a defective recessive allele is frequent, then

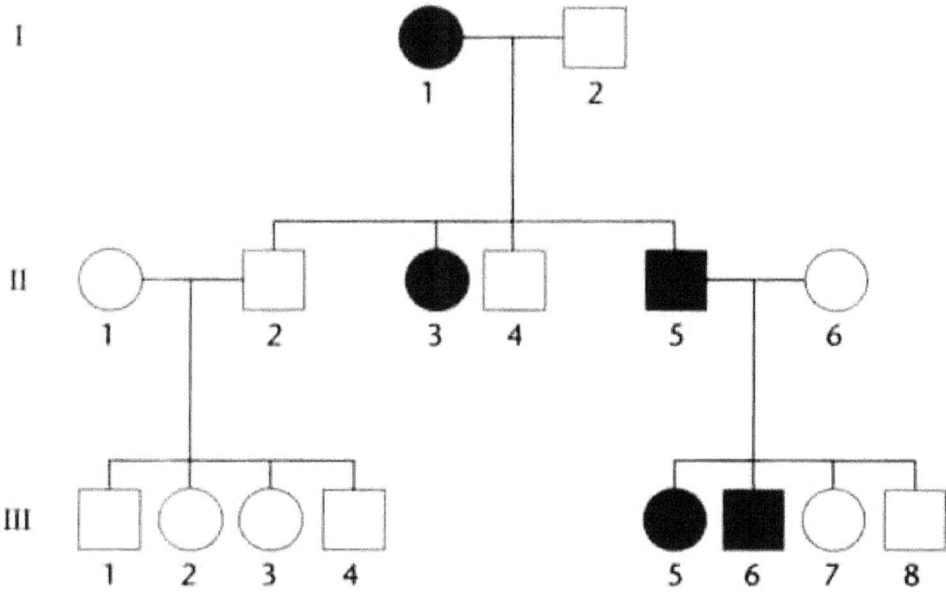

Figure 3.16: A Sample Pedigree with Features of an Autosomal Dominant.

Table 3.5: Catalog of some Human Autosomal Recessive Disorders

Disorder	*Prevalence*	*Description*
Cystic fibrosis	1/2,500	Multisystem; defective chloride transport in epithelial tissue; blockage of Ductule and small airways; severe lung disease; pancreatic insufficiency; chronic inflammation of the nasal sinuses (sinusitis); infertility
Lamellar ichthyosis	1/25,000	Disorder of the skin; disfigurement; large scales; variable redness
Wilson disease	1/40,000	Chronic liver disease; progressive neurological impairment; accumulation of copper in the liver, brain, and other tissues
Gaucher disease type 1	1/50,000	Increased blood cell breakdown in the spleen (hypersplenism); enlargement of the liver and spleen (hepatosplenomegaly); fragile bonest
Friedreich ataxia	1/50,000	Onset at puberty; inability to coordinate muscles for voluntary movement (ataxia); inability to speak (dysarthria); Muscular Atrophy
Childhoodspinal muscular atrophy	1/10,000	Genetically heterogeneous; variable; degeneration of the anterior horns of the spinal cord; weakness and wasting of the muscles; often fatal by 20 years of age or earlier
Phenylketonuria	1/10,000	Deficiency of liver enzyme phenylalanine hydroxylase; brain damage; mental retardation; excess phenylpyruvate in the urine; accumulation of phenylalanine in the blood
β- Thalassemia	1/20,000	Severe depletion of red blood cells (anemia); enlarged spleen (splenomegaly); bone deformities
Galactokinase deficiency	1/40,000	Inability to utilize galactose; cataracts; mild mental retardation
α1 – Antitrypsin deficiency	1/3,500	Breathlessness (emphysema); cirrhosis of the liver

homozygous recessive individuals have a high probability of marrying heterozygotes. In these cases, an autosomal recessive condition may mimic an autosomal dominant pattern. Usually, the nature of the mode of inheritance becomes apparent after pedigrees from many different families with the same condition are assembled and examined. Males and females are equally affected each subsequent sib of an affected child has 1 and 4 chance of being affected I.

3.13.3 X-linked Onheritance

In fertile human males, half the sperm carry a Y chromosome and the other half an X chromosome. All unfertilized eggs of a female receive a single X chromosome. Consequently, within a population, half of the next generation will be XX (females) and the other half will be XY (males) (**Figure 3.18**). The segregation of the sex chromosomes during

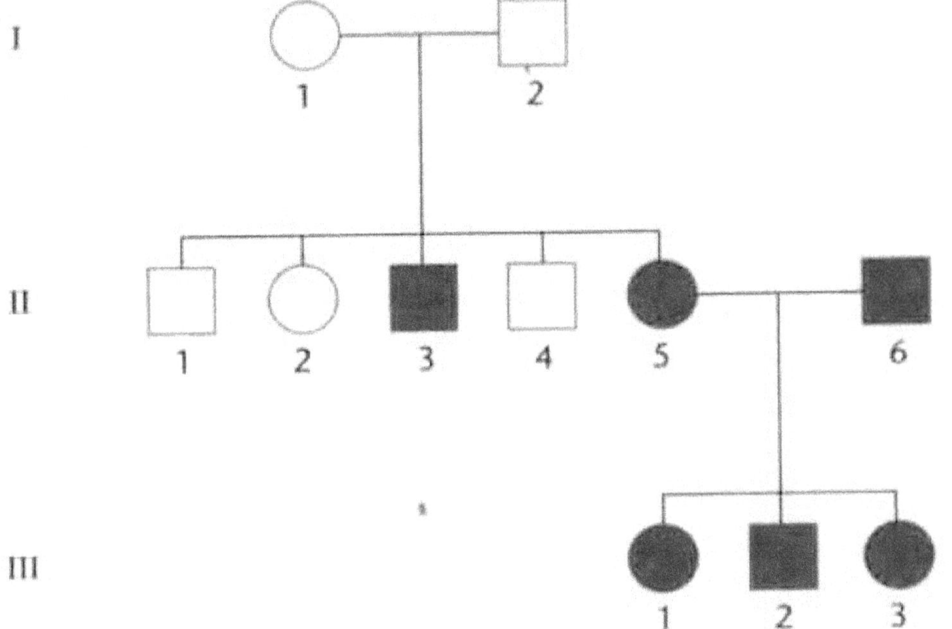

Figure 3.17: A Sample Pedigree with Features of an Autosomal Recessive Disorder.

meiosis is an effective way to maintain a sex ratio close to 1:1 from one generation to the next. With this sex determination system, all sons inherit a Y chromosome from their fathers. In humans, an important gene for determining testis formation (SRY, sex determining region of the Y chromosome) is located at chromosome Yp11.3 position.

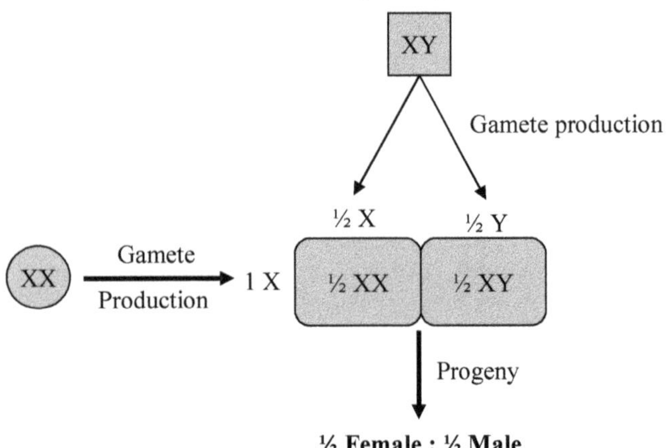

Figure 3.18: Distribution of Sex Chromosomes and Determination of the Sex Ratio in Humans.

Except for two segments (at least), which are designated pseudo autosomal regions (**Figure 3.19**), the sex chromosomes do not share much chromosomal material. The pseudo autosomal regions are locations where synapsis and crossing over occur between X and Y chromosomes. Synapsis of the X and Y chromosome ensures the correct segregation of the sex chromosomes during meiosis. Both synapsis and crossing over are more common between the Xp-Yp (1-1) than the Xq – Yq (2-2) pseudo autosomal regions.

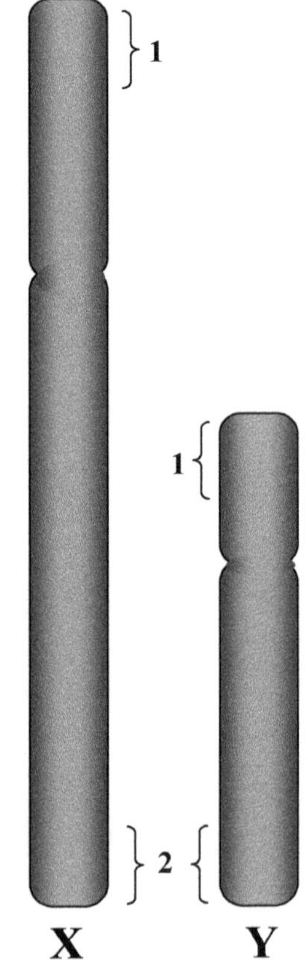

Figure 3.19: Human Sex Chromosomes.

The numbered brackets mark the approximate locations of two pseudo autosomal regions. The Xp–Yp pseudo autosomal regions (1–1) synapse more frequently than the Xq–Yq pseudo autosomal regions (2–2).

Approximately 23 genes have been localized to the non-pseudo autosomal portion of the Y-chromosome. These Y– linked genes are called holandric genes. The genetic makeup of the X chromosome is well known. More than 285 X-liked genes have been identified.

A single X-linked recessive allele is sufficient to produce a phenotype in human males, because there is little homologous genetic information shared between the X and Y chromosomes. Consequently, in these cases, there is virtually no chance of a normal dominant allele masking the effect of a recessive allele. The dominant or recessive effect of an X-linked gene is determined by the phenotype in females because they have two X chromosomes. In pedigrees with an X-linked dominant condition, each generation usually has one affected individual. All daughters of affected males are affected, both sons and daughters of an affected heterozygous female may be affected, and generally, twice as many females as males are affected (**Figure 3.20**). In pedigrees with X-linked recessive disorders, all sons of an affected mother are affected, affected fathers never transmit the trait to their sons, unaffected parents may have affected offspring, and generally, there are more affected males than females (**Figure 3.21**).

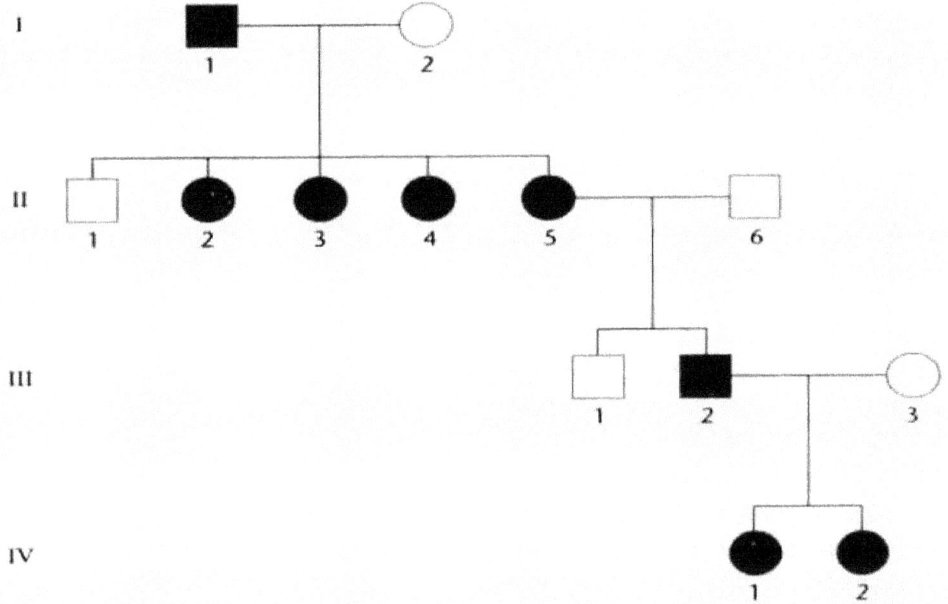

Figure 3.20: A Sample Pedigree with Features of an X-linked Dominant Inherited Disorder.

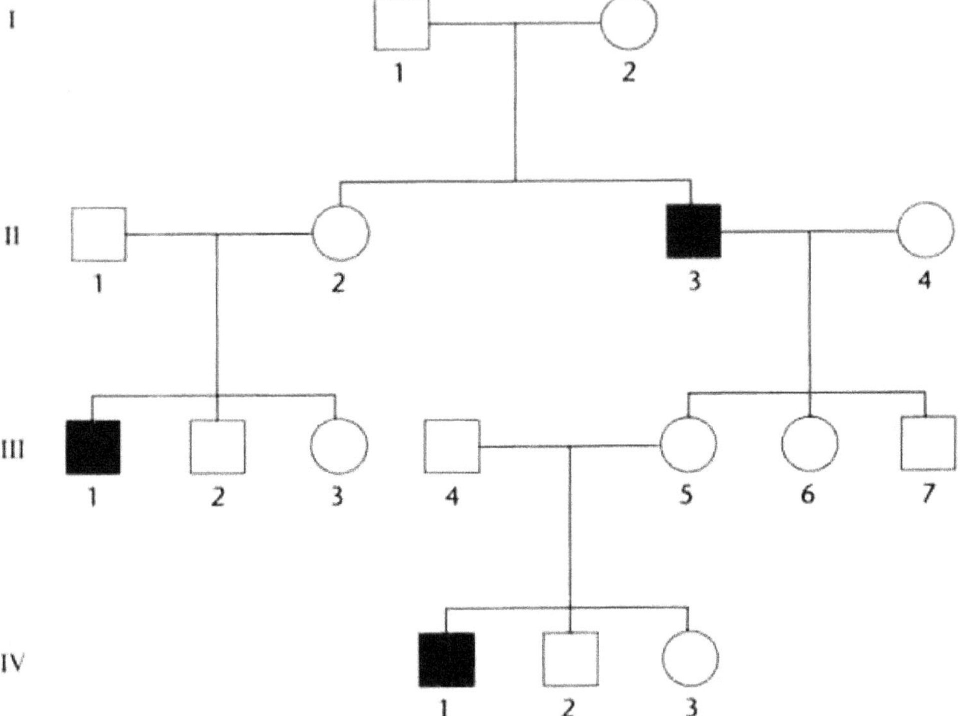

Figure 3.21: A Sample Pedigree with Features of an X-linked Recessive Inherited Disorder.

3.13.4 Example of X-linkage in Humans

In humans, many genes and the traits controlled by them are recognized as being linked to the X chromosome. These X –linked traits can be easily identified in pedigrees, characterized by a crisscross pattern of inheritance. A pedigree for one form of human color blindness is shown in **Figure 3.22**. The mother in generation I pass the trait on to all her sons, but to none of her daughters. If the offspring in generation II have children by normal individuals, the color –blind sons will produce all normal male and female offspring (III-1, 2, and 3); the normal – vision daughters will produce normal –vision female offspring (III-4, 6 and 7), as well as color blind (III-8) and normal –vision (III-5) male offspring.

Many X–linked human genes have not been identified, as shown in **Table 3.6**. For example, the genes controlling two forms of hemophilia and two forms of muscular dystrophy are located on the X chromosome. In addition, numerous genes whose expression yields enzymes are X–linked. Glucose-6-phosphate dehydrogenase and hypoxanthine–guanine– phosphoribosyl transferase are two examples. In the latter case, the severe Lesch–Nyhan syndrome results from the mutant form of the X-linked gene product.

Because of the way in which X-linked genes are transmitted under unusual circumstances may be associated with recessive X–linked disorders in comparison to recessive autosomal disorders. For example, if an X–linked disorder is lethal for the affected individual prior to reproductive maturation,

Table 3.6: Human X-linked Traits

Condition	*Characteristics*
Color blindness, deutan type	Insensitivity to green light
Color blindness, protan type	Insensitivity to red light
Fabry's disease	Deficiency of galactosidase A; heart and kidney defects, early death
G-6-PD deficiency	Deficiency of glucose-6-phosphate dehydrogenase; severe anemic reaction following intake of primaquines in drugs and certain foods, including fava beans
Hemophilia A	Classical form of clotting deficiency; deficiency of clotting factor VIII
Hemophilia B	Christmas disease; deficiency of clotting factor IX.
Hunter syndrome	Mucopolysaccharide storage disease resulting from iduronate sulfatase enzyme deficiency; short stature, claw like fingers, coarse facial features, slow mental deterioration, and deafness
Lchthyosis	Deficiency of steroid sulfatase enzyme; scaly dry skin, particularly on extremities
Lesch-Nyhan syndrome	Deficiency to hypoxanthine –guanine phosphoribosyl transferase enzyme (HPRT) leading to motor and mental retardation, self mutilation, and early death
Muscular dystrophy	Progressive, life-shortening disorder characterized by muscle degeneration and weakness; (Duchene type) sometimes associated with mental retardation; deficiency of the protein dystrophin

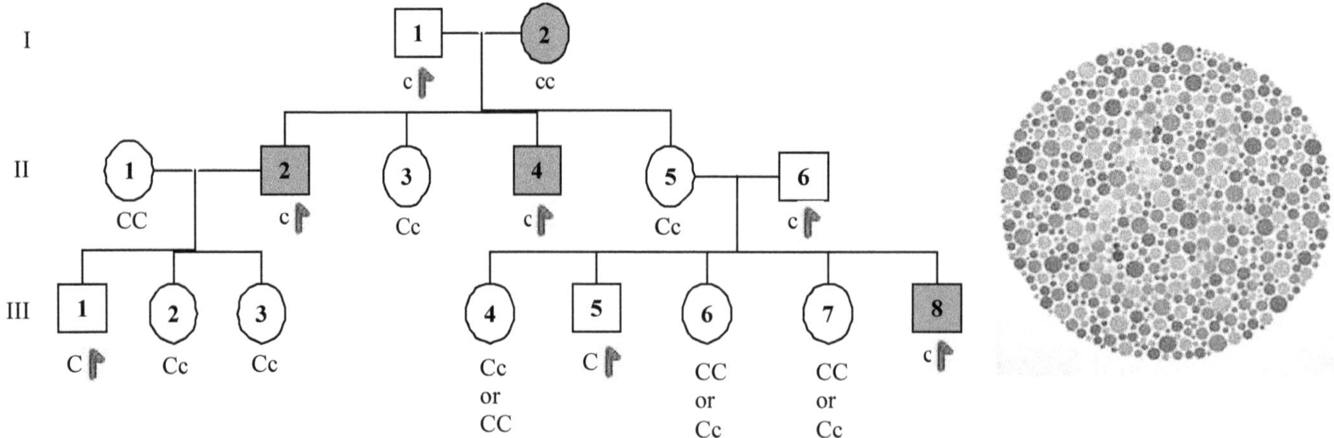

Symbols

C = normal vision; c = color blindness;

= Y chromosome

Figure 3.22: A Human Pedigree of the X-linked Color-Blindness Trait.

The most portable genotypes of everyone in the pedigree. The photograph is of an Ishihara color-blindness chart. Red-green color-blind individuals see a 3 rather than the 8 visualized by those with normal vision.

the disorder occurs exclusively in males. This is because the only source of the lethal allele in the population is heterozygous females who are "carriers" and do not express the disorder. They pass the allele to half of their sons, who develop the disorder because they are hemizygous but rarely reproduce. Heterozygous females also pass the allele to half of their daughters, who like their mothers, become carriers but do not develop the disorder. Examples of such an X–linked disorder in humans include the Duchene form of muscular dystrophy (DMD) and Lesch-Nyhan disease. DMD has an onset prior to age 6 and is often lethal around 20. Affected males are unable to reproduce.

3.14 Lesch–Nyhan Syndrome: The Molecular Basis of a Rare X-linked Recessive Disorder

Lesch–Nyhan syndrome (LNS) is a devastating disease that is first apparent in infants at age 3 to 6 months when orange particles (sometimes referred to as orange sand) appear in the urine and discolor the affected infant's diaper. The urinary stones contain urate crystals which harbor many future difficulties that lead to an early death. It is found in males and is the result of the complete or nearly complete loss of activity of a critical enzyme, hypoxanthine-guanine phosphoribosyl transferase (HPRT). This enzyme imparts the ability to metabolically recycle purines, one of the two major types of nitrogenous bases that make up nucleotides in DNA. While purines adenine and guanine can be synthesized from basic chemical components, mammals have evolved the ability to extract them from DNA that is being degraded, recovering them in the form of the purine hypoxanthine. Under the direction of HPRT< hypoxanthine can be converted back to adenine and guanine-containing nucleotides. When this mechanism fails because of mutation, the excess hypoxanthine is converted to uric acid, which accumulates well beyond the body's ability to excrete it.

This metabolic or biochemical disorder shows different effects severe form is mental retardation, seizures, and aggressive, uncontrolled spastic movements (resembling cerebral palsy) that include self-mutilation of the fingers and lips. Patients need 24–hour care throughout their lives and almost always die before the 3rd decade of their life mostly at this time kidney failure occurs.

The gene involved in LNS is located on the long arm of the X chromosome and consists of 44,000 base pairs (44kb). However, the HPRT gene product has only 654 base pairs to encode it. The gene contains 9 exons and 8 introns. Mice have a nearly identical gene that is 95 per cent homologous to their human counterpart.

In normal individuals the enzyme is ubiquitous in tissues throughout the body, however, found in higher concentrations in the basal ganglia of brain cells. The behavioral phenotype of the LNS patients depends upon the concentration or presence or absence of the enzyme. There is no cure for LNS patients. Because the responsible gene is recessive and X-linked and since affected males never reproduce, females, while they can be carriers of the mutant gene, never become homozygous and never develop LNS. To summarize X-linked recessive; X-linked dominant and Y-linked we should keep the following points in mind.

3.15 X-linked Recessive

Affected boys may have affected maternal uncles, never transmitted from father to son, affects males, females are carriers, affected males in a pedigree are linked through females and not through unaffected males. Subsequent brothers of affected; sisters are not affected but have a 1 in 2 risks of being carriers.

3.16 X-linked Dominant

Quite like autosomal dominant except that all daughters and no sons of an affected father are affected. Conditions are often milder.

3.17 Y-linked

All sons of affected father are affected therefore it affects only males.

3.18 X-linked Inheritance with Cellular Interference

There is an X linked syndrome known as Craniofrontonasal syndrome. In this syndrome, heterozygous females are more severely affected as compared to hemizygous males. This is caused by a loss of function mutation in the EFNB1 gene that encodes the ephrin cell signaling molecule. Ephrin on a cell surface results in repulsive interactions with cells that carry a matching ephrin receptor. In females, there are some of the cells lacking ephrin hence correct tissue boundaries are not maintained. In males, if this mutation is there other members of the ephrin gene family take over.

3.19 Sex Bias

This is because of physiological or anatomical differences between the two sexes. For example, mutations in the BRACA2 gene on chromosome 13 causes ovarian cancer in females. These conditions are called sex limited.

3.20 In Sex-limited and Sex-influenced Inheritance, an Individuals' Sex Influences the Phenotype

Some time the phenotype is absolutely limited to one sex. Such types of characters are called sex-limited. In humans, the well-cited example is of baldness in males. The pattern of baldness in humans, where the hair is very thin or absent on the top of the head, is inherited in the following way as demonstrated in **Table 3.7**.

Table 3.7: Sex Limited Traits

Genotype	*Phenotype*	
	♀	♂
BB	Bald	Bald
Bb	Not bald	Bald
Bb	Not bald	Not bald

Even though females can display pattern baldness, this phenotype is much more prevalent in males. When females inherit the BB genotype, the phenotype is much less pronounced than in males and is expressed later in life.

Rett syndrome (OMIM 312750) if males are affected generally the pregnancy is lost in the early gestational period. Hence mostly females with this disease were seen in some males are survived but their phenotype is so different that they were not identified till molecular tests for MECP2 were not available. This gene is responsible for Rett syndrome.

3.21 Mitochondrial Inheritance

A vertical pedigree pattern, children of affected men are never affected, all children of affected women may be affected. However, mitochondrial conditions are typically extremely variable even within a family. Mitochondria are derived from the egg.

3.22 Genetic Heterogeneity

A genetic disorder, which may appear to be a single entity, is frequently found to be genetically heterogeneous on closer analysis. That is, it may include phenotypes that are similar but are determined by different genotypes. Genetic heterogeneity may be the result of different mutations at the same locus (allelic heterogeneity), mutations at different loci (locus heterogeneity), or both.

3.23 Locus Heterogeneity

When the same disease phenotype can be caused by a mutation in different loci (genotype) it is termed as locus heterogeneity osteogenesis imperfect a type-2 can be caused by mutation either in collagen gene is present on Chr.17 or Chr.7. Another disease retinitis pigmentosa, a common cause of visual impairment due to photoreceptor degeneration is associated with abnormal pigment distribution in the retina, has long been known to occur in autosomal dominant, autosomal recessive, and X-linked forms. Recent DNA analysis has shown that there are at least 3 X-linked forms, 12 autosomal dominant forms, and five autosomal recessive forms of retinitis pigmentosa, and there are many genetic diseases that can manifest with retinitis pigmentosa in conjunction with other defects. Autism is another disorder with locus heterogeneity, and multiple genome screens have been performed to identify potential linkage regions with no consistent candidate gene yet. However, a recent study using extended autism families identified a novel linkage peak on chromosome 12q, revealing substantial locus heterogeneity in autism. Certain diseases can have different causes, known as locus heterogeneity. For example, osteogenesis imperfecta type-2 can be caused by mutations in either the collagen gene on Chr.17 or Chr.7. Similarly, retinitis pigmentosa, a common cause of visual impairment, can be caused by changes in genes that control cells in the retina. These changed genes are passed down from parents to children. RP is linked to many different genes and can be inherited in different ways.

3.24 Allelic Heterogeneity

Allelic heterogeneity may result in clinical variation. Many loci own more than one mutant allele, and at a given locus, there may be quite a few or many mutations. Sometimes, these different mutations clinically cause similar disorders. In other cases, different mutant alleles at the same locus result in very different clinical presentations. For example, some mutations in the RET gene, which encodes a receptor tyrosine kinase, can cause dominantly inherited failure of development of colonic ganglia, leading to defective colonic motility and severe constipation (Hirschsprung disease), whereas other mutations in the same gene result in dominantly inherited cancer of the thyroid and adrenal glands (multiple endocrine neoplasia type IIa and IIb). The third group of mutations in RET causes both Hirschsprung disease and multiple endocrine neoplasia in the same individuals.

3.25 Atypical Patterns of Inheritance

In genetic medicine, single-gene disorders typically adhere to the principles of Mendelian inheritance. However, there are exceptions to these rules that must be taken into account when analyzing mutations or observing unusual disorders. These variations from the typical pattern can be caused by allelic heterogeneity, genetic heterogeneity, locus heterogeneity, or linkage.

When presented with a patient who exhibits unusual symptoms, medical professionals will compare those symptoms to known syndromes. If a match is not found, further research will be conducted to determine whether the condition is inherited or caused by environmental factors.

If the patient's immediate family members also have the condition, medical professionals will create a pedigree to study the genetic inheritance pattern. However, a single pedigree may not provide sufficient information to deduce the mode of inheritance unless it is quite large with many affected individuals. In these cases, studying additional pedigrees can provide more information. Medical professionals must look for physiological and physical differences to distinguish between the effects of different loci as genetic heterogeneity is a common issue.

Incomplete penetrance is also a confounding factor in pedigree analysis. Sometimes, even if a person has a specific genotype, the expected phenotype may not be produced. Penetrance is a measure of the frequency of occurrence of the phenotype when a particular genotype or allele is present. Incomplete penetrance occurs when the penetrance of a genotype is less than 100 per cent. The age of onset of a disorder can also vary over time, causing individuals to be misclassified in pedigrees.

Sometimes, people in the same family can have different physical characteristics or medical conditions even if they have the same genetic makeup. This is called phenotypic variation or clinical heterogeneity. Osteogenesis imperfecta (OI) is a condition that can cause brittle bones, fragile skin, loose joints, deafness, soft teeth, and blue tinted sclera. Even within the same family, some members may only have blue sclera while others may have severe physical limitations due to brittle bones.

When trying to understand how diseases are inherited through family trees, there are many complex factors to consider. Incomplete penetrance can happen when a gene product from a different gene location cancels out the effects of a disease-causing gene. Age-dependent penetrance happens when tissue damage builds up over time due to a changed gene product. In these cases, it may not be clear that someone has a disease until a certain level of damage or cell death has happened. Clinical heterogeneity can also happen if other genes change how a disease-causing gene works. Even though these factors can be complicated, it's important to think about them when interpreting how diseases run in families.

3.26 Genotype-Phenotype Correlation

In the field of genetics, the interaction between genes or genotypes can result in a particular phenotype. However, the expected phenotype may not always manifest even if the genotype is present. This can occur due to various factors like the environment, chance, and modifier genes.

3.27 Penetrance and Expressivity

Penetrance and expressivity are commonly used terms in genetic analysis to explain why some genetic traits or illnesses may not be expressed even with the presence of the genetic code. Penetrance refers to the frequency of gene expression, presented as a percentage of the population with the gene and corresponding phenotype. On the other hand, expressivity refers to the extent of gene expression in an individual, also expressed as a percentage.

3.28 Difference between Penetrance and Expressivity

While penetrance and expressivity may seem similar, they differ in that penetrance relates to gene expression frequency in a population, while expressivity relates to the degree of expression in an individual. Both concepts can vary depending on factors such as age, genetic makeup, and environment.

- It's crucial to understand penetrance and expressivity in order to fully grasp genetic inheritance. Penetrance refers to how often a gene is expressed in a population, while expressivity relates to how strongly a gene is expressed in an individual. These concepts can be influenced by various factors and are vital to understanding genetic inheritance.
- For instance, retinoblastoma, a type of eye cancer, is caused by a dominant mutation of one gene, but only 75 per cent of people with the mutant allele develop the disease. This is an example of incomplete penetrance. Additionally, expressivity refers to the severity of a particular genotype in a phenotype and can also vary.
- Cystic fibrosis is a disease inherited recessively from mutations in the cystic fibrosis transmembrane conductance regulator (CFTR) gene. Patients with cystic fibrosis exhibit individual differences, and the severity of the disease in the lungs can differ among siblings who carry the same CFTR mutations. This shows how the environment and other non-CFTR genetic factors, called modifier genes, can affect the phenotype. Modifier genes change the phenotypes produced by the alleles

of other genes, such as inflammation, airway reactivity, and alternative ion channels.

- ✰ Understanding the complexities of penetrance and expressivity at the molecular level can be challenging as variable phenotypes can be caused by various factors. Nonetheless, having a thorough understanding of these concepts is crucial in comprehending genetic inheritance. Modifier genes
- ✰ Environmental factors
- ✰ Allelic variation
- ✰ Complex genetic and environmental interactions

In most cases in which a particular genotype is inherited, it is not fully known why the same allele can cause subtly different or profoundly different phenotypes. In some cases, however, there is genetic evidence that modifier genes influence phenotypic variation. Modifier genes can affect penetrance, dominance, and expressivity. A genetic modifier, when expressed, can alter the expression of another gene. Modifier genes can affect transcription and alter the immediate gene transcript expression, or they can affect phenotypes at other levels of the organization by altering phenotypes at the cellular or organismal level.

Some examples of modifier genes identified in mice and humans, along with their modifier effects and phenotypic consequences. As you can see from the table, many more modifiers have been identified in mice than in humans because of the ability to perform gene targeting experiments on inbred mouse strains. For example, an unknown modifier associated with specific mouse genetic backgrounds can alter the penetrance of an allele (called disorganization), which causes birth defects. Likewise, another unknown modifier that is associated with specific mouse genetic backgrounds affects the expressivity of an allele resulting in mice with different tail lengths. A third modifier gene, named dsu, suppresses the expression of several different coat color alleles in mice, including ashen, leaden, ruby-eye, and ruby-eye-2. To better understand the role of modifier genes in humans, in one of the study 141 members of a family afflicted with isolated deafness (*i.e.*, deafness was the only clinical finding in this family), which is caused by the DFNB26 gene. The researchers noticed that the disease was incompletely penetrant. While most individuals who were homozygous for the gene were deaf, seven homozygous family members had normal hearing. Through linkage studies, the scientists identified a dominant modifier, called DFNM1, which suppressed deafness in homozygous individuals. It is not yet known how this modifier gene suppresses deafness. It may suppress the mutant gene's expression or prevent deafness through another mechanism.

3.29 Mosaicism

If a person's body carries two or more genetically different cell lines it is called a mosaic. During the process of cell division (mitosis) every cell should get a complete and identical set of genes. However, if some cell carries the mutated gene or abnormal chromosome and these cells tend to grow. Sometimes the mutation arises sufficiently early in embryonic development and the mutant line makes up a significant part of the whole body. The person may show features of the disease with a milder phenotype or with a patchy distribution reflecting the distribution of mutant cells. In case the mutation affects the germline (sperm or egg or their progenitors).

Mosaicism causes problems in pedigree analysis. Let us suppose a person is all normal but produces children who mimic dominant or X-linked disease this shows that there is mosaicism for the mutant gene in the egg (in the mother) or the sperm (in father). One of the examples is neurofibromatosis type 2 (NF2), NF2 is a rare genetic condition that is primarily characterized by benign tumors of the nerves that transmit sound impulses from the inner ears to the brain. The pattern of the disease is autosomal dominant. If a man or women has germline mosaicism for NF2, they have a mixture of sperm or egg cell some having the mutant NF2 while other as normal NF2. If the mother's 50 per cent egg cell contains a mutant gene and 50 per cent of children receive this gene, they will be affected. In such a type of inheritance, it is difficult to assess the exact inheritance pattern as we do not know what per cent of cells have the mutant genes.

3.30 The Environment can Affect the Phenotypic Expression of a Genotype

Temperature can have a visible effect on phenotype. For example, temperature influences the unique coat color pattern of Siamese cats. These cats are homozygous for one of the multiple alleles of a gene that encodes an enzyme catalyzing the production of the dark pigment melanin. Interestingly the enzyme generated by the variant "Siamese" allele does not function at the cat's normal body temperature. It becomes active only at the lower temperatures where it promotes the production of melanin, which darkens the cat's ears, nose, paws, and tail. The enzyme is temperature sensitive. Under the normal environmental conditions in temperate climates, the Siamese phenotype does not express different phenotypes. Temperature can also affect survivability. Drosophila melanogaster develops and multiplies normally at temperatures between 180C and 290C; but if the temperature rises beyond normal limits for a short time, fruit flies get reversibly paralyzed, and if the high temperature persists for more than a few hours, they die. These insects carry a temperature-sensitive allele of the shibire gene, which encodes a protein essential for nerve cell transmission, among other functions. This type of allele is known as conditional lethal as it is lethal only under certain conditions. The range of temperatures under which the insects remain viable is the permissive range; the lethal temperatures above which they cannot survive. Thus, at one temperature the allele gives rise to a phenotype that is indistinguishable from the wild type, while at another temperature, the same allele generates a mutant phenotype (lethality). Flies with the wild type shibire allele are viable even at the higher temperatures.

In Drosophila melanogaster an allele named 1(1)su(f)ts76a, is temperature –sensitive and leads to lethality, female sterility, and abnormal bristle formation at 29 degrees C it closely resembles two other conditional alleles of su(f), 1(1)su(f)ts67g and 1(1)ts726. Female sterility at 29 degrees C is characterized by both disorganized egg chambers in the ovarioles and also chorion-deficient oocytes. Both of these abnormalities may be the result of premature follicle cell death. The observations on 1(1)su(f)ts76a are consistent with the proposal that the similar allele, 1(1)ts726, is a cell lethal mutation specifically affecting mitotically active cells.

3.31 Genetic Background: Suppression and Position Effects

Though it is difficult to assess the specific effect of the genetic background and the expression of a gene responsible for determining a potential phenotype, two effects of genetic background have been well characterized.

The expression of other genes throughout the genome may influence the phenotype produced by a gene. One of the examples is genetic suppression. Mutant genes such as suppressor of sable (su-s), suppressor of forked (su-f), and suppressor of Hairy-wing (suHw) in Drosophila completely or partially restore the normal phenotype in an organism that is homozygous (or hemizygous) for the sable, forked, and Hairy-wing mutations, respectively. The flies' hemizygous for both forked (a bristle mutation) and su-f have normal bristles. Let us suppose the suppressor gene causes the complete reversal of the expected phenotypic expression of the original mutation. Suppressor genes act like modifier genes modifying the effect of the primary gene effects.

Second, the physical location of a gene in relation to other genetic material may influence its expression. This is known as the position effect. For example, if a region of a chromosome is relocated or rearranged due to translocation or inversion the normal expression of the other genes located on the chromosome may get modified. This is particularly true if the gene is relocated to or near certain areas of the chromosome that are prematurely condensed and genetically inert, referred to as heterochromatin.

An example of a position effect involves female Drosophila heterozygous for the X-linked recessive eye-color mutant white (w). The w+/w genotype causes a wild-type brick red eye color. However, if the region of the X chromosome containing the wild-type w +allele is translocated so that it is close to a heterochromatic region, the expression of the w +allele is modified. Instead of red color, the eyes are variegated or mottled with red and white patches. Therefore, following translocation, the dominant effect of the normal w + allele is intermittent. A similar position effect is produced if a heterochromatic region is relocated next to the white locus on the X chromosome. It appears that heterochromatic regions may inhibit the expression of adjacent genes. A similar situation is seen in other organisms providing proof that alteration of the normal arrangement of genetic information can modify its expression.

Phenotypic changes may occur due to exposure to chemicals or other environmental agents that are like those caused by mutant alleles of specific genes. This type of phenotype is labeled as phenocopy. Phenocopies are not heritable because they do not arise from a change in a gene. In humans, ingestion of the sedative thalidomide by pregnant women in the early 1960s produced a phenocopy of a rare dominant trait called phocomelia. By disrupting limb development in otherwise normal fetuses, the drug mimicked the effect of the phocomelia-causing mutation. When this became evident, thalidomide was withdrawn from the market.

Some types of environmental change may have a positive effect on an organism's survivability, for *e.g.*, children born with the genetically determined disease known as phenylketonuria, or PKU will develop a range of neurological problems, including convulsive seizures and mental retardation unless they are put on a special diet. Newborn screening can detect this defect. The health care practitioner prescribes a protective diet for such newborns. This diet excludes phenylalanine; the diet must also provide enough calories to prevent the infant's body from breaking down its own proteins. This can be considered as a simple change in diet (environment) that has a protective effect although somewhat unpleasant because it is highly limiting, which now enables many PKU infants to develop into healthy adults.

Finally, two of the top killer diseases in the United States cardiovascular disease and lung cancer also illustrate how the environment can alter phenotype by influencing both expressivity and penetrance. People may have a genetic predisposition to heart disease which is triggered by environmental factors like diet *etc.* Similarly, some people are born genetically prone to lung cancer, but whether they get it (penetrance) is strongly determined by whether or not they choose to smoke. Thus, the environment including temperature, diet, *etc.* interacts with the genotypes and results in the final phenotype.

The biological environment can also influence the phenotypic expression of genes. The pattern of baldness in humans is a well-known example here the biological factor is gender. Premature pattern baldness is due to an allele that is expressed differently in males and females. In males, both homozygote and heterozygote for this allele develop bald patches, whereas in females, only the homozygotes show a tendency to become bald, and this is usually limited to general thinning of the hair. The expression of this allele is probably triggered by the male hormone testosterone. Females produce much less of this hormone and are therefore seldom at risk to develop bald patches. The sex influenced the nature of pattern baldness shows that biological factors can control the expression of genes.

3.32 Nutritional Effects

In microorganisms, mutations that prevent the synthesis of nutrient molecules are quite common, such as when an enzyme essential to a biosynthetic pathway becomes inactive. A microorganism bearing such a mutation is called an auxotroph. If the product of a biochemical pathway can no longer be synthesized, and if that molecule is essential to normal growth and development, the mutation prevents growth and may be lethal. For example, if the bread mold Neurospora can no longer synthesize the amino acid leucine, proteins cannot be synthesized. If leucine is present in the growth medium, the detrimental effect can be eliminated. Nutritional mutants are crucial for genetic studies in bacteria and served as the basis for George Beadle and Edward Tatum's proposal, in the early 1940s, that one gene function to produce one enzyme.

A slightly different set of circumstances exists in humans. The presence or absence of certain dietary substances, which everyone takes without harm, can adversely affect individuals with abnormal genetic constitutions. Often, a mutation may prevent an individual from metabolizing some substance commonly found in normal diets. For example, those afflicted with the genetic disorder phenylketonuria cannot metabolize the amino acid phenylalanine. Those with galactosemia cannot metabolize galactose. If the diet is modified the correction is possible.

The common case of lactose intolerance, in which individuals are intolerant of the milk sugar lactose, illustrates the general principles involved. Lactose is disaccharides consisting of a molecule of glucose linked to a molecule of galactose and constitutes 7 per cent of human milk and 4 per cent of cow's milk. To metabolize lactose, humans require the enzyme lactase, which cleaves the disaccharide. Adequate amounts of lactase are produced during the first few years after birth. However, in many people, the level of this enzyme soon drops drastically. As adults, these individuals become intolerant of milk. The major phenotypic effect involves severe intestinal diarrhea, flatulence, and abdominal cramps. This condition is particularly prevalent in (although not limited to) Eskimos, Africans, Asians, and Americans with these heritages. In some of these cultures, milk is converted to cheese, butter, and yogurt, where the amount of lactose is reduced significantly, and the adverse effects can be reduced.

3.33 Single Genes can have Multiple effects

A single gene may have multiple phenotypic effects, which is not an uncommon phenomenon. This is referred to as pleiotropy some of the examples are cited below: Marfan syndrome, a human malady resulting from an autosomal dominant mutation in the gene encoding the connective tissue protein fibrillin. Fibrillin is important to the structural integrity of the lens of the eye, to the lining of vessels such as the aorta, and to bones, among others. Therefore, the phenotype is associated with Marfan syndrome includes lens dislocation, increased risk of aortic aneurism, and lengthened long bones in limbs. This disorder is of historical interest in that speculation is there suggesting that Abraham Lincoln was afflicted.

A second example involves another human autosomal dominant disorder. It is an autosomal dominant condition caused by a deficiency of protoporphyrinogen oxidase (chromosome 14). It is relatively rare in most parts of the world except for South Africa where it is common (1 in 300 individuals) among the whites and others. The high South African prevalence arises from the "founder" effect where many of the cases can be traced to a pair of early settlers who immigrated to South Africa from Holland in 1688. Porphyria variegate afflicted individuals can not adequately metabolize the porphyrin component of hemoglobin when this respiratory pigment is broken down as red blood cells are replaced. The accumulation of excess porphyrins is immediately evident in the urine, which takes on a deep red color. However, this phenotypic characteristic is merely diagnostic. The severe aspects of the phenotype are due to the toxic nature of the buildup of porphyrins found in the brain. Phenotypically patients develop abdominal pain, muscular weakness, fever, a high pulse, insomnia, headaches, vision problems (that can lead to blindness), delirium, and ultimately convulsions. Like Marfan syndrome, porphyria variegate is also of historical significance. King George III, King of England during the American Revolution, is believed to have suffered from episodes involving all of the above symptoms. He ultimately became blind and senile prior to his death.

3.34 Genetic Expression Onset

It is not essential that all genetic traits appear at the time of birth. One must keep in mind that the prenatal, infant, pre-adult and adult phases require different genetic information. The Tay-Sachs disease, inherited as an autosomal recessive, is a lethal lipid-metabolism disease involving an abnormal enzyme, hexosaminidase A. Newborns appear phenotypically normal for the first few months. Then, show developmental delay paralysis and blindness and finally die.

Overproduction and accumulation of uric acid, a waste product of normal chemical processes that are found in blood and urine causes Lesch-Nyhan syndrome. These patients have gouty arthritis (arthritis caused by an accumulation of uric acid in the joints), kidney stones, and bladder stones. There are nervous system associated problems that may be linked to behavioral disturbances. There exist jerking movements which are all involuntary muscle movements. Patients with Lesch-Nyhan syndrome is generally not able to walk, require assistance for sitting and are generally wheelchair-bound. Self-injury, including biting and headbanging, is the most common. It is inherited as an X-linked recessive disease. The disorder is due to a mutation in the gene encoding hypoxanthine-guanine phosphoribosyl transferase (HPRT). Newborns are normal for six to eight months prior to the onset of the first symptoms. Duchene muscular dystrophy (DMD) is an X-linked recessive disorder associated with progressive muscular wasting. It is not usually diagnosed until the age of 3 to 5 years. It is fatal in the early twenties.

Huntington's disease is the most variable of all inherited human disorders as far as the age of one set is considered of age. Inherited as an autosomal dominant, Huntington's disease affects the frontal lobes of the cerebral cortex, where progressive cell death occurs over a period of more than a decade. Brain deterioration results in spastic uncontrolled movements, intellectual, emotional deterioration, and ultimately death. The disease is reported at all ages however, it occurs most frequently between 30 and 50 years of age, with a mean onset age of 38.

The critical expression of normal genes varies throughout the life cycle of organisms, including humans. Gene products may play more essential roles at certain times. Further, it is likely that the internal physiological environment of an organism changes with age.

3.35 Genetic Anticipation

In some genetic conditions, the signs and symptoms tend to become more severe and may occur at an earlier age as the disorder is passed from one generation to the next. This is called anticipation. Anticipation is most often seen in Huntington's disease, myotonic dystrophy, and fragile X syndrome, all these are disorders of the nervous system.

Anticipation is associated with a mutation called trinucleotide repeat expansion. A trinucleotide repeat is a sequence of three DNA building blocks (nucleotides) that are repeated a number of times in a row. DNA becomes unstable due to these triplet repeats, therefore, cause errors at the time of cell division. The repeat number may change as the gene is passed from parent to child. This condition is called trinucleotide repeat expansion. In some cases, the trinucleotide repeats keep on expanding until the gene stops functioning normally. This expansion causes the features of some disorders to become more severe with each successive generation.

In Huntington's disease, it only occurs when a male transmits the disease while the opposite is true for congenital myotonic dystrophy (maternal effect). Anticipation is due to trinucleotide repeat expansions, which may be in the coding region (Huntington's disease) or in the untranslated region (Fragile X syndrome). The opposite of anticipation is also true and is reported in congenital myotonic dystrophy. This is called a reverse mutation, negative expansion, or contraction. A sex-bias in meiotic instability has been reported in Friedreich's ataxia: paternally transmitted alleles tend to decrease in a linear way that depends on the paternal expansion size, whereas maternal alleles can either increase or decrease.

3.36 Occurrence of the Disease for the First Time in the Family

1. This is the first-de Novo- mutation (in a disease for which the mutation has a very high rate as in Rett syndrome; also occurs in congenital adrenal hyperplasia, achondroplasia - seven-eight of the mutations are new due to methylated CG dinucleotides in FGFR3). De novo microdeletion or point mutations in the transcriptional coactivator CREB-binding protein (CBP) on 16p13.3 result in Rubinstein-Taybi syndrome with a recurrence risk of 0.1 per cent in sibs. The mutation rate in the NF1 gene is one of the highest known in humans; with approximately 50 per cent of all NF1 patients having novel mutations. About one-quarter of affected individuals with Marfan syndrome have new mutations; thus, a paternal age effect is present in sporadic cases as in all genetic syndromes with a high rate of new mutations. The new mutation rate ranges between 10^{-4} to 10^{-7} with a median of 10^{-6}.
2. Both parents are healthy carriers of a recessive mutation means mutation occurs commonly in the population.
3. If the gene is segregated in previous generations with no diseased individuals, the penetrance may have been low in earlier generations (acute intermittent porphyria is an example of low penetrance genetic disease). If it is an X-linked recessive disease and all the children so far are girls, the disease appears in the first male child.
4. It may be an adult-onset disease with genetic anticipation but the carriers of the gene in previous generations did not live long enough to express the disease. Huntington's disease is an example of a late-onset genetic disease with anticipation. A similar situation may arise due to the parent of origin effect. Differences in gene expression according to the parent from whom the gene was derived occur in Huntington's disease and in myotonic dystrophy and might be due to a difference in methylation of the genes in the two sexes or unknown modifier genes acting in a sex-limited fashion.

3.37 Genomic (Parental) Imprinting

When the origin of the gene is either from the father or mother it is referred to as genomic imprinting. The term genomic imprinting is restricted to mono-allelic gene expression or it may be due to inactivation of the maternal or paternal allele of a given locus. A hypothetical example is shown in **Figure 3.23**.

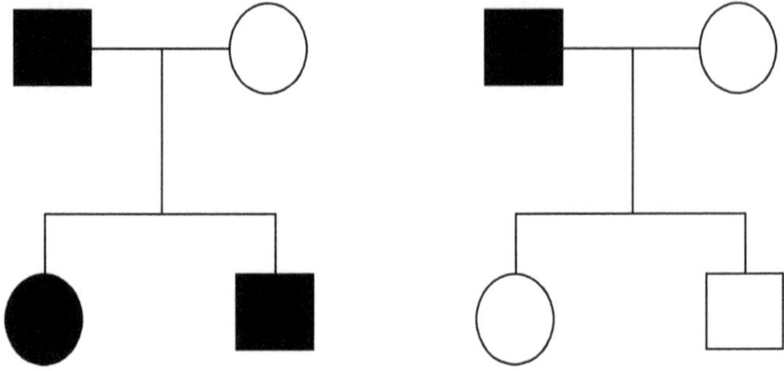

Figure 3.23: Hypothetical Imprinted Genes Responsible for Body Color (LEFT).

The black pigment is maternally imprinted which indicates that the maternal allele is inactivated. Mattings between a male who possesses the allele for pigment and a female who possesses the allele for no pigment produces offspring that show only the pigmented phenotype. In this case, the mother's allele is imprinted and inactivated in the offspring. Therefore, the only actively expressing allele is the father's pigmented allele, which is not imprinted in the offspring. In this example, the pigmented gene is paternally imprinted. The father allele is imprinted and inactivated in the offspring. Therefore, the only actively expressing allele is the mother's no pigment allele, which is not imprinted in the offspring.

Because there is a monoallelic parent-of-origin specific expression from an imprinted autosomal locus, genomic imprinting is counter to classical Mendelian genetic theory, which states that there is an equal inheritance of parental traits. An imprinted mutant allele would appear to be recessive when inherited from one sex, because it would be inactive (and consequently invisible), whereas it would be the only active allele when inherited from the other sex and, therefore, appear to be dominant (**Figure 3.24**).

A darkened body indicates an individual that is mutant for the hypothetical imprinted locus. A cross is used to indicate the imprinted/inactive allele. Both parents are homozygous for the normal allele at the imprinted locus. Although only one allele is active (the paternal copy) in the offspring produced from these parents, it must be a normal allele and therefore all offspring will have a normal phenotype. The mother is a homozygous mutant at the imprinted locus, and the father is normal. Since this hypothetical locus is maternally imprinted, the maternal mutant copy will be inactivated in their offspring and the

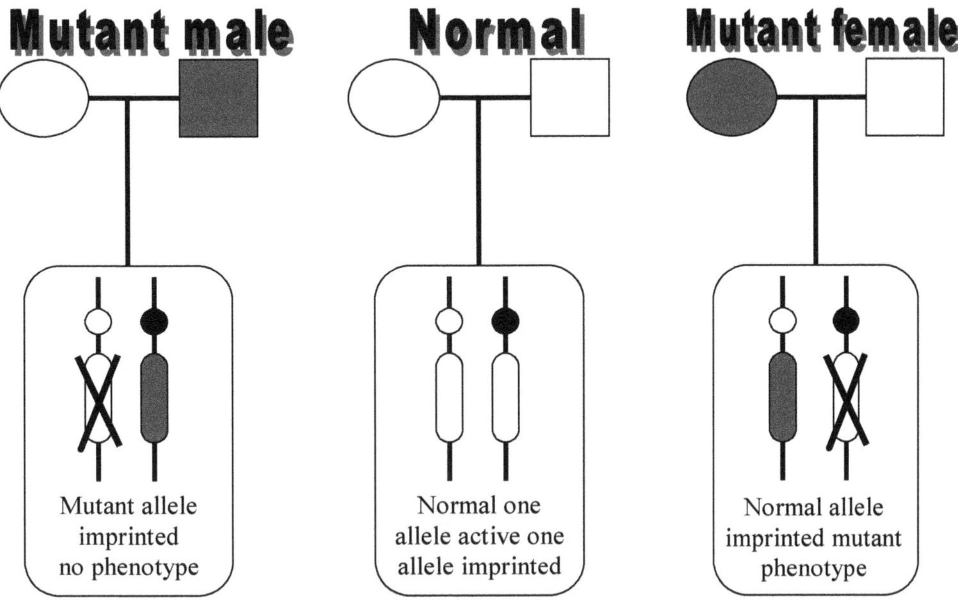

Figure 3.24: Schematic Representation of the Phenotypic Effects of Maternal Imprinting of a Mutant Allele.

paternal normal copy will be the only active allele. The offspring will be phenotypically normal, and the mutant allele will appear to be a recessive mutation. The mother is homozygous normal at the imprinted locus, and the father is a homozygous mutant. The maternal normal allele is imprinted and inactivated in the offspring of these parents. The only allele that is active in the mutant paternal copy. Therefore, all offspring produced from these parents will display the mutant phenotype, and the mutant allele will appear to be a dominant mutation.

As stated previously, the genome imprinting phenomenon has been most intensively studied in mammals. One type of epigenetic modification in mammals involves the addition of a methyl group to position 5 of cytosine and occurs most frequently at CpG dinucleotides. This type of DNA modification is significant because, in general, methylated DNA sequences are transcriptionally inactive. Furthermore, the methylation of DNA residues can be stably preserved through the replication process by the action of maintenance methylases, which use the replicated hem methylated DNA as a template It is, therefore, suggested that DNA methylation may be the epigenetic marking responsible for the imprinting phenomenon in mammals, or at least reflects the imprinted state of a locus produced during gametogenesis. It is logical to assume that the epigenetic differences observed in the paternal and maternal genetic contributions would be established during the only time when the two genomes are segregated; *i.e.*, during gametogenesis. If methylase differences exist in the gametes, one would expect to find different DNA methylation patterns in sperm and eggs. These differences should be perpetuated at least through early development when the final somatic tissue methylation pattern is established. It has been demonstrated in mice that sperm DNA is more methylated than oocyte DNA. In the mouse, a period of overall demethylation has been observed from fertilization, lasting until approximately the blastocyst stage, followed by extensive de novo methylation. This period begins at gastrulation and ultimately produces methylation levels higher than those observed in either of the gametes. Similar changes in the methylation phenotype of a variety of transgenes in mice have been observed during development. In **Table 3.8** some of the mammalian imprinted genes are shown.

Table 3.8: Summary of Mammalian Endogenous Imprinted loci

Imprinted loci	*Active allele*
Igt2	Paternal
Igt2r	Maternal
H19	Maternal
Snrp	Paternal
Ins1 and 2	Paternal
U2afbp-rs	Paternal
Zn1127	Paternal
Par1	Paternal
Par 5	Paternal
Xist	Paternal
Mas	Paternal
IPW	Paternal
WT1	Maternal

Imprinted loci	Active allele
Mash-2	Maternal
p57(KIP2)	Maternal
Pef1/MEST	Paternal
ApoE	Paternal
Pk3	Paternal
Pplcy	Paternal

The estimated imprinted genes in the mammalian genome range from less than 100 to greater than 200. Like the imprinted transgenes, these endogenous genes show high levels of DNA methylation at the imprinted/inactive allele. At present, no consensus sequences for imprinted loci have been identified at/or surrounding these imprinted genes. However, the imprint imposed at these loci is a very specific and localized process. The imprinted genes H19 (maternal) and Igf2 (paternal) are tightly linked to each other and localized to the same region of mouse chromosome 7 but are imprinted in the opposite direction.

The pronuclear transplantation experiments on mice have unequivocally demonstrated that there is an absolute requirement for a genetic contribution from both sexes in order for development to proceed normally, *i.e.*, maternal and paternal contributions are not equivalent. Embryos containing only maternal contributions develop minimal extra-embryonic tissues (trophectoderm), whereas a poorly developed embryo proper is characteristic of embryos containing only paternal genomes. The importance of the mammalian embryo of both maternal and paternal genomes is most apparent in the very low viability of mammalian parthenogenesis.

The timing of the developmental arrest and the phenotype of parthenogenetic mouse embryos is highly variable and is most dependent on the genetic background of the embryos. Parthenogenetic embryos produced from one inbred strain of mice may only develop to blastula or gastrula stages, whereas parthenogenetic embryos produced from another strain of mice might develop to the 40-somite stage. Understanding of the role of genomic imprinting in development has been hampered by the lack of a distinct/consistent embryonic phenotype in parthenogenetic embryos and the lack of a precise role for the presently known imprinted genes in development.

Considerable interest has been generated by the realization that the penetrance and severity of many complex human diseases can be affected by the sex of the parent contributing the allele responsible for the disease. The parent of origin effects associated with a disease suggests that imprinting may be involved. Some of these putative imprinted diseases include Huntington's disease, cystic fibrosis, Prader-Willi, and Angelman syndromes. Parent of origin effects is also seen for birth defects such as spina bifida and certain cancers. The role of imprinting in the development of cancers such as Wilm's tumor, rhabdomyosarcoma, and osteosarcomas has been particularly well studied. By examining DNAs collected from tumors and normal tissues, it has been revealed that the development and proliferation of cancerous tumors may be associated with a loss of imprinting at several different loci. Two examples of these imprinted loci are Igf2 (insulin-like growth factor), which is involved in beta-cell tumorigenesis, hepatoblastoma, Wilm's tumor, and rhabdomyosarcoma, and the gene H19, which has been shown to be involved in hepatoblastoma, lung cancer, Wilm's tumor and bladder carcinoma The development of these cancers is thought to be due to the relaxation of the normal monoallelic expression or 'imprinted' status of these genes resulting from the loss of DNA methylation at the imprinted allele.

There are different theories regarding the evolutionary origins/advantages of genomic imprinting. Few of these theories have been able to satisfy all the disparate information from our present understanding of the phenomenon, such as i) broad phylogenetic distribution, ii) the different types of imprinted genes, and iii) both paternal and maternal imprinting. The general way the kinds of influences that could produce imprinting and suggest several roles for imprinting: (1) imprinted and non-imprinted alleles of a locus may confer different phenotypes; (2) modifiers of imprinting (so-called "imprinted genes") may have pleiotropic effects such that they are selected for reasons other than their action on the imprinted locus; and (3) imprinting could have co-evolved with other traits. It has been suggested that the function of genomic imprinting is to produce functional haploid gene sets in order for the cell's regulatory machinery to be able to 'fine-tune' itself on just a single gene copy. It was further implied that imprinting would also serve to prevent possible detrimental crosstalk between maternal and paternal transcripts at developmentally important regulatory loci. Genome imprinting as a mechanism of gene regulation originated from the use of DNA methylation as a host defense mechanism. The same processes that inactivate incoming viruses in oocytes might also result in the inactivation of some of the organism's own (maternal) loci. Imprinting acts as a mechanism of surveillance for chromosome loss and thus protects against cancers and mono- and trisomic embryos. If imprinted loci are scattered throughout the genome, loss of a chromosome or portion of a chromosome would result in the absence of an active copy of an allele producing probable detrimental effects. Other theories are based solely on genomic imprinting as it occurs in mammals. It has been seen that more specifically, androgenetic embryos show poorly developed embryos and excessive trophectoderm (extraembryonic supportive tissue), whereas gynogenetic embryos have well-developed embryos and poorly developed trophectoderm. These "imprinting" phenotypes of andro- and gynogenetic embryos seem to suggest the possibility of the male using imprinting to maximize maternal input to the embryo. Finally, imprinting in mammals may be a means of protecting the female from ovarian germ cell tumors. Imprinting would prevent the development of malignant trophoblast disease because of the viability of unfertilized eggs and the inability of parthenogenetically activated oocytes to implant and develop.

4

Nucleic Acids as Genetic Material

DNA was first discovered in the late 19th century, but it wasn't until the mid-20th century that it was recognized as genetic material. Two experiments were conducted to prove this, which are now considered classics. Initially, scientists believed proteins were genetic material due to their abundance in cells, constituting 50 per cent of the dry weight, although the amount varies between cells. Proteins comprise 20 amino acids, making them more complex than nucleic acids, which seemed too simple to carry such complex information. However, proteins do not meet the criteria necessary for genetic material.

1. DNA stable form must contain information about an organism's cell structure, function, development, and reproduction.
2. It must replicate accurately so that daughter cells have similar genetic information to parental cells.
3. It must be capable of change. Without change, organisms will be incapable of variation and adaptation. Hence, no evolution will occur. A Swiss biochemist, Friedrich Miescher, isolated a substance from pus cells found in bandages and did chemical tests, which indicated that it contained carbon, hydrogen, oxygen, nitrogen, and phosphorus. A phosphate group is not found in proteins. Miescher then searched for the same substance in other sources and found it in the nucleus of all the samples; therefore, he named it nuclein. In the early 1900s, experiments demonstrated that chromosomes carry hereditary information. Further experimentation in the 1920s showed that genetic materials are a class of compounds that include DNA and RNA.

4.1 Discovery of Transformation in Bacteria

While serving as a British medical officer, Frederick Griffith made a significant discovery about bacterial transformation - a type of recombination that occurs in bacteria. Although he couldn't prove that DNA was involved in the process, his experiments laid the foundation for future discoveries.

Table 4.1 shows that pneumococci display different phenotypes, resulting in variability. Griffith's transformation demonstration involved two key phenotypic characteristics - the presence or absence of a polysaccharide capsule surrounding bacterial cells and the type of capsule determined by the specific molecular composition of the polysaccharides. Pneumococci with capsules form smooth, large colonies and are categorized as Type S when grown

Table 4.1: Characteristics of *Streptococcus pneumoniae* Strains when grown on Blood Agar Medium

	Colony Morphology				*Reaction with antiserum Prepared against*	
Type	Appearance	Size	Capsule	Virulence	Type 115	Type 115
IIRa	Rough	Small	Absent	Avirulent	None	None
IIS	Smooth	Large	Present	Virulent	Agglutination	None
IIIR a	Rough	Small	Absent	Avirulent	None	None
IIIS	Smooth	Large	Present	Virulent	None	Agglutination

on blood agar in Petri dishes. These encapsulated pneumococci are pathogenic, causing pneumonia in mammals such as mice and humans. Virulent Type S pneumococci mutate to a nonpathogenic form without a polysaccharide capsule at a frequency of approximately one per 107 cells. These non-encapsulated, virulent pneumococci form small, rough-surfaced colonies. They are designated as Type R. The polysaccharide capsule makes bacterial cells impervious to destruction by white blood cells, making them virulent. When a capsule is present, it can be of several antigenic types depending on the specific molecular composition of the polysaccharides and, ultimately, the cell's genotype.

Through his experiments with pneumococci in mice, Griffith discovered something remarkable. When he injected a combination of heat-killed Type IIIS pneumococci and live Type IIR pneumococci, the mice developed numerous cases of pneumonia. What's interesting is that the dead Type IIIS cells transformed the live Type IIR cells into Type IIIS cells. What caused this transformation is not solely due to mutation. Even harmless Type IIR cells became dangerous Type IIIS cells. The evidence suggests that something within the dead Type IIIS cells was responsible for this conversion. Type R cells don't possess capsules but carry genes that allow them to produce a specific type of capsule (Type II or III) once the capsule formation block is removed. Type IIR cells come from Type II cells and can mutate back to encapsulated Type S cells. The capsules produced are of Type II.

Different capsule types can be distinguished immunologically. When Type II cells are introduced into a rabbit's bloodstream, the immune system produces antibodies specifically targeting Type II cells. These Type II antibodies can clump together Type II pneumococci but not Type I or Type III pneumococci. Additionally, experiments have shown that transformation, first described by Griffith, does not require a living host. It can occur in a test tube when live Type IIR cells are cultivated with dead Type IIIS cells or extracts of Type IIIS cells. Griffith's experiments demonstrated that the offspring inherited the Type IIIS phenotype of the transformed cells due to a permanent inherited alteration in the cells' genotype. This transformation is vital in determining the chemical basis of heredity in pneumococcus.

Griffith's unexpected discovery was that if the injected heat-killed Type IIIS pneumococci (virulent when alive) plus live Type IIR pneumococci (avirulent) into mice, many of the mice succumbed to pneumonia, and live Type IIIS cells were recovered from the carcasses **(Figure 4.1)**.

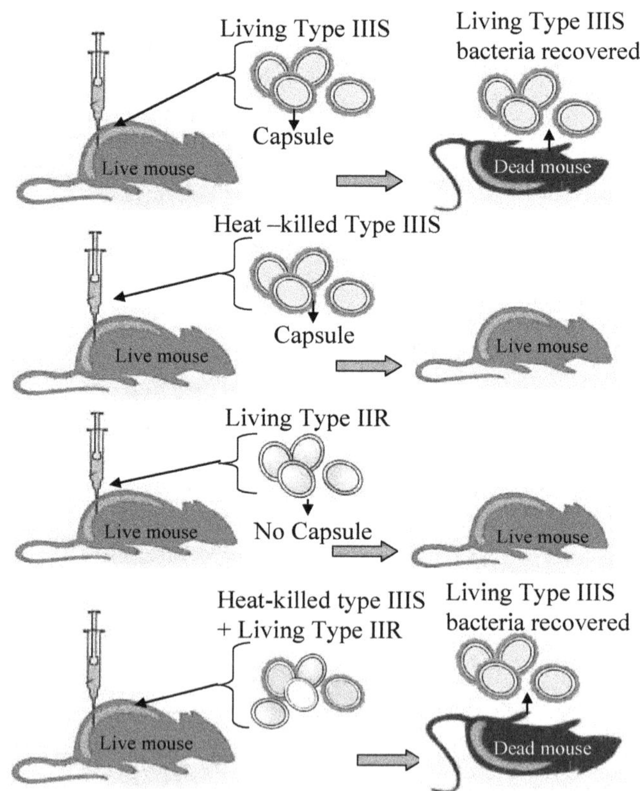

Figure 4.1: Transformation in *Streptococcus pneumoniae* in Griffith Experiments.

When mice were injected with heat-killed Type IIIS pneumococci alone, none of the mice died. The observed virulence was therefore not due to a few types in S type cells that survived the heat treatment. The live pathogenic pneumococci recovered from the carcasses had Type III polysaccharide capsules. This result is important because non encapsulated Type

R cells can mutate back to encapsulated Type S cells. However, when such a mutation occurs in a Type IIR cell, the resulting cell will become Type IIS, not Type IIIS. Thus, the transformation of a virulent Type IIR cells to virulent Type IIIS cells cannot be explained by mutation. Instead, some component of the dead Type IIIS cells (the "transforming principle") must have converted living Type IIR cells to Type IIIS.

4.2 Proof that DNA mediates Transformation

In 1944, Oswald Avery, Colin MacLeod, and Maclyn McCarty conducted groundbreaking research that showed that DNA, not protein or RNA, was a genetic material. They focused on Streptococcus pneumoniae and the "transforming principle" that caused the transformation. They discovered that when highly purified DNA from Type IIIS pneumococci were mixed with Type IIR pneumococci, some pneumococci transformed into Type IIIS (**Figure 4.2**).

To confirm that DNA was the transforming principle, Avery's team used enzymes to break down DNA, RNA, and protein. Only DNase treatment impacted the transforming activity of the DNA preparation, removing all transforming activity. Initially, the mechanism behind this phenomenon wasn't clear, but it's now known that during transformation, the segment of DNA in the pneumococcus chromosome containing the genetic information for a Type III capsule is physically inserted into the chromosome of the Type IIR recipient cell.

4.3 The Hershey Chase Bacteriophage Experiments

Hershey and Martha Chase, a lab technician, conducted an experiment to investigate whether proteins carry genetic material. They transferred proteins and DNA between a virus and its host using the T2 bacteriophage as the vehicle for delivery. The T2 bacteriophage is a bacterial virus that has a simple structure consisting of a protein-based outer wall and a DNA core, making it an ideal research candidate. When the phage injects its genetic material into a bacterium, its protein shell remains attached to the host. The virus then takes over the bacterium's reproductive mechanisms and uses them to replicate itself, ultimately destroying the host in the process.

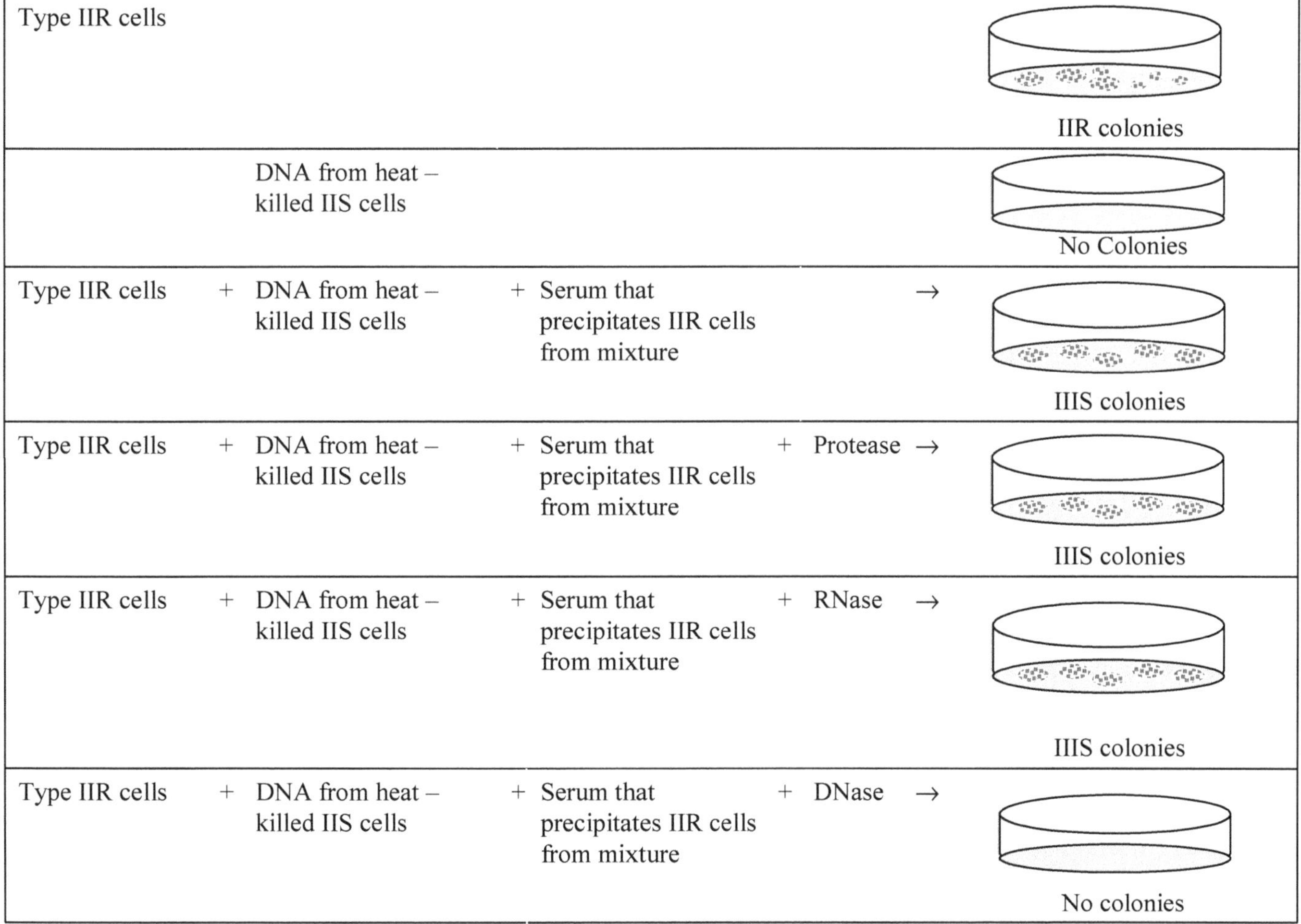

Figure 4.2: DNA is the Transforming Molecule Proved by Avery, Macleod, and McCarty.

Although the protein shell stays outside, researchers believe that the protein transfers from the bacterium to the virus during attachment.

Alfred Hershey and Martha Chase experimented using a radioactive isotope of phosphorus called Phosphorus-32 to label T2 phage DNA. They tracked the location of DNA and protein by measuring the concentration of radiation since DNA has more phosphorus than protein. After infecting E. coli with the tagged phages, they used a blender to separate the protein from the host. They then isolated the bacterium from the phages and protein via high-speed centrifugation (**Figure 4.3**).

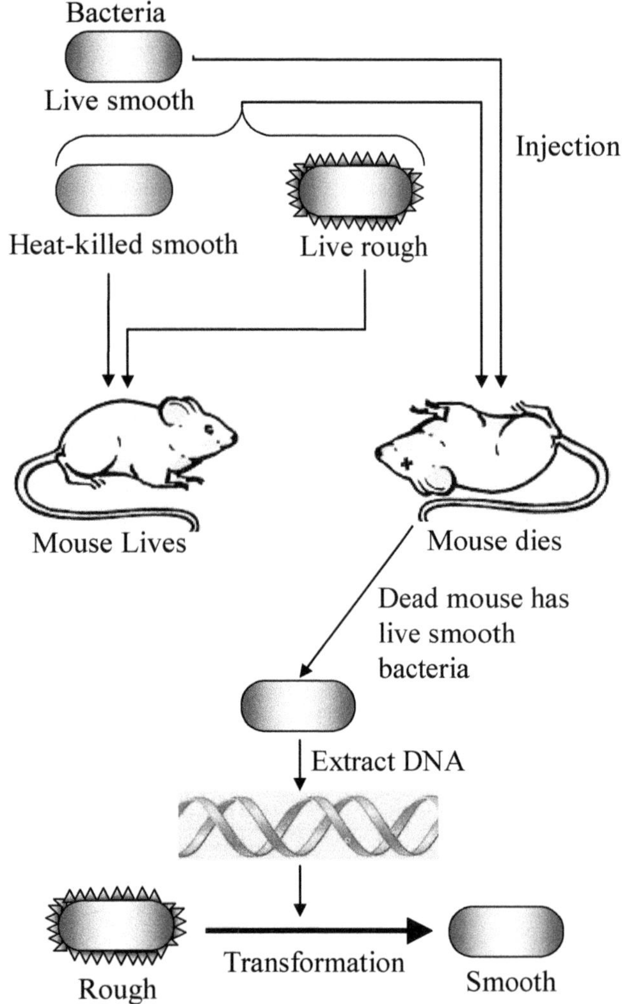

Figure 4.3: Avery Griffith's Macleod and McCarty Combined Data to Prove that DNA is the Transforming Factor from *Streptococcus pneumoniae.*

After separation, the scientists evaluated the radiation concentrations in the E. coli cells and protein shells. They found that radioactive phosphorus was present in the bacterial sample, indicating that DNA had been transferred from the bacteriophage to the host organism. They also found that the virus replicated in each host cell, even though the protein shell detached during phage reproduction. This suggests that the protein shell is unnecessary for the replication process after the initial insertion of genetic material.

Hershey and Chase used sulfur to tag the proteins for their second experiment. They infected E. coli cells, separated the shells from the host, and tested for sulfur. Similarly, sulfur was only present in the protein shells, not the bacteria. However, the phage's genetic material was still replicated despite removing the protein shell from the bacteria.

Hershey examined the offspring of the phages and found that the young bacteriophages had DNA labeled with phosphorus, but their protein lacked any radioactivity. Hershey's experiments showed that DNA, not protein, is the source of genetic material. Their research was published in The Journal of General Physiology as "Independent Functions of Viral Protein and Nucleic Acid in Growth of Bacteriophage." This marked a significant milestone in scientific research, and scientists worldwide sought details on the experiments. The work of Alfred Hershey and Martha Chase laid the foundation for genetic analysis, eventually leading to Watson and Crick's success in determining the structure of DNA.

Unfortunately, when live smooth bacteria are injected into mice, it leads to pneumonia and eventual death. However, treating the bacteria beforehand can eliminate the risk of disease. Interestingly, co-injecting non-virulent bacterial strains with heat-treated virulent bacteria can transform them and cause infection.

Hershey and Chase's evidence that phage T2's genetic material is DNA had a minor error. Alongside the DNA, a significant amount of 35S (protein) was also found to be injected into host cells. Nevertheless, these experiments provided conclusive evidence that phage proteins contain genetic information. Recently, scientists have demonstrated that the genetic material of these bacterial viruses is indeed DNA, not protein (as shown in **Figure 4.4**), by infecting pure phage DNA into protoplasts of E. coli. These experiments resulted in the production of typical infective progeny phages, also known as transfection experiments.

4.4 Structure of Nucleic Acids

4.4.1 DNA Structure

The hereditary material is DNA in most of the organisms. However, several viruses use RNA (ribonucleic acid) as the building block for their genome. DNA and RNA are polymeric molecules made up of linear chains of subunits called nucleotides. Each nucleotide has three parts:

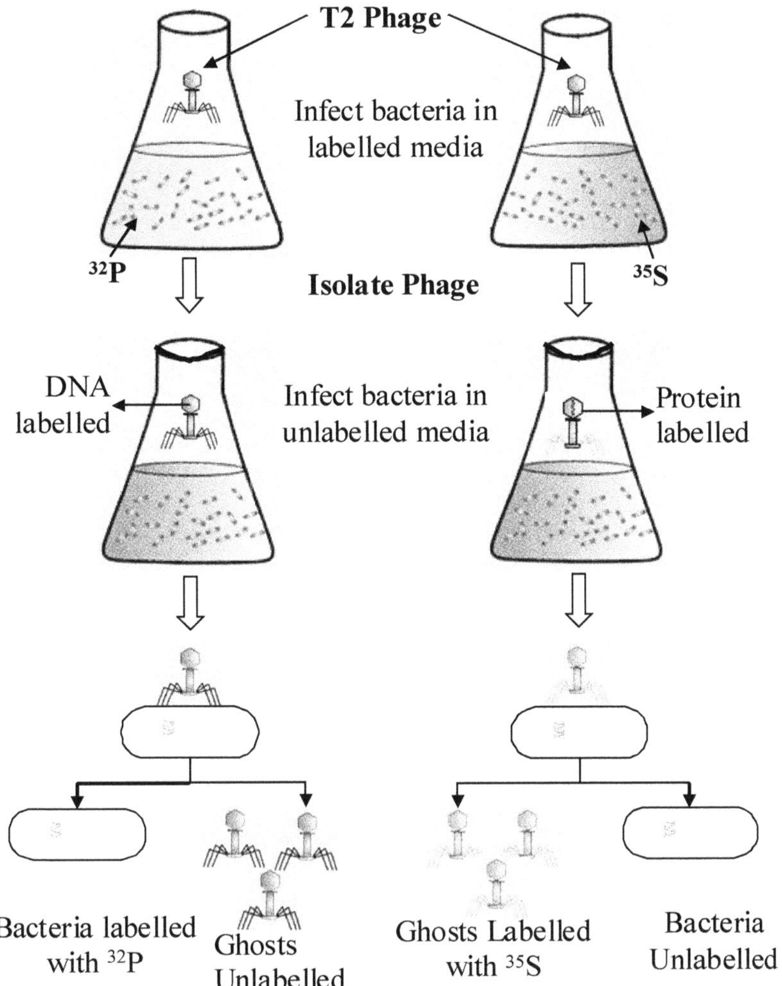

Figure 4.4: Hershey-Chase Blender Experiment to Show that Nucleic Acid was the Genetic Material.

a nitrogenous base, a five-carbon atom sugar, and a phosphate group (**Figure 4.5**).

When you combine base and sugar, nucleosides are formed. If you add phosphate groups to these nucleosides, you get nucleotides. The nucleotides in DNA are called deoxyribonucleotides, and they contain two sugar molecules called deoxyribose. The ribonucleotides in RNA contain the sugar ribose. Nucleotide bases can be either double-ringed purine or single-ringed pyrimidine. DNA and RNA have two purine-containing nucleotides (adenine and guanine) and two pyrimidine-containing nucleotides (cytosine and uracil for RNA and thymine for DNA).

Figure 4.5 shows the numbering system used to identify nucleotides in this text. Pyrimidine and purine rings have numbered carbon and nitrogen atoms, with numbers 1-6 and 1-9, respectively. The carbon atoms of the sugar ring are numbered 1′ to 5′. The 2′-deoxyribose is attached to the 2′ carbon of the sugar ring and lacks a hydroxyl group. In DNA and RNA, sugar-phosphate bonds connect individual nucleotides, with the hydroxyl group on the 3′ carbon of one nucleotide bonding with the phosphate group on the 5′ carbon of another nucleotide (**Figure 4.6**). Dinucleotides, trinucleotides, and polynucleotides are formed when two, three, and a long chain of nucleotides are connected, respectively.

The guanine sugar ring contains phosphates categorized as alpha, beta, or gamma. When forming dinucleotides, the alpha and beta phosphates represented by pyrophosphate are lost, and the phosphodiester bond links the 3′ hydroxyl to the phosphate on the 5′ carbon atom of the sugar.

Erwin Chargaff, a professional chemist, analyzed the chemical composition of nucleotides. He subjected DNA from various organisms to solid acid treatment, hydrolyzing it before separating the nucleotides through paper chromatography.

To his surprise, his experiments revealed that the relative ratios of the four bases were not equal, nor were they random. The number of adenine (A) residues in all DNA samples was identical to the

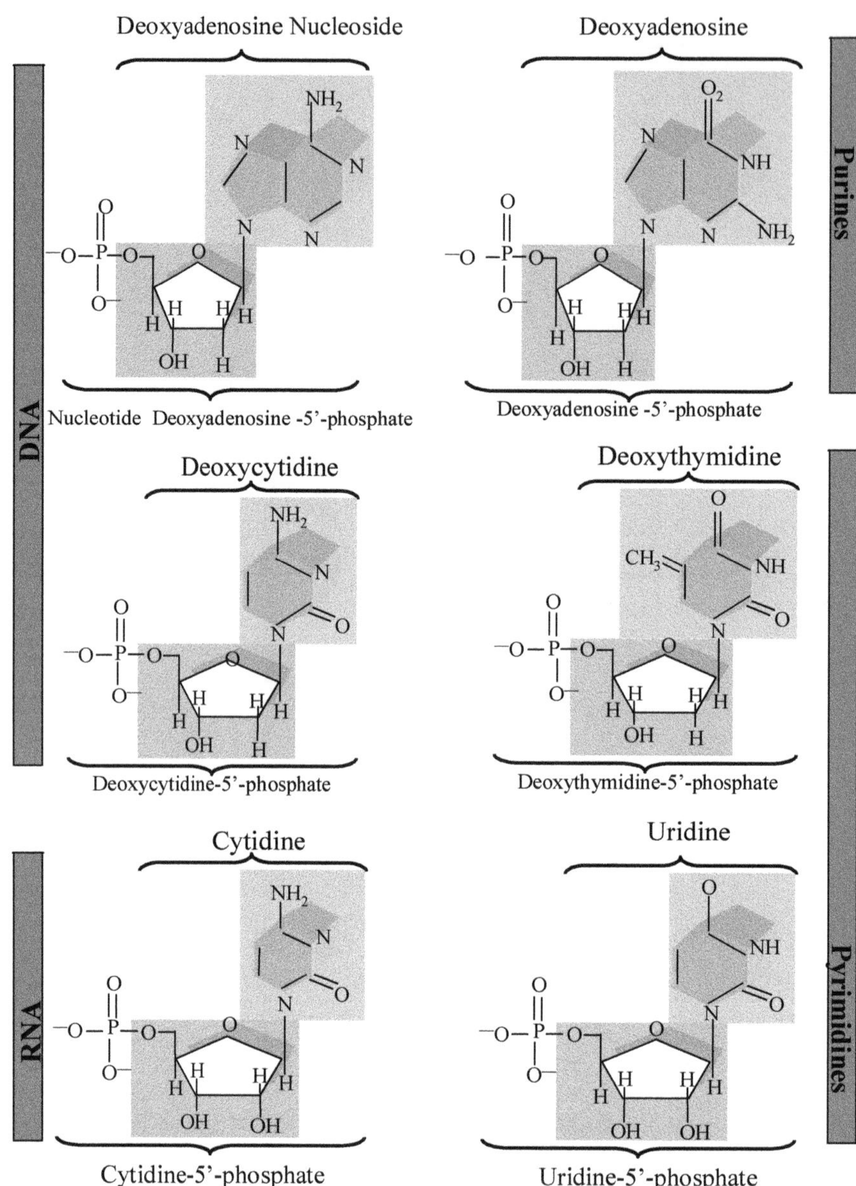

Figure 4.5: Purines and Pyrimidines of Nucleic Acids.

number of thymine (T) residues, while the number of guanines (G) residues was equal to the number of cytosine (C) residues (**Table 4.2**). In summary, the Chargaff rule can be expressed as A = T and G = C, with the sum of purines equaling the sum of pyrimidines. However, the percentage of (C + G) does not necessarily equal that of (A + T).

This specific arrangement is responsible for genetic specificity. The significance of the A = T and G = C relationships became fully apparent after the three-dimensional structure of DNA was solved. A always pairs with T, and G always pairs with C.

Scientists have long been fascinated by using X-ray diffraction to explore the structure of DNA. This method involves directing X-rays at a regular arrangement of molecules, which may be either crystals or fibers. X-rays diffract and create spots on X-ray film when interacting with an atom. By analyzing these diffraction patterns, scientists can gain insights into the shape and structure of the molecules in the arrangement. In 1938, William Astbury used this technique to examine DNA fibers and discovered a repeating unit of 0.34 nm within the DNA. Later, Rosalind Franklin improved upon this X-ray data from highly purified DNA samples obtained between 1950 and 1953. Her work confirmed the 0.34 nm periodicity and suggested that the structure of DNA was some helix. While Franklin did not propose a model for the design of DNA, Linus Pauling and Robert Corey utilized Franklin's data, among others, to suggest that DNA was a triple helix with phosphates near the center of the axis and bases on the outside.

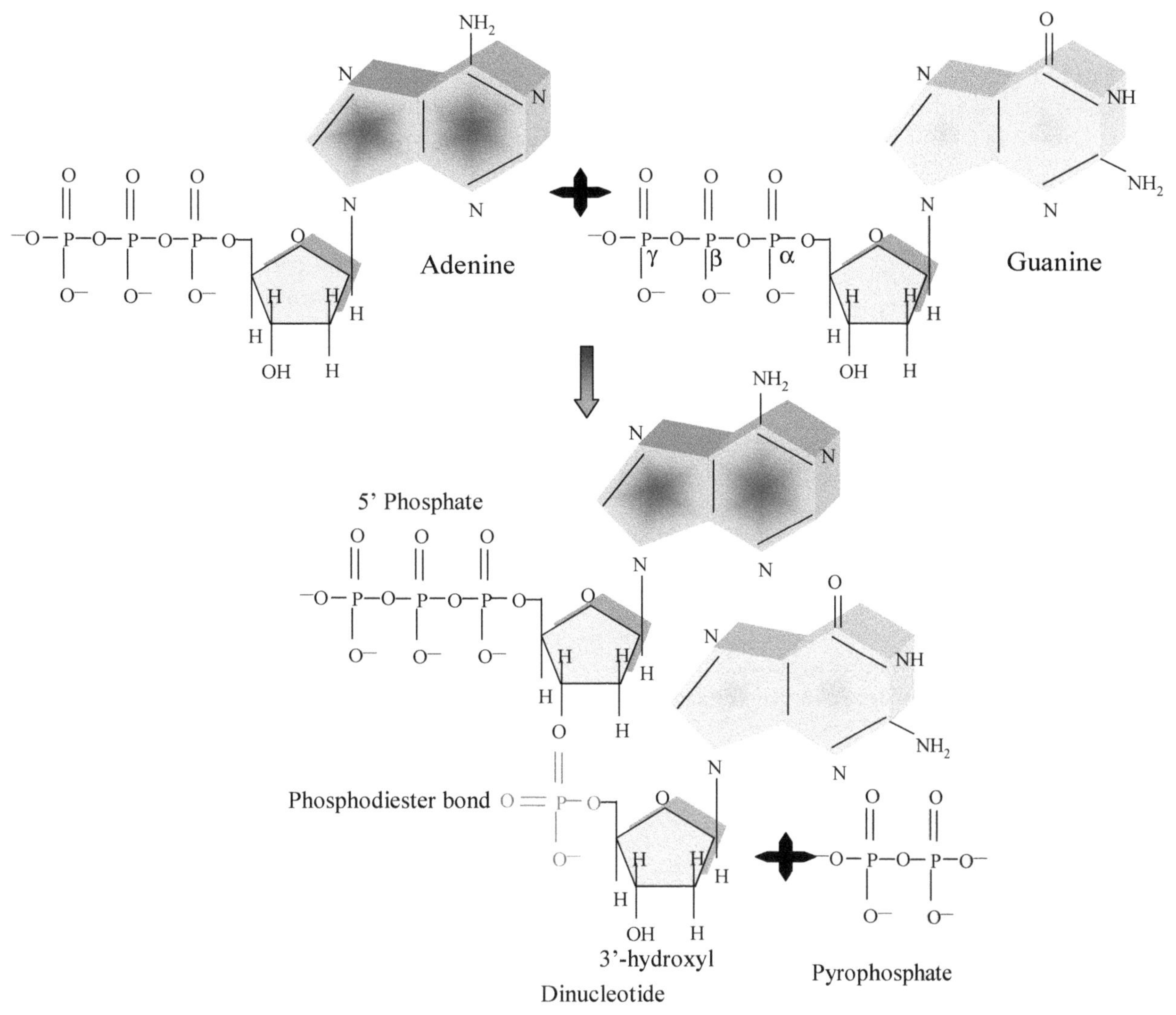

Figure 4.6: Joining of Adenine and Guanine.

Table 4.2: Chargaff's Rules Applied to different Species

Organism	*A to G*	*T to C*	*A to T*	*G to C*	*Purines: Pyrimidine's*
Ox	1.29	1.43	1.04	1	1.1
Human	1.59	1.75	1	1	1
Hen	1.45	1.29	1.06	0.91	0.99
Salmon	1.43	1.43	1.02	1.02	1.02
Sea urchin	1.83	1.80	1.02	1.00	1.01
Wheat	1.22	1.18	1	0.97	0.99
Yeast	1.67	1.92	1.03	1.2	1
Hemophilus influenzae	1.75	1.54	1.06	0.93	1.01
Escherichia coli	1.05	0.95	1.09	0.99	1
Serratia marcescens	0.76	0.63	1.03	0.85	0.92
Bacillus schatz	0.68	0.85	1.07	0.9	0.96

4.4.2 The Double Helix

During the 1950s, the structure of DNA was a subject of much debate among scientists. Despite having a good understanding of its primary structure, its secondary structure remained a mystery. It was Rosalind Franklin, while working at King's College in London, who suggested that DNA had a long, helical structure with uniform thickness using X-ray diffraction. Later, Francis Crick and James Watson, researchers at Cambridge University, proposed a double-stranded helical model based on hydrogen-bonded base-pairing interactions, which matched most of the known facts and was confirmed by future discoveries.

In 1953, Watson and Crick attempted to build molecular models of DNA but found that the structure described by Pauling-Corey needed to be corrected due to the proximity of some atoms. Using X-ray diffraction results and the Chargaff rules, they proposed the famous double-helix model, which was later updated due to high-resolution X-ray diffraction.

The DNA structure consists of two long polynucleotide chains coiled around an axis to form a right-handed double helix with clockwise turns. The two chains run in opposite directions and have a specific orientation. They have flat bases located perpendicular to the axis and stacked on each other with a 0.34 nm distance between them inside the helix. Hydrogen bonds pair the nitrogenous bases of opposite strands. Each complete turn of the helix is 3.4 nm long, about ten bases from each strand (10.4bp) from one complete turn of the helix. The double-helical structure has larger major and smaller minor grooves, and its diameter measures approximately 2nm. The nitrogenous bases at the center of the helix occupy the most significant position.

4.4.3 Base Pairing

The pairing of the nitrogenous bases in the center of the helix is the most significant feature of the model by Watson and Crick; still, other factors are essential to understanding the double helix.

4.4.4 The Antiparallel Helix

The structure of the DNA molecule consists of two polypeptide chains that run in opposite directions. One chain moves from 5′ to 3′, while the other moves from 3′ to 5′, as shown in **Figure 4.7**. The labels 5′ and 3′ refer to the numbering system of the sugar ring DNA sequences. To begin a single DNA chain, a free phosphate group attaches to the 5′ carbon of a deoxyribose ring. Additional nucleotides are linked to the chain through phosphodiester bonds,

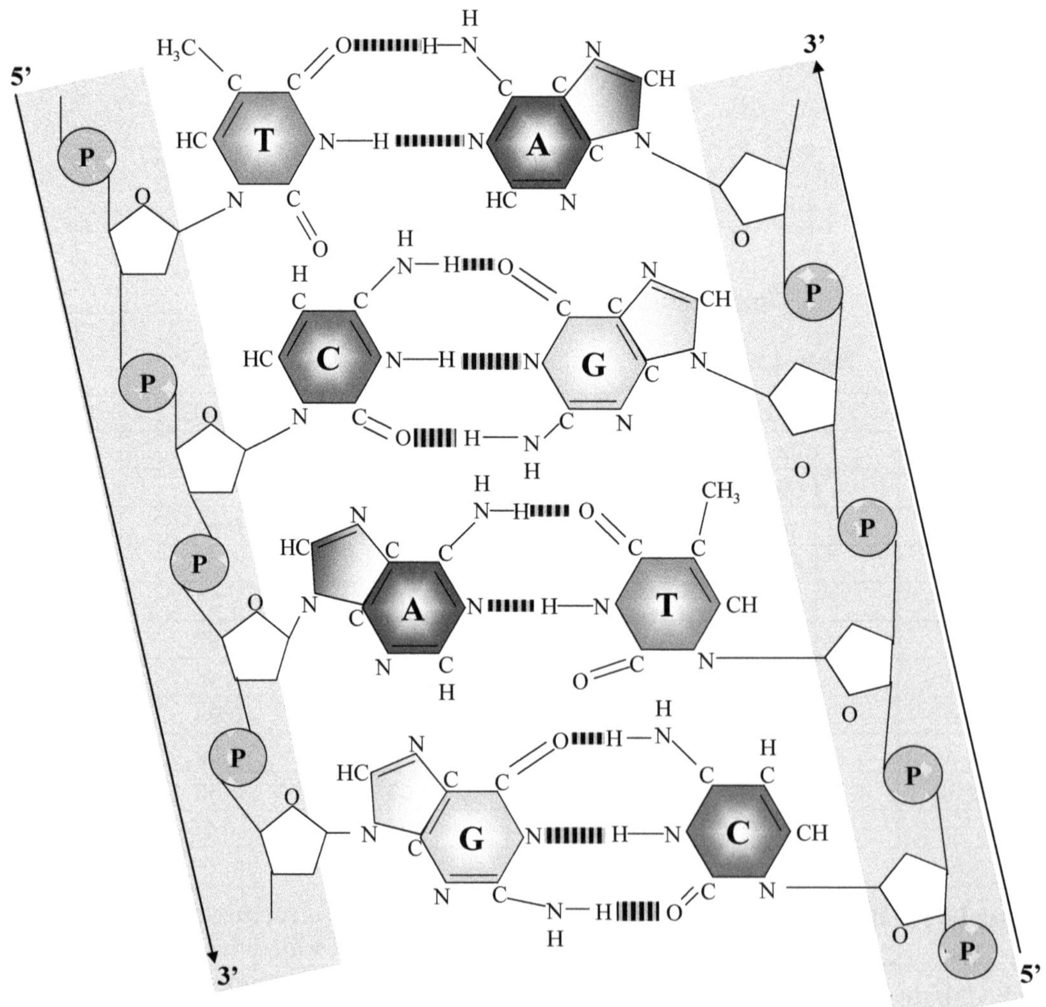

Figure 4.7 DNA Base Pairing and Complementation.

which connect the 3′ carbon atom's hydroxyl group of one sugar to the phosphate on the 5′ carbon of an adjacent sugar. The chain ends with a free hydroxyl group attached to the 3′ carbon atom of the last sugar. It's important to note that these details often need to be more apparent when viewing the molecule's structure.

4.5 Structure of the Nucleotides

The DNA molecule is made up of four subunits which are called nucleotides. Each nucleotide has five-carbon atoms designated as 1′, 2′, 3′, 4′, 5′, which are otherwise called one prime two prime, and three prime, *etc.* Sugar, deoxyribose, with a nitrogenous base attached to one end and a phosphate group attached to the other from the ring. The primers distinguish the carbons of the sugar from the numbered carbon and nitrogen in the nitrogenous base. The nitrogenous base is attached to the l′ carbon of the sugar and a phosphate group is attached to the 5′ carbon. An oxygen atom that is usually present on the 2′ carbon in the ribose sugar is absent in DNA, hence the prefix deoxy in the term deoxyribonucleic acid. The four nucleotides have complicated chemical names: deoxythymidine 5′-monophosphate (dTMP), deoxycytidine 5′-monophosphate (dCMP), deoxyadenosine 5′-monophosphate (dAMP), and deoxyguanosine 5′ monophosphate (dGMP). However, for simplicity, each nucleotide in a DNA molecule is usually referred to by the name of its base *i.e.*, thymine, cytosine, adenine, and guanine, these are referred to as T, C, A, or G. C and T, these are also called pyrimidine's and are like each other in structure, each having a single six-membered ring in the nitrogenous base. The other two nucleotides, A and G, are purines, and are like each other in the structure; each purine has a nitrogenous base composed of two rings, in this case, the six-membered ring gets fused to a five-membered ring (**Figure 4.8**).

4.6 Base Pairs and Stacking

A computer-generated model of high-resolution X-ray crystallographic diffraction data has shown that the bases of both DNA chains are flat structures that lie approximately perpendicular to the helical axis. All the bases themselves are stacked upon each other. All the base pairs are not perpendicular to the helical axis, and some show propeller twist, as a result, the purine and pyrimidine pair are not lying flat but are twisted with respect to each other, like the blades of

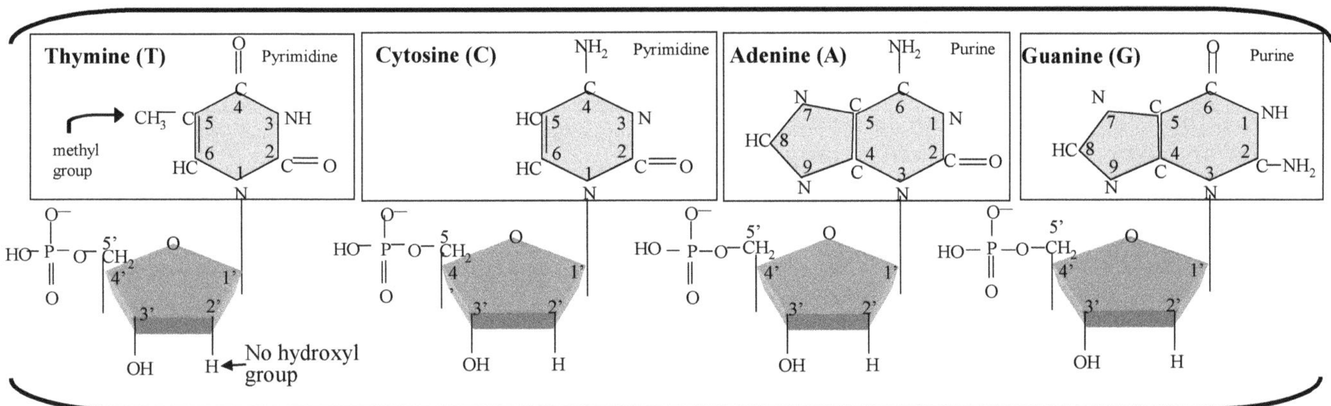

Deoxyribonucleotides : subunits of DNA

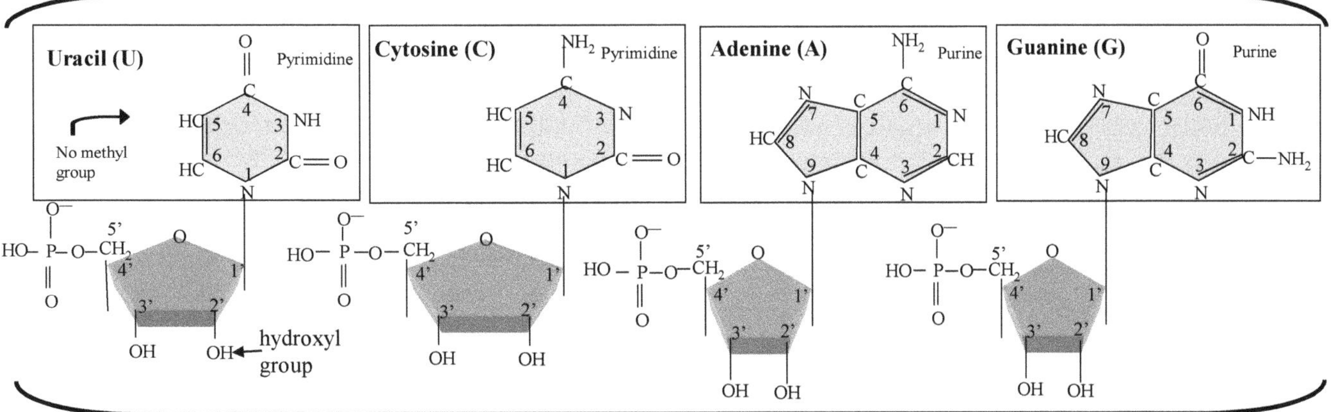

Ribonucleotides : subunits of RNA

Figure 4.8: The Four Nucleotide Subunits of DNA and RNA.

a propeller. The pairing of a purine (A or G) with a pyrimidine (T or C) within the helix is important for the integrity of the helix. The constant length of the purine – pyrimidine pairing may be disrupted if purine – purine is too long or pyrimidine – pyrimidine is too small pairings occur. The purine–pyrimidine pairs are complementary to each other further the two strands of a single DNA molecule are also complementary to one another. Thus, if the sequence 5′-ATGATCAGTACG-3′ occurs on one strand of the DNA, the other strand must have the sequence 5′–CGTACTGATCAT-3′. These two sequences are complementary to each other:

Strand one: 5′– ATGATCAGTACG–3′

| || || ||||| ||

Strand two: 3′–TACTAGTCATGC–5′

The pairing of two DNA strands is very specific. Precise matches between two DNA strands, like those shown above, are highly stable and they readily form helices. In **Table 4.1** the ratios of individual nucleotides isolated from DNA of various sources is shown.

The sugar-phosphate backbone provides a barrier in reading the DNA base sequence. The grooves have their own importance as they provide a mechanism by which the bases can be distinguished from one another. There are major and minor grooves. These are ~ 0.12nm major groove floor is composed of nitrogen and oxygen atoms which belong to the unique position of each base pair. Minor groove contains nitrogen and oxygen atoms that are often common to either purines or pyrimidines.

One of the most common ways in which proteins can recognize specific DNA sequences is by the insertion of a protein α –helix, which is a protein secondary structure motif in which a right-handed helix is formed by amino acids on a polypeptide chain. Each amino acid in the helix occupies a vertical distance of 0.15 nm, and there are 3.6 amino acid residues per turn of the helix. The diameter of the polypeptide backbone in an α – helix is ~ 0.5nm; however, the amino acid side chains project away from the helical axis. A protein α – helix can fit almost exactly into the major groove of double-stranded DNA. The amino acid side chains that project away from the a-helix are able to form hydrogen bonds with the DNA bases in the major groove. This is described as a DNA–protein interaction.

4.7 Role of Hydrogen Bonds in DNA

A number of factors are responsible for the stability of the DNA double helix structure among them hydrogen bonds internal and external hydrogen bonds stabilize the DNA molecule. The two strands of DNA stay together by H bonds that occur between complementary nucleotide base pairs. There are approximately 2500 or so hydrogen bonds that hold together every kilobase of DNA hence providing extraordinary stability to the helix. Two hydrogen bonds occur between the adenosine and the thymine base pairs, and between the cytosine and the guanine, there are three.

While each hydrogen bond is extremely weak (compared to a covalent bond for example), the millions of H-bonds together represent an extremely strong force that keeps the two DNA strands together. In addition, other groups of the base rings (polar groups) can form external hydrogen bonds with surrounding water that give the molecule extra stability. If the hydrogen is bonded to an oxygen or nitrogen atom the sharing of the electrons is not quite equal. The bond becomes slightly polarized. This means that the hydrogen atom has a slight positive charge while the oxygen or nitrogen atoms are slightly negative. If two molecules with this polarization are close to one another they can 'hydrogen bond'. The positive hydrogen atom of the first molecule is attracted to the negative oxygen or nitrogen atom of the second molecule. This bond is weaker than the covalent bond but is the attraction that gives DNA its shape. In general terms, a chemical bond is an attractive force that holds atoms together. The spontaneous formation of a bond between two free atoms involves the release of internal energy of the unbonded atoms and its conversion to another energy form, *e.g.* heat. The strength of a particular bond is measured by the amount of energy released upon the formation of the bond. The stronger the bond, the more energy is released. The change in energy (ΔG) that accompanies bond formation is used to describe the strength of a bond. The value of ΔG is calculated according to the following equation:

$$\Delta G = -RT \text{ in Keq}$$

where R is the universal gas constant (8.314 J K-I mol-I), T is the absolute temperature and In Keq is the natural log of the equilibrium constant between the bonded and the unbonded forms of the molecule. Covalent bonds are short (0.095 nm) and are very strong, with ΔG values in the range of -100 to -500 kJ mol-I. Hydrogen bonds, on the other hand, are longer (approximately 0.3 nm) and are much weaker, with ΔG values in the range of -10 to -30 kJ mol-1.

As found in the double helix, adenine forms two hydrogen bonds with thymine, and cytosine forms three hydrogen bonds with guanine.

4.8 Reversible Denaturing of DNA

DNA molecules are extremely stable and in dehydrated form can survive many years. DNA has been isolated from Egyptian mummies which are approximately 3000 years old. The double helix is still intact in these mummies, but the DNA is broken into small fragments of nucleotides joined as a contiguous unit, the DNA is composed of short, 500-1000 base pair (bp) fragments.

A major consequence of the non-covalent forces that hold the double helix together is that the two constituent strands of DNA may be separated (denatured) simply by heating. The thermal energy of heating a DNA sample will break the relatively weak hydrogen bonds connecting the two strands of the helix but do not alter the covalent linkages that hold each strand together. The separation of the two DNA strands in a helix results in a change in the physical properties of DNA. One of these changes is the way in which DNA absorbs UV light. DNA absorbs light at a wavelength of 260 nm due to the presence of alternating single and double bonds in the DNA bases. Interestingly, each of the four nucleotide bases has a slightly different absorption spectrum, and the overall absorption spectrum of DNA is the average of the different absorption spectra of four nucleotides. A pure DNA solution appears transparent to the eye, and absorption does not become measurable until approximately 320 nm. Moving further into the UV region, there is an absorption peak at about 260 nm, followed by a dip between 220 and 230, and then the solution becomes essentially opaque in the UV light. A solution of double-stranded, native DNA, with a concentration of 0.05 mg/mL, has an absorbance (or optical density, OD) of about 1.0 at the 260 nm peak. When a DNA helix is denatured, it becomes single-stranded, and the absorbance increases. The same 0.05 mg mL double-stranded DNA solution will absorb 1.5 when it is in the single-stranded form. This type of absorbance data is often expressed as

1.0 A260 nm double-stranded DNA = 50 μg/mL

1.0 A 260 nm single-stranded DNA = 33 μg/mL

There is an increase in the absorbance as double-stranded DNA converts into this phenomenon single-stranded, which is known as the hyperchromic effect, which shows that there is an interaction between the electronic dipoles in the stacked bases of the double helix. The stacking of the base pairs in double stranded DNA has the effect of dampening the absorption of individual nucleotides due to the competing interactions of adjacent base pairs in the stack. In the single-stranded form, there is no such competition hence single-stranded DNA absorbs light to a greater extent than if it was double-stranded.

Upon heating, the DNA strands separate, and the process is called melting. Since AT base pairs have only two hydrogen bonds that hold them, therefore, the regions of DNA with a high level of A and T residues separate first. As the temperature continues to increase, more and more of the DNA will become single-stranded and finally, complete strand separation occurs. When there is rise of temp 50 per cent DNA converts into a single strand this temperature is called melting temperature (Tm). The Tm of all DNA molecules is not the same; it depends upon the length of DNA, and the proportion of GC and AT base pairs that a DNA molecule contains. Generally, short DNA molecules have low Tm values, as do those that contain a high proportion of AT base pairs. The Tm can be calculated according to the following equation:

$$Tm = 16.5(\log[Na+]) + O.4l(\text{per cent GC}) + 81.5\ ^{o}C$$

where [Na+] is the concentration of sodium ions in the DNA solution and (per cent GC) is the percentage of GC residues in the duplex. Thus, for a DNA molecule that contains 40 per cent GC residues (typical for a mammalian genome) in a solution of 50mM NaCI, the melting temperature is 76.40C. Short DNA molecules (15-30 bases long) are often used in genetic engineering experiments. For short regions of complementation, the equation above breaks down, and the following estimation (sometimes called the Wallace rule) is used to determine the Tm:

$$Tm = 2(AT) + 4(GC)$$

Therefore, an oligonucleotide of 20 bases of single-stranded DNA containing five A residues, six T residues, three C resides, and six G residues will have an annealing temperature to its complementary DNA sequence of approximately 2(11) + 4(9) = 58 DC. The importance of annealing temperature is important when polymerase chain reaction (PCR) or nucleic acid hybridization is carried out.

The formation of single-stranded DNA is entirely a reversible process. Single-stranded DNA made by heating a duplex will reform when the temperature is reduced. As the temperature falls, the thermal energy is reduced, and random collisions between the complementary strands will result in re-association. Providing that the temperature is reduced relatively slowly (1-2 DC per second), the complete duplex formation will result. The two complementary single-stranded DNA molecules will come together and reform the exact duplex that was present before the sample was heated. If the temperature drops rapidly correct base pairing will not occur.

4.9 Structure of DNA in the Cell

The genome of all organisms contains different types of nucleic acids. The viruses may be double-stranded, or they can be single-stranded DNA, or they may contain only RNA. Eukaryotes are composed of double-stranded DNA molecules. Interestingly prokaryotes have a circular DNA molecule. This ring is formed when the 5′ and 3′ ends are joined to each other. Plasmids are found in prokaryotic cells and are considered extra-chromosomal DNA molecules. They play a pivotal role in the understanding of the advances in molecular biology and genetic engineering. Plasmid DNA may be closed circles of either single-stranded or double-stranded DNA.

The DNA molecule has undergone electron microscopy and it has been indicated that DNA is more string-like than a rod-like structure and wraps around itself to form a variety of irregular structures. A vast array of proteins is found inside the eukaryotic cell. These are required for DNA replication, its transcription into RNA, and its packaging within the cell. DNA is wrapped around these proteins or bends the DNA molecule. For short linear DNA molecules, bending is not a major problem. If some kind of stress is given one part of a DNA molecule results in twisting of the double helix which can be relieved by untwisting another fragment of the DNA. Linear DNA allows free rotation of the DNA strands with free ends. Stress does not influence the circular DNA due to a lack of free ends, as it cannot simply untwist to counteract a twist elsewhere in the molecule. The formation of DNA supercoils is due to the twisting of the DNA molecules (**Figure 4.9**). The formation of the DNA supercoiling is shown in Figure 4.9.

Super coiling can only be introduced into or released from a closed-circular DNA molecule by breaking at least one of the phosphodiester backbones. This results in the double helix turning in the linear molecule of DNA which can explain the level of supercoiling. The supercoiling further illustrates the linking number (LK). If the supercoils are introduced into the DNA molecule it reduces or increases the number of helical turns and gets fascinated when the circular DNA is formed where the linking numbers are distorted (**Figure 4.9**).

It has been recently discovered that on rare occasions, DNA takes on a different shape than its typical helix structure. This means that 'multiple layers' of information are stored in the genetic code. The DNA molecule can bend and flex, something like a rope ladder, but throughout these gyrations, its building blocks–called bases–remain paired up just the way they were initially described by James Watson and Francis Crick, who proposed the spiral-staircase-like structure in 1953. There are transient, alternative forms in which some steps on the stairway come apart and resemble stable systems other than the typical Watson-Crick base pairs. NMR has revealed that when there occurs a chemical shift of these alternative forms, there is an orientation in which certain bases are flipped at 180 degrees.

4.10 The Eukaryotic Nucleosome

The human genome has a massive amount of DNA, which is roughly around 3.2 x 109 base pairs. All humans are diploid organisms. Most of our cells' total amount of DNA is nearly 6.4 x 109 base pairs. At 0.33 nm per base pair, this corresponds to an overall length of about 2.1 m. Such a large DNA molecule is highly compact. It is related to many proteins that result in the wrapping of DNA into nucleosomes. During interphase, the genetic material (together with its associated proteins) is uncoiled to some degree and dispersed throughout the nucleus as chromatin. When mitosis begins, the chromatin condenses to a large extent, and at the time of prophase, it is dense so that it is accepted as chromosomes. It represents a reduction in length up to 10000-fold. Electron microscopic observations illustrate that chromatin fibers are like linear arrays of spherical particles. The particles are found regularly along the axis of a chromatin strand and look like beads on a string, which are known as the nucleosome.

The proteins related to DNA in chromatin are divided into primary, positively charged histones, which are essential and are less positively charged

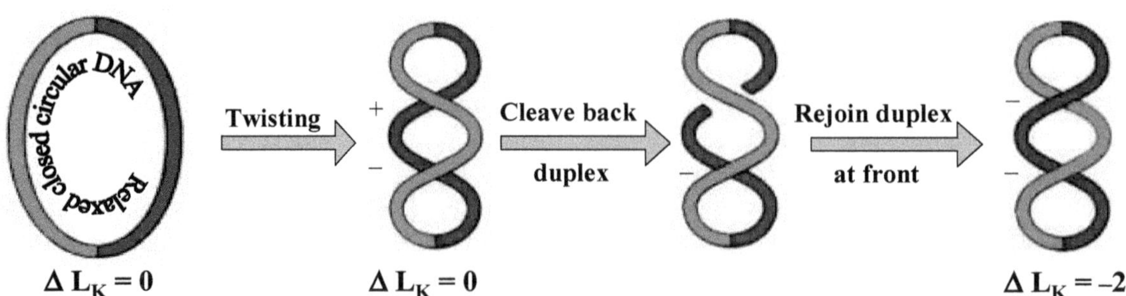

Figure 4.9: DNA Supercoiling.

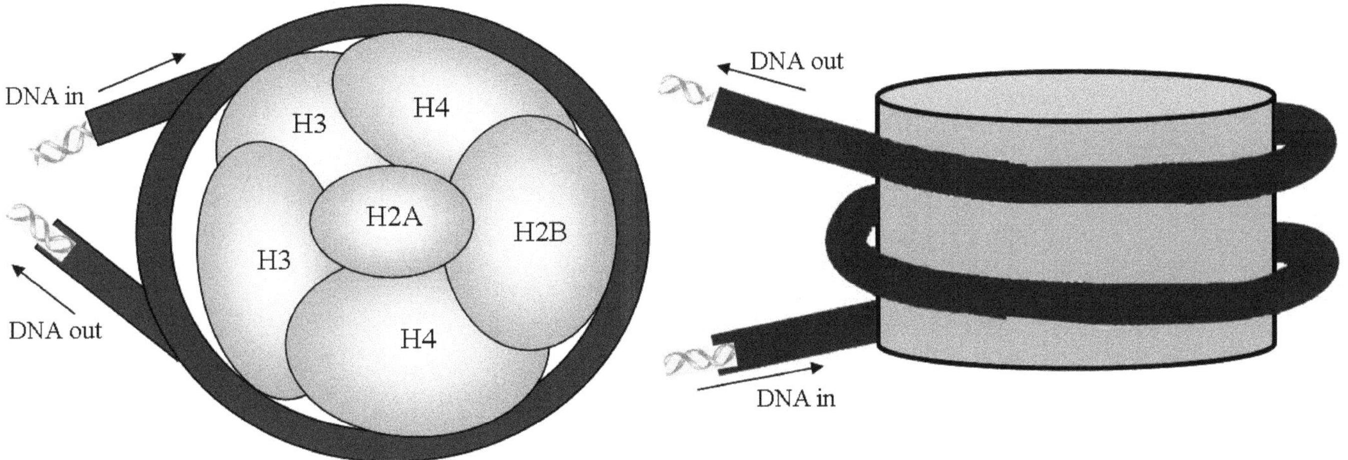

Figure 4.10: DNA Wraps around the Core Nucleosome.

non-histones. Histones play the most essential structural role in the proteins associated with DNA. Histones contain lysine and arginine, making it possible for them to bind through electrostatic exchanges to the negatively charged phosphate groups of the DNA nucleotides. There are five diverse types of histone proteins called – H1, H2A, H2B, H3, and H4. A nucleosome core particle consists of two copies each of histones H2A, H2B, H3, and H4 to form a histone octamer around which ~150 base pairs of DNA are wrapped in a left-handed superhelix, which completes about 1.7 turns per nucleosome (**Figure 4.10**). Richmond and colleagues in 1997 were able to solve the X-ray crystal structure of the nucleosome-DNA complex at a higher resolution. It is possible to see the DNA helix as it encircles the histone octamer. The amino-terminal ends of some of the histones overhang from the octamer and project away from the nucleosome. The amino-terminal tails contain the adjacent nucleosomes to generate nucleosome-nucleosome associations. Histone tails play an important role in gene expression and regulation.

The 2 nm DNA double helix is 10 nm in diameter and is coiled into the nucleosome core. Roughly 200 base pairs of DNA link to each core particle to form the 'beads on a string', which can be visualized microscopically. Histone H1, which is not part of the core octamer, may be located at the site where DNA enters and leaves the nucleosome and may act to seal the DNA around the nucleosome (**Figure 4.11**).

The nucleosome configuration is an exciting step; the DNA is shortened to about one-third of its new length in the nucleus, however, chromatin does not survive in this comprehensive form. Instead, the 10 nm chromatin fiber is further packed into a thicker 30 nm fiber, which is labeled as a solenoid. Numerous nucleosomes are packed closely; however, the precise orientation and details of the structure are not clear. It has recently been recommended that the 30 nm fiber might adopt a compact helical zigzag pattern with about four nucleosomes per 10 nm. The second level of packaging is generated after the formation of the overall length of the DNA *i.e.*, 30nm fiber.

A series of loop structures are formed by 30nm fiber; this condenses and creates the structure of the chromatin fiber. The fibers are then coiled along the chromosome arms that include a chromatid; this form is the metaphase chromosome. The tight packing of the DNA into a chromosome presents a mammoth challenge to both the replication of DNA and its transcription.

In the overall transition from a fully extended DNA helix to the very condensed status of the mitotic chromosome, a packaging ratio of about 500:1 should be achieved.

4.11 Different DNA Structures

There are different structures of DNA named as A-, B-, and Z-DNA (**Figure 4.12**). In **Table 4.3** the different forms of DNA are illustrated.

4.11.1 A-DNA and B-DNA

The 10.0 base pairs are present in the A DNA and B DNA. In these forms there exit 3600 turns. The bases are far away from the perpendicular to the helix axis (which would be 0°) by 13° in A-DNA and 20° in B-DNA.

B-DNA is found under humid conditions. B-DNA is a double-helical structure analyzed in hydrated fibers outside the cell. However, the A-DNA is seen only in non-humid conditions. The A-DNA double helix is small and wide with a narrow and very deep major groove and a wide, shallow minor groove. The B-DNA double helix is thinner and longer for a similar number of base pairs than A-DNA, with a wide major

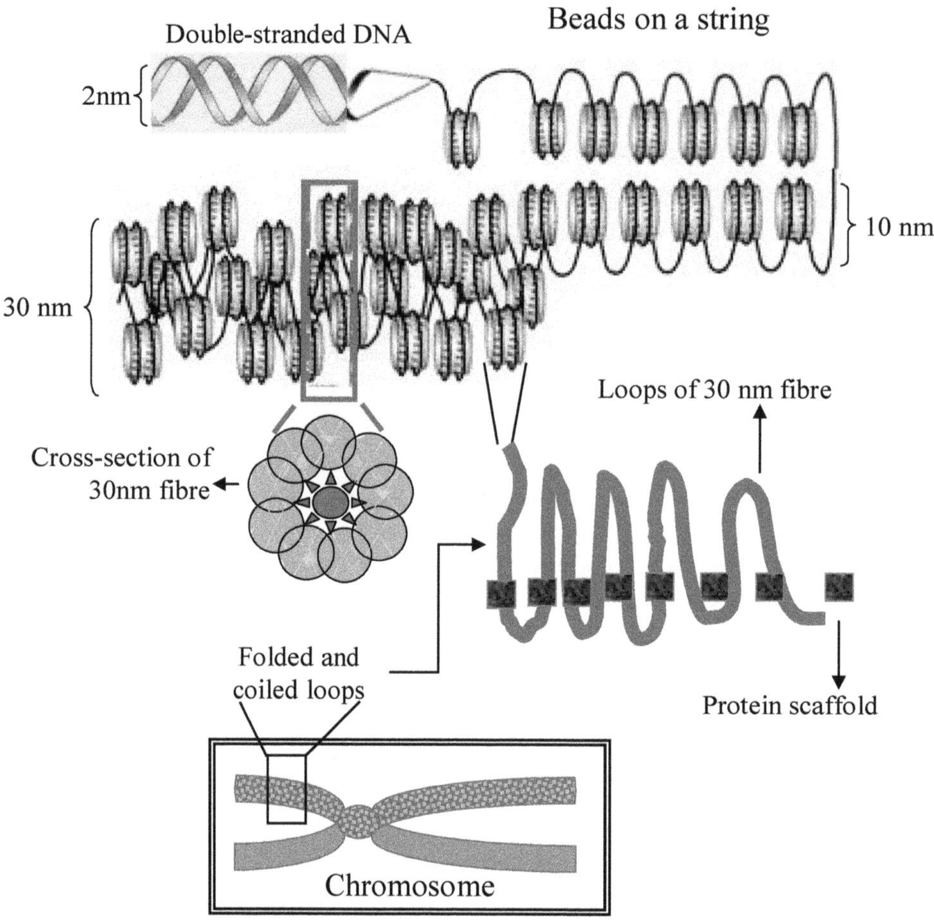

Figure 4.11: Packaging of DNA into the Eukaryotic Chromosome.

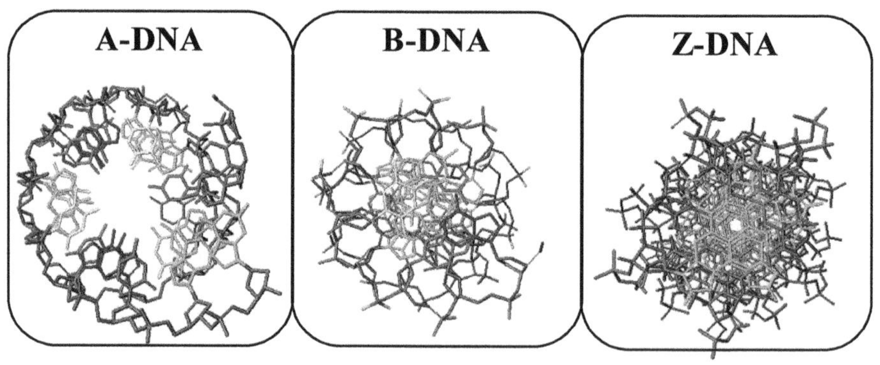

	A-DNA	B-DNA	Z-DNA
Helix diameter	2.6 nm	2.4nm	1.8 nm
Helix rotation	Right-handed	Right-handed	Left-handed
Rotation per bp	2.5 nm	3.4 nm	4.5 nm
bp per helical turn	11	10.4	12

Figure 4.12: Different forms of DNA. A-DNA, B-DNA and Z-DNA.

groove and a narrow minor groove; both grooves are of similar depths.

4.11.2 Z-DNA

Z-DNA is a left-handed helix with a zigzag sugar-phosphate backbone. (The latter property gave this form of DNA its "Z" title.) Z-DNA is found when salt conditions are high in the initial phase. This DNA form is stabilized under low salt conditions. Z-DNA has 12.0 base pairs per complete helical turn. The bases are tilted away from the right angles to the helix axis by 8.8 degrees. The Z-DNA helix is thin and elongated with a deep, minor groove. The major groove is very near to the surface of the helix, so it is not distinct. Different features of different DNAs are shown in **Table 4.3**.

Table 4.3: Properties of A-DNA, B-DNA, and Z-DNA

Property	*A-DNA*	*B-DNA*	*Z-DNA*
Major groove	Enormously narrow and very deep	Wide and of intermediate depth	Flattened out on helix surface
Minor groove	Very wide and shallow	Narrow and of intermediate depth	Extremely narrow and very deep
Helix axis location	Major groove	Through base pairs	Minor groove
Helix diameter	2.2nm	2.0nm	1.8nm
Overall morphology	Short and wide	Longer and thinner	Elongated and thin
Base pairs per helix turn	11.0	10.4	12.0
Helix direction	Right-handed	Right-handed	Left-handed

4.12 The RNA Molecule

4.12.1 The discovery of RNA as viral genetic material

RNA is a genetic material that was discovered by Heinz-Fraenkel-Conrat in 1957. They conducted an experiment on tobacco mosaic virus (TMV). This is a small virus and is made up of a single molecule of RNA covered by a protein coat. The protein coat differs in different viruses, which can be distinguished based on this. Fraenkel-Conrat and colleagues treated TMV particles of two different strains with chemicals that removed the protein coats of the viruses from the RNA molecules. Further, they separated the proteins from the RNA and mixed the proteins of one strain with the RNA molecules, resulting in complete, infective viruses. These are composed of proteins from one strain and RNA from the other strain. The progeny viruses were always genotypically and phenotypically the same as the parent strain (**Figure 4.13**). This showed that the genetic information of TMV is stored in RNA and not in proteins. The tobacco leaves were infected with re-constituted mixed viruses.

There exists a high degree of similarity between RNA and DNA. It is composed of ribonucleotides that differ only somewhat from the deoxyribonucleotide counterparts of DNA. Ribonucleotide present in RNA has carbon ribose sugars, whereas the deoxyribonucleotides has a nitrogenous base attached to their first carbon and a phosphate group attached to their 5′ carbon. Ribonucleotides, however, have a hydroxyl group attached to the 2′ carbon, whereas deoxyribonucleotides have the only hydrogen at this position. Like DNA, RNA consists of four nucleotides that differ from each other only in the composition of the nitrogenous base. Three of the nitrogenous bases are impossible to differentiate from three of the bases in DNA-cytosine, adenine, and guanine. The fourth nitrogenous base in RNA, uracil (U), is very similar to thymine in DNA. Thymine has a methyl group attached to carbon number 5, but this methyl group is not present in uracil.

The DNA and RNA from the phosphate group at the 5′ end of one nucleotide attach to the 3′ carbon of another nucleotide, as illustrated in **Figure 4.14**. We call the attachments phosphodiester bonds as there are two ester linkages, one on either side of the phosphorus atom. These phosphodiester bonds link one nucleotide to the next to form a chain of nucleotides called a single strand of DNA or RNA. The 5′ carbons of each nucleotide in a single strand point towards the same direction. At one end a strand has a 5′ carbon with a phosphate group attached, and the other end terminates with a 3′ carbon attached

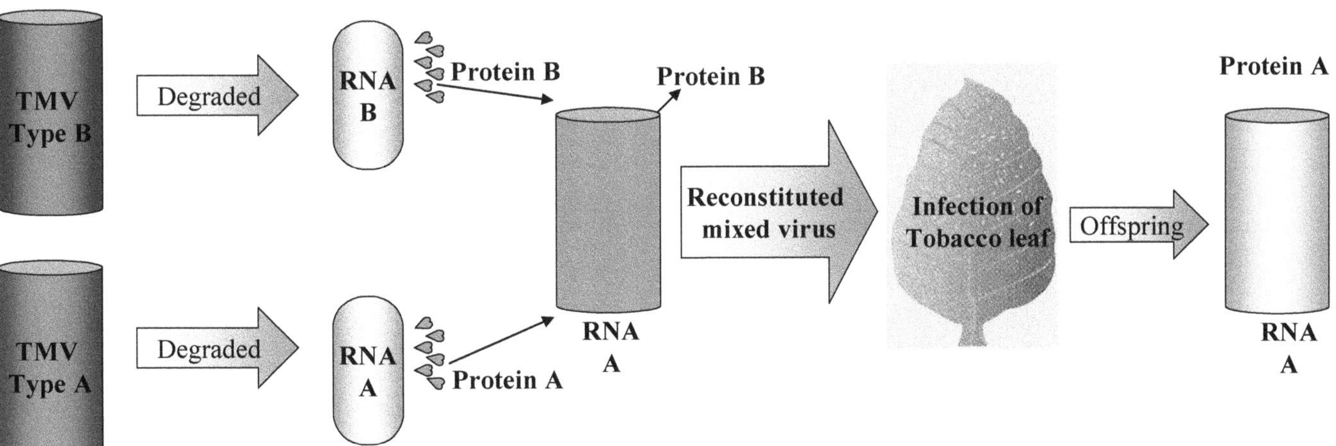

Figure 4.13: Genetic Material of Tobacco Mosaic Virus (TMV) is RNA, not Protein.

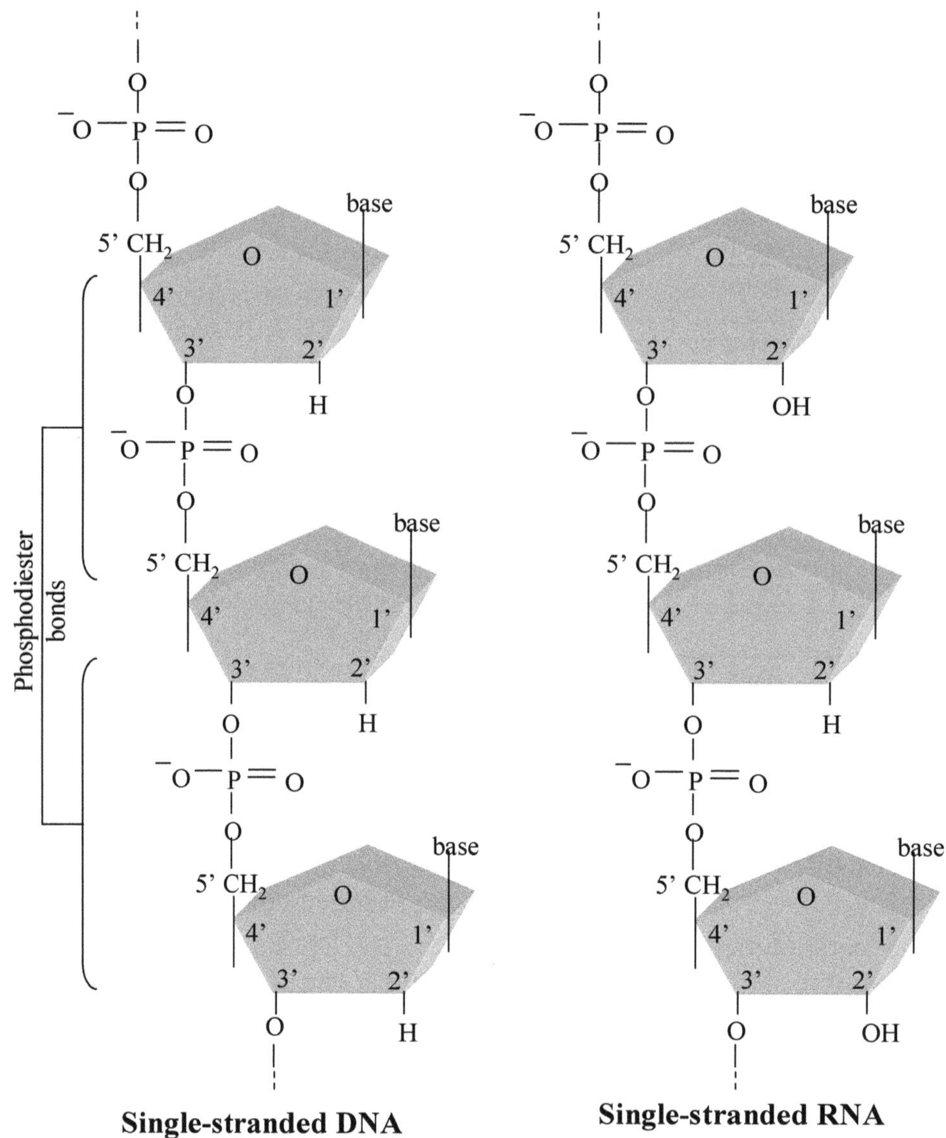

Figure 4.14: Phosphodiester Bonds in Single Strands of (a) DNA and (b) RNA.

to a hydroxyl group. DNA and RNA also exist in double-stranded forms in which two single strands of nucleotides wind around each other to form a double helix. Most DNA molecules are double-stranded, whereas most RNA molecules are single-stranded. The nucleotides T, C, A, G, and U attach to each other with phosphodiester bonds to form the sugar-phosphate backbones of nucleic acids.

4.12.2 Types of RNA

There are several types of RNA present in the eukaryotic cells.

4.12.2.1 Messenger RNA (mRNA)

RNA is synthesized in the nucleus and are transferred to act in the cytoplasm before export occurs naked RNA is complexed with proteins, which are known as the heterogeneous nuclear ribonucleoprotein particles (hnRNP's), and the proteins (other than cap-binding proteins and poly [A] binding protein) are called hnRNP proteins.

There are approximately 20 different hnRNP proteins in each human cell. Some assist in modifying the pre-mRNA to make synthesis easier. The properly processed mRNAs get exported. If mRNAs are not correct, then defective proteins would be formed. There are about 3,000 to 15,000 different species of mRNA. Variation is seen in mRNAs at the size, sequence, and secondary structure and folding takes place according to the above properties. Various mRNAs in a cell share not many characteristic features, which may act as an export signal. However, they share many proteins, so one of the suggested mechanisms for signaling export is that one or more hnRNP proteins might contain nuclear export signals (NESs). According to this model, proteins with NESs would bind to the mRNA, and a transport receptor would differentiate the NE signals in the protein. At least some hnRNP proteins have NESs, but others do not. The export of some mRNAs is modulated by signals found in mRNA-binding proteins that are not hnRNP proteins.

Although cellular mRNAs are generally not exported until all RNA handling has been finished, some viral RNAs are exported without removing introns. Human immunodeficiency virus – 1 (HIV-1) RNA is reasonably well understood. Like all retroviruses, HIV-1 produces a set of overlapping mRNAs. Some mRNA is spliced. However, full-length, non-spliced viral mRNA must be exported because this is the form of viral RNA used to produce new infectious viral particles, and it also functions as an mRNA to produce some HIV proteins. Protein transport is handled by the synchronized in response to external signals and intracellular requirements, and mRNA export can be regulated. The best example of regulated mRNA export occurs as a part of the response of cells to stresses such as heat, *etc.*, which is referred to as heat shock response. Heat shock response is due to polyadenylated RNAs not being exported and accumulating in the nucleus. The cell does not control the expression of heat shock genes, which encode proteins that protect the cell from damage. Some mRNA exports are prohibited, but the competent export of heat shock mRNAs is allowable. The exact mechanism by which this occurs is not precise. It is unclear how mRNA is exported by heat shock; it does not need new protein synthesis and can be detected within a few minutes after heat shock. These observations suggest that a signal transduction pathway is induced. This may modify the nuclear transport machinery to selectively export stress reaction mRNAs or alter mRNA biogenesis so that only stress reaction mRNAs are correctly packaged and recognized for export.

Transcription and processing of mRNAs occur in the nucleus, which requires hnRNPs to move from transcription sites to NPCs at the nuclear margin. Sites of transcription are also the sites of most mRNA processing after which 5′ capping occurs as soon as the RNA is long enough to emerge from the RNA polymerase holoenzyme; splicing begins during transcription, and 3′ ends are generated by cleavage of the RNA while it is still a growing chain. In the case of defective splicing, mRNAs are not exported.

It is thought that hnRNPs disperse from sites of transcription and processing through the interchromosomal spaces to nuclear pore complexes and are then exported from the nucleus.

Exporting mRNA into the cytoplasm is quite a complex process compared to protein transport across the nuclear membrane. Many novel factors are essential during the export of mRNA. The mRNA export process is synchronized with transcription and mRNA processing to export only fully processed mRNAs. In fact, some of the proteins required for mRNA export also take part in the processing events that must be completed correctly for the mRNA export.

The mRNA's 5′ cap structure is not necessary for mRNA export but stimulates the process. In contrast, splicing and 3′ processing are coupled to export. Splicing and polyadenylation factors begin to relate to mRNAs during transcription. Once the splicing is completed, the protein complex is called the exon junctional complex (EJC), which remains close to the splice junction in metazoans. The occurrence of this complex may be measured as a signal that the mRNA has been spliced.

4.12.2.2 snRNAs are Exported, Modified, Assembled into Complexes, and Imported

Small nuclear riobo nucleoprotein particles (snRNPs) are RNA protein complexes and other nuclear RNA processing. The RNAs found in snRNPs, the U snRNAs, are synthesized in the nucleus. However, in metazoan cells, the configuration of functional snRNPs needs U snRNA export, alteration in the cytoplasm, association with proteins in the cytoplasm, and import of the U snRNP complex. The imported U snRNP complex then undergoes final gathering into the snRNP.

Most U snRNAs are composed of RNA polymerase II. Like mRNAs, which are also RNA polymerase II products, U snRNAs have monomethylated 5′ caps. Still, they are unlike those of mRNAs as they do not have a poly(A) tail, and their synthesis uses a unique signal for 3′ processing that results into RNA that is not polyadenylated. The U snRNA cap is a key signal for its export. When the RNA has entered the cytoplasm, its cap is methylated to become a trimethyl-G cap, and the RNA is assembled into an RNA-protein complex by interacting with a set of proteins called the Sm proteins. The trimethyl cap contributes to the splicing process in which U-snRNPs function and the proteins help to create the essential overall three-dimensional structure of each snRNP. These complexes are called U snRNPs. They are imported into the nucleus by a transport receptor consisting of an adaptor (snurportin) and importin β. The cap-binding and Sm proteins provide dual signals for import. U snRNAs do not come out from the nucleus into budding yeast. In yeast, U snRNPs are assembled in the nucleus from RNAs and imported proteins miRNA plays an important role in the regulation of gene expression. miRNA is only 21 – 22 nucleotides long and is seen in multicellular organisms be it plant or humans. miRNA represents 4 per cent of the genome they play an important role in the regulation of varied pathways, including development, differentiation, and apoptosis (programmed cell death), organogenesis, and cell proliferation. They act as binding to target mRNAs in the cytoplasm, in some cases blocking their translation and in others promoting their turnover.

miRs are products of RNA polymerase II transcription, and in animals, they undergo the processing of a large precursor by RNase III-like enzyme complexes. Some genes encoding miRs are found in regions between protein-coding genes. Others are situated within the intronic regions of protein-coding genes. Often, miR precursors contain multiple miRs.

miRs undergo many processing steps. The enzyme Drosha in the nucleus along with the Pasha protein processes the large precursor to yield smaller hairpin precursors, called pre-miRs, each having a single miRs and flanking sequence and likely to assume a hairpin pattern. These precursors are exported by exportin -5, in a manner mechanistically like the export of tRNAs by exportin –t. Once in the cytoplasm, another RNase III-like enzyme, Dicer, interacts with other factors to process the hairpin and yield the functional miR, which is associated with multiple proteins to form the RNA-induced silencing complex (RISC). This then interacts with target mRNAs.

4.12.2.3 Ribosomal RNA (rRNA)

Ribosomal RNA is involved in constructing ribosome machinery for synthesizing proteins by translating mRNA. Ribosomes comprise large complexes constituting two subunits with about 80 proteins and four ribosomal RNAs (rRNAs). The subunits are assembled separately in the nucleolus and reach the cytoplasm for final assembly. The ribosomal subunits are among the largest complexes transported through NPCs, at the upper limit of what the nuclear pore channel can have. Because these subunits are large, other macromolecules may be excluded from transport when a ribosomal subunit goes through the channel.

The assembly and export of the two ribosomal subunits are involved in highly interactive processes and depend heavily on nuclear transport. Up to 50 per cent of all nuclear transport may be occupied in ribosome biogenesis. The ribosomal components are among the most profuse proteins and RNAs in the cell. Ribosomal proteins are synthesized in the cytoplasm, imported into the nucleus, and enter the nucleolus. There, they interact with rRNA precursors. Following proper assembly, each subunit is exported through the NPC. Subunits appear to remain in the nucleolus until they are correctly pre-assembled for export, but what signals are released from the nucleolus into the nucleoplasm is unknown. In the cytoplasm, maturation of the subunits is accomplished, and subunits can then act together with translation initiation factors, tRNAs, and mRNAs to form mature translating ribosomes. As for other exported macromolecules, the export of ribosomal subunits is receptor mediated.

The large (60S) ribosome subunits need multiple receptors. One of these receptors is Crm1, which interacts with the 60S subunit through an adaptor protein called Nmd3. The 60S subunit also requires two other export receptors: the Mtr2/Mex67 heterodimer that plays a vital role in mRNA export and another protein. The export of the small 40S subunit also needs Crm1. Although it will likely interact with the 40S subunit through an adapter, Nmd3 is unnecessary for small subunit export, and an adapter has not yet been identified.

4.12.2.4 Transfer RNA (tRNA)

The ribonucleic acid (RNA) that is directly in use for the translation of the sequence of nucleotides in messenger RNA to amino acid sequences for the building of proteins is recognized as the transfer RNA usually called tRNA. The production of the tRNA itself is aimed at by the DNA in the cell that provides a pattern to produce RNA by "transcription". tRNA is commonly called the cloverleaf structure of RNA. It is a two-dimensional projection of the form of the molecule. Its actual three-dimensional shape is more complex, but this is a common way to depict it to highlight its function of binding an amino acid to one end parallel to the anticodon on the opposite end. This anticodon binds to a codon consisting of three nitrogenous bases which specify an amino acid according to the genetic code (**Figure 4.15**). The tRNA has been analyzed for a number of yeast and its structure was the culmination of seven years of work by Robert Holley of Cornell University. When the messenger RNA blueprint for a protein reaches a ribosome for the process of building a protein by translation of that blueprint, tRNA molecules with all the required amino acids must be there for the process to proceed. Since most proteins use all twenty amino acids, all must be accessible, attached to suitable tRNA molecules. In this example, the tRNA has bonded to the codon GCC on the mRNA, GCC being one of the codons corresponding to alanine. This leads to the assignment of alanine in the proper place in

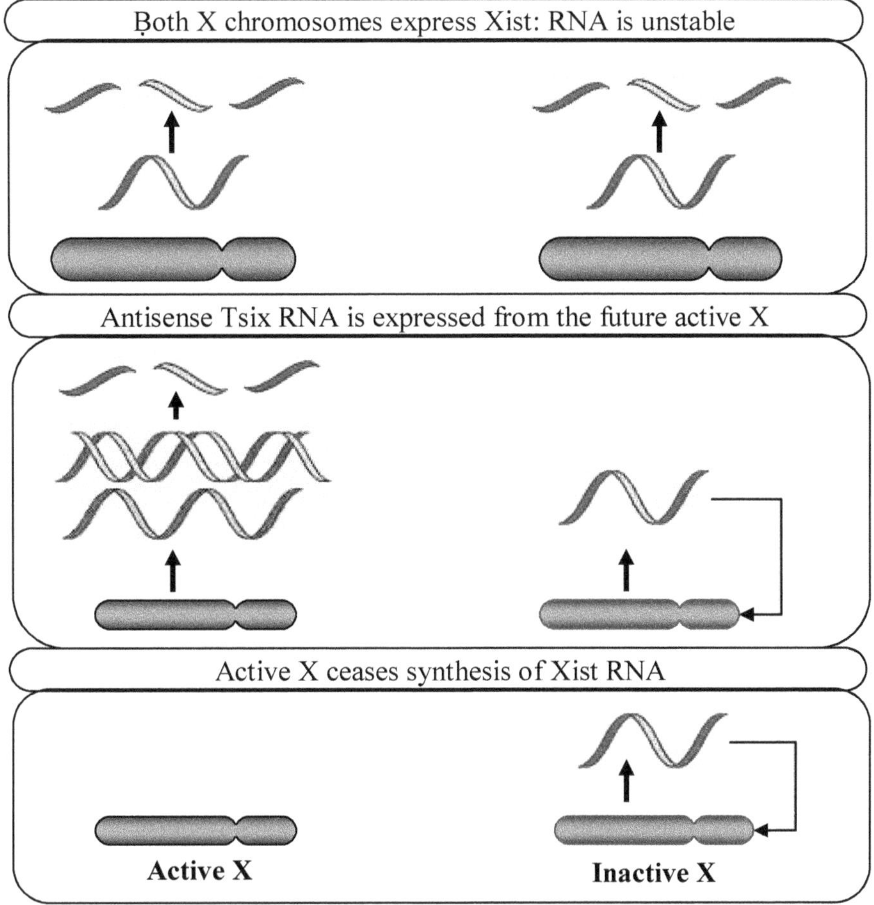

Figure 4.15: X inactivation Involves the Stabilization of XIST RNA, which Coats the Inactive Chromosome. Tsix prevents XIST expression on the future active X.

the growing polypeptide chain that is to become a protein.

Many forms of tRNA have roughly the same size and shape, varying from about 73 to 93 nucleotides. Besides the usual bases A, U, G, and C, all have a significant number of modified bases, which are thought to have been formed by alteration after the transcription. All tRNAs have sequences of nucleotides that are complementary to other parts of the molecule and base-pair to form the five arms of the tRNA. Four of the arms are reliable, but the variable arm can range from 4 to 21 nucleotides.

The exportin-t binds tRNA directly in the presence of Ran-GTP inside the nucleus and forms a trimeric tRNA-exportin Ran-GTP complex. This complex is like those formed between other exportins and NES-containing proteins. The trimeric complex passes through the nuclear pore to the cytoplasm. After Ran-GAP stimulates GTP hydrolysis by Ran, the complex dissociates and releases tRNA.

The mRNA's 5′ cap structure is not necessary for mRNA export but stimulates the process. In contrast, splicing and 3′ processing are coupled to export. Splicing and polyadenylation factors begin to relate with mRNAs at the time of transcription. Once the splicing is completed, the protein complex is called exon junctional complex (EJC) which remains close to the splice junction in metazoans. The occurrence of this complex may be measured as a signal that the mRNA has been spliced.

4.12.2.5 XIST RNA

The X chromosome has more than 1,000 genes that are necessary for appropriate development and cell viability. However, females carry two copies of the X chromosome, resulting in a potentially toxic double dose of X-linked genes. To rectify this imbalance, mammalian females have evolved a unique mechanism of dosage compensation distinct from that used by organisms such as flies and worms. The machinery is, by the process called X-chromosome inactivation (XCI), female mammals transcriptionally silence one of their two Xs in a complex and highly coordinated manner. This was developed by Lyon in 1961 hence also called the Lyon hypothesis. The inactivated X chromosome then forms a compact structure called a Barr body, and it is stably maintained in a silent state. A prime example of X inactivation is in the coat-color patterning of tortoiseshell or calico cats. In cats, the fur pigmentation gene is X-linked, and depending on which copy of the X chromosome each cell chooses to leave active, either an orange or black coat color results. X inactivation only occurs in cells with multiple X chromosomes, which explains why almost all calico cats are female. X inactivation exists in two different forms: random and imprinted. Although both forms utilize the same RNAs and silencing enzymes, they differ in terms of their developmental timing and mechanism of action.

XIST codes for an RNA that lacks open reading frames. The XIST RNA "coats" the X chromosome, from which it is synthesized, suggesting that it has a structural role. Prior to X-inactivation, it is synthesized by both female X chromosomes. Following inactivation, the RNA is found only on the inactive X chromosome. The transcription rate remains the same before and after inactivation, so the transition depends on posttranscriptional events. An antisense RNA generated from XIST, Tsix, plays a role in regulating the stability of XIST. Tsix is active on the future active X but downregulated on the future inactive X; this regulation allows the persistence of XIST on the inactive X and thus leads to silencing (**Figure 4.15**).

Previous to X-inactivation, XIST RNA decays with a half-life of ~2Hr. X-inactivation is mediated by stabilizing the XIST RNA on the inactive X chromosome. The XIST RNA reveals an intermittent distribution along the X chromosome, suggesting that association with proteins to form particulate structures may be the means of stabilization. The precise mode of action of XIST RNA is limited to dispersal in cis along the chromosome. Accumulation of XIST on the future inactive X results in the exclusion of transcription machinery (such as RNA polymerase II) and leads to a series of chromosome-wide histone modifications, including H4 deacetylation and precise methylation of both H3 and H4. Late in the process, an inactive X-specific histone variant, macro H2A, is incorporated into the chromatin, and promoter DNA is methylated. At this point, the heterochromatic of the inactive X is stable, and XIST is not necessary to uphold the silent state.

5

DNA Replication and Recombination

The most significant aspect of Watson and Crick's discovery of DNA structure was not that it provided scientists with a three-dimensional model of this molecule but that this structure revealed how DNA was replicated. As noted in their 1953 paper, Watson and Crick strongly suspected that the specific base pairings within the DNA double helix existed to ensure a controlled system of DNA replication. However, it took several years of subsequent study, including a classic 1958 experiment by American geneticists Matthew Meselson and Franklin Stahl, to understand the exact relationship between DNA structure and replication. Replication is an essential step in the precise function of DNA. Around 3x 109 (3 billion) base pairs are found in 23 pairs of chromosomes. If an error occurs in one in a million, it will create 3000 errors during each replication cycle of the genome.

The base pairing in DNA reveals that if two strands of a DNA molecule are separated, they can serve as a template for forming a complementary strand by bringing in individual nucleotides to base-pair with their complementary base on the template and collectively joining the new nucleotides. Thus, each DNA molecule after replication will have one of the original strands plus one newly synthesized strand. This model of DNA replication is stated to be semi-conservative.

Initially, three models of DNA replication were projected. These were conservative, dispersive, and semi-conservative replication models. Conservative replication projected that one DNA molecule has a newly synthesized DNA after replication while the other molecule is entirely original DNA. Dispersive replication recommended that each DNA molecule after replication might consist of a part of the new and old DNA, which is interspersed. The conservative, semi-conservative, and dispersive models are shown in **Figure 5.1**.

To validate various models of replication, Meselson and Stahl carried out different experiments, as shown below.

5.1 The Meselson-Stahl Experiment

In 1958, M. S. Meselson and F. W. Stahl published a paper explaining an experiment showing strong evidence that semiconservative replication is the model bacterial cells use to make new DNA molecules. They grew E. coli cells for several generations in a medium containing 15NH4Cl (ammonium chloride), the only nitrogen source. A "heavy" isotope of nitrogen, 15N has one more neutron than the naturally occurring 14N isotope; thus, molecules containing 15N are denser than those containing 14N. Unlike radioactive isotopes, 15N is stable; after several generations, all nitrogen-containing molecules, including the nitrogenous bases of DNA, had the heavier isotopes in the E. coli cells. Their success was because DNA having 15N can be identified from DNA containing 14N. They used sedimentation equilibrium centrifugation or density gradient centrifugation. Samples were enforced to centrifugation through a density gradient of a heavy metal salt, such as cesium chloride. The principle

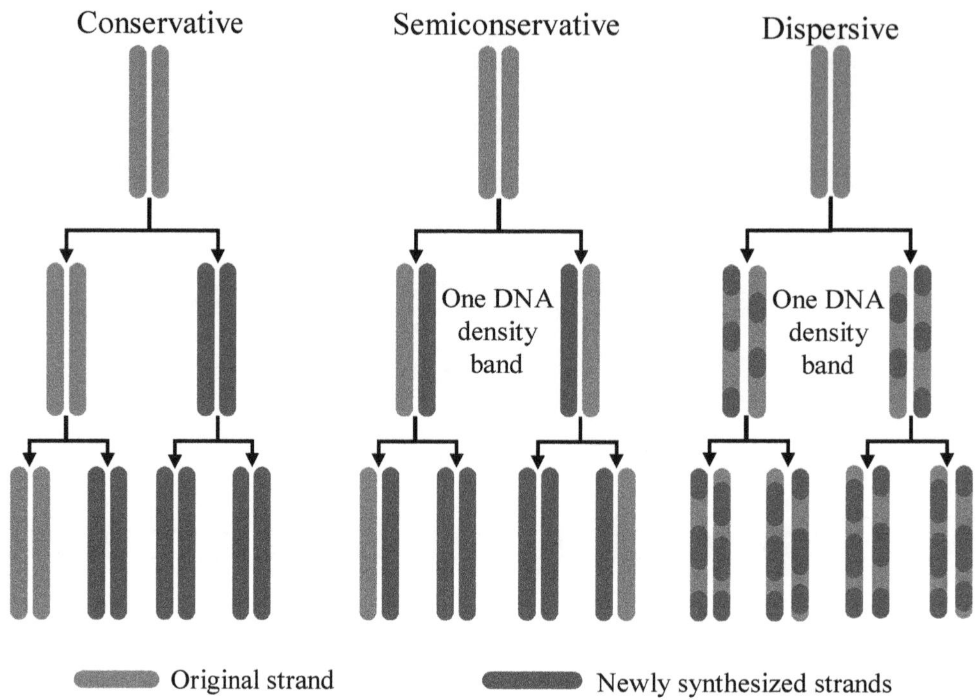

Figure 5.1: Three Proposed DNA Replication Mechanisms, *i.e.*, Conservative, Semi-conservative, and Dispersive Models.

is that molecules of DNA will reach equilibrium when their density equals the density of the gradient medium. In this case, 15N-DNA will reach this point at a position closer to the bottom of the tube than will 14N-DNA.

In this experiment (**Figure 5.2**), uniformly labeled 15N cells were transferred to a medium containing only 14NH4Cl. All "new" synthesis of DNA during replication contained only the lighter isotope of nitrogen. The time of transfer to the new medium was taken as time zero (t = 0). Replication was over after several generations after each replication cells were removed. DNA was isolated from each sample and centrifuged using sedimentation equilibrium centrifugation.

After one generation, the isolated DNA was present in only a single band of intermediate density. This was the expected result for semi-conservative replication in which each replicated molecule was composed of one new 14N-strand and one old 15N-strand. These results do not match with the conservative replication, in which two distinct bands should occur. Hence this model may be rejected.

After two cell divisions, DNA samples showed two density bands-one intermediate band and the other lighter band corresponding to the 14N position in the gradient. Similar results were even seen in the third generation. This illustrated that the replication is semi-conservative.

However, one should not forget dispersive replication, which explains the intermediate density of DNA. However, Meselson and Stahl ruled out this mode of replication based on the two observations. (1) After the first generation of replication in a 14N-containing medium, they were able to isolate the hybrid molecule, which they heated denatured upon heating a duplex got separated into a single strand. When the densities of the single strand of the hybrid were determined, they revealed either a 15N profile or a 14N profile, but no intermediate density was seen.

These results showed that there exists a dispersive mode of replication but in support of a semi-conservative mode of replication. If the following replication was dispersive, all generations after t = 0 should show DNA of an intermediate density. However, in the actual experiment, it was not seen. The Meselson-Stahl experiment provided conclusive support for bacterial semi-conservative replication and ruled out both models.

5.2 Semi-Conservative Replication in Eukaryotes

In 1958, the Meselson-Stahl experiment proved that DNA replication was semi-conservative in bacteria. In the early 1960s, Herbert Taylor did a root-tip experiment to demonstrate that a eukaryotic cell replicated by semi-conservative replication. They used the root tips of the broad bean, which has many dividing cells. Replication was monitored

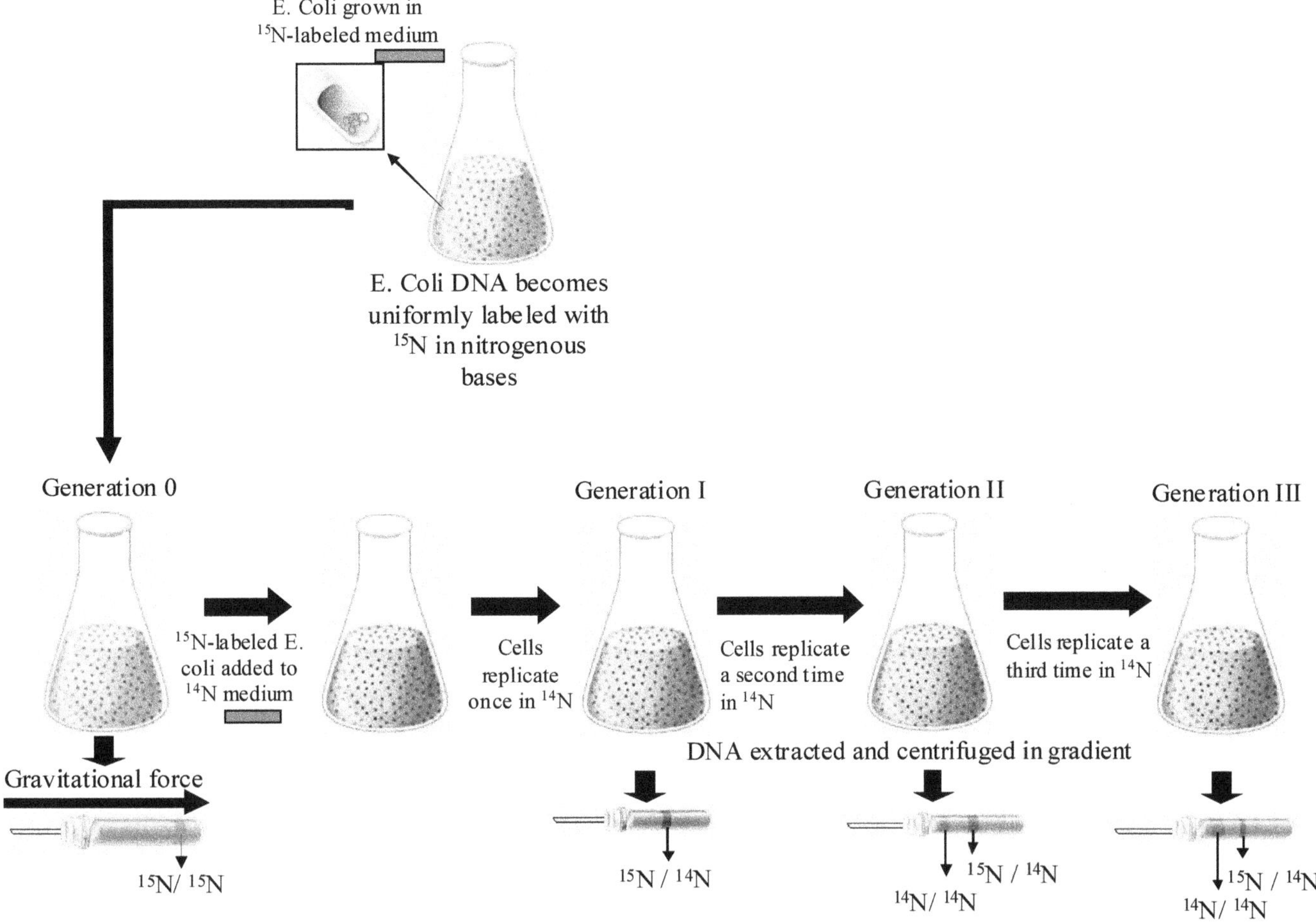

Figure 5.2: Meselson-Stahl Experiments.

by labeling DNA with 3H-thymidine, followed by auto-radiography. Autoradiography helped locate the newly synthesized DNA inside the cell. They continued this experiment, arrested the cultures at metaphase by adding colchicine, and then examined the chromosomes by autoradiography. They detected the radioactive thymidine, and it was found only in association with chromatids with newly synthesized DNA. These results are in favor of the semi-conservative mode of replication. Both sister chromatids show presence after the first replication cycle radioactivity, revealing that each chromatid contains one newly synthesized DNA strand with radioactivity and one old unlabeled strand. In the second replication cycle, which is carried out in the unlabeled medium, out of the two sister chromatids of each chromosome, only one should be radioactive, as one of the parent strands is unlabeled. There may be exceptions of sister chromatid exchanges. In some cases, this result was observed. These experiments proved the semiconservative model.

Although the semi-conservative nature of DNA replication has been confirmed, many questions about replication remained unanswered. These questions were at what specific site on the chromosome the replication is initiated, at random sites or multiple locations. The answer depends on the nature of the organism under study. For instance, bacteria have a single specific origin of replication; it begins at the same point on the chromosome every time. The replicon of E. coli, which is only 4.2Mb genome and the replication site, is called OriC. It is a nine-base-pair (bp) sequence and is repeated four times within the replication. There is clear evidence showing the origin and direction of replication. John Cairns was able to track the replication in E. coli using radioactive precursors of DNA synthesis followed by radioactivity. He showed that in E. coli, there is a single oriC, from where the replication is initiated. DNA synthesis in bacteriophages and bacteria originates simultaneously; the entire chromosome constitutes one replicon. Single origin is characteristic of bacteria, which have only one circular chromosome. Still, other results again rely on autoradiography, showing that replication is bidirectional, moving away from oriC in both directions. This results in two replication forks that migrate farther and farther apart as replication proceeds. These forks eventually merge as semi-conservative replication of the entire chromosome is completed at a termination region called ter. Still, other results, again on autoradiography, demonstrated that replication is in both directions moving away from oriC in both directions. The DNA synthesis in bacteria involves five polymerases and some other enzymes.

A particular protein called DnaA coded by the gene dnaA allows the unwinding of the helix. Several subunits of the DNA A protein bind to each of several 9mers. This step facilitates the subsequent binding of DnaB and DnaC proteins that further open and destabilize the helix. Proteins such as these, which require the energy usually supplied by the hydrolysis of ATP to break hydrogen bonds and denature the double helix, are called helicases. Other proteins, called single-stranded binding proteins (SSBPs), stabilize this open conformation **Figure 5.3**.

5.3 DNA Polymerase I

Arthur Kornberg and colleagues in 1957 first reported the role of enzymes in DNA. They isolated an enzyme from E. coli that could direct DNA synthesis in a cell-free (in vitro) system. The enzyme is now known as DNA polymerase I, as it was the first of many other similar enzymes to be isolated. Depiction of the activity of this enzyme in vitro discovered that it had specific necessities for its activity. It required 5′-triphosphate forms of the four nucleotides, which are already present in the DNA. Pre-existing DNA has two functions. (1) One serves as a template for the synthesis of the new DNA, and the other serves as a primer for DNA synthesis. It is now known that DNA polymerase I cannot initiate DNA synthesis without having a free 3′-OH to add a new nucleotide to DNA synthesis; therefore, it needs a primer, a preexisting piece of nucleic acid to serve as an initiator of DNA synthesis. DNA polymerase I synthesizes DNA by forming a bond between the 5′ phosphate of the incoming nucleotide and the 3′ OH group of the nucleotide at the end of the growing DNA chain. This demonstrated that DNA chain synthesis occurs from the 5′ to 3′ direction.

DNA polymerase I has polymerase activity and two other enzymatic activities, both of which are exonuclease activities. Exonucleases digest DNA by cleaving phosphodiester bonds, which results in the chewing of the nucleotides from the end of the DNA chain. DNA polymerase I has 3′ to 5′ exonuclease activity, degrading DNA in a direction opposite to the synthesis. This provides the enzyme with a proofreading function. If a wrong nucleotide gets inserted into a growing chain, the enzyme can digest it with the 3′ to 5′ exonuclease activity and insert the correct nucleotide. Kornberg determined that there were two significant requirements for in vitro DNA synthesis under the direction of DNA polymerase I, *i.e.*, all four deoxyribonucleoside triphosphates (dNTPs) and template DNA.

If these four deoxyribonucleoside triphosphates are absent, there will be no synthesis. If derivatives of these precursor molecules other than the nucleoside triphosphate were used (nucleotides or nucleoside diphosphates), the synthesis would also not occur. If no template DNA is added, synthesis of DNA will occur but will be significantly reduced (**Figure 5.4**).

How each nucleotide is added to the growing chain is the function of the specificity of DNA polymerase I, which contains 928 amino acids. The precursor dNTP has the three phosphate groups attached to deoxyribose's 5′ -carbon. As the two terminal phosphates are cleaved during synthesis, the remaining phosphate groups are connected to the 5′ –carbon and are covalently linked to the 3′ -OH group of the deoxyribose to which it is added. Thus, chain

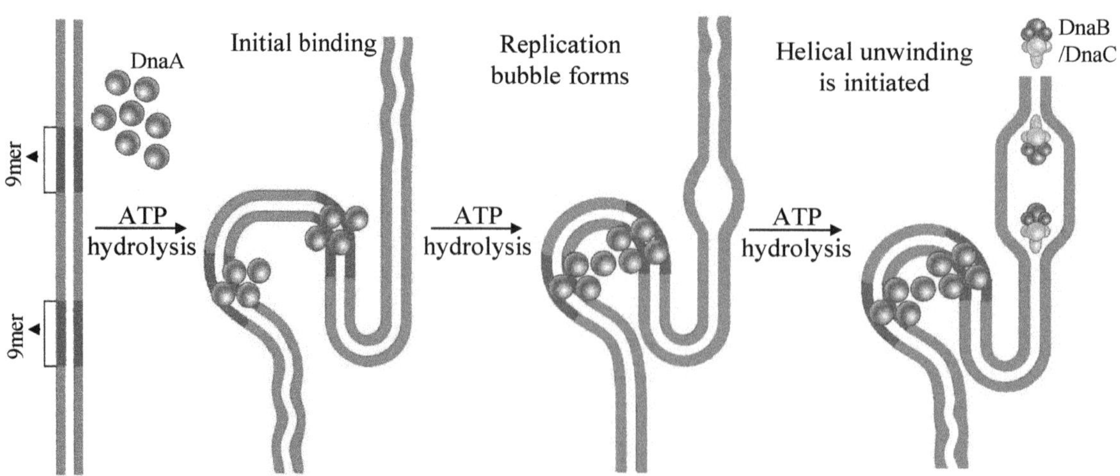

Figure 5.3: Helical Unwinding of DNA.
Three proteins called Dna A, B and C proteins are involved at the time of unwinding.

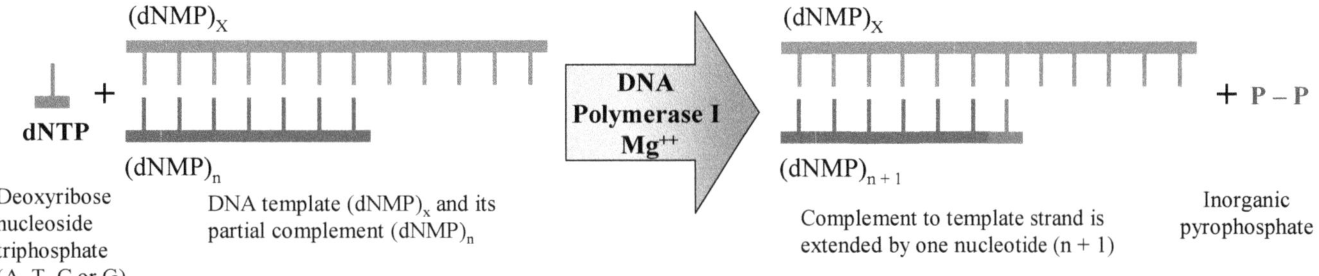

Figure 5.4: DNA Polymerase I Catalyzes the Chemical Reaction.

elongation occurs in the 5′ to 3′ direction by adding one nucleotide to the growing 3′ end at a time. Each step provides a newly exposed 3′-OH group, which can participate in the next edition of a nucleotide as DNA synthesis proceeds **Figure 5.5**.

The base composition of the DNA template is different in different organisms. The composition of the template and product among triphosphate (T2), *E. coli,* and calf are shown in **Table 5.1**.

5.4 Synthesis of Biologically Active DNA

Many experiments have proved that DNA replication is by semi-conservative method but still people were not fully convinced that DNA polymerase I was the enzyme that helps in replication of DNA within cells (in vivo). The doubt arose because in vitro synthesis was much slower than in vivo synthesis, the enzyme was much more effectively replicating single-stranded DNA than double-stranded DNA, and the enzyme appeared to degrade DNA and synthesize, showing that the enzyme contains exonuclease activity. A further experiment was carefully designed on a small bacteriophage qxI 74. Each newly synthesized strand was distinguished and isolated from the template strand. The experiment protocol dictated that these

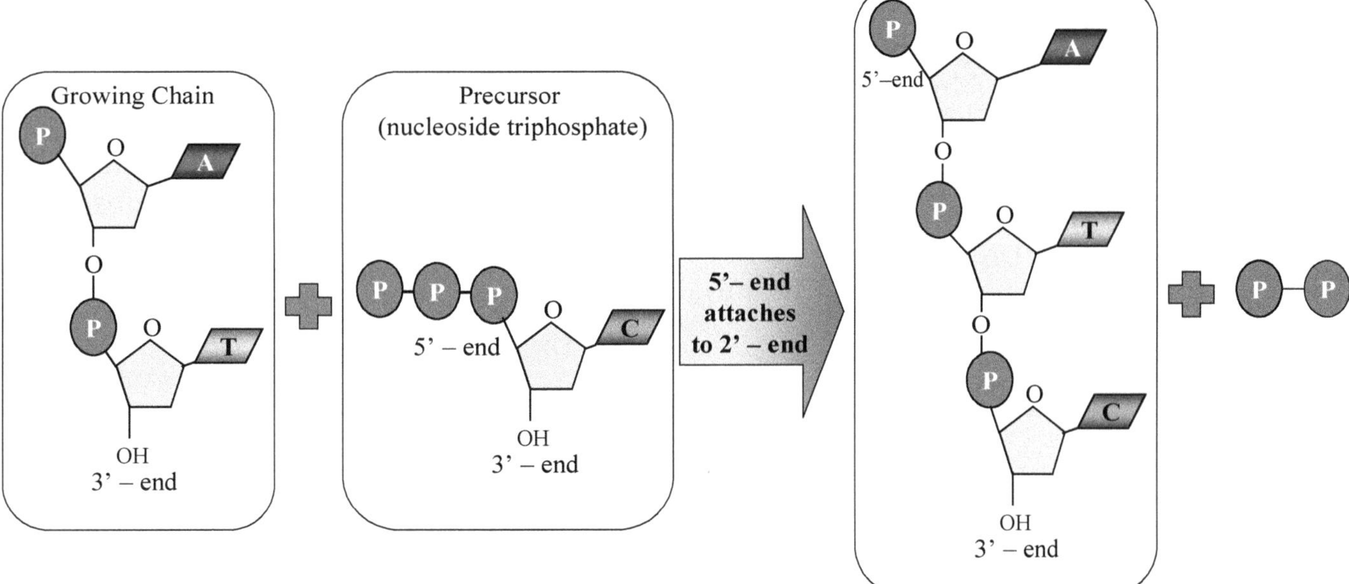

Figure 5.5: Synthesis of DNA from 5' to 3' end.

Table 5.1: Base Composition of the DNA Template

Organism	*Template or Product*	*Per cent A*	*Per cent T*	*Per cent 6*	*Per cent C*
Triphosphate (T2)	Template	32.7	33.0	16.8	17.5
	Product	33.2	32.1	17.2	17.5
E. coli	Template	25.0	24.3	24.5	26.2
	Product	26.1	25.1	24.3	24.5
Calf	Template	28.9	26.7	22.8	21.6
	Product	28.7	27.7	21.8	21.8

(+) strands must have been synthesized in vitro under the direction of Kornberg's enzyme. The critical test of biological activity was the process of transfection, in which newly synthesized (+) strands were added to bacterial protoplasts (bacterial cells minus their cell wall). After infection by the synthetic DNA, mature phages were produced- the synthetic DNA had successfully directed reproduction.

5.5 DNA Polymerase II, III, IV, and V

The general mechanism for DNA replication in eukaryotes is like that in prokaryotes, but the enzymes differ. Most enzymes responsible for DNA replication in yeast and mammals have now been characterized. There are five known DNA polymerases in mammals, called α, β, δ, e, and g. DNA polymerases and δ are both essential for replication and are probably the DNA polymerases responsible for most of the DNA synthesis. There is strong evidence that α is responsible for the synthesis of the lagging strand and δ is responsible for the synthesis of the leading strand.

DNA polymerase α also functions as a primase, synthesizing RNA primers on a DNA template to initiate replication (**Figure 5.6**). The association's two minor subunits (48 and 58 kD*) may act independently as a primase (**Figure 5.6**). They may also act as a primase as part of DNA polymerase α (Figure 5.6)–the 180 and 70 kD subunits of DNA polymerase α function in DNA replication. The combined DNA polymerase and primase activities of DNA polymerase α constitute strong evidence that this enzyme synthesizes the Okazaki fragments that become the lagging strands. DNA polymerase δ has no primase activity. DNA polymerases e and β help the damaged DNA to get repaired. The fifth DNA polymerase in eukaryotes g is found only in mitochondria. This enzyme synthesizes and proofreads the organelle's DNA.

After a primer is synthesized on a strand of DNA and the DNA strands unwind, synthesis and elongation can proceed in only one direction. As previously mentioned, DNA polymerase can only be added to the 3′ end, so the 5′ end of the primer remains unaltered. Consequently, synthesis proceeds immediately only along the so-called leading strand. This immediate replication is known as continuous replication.

The other strand (in the 5′ direction from the primer) is called the lagging strand, and replication along it is called discontinuous replication. The double helix must unwind before another primer can be synthesized further on the lagging strand. Synthesis can then occur from the 3′ end of that new primer. Next, the double helix unwinds a bit more, and another spurt of replication proceeds. As a result, replication along the lagging strand can only proceed in short, discontinuous spurts. DNA polymerases δ, e, and g have 3′→5′ exonuclease activities, indicating that they are capable of proofreading. None of the eukaryotic DNA polymerases have a 5′→3′ exonuclease activity, like that of DNA polymerase I in E. coli. Instead, several other eukaryote enzymes remove nucleotides in the 5′→3′ direction. A model for simultaneous synthesis of both DNA strands in eukaryotes has been devised. It is like the prokaryotic model. The eukaryotic model is based on the observation that DNA polymerases α and δ associate with each other in a single complex. This model α synthesizes the lagging strand while δ synthesizes

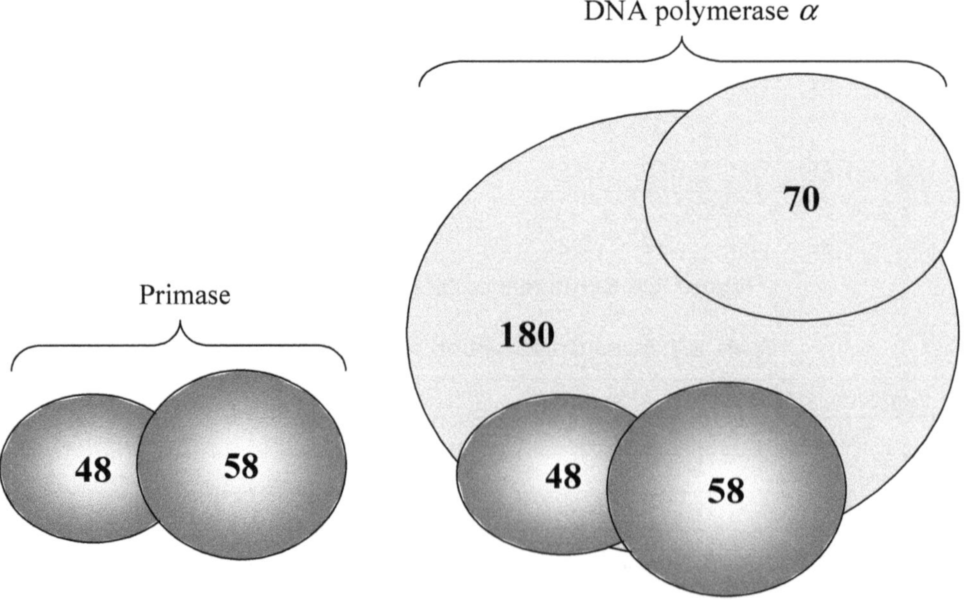

Figure 5.6: DNA Polymerase α among Mammals.

the leading strand simultaneously. The fragments of newly synthesized DNA along the lagging strand are called Okazaki fragments, named in honor of their discoverer, Japanese molecular biologist Reiji Okazaki. Okazaki and his colleagues made their discovery by conducting what is known as a pulse-chase experiment, which involved exposing replicating DNA to a short "pulse" of isotope-labeled nucleotides and then varying the length of time that the cells would be exposed to non-labeled nucleotides. The labeled nucleotides were incorporated into growing DNA molecules only during the initial few seconds of the pulse; after that, only non-labeled nucleotides were incorporated during the chase. The scientists then centrifuged the newly synthesized DNA and observed that the shorter chases resulted in most of the radioactivity appearing in "slow" DNA. The sedimentation rate was determined by size: smaller fragments precipitated more slowly than larger fragments because of their lighter weight. As the investigators increased the length of the chases, radioactivity in the "fast" DNA increased with little or no increase of radioactivity in the slow DNA. The researchers correctly interpreted these observations to mean that, with short chases, only very small fragments of DNA were being synthesized along the lagging strand. As the chases increased in length, giving DNA more time to replicate, the lagging strand fragments started integrating into longer, heavier, hence allowing faster sedimentation DNA strands. Scientists know that the Okazaki fragments of bacterial DNA are typically between 1,000 and 2,000 nucleotides long, whereas in eukaryotic cells, they are only about 100 to 200 nucleotides long.

A mutant strain of E. coli that was deficient in polymerase I activity still duplicated its DNA and successfully reproduced. However, DNA repair was not possible. The mutant strain is susceptible to ultraviolet light (UV) and radiation; this damages DNA and is mutagenic. Nonmutant bacteria can repair a great deal of UV-induced damage.

These observations revealed the following conclusions:

1. At least one other enzyme that is accountable for replicating DNA in vivo is also there in E. coli cells.
2. DNA polymerase may serve a secondary function in vivo. Kornberg and others now believe this function is critical to DNA synthesis's fidelity. Still, the enzyme does not synthesize the entire complementary strand during replication. To date, four other unique DNA polymerases have been isolated from cells lacking polymerase I activity and from normal cells that contain polymerase I. **Table 5.2** shows several characteristics of DNA polymerase I with DNA polymerase n and Ill. While none of the three can initiate DNA synthesis on a template, all three can elongate the existing DNA strand, called a primer.

Table 5.2: Polymerases I, II, and III of Bacterial DNA

Properties	*I*	*II*	*III*
Initiation of chain synthesis			
5' -3' polymerization	+	+	+
3' -5' exonuclease activity	+	+	+
5' -3' exonuclease activity	+		
Molecules of polymerase/cell	400	?	15

DNA polymerase I also demonstrates 5′ to 3′ exonuclease activity. This activity allows the enzyme to excise nucleotides, starting at the end at which synthesis begins and proceeding in the same direction of synthesis. Two final observations probably explain why Kornberg isolated polymerase I and not polymerase III: polymerase I is present in more significant amounts than polymerase III, more so it is more stable.

Polymerase III is the enzyme responsible for the 5′ to 3′ polymerization essential to in vivo replication. Its 3′ to 5′ exonuclease activity also provides a proofreading function that is activated when it inserts an incorrect nucleotide. The synthesis stops, and the polymerase "reverse course," excising, enabling it to get excised off incorrect nucleotide. Then, it proceeds back in the 5′ to 3′ direction, synthesizing the complement of the template strand. Polymerase I is believed to be responsible for removing the primer and for the synthesis that fills the gaps produced during synthesis. Its exonuclease activities allow for its participation in DNA repair. Polymerase II, as well as polymerase IV and V, is involved in various aspects of the repair of DNA that external forces have damaged. Polymerase II is encoded by a gene activated by disruption of DNA synthesis at the replication fork.

DNA polymerase III molecules and their active form are called a holoenzyme, which consists of a dimer containing ten different polypeptide subunits (**Table 5.3**) with a molecular weight of 900,000 Da. The largest subunit, a, has a molecular weight of 140,000 Da and, along with subunits and δ, constitutes the core enzyme responsible for the polymerization activity. The α subunit is responsible for nucleotide polymerization on the template strands, whereas the subunit of the core enzyme possesses the 3′ to 5′ exonuclease activity.

Table 5.3: DNA Polymerase III Holoenzyme Subunits

Subunit	*Function*	*Groupings*
A	5' – 3' polymerization	Core enzyme: elongates poly-nucleotide chain and proofreads
ε	3' – 5' exonuclease	
θ	Core assembly	
γ δ δ' χ Ψ	Loads enzyme on template (serves as clamp loader)	γ complex
β	Sliding clamp structure (processivity factor)	
τ	Dimerizes core complex	

The second group of five subunits ('Y, S, S', X, and 1/1) forms what is called the 'Y complex, which is involved in "loading" the enzyme onto the template at the replication fork. This enzymatic function requires energy and is dependent on ATP hydrolysis. The β-subunit is a clamp that prevents the core enzyme from falling off the template during polymerization. Finally, the T subunit functions to dimerize two core polymerases, facilitating the simultaneous synthesis of both strands of the helix at the replication fork. The holoenzyme and several other proteins at the replication fork together form a huge complex (nearly as large as a ribosome) known as the replisome.

5.6 Many Complex Processes are Involved in DNA Replication

DNA replication is semi-conservative and bidirectional in bacteria and viruses, along with a single replication. Bidirectional synthesis creates two replication forks that move in opposite directions away from the origin of synthesis. The following mechanisms are followed.

1. A mechanism must exist by which the helix undergoes local unwinding and is further stabilized; in this configuration the synthesis proceeds along both strands.
2. As unwinding and subsequent DNA synthesis proceed, increased coiling creates tension further down the helix, which must be reduced.
3. A primer of some sort must be synthesized so that polymerization can commence under the direction of DNA polymerase. Surprisingly, RNA and not DNA serves as the primer.
4. Once the RNA primers are synthesized, DNA polymerase III starts to synthesize the DNA complement of both strands of the parent molecule. Because the two strands are anti-parallel to one another, continuous synthesis in the direction that the replication fork moves is possible along only one of the two strands. On the other strand, synthesis is discontinuous in the opposite direction.
5. The RNA primers must be removed before the completion of replication. The temporarily created gaps must be filled with DNA complementary to the template at each location.
6. The newly synthesized DNA strand that fills each temporary gap must be joined to the adjacent strand of DNA.
7. While DNA polymerases accurately insert complementary bases during replication, they are not perfect, and occasionally, incorrect bases are added to the growing strand. A proofreading mechanism that corrects errors is an integral process during DNA synthesis.

As for unwinding proceeds, a coiling tension is created ahead of the replication fork, often producing supercoiling. In circular molecules, supercoiling may take the form of added twists and turns of the DNA. Supercoiling can be relaxed by DNA gyrase, a member of a larger group of enzymes called DNA topoisomerases. The gyrase makes either single- or double-stranded "cuts" and catalyzes localized movements that unfold the twists and knots created during supercoiling. Finally, strands are resealed, for this energy is provided by ATP hydrolysis. The DNA polymerase and other enzymes together are called replisomes. Initial binding of many monomers of DNA occurs at DNA sites containing repeating sequences of 9 nucleotides, called 9mers. Not illustrated are 13mers, which are also involved.

5.7 RNA Primer and DNA Synthesis

A short segment of RNA (about 5 to 15 nucleotides long), complementary to DNA, is synthesized first on the DNA template. Synthesis of the RNA is directed by a form of RNA polymerase called primase, which does not require a 3' end to be free to start the synthesis. This short segment of RNA that DNA polymerase III begins to add 5' -deoxyribonucleotides, initiating DNA synthesis (**Figure 5.7**). The RNA primer gets separated and replaced with DNA. This takes place with the help of DNA polymerase I. Recognized in viruses, bacteria, and several eukaryotic organisms, RNA priming is a universal phenomenon during the initiation of DNA synthesis.

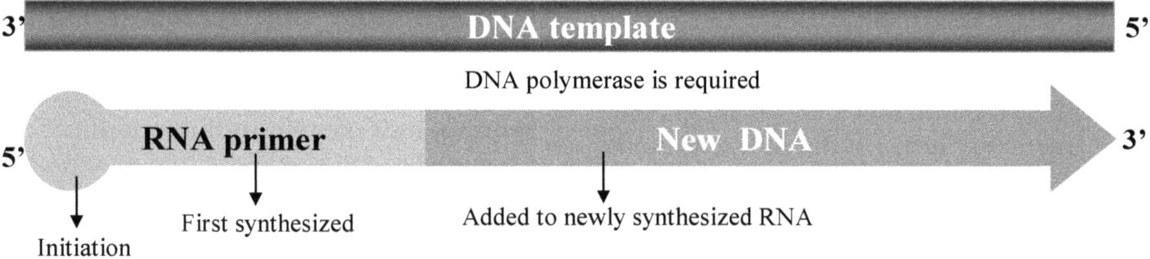

Figure 5.7: Initiation of DNA Syntheses.

5.8 Anti-parallel Strands Require Continuous and Discontinuous DNA Synthesis

When bacteriophage DNA is replicated in E. coli, some of the newly formed DNA, which is bonded to the hydrogen, forms the template strand and is present as small fragments containing 1000-2000 nucleotides. RNA primers are part of such a fragment. The discontinuous synthesis of DNA requires enzymes that remove both the RNA primer and unite the Okazaki fragments into the lagging strand. As we have noted, DNA polymerase I removes the primer and replaces the missing nucleotides. Joining the fragments appears to be the work of DNA ligase, which can catalyze the formation of the phosphodiester bond that seals the nick between the discontinuously synthesized strands. The evidence that DNA ligase performs this function during DNA synthesis is strengthened by observing a ligase-deficient mutant strain (lig) of E. coli, where many un-joined Okazaki fragments accumulate.

5.9 Role of Polymerase III

Once the strands unwind and the replication fork progresses down the helix, only one strand can serve as a template for continuous DNA synthesis. This DNA strand is called the leading strand. As the fork progresses, many initiation points are necessary on the opposite DNA template, resulting in discontinuous DNA synthesis.

It has been suggested that both DNA strands can be replicated simultaneously. The lagging strand forms a loop, nucleotide polymerization can occur on both template strands under the direction of a dimer of the enzyme. After the synthesis of 100 to 200 base pairs, the monomer of the enzyme on the lagging strand will encounter a completed Okazaki fragment, at this point, it releases the lagging strand. A new loop is then generated with the lagging template strand, and the process is repeated. Looping inverts the orientation of the template, but not the direction of actual synthesis on the lagging strand, which is always in the 5′ to 3′ direction.

The holoenzyme facilitates synthesis at the replication fork is a dimer of the β subunit that forms a clamp-like structure around the newly formed DNA duplex. This β-subunit clamp prevents the core enzyme from falling off the template as polymerization proceeds. The- subunit dimer is often called a sliding clamp because the entire holoenzyme moves along the parent duplex, advancing the replication fork.

DNA replication is prone to errors, but these are corrected, and a new strand is precisely complementary to the template strand at each nucleotide position is formed. Although the action of DNA polymerases is very accurate, synthesis is not perfect, and a non-complementary nucleotide is occasionally inserted erroneously. The DNA polymerases all possess 3′ to 5′ exonuclease activity to compensate for inaccuracies. This property allows them to detect and excise a mismatched nucleotide (in the 3′ -5′ direction). The mismatched nucleotide is removed; 5′ to 3′ synthesis can again proceed. This process is called proofreading, which increases the fidelity of synthesis by a factor of about 100. In the case of the holoenzyme form of DNA polymerase, the epsilon (E) subunit is directly involved in the proofreading step. In strains of E. coli with a mutation that has rendered the E subunit nonfunctional, the error rate (the mutation rate) during DNA synthesis is increased substantially.

The DNA replication is understandable by a coherent model. At the advancing fork, a helicase unwinds the double helix. Once unwound, single-stranded binding proteins associated with the strands prevent the helix's reformation. In advance of the replication fork, DNA gyrase diminishes the tension created as the helix supercoils. Each half of the dimeric polymerase is a core enzyme bound to one of the template strands by a β-subunit sliding clamp. Continuous synthesis occurs on the leading strand, while the lagging strand must loop around for simultaneous synthesis to appear on both strands. It is essential for the replication of the lagging strand, and it is the action of DNA polymerase I and DNA ligase to replace the RNA primer with DNA and join the Okazaki fragments, respectively. Many genes' controls replication. If mutations occur, they interrupt or seriously impair some aspect of replication; ligase-deficient and proofreading-deficient mutations may be present. These mutations are lethal genetic analysis that frequently uses conditional mutations expressed under one condition. For example, a temperature-sensitive mutation may not be described at a particular permissive temperature. The mutant phenotype is described and can be studied when mutant cells are grown at a restrictive temperature.

As shown in **Table 5.4**, a variety of genes in E. coli have the subunits of the DNA polymerases encoding products, which are involved in the specification of the origin of synthesis, helix unwinding, and stabilization, initiation, and priming, relaxation of supercoiling, repair, and ligation. A newly discovered large group of genes reveals the complexity of the replication process, even in the relatively simple prokaryote. Given the enormous quantity of DNA that must be unerringly replicated quickly, this level of complexity is not unexpected.

Table 5.4: Products of *E. coli* Mutant Genes

Mutant Gene	*Enzyme*
polA	DNA polymerase I
polB	DNA polymerase II
dnaE, N, Q, X, Z	DNA polymerase III subunits primase
dnaG	Initiation
dnaA, I, P dnaB,C	Helicase at *oriC*
oriC	Origin of replication
gyrA, B	Gyrase subunits
Lig	Ligase
Rep	Helicase
Ssb	Single-stranded binding proteins
rpoB	RNA polymerase subunit

Eukaryotic DNA synthesis is more complex as compared to prokaryotic DNA replication. In both systems, double-stranded DNA is unwound at replication origins, replication forks are formed, and bidirectional DNA synthesis creates leading and lagging strands from templates under the direction of DNA polymerase. Eukaryotic polymerases have the exact fundamental requirements for DNA synthesis as do bacterial systems: four deoxyribonucleoside triphosphates, a template, and a primer. However, eukaryotic cells contain much more DNA per cell, and eukaryotic chromosomes are linear rather than circular because this DNA is complex with proteins; eukaryotes face many complexities not encountered by bacteria. Thus, the process of DNA synthesis is more complicated in eukaryotes and more challenging to study.

5.10 Multiple Replication Origins

The most apparent difference between eukaryotic and prokaryotic DNA replication is that eukaryotic chromosomes contain multiple replication origins, in contrast to the single site that is part of the E. coli chromosome. Multiple origins, visible under the electron microscope as "replication bubbles" that form as the helix opens, each provide two potential replication forks. Multiple origins are essential so that the entire genome of a typical eukaryote can be completed in a reasonable time. Eukaryotes have a larger quantity of DNA than bacteria, *e.g.*, yeast has four times as much DNA, and Drosophila has 100 times as much as E. coli. The rate of synthesis by eukaryotic DNA polymerase is much slower-only about 2000 nucleotides per minute, a rate 25 times less than the comparable bacterial enzyme. Under these conditions, replication from a single origin of a typical eukaryotic chromosome might take days to complete. However, entire eukaryotic genomes are usually replicated in a few hours.

Insights are now available concerning the molecular nature of the multiple origins and the initiation of DNA synthesis at these sites. Most of the information has been taken from the study of yeast (*e.g.*, Saccharomyces cerevisiae), which contain approximately 250-400 replicons per genome; subsequent studies have used mammalian cells, which have as many as 25,000 replicons. The origins of yeast have been isolated and are called autonomously replicating sequences (ARSs). They consist of a unit containing a consensus sequence of II base pairs flanked by other short sequences involved in efficient initiation. DNA synthesis is restricted to the S phase of the eukaryotic cell cycle. Research has shown that many origins are not activated at once; clusters of 20-80 adjacent replicons are activated sequentially throughout the S phase until all DNA is replicated.

How the polymerase finds the ARSs among so much DNA is an obvious recognition problem. The solution has a mechanism that is initiated before the S phase. During the G1 phase of the cell cycle, all ARSs are initially bound by a group of specific proteins (six in yeast), forming what is called ORC (the origin recognition complex). Mutations in the ARSs or any other gene encoding these proteins of the ORC abolish or reduce the initiation of DNA synthesis. As these recognition complexes are found in the G1, synthesis is not initiated at these sites until the S phase; other proteins must still be involved in the initiation signal. The most important of these proteins are specific kinases, critical enzymes involved in phosphorylation; these are the integral parts of cell cycle control. When bound with ORC, a pre-replication complex is formed accessible to DNA polymerase. After these kinases are activated, they serve to complete the initiation complex, directing localized unwinding and triggering DNA synthesis. Activation also inhibits the reformation of the pre-replication complexes once DNA synthesis has been achieved at each replicon. This is a crucial mechanism, as it distinguishes segments of DNA that have completed replication from segments of un-replicated DNA, thus maintaining orderly and efficient replication. This ensures that replication only occurs once along each stretch of DNA during each cell cycle.

5.11 Eukaryotic DNA Polymerases

The most multifaceted aspect of eukaryotic replication is the collection of polymerases required for DNA synthesis. However, only four are involved in DNA replication, while the remainder is involved in repair processes. For the polymerases to have access to DNA, the topology of the helix should first be modified. As synthesis is triggered at each origin site, the double strands are opened within an AT-rich region, allowing the entry of a helicase enzyme that proceeds to unwind the double-stranded DNA further. Before polymerases can begin synthesis, histone proteins are complex with the DNA and must be stripped away or otherwise modified. As DNA synthesis then proceeds, histones get re-associated with the newly formed duplexes, reestablishing the characteristic nucleosome pattern. In eukaryotes, the synthesis of new histone proteins is tightly coupled to DNA synthesis during the S phase of the cell cycle.

The terminology and characteristics of six different polymerases are summarized in **Table 5.5**. Three of these (pol α,δ, and ε) are essential to nuclear DNA replication in eukaryotic cells. Two (pol β and ζ are thought to be involved in DNA repair. The sixth form (pol γ) is involved in synthesizing mitochondrial DNA. Its replication function is presumably limited to that organelle even though a nuclear gene encodes it. All but one of the six enzyme forms (the P form) consists of multiple subunits. Various subunits perform different functions at the time of replication.

Pol α and δ are the primary forms of the enzyme involved in initiation and elongation during nuclear DNA synthesis. Two of the four subunits of the Pol α enzyme function in synthesizing the RNA primers during the initiation of synthesis of both the leading and lagging strands. After the addition of about ten ribonucleotides, another subunit helps in further elongation of the RNA primer by adding 20-30 complementary deoxyribonucleotides. Pol α is said to possess low processivity, a term that essentially reflects the length of DNA is synthesized by an enzyme before it dissociates from the template. Once the primer is in place, an event known as polymerase switching occurs, whereby Pol α dissociates from the template and is replaced by Pol δ. This form of the enzyme possesses high processivity and elongates the leading and lagging strands. It also includes 3′ to 5′ exonuclease activity, which allows it to proofread. Pol ε, the third essential form, possesses the same general characteristics as Pol δ, but it is believed to operate under different cellular conditions or to be restricted to the synthesis the lagging strand. In yeast, mutations that render Pol ε inactive are lethal, attesting to the essential functions at the time of replication.

The process described applies to both leading and lagging strand synthesis. At both the strands, RNA primers must be replaced with DNA. On the lagging strand, the Okazaki fragments, which are about ten times smaller (100-150 nucleotides) in eukaryotes than in prokaryotes, must get ligated.

To accommodate the increased number of replicons, eukaryotic cells contain many more DNA polymerase molecules than bacteria. While E. coli has about 15 copies of DNA polymerases per cell, there may be up to 50,000 copies of the form of DNA polymerase in animal cells. As has been pointed out, the presence of a greater number of smaller replicons in eukaryotes compared with bacteria compensates for the slower rate of DNA synthesis in eukaryotes. E. coli requires 20 to 40 minutes to replicate its chromosome, while Drosophila, with 40 times more DNA, is known during embryonic cell divisions to accomplish the same task in only 3 minutes.

During replication, complications may arise at the linear ends of the chromosome. The ends of the chromosomes, called telomeres, contain long stretches of start repeating sequences of DNA. These sequences

Table 5.5: Eukaryotic DNA Polymerases

Location	*Polymerase*					
	A	*B*	δ	ε	γ	*Z*
	Nucleus	*Nucleus*	*Nucleus*	*Nucleus*	*Mitochondria*	*Nucleus*
3'-5' Exonuclease Activity	No	No	Yes	Yes	Yes	No
Essential to Nuclear Replication?	Yes	No	Yes	Yes	No	No

are attached to the telomeric proteins. Due to this, specificity and stability are maintained.

Telomeres are essential because the double-stranded "ends" of DNA molecules at the termini of chromosomes potentially resemble what are called double-stranded breaks (DSBs) that can occur if the chromosome should be fragmented internally. In such a case, the resulting double-stranded ends can fuse. If this fusion does not happen, there may be degradation. Thus, telomeres are believed to protect the ends of intact eukaryotic chromosomes from such a fate.

Now, consider semi-conservative replication near the end of a double-stranded DNA molecule. As the synthesis of the leading strand usually proceeds to the end, a difficulty arises on the lagging strand once the RNA primer is removed (**Figure 5.8).** Usually, the newly created gap would be filled by adding a nucleotide to the existing 3′ -OH group provided during discontinuous synthesis (**Figure 5.8**); however, since it is the end of the DNA molecule's end, no strand is to provide the 3′ -OH group.

As a result, a gap remains on the lagging strand during each successive round of synthesis that will shorten the end of the chromosome by the length of the RNA primer. Because this is potentially such a significant problem, however, it is thought that molecular evolution and subsequently be conserved and shared by almost all eukaryotes. As pointed out above, telomeric DNA in eukaryotes consists of many short, repeated nucleotide sequences. For example, the template producing the lagging strand at the end of the Tetrahymena chromosome contains many repeats of the sequence 5′ TTGGGG-3′.

As shown in **Figure** 5.9, telomerase adds several repeats of this six-nucleotide sequence to the 3′ end of the lagging strand (using 5′-3′ synthesis), thus preventing chromosome shortening. Repeats fold

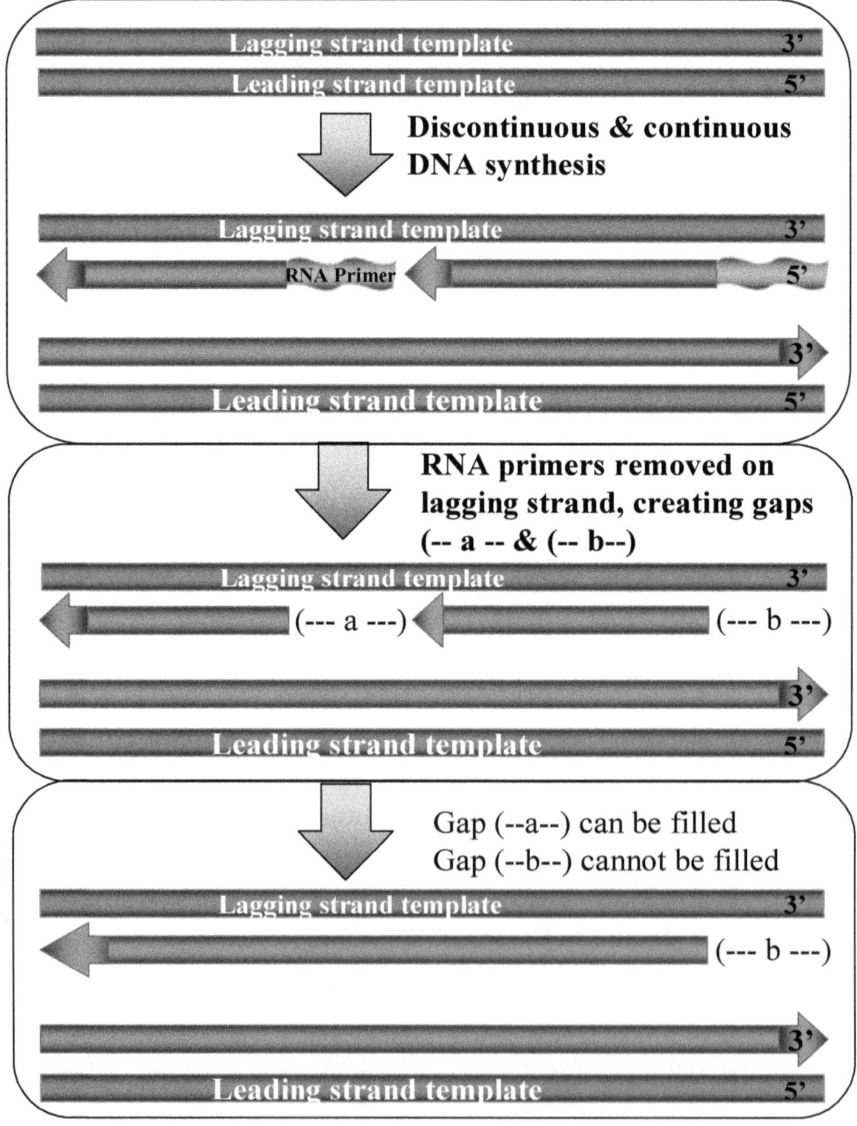

Figure 5.8: Problems Involved at the Time of the Replication of the Ends of Linear Chromosomes. A gap (- -b- -) is left following synthesis on the lagging strand.

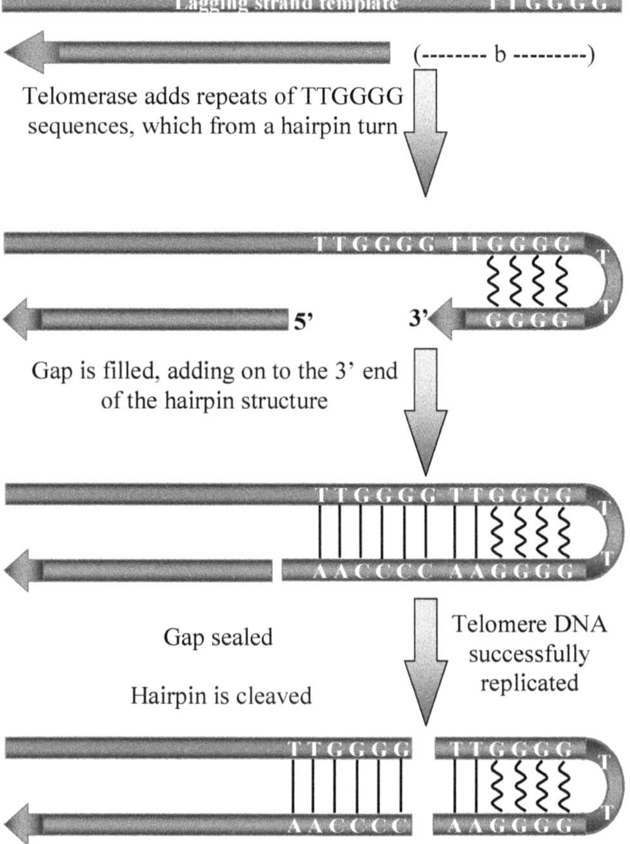

Figure 5.9: The Enzyme Telomerase Directs Synthesis of the TTGGGG Sequences Lead to the Extension of the Lagging Strand.

back on themselves, forming a "hairpin loop" that stabilizes by hydrogen bonding between opposite guanine residues (G=G), creating a free 3′ -OH group on the end of the hairpin that can serve as a substrate for DNA polymerase I, this allows to fill the gap that otherwise shortens the chromosome subsequently. The hairpin loop can be cleaved off at its terminus, and the potential loss of DNA in each subsequent replication cycle is averted.

The enzyme is highly unusual because it is a ribonucleoprotein, constituting within its molecular structure a short piece of RNA essential to its catalytic activity. The RNA component serves as a "guide" and a template for synthesizing its DNA complement, a process called reverse transcription. In Tetrahymena, the RNA contains several repeats of the sequence 5′-CCCCAA-3′, complementing the repeating telomeric DNA sequence that must be synthesized (5′ - TTGGGG-3′).

Reverse transcription of this overlapping sequence, which synthesizes DNA on an RNA template, leads to the extension of the lagging strand, as depicted in **Figure 5.10**. It is believed that the enzyme is then translocated toward the end of the lagging strand and the sequences of events are repeated. In yeast, of which such synthesis has been investigated in detail, specific proteins are responsible for bringing the synthesis to an end, thus finding out the number of repeating units added to the end of the lagging strand. Similar proteins no doubt function in the same way in other eukaryotes.

Analogous enzyme functions have now been found in almost all eukaryotes studied. In humans, the telomeric DNA sequence that creates the lagging strand and is repeated is 5′-TTAGGG-3′, differing from Tetrahymena by only one nucleotide.

Telomeric DNA sequences have been highly conserved throughout evolution, reflecting the critical function of telomeres. The telomere shortening is linked to a molecular mechanism involved in cellular aging. In fact, in most eukaryotic somatic cells, telomerase is not active and thus, with each cell division, the telomeres of each chromosome shorten. After many divisions, the telomere is compressed to a greater extent, and the cell loses the capacity for further division. Malignant cells, on the other hand, maintain telomerase activity and hence keep on dividing.

5.12 DNA Recombination, like DNA Replication, is directed by Specific Enzymes

Several models have been projected to show homologous recombination and share certain standard features. All models are based on the proposals put forth independently by Robin Holliday and Harold L. K. Whitehouse in 1964. They also depend on the complementary between DNA strands for exchange precision. Each model is dependent on a series of enzymatic processes to accomplish genetic recombination.

One such model is shown in **Figure 5.10**. It begins with two paired DNA duplexes or homologs [1st step], in each of which, an endonuclease introduces a single-stranded nick at an identical position [2nd step]. The ends of the strands produced by these cuts are then displaced and subsequently paired with their complements on the opposite duplex [3rd step]. A ligase then seals the loose ends [4th step], creating hybrid duplexes called heteroduplex DNA molecules. The exchange creates a cross-bridge structure. The position of this cross-bridge can then move down the chromosome by a process referred to as branch migration [5th step], which occurs because of a zipper-like action as hydrogen bonds are broken and then reformed between complementary bases of the displaced strands of each duplex. This migration

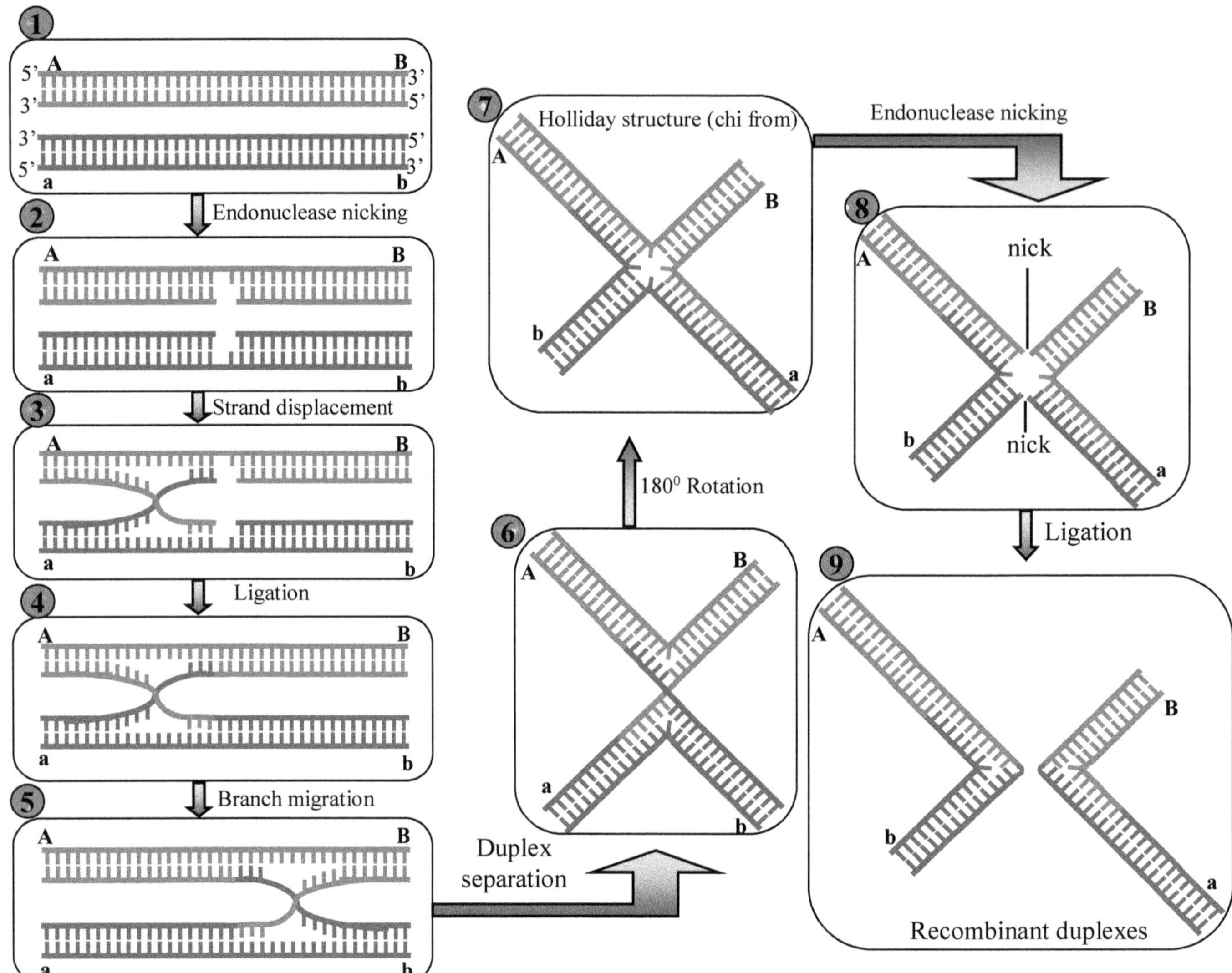

Figure 5.10: Genetic Recombination.

yields an increased length of heteroduplex DNA on both homologs.

If the duplexes now separate [6th step] and the bottom portions rotate 1800 [7th step], an intermediate planar structure called X form- the characteristic Holliday structure- is created. If the two strands on opposite homologs previously uninvolved in the exchange are now nicked by an endonuclease [8th step] and ligation occurs [9th step], recombinant duplexes are created. The arrangement of alleles is altered because of recombination.

Evidence supporting this model includes the electron microscopic visualization of X-form planar molecules from bacteria in which four duplex arms are joined at a single exchange point [**Figure 5.10** (7th step)]. Significant evidence comes from the discovery of the RecA protein in E. coli. The molecule promotes the exchange of reciprocal single-stranded DNA molecules, as occurs in the 3rd step of the model. RecA also enhances the hydrogen-bond formation during strand displacement, thus initiating heteroduplex formation. Many other enzymes are essential for the nicking and ligation process. These have also been found and investigated. The products of the recB, recC, and recD genes are thought to be involved in the nicking and unwinding of DNA. There may be different mutations that may prevent genetic recombination. These have been discovered in viruses and bacteria. These mutations represent genes whose products play an essential role in this process.

5.13 Gene Conversion is the Result of DNA Recombination

Gene conversion, one of the two mechanisms of homologous recombination, involves the unidirectional transfer of genetic material from a 'donor' sequence to a highly homologous 'acceptor'. Considerable progress has been made in understanding the molecular mechanisms that underlie gene conversion, its formative role in human genome evolution, and its implications for a human inherited disease. It was first reported that yeast gene conversion is characterized by a nonreciprocal genetic exchange between two closely linked genes. For example, suppose we have to cross two Neurospora strains. In that case, each bearing a separate mutation (a+ X +b), reciprocal recombination between the genes would yield spore pairs of the + + and the ab genotypes. However, a nonreciprocal exchange delivers one team without the other. Working with pyridoxine mutants, Mitchell observed several asci-containing spore pairs with the + + genotype but not the reciprocal product (ab). Because the frequency of these events was higher than the predicted mutation rate and consequently could not be accounted for by mutation, they were called "gene conversions." They were so named because it appeared that one allele had somehow been "converted" to another in which genetic exchange had also occurred.

Gene conversion is now considered to be a consequence of the process of DNA recombination. Two mechanisms for gene conversion and homologous recombination have been shown. The double-strand break repair model A double-strand break is expanded to a gap, which is then repaired by copying a homologous sequence. The gene conversion is often accompanied by a crossing-over of the flanking sequences as shown in **Figure 5.11,** which is the successive half crossing-over model. Half crossing-over leaves one recombinant duplex and one or two end(s) out of two parental duplexes. The resulting ends are, in turn, recombinogenic. A successive round of the half crossing-over mechanism explains why apparent plasmid gene conversion in the RecF pathway of E. coli is not accompanied by crossing-over. This model can explain chromosomal gene conversion if one assumes that the donor is first replicated. Gene conversion during mating-type switching in yeast, antigenic variation in unicellular microorganisms, and chromosomal gene conversion in mammalian somatic cells are explained by this model. Distinguishing between these two mechanisms is essential in understanding recombination in yeast and mammalian cells and its application to gene targeting.

Suppose that the G a"C pair on one of the two homologs was responsible for the mutant allele, while the A =T pair was part of the wild-type gene sequence on the other homolog. Conversion of the Ga"C pair to A =T would change the mutant allele to wild type, just as Mitchell initially observed (**Figure 5.11**).

Gene conversion is necessary in the context of evolution; the availability of complete genome sequences from multiple species has greatly facilitated studies of the role of gene conversion in creating specific multigene families. The use of population genetic data to make inferences about the impact of gene conversion at the genome level is gaining popularity. These approaches, when combined with new or better statistical methods, are probable to provide better estimates of the occurrence and spatial allocation of gene-conversion actions along with human chromosomes. Further incorporation of data from genome-wide studies, sperm typing, and disease analysis should give a better portrait of the impact that gene conversion shows the evolution of human and other mammalian genomes certainly, the spectrum of gene-conversion events result in human disease.

5.14 Summary

When a cell divides, it is important that each daughter cell receives an identical copy of the DNA. This is accomplished by the process of DNA replication. The replication of DNA occurs before the cell begins to divide into two separate cells.

The discovery and characterization of the structure of the double helix provided a hint as to how DNA is copied. Recall that adenine nucleotides pair with thymine nucleotides and cytosine with guanine and that DNA is double-stranded. This means that the two strands are complementary to each other. For example, a strand of DNA with a nucleotide sequence of AGTCATGA will have a complementary

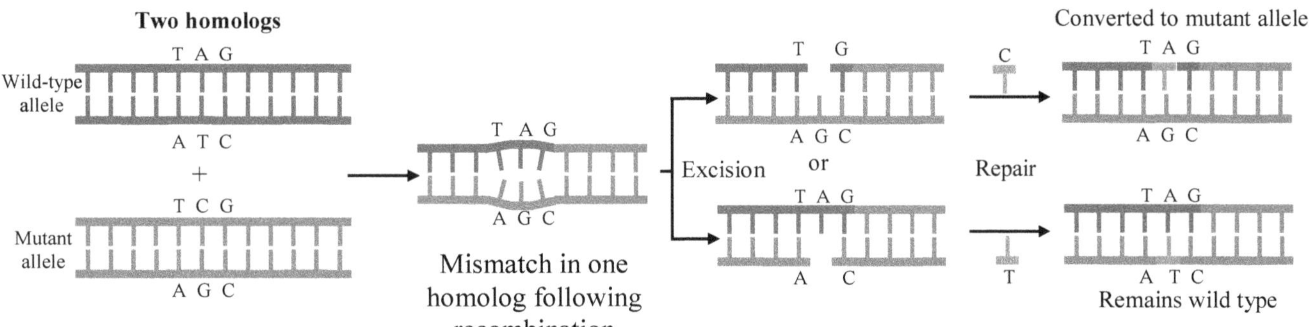

Figure 5.11: Gene Conversion Mechanisms.

strand with the sequence TCAGTACT. Because of the complementarity of the two strands, having one strand makes it possible to recreate the other strand. This model for replication suggests that the two strands of the double helix separate during replication, and each strand serves as a template from which the new complementary strand is copied.

Gray indicates the original DNA strands, and blue shows newly synthesized DNA. During DNA replication, each of the two strands that make up the double helix serves as a template from which new strands are copied. The new strand will complement the parental or "old" strand. Each new double strand comprises one parental strand and one new daughter strand. This is known as semiconservative replication. When two DNA copies are formed, they have an identical sequence of nucleotide bases and are divided equally into two daughter cells. Many other aspects have been touched on in this chapter.

6

Transcription

DNA that makes up the human genome contains different inherited units called genes. Each gene encodes a unique protein that performs a specialized function in the cell. The human genome contains more than 20,000 genes. Cells use the two-step process of transcription and translation to read each gene and produce the string of amino acids that make up a protein. The basic rules for translating a gene into a protein are laid out in the Universal Genetic Code.

DNA transcription is the 1st step of new protein synthesis. Three types of RNA help in the production of new proteins: messenger RNA (mRNA), ribosomal RNA (rRNA), and transfer RNA (tRNA). Production occurs in the nucleus, and a set of genes forms numerous enzymatic complexes. Once the RNA is synthesized, it leaves the nucleus in the cytoplasm, where it is triggered during translation. The chromosomes get and de-condensed in the nearby regions. This helps the genes to be read at the transcription site. Very few genes are in an active state. Two types of genes have been distinguished: (i) housekeeping genes constitutively expressed at a relatively constant level across many or all known conditions. (ii) genes that encode for exact products of importance for cell differentiation. Both extrinsic and intrinsic signals often determine the activity of these genes.

6.1 Transcription

The transcription process can be visualized using electron microscopy. It was first observed using this method in 1970. DNA molecules look like "trunks," with many branches, and these are RNA "branches." Sometimes, enzymes that degrade DNA and RNA are added to the molecules. The presence of RNA is lost. DNA is double-stranded, but only a single strand is required, which serves as a template and is used for transcription at any given time. The template strands are noncoding, while the non-template strand is called the coding strand because its sequence will be the same as that of the new RNA molecule. In most organisms, the strand of DNA that serves as the template for one gene may be the non-template strand for other genes within the same chromosome.

Polymerases are large enzymes composed of approximately a dozen subunits, and when active on DNA, they are also typically complex with other factors. In many cases, these factors signal which gene is to be transcribed. Three different types of RNA polymerase exist in eukaryotic cells, whereas bacteria have only one. In eukaryotes, RNA pol I transcribe the genes that encode most of the ribosomal RNAs (rRNAs), and RNA pol III transcribes the genes for a tiny rRNA, plus the transfer RNAs that play a vital role in the translation process, as well as other small regulatory RNA molecules. Thus, RNA pol II transcribes the messenger RNAs, which serve as the templates to produce protein molecules.

6.1.1 (a) Initiation

Initiation is the process of copying genetic information from DNA to RNA, which results in a specific protein formation. Humans have approximately 21 thousand genes, but only a few are expressed in a particular cell at an exact time. The expression of genes depends upon several proteins to be produced. One of the essential proteins is RNA polymerase; it gets bound to double-stranded DNA at the time of gene copying. RNA polymerase recognizes explicitly and binds to promoters homologous to the TATA box. These are specialized DNA sequences near the beginning of a gene where transcription may start. All promoters contain two characteristic short sequences, only 6-10 nucleotides that help in RNA polymerase binding. These short sequences are almost similar in different promoters. In bacteria, the complete RNA polymerase (the holoenzyme) consists of a core enzyme plus an initiation of transcription (IT) subunit, which is involved at the time of initiation. The IT subunit is an essential component as it reduces RNA polymerase's general affinity for DNA but simultaneously increases RNA polymerase's affinity for the promoter. As a result, the RNA polymerase holoenzyme can attach to the promoter and bind tightly to it, forming the promoter complex.

The RNA polymerase unwinds a part of the double helix, exposing unpaired bases on the strand that may serve as the template. This portion is known as the transcription bubble. This bubble is approximately 12–14bp. The enzyme can recognize the template strand and identify two nucleotides that initiate copying. RNA polymerase, which aligns the first two ribonucleotides of the new RNA, occurs at the 5′ end of the final RNA product. The end of the gene transcribed into the 5′ end of the mRNA is called the 5′ end of the gene. The mRNA grows in the 5′-to-3′ direction because RNA polymerase adds nucleotides at the 3′ end of the growing RNA chain. After aligning the first two ribonucleotides, RNA polymerase catalyzes the formation of a phosphodiester bond between them, resulting in a diribonucleotide. Soon after that, the RNA polymerase releases the initiation of the transcription subunit (**Figure 6.1 (a)**).

The initiation can be summarized in the following manner:

1. First, the ribosome dissociates into its two constituent subunits, called the 40S and 60S subunits. The subunits are prevented from spontaneous re-association by binding of initiation factor eIF-3 to the small ribosomal subunit.
2. Initiation factor eIF-2 is a GTP-binding protein that recognizes explicitly initiator tRNA (Met-tRNA, Met) forming a ternary complex, eIF2-GTP-Met-tRNA, Met.
3. The ternary complex binds to the 40S ribosomal subunit.
4. The initiation factor eIF-4F recognizes the cap structure at the 5′end of the messenger RNA. This factor has a subunit that specifically interacts with the cap structure. The initiation factor binding guides the activated 40S subunit, containing the initiator tRNA, to the 5′end of the mRNA. This type of initiation is known as cap-dependent and the most common type. The initiation can also occur through an alternative cap-independent initiation mechanism.
5. The mRNA-bound subunit travels along the 5′-untranslated end of the mRNA until it reaches the first AUG codon, which will serve as the start codon for the translation process. This process, known as scanning, requires energy (ATP) and additional initiation factors.
6. When the activated 40S subunit has reached the start codon, the 60S subunit binds to the 40S subunit. This reaction requires an additional initiation factor, eIF-5, which hydrolyses the eIF-2-bound GTP, releasing the initiation factors from the ribosome. As a result of the initiation process, the initiator Met-tRNA becomes positioned in the ribosomal P-site.

6.1.2 (b) Elongation

With the primer as the starting point for the leading strand, a new DNA strand grows one base at a time. The existing strand is a template for the new strand. For example, if the next base on the existing strand is an A, the new strand receives a T. The enzyme DNA polymerase controls elongation, which can occur only in the leading direction. The lagging strand unwinds in small sections that DNA polymerase replicates in the ultimate guide. The resulting small "Okazaki fragments" can contain 1,000 to 2,000 bases in bacteria, but eukaryotic organisms having cells with nuclei have fragments of only 100 to 200 bases. The fragments terminate in an RNA primer that is subsequently removed so that enzymes can stitch the fragments into an elongating strand, and the ribosome moves along the mRNA in the 5′-to-3′direction, which requires the elongation factor G, in a process called translocation. The tRNA that corresponds to the second codon can then bind to the A site, a step that requires elongation factors (in E. coli, these are called EF-Tu and EF-Ts), as well as guanosine triphosphate (GTP) as an energy source for the process. Upon binding the tRNA-amino acid complex in the A site, GTP is cleaved to form guanosine diphosphate (GDP), then released along with EF-Tu to be recycled by EF-Ts for the next round. Next, peptide bonds between the now adjacent first and second amino acids are formed through a peptidyl transferase activity. For many years, it was thought that an enzyme catalyzed this step, but recent evidence indicates that the transferase activity is a catalytic function of rRNA. After the peptide bond is formed, the ribosome shifts or translocate again, thus causing the tRNA to occupy the E site. The tRNA is then released to the cytoplasm to pick up another amino acid. In addition, the A site is now empty and ready to receive the tRNA for the next codon. This process is repeated until all the codons in the mRNA have been read by tRNA molecules and the amino acids attached to the tRNAs have been linked together in the growing polypeptide chain in the appropriate order. At this point, the translation must be terminated, and the nascent protein must be released from the mRNA and ribosome.

Once an RNA polymerase is moved from the promoter, other RNA polymerase molecules can proceed to start the transcription. Many RNA polymerases can act simultaneously (**Figure 6.1 (b)**).

6.1.3 (c) Termination

Termination means the end of transcription. There are two types of terminators. These are basically DNA sequences termination bubble collapses when the last base is added. This is an intrinsic mechanism called a terminator, which causes the RNA polymerase core enzyme to terminate transcription on its own; some extrinsic terminators require proteins other than RNA polymerase-particularly a polypeptide is known as rho-to bring the termination. These terminators are very specific and have specific sequences in the mRNA; they are transcribed from particular DNA regions and start the transcription termination. Terminators often form hairpin loops in which nucleotides within the mRNA pair with nearby complementary nucleotides are situated. After termination, RNA polymerase and a completed RNA chain are released from the DNA (**Figure 6.1 (c)**). It takes place both in Prokaryotes and Eukaryotes.

6.1.4 (d) Primary Transcript

RNA created after the action of the RNA polymerase is known as the primary transcript. The bases in the primary transcript complement the bases between the initiation and termination. Start codons code for all the amino acids in the polypeptide to be built, and it carries these bases (**Figure 6.1 (d)**).

In prokaryotes, the primary transcript degrades rapidly to give mature products (rRNA and tRNA). DNA-dependent RNA polymerases and RNA polymerases are responsible for the transcription. These enzymes allow only one nucleotide to be incorporated into the RNA strand. This RNA strand now becomes complementary to the DNA strand, which is known as the template. The initial step in the synthesis of RNA is a specific polymerase associated with the DNA template. RNA polymerases cannot recognize promoters independently; for this, the help of extra proteins is compulsory. These proteins are called transcription factors. If there are no transcription factors, initiation will not take place. The polymerase moves along the template DNA strand toward its 5′ end (3′ → 5′ direction). As the polymerase moves, the DNA gets unwound temporarily, and the polymerase assembles as a complementary strand of RNA that grows from its 5′ terminus in a 3′ direction (**Figure 6.2a, b and in Figure 6.2a.c**). Hydrolysis of ribonucleotide triphosphate precursors (NPS) generates nucleoside monophosphate.

They are polymerized into a covalent chain. Reactions are essential for synthesizing nucleic acids (and proteins) different from those of intermediary metabolism. The other enzyme, a pyrophosphatase, in which the pyrophosphate (PPj) is formed in the first reaction, is hydrolyzed to inorganic phosphate. As the polymerase moves along the DNA template, it incorporates complementary nucleotides into the growing RNA chain. A nucleotide is inserted into the RNA strand if it can form a proper base pair with the nucleotide in the DNA strand, which is getting transcribed.

The incoming adenosine 5′-triphosphate pairs with the thymine-containing nucleotide of the template. Once the polymerase moves through the DNA, the double helix is formed. As a result, the RNA chain with its template is a DNA-RNA hybrid. A bacterial RNA polymerase can incorporate 50 to 100 nucleotides into a growing RNA molecule per second, and most genes in a cell get transcribed side by side by side by side by various polymerases. RNA polymerases can form long RNAs. Which causes the enzyme to remain attached to the DNA over long stretches?

6.2 Transcription in Prokaryotes

A single RNA polymerase constituted of four subunits is present in E. coli. This includes the core enzyme. The core enzyme can be purified from the bacterial cells and added to a solution of bacterial DNA molecules and ribonucleoside triphosphates; the enzyme will bind to the DNA and start the synthesis of RNA. There exists an accessory polypeptide called sigma factor (σ), which is added to the RNA polymerase before it attaches to DNA; transcription starts at selected locations and increases the affinity for promoter sites in DNA but decreases its relationship to DNA in general (**Figure 6.3**).

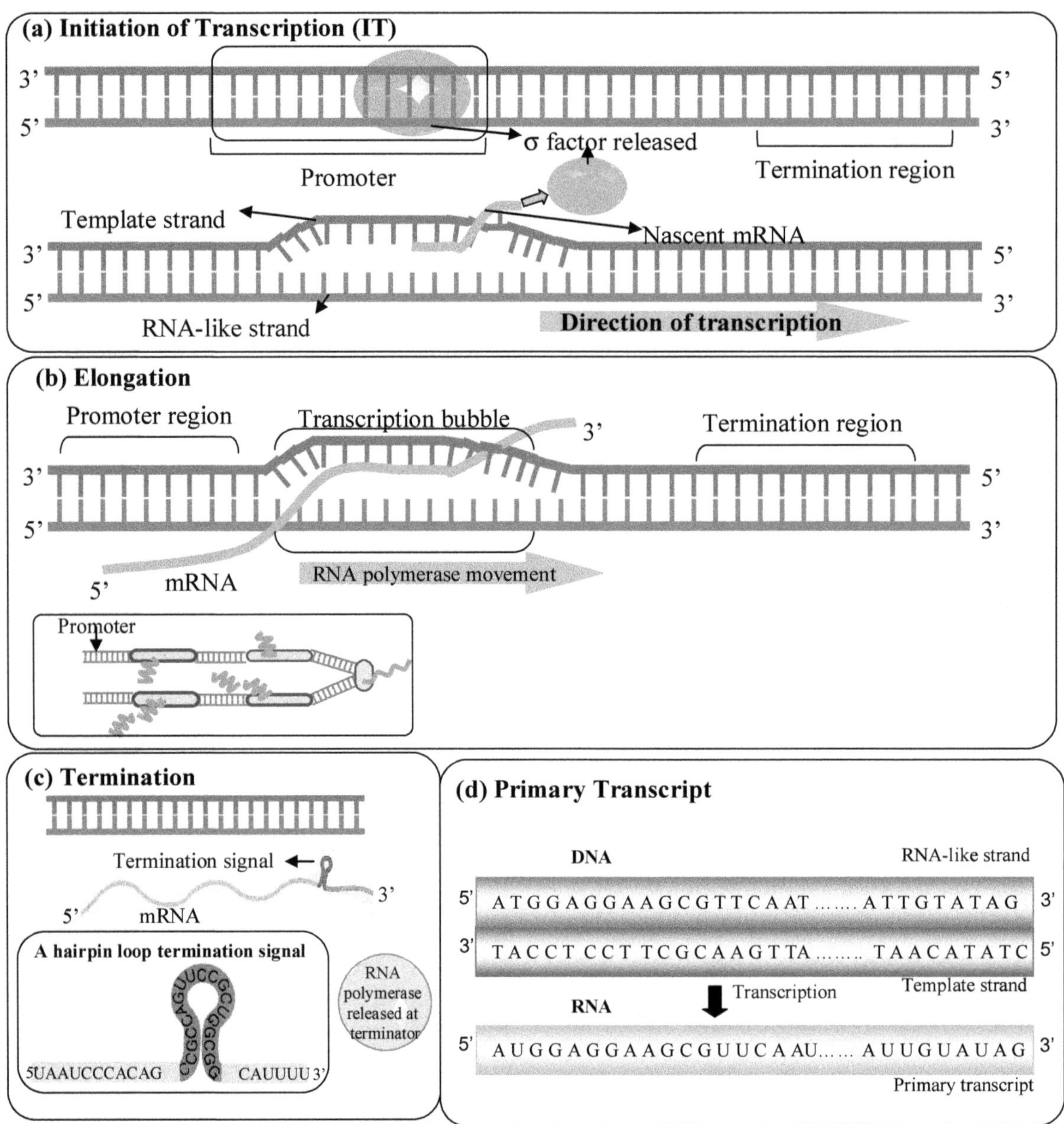

Figure 6.1: RNA Polymerase and its Involvement in Transcription.

The RNA polymerase gets bound to the promoter and separates (or melts) the two DNA strands in the region surrounding the start site, which makes the template strand accessible to the enzyme's active site. Transcription initiation appears to be difficult because an RNA polymerase typically makes several unsuccessful attempts to assemble an RNA transcript. Approximately ten nucleotides get successfully incorporated into a growing transcript; the enzyme undergoes a significant change in conformation and releases the sigma factor and the promoter DNA. Typical promoters vary from 20 to 200 bases long. The length of the supporters is variable, hence binding strength with RNA polymerases. However, it goes. Conformational changes result in the stabilization of the polymerase, which converts the enzyme into an elongation complex. This is the final step.

Bacterial promoters are found before the initiation site of RNA synthesis (**Figure 6.4**). The nucleotide at which transcription starts is called + 1 upstream from (toward the 3′ end) and the preceding nucleotide -1 downstream (toward the 5′ end). The two short stretches of DNA of the bacterial gene are similar between different genes. One of these stretches is approximately 35 bases upstream from the initiation site and typically occurs as the sequence TTGACA (**Figure 6.4**). This TTGACA sequence is called a consensus sequence. This is located approximately 35bp upstream from the transcription start site +1 and TATAAT, centered at ten base pairs upstream from

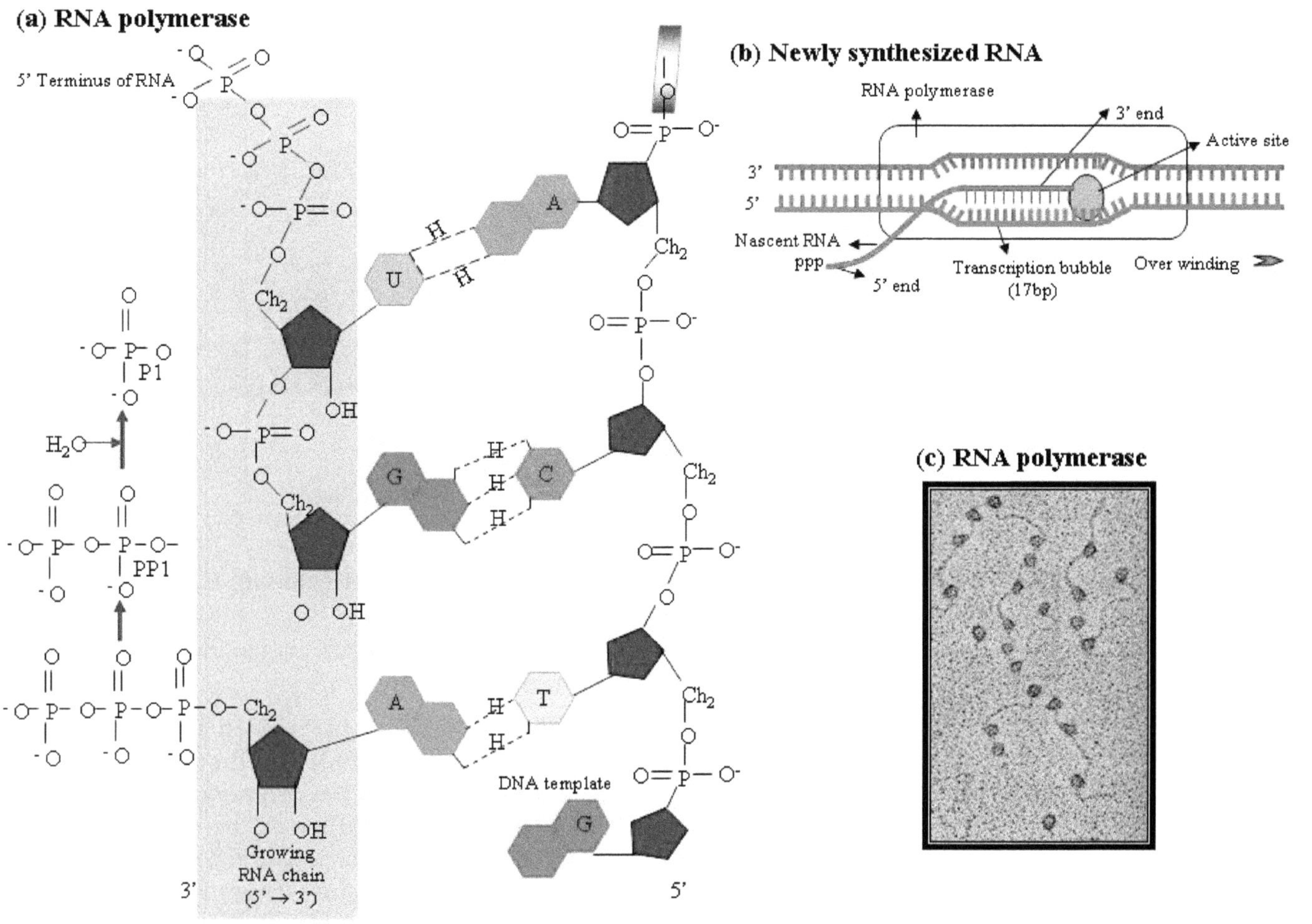

(a) RNA polymerase

5' Terminus of RNA

P1

PP1

Growing RNA chain (5' → 3')

3'

DNA template

5'

(b) Newly synthesized RNA

RNA polymerase

3' end

Active site

3'

5'

Nascent RNA

ppp

5' end

Transcription bubble (17bp)

Over winding

(c) RNA polymerase

Figure 6.2: Transcription and Chain Elongation.

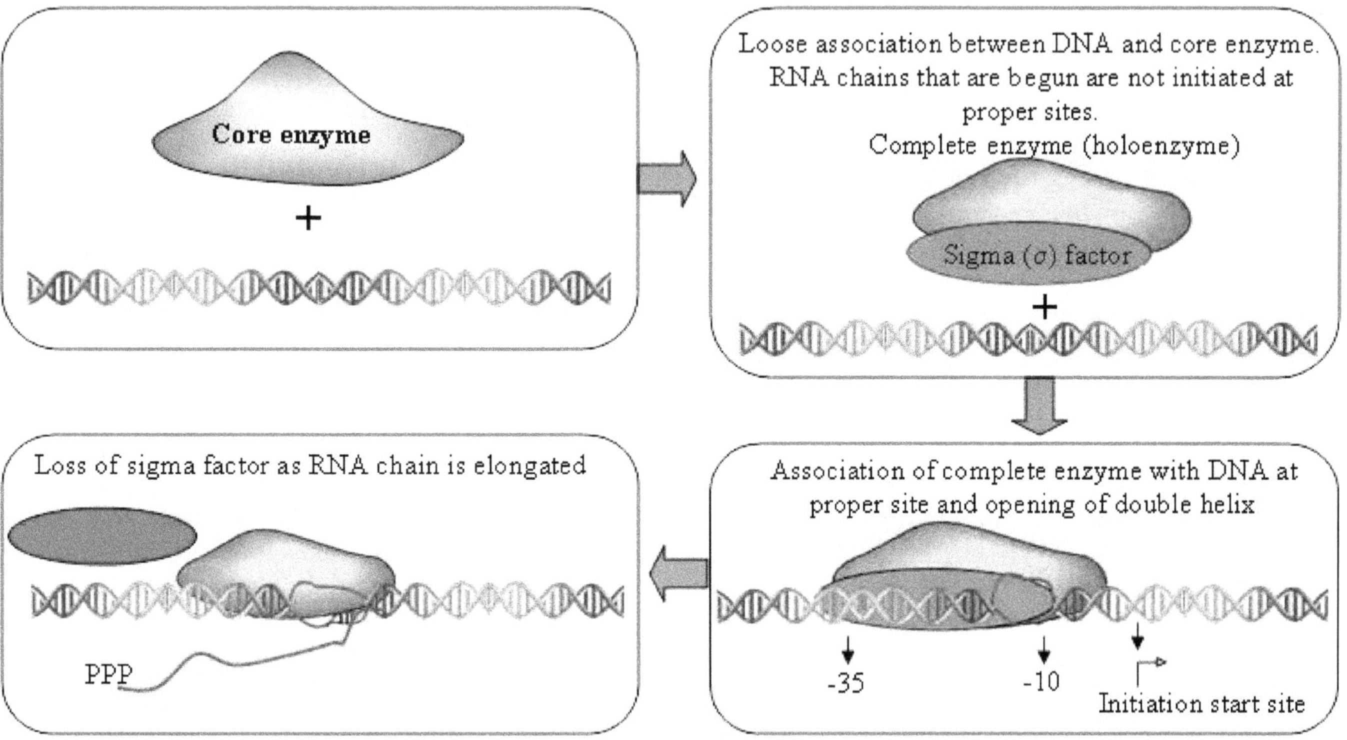

Figure 6.3: Initiation of Transcription in Prokaryotes.

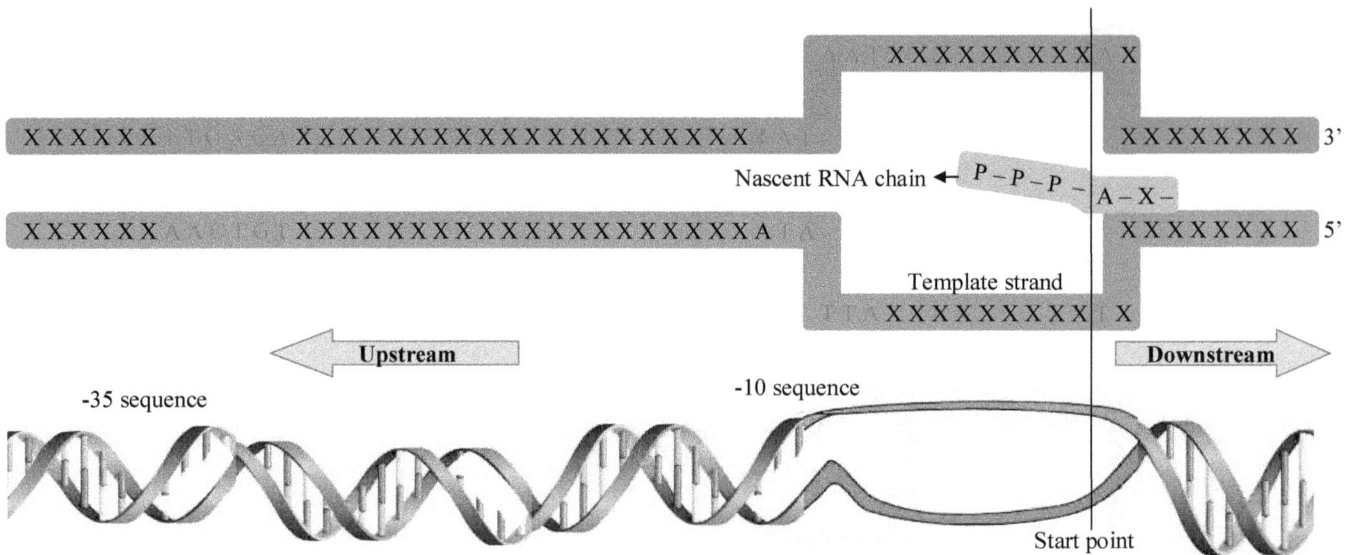

Figure 6.4: Basic Elements of a Promoter Region in the DNA of the Bacterium *E. coli.*

the +1 site. The -10 sequences, called the TATA box, are similar to sequences found in many eukaryotic promoters. The position of the promoter sequences determines where and on which strand the RNA polymerase begins synthesis; more so, A or G is often the first nucleotide in the transcript.

This site in the promoter region is known as the -35 region, which is identified by the factors that recognize different versions of the -35 sequence. One of them is J'70, the "housekeeping" CJ' factor because it initiates transcription in most genes. The J' factors initiate transcription of fewer specific genes contributing to the typical response. For example, when E. coli cells are subjected to a sudden rise in temperature, a new J' factor is synthesized that recognizes a different promoter sequence, leading to the coordinated transcription of a battery of heat-shock genes. The polymerase stops transcription when it reaches termination sequences. The polymerase enzyme disassociates from the DNA, releasing the newly synthesized RNA. Two types of termination events are known: (1) those that depend upon only the transcription termination sequence in the DNA and (2) those that require the presence of a termination protein in addition to the transcription termination sequence.

6.3 Transcription and RNA Processing in Eukaryotic Cells

The eukaryotic cells have three discrete RNA polymerases that contrast with bacterial cells. These are called polymerase I, II, and III, with different functions illustrated below. RNA polymerase I transcribe two large RNA components of the ribosomes; RNA polymerase – II is used in the protein-coding genes, while RNA polymerase – III transcribes and transfers to RNA.

The initiation starts with transcription by an enzyme called RNA polymerase II at specific DNA sites known as the promoter site constituting DNA sequences, followed by elongation and termination, and by various RNA processing events that finally produce a messenger RNA that goes out of the nucleus for translation. All these steps can be precisely regulated; much of the gene expression is controlled at the level of transcription initiation. Thus, understanding gene regulation implies deciphering how RNA polymerase II is directed to the genes that must be transcribed at the right time and the proper level in particular cell types.

6.4 Basal and Activated Transcription

Basal transcription usually means constitutive transcription, primarily at a low level. When activated, it is enhanced and called activated transcription. Every gene has a specific transcription start point in eukaryotes. Additional factors called general transcription factors (GTFs) are necessary. GTP's interact directly or indirectly with the RNA polymerase II at the gene promoter. GTF is required to position the RNA polymerase II at the gene promoter, help in meeting the two DNA strands, and enable the enzyme to initiate and elongate transcription. They contain TATA-binding proteins (TBP), TFIIA, TFIIB, TFIIE, TFIIF, and TFIIH, contributing different activities necessary for the transcription process. For example, TBP recognizes the TATA box, an AT-rich short sequence characteristic of many polymerase II promoters. TFIIH is a protein complex containing a helicase that unwinds DNA and a kinase that phosphorylates the RNA polymerase's C-terminal domain (CTD). This modification is necessary to release the polymerase from the promoter. In the presence of these general transcription factors, purified RNA polymerase II, incubated with purified DNA, can specifically bind a promoter sequence to initiate the transcription in vitro (Figure 6.5a, b). In vitro, basal transcription can be increased by adding DNA-binding proteins that interact with specific sequences usually situated near the upstream of the promoter region recognized by RNA polymerase II in association with the GTFs. These sequence-specific proteins, or transcriptional activators, are responsible for the activated transcription level. Under the in vitro conditions, activated transcription is generally only a few times higher than the basal level and requires additional factors such as the TBP-associated factors TAFs.

Additional factors are required for specific transcription initiation and activation in vivo on a DNA template packaged into nucleosomes (cylinder-shaped histone octamer complexes with DNA wrapped around them). There is a Mediator complex and additional factors that interact with Pol II; there are nucleosome remodeling factors SWI/SNF and SAGA, and the sequence-specific transcriptional activators, which are typically composed of a DNA-binding domain (DBD) and an activation domain (AD) that can potentially interact with all the above factors to stimulate transcription. Arrows indicate (Figure 6.5) potential molecular interactions, while the bent arrow on the DNA outline shows specific transcription initiation. The promoter region is where the transcription complex binds and initiation starts; transcriptional activators have the binding site.

In eukaryotic cells, DNA is tightly packaged into so-called nucleosomes. The Histone octamer complexes with DNA wrapped around them resulted in decreased accessibility to DNA of the RNA polymerase and the GTFs compared to the situation in vitro. Initiation and elongation in vivo are more complex and require many more factors (**Figure 6.5c**). The transcription complex that contains the RNA polymerase II, on its own, is usually unable to recognize a promoter sequence in vivo, in contrast to its ability to do so on naked DNA in vitro. Basal transcription hardly occurs in vivo. Nucleosome DNA is not accessible to the transcription complex. The addition of sequence-specific transcriptional activators binding upstream of the promoter activated transcription of the nucleosome DNA in vitro, thus indicating that this type of transcription factor can dilute the nucleosome's inhibitory effect. The level of transcription can be achieved gradually, from very low to very high, by changing the state of the DNA template (from tight nucleosome to naked DNA), the number and type of transcription factors helping the polymerase, the number and strength of transcriptional activators, and the number of their binding sites on DNA. Thus, the transcription level of any given gene in an organism will be determined by all these parameters applied to that gene at its chromosomal location.

6.5 Activators and Enhancers

Enhancers generally have multiple clustered binding sites for transcription factors. Cooperative binding of multiple activators to nearby sites in an enhancer forms a multi-protein complex called an enhanceosome. Assembly of enhancing some often requires small proteins that bind to the DNA minor groove and bend the DNA sharply, allowing proteins on either side of the bend to interact more readily.

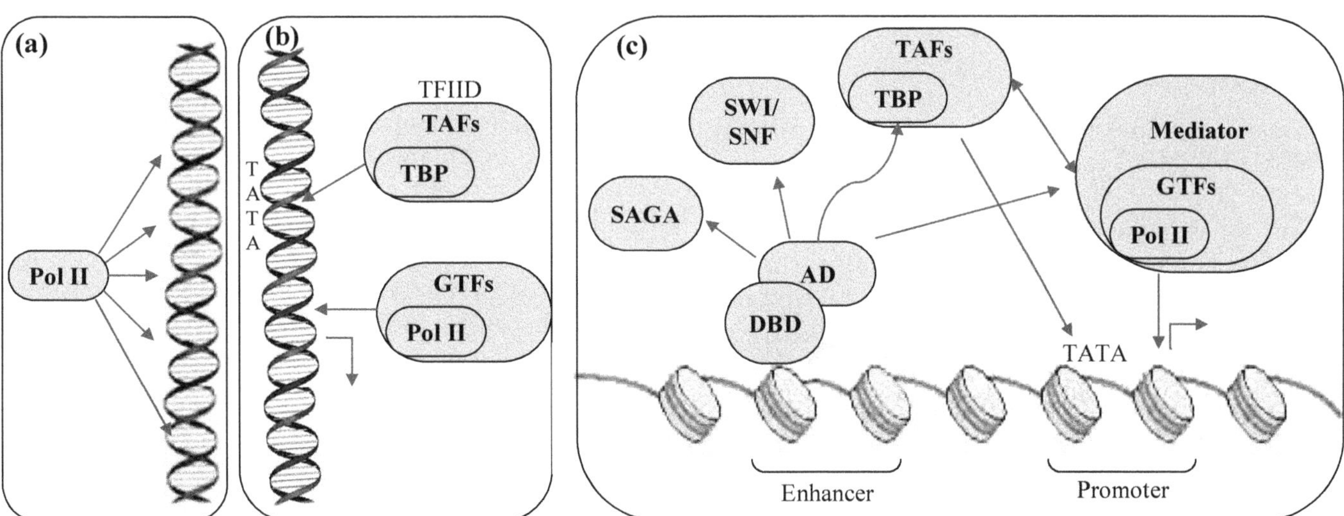

Figure 6.5: Different Transcription Factors for Specific Promoter Binding and Transcription Initiation through RNA Polymerase II.

Activators are generally modular proteins containing a single DNA-binding domain and one or few activation domains; different environments commonly are linked through flexible polypeptide regions. This may help activation domains in other activators act together even when their DNA-binding fields are bound to sites separated by tens of base pairs.

Cis-and-transacting elements activate the initiation of DNA transcription by RNA polymerase II. As the name suggests, cis-acting elements are DNA sequences that can influence gene transcription. Functionally, they can be subdivided into promoter and enhancer elements, both of which are necessary for a gene's specific and activated expression (Figure 6.5c). Trans-acting elements are proteins (transcriptional activators) that bind specific Cis-acting elements to activate transcription. Cis-acting elements are harmful when they attach to the transcriptional repressors. These allow down-regulation of the expression of the genes.

6.6 Enhancers

Enhancers enhance gene expression, which in other ways means gene activation. These are important as they are responsible for metazoans' tissue and time-specific transcription activation. There are specific sequences in the enhancers that can bind different transcriptional activators. One characteristic of these regulatory elements is that they can activate transcription from several thousands of base pairs upstream or downstream of a gene, independently of their orientation. Gene regulatory sequences in yeast are functionally equivalent to the metazoan enhancers and are called upstream activating sequences (UASs). Yeast UASs show a much lower degree of flexibility than their metazoan counterparts but can nevertheless activate transcription from a distance up to 1200 base pairs approximately from the promoter.

6.7 Activators

Transcriptional activators in eukaryotic cells are modular proteins typically made up of a sequence-specific DNA binding domain and an activating region (or domain). It is generally accepted that activators interact through their activating parts with components of the RNA polymerase II transcriptional apparatus that bind to the promoters. Some activators can also interact with factors that modify or remodel nucleosomes (chromatin). According to their amino acid activation, domains have been classified as acidic (*e.g.*, the viral VP16 and the yeast Gal4), glutamine-rich (*e.g.*, SP1), or proline-rich (*e.g.*, CTF). The binding of the activator to DNA serves to locate it to a position from which the activation domain can activate transcription. DNA binding per se is not sufficient for activating transcription. In higher eukaryotes, many genes are activated not by just one but by several activators. The use of several, and often different, activators to activate one gene establishes a combinatorial control that allows differential regulation of many genes with a relatively small number of transcriptional activators A wide variety of mechanisms has been devised in nature to regulate proteins. One mechanism to silence an activator is keeping it outside the nucleus, away from the target genes. This mode of activator control is seen in the case of NFkB. This activator is held outside the nucleus by its inhibitor IkB. Through phosphorylation of IkB upon specific signals, NFkB dissociates and enters the nucleus, where it binds and activates its target genes. Another regulatory mechanism is to mask and inactivate the activation domain by another protein. The activator Gal4, for example, can be inactivated by Gal80, a protein that binds to the activation domain of Gal4 and interferes with protein interactions necessary for its activation function. It has been shown that individual activators can also be shut down immediately after they have stimulated transcription from a promoter. For example, such a mechanism shuts down the yeast transcriptional activator Gcn4. Once bound to DNA, activators contact the transcriptional apparatus binding to the promoter. Each time a Gcn4 protein gets the transcriptional machinery to activate transcription, the kinase Srb10, which is a general component of the transcription complex, phosphorylates Gcn4 and signals it for destruction by ubiquitin-mediated proteolysis. The ability of the transcriptional machinery to mark activators for destruction provides an efficient mechanism to limit the number of times a DNA-bound activator can activate transcription from a gene promoter.

6.8 Post-recruitment Steps of Transcription Activation

A stable initiation complex must be created by promoter clearance, and elongation produces an RNA transcript. It is now widely accepted that modification of the C-terminal domain (CTD) of the largest subunit of polymerase II by phosphorylation provides at least part of the signal for the switch from initiation to elongation. Concurrent with the CTD phosphorylation, probably due to this modification, there is an exchange of factors associated with the polymerase, which marks the elongating complex and distinguishes it from the initiation complex. While Pol II molecules of the promoter-bound initiation complex lacking phosphate on their CTDs are associated with the so-called mediator complex, which is contact sites for transcriptional activators, Pol II molecules with phosphorylated CTDs are found associated with DNA sequences downstream of promoters and with different protein complexes, such as the elongator complex and various RNA processing factors. Indeed, as transcription elongation proceeds under the control of so-called elongating factors, RNA capping, splicing, and ultimately polyadenylation are carried out by factors that have been found associated with the phosphorylated Pol II.

Transcription activation by sequence-specific activators in eukaryotes involves the recruitment of different factors with different activities that can affect the chromatin structure, causing the formation of the Pol II initiation complex at promoters and influence the efficiency of promoter clearance and elongation. Thus, all these complex interactions of activators result in conferring the specificity and competence required for correct and productive transcription of specific genes to the polymerase.

6.9 Ribosomal RNAs

Ribosomes are molecules in which protein synthesis takes place. Ribosomal RNA (rRNA) is an assembly of amino acids that can synthesize proteins. A tiny particulate structure located in the cytoplasm of the cell, which is located outside the nucleus, the ribosome is composed of two subunits, one larger and the other smaller than the other. These come together during polypeptide synthesis to form a mature ribosome. The active site is on the large unit of the ribosome. Ribosomal RNA molecules can catalyze its chemical reactions. Ribosomes are more than 80 per cent of the RNA in most of the cells. The human genome has five rDNA clusters located on different chromosomes.

The bulk of a nucleolus, which contains many rDNA, is composed of ribosomal subunits that give the nucleolus a granular appearance. Embedded within this granular mass are one or more rounded cores consisting primarily of fibrillar material.

6.10 Synthesizing the rRNA Precursor

The 70S ribosome found in the prokaryotes is composed of a 30S subunit and a 50S subunit. The former contains 16S rRNA, and the latter 23S and 5S rRNAs. In bacterial genes, the sequences specifying 16S, 23S, and sometimes also 5S RNA are arranged in a series of mRNA, which is transcribed from DNA as a 30S unit (p30S). Experimental evidence indicates that the 30S unit has the 16S component at the 5′ end and the 23S part is at the 3′ end, with spacer units between the two components. 5S rRNA is transcribed at or near the 3′ end in some prokaryotes. At the time of processing, the p30S transcriptional unit is cleaved within the spacer segment by RNase III into 25S and 18S segments. These are then shortened to p23S and pl6S segments correspondingly by RNase III. Secondary trimming of these intermediates produces the final size, 23S, and 16S, respectively.

The enzyme implicated in secondary trimming is probably a ribonuclease, designated as RNase M or maturase. In E. coli, trimming of the pl6s component involves the removal of approximately 200 nucleotides from both 3′ and 5′ ends. The final cleavages occur when the rRNAs are associated with structural proteins of ribosomes.

Some nucleoside modifications occur in the p30S component; cleavage also causes changes. In E. coli, the precursors of 5S rRNA are only a few nucleotides bigger than mature 5S rRNA. In Bacillus, there are two types of more extensive precursors. One, designated as p5A, is 180 nucleotides long, while the other, P5B′ is 140-150 nucleotides long. Both are apparently transcribed on different 5S genes.

Cleavage of p5A by the enzyme RNase M5 splits it into three components: a 5′ terminal piece of about 20 nucleotides, a mature 5S rRNA of 118 nucleotides, and a 3′ terminal part of about 40 nucleotides. The 5′ and 3′ terminal elements are broken down to their mononucleotides by an exonuclease, which serves a scavenger function.

Mature 5S rRNA is resistant to this enzyme. The precursor p5B gets split by RNase M5 into a 22 nucleotide 5′ terminal piece, a mature 58 mRNA, and some smaller 3′ terminal pieces. In eukaryotes, the 80S ribosome consists of 40S and 60S subunits. The 40S subunits contain 16-18S rRNA, and the 60S subunits 25-28S rRNA, 5.8S RNA, and 5S rRNA.

The genes coding for 16-18S rRNA and 25S-28S rRNA are arranged in clusters of hundreds to thousands of copies. In eukaryotes, the transcribed rRNA is 45S rRNA in mammals and 36-38S rRNA in lower eukaryotes, with molecular weights of 4.5 million and 2.6-2.8 million, respectively.

6.11 Processing the rRNA Precursor

Eukaryotic ribosomes have four distinct ribosomal RNAs, three in the large subunit and the other in the small subunit. In humans, the large subunit comprises 28S, 5.8S, and 5S RNA molecules, and the small subunit contains an 18S RNA molecule. Three of these rRNAs (the 28S, 18S, and 5.8S) are carved by various nucleases from a single primary transcript (called the pre-rRNA). The 5S rRNA is synthesized from a separate RNA precursor outside the nucleolus.

Compared with other RNA transcripts, two of the peculiarities of the pre-rRNA are many methylated nucleotides and pseudouridine residues. As soon as the human pre-rRNA precursor undergoes first cleavage, more than 100 methyl groups are added to the ribose groups in the molecule, and approximately 95 of its uridine residues are chemically converted to pseudouridine (**Figure 6.6a**). All these modifications occur after the nucleotides are incorporated post-transcriptionally into the nascent RNA. The function of the methyl groups and pseudouridines is unclear. It is speculated that these modified nucleotides may protect parts of the pre-rRNA from enzymatic cleavage, promote folding of rRNAs into three-dimensional structures, and/or promote interactions of rRNAs with other molecules.

Because rRNAs are so heavily methylated, their synthesis can be followed by incubating cells with radioactively labeled methionine. This compound serves in most of the cells as a methyl group donor.

The pathways in the processing of rRNA primary transcript are indicated in **Figure 6.7**.

Electron microscopy shows total transcription. The first RNP particle attaches to a pre-rRNA transcript and contains the U3 snoRNA, which binds to the 5′ terminus of the precursor molecule. Different snRNPs help in enzymatic cleavage.

6.12 Synthesis and Processing of the 5S rRNA

A 5S rRNA is about 120 nucleotides long. In eukaryotes, the 5S rRNA molecules are encoded by many similar genes. The 5S rRNA genes are arranged in a tandem array, with the transcribed portions alternating with non-transcribed spacers to form repeating units analogous to the repeating units encoding the 45S rRNA precursor. RNA polymerase III transcribes the 5S rRNA genes. The 5′ end of the primary transcript is identical to that of the mature 5S rRNA, but the 3′ end usually has extra nucleotides that are removed during processing. RNA polymerase III is unusual among the three polymerases, which bind to the promoter site located within the transcribed portion of the gene rather than a site upstream from the gene's 5′ flanking region.

6.13 Transfer RNAs

There are approximately 50 different species of transfer RNA in plant and animal cells, each encoded by a DNA sequence repeated several times within the genome. Yeast cells have a total of about 275 tRNA genes, fruit flies about 850, and humans about 1,300. Transfer RNAs are synthesized from genes that are found in small clusters scattered throughout the genome (**Figure 6.8**).

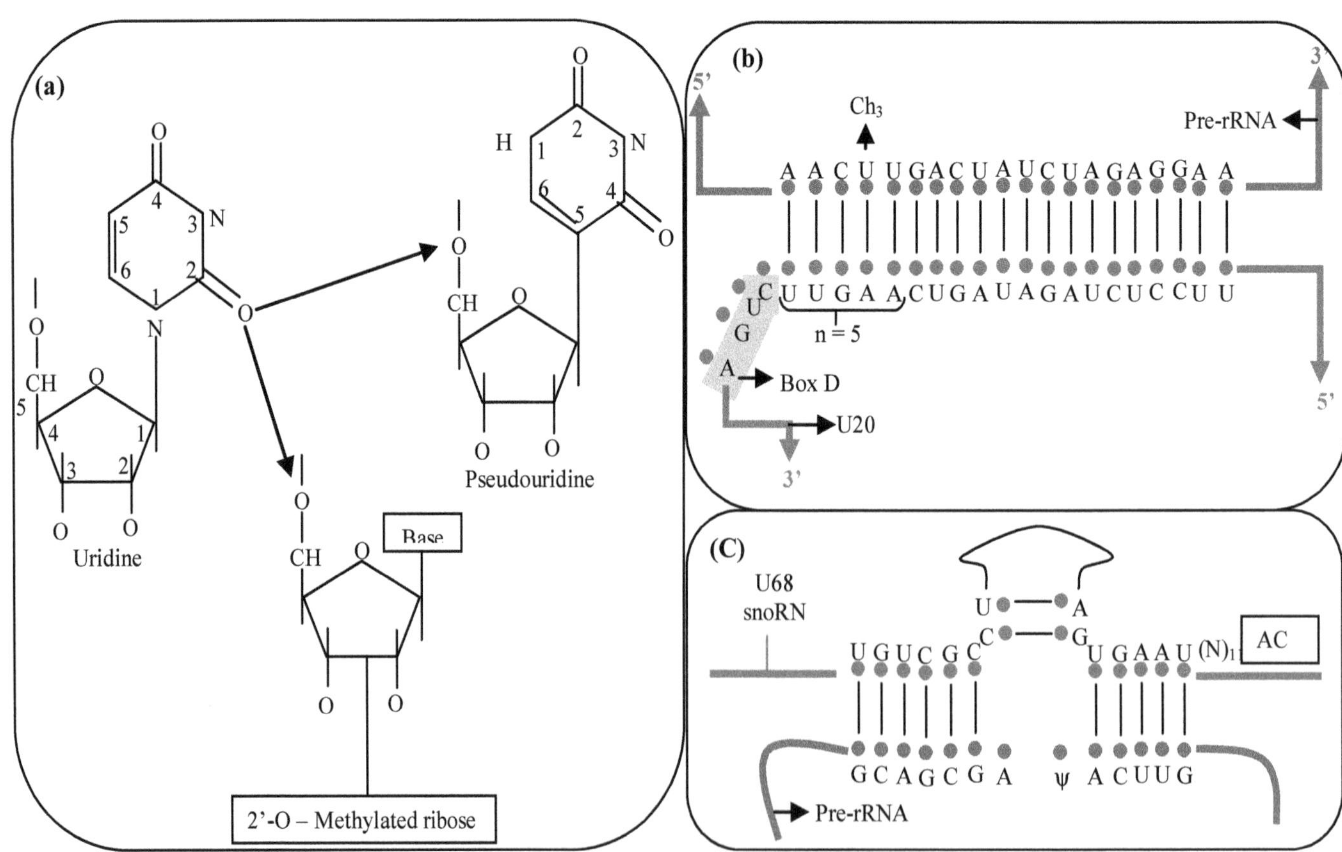

Figure 6.6: Pre-rRNA Modifications.

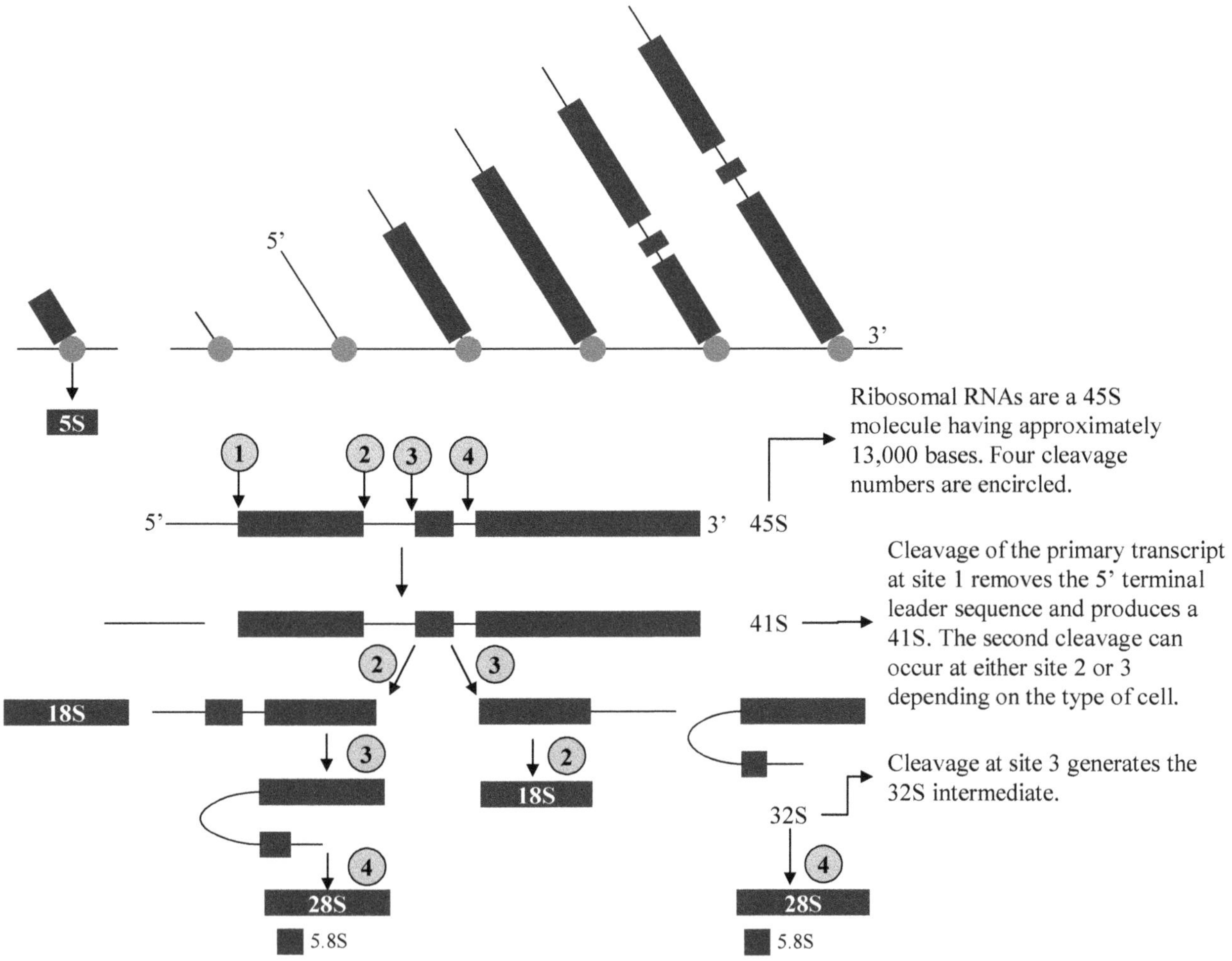

5'
3'
5S
1
2
3
4
5'
3'
45S
Ribosomal RNAs are a 45S molecule having approximately 13,000 bases. Four cleavage numbers are encircled.
Cleavage of the primary transcript at site 1 removes the 5' terminal leader sequence and produces a 41S. The second cleavage can occur at either site 2 or 3 depending on the type of cell.
41S
18S
18S
Cleavage at site 3 generates the 32S intermediate.
32S
28S
5.8S
28S
5.8S

Figure 6.7: Processing of Mammalian Ribosomal RNA.

1 kb
2 kb
3 kb
Tyr
Met-A
Met-B
Leu
Lys
Phe
Asn
Ala

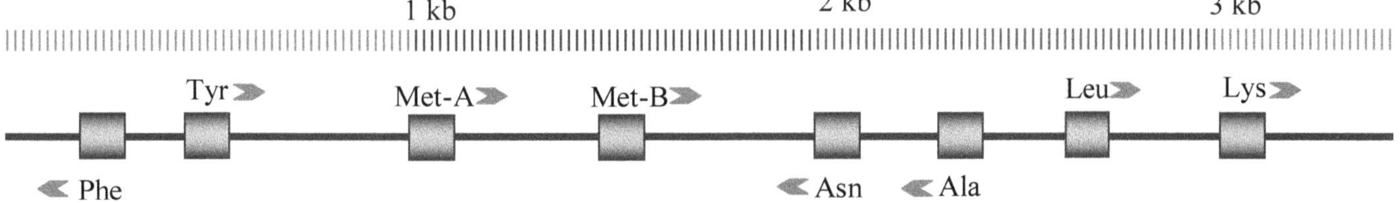

Figure 6.8: Arrangement of Genes that Code for Transfer RNAs in *Xenopus*.

Like the 5S rRNA, tRNAs are also transcribed through RNA polymerase III transcribe tRNA and the promoter sequence lies within the coding region of the gene and not in the 5′ flanking region. The primary transcript of a transfer RNA molecule is larger than the final product, and pieces on both the 5′ and 3′ of the precursor tRNA (and a small interior piece in some cases) should be trimmed away. One of the enzymes involved in pre-tRNA processing is an endonuclease called ribonuclease P. It is present in bacterial and eukaryotic cells and consists of RNA and protein subunits.

6.14 Messenger RNAs

mRNA has the following characteristic features. (1) They have large molecular weights (up to about 80S, or 50,000 nucleotides); (2) as a group, they are represented by RNAs of diverse (heterogeneous) nucleotide sequences; and (3) they are found only in the nucleus. These RNAs are referred to as heterogeneous nuclear RNAs (hnRNAs). When cells that are incubated in [3H] uridine or [32P] phosphate for a brief pulse are placed into an unlabeled medium, and a constant watch is kept for an hour or more before they are killed. RNA is extracted, the amount of radioactivity in the large nuclear RNAs drops sharply and appears instead in much smaller mRNAs found in the cytoplasm. These early experiments suggested that the large, rapidly labeled hnRNAs were precursors to the smaller cytoplasmic mRNAs.

RNA in the cell is present in two forms, these are 18S and 28 rRNA.

mRNA is present in a significant fraction of the RNA in the cell. Thus, even though the mRNAs and their heterogeneous nuclear RNA precursors constitute only a tiny percentage of the total RNA of most eukaryotic cells, they contain a large portion of the RNA that is being synthesized by that cell at any moment. hnRNA and mRNAs are degraded after relatively brief periods. This is particularly true of the snRNAs, which have only a few minutes of half-lives. In contrast, the rRNAs and tRNAs have half-lives that are measured in days or weeks and thus gradually accumulate to become the predominant species in the cell. The half-lives of mRNAs vary depending on which species mRNA belongs to. It differs from 15 minutes to a few days.

The presence of intervening sequences in nonviral cellular genes was first reported in 1977 by Alec Jeffreys and Richard Flavell of the University of Amsterdam. These researchers discovered an intervening sequence of approximately 600 bases located directly within a part of the globin gene that codes for the amino acid sequence of the globin polypeptide (**Figure 6.9**). It has been seen that there are intervening sequences existing in other genes. The presence of genes with intervening sequences are called split genes-is the rule, not an exception. Those parts of a split gene that contribute to the mature RNA product are called exons, whereas the intervening sequences are called introns. Split genes are widespread among eukaryotes, both in the simpler eukaryotes, such as yeast and protists, and in all living beings; the size and structure of introns in eukaryotes are simpler, less in number, and smaller in size as compared to plants. Introns are seen in all types of genes, including those that code for tRNAs, rRNAs as well and mRNAs.

Cells may produce a primary transcript that corresponds to the entire transcription unit, and those portions of the RNA corresponding to the introns in the DNA are somehow removed. If this holds then the segments corresponding to the introns should be present in the primary transcript.

The mRNA formation in eukaryotic cells occurs by removing internal sequences of ri- bonucleotides from a much larger pre-mRNA. Many steps are involved in this process, as described below.

6.15 The Processing of Eukaryotic Messenger RNAs

The nucleus is the center where mature mRNA is produced and then transported to the cytoplasm. The conversion process required the addition of a methylated cap at the 5′ end and a Poly-A Tail at the 3′ end are added.

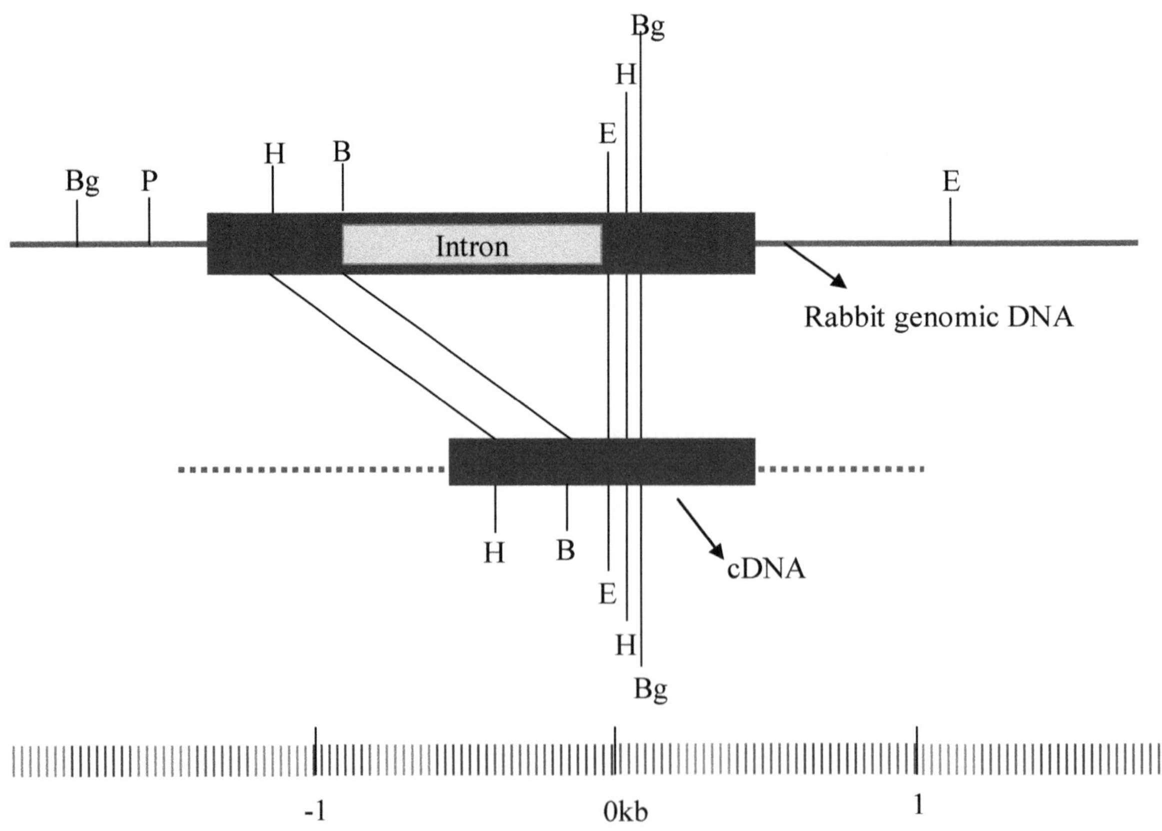

Figure 6.9: Discovery of Introns in a Eukaryotic Gene.

The 52 end of eukaryotic mRNA is capped with a guanine nucleotide (which is methylated, forming 7-methyl guanosine). The cap (52 -G) is added to the mRNA after transcription. A nuclear enzyme, guanylyl transferase, catalyzes the addition of 52 G. The cap is linked to the 52 terminus of the mRNA through an unusual 52 -triphosphate linkage. The 52 cap is formed by condensation of a molecule of GTP with the triphosphate at the 52 end of the transcript. The guanine is methylated at N-7 to form 7-methylguanosine. Additional methyl groups are added to the 22 hydroxyls (-OH) of the first and second nucleotides adjacent to the cap. The methyl groups are derived from S Adenosyl L Methionine. At the 32 end, most eukaryotic mRNAs have a string of 80-250 adenylate residues called the poly(A) tail.

6.16 Addition of Poly-A Tail to mRNA

1. Polyadenylation is the post-translational modification. Following the transcription of mRNA, poly-A polymerase adds the poly-A tail.
2. RNA polymerase II, which makes the precursor to mRNA called hnRNA, runs along with the DNA and synthesizes the precursor to mRNA.
3. AAUAAA sequence in the precursor mRNA (or hn RNA) signals an endonuclease to cleave between 10 and 30 nucleotides down this signal, and then Poly A Polymerase, using ATP as the source of A, adds the long A chain at the 3′ end. This AAUAAA sequence is required, but not sufficient, for poly-A addition to the 3′end.
4. As the poly-A mRNA gets used more and more, the poly-A tail gets smaller. If the mRNA synthesis is shut down, one will notice that the poly-A tail gets smaller as the mRNA is used more and more.

The 3′ sequence is after the poly-A addition sequence, it is cleaved off from the main mRNA molecule 10 to 30 nucleotides downstream from the poly-A addition site.

6.17 5' Cap of Eukaryotic mRNA

The cap is a post-transcriptional modification present in all eukaryotic mRNA. It is added by enzymes following transcription. Cap synthesis takes place in the nucleus.

1. Cap found on both poly A+ and poly-A- mRNA.
2. Cap recognized by a ribosome that the RNA is an mRNA.
3. It has a 5′ to 5′ linkage in it.
4. The cap is 7-methylated guanine that is on the 5′ end of the mRNA - it is this 7-methylguanine that is connected through three phosphodiester linkages via a 5′-5′ linkage.
5. RNA polymerase and DNA polymerase cannot make a 5′ to 5′ linkage (only 5′ to 3′) so we know that the cap structure is made after the RNA/DNA polymerase is finished with that region of the molecule.
6. mRNA is a negatively charged molecule. Since the cap formation causes the 7 nitrogen to have a positive charge, the cap sticks out due to its positive charge and is easily recognizable by the ribosome.
7. Only RNA polymerase II products are "capped" - RNA polymerase I and III products are not capped.
8. All Eukaryotic mRNAs are capped. There are three different types of caps - cap 0, cap 1, and cap 2. All three caps have the 7 methylations of guanine. The significant difference between the caps is in the other methylations: Cap 0 has only the 7-methylguanosine site methylation. Cap 1 has the same 7 methylations and methylation on the ribose of the first nucleotide after the cap. Cap 2, the most popular type of cap, has the 7-methyl guanine and the ribose of the first two nucleotides after the cap is methylated.

6.18 Mechanism of Cap Synthesis

RNA Polymerase II starts synthesizing hnRNA (the precursor to mRNA), as shown below.

pppG + pppC → pppGpC + ppi (Enzyme RNA polymerase II)

(the pyrophosphate is hydrolyzed to drive the reaction to the right.)

This is simply RNA synthesis. The pppG is GTP, with three phosphates on the 5′ end of the RNA. If uracil is added, then the reaction would be: pppGpC + pppU → pppGpCpU + ppi (Enzyme RNA polymerase II) again the pyrophosphate is hydrolyzed to drive the reaction to the right.

The RNA chain is being extended upon the addition of UTP. This is the continuation of RNA synthesis. After synthesizing a few ribonucleotides, cap formation begins. RNA synthesis is not yet finished but the cap formation begins. The cap is synthesized before the polyadenylated tail is formed. A specific phosphatase comes along and cleaves the terminal phosphate from the 5′G. This will make the 5′ end a diphosphate instead of a triphosphate: pppGpCpU... → ppGpCpU... + pppG → GpppGpCpU. + ppi →

m7GpppGpCpA ... The enzyme used is Enzyme 1: Phosphatase, Enzyme 2: Guanyl transferase, Enzyme 3, 4, 5, *etc.*: SAM (S adenosyl – L- methionine) While the first reaction occurs, RNA polymerase II continues along, synthesizing the rest of the RNA chain.

GTP added by guanyl transferase enzyme in the unusual 5′ to 5′ linkage, and the pyrophosphate is given off. The hydrolysis of the pyrophosphate completes the reaction. We now have the 5′ to 5′ linkage.

Methyl groups are added on by S-adenosyl-L-methionine (SAM), the first methyl group added to get 7-methyl guanine. Methyl groups are added one at a time, starting at the guanine and then moving to the terminal nucleotide's ribose, and finally to the penultimate sugar (all depending on whether it is a cap 0, cap 1, or cap 2.)

The nucleotide at the 5′ end of a eukaryotic mRNA is a G in reverse orientation from the rest of the molecule; it is connected through a triphosphate linkage to the first nucleotide in the primary transcript. A special capping enzyme adds it to the primary transcript after polymerization of the transcript's first few nucleotides. Enzymes are known as methyltransferases, then add methyl (-CH3) groups to the backward G and one or more of the succeeding nucleotides in the RNA, forming a so-called methylated cap (**Figure 6.10**). It is important to step for the translation of the mRNA into protein.

Like the 5′ methylated cap, the 3′ end of most eukaryotic mRNAs is not directly encoded by the gene. The mRNAs at the 3′ end consist of 100-200 in most of the eukaryotes. This is called a poly-A tail (**Figure 6.11**).

The ribonuclease cleaves the primary transcript. Under the first step, it forms a new 3′ end; upon the sequence AAUAAA, which is found in poly-A-containing mRNAs 11-30 nucleotides upstream of the position where the tail is added.

poly-A polymerase gets added in the next step, and the 3′ end gets exposed due to cleavage. The function of poly-A is controversial, but it is thought that it may have three functions: aiding mRNA transport out of the nucleus into the cytoplasm stabilizing mRNAs in the cytoplasm.

6.19 RNA Splicing Removes Introns from the Primary Transcript

Most introns start from the sequence GU and end with the sequence AG (in the 5′ to 3′ direction).

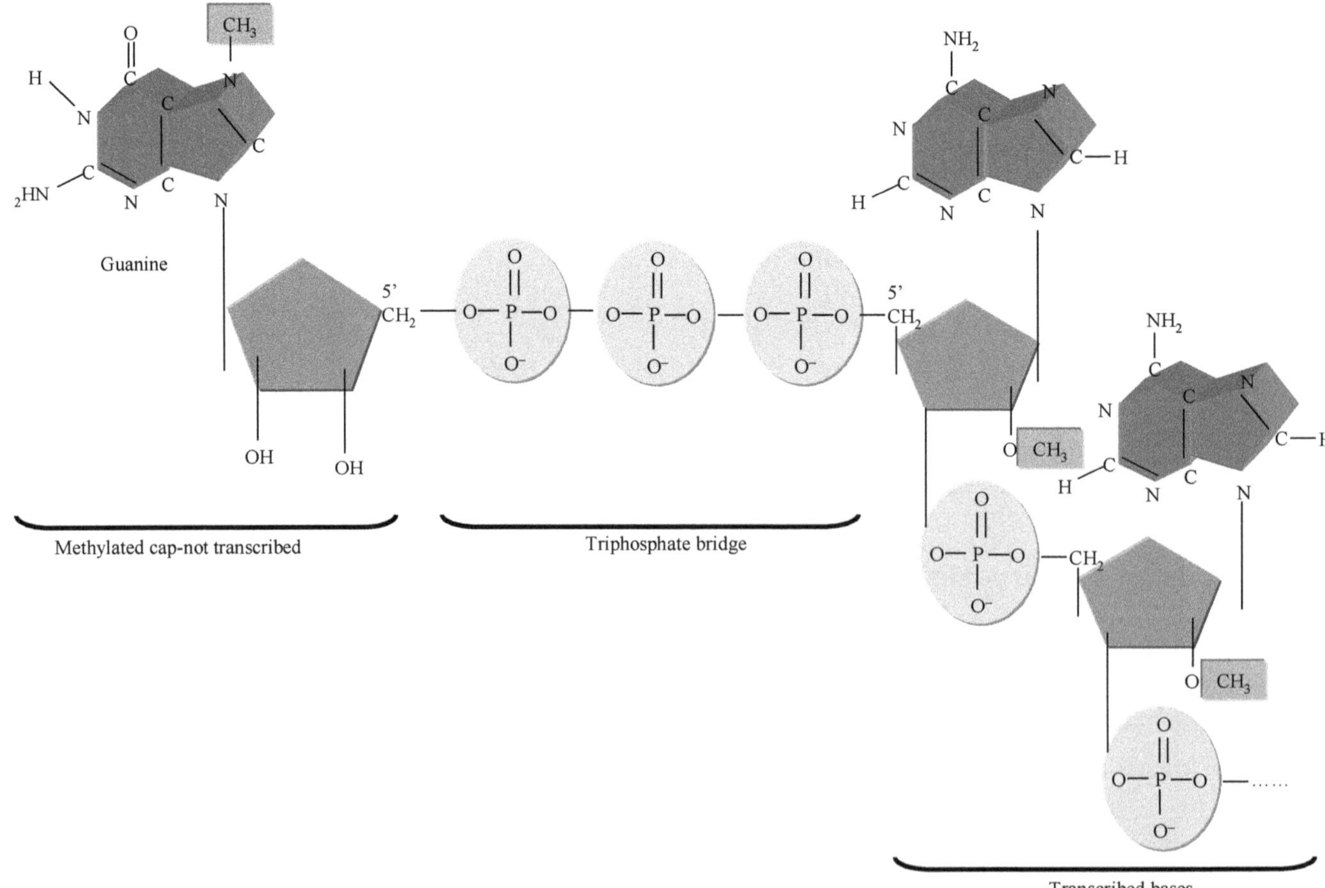

Figure 6.10: Presence of the Methylated Cap at the 5' end of Eukaryotic mRNAs.

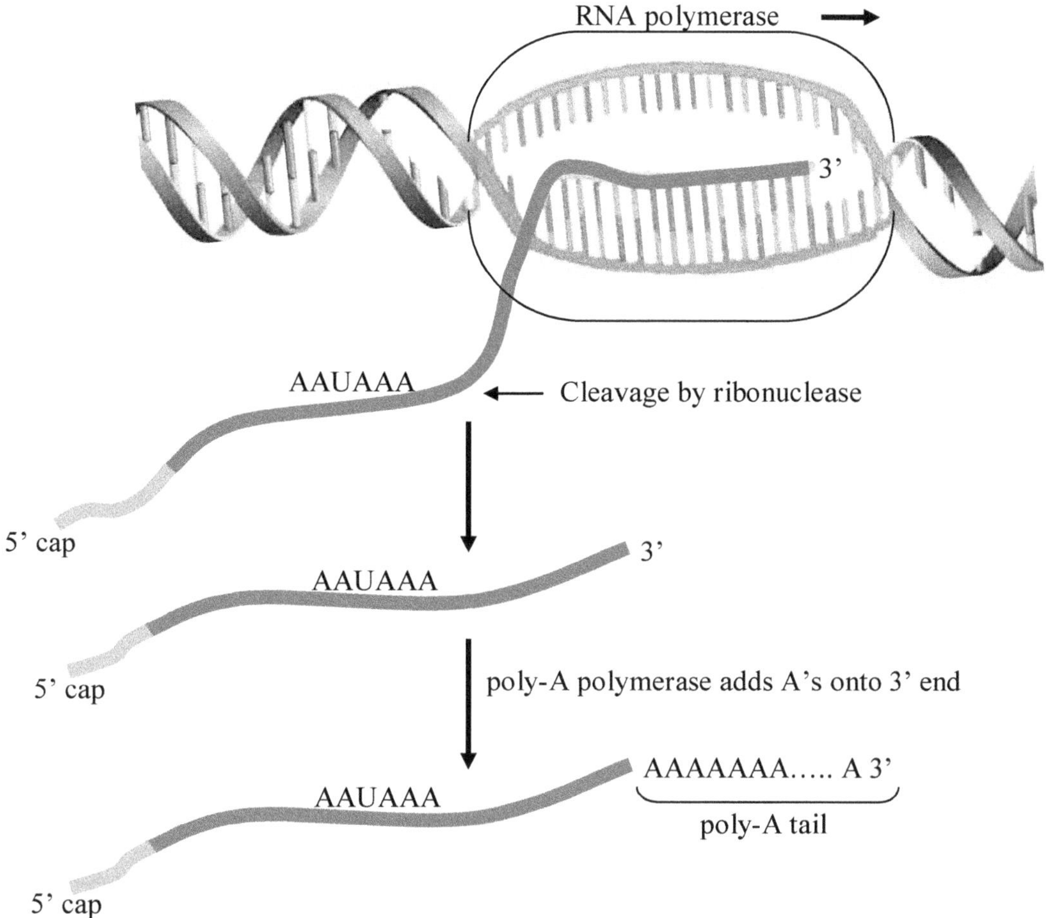

Figure 6.11: RNA Processing, which Adds a Tail to the 3' end of Eukaryotic mRNAs.

Depending upon their function, they are called splice donor and splice acceptor sites. The sequences at the two sites are insufficient to signal an intron's presence. The branch site is located 20 - 50 bases upstream of the acceptor site. The consensus sequence of the branch site is "CU (A/G) A(C/U)," where A is conserved in all genes. In over 60 per cent of cases, the exon sequence is (A/C) AG at the donor site and G at the acceptor site (**Figure 6.12**).

The introns are classified into the group –I and group –II, group –I introns are seen in nuclear, mitochondrial, and chloroplast rRNA genes, group-II are also seen in mitochondrial and chloroplast mRNA genes. Most of them have the self-splicing mechanism which means no protein factors are required for the intron to be accurately and efficiently spliced out.

Group, I introns require an external guanosine nucleotide that contains 3′-OH as a cofactor. The resultant 3′-OH at the 3′ end of the 5′ exon then attacks the 5′ nucleotide of the 3′ exon releasing the intron and covalently attaching the two exons together. The 3′ end of the 5′ exon is termed the splice donor site and the 5′ end of the 3′ exon is termed the splice acceptor site (**Figure 6.13**).

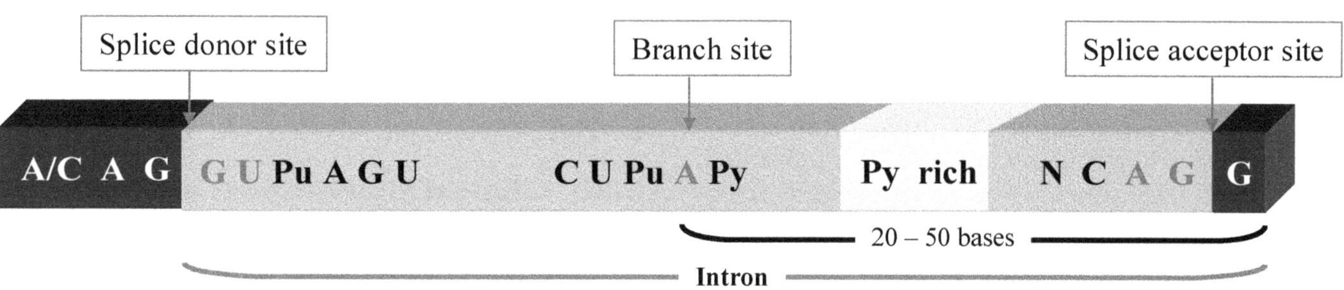

Figure 6.12: The Consensus Sequence for Splicing.
Pu = A or G; Py = C or U.

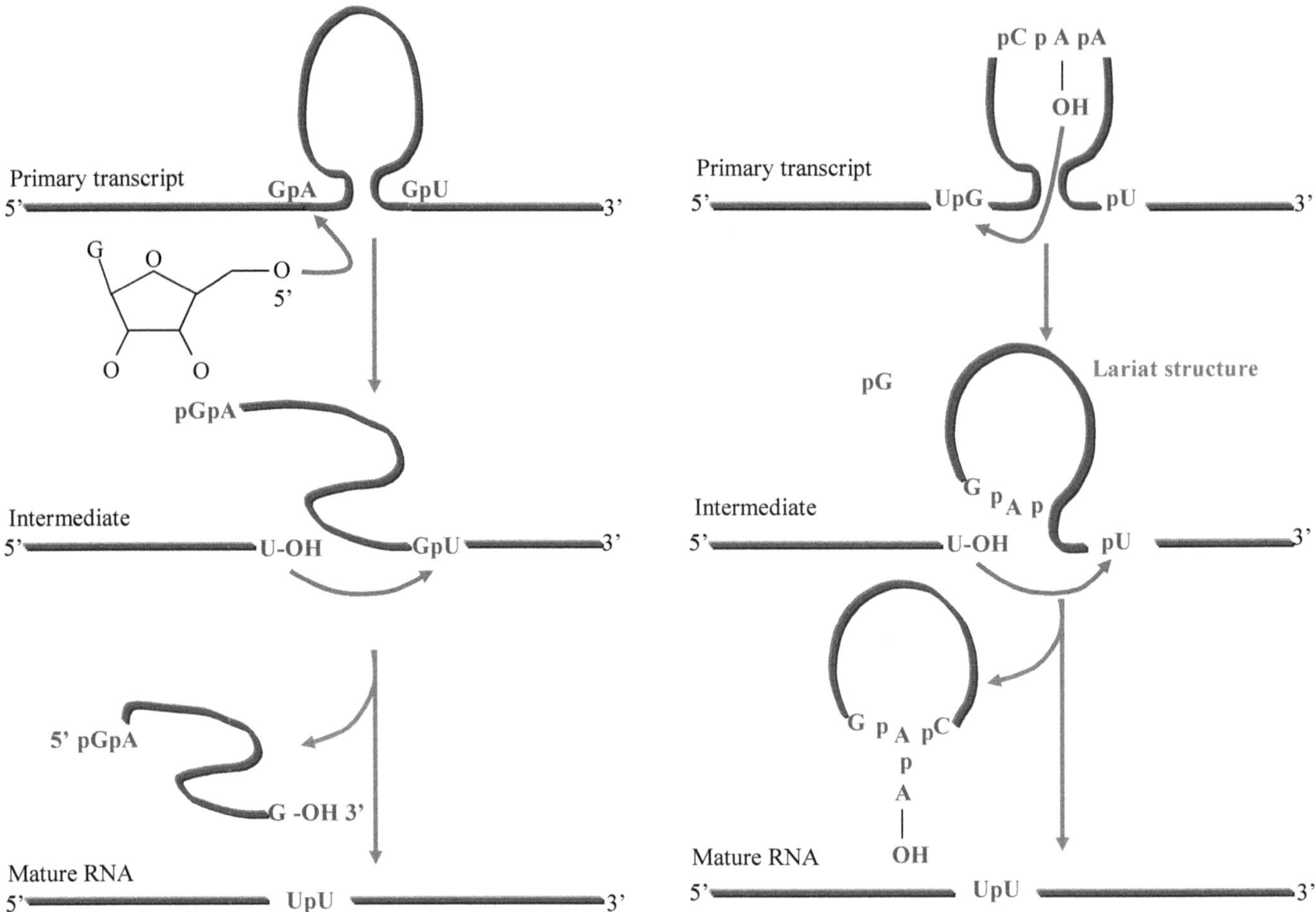

Figure 6.13: The Splice Donor Site and the 5' end of the 3' exon.

6.19.1 Group 1 Introns Splicing

Group II introns are spliced similarly, except that instead of an external nucleophile the 2′-OH of an adenine residue within the intron is the nucleophile. This residue attacks the 3′ nucleotide of the 5′ exon, forming an internal loop called a lariat structure. The 3′ end of the 5′ exon then attacks the 5′ end of the 3′ exon as in group I splicing, releasing the intron and covalently attaching the two exons.

6.19.2 Group II Introns Splicing

The third class of introns is also the most significant class found in nuclear mRNAs. This class of introns undergoes a splicing reaction like group II introns in that an internal lariat structure is formed. However, the splicing is catalyzed by specialized RNA-protein complexes called small nuclear ribonucleoprotein particles (snRNPs, pronounced "snurps"). The RNAs found in snRNPs are identified as U1, U2, U4, U5, and U6. The genes encoding these snRNAs are highly conserved in vertebrates and insects and are also found in yeast and slime molds, indicating their importance.

Analysis of many mRNA genes has led to the identification of highly conserved consensus sequences at the 5′ and 3′ ends of essentially all mRNA introns (**Figure 6.14**).

The U1 RNA has sequences that are complementary to sequences near the 5′ end of the intron. The binding of U1 RNA distinguishes the GU at the 5′ end of the intron from other randomly placed GU sequences in mRNAs. The U2 RNA also recognizes sequences in the intron, in this case near the 3′ end. The addition of U4, U5, and U6 RNAs forms a complex identified as the spliceosome that then removes the intron and joins the two exons together.

The fourth class of introns is those found in certain tRNAs. These introns are spliced by a specific splicing endonuclease that utilizes the energy of ATP hydrolysis to catalyze intron removal and ligation of the two exons together.

6.20 Clinical Significances of Alternative and Aberrant Splicing

As cells always conserve energy splicing is seen as a waste of energy. However, one must not

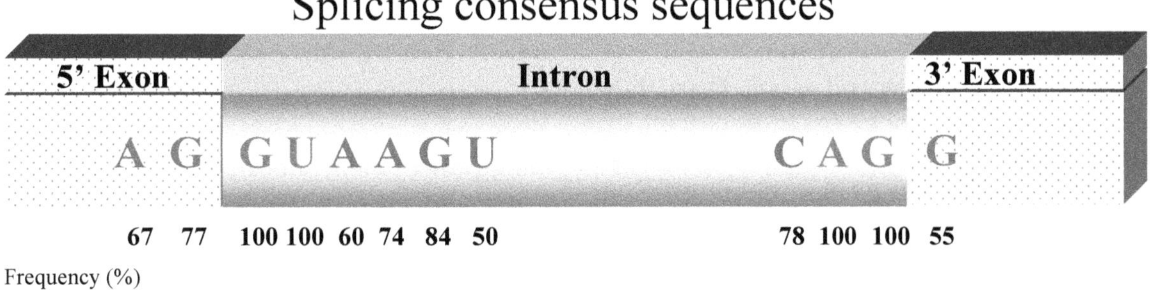

Figure 6.14: Conserved Consensus Sequences at 5' and 3' ends.

forget that introns save genetic makeup, which can get disturbed by outside influences like chemicals, radiation, *etc.* Alternate splicing is also provided.

This process of alternative splicing has been identified to occur in the primary transcripts from at least 40 different genes. Depending upon the site of transcription, the calcitonin gene yields an RNA that synthesizes calcitonin (thyroid) or calcitonin-gene-related peptide (CGRP, brain). The alternative splicing that occurs in the -tropomyosin transcript is even more complex. There are approximately eight different alternatively spliced tropomyosin mRNAs. If there are splicing abnormalities, these can lead to various disease states. Some of the examples are β Thalassemia due to globin primary transcript abnormalities.

Connective tissue diseases exhibit humoral autoantibodies that recognize cellular RNA-protein complexes.

Many eukaryotic genes are longer than the corresponding mRNA. mRNA of the human gene for dystrophin is shown in (**Figure 6.15**). Abnormalities in the dystrophin gene underlie the genetic disorder of Duchenne muscular dystrophy (DMD). The dystrophin gene is 2.5 million nucleotides – or 2500 kilobases (kb) – long, whereas the corresponding mRNA is approximately 14,000 nucleotides, or 14kb, in length. The gene contains DNA sequences that are not present in mature mRNA. Those regions of the gene that end up in the mature mRNA are scattered throughout the 2500 kb of DNA.

Sequences found in both a gene's DNA and the mature messenger RNA are called exons (for "expressed regions"). Some of the exons or parts of exons near the beginning and end of a gene may not carry amino acid specifying codons. These exons instead encode un-translated regions at the 5′ and 3′ ends of the mRNA that play an essential role in regulating the efficiency of mRNA translation into protein.

Exons also vary in size from 50 to a few thousand nucleotide pairs, while introns range from 50 to over 100,000 nucleotide pairs (in the DMD gene, the mean intron length is 35kb, but one intron is an impressive 400 km long). Introns do not appear and hence do not encode polypeptides. In humans, there are very few genes that do not contain introns.

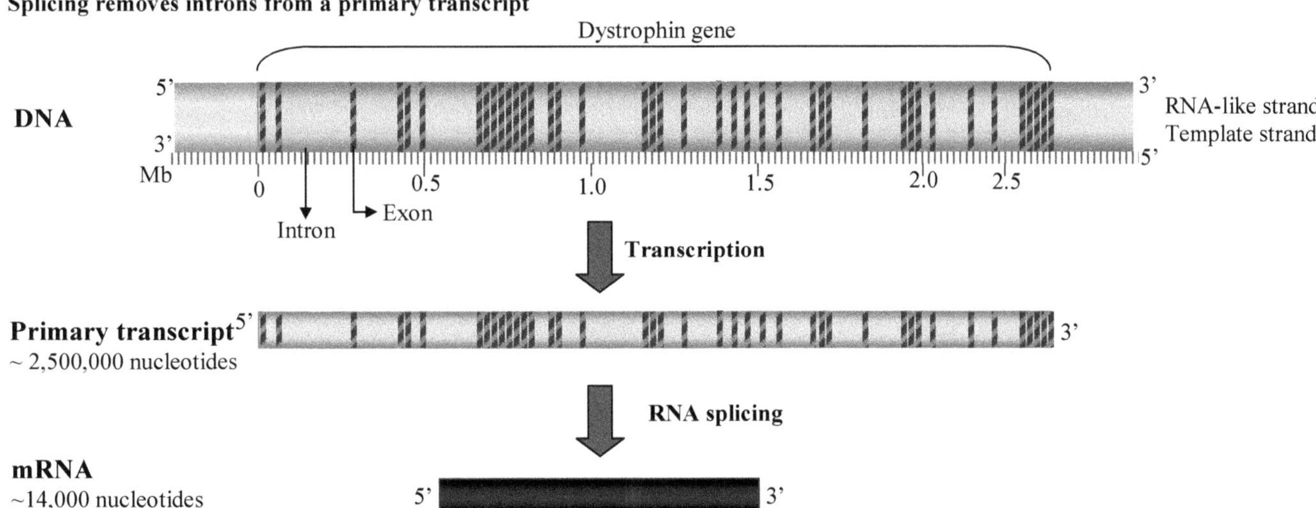

Figure 6.15: The Human Dystrophin Genes where RNA Splicing Plays a Critical Role.

How do cells make a mature mRNA from a gene whose coding sequences are interrupted by introns? The answer is that cells first make a primary transcript containing all a gene's introns and exons. Then, they remove the introns from the primary transcript by RNA splicing. This process deletes introns and joins together successive exons to form a mature mRNA consisting only of exons (**Figure 6.15**). Because the first and last exons of the primary transcript become the 5′ and 3′ ends of the mRNA, while all intervening introns are spliced out, a gene must have one more exon than introns. To construct the mature mRNA, splicing must be remarkably precise. For example, if an intron lies within a codon, splicing must remove the intron and reconstitute the codon without disrupting the reading frame of the mRNA.

Figure 6.16 illustrates the details of RNA splicing. Three types of short sequences are splice donors, splice acceptors, and branch sites. These help to ensure the specificity of splicing. During splicing, there occur two sequential cuts in the primary transcripts. The first cut is at the splice –donor site, at the 5′ end of the intron. After this first cut, the new 5′ end of the intron attaches, via a novel 2′ – 5′ phosphodiester bond, to an A at the branch site located within the intron, forming a so-called lariat structure. The second cut is at the splice –acceptor site, at the 3′ end of the intron; this cut removes the intron. The intron which is not required gets degraded, and the precise splicing of adjacent exons completes the process of intron removal.

Spliceosomes are required at the time of splicing. The spliceosome comprises four subunits known as small nuclear ribonucleoproteins, or snRNPs (pronounced "snurps"). Each snRNP contains one or two small nuclear RNAs (snRNAs) 100 – 300 nucleotides long. Certain snRNAs can base pair with the splice donor and splice acceptor sequences in the primary transcript, so these snRNAs are particularly important in bringing together the two exons that flank an intron.

The spliceosome structure is highly complex; some primary transcripts can splice themselves without a spliceosome or any other factor. These rare primary transcripts act as ribozymes.

The function of introns is not known. They may act like building blocks. They allow the shuffling of exons to make new genes, which appear to have played a vital role in the evolution of complex organisms. The exon –the as-module proposal is attractive because it is easy to understand the selective advantage of the potential for exon shuffling. However, there is no proof of introns functional role (**Figure 6.17**).

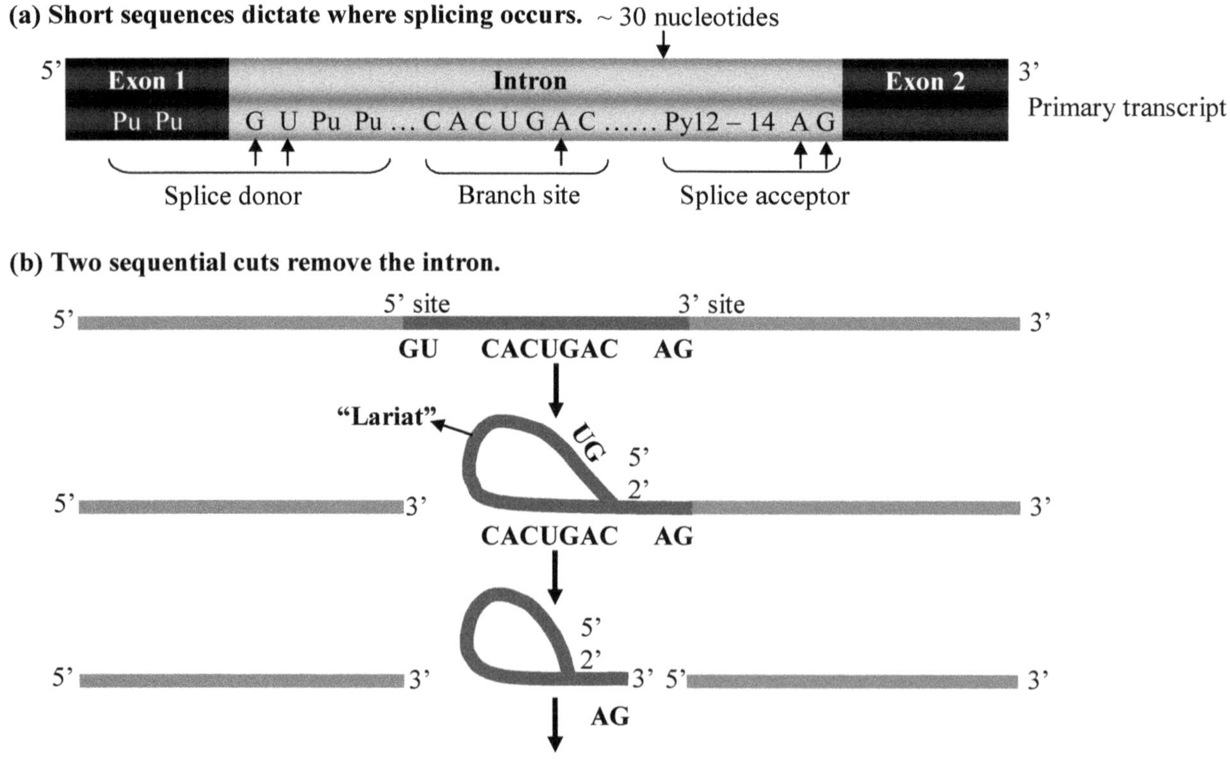

Figure 6.16: RNA Processing Splices out introns, Resulting in the Joining of Adjacent Exons.

Spliceosome components

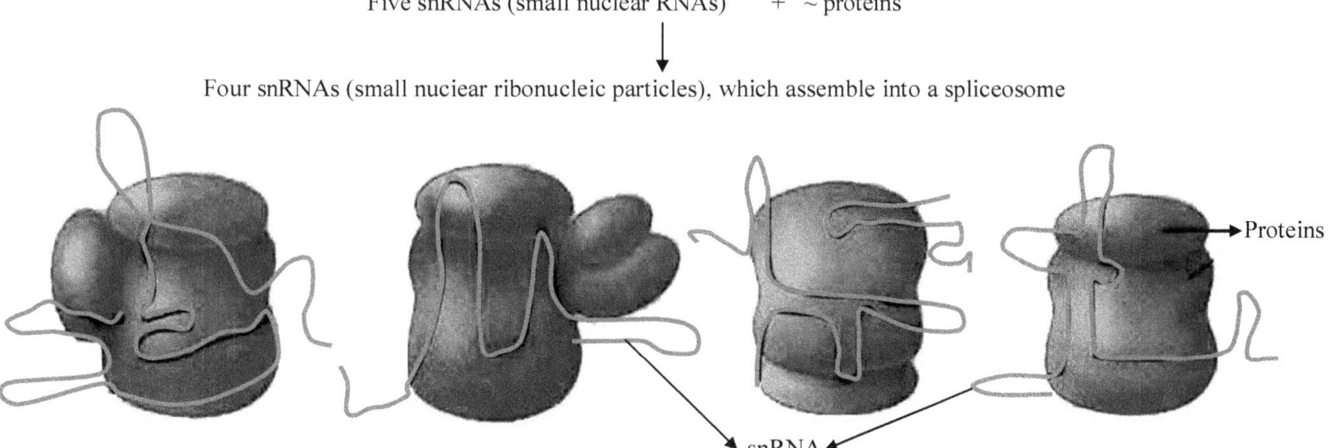

Figure 6.17: Splicing is Catalyzed by the Spliceosome.

6.21 Evolutionary Implications of Split Genes and RNA Splicing

Biological evolution depends upon catalyzing chemical reactions through RNAs. RNAs can store genetic information and can also catalyze the reaction they serve as genetic material, and they catalyze chemical reactions. RNA has the capability of replicating itself.

Introns are capable of self-splicing; their presence in the middle of genes would not have caused a problem. Sequences corresponding to introns would have excised themselves from the primary RNA transcripts. Over time, the catalytic portions of introns moved from the interior of protein-coding genes to separate locations in the genome. The RNAs encoded by these "new" genes continued participating in the splicing process. Over additional time, the RNAs encoded by these genes evolved into the snRNAs, and their catalytic activity became dependent on proteins. Together, the snRNAs and proteins evolved into the snRNP components of the spliceosome. Although the presence of introns may have created an added burden for cells because they had to remove these intervening sequences from their transcripts, introns are not without their virtues. Many primary transcripts can be processed by two or more pathways so that a sequence that acts as an intron in one pathway becomes an exon in an alternate pathway. As a result of this process, called alternate splicing, the same gene can code for more than one polypeptide.

It was also discovered that certain RNAs-namely, the snoRNAs required for rRNA processing, are encoded by introns rather than exons. Most of these snoRNAs are contained within the introns of genes that code for specific polypeptides involved in ribosome assembly and function. When these genes are transcribed, and the introns are removed from the primary transcripts, rather than being discarded, some of these introns are processed into snoRNAs. Several genes have been discovered in which the role of introns and exons is essentially reversed. These genes are transcribed into primary transcripts whose introns are processed into snoRNAs, while the exons are degraded without ever giving rise to an mRNA. The presence of introns is also thought to have had a major impact on biological evolution. Various proteins have homologous parts. These are encoded by genes that are almost certainly composites made up of parts of other genes. The movement of genetic "modules" among different genes-a process called exon shuffling-is greatly facilitated by the presence of introns, which act like inert spacer elements between exons. Genetic rearrangements require breaks in DNA molecules, which can occur within introns without introducing mutations that might impair the organism. Over long periods, exons can be shuffled independently in various ways, allowing a large number of combinations in search of new and helpful coding sequences. As a result of exon shuffling, evolution need not occur only by the slow accumulation of point mutations. Still, it might also move ahead by "quantum leaps" with new proteins appearing within a single generation.

Transcription means the synthesis of new proteins. DNA is transcribed to RNA to produce mRNA (messenger = Ribo Nucleic Acid), rRNA (ribosomal RNA), or tRNA (transport RNA). This happens in the nucleus by means of enzymatic complexes produced by specific genes. The RNA is further transported outside the nucleus to the cytoplasm, where it becomes active in the translation (the actual synthesis of proteins). During transcription, the chromosomes are locally de-condensed so that the genes present at this site can be read. Two types of genes are involved: 1) so-called "housekeeping genes" that are continually read and encoded for products that are necessary for the metabolism and existence of a cell, and 2) genes that encode for specific products of importance for cell differentiation. The activity of these genes (yes or no in the state of transcription) is often determined by a confluence of external factors and internal cellular signals, a longer time, and increasing the efficiency of the initial translation steps.

7

The Genetic Code

The code consists of the sequence of nucleotides in the stretch of DNA molecule. A specific part of the code is transcribed into a sequence of nucleotides in the messenger RNA (mRNA). The mRNA moves from the nucleus to the cytoplasm, translating it into protein by the ribosomes. A codon consists of three uninterrupted nucleotides in the mRNA codes for each amino acid in the protein. An alteration in the code may cause the incorrect sequence of the amino acids in the protein; this alteration is known as the mutation. Among many scientists, three scientists, Marshall Warren Nirenberg, Har Gobind Khorana, and Robert William Holley, were interested to know how nucleotides code the DNA and, further, how these can be translated into the 20-letter alphabet of amino acids, the building blocks that makeup proteins. They tagged the eukaryotic cell amino acids with radioactivity. It was found that the protein synthesis takes place in the cytoplasm, and the radioactive amino acids are incorporated in the polypeptides. From this discovery, it was clear that the survival of an intermediate molecule (RNA) made in the nucleus is competent enough to send DNA sequence information to the cytoplasm, where it can direct protein synthesis. RNA has an excellent potential for base pairing with a strand of DNA. Hence, the cellular machinery copies a DNA strand into a complementary RNA strand. It is just like DNA-to-DNA copying occurs during DNA replication. Subsequent studies in eukaryotes on incorporating radioactive uracil (a base found only in RNA) into the molecules of RNA showed that although the molecules are synthesized in the nucleus, at least some of them move to the cytoplasm. mRNA is a molecule that goes into the cytoplasm (**Figure 7.1**). They stay in the nucleus from the transcription of DNA sequence information through base pairing, go into the cytoplasm after processing, and then resolve the proper order of amino acids during protein synthesis.

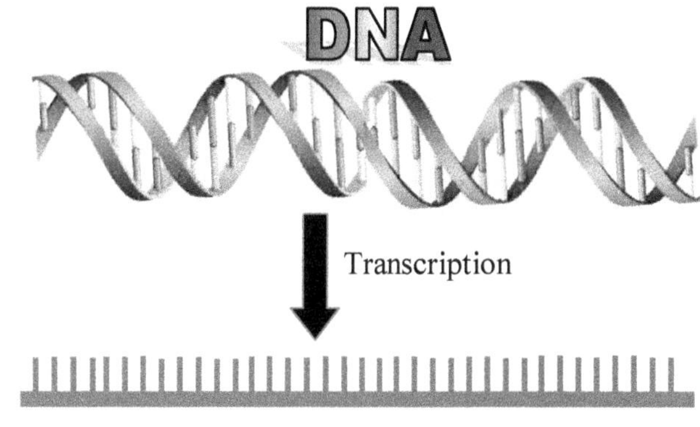

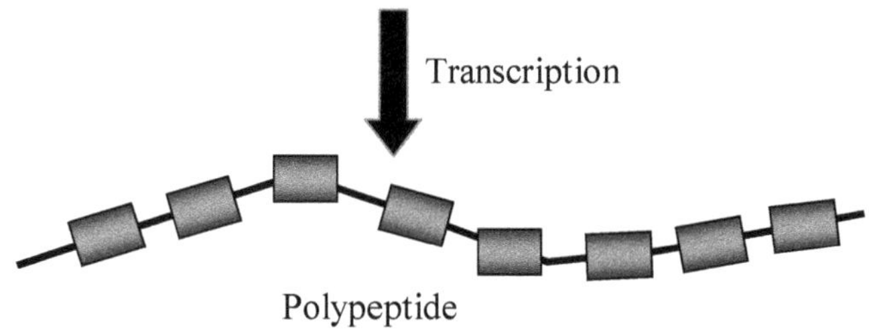

Figure 7.1: Gene Expression: Constitute DNA to mRNA and mRNA to Polypeptides.

It was difficult to know which RNA would use which codon during translation. Two critical discoveries showed the significance of the genetic code. In the first experiment, cellular extracts with mRNA were obtained and polypeptides were synthesized in a test tube. They labeled these extracts as "in vitro translational systems." In the second experiment, the synthesis of artificial mRNAs was carried out which contained only a few codons of known composition. When added to in vitro translational systems, these simple, synthetic mRNAs directed the formation of very simple polypeptides.

In 1961, Marshall Nirenberg and Heinrich Matthaei added a synthetic poly-U, which was 5′. UUUUUUUUUUUU. 3′ mRNA to a cell-free translational system derived from *E. coli.* With the poly-U mRNA, phenylalanine (Phe) was the only amino acid incorporated into the resulting polypeptide (**Figure 7.2a**). Because UUU is the only possible triplet in poly U, UUU must be a codon for phenylalanine. Similarly, Nirenberg and Matthaei showed that CCC encodes proline (Pro), AAA is a codon for lysine (Lys), and GGG encodes for glycine (Gly) (**Figure 7.2b**).

Dr. Har Gobind Khorana constructed mRNAs with repeating dinucleotides, such as poly-U (5′. UCUCUCUC. 3′), repeating trinucleotides, such as poly-UUC, and repeating tetra nucleotides, such as poly-UAUC, and used them to direct the synthesis of slightly more complex polypeptides. As **Figure 7.2b** shows, the results limited the coding possibilities; however, some ambiguities remained. For example, poly-UC encodes the polypeptide N. Ser-Leu-Ser-Leu-Ser-Leu. C where serine and leucine alternate with each other. The mRNA contains only two different codons (5′ UCU 3′ and 5′ CUC 3′); it is not obvious which corresponds to serine and which to leucine.

With this knowledge, scientists hypothesized that one would know not only how RNA translates messages from DNA to build proteins but also how one would be able to read the entire genetic code of living organisms. Nirenberg's team extended its experiments with synthetic RNA. During this period, they discovered that AAA (three adenosines) was the code word for the amino acid lysine, and CCC (three cytosines) was the code word for proline. GGG (three guanines) did not work as a messenger at all. They also discovered that by replacing one or two units of a triplet with other nucleotides, they could direct the production of other amino acids. They found, for example, that a synthetic RNA composed of one

(a) Poly-U mRNA encodes polyphenylalanine.

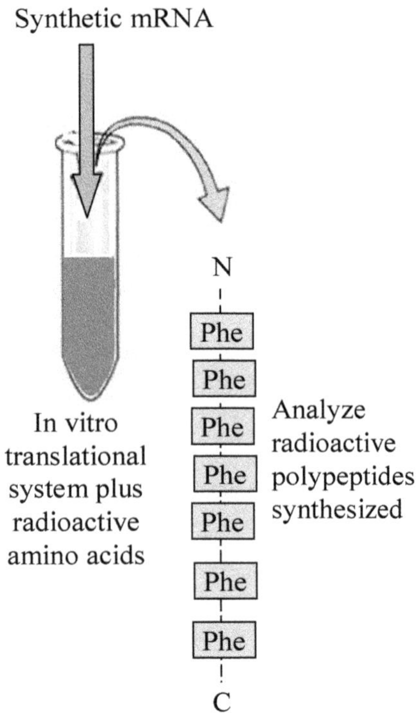

Synthetic mRNA

N

Phe

Phe

Phe

Phe

Phe

Phe

Phe

C

In vitro translational system plus radioactive amino acids

Analyze radioactive polypeptides synthesized

(b) Analyzing the coding possibilities.

Synthetic mRNA		Polypeptides synthesized
Poly-U	UUUU	Phe-Phe-Phe.....
Poly-C	CCCC	Pro-Pro-Pro ....
Poly-A	AAAA	Lys-Lys-Lys ....
Poly-G	GGGG	Gly-Gly-Gly ...
Repeating dinucleotides		**Polypeptides with alternating amino acids**
Poly-UC	UCUC	Ser-Leu-Ser-Leu ...
Poly-AG	AGAG ...	Arg-Glu-Arg-Glu ...
Poly-UG	UGUG	Cys-Val-Cys-Val ...
Poly-AC	ACAC ...	Thr-His-Thr-His ...
Repeating trinucleotides		**Three polypeptides each with one amino acid**
Poly-UUC	UUCUUCUUC ...	Phe-Phe .... and Ser-Ser ... and Leu-Leu ..
Poly-AAG	AAGAAGAAG ...	Lys-Lys ... and Arg-Arg ... and Glu-Glu ...
Poly-UUG	UUGUUGUUG ...	Leu-Leu ... and Cys-Cys ... and Val-Val...
Poly-UAC	UACUACUAC ...	Tyr-Tyr ... and Thr-Thr ... and Leu-Leu ...
Repeating tetranucleotides		**Polypeptides with repeating units of four amino acids**
Poly-UAUC	UAUCUAUC	Tyr-Leu-Ser-Ile-Tyr-Leu-Ser-Ile ...
Poly-UUAC	UUACUUAC	Leu-Leu-Thr-Tyr-Leu-Leu-Thr-Tyr ...
Poly-GUAA	GUAAGUAA ...	None
Poly-GAUA	GAUAGAUA ...	None

Figure 7.2: Different Ways Coding can take place.

unit of guanine (G) added to two units of uracil (UU) directed that valine be added to a developing amino acid chain. In Nirenberg's shorthand method, the code word for valine was GUU. In the late 1950s, the biochemist Sydney Brenner coined the term "codon" to describe the fundamental units engaged in protein synthesis, even though the units had yet to be fully determined. Francis Crick popularized the term in 1959. After 1962, Nirenberg began to use "codon" to characterize the three-letter RNA code words. With one of each of the four nucleotides occupying a place in a three-letter codon arrangement, Nirenberg quickly deduced that there were 64 possible combinations (4 x 4 x 4) of three-letter codons. In 1964 and 1965, Nirenberg's postdoctoral researcher, Philip Leder, developed a sophisticated filtration machine that helped the team determine the order of the nucleotides in the codons. This development speeded up the process of assigning code words to amino acids. By 1966, Nirenberg announced that he had deciphered the sixty-four RNA codons for all twenty amino acids. This remarkable personal and scientific accomplishment held great significance, not only for Nirenberg but also for the history of modern science. At a 1966 conference, Geoffrey Zubay, a professor of biology at Columbia University, remarked that Francis [Crick] predicted the entire code would be solved in [1961]. It has taken a few years longer than that and the pace; in a 1967 talk, Nirenberg characterized messenger RNA as a "robot" whose purpose was to obey the commands of DNA and carry out vital genetic instructions. George and his wife Muriel Beadle wrote in 1966, "The deciphering of the DNA code has revealed our possession of a language much older than hieroglyphics, a language as old as life itself, a language that is the most living language of all even if its letters are invisible and its words are buried deep in the cells of our bodies." In

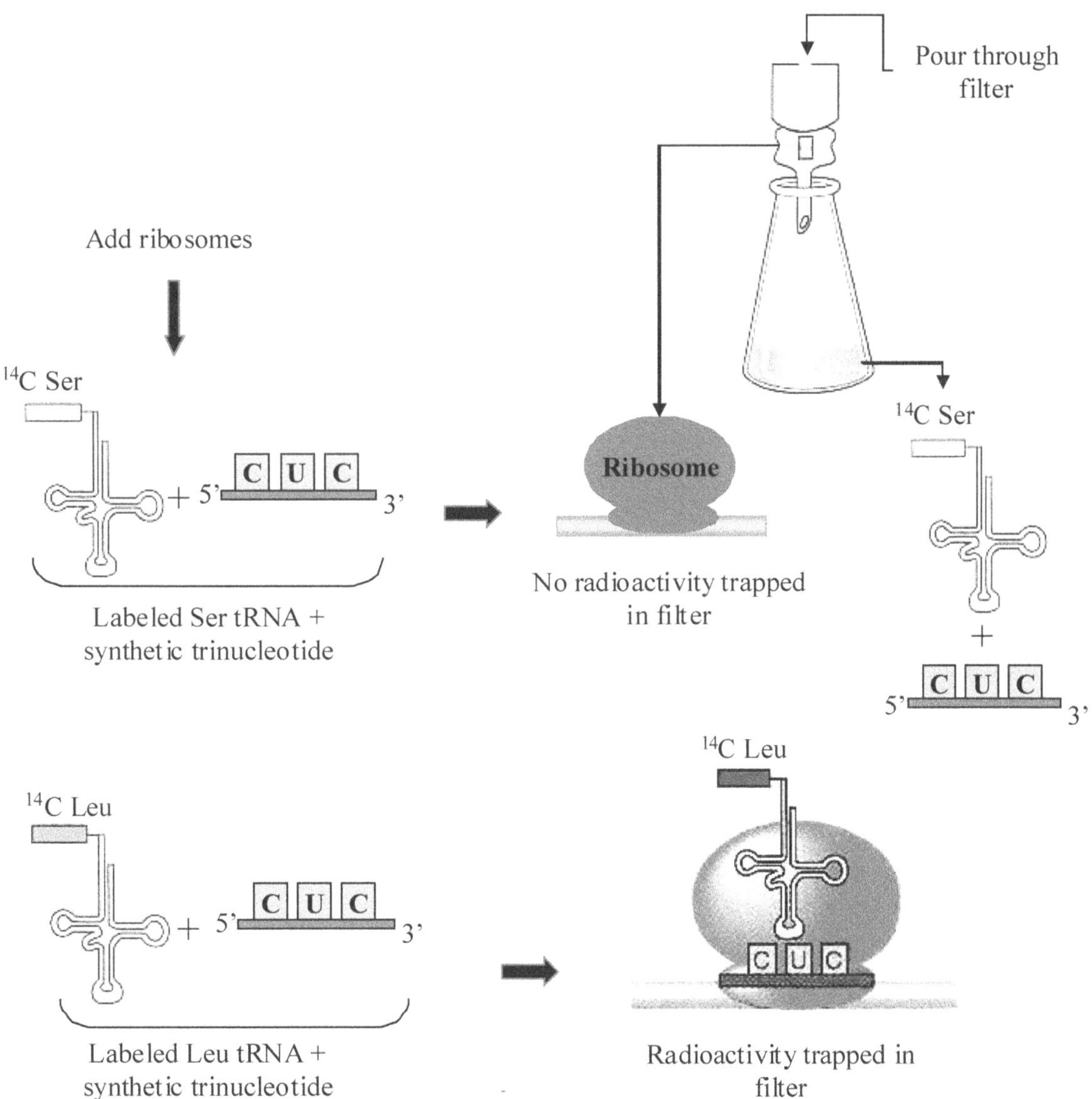

Figure 7.3: Cracking the Genetic Code with mini-mRNAs.
The 5'-to-3' direction in mRNA corresponds to the N-Terminal-to-C-Terminal direction in the polypeptide.

October 1968, Nirenberg received the news that he had won the Nobel Prize in Medicine or Physiology, an honor he shared with Robert W. Holley and Har Gobind Khorana for their collective efforts in deciphering different aspects of the genetic code. They filtered the mixture of synthetic mRNAs and translational systems containing a tRNA attached radioactively labeled amino acid (**Figure 7.3**). tRNAs have a property that if it is carrying an amino acid, it can pass through the filter paper.

If a tRNA carrying an amino acid binds to a ribosome, it is not able to pass through the filter. Still, it will get stuck to the filter because it is larger and more so is comprised of a complex of the ribosome, amino-acid-carrying tRNA and also small mRNA. Nirenberg and Leder used this approach to see which small mRNA caused the entrapment of which radioactively labeled amino acid in the filter; they knew from Khorana's earlier work that CUC encoded either serine or leucine. When they added the synthetic triplet CUC to an in vitro system where the radioactive amino acid was serine, this tRNA-attached amino acid passed through the filter, and the filter thus emitted no radiation. But when they added the same triplet to a system where the radioactive amino acid was leucine, the filter lit up with radioactivity, indicating that the radioactively tagged leucine was attached to a tRNA got bound to the ribosome mRNA complex and hence was present on the filter paper. CUC thus encodes leucine and not serine.

Nirenberg and Leder used this technique to determine all the codon-amino acid correspondence in the genetic code. Next, they added six-nucleotide-long 5′ AAAUUU 3′ to an in vitro translational system; the product was N-Lys-Phe-C, but no N-Phe-Lys-C. Since AAA is the codon for lysine and UUU is the codon for phenylalanine, the codon near the 5′ end of the mRNA encoded the amino acid in proximity to the N terminus of the corresponding polypeptide. Similarly, the codon close to the 3′ end of the mRNA encoded the amino acid nearest to the C terminus of the resulting polypeptide.

The next step was to understand how the polarities of the macromolecules participating in gene expression relate to each other, we know that the gene is a segment of a DNA which is the double helix, but one of the two strands serves as a template for the mRNA is the template strand. The other strand is the RNA-like strand because it has the same polarity and sequence (**Figure 7.4**).

7.1 Nonsense Codons Cause Termination of the Polypeptide Chain

Three triplets-UAA, UAG, and UGA-do were found to not communicate to any amino acids. When these codons come, they stop the translation. As an example of how investigators established this fact, let us consider the case of poly-GUAA. This mRNA will not generate a long polypeptide because, in all reading frames, it has the stop codon UAA. A tRNA represents none of the termination codons. They function entirely differently from other codons and are recognized directly by protein factors. The stop codons that stop translation are called nonsense codons. The UAA is named the ocher codon, UAG is also called an amber codon, and UGA is the opal codon. The basis of this terminology is based on the last name of one of the early investigators, Bernstein-which means "amber" in German; ocher and opal derive from their similarity with amber as semiprecious materials. During this process, 64 possible codons were formed in **Figure 7.5**. The decoder chart shows a sequence for the 64 possible codons in the mRNA. The codon assignments in Figure 7.5 are universal, regardless of the source of a cell. The primary exceptions to the universality of the genetic code occur in the codons of mitochondrial mRNAs. For example, in human mitochondria, UGA

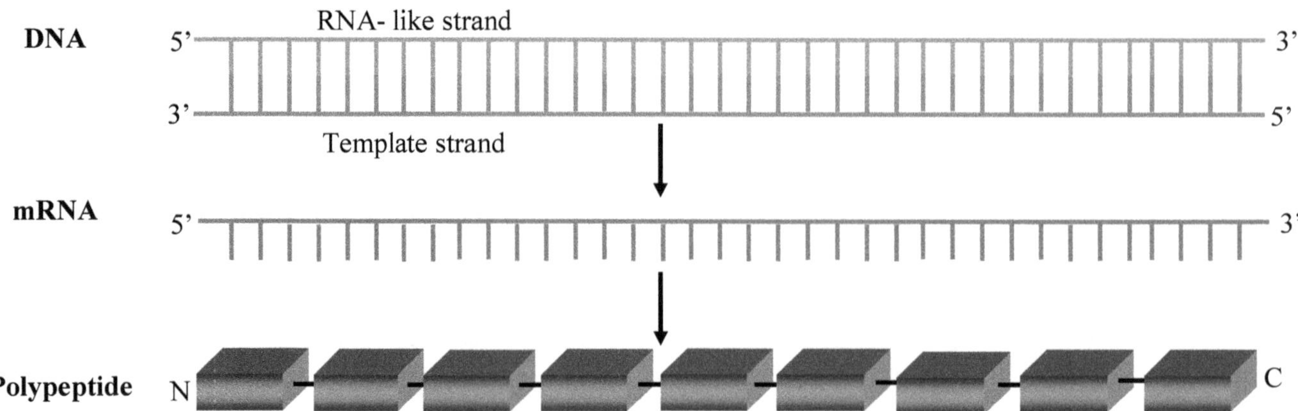

Figure 7.4: mRNA and Polypeptide Synthesis.

First letter	Second letter: U	Second letter: C	Second letter: A	Second letter: G	Third letter
U	UUU Phenylalanine	UCU Serine	UAU Tyrosine	UGU Cysteine	U
U	UUC Phenylalanine	UCC Serine	UAC Tyrosine	UGC Cysteine	C
U	UUA Leucine	UCA Serine	UAA Stop	UGA Stop	A
U	UUG Leucine	UCG Serine	UAG Stop	UGG Tryptophan	G
C	CUU Leucine	CCU Proline	CAU Histidine	CGU Arginine	U
C	CUC Leucine	CCC Proline	CAC Histidine	CGC Arginine	C
C	CUA Leucine	CCA Proline	CAA Glutamine	CGA Arginine	A
C	CUG Leucine	CCG Proline	CAG Glutamine	CGG Arginine	G
A	AUU Isoleucine	ACU Threonine	AAU Asparagine	AGU Serine	U
A	AUC Isoleucine	ACC Threonine	AAC Asparagine	AGC Serine	C
A	AUA Isoleucine	ACA Threonine	AAA Lysine	AGA Arginine	A
A	AUG (start) Methionine	ACG Threonine	AAG Lysine	AGG Arginine	G
G	GUU Valine	GCU Alanine	GAU Aspartic acid	GGU Glycine	U
G	GUC Valine	GCC Alanine	GAC Aspartic acid	GGC Glycine	C
G	GUA Valine	GCA Alanine	GAA Glutamic acid	GGA Glycine	A
G	GUG Valine	GCG Alanine	GAG Glutamic acid	GGG Glycine	G

Figure 7.5: Genetic Code Depicted as a Universal Decoder Chart, which Shows 64 Possible mRNA and Codons that Code the Corresponding Amino Acid.

is read as tryptophan rather than stop codon, AUA is read as methionine rather than isoleucine, and AGA and AGG are read as a stop rather than arginine.

The codon chart in **Figure 7.5** indicates that the amino acid assignments are non-random; the codon boxes for a specific amino acid tend to cluster within a particular portion of the chart. Clustering revealed the similarity in a codon for the same amino acid. Due to this similarity in codon sequence, spontaneous mutations causing single base changes in a gene often will not change the corresponding protein's amino acid sequence. The "safeguard" aspect of the code goes beyond its degeneracy. Codon assignments are such that similar amino acids tend to be specified by similar codons. Various hydrophobic amino acids (depicted as brown boxes in **Figure 7.5**) are clustered in the first two columns of the chart. This results in a mutation that further causes base substitution in one of these codons, which is likely to substitute one hydrophobic residue for another. Also, the most significant similarities between amino acid-related codons occur in the first two nucleotides of the triplet. In contrast, the more similarities between amino acid-related codons occur in the first two nucleotides of the triplet, the more significant variability is seen in the third nucleotide. For example, glycine is encoded by four codons, all of which begin with the nucleotide GG.

7.2 How to Verify the Code

Most mutagens change into a single nucleotide in a codon. As a result, most missense mutations that change the identity of a single amino acid should be single-nucleotide substitutions, and analyses of these substitutions should conform to the code. There are two trp- trp-auxotrophic mutations in the E. coli tryptophan synthetase gene that produced two different amino acids (arginine, or Arg, and glutamic acid, or Glu) at the same position-amino acid 211-in the polypeptide chain (**Figure 7.6a**). According to the code, these mutations could have resulted from single-base changes in the GGA codon that usually inserts glycine (Gly) at position 211 (**Figure 7.6b**).

7.3 Experimental Pieces of Evidence for the Function of Genetic Code

It has been proved that there exist different properties of the genetic code. The genetic code is non-overlapping; three bases encode an amino acid. These triplets are termed codons. The code is read from a fixed starting point and continues to the end of the coding sequence. We know this because a single frameshift mutation anywhere in the coding sequence alters the codon alignment for the rest of the sequence, and the code is degenerate in that some amino acids are specified by more than one codon.

(a) Altered amino acids in trp- mutations and trp+ revertants

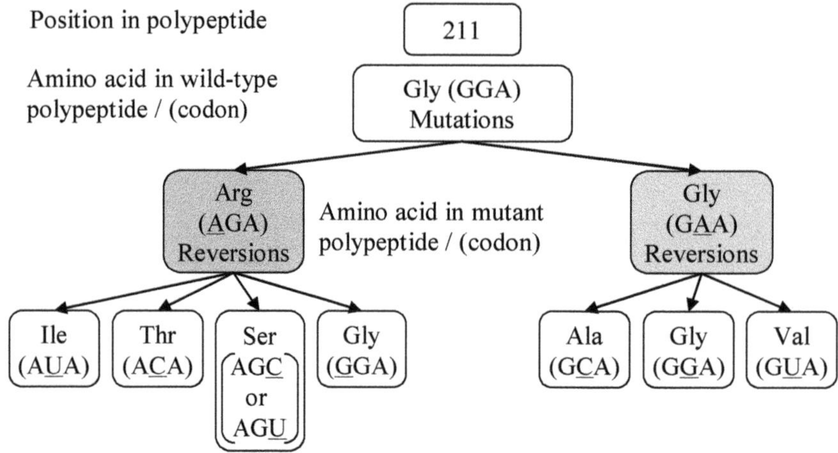

Upon single-base substitutions the amino-acid substitutions of *trp-* mutations and *trp+* revertants are produced

(b) Amino acid alterations that accompany intragenic suppression

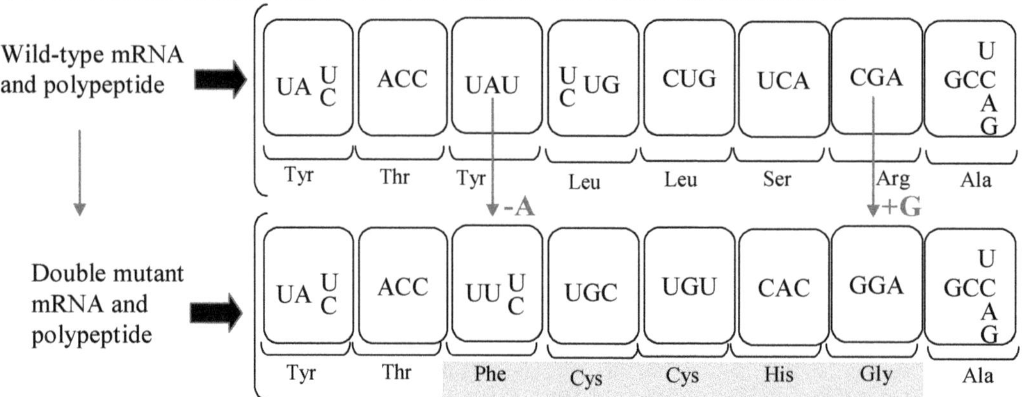

The genetic code predicts the amino-acid alterations *(yellow)* that would arise from single-base-pair deletions and suppressing insertions.

Figure 7.6: Verification of the Genetic Code.

Although it may not be necessary to present experimental evidence which proved the validity of these properties, it may be useful to explain the meaning of these properties of the genetic code listed above.

7.4 The Code is Triplet

As earlier outlined, singlet and doublet codes are not enough to code for 20 amino acids; it was pointed out that triplet code is the minimum requirement. But it could be a quadruplet code or of a higher order. As pointed out above, in a triplet code of 64 codons, there is an excess of 44 codons; therefore, more than one codon is present for the same amino acid. This excess will be still more significant if more than three-letter words are used. In the quadruplet code, there will be 4x4x4x4 = 256 possible words.

If the groups in the code contain one or two nucleotides, it would have 4 + 16 = 20 groups and should account for all the amino acids, but there would be nothing left over to signify the pause needed to denote where one group finishes and the next starts. Groups of three nucleotides in a row would provide 43 = 64 different triplet combinations, which are sufficient to code for all the amino acids. If the code consisted of doublets and triplets, a signal denoting pause would again be required. But a triplets-only code would need any symbol for the "pause" if the mechanism for counting to three and distinguishing among successive triplets were very reliable.

Although this reasoning- explaining the unknown in terms of the known by looking for the simplest possible explanation- generates a theory, it does not prove it. Each nucleotide triplet is called a codon. Each codon, designated by the bases defining its three nucleotides, specifies one amino acid. For example, GAA is a codon for glutamic acid (Glu), and GUU is a codon for valine (Val). Because the code comes into play only during the translation part of gene expression, that is, during the decoding of messenger RNA to the polypeptide, geneticists usually present the code in the RNA dialect of A, G, C, and U, as depicted in **Figure 7.7**.

Although techniques for determining both nucleotide and amino-acid sequence are available today, these were not available when researchers were trying to crack the genetic code in the 1950s and 1960s. They established a polypeptide's amino-acid sequence but not the nucleotide sequence of DNA or RNA. Because of their inability to read the nucleotide sequence, they used various genetic and biochemical techniques to learn about the code. They began by examining how different mutations in a single gene affected the amino-acid sequence of the gene's polypeptide product. In this way, they could use the specific mutations to understand the normal, *i.e.*, the relationship between the genes and polypeptides.

Hence, messenger RNA is the segment of nucleotide that translates into protein. This demonstrates that there is a similarity between the information contained in the mRNA, which is composed of a specific sequence of nucleotides, and the information contained in polypeptides, which is composed of a specific sequence of amino acids. In other words, mRNA and the polypeptides are collinear, which means that the 5′ → 3′ linear sequence of nucleotides in a segment of mRNA corresponds directly to the amino–carboxyl linear sequence of amino acids in the polypeptides.

Proteins are complicated with three-dimensional structures. If unfolded and stretched out from the N terminus to the C terminus, proteins have a one-dimensional, linear structure-specific sequence of amino acids. Suppose the information in a gene and its corresponding protein are collinear. In that case, the consecutive order of bases in the DNA from the beginning to the end will stipulate the consecutive order of amino acids from one end to the other of the outstretched protein. This implies that a gene and its protein product have definite polarities with an invariant relation to each other. Collinearity was confirmed by examining the results of E. coli by Yanofsky in 1960.

7.5 The Code is Non-overlapping

Non-overlapping code means that a base in an mRNA is not used for two different codons. An overlapping code can mean coding for four amino acids from six bases. In actual practice, six bases code for not more than two amino acids. However, in overlapping genes described in the organization of genetic material, three split genes, overlapping genes, and pseudogenes, it is shown that the same base can

Second letter

First letter	U		C		A		G		Third letter
U	UUU	Phenylalanine	UCU	Serine	UAU	Tyrosine	UGU	Cysteine	U
	UUC	Phenylalanine	UCC	Serine	UAC	Tyrosine	UGC	Cysteine	C
	UUA	Leucine	UCA	Serine	UAA	Stop	UGA	Stop	A
	UUG	Leucine	UCG	Serine	UAG	Stop	UGG	Tryptophan	G
C	CUU	Leucine	CCU	Proline	CAU	Histidine	CGU	Arginine	U
	CUC	Leucine	CCC	Proline	CAC	Histidine	CGC	Arginine	C
	CUA	Leucine	CCA	Proline	CAA	Glutamine	CGA	Arginine	A
	CUG	Leucine	CCG	Proline	CAG	Glutamine	CGG	Arginine	G
A	AUU	Isoleucine	ACU	Threonine	AAU	Asparagine	AGU	Serine	U
	AUC	Isoleucine	ACC	Threonine	AAC	Asparagine	AGC	Serine	C
	AUA	Isoleucine	ACA	Threonine	AAA	Lysine	AGA	Arginine	A
	AUG	(start) Methionine	ACG	Threonine	AAG	Lysine	AGG	Arginine	G
G	GUU	Valine	GCU	Alanine	GAU	Aspartic acid	GGU	Glycine	U
	GUC	Valine	GCC	Alanine	GAC	Aspartic acid	GGC	Glycine	C
	GUA	Valine	GCA	Alanine	GAA	Glutamic acid	GGA	Glycine	A
	GUG	Valine	GCG	Alanine	GAG	Glutamic acid	GGG	Glycine	G

Figure 7.7: Sixty-one Codons Represent the 20 Amino Acids, and 3 Codons Signify Stop Codons. In total there are 64 codons.

be used for different codons, but only to varying occasions in time and/or space, so that the same base cannot be used for two different codons during the synthesis of the same protein. This is precisely what we mean by non-overlapping code.

Table 7.1: The Discrepancy between an Overlapping and Non-overlapping Genetic Code

Base Sequence	*Codons*	
Original sequence.... AGCATCG...	Overlapping code.... AGC, GCA, CAT, ATC, TCG, ...	Non-overlapping code...., AGC, ATC, ...
Sequence after single-base substitution.... AGAATCG ...	Overlapping code...., AGA, GAA, AAT, ATC, TCG, ...	Non-overlapping code...., AGA, ATC, ...

Mutant proteins, such as the mutant hemoglobin responsible for sickle cell anemia, *etc.*, have provided much information about genetic code's overlapping or non-overlapping nature. The mutant protein contained a single amino acid substitution in sickle cell anemia. If the code were overlapping, a change in one of the base pairs in the DNA would be expected to affect three consecutive codons (**Table 7.1**), resulting in three successive amino acids in the equivalent polypeptide. If, however, the code is non-overlapping and each nucleotide is part of only one codon, then only one amino acid replacement would be expected. This indicated that the code was non-overlapping.

7.6 The Genetic Code is a Triplet

7.6.1 A Codon is Composed of more than one Nucleotide

It has been demonstrated that different point mutations (changes in only one nucleotide pair) may affect the same amino acid, as shown in **Figure 7.8,** mutation 23 changed the glycine (Gly) at position 211 of the wild-type polypeptide chain to arginine (Arg). In contrast, mutation 46 yielded the same position of glutamic acid (Glu). In another example, mutation 78 changed the glycine at position 234 to cysteine (Cys), while mutation 58 produced aspartic acid (Asp) at the same position. In both cases, Yanofsky showed recombination could occur between the two mutations that changed the identity of the same amino acid, and such recombination would produce a wild-type tryptophan synthetase gene (**Figure 7.8**).

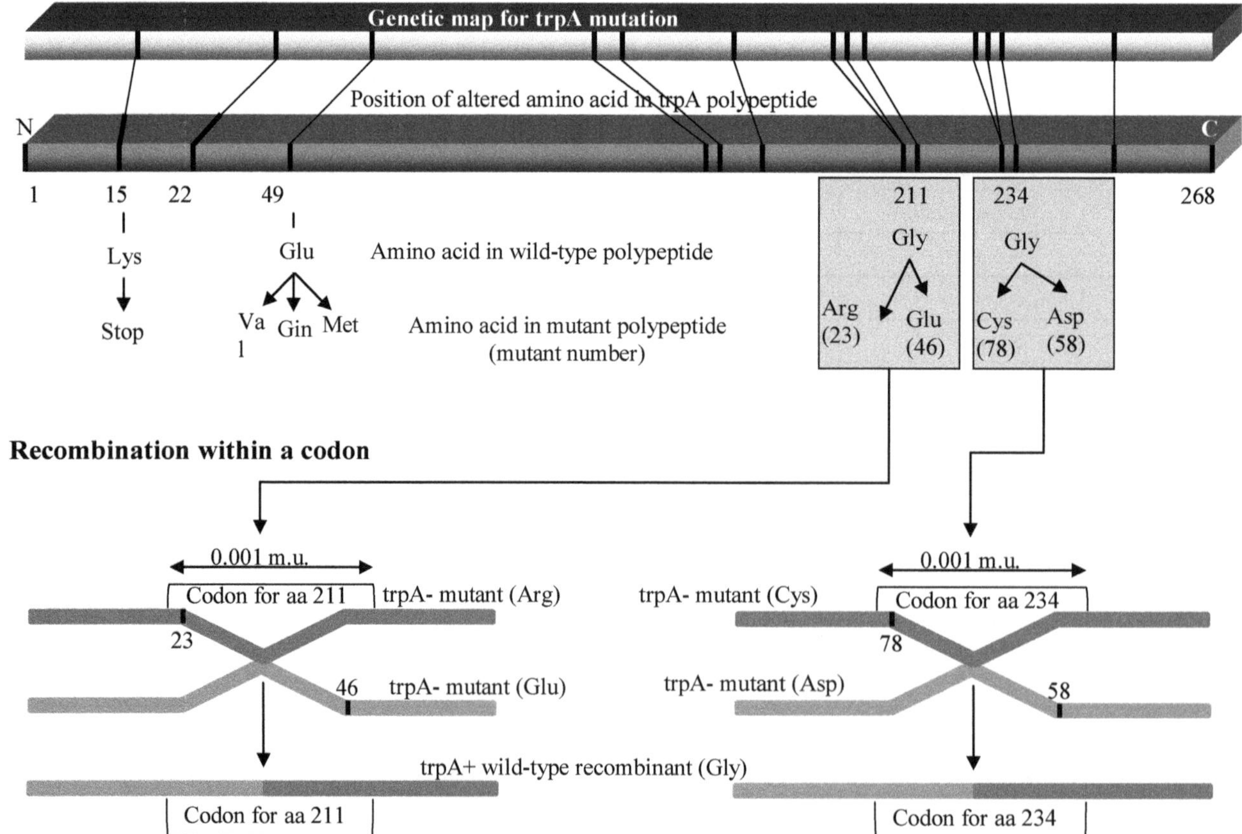

Figure 7.8: Mutations Result into Collinear Genes with the Sequence of Amino Acids in the Encoded Polypeptide.

Because the smallest unit of recombination is the base pair, two mutations capable of recombination, in this case, in the same codon because they affect the same amino acid, must be indifferent (although nearby) nucleotides. Thus, a codon contains more than one nucleotide.

As **Figure 7.8** illustrates, each point mutation in the tryptophan synthetase gene characterized by Yanofsky alters the identity of only a single amino acid. This is also true of the point mutations examined in many other genes, such as the human genes for rhodopsin and hemoglobin. Since point mutations change only a single nucleotide pair and most point mutations affect only a single amino acid in a polypeptide, each nucleotide in a gene must influence the identity of only a single amino acid. On the contrary, if a nucleotide were part of more than one codon, a mutation in the nucleotide would affect more than one amino acid.

Each mutation changes only a single amino acid, so each nucleotide is part of only a single codon. Codons must include two or more base pairs. When two mutant strains with different amino acids at the same position were crossed, recombination could produce a wild-type allele.

Although the most efficient code allowing four nucleotides to specify 20 amino acids requires three nucleotides per codon, more complicated scenarios are still possible. But in 1955 Francis Crick and Sydney Brenner obtained convincing evidence for the triplet nature of the genetic code by studying mutations in the bacteriophage T4 rllB gene (**Figure 7.9**). They induced mutations with proflavine, an intercalating mutagen that can insert itself between the paired bases placed in the center of the DNA molecule (**Figure 7.9a**). They assumed that proflavine would act like other mutagens, causing single-base substitutions. If this were true, it would be possible

(a) The mutagen proflavin can insert between two base pairs.

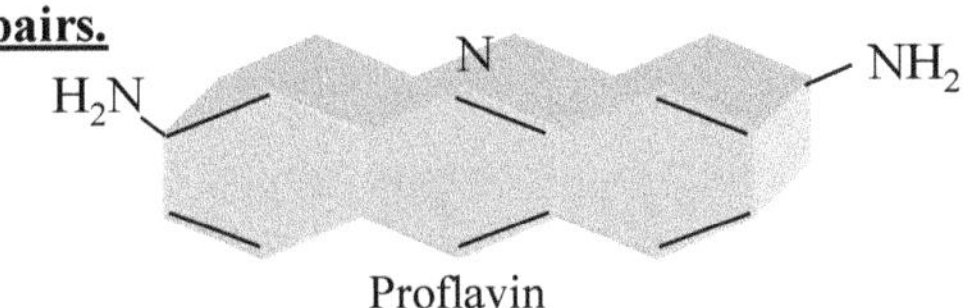

Proflavin

The mutagen proflavin slips between adjacent base pairs, eventually causing a deletion or insertion.

(b) Consequences of exposure to proflavin.

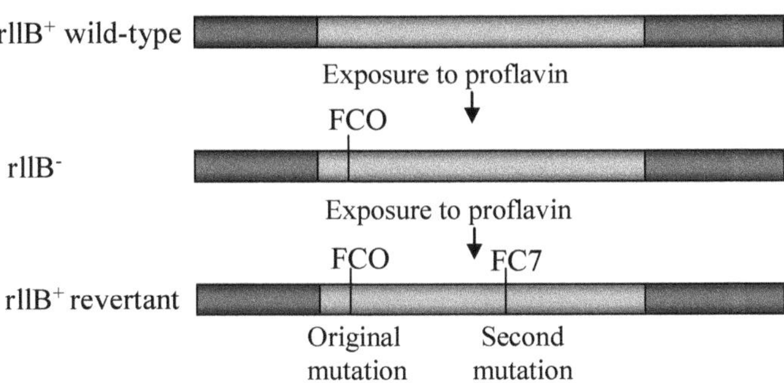

Treatment with proflavin produces a mutation at one site (FCO). A second proflavin exposure results in a second mutation (FC7) within the same gene, which suppresses FCO.

(c) rllB+ revertant X wild type yields rllB- recombinants.

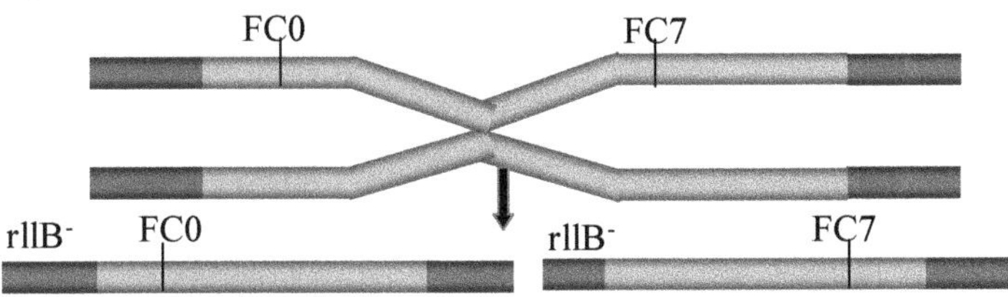

When the revertant is crossed with a wild-type strain, crossing-over separates the two *rllB-* mutations FCO and FC7. The reversion to an *rllB+* phenotype was thus the result of intragenic suppression.

(d) Different sets of mutations generate either a mutant or a normal phenotype.

Proflavin- induced mutations (+) insertion (-) deletion	Phenotype
– Or +	Mutant
– – Or + +	Mutant
– – – – Or – – – – – or + + + + Or + + + + +	Mutant
– +	Wild type
– – – or – – – – – – or + + + Or + + + + + +	Wild type

Evidence for a triplet code.

Figure 7.9: Frameshift Mutations in the Bacteriophage T4 *rllB* gene Showed that Codons Consist of Three Nucleotides based on Frame Shift Mutations.

to generate revertants through treatment with other mutagens that might restore the wild-type DNA sequence. Surprisingly, genes with proflavine-induced mutations did not revert to wild type upon treatment with other mutagens known to cause nucleotide substitutions. Only further exposure to proflavine caused proflavine-induced mutations to revert to wild type (**Figure 7.9b**).

A single (-) or a single (+) mutation destroyed the function of the rIIB gene and produced a rIIB- phage. Similarly, any gene with two base changes of the same sign (- - or + +), or with four or five insertions or deletions of the same sign (for example, + + + +), also generated a mutant phenotype. However, genes containing three or more than three mutations of the same sign (for example, + + + or - - - - - -), as well as genes containing a (+ -) pair of mutations, generated rIIB+ wild-type individuals. The intragenic suppression allowed the reconstitution of the reading frame, thereby restoring the lost or aberrant genetic function produced by other frameshift mutations in the gene.

In **Figure 7.10a**, intragenic suppression is shown, which occurs only if, in the portion in between two frameshift mutations of opposite sign, a gene still dictates the appearance of amino acids if these amino acids are not the same as those appearing in the normal protein.

Although the genetic experiments just described enabled remarkably preexisting insights about the nature of the genetic code, they did not make it possible to assign particular codons to their corresponding amino acids. This awaited the discovery of messenger RNA and the development of techniques for synthesizing simple messenger RNA molecules that researchers could use to manufacture simple proteins in the test tube.

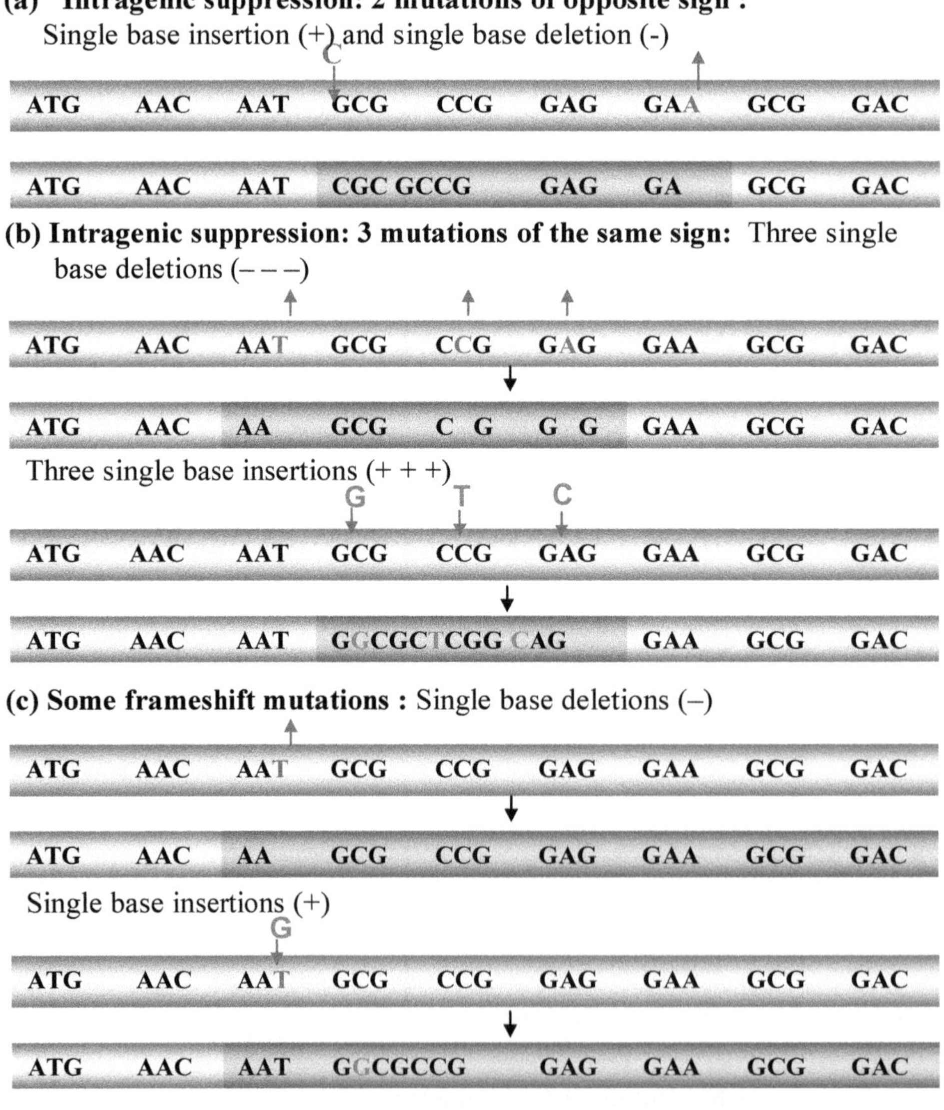

Figure 7.10: Codons Consist of Three Nucleotides.

The relatedness of all living organisms is also known from the comparisons of genes with similar functions in different organisms. For example, there is a striking similarity between the genes for many proteins in bacteria, yeast, plants, worms, flies, mice, and humans. Moreover, it is often possible to place a gene from one organism into the genome of a very different organism and see how it functions normally in the new environment. Recently, some of the mice models have been created for HIV infection to test the efficiency of various drugs that can be tried on humans. Human genes that help to regulate cell division, for example, can replace related genes in yeast and enable the yeast cells to function normally.

One of the most striking examples of relatedness at this level of biological information was uncovered in studies of eye development. Both insects and vertebrates (including humans) have eyes, but they are of very different types. Biologists have assumed that the evolution of eyes occurred independently in the lineages leading to present-day insects and present-day vertebrates. The eye is an example of convergent evolution, that is, of evolution in which structurally unrelated but functionally analogous organs emerge in different species due to natural selection. Studies of a gene called the Pax6 gene have turned this view upside down.

Mutations in the Pax6 gene lead to a failure of eye development in both humans (with a condition known as aniridia), and mice, and molecular studies have suggested that Pax6 might play a central role in the initiation of eye development in all vertebrates.

Remarkably, when the human Pax6 gene is expressed in cells along the surface of the fruit fly body, it induces numerous little eyes to develop there. This observation revealed that there was a single origin of the eye in an ancestor common to flies and humans and that, after 600 million years of divergent evolution, both vertebrates and insects still share the same main control switch for initiating eye development. Studies of many other genes have shown that the entire developmental program of flies and humans uses many of the same genes. These genes have duplicated and evolved for divergent functions but still retain their ancestral markings.

7.7 Translation in Mitochondria Reveals that the Genetic Code is not Universal

The mtDNA carries its own rRNA and tRNA genes and has its translational apparatus. The mitochondrial translation is quite unlike the cytoplasmic translation of mRNAs transcribed from nuclear genes in eukaryotes. Many aspects of the mitochondrial translational system resemble translation in prokaryotes. Mitochondria also initiates translation, just like in bacteria. Moreover, bacterial translation inhibitors, such as chloramphenicol and erythromycin, which do not affect eukaryotic cytoplasmic protein synthesis, are potent inhibitors of mitochondrial protein synthesis.

The genetic code is almost, but not quite, universal. The mtDNA sequences of tRNAs and protein encoding genes in several species cannot explain the sequences of the resulting proteins in terms of the "universal" code. For example, in human mtDNA, the codon VGA specifies tryptophan rather than stop codon (as in the standard genetic code); AGG and AGA specify stop instead of arginine, and AVA specifies methionine rather than isoleucine (**Table 7.2**). No single mitochondrial genetic code functions in all organisms and higher plants' mitochondria use the universal code. Moreover, while an f-Met-tRNA usually initiates translation in mitochondria by reading AVG or AVA, other triplets, which do not specify methionine, often mark the site of initiation. The genetic codes of mitochondria probably diverged from the universal code by a series of mutations occurring some time after the organelles became established components of eukaryotic cells. Mitochondrial Genetic code reveal Chloroplast DNA sizes as depicted in **Table 7.3**

Table 7.2: Mitochondria Genetic Code Variation

Characteristic	*Universal code*	*mtDNA Code*
Number of tRNAs	32	22
UGG	Trp	Trp
UGA	Stop	Trp
AGG	Arg	Stop
AGA	Arg	Stop
AUG	Met	Met
AUA	Ile	Met

Table 7.3 Chloroplast DNA Sizes

Organism	*Size (kb)*
Chlamydomonas reinhardtii	196
Marchantia (liverwort)	121
Nicotiana tabacum (tobacco)	156
Oryza sativa (rice)	135

7.8 Altered Genetic Code

Since elucidating the genetic code in the 1960s, 24 alterations in codon identity have been recorded in prokaryotic and eukaryotic translation systems. These alterations involve a redefinition of the identity of both sense and nonsense codons and codon un-assignment (codons vanished from genomes). Artificial expansions of the genetic code to incorporate non-natural amino acids and natural incorporation of selenocysteine and pyrrolysine (22nd amino acid) have also been reported. Sec is incorporated in both prokaryotic and eukaryotic selenoproteins through reprogramming of UGA stop codons by novel translation elongation factors (selenoprotein translation factor B prokaryotes, elongation factor [EF]-Sec, and selenium-binding protein 2 eukaryotes), a new tRNA (tRNASec), and a Sec mRNA insertion element. L-pyrrolysine insertion occurs in the archaeon Methanosarcina barkeri through reprogramming of the UAG stop codon by a pyrrolysine insertion sequence in the methylamine methyltransferase mRNA. The flexibility of the genetic code is further exemplified by the absence of glutamine and asparagine aminoacyl-tRNA synthetases in several mitochondria and archaeal and bacterial species. In those cases, aminoacylation of tRNAGln and tRNAAsn is accomplished by an ATP-dependent transamidation reaction on discharged Glu-tRNA Gln and Asp-tRNAAsn.

Methanococcus jannaschii, Methanopyrus kandleri, and Methanothermobacter thermoautotrophicus all lack canonical cysteinyl-tRNA synthetases and charge tRNACys with the intermediate substrate O-phosphoseryl (Sep), using the enzyme Sep-tRNA synthetase. Sep-tRNACys is then converted to Cys-tRNA Cys by Sep-tRNA: Cys-tRNA synthetase.

The mtDNA genetic code is simplified such that a modified U in the tRNA "wobble" position can read all four codons in a codon family (that is, UUU, UUC, UUA, UUG). An unmodified U can read both purines and G can read both pyrimidines. Tryptophan tRNA has a U in the wobble position so that it will read both the traditional UGG codon and the associated UGA stop codon as tryptophan. Similarly, the methionine codon reads both AUG and the associated AUA as methionine. Finally, one of the two arginine tRNAs was deleted so that two of the six arginine codons (AGG and AGA) now function as stop codons.

A close look at the chloroplast genome of the M. polymorpha (liverwort), the first cpDNA was fully sequenced and illustrated the details of cpDNA structure and function. Liverwort cpDNA contains 92 protein-encoding genes and 36 genes for tRNAs and rRNAs, which have more genes than are carried by the mitochondrial genomes of much larger plants. The rRNA genes of the chloroplast encode four ribosomal RNAs. These are 16S, 23S, 4.5S, and 5S. These are found in eubacteria. Another similarity with the bacterial genome is that some functionally related cpDNA genes are organized in clusters that resemble bacterial operons.

The 92 polypeptides encoded by M. polymorpha cpDNA account for only a small fraction of the proteins active in chloroplasts. However, cpDNA encodes proteins for a wider range of processes than does mtDNA. The cDNA-encoded proteins include many of the molecules that carry out photosynthetic electron transport and other aspects of photosynthesis, as well as RNA polymerase, translation factors, ribosomal proteins, and other molecules active in chloroplast gene expression. The RNA polymerase of chloroplasts is like the multi subunit bacterial RNA polymerases. Inhibitors of bacterial translation, such as chloramphenicol and streptomycin, inhibit translation in chloroplasts, as they do in mitochondria.

The function of some chloroplast genes is not yet known. Because most of these unidentified reading frames (URLs) are well conserved in many cpDNAs and have counterparts in bacteria, they are likely key elements in chloroplast gene expression and organelle function.

7.9 Codon Usage Bias

Codon usage bias (or simply codon bias) is a phenomenon of non-uniform usage of codons during the translation of genes to proteins. Specific codons are used more often than alternate synonymous ones in genes with such bias. Such codons are often referred to as optimized codons or preferred codons. The balance between mutation and translational selection manifests the extent of codon bias. The bias is most prominent in certain types of genes, generally in highly expressed genes. Variation of codon optimization among genes provides differential efficiency as well as accuracy in the translation of genes. The selection associated with translational efficiency/accuracy is often termed 'translation selection'.

The study of codon bias is gaining attention with the advent of whole-genome sequencing of numerous organisms. Molecular evolutionary investigations suggest that codon usage bias varies between and between genomes and may be relevant to understanding genome evolution among related species. In the post-genomic era, studies on codon bias are now directed toward analyzing the whole coding genome rather than specific genes or sets of genes to understand the translational selection of genomes better. Various factors such as expression level, gene length, composition bias (per cent G+C content and GC skew), recombination rates, and RNA stability, among others, are known to influence codon bias. Genome-wide investigations of codon bias patterns, their causes and consequences, and the identification of selective forces that shape their evolution are significant to studies of genome biology. Apart from non-random usages of synonymous codons, the codon context sequences are also essential features of genes as these sequences correspond to the A and P sites of the ribosome during translation. The A site binds to an aminoacylated tRNA, and the P site binds to a peptidyl-tRNA (a tRNA bound to the synthesized peptide). Thus, a codon context occupies these two sites of an active ribosome at any given time. Hence, optimization of codon context sequence may profoundly affect the translation selection of genes. Analyses of codon usage patterns in insects are fascinating for several reasons. The exceptional diversity of insects within the animal kingdom makes insects suitable for studying codon bias patterns at variable evolutionary time scales. Moreover, the recent expansion of genome sequencing of different insect species complexes (such as the twelve Drosophila species, several species of ants, honeybees, wasps, and multiple mosquitoes) also provides an exceptional opportunity for studying the evolution of codon bias within specific families and genera. In this context, one of the primary objectives of this work is to understand the patterns of codon usage bias among dipteran and hymenopteran sequenced genomes. The Diptera and Hymenoptera orders include most insect genomes sequenced so far. Although studies have been conducted on codon usage bias within the Drosophila genus or between a few other insect species, a comprehensive analysis of codon bias patterns between dipteran and hymenopteran sequenced genomes is lacking.

It has been reported that there is a lower rate of evolutionary change at synonymous sites relative to that of pseudogenes in **Table 7.4**. This observation suggests that synonymous substitutions are not completely neutral from the perspective of natural selection and that some triple codons may be favored over others. This hypothesis is reinforced by the finding that synonymous codons are not used equally throughout the coding sequences of many organisms. For example, the redundancy of the genetic code allows six different codons (UUA, UUG, CUU, CUC, CUA, and CUG) to specify the amino acid leucine, but 60 per cent of the leucine codons found in Escherichia coli are CUG and 80 per cent in yeast are UUG. Since the alternative synonymous codons specify the same amino acid, selection must favor some synonymous codons over others, or they would all be equally used. Some synonymous codons pair with different tRNAs that carry the same amino acid. Therefore, a mutation to a synonymous codon does not change the amino acid but may change the tRNA used by a ribosome during translation. Studies of the different tRNAs reveal that within a cell, the amounts of the acceptor tRNAs (different tRNAs that accept the same amino acid) differ, and the most abundant tRNAs are those that pair with the most frequently used codons. Selection may favor one synonymous codon over another because the tRNA for the codon is more productive, and the translation of mRNAs that contain that codon is more efficient. Alternatively, the bonding energy between codon and anticodon of synonymous codons may differ slightly because differences in translation efficiency and bonding energy appear to be subject to natural selection. This is especially true in genes that are expressed at high levels and in an organism with short generation times and large population sizes, such as bacteria, yeast, and fruit flies, where codon usage is most apparent.

Table 7.4: Evolutionary Change in DNA Sequences of Mammalian Genes

Sequence	*Nucleotide Substitutions per Site per Year (x 10^{-9})*
Functional genes	
5' Flanking region	2.36
Leader	1.74
Coding sequence, synonymous	4.65
Coding sequence, non-synonymous	0.88
Intron	3.70
Trailer	1.88
3' Flanking region	4.46
Pseudogenes	4.85

There exists a propensity to use preferred rather than less-used codons. It is pronounced within the genes of bacterial organisms, which are expressed at the highest levels. In fact, a gene's loyalty to an organism's codon usage bias has proven to be one of the most dependable indicators of the relative amount of expression of the gene during the organism's lifetime. Hence, highly expressed genes should also be expected to use energetically inexpensive amino acids instead of energetically expensive amino acids wherever possible. The energy of synthesizing each of the 20 different amino acids varies greatly, with glycine requiring an investment of only 11.7 adenosine triphosphate (ATP) equivalents and tryptophan requiring an average of 78.3 ATP equivalents. Some places within a protein's primary sequence exist where only a tryptophan will do. Still, wherever it is possible to substitute glycine (or delete the residue altogether), natural selection should favor the change – especially in highly expressed genes where the cost savings would have the most significant effect. Analyses of the hundreds of completely sequenced prokaryotic genomes suggest that this is the case.

8

Mutation

Any modification in the DNA sequence is called a mutation. Mutations result in variation, which may be neutral or harmful, resulting in a lethal effect. Variation is visible in the form of normal or wild-type alleles in the population, which are switched to variant alleles once the mutation occurs. The variant allele could be neutral (having no effect or harmful/ lethal or beneficial. A beneficial variant gets selected in the population. The lethal variants are removed during natural selection.

There is another term called polymorphism, defined as the least common allele that must have a frequency of one per cent or more in the population. If the gene frequency in the population is less than 1 per cent, the gene is not polymorphic. Some mutations cause diseases that reduce fitness. However, if the new mutation is neutral or beneficial, it will exist and survive in the population. After several generations are in the process of evolution, its frequency may change.

Lethal mutations may be responsible for the disease, but polymorphic sequence variation influences disease susceptibility and may influence drug responses. Some polymorphic alleles provide a selective advantage; the most quoted example is sickle cell disease. Among Caucasians, it is not a frequently found variant of the beta-globin gene that results in a severely debilitating blood disorder called sickle cell anemia. However, the same allele is polymorphic in certain parts of Africa because it confers resistance to the blood-borne parasite that causes malaria.

8.1 Types of Mutation

Any change in the DNA in the form of deletion, single base-pair substitution, or insertion (point mutation) alters the function of the DNA. It is important to note that change occurs in the exonic sequences; it can also happen in the intronic sequence and may affect the expression of the gene. Mutation can be in the somatic cells or the germ cells. When it is in the somatic cells, it is called somatic mutation. If it is in the germ cells, it is called germline mutation. Mutation can be spontaneous, where no causative agent is required. Generally, these mutations occur at the time of DNA replication. The rate of spontaneous mutation varies. If the size of the gene is large, it mutates more frequently. The occurrence of mutation is called a mutation event, and the actual occurrence is called mutation frequency. A study of the five coat color loci in mice revealed that the mutation rate ranged from 2 x10-6 to 40 x10-6 per gamete per gene. Data from several studies on eukaryotic organisms shows that the spontaneous mutation rate is generally 2-12 x10-6 mutations per gamete per gene. Given that the human genome contains approximately 21 thousand genes, we can predict that 1-5 human gametes would have a mutation in some gene. The normal gene is called the wild-type gene. Any change in the wild-type gene is a forward mutation; any change back to the wild-type is a reverse mutation. A mutated gene is called a mutant. If the change occurs in chromosomal number, deletion of the chromosome, or any other change, it is called a chromosomal mutation.

8.2 How do DNA Changes affect the Phenotype?

Point mutation could be due to base substitutions based on additions or deletions. Base substitution means one base pair is substituted for another. These can be transitions and transversions. In the transition category, purine is replaced by purine A→G or G→A, pyrimidine is replaced by pyrimidine, while in transversion, pyrimidine is replaced by purine C→A, C→G, T→A, T→G. The consequences of these mutations could be the degeneracy of the code or the existence of translation–termination codons. The outcome could be a silent substitution, where one codon for an amino acid is converted into another codon for the same amino acid. There could be a missense mutation: the codon for one amino acid is replaced by a codon of the other amino acid. In nonsense mutation, the codon for one amino acid is replaced by a stop codon. The severity of these mutations differs; for example, in missense mutation, there is the substitution of a chemically similar amino acid (synonymous substitution); This mutation results in a less severe effect on the protein structure and function. The result is more severe if chemically different amino acids are substituted (nonsynonymous mutations). Sometimes, mutations are closer to the active protein site and may lead to a lack of functional mutations. If mutations occur in less crucial areas, they may not have any effect.

Nonsense mutations have considerable effect as they lead to the premature termination of translation. Thus, the protein function may be affected. Sometimes simple addition or deletion may result in adverse consequences as the sequence of mRNA is read by the translational apparatus in a group of three base pairs (codons). This addition or deletion may cause a shift in the reading frame resulting in frameshift mutations lethal mutations conditional mutations. In condition mutations, mutant allele causes a mutant phenotype in certain environmental conditions only. Some of the Drosophila mutations are heat sensitive lethal. The heterozygous (H+/H) are wild type, and they survive at 200C (permissive condition) but die at higher temperatures, *i.e.*, 300C (restrictive conditions). Conditional mutations allow determining the sensitive period at which specific time the gene acts. The organism can be shifted to different conditions.

8.3 Biochemical Mutations

Microorganisms are prototrophic. They can exist in a substrate of simple inorganic salts, and an energy source, such as a growth medium called a minimal medium. However, when the mutation occurs, mutants can grow only when supplied with some additional nutrients. These are called autotrophic. One fungus (mutant) class grows only when provided with nitrogenous base adenine. These are called ad mutants. Among humans, different biochemical disorders of inborn errors of metabolism are seen as due to biochemical mutations.

8.4 Loss of Function Mutation

In a diploid cell, there are two wild-type alleles of a gene, both producing normal gene products. Let us suppose that one of the wild-type alleles undergoes mutation, but the other wild-type allele can produce the normal gene product. In such cases of loss of function, the mutation is recessive. Some loss of functional mutations are dominant.

8.5 Gain of Function Mutation

Sometimes mutation adds a new function to the gene, and it is expressed phenotypically. Most of these mutations are dominant. These are in the regulatory regions of the gene and are known as the gain of function mutations.

8.6 How Common are Mutations?

The behavior patterns of an organism, for example, the mating behavior or circadian rhythms of animals, can be changed due to mutation. The primary effect of behavioral mutations is often difficult to analyze. For example, the mating behavior of a fruit fly may be impaired if it cannot beat its wings. However, the mutation can be in flight muscles or the nerves leading to the brain where the nerve impulses that initiate wing movements originate.

Mutations may affect the regulation of genes. If a mutation occurs in the lac operon, a regulatory gene, it can produce a product that controls the transcription of other genes. In some cases, a region of DNA either close to or far away from a gene may also modulate its activity. In either case, a mutation in a regulatory gene or a gene control region can disrupt normal regulatory processes and permanently activate or inactivate a gene. The regulatory mutations are an excellent source for understanding genetic regulation.

Sometimes mutation may interrupt a process that is essential to the survival of the organism; hence lethal mutation. For example, a mutant bacterium that has lost the ability to synthesize an essential amino acid will cause it to grow and eventually will die when placed in a medium lacking that amino acid. Inherited human biochemical disorders are good examples of lethal mutations, *e.g.*, Tay-Sachs disease and Huntington's disease, which are caused by mutations that result in lethality but at different points in the life cycle of humans.

Mutations are rare events. The rate can be calculated by knowing the mutation frequency over time. This could be an organismal generation, cell division, or generational span. The rate of spontaneous mutation in various organisms reveals that it is exceedingly low for all organisms and varies considerably between different organisms. Among species, the spontaneous mutation rate varies from gene to gene.

Viral and bacterial genes undergo spontaneous mutation on an average of 1 in 100 million (10-8) per cell division. Neurospora exhibits a similar rate, but maize, Drosophila, and humans demonstrate a rate of several orders of magnitude higher. The genes studied in these groups average between 1/1,000,000 and 1/100,000 (10-6 and 10-5) mutations per gamete formed. Mouse genes have a higher spontaneous mutation rate -1/100,000 to 1/10,000 (10-5 to 10-4). It is not clear why such a significant variation occurs in mutation rates. The divergence between organisms might reflect the relative efficiency of their DNA proofreading and repair systems.

The frameshift mutation results in a complete loss of standard protein structure and function. It is very important to keep in mind where the point mutation occurs. Suppose the mutation occurs at splice donor or splice acceptor sites for introns in eukaryotic pre-mRNAs, RNA polymerase–binding sites in promoters, and 16S rRNA-bindings sites (shine-Delgar no sequences) upstream of the translation–initiation sites of prokaryotic mRNAs. The regulatory mutations affect the amount of protein in the gene.

8.7 Somatic and Germinal Mutations

If a change occurs in somatic cells, it is a somatic mutation. If it occurs in germinal cells, it is a germinal mutation. In the case of somatic mutations, it appears in single-cell and then descends into different progenitor cells called a clone. These cells stay together; hence the phenotypically mutant cells are called the mutant sector. If the mutation occurs in the dividing cell, it may affect the phenotype, but if it occurs in a postmitotic cell, there is no effect on the phenotype. In the case of malignancies, the mutation occurs in proto-oncogenes, and cell division is uncontrolled division, resulting in a cluster of cells called a tumor.

Suppose the mutant sex cell participates in fertilization. In that case, the mutation will be passed on to the next generation, *e.g.*, X-linked hemophilia mutation in European royal families is thought to have arisen in the germ cells of Queen Victoria or one of her parents. The mutation was expressed only in her male descendants.

Further, one must keep in mind various ways of categorizing mutations often overlap a morphological mutation.

8.8 Harmful Mutations in Humans

Most of the mutations may occur in the portion of the genome that does not contain genes. These are neutral mutations as they do not affect gene products. While this information is of great interest to geneticists, perhaps a more significant piece of information regarding spontaneous mutations is the rate at which deleterious mutations occur. It is difficult to determine the rate of mutation at the nucleotide level. After the sequencing became popular accurate estimates are possible. The rate of deleterious mutations in humans is surprisingly high.

8.9 DNA Replication Errors cause Mutations

The DNA molecule is not stable in the typical cellular environment. Every base pair in the DNA helix has a certain probability of undergoing mutation. This is also essential for evolution to proceed. However, whenever there is a damaging mutation the DNA repair mechanism also occurs to correct the mutations (**Figure 8.1**).

The chemistry of the cellular environment itself is a significant source of damage. DNA is subject to spontaneous water-mediated reactions that include deamination of cytosine, guanine, and adenine, which converts these to uracil, xanthine, and hypoxanthine respectively. Cytosine deamination is shown in **Figure 8.2**, is the most frequent, and estimated at ~100 – 500 events/cell/day in a mammalian cell. Since uracil, xanthine, and hypoxanthine are normally not present in DNA, repair machinery preferentially removes these bases.

Hydrolysis of the N-glycosyl bond (which links the base to the sugar) also occurs spontaneously in the cellular environment, most frequently when the base is in purine (~10,000 events/cell/day), shown in **Figure 8.3**. Loss of pyrimidine is less common (~500 events/cell/day). Cleavage of the N-glycosyl bond thus results in a basic site (a site lacking a base), also known as an AP (for apurinic/apyrimidinic) site.

Another significant source of damage results from the reactive oxygen species (ROS), which are generated both by normal cellular metabolism (endogenous source) as well as resulting from ionizing radiation (exogenous source). However, cells have inbuilt mechanisms to prevent damage from ROS production, including enzymes that convert reactive species such as superoxide and hydroxyl radicals to less reactive products. Estimating how

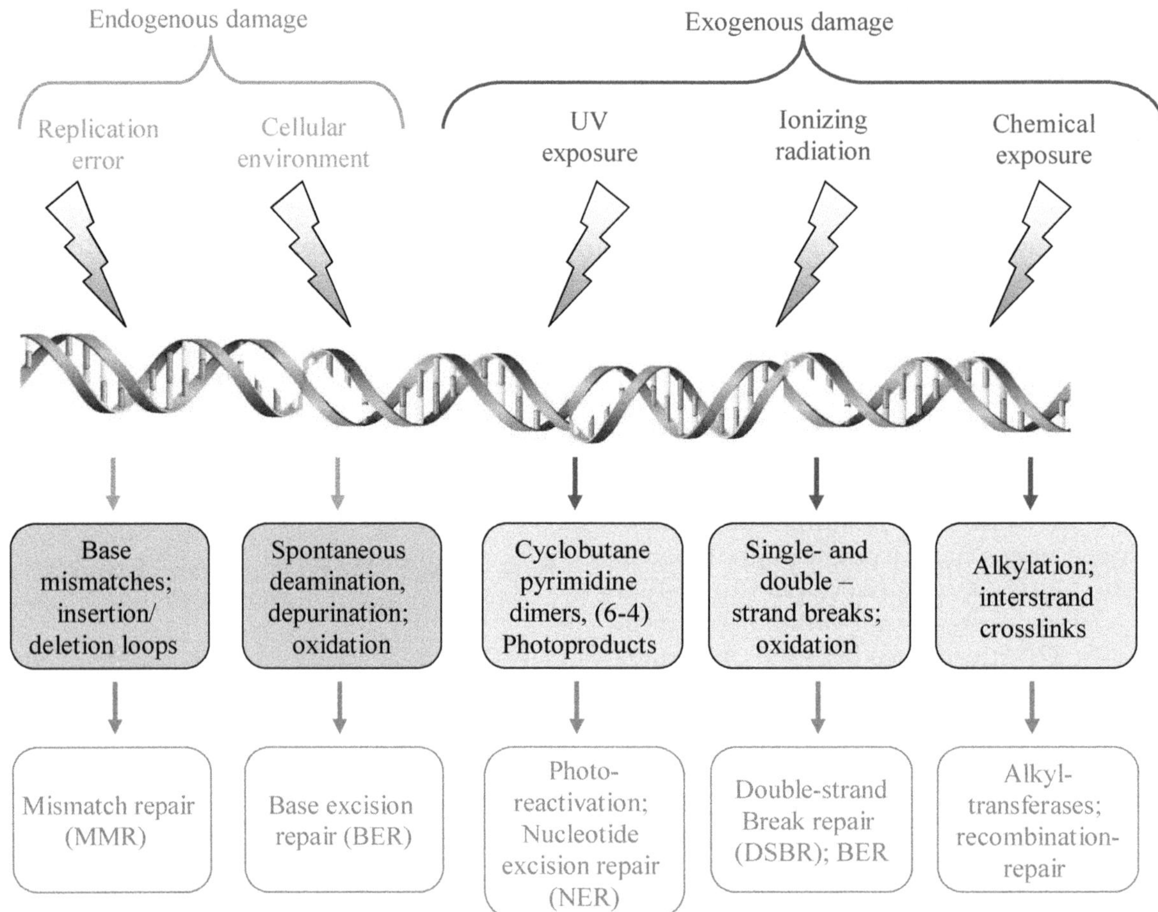

Figure 8.1: DNA once Damaged can be Repaired through Specialized Repair Mechanisms.

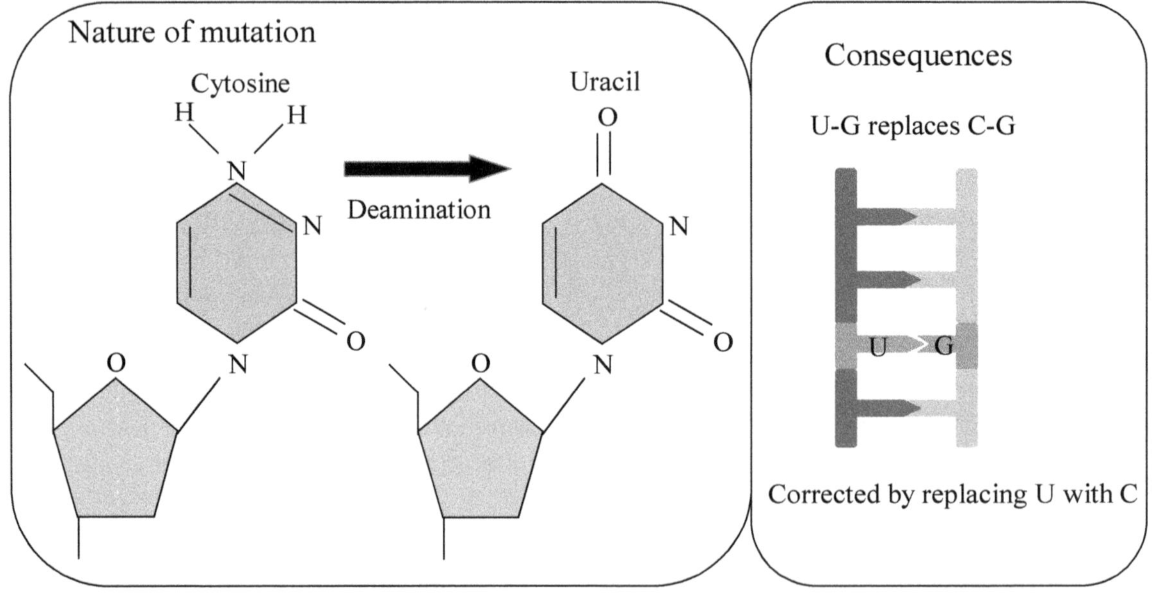

Figure 8.2: Spontaneous Deamination of Cytosine Creates a U : G Base Pair.

much damage from endogenous ROS production occurs in a healthy cell is difficult. Endogenous ROS production typically increases during cellular aging because of decreased electron transport chain efficiency. Whether produced by internal chemistry or exogenous ionizing radiation, hydroxyl radicals can result in almost 80 different kinds of specific base damage. One of the most common, 8-oxoguanine (8-oxoG), is shown in **Figure 8.4**. 8-oxoG is capable of base pairing with both cytosine and adenine; 8-oxoG:

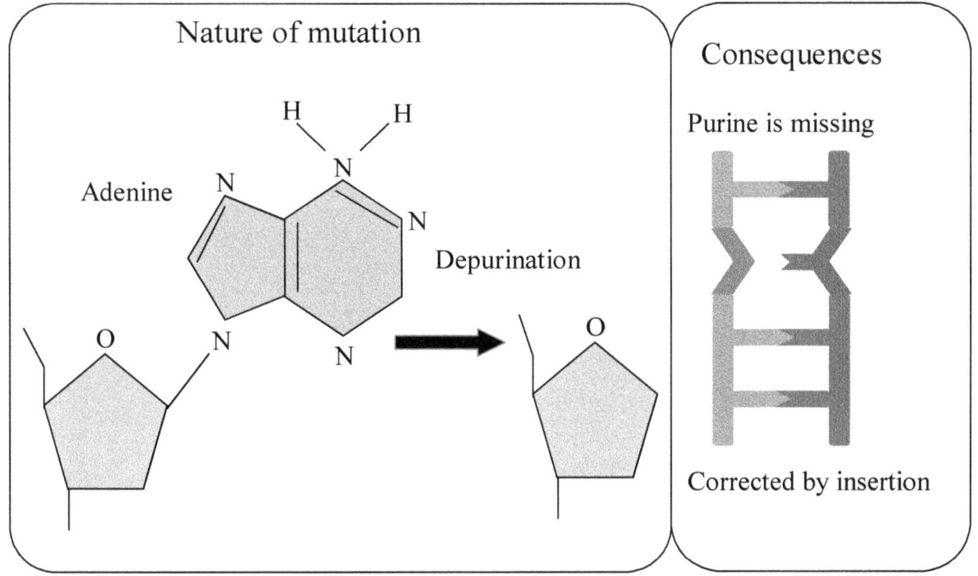

Figure 8.3: Spontaneous Depurination Creates a Basic Site which is at Lower Rate.

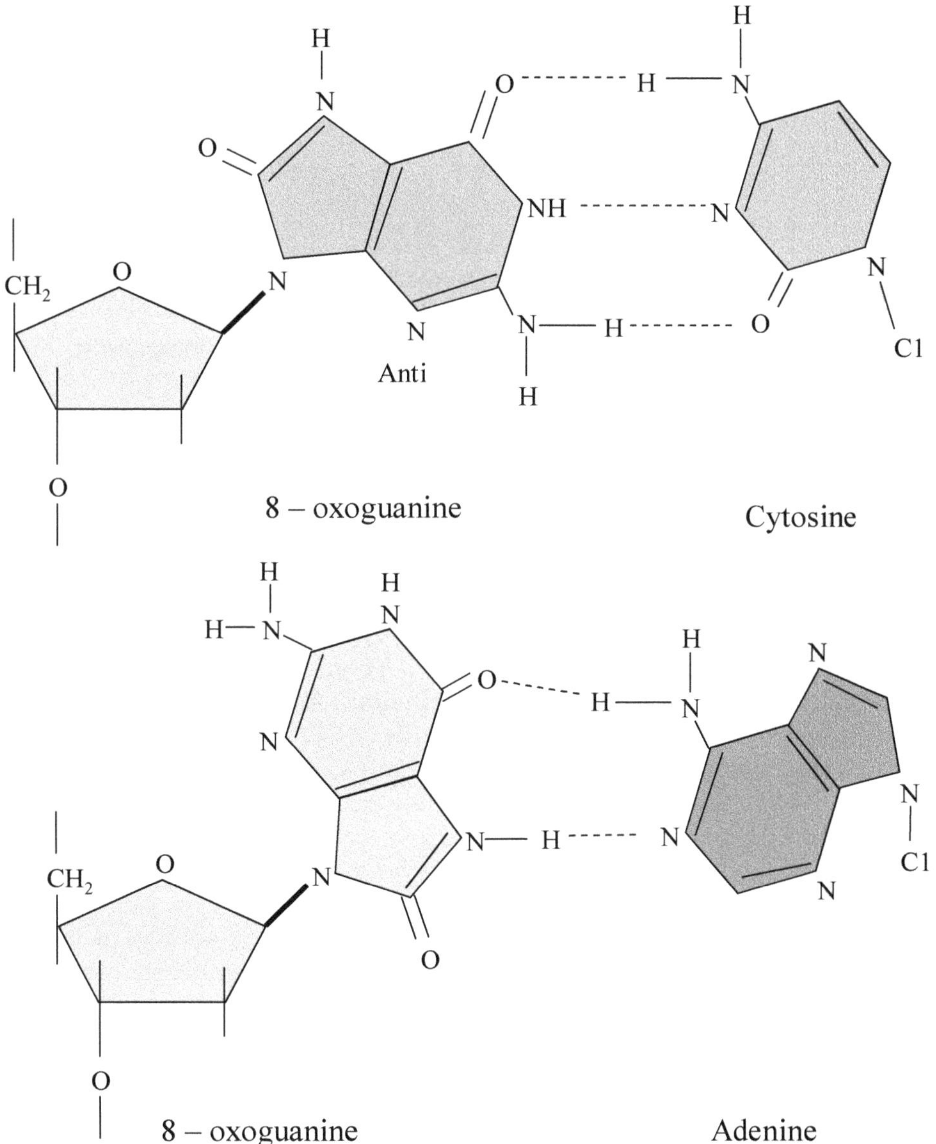

Figure 8.4: Oxidative Damage Causes a Variety of DNA Tensions; some (such as 8 -oxo guanine) cause base mispairing and/or structural distortions of the DNA.

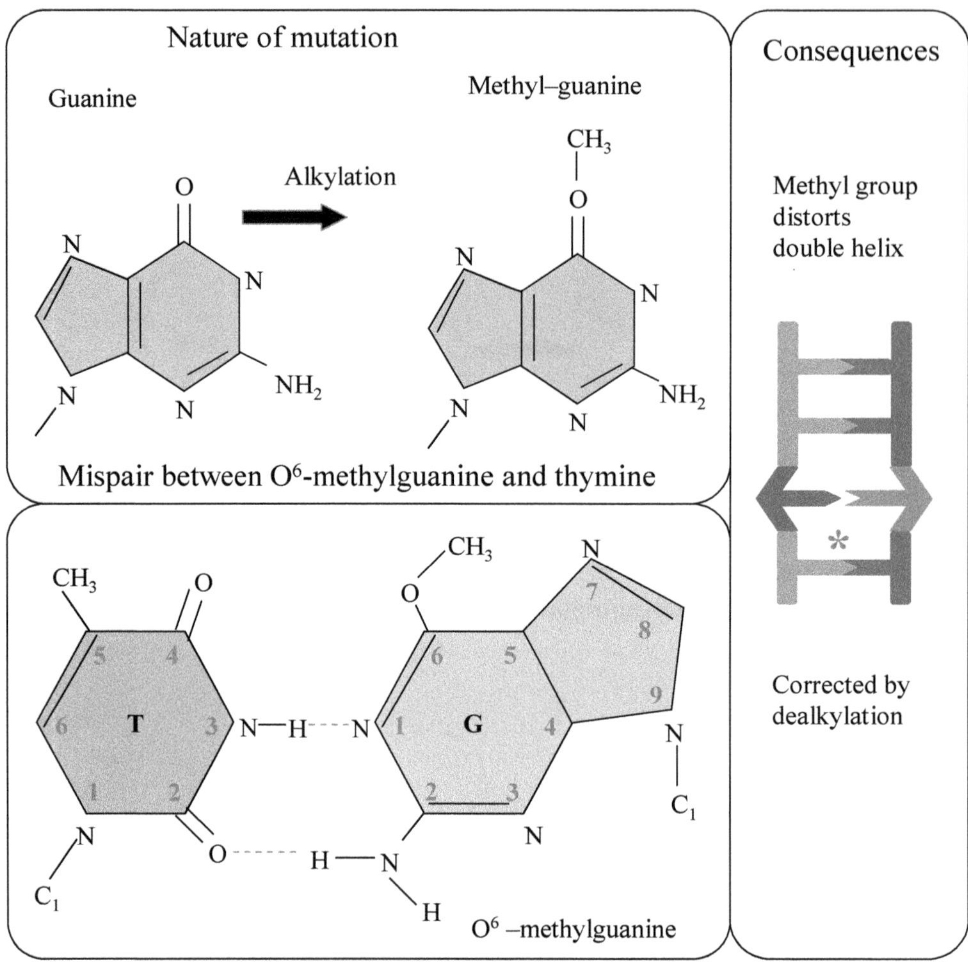

Figure 8.5: Distortions of the Double Helix Cause Base Mispairing.

A base pairs can be converted to T : A base pairs during replication if the damage is uncorrected.

Exogenous sources of damage are far more varied. These are radiation chemical mutagens (alkylation) that attack DNA. Alkylation can occur at most of the nitrogen and oxygen atoms in the bases (except the nitrogen linked to deoxyribose), as well as the oxygen atoms in the phosphate groups that are not involved in phosphodiester bonds. **Figure 8.5** shows the structure of 06-methyl guanine, a type of product created by alkylating agents such as methyl methane sulfonate (MMS). While 06 methyl guanine is not a significant product (the N -7 position of guanine is the most frequent methylation target), it is a significant one as its base pairs with T. Thus, like 8-oxo G, described previously, this lesion can result in the conversion to an A: T base pair if it is not corrected before replication.

Polycyclic aromatic hydrocarbons (PAHs) are a diverse family of molecules containing two or more fused aromatic rings; some members of this family are inadvertently converted into alkylating epoxides by the action of cytochrome P450 enzymes. Cytochrome P450 also converts aflatoxins into reactive epoxides. Aflatoxins are a class of carcinogens produced by fungi in the Aspergillus genus, which grow on grains like peanuts, rice, and corn and constitute a serious public health threat. **Figure 8.6** shows the result of an aflatoxin epoxide attack on guanine, which creates a lesion leading to replication errors.

There are two reactive sites in a few alkylating agents; hence, they can form two covalent links with DNA. This can occur (within the same strand) or across the double helix cross-links. One of the first agents recognized to create interstrand cross-links was bis [2-chloroethyl] methylamine, more infamously known as the chemical warfare agent nitrogen mustard gas. This molecule and the cross-link it forms are shown in **Figure 8.7**. Despite its inauspicious beginnings, nitrogen mustard gas was the first of several crosslinking compounds that have proven to be useful chemotherapeutic drugs, including the simple molecule cisplatin (also shown in **Figure 8.7**). However, these drugs do have serious side effects due to their general cytotoxicity.

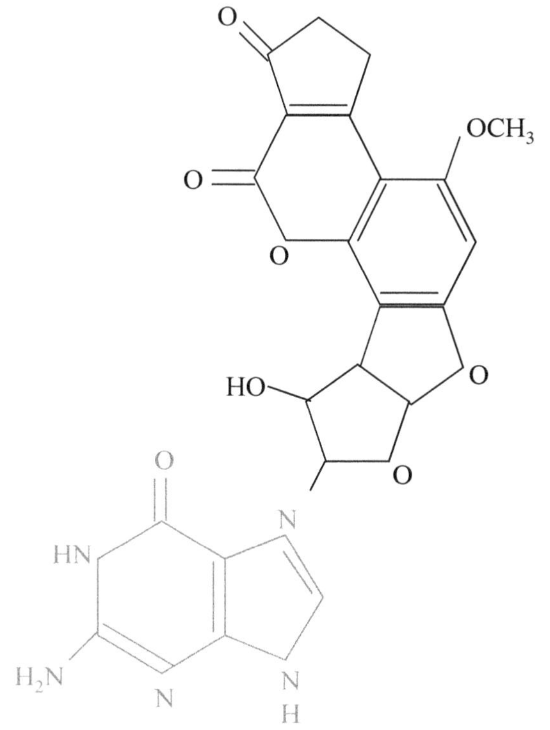

Aflatoxin attached to the N7 position of the guanine base, which is shown in red

Figure 8.6: Environmental Agents get Converted into *Alkylating* Agents through Cytochrome p450.

8.10 Induced Mutations

Mutations that result due to the influence of an external factor are called induced mutations. Induced mutations can be due to natural or artificial agents, called mutagenic agents. For example, radiation from cosmic and mineral sources and ultraviolet radiation from the sun can induce mutations. The earliest demonstration of the artificial induction of mutations occurred in 1927 when Hermann. Muller reported that X-rays could cause mutations in Drosophila. In 1928, Lewis J. Stadler reported that X-rays had the same effect on barley. In addition to various forms of radiation, numerous natural and man-made chemical agents are also mutagenic. The first report of the mutagenic action of a chemical came to light in 1942 by Charlotte Auerbach, who demonstrated that the nitrogen component of poisonous mustard gas used in World Wars I and II could cause cell mutations. Since then, many other mutagenic chemicals have been identified.

It is possible to distinguish chemical mutagens by their mode of action and these chemicals structurally resemble purines and pyrimidines and may be incorporated into DNA in place of the normal bases during DNA replication: bromouracil (BU) –artificially created compound extensively used in research. It resembles thymine (has a Br atom instead of methyl group) and can be incorporated into DNA, and it pairs with A-like thymine. It has a higher likelihood for tautomerization to the enol form (BU*) aminopurine -adenine analog, which can pair with T or (less well) with C; causes A: T to G: C or G: C to A: T transitions. Base analogs cause transitions, as do spontaneous tautomerization events. Other mutagenic base analogs are 2 amino purines (2- AP),

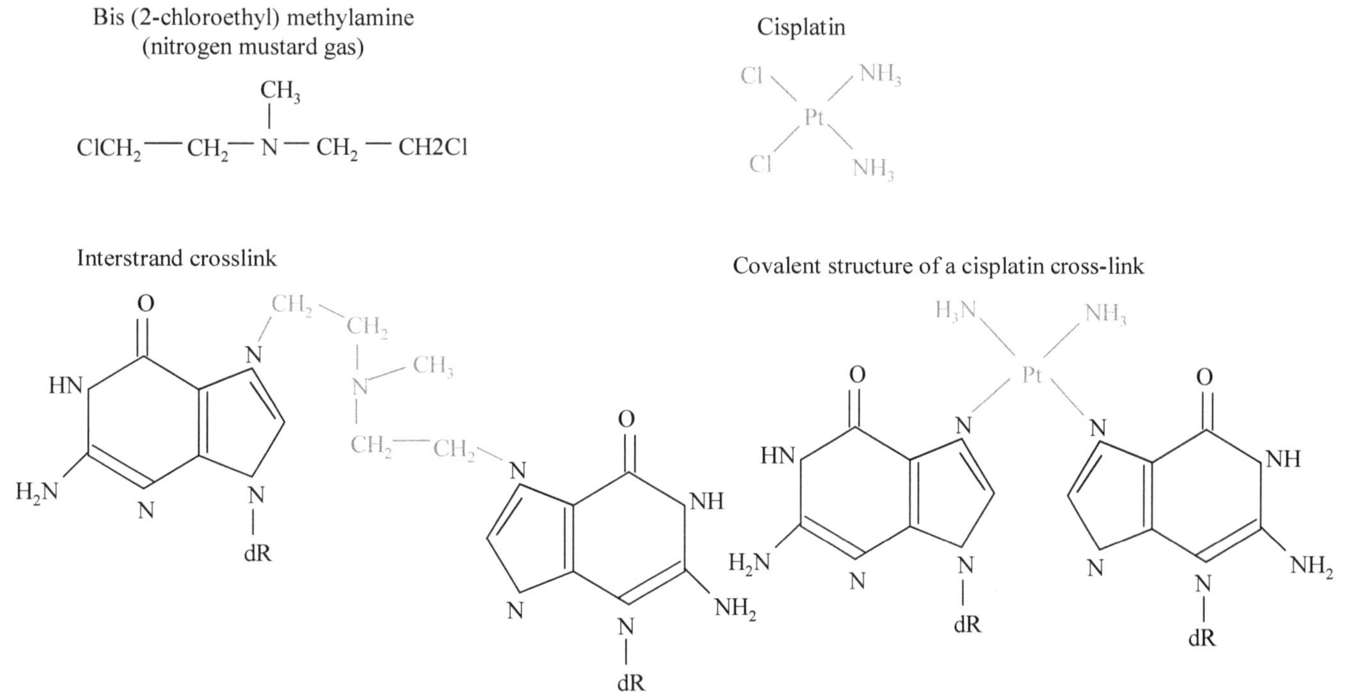

Bis (2-chloroethyl) methylamine
(nitrogen mustard gas)

$ClCH_2—CH_2—N(CH_3)—CH_2—CH2Cl$

Cisplatin

Interstrand crosslink

Covalent structure of a cisplatin cross-link

Figure 8.7: Nitrogen Mustard Gas and Cisplatin are Cross-linking Agents (Chemotherapeutic drugs).

which can act as an analog of adenine. It can base pair with cytosine A= T to G a" C following replication.

8.11 Chemicals that Alter Structure and Pairing Properties of Bases

There are many mutagens; some well-known examples are nitrous acid–formed by the digestion of nitrites (preservatives) in foods. It causes C to U, C to T, and A to hypoxanthine deamination's. The alkylated base may degrade to yield a baseless site, which is mutagenic and can undergo recombination or mispair to result in mutations upon DNA replication. Intercalating agents like acridine orange, proflavine, and ethidium bromide are also mutagenic. All are flat, multiple-ring molecules that interact with DNA bases and insert between them. This insertion causes a frameshift mutation (**Figure 8.8**).

8.12 Agents Altering DNA Structure

Large molecules that bind to bases in DNA and cause them to be noncoding–we refer to them as "bulky" lesions (*e.g.*, NAAAF) causing Intra and inter-strand crosslinks (*e.g.*, psoralens found in some vegetables and used in treatments of some skin conditions) chemicals causing DNA strand breaks (*e.g.*, peroxides) (**Figure 8.9**). In **table 8.1** different alkalating agents have been listed

8.13 Acridine Dyes and Frame Shift Mutations

Frameshift mutations are caused by adding or removing one or more base pairs in the polynucleotide sequence of the gene. Acridine dyes are like nitrogenous base pairs and intercalate, or wedge, between the purines and pyrimidines of intact DNA. Intercalation introduces a conformational change in the DNA helix and causes deletions and insertions that create frameshift mutations (**Figure 8.10**).

The resultant frameshift mutations are generated at gaps produced during replication, repair, or recombination. During these events, there is the possibility of slippage and improper base pairing of one strand with the other.

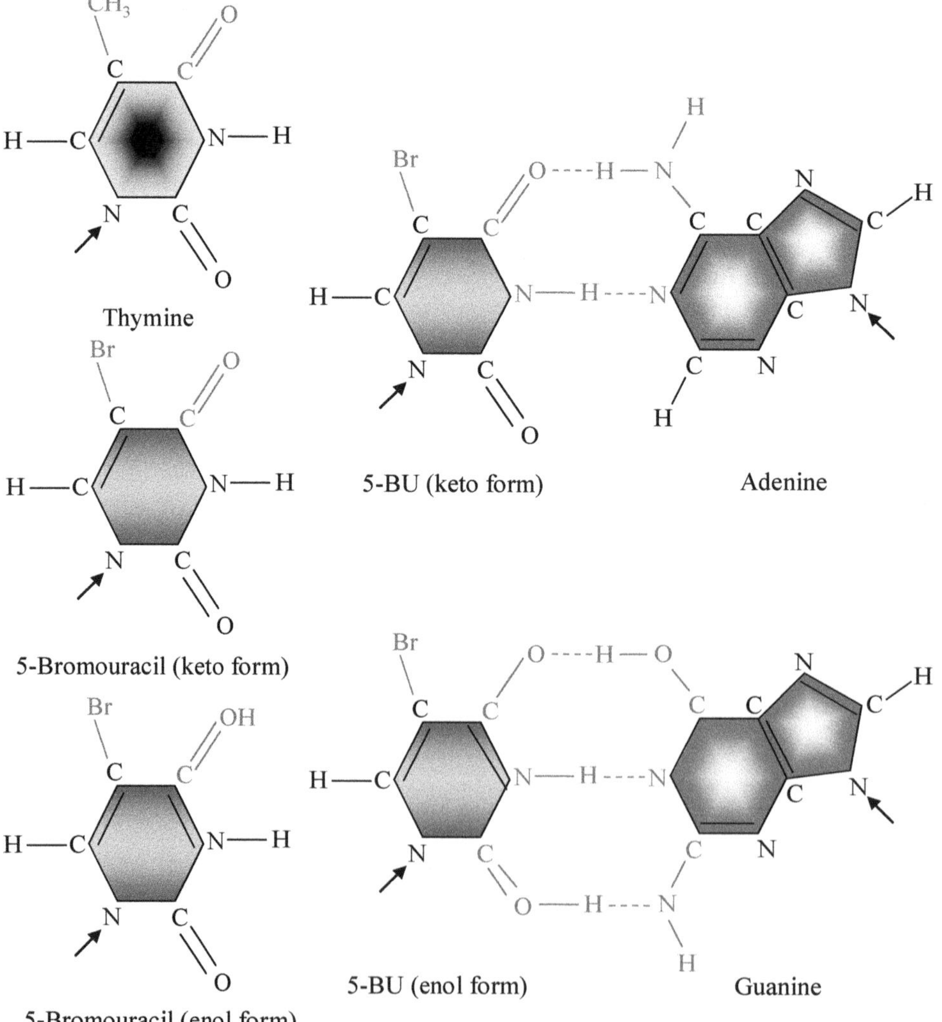

Figure 8.8: Structure of Thymine 5' Bromouracil (keto and enol form) Adenine and Guanine.

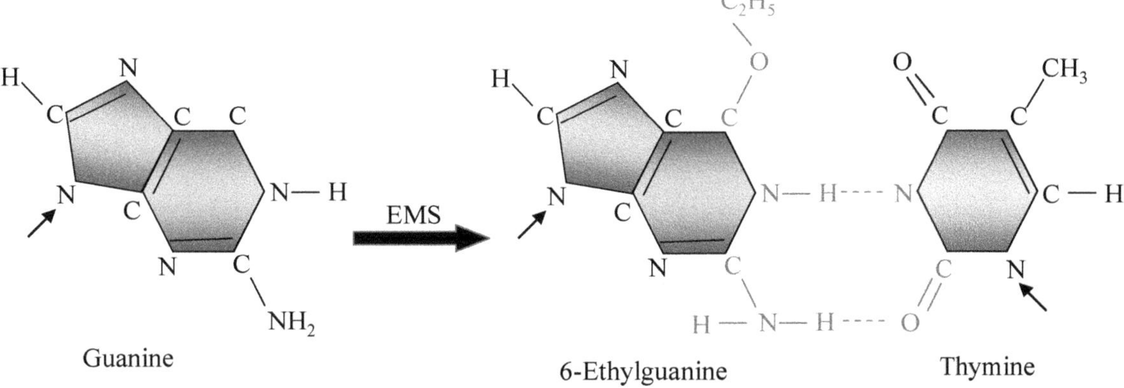

Figure 8.9: Conversion of Guanine to 6-ethylquanine by the Alkylating Agent Ethyl Methane Sulfonate (EMS)–the 6-ethylquanine Base Pairs with Thymine.

Table 8.1: Different Alkylating Agents

Common Name or Symbol	*Chemical Name*	*Chemical Structure*
Mustard gas (sulfur)	Di-(2-chloroethyl) sulfide	$Cl - CH_2 - CH_2 - S - CH_2 - CH_2 - Cl$
EMS	Ethyl methane sulfonate	$CH_3 - CH_2 - O - S(=O)(=O) - CH_3$ (O above and below S, double bonds)
EES	Ethylethane sulfonate	$CH_3 - CH_2 - O - S(=O)(=O) - CH_2 - CH_3$ (O above and below S, double bonds)

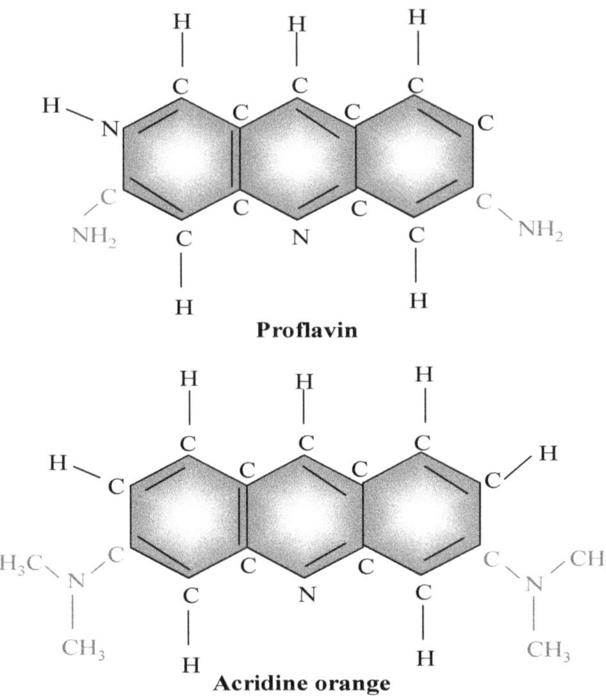

Figure 8.10: The Proflavine and Acridine Orange Intercalate into DNA.

8.14 Ultraviolet Light and Thymine Dimmers

Ultraviolet light causes mutation. The significant effect of UV radiation is the creation of pyrimidine dimmers, particularly those involving two thymine residues. While cytosine-cytosine and thymine–cytosine dimmers may also be formed, they are less prevalent than thymine dimmers. The dimmers distort DNA conformation and inhibit normal replication. As a result, errors can be inserted in the base sequence of DNA during replication.

8.15 Ionizing Radiation

The ionizing radiation can destroy biological organisms and result in DNA damage.

Extensive doses of ionizing radiation have been shown to have a mutating effect on future generations of individuals receiving the dose of ionizing radiation. Examples of ionizing radiation are energetic Beta particles, neutrons, alpha particles, and energetic photons (UV and above). The energy required to ionize an atom or molecule may vary widely. X-rays

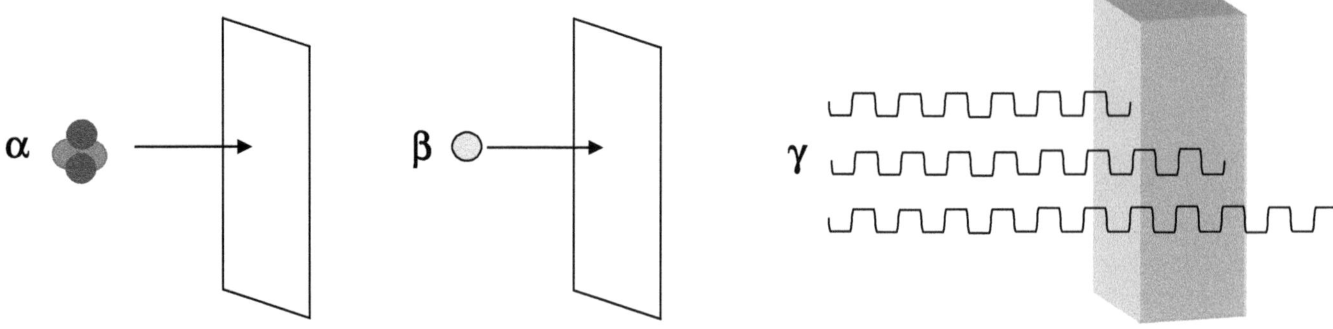

Figure 8.11: Ionizing Radiation.

and gamma rays ionize almost any molecule or atom; far ultraviolet, near ultraviolet, and visible light are ionizing to very few molecules; microwaves and radio waves are non-ionizing radiation (**Figure 8.11**).

Ionizing radiation is produced by radioactive decay, nuclear fission, and nuclear fusion, by extremely hot objects (the hot sun, *e.g.*, produces ultraviolet), and by particle accelerators that may produce fast electrons or protons or bremsstrahlung or synchrotron radiation.

8.16 Radiation Levels

The associations between ionizing radiation exposure and cancer development due to somatic mutation are primarily based on populations exposed to relatively high levels of ionizing radiation, such as Japanese atomic bomb survivors and recipients of selected diagnostic or therapeutic medical procedures.

Cancers like leukemia, thyroid, breast colon, liver, lung esophagus, ovarian multiple myeloma, stomach cancers, *etc.* United States Department of Health and Human Services literature suggests a possible association between ionizing radiation exposure and prostate, nasal cavity/sinuses, pharyngeal and laryngeal, and pancreatic cancer. The period between radiation exposure and cancer detection is known as the latent period. Studies of occupational workers exposed to chronic low radiation levels above normal background have provided mixed evidence regarding cancer and transgenerational effects. Cancer results, although uncertain, are consistent with estimates of risk based on atomic bomb survivors and suggest that these workers do have a slight increase in the probability of developing leukemia and other cancers.

Adaptive mutation revolves around the controversial idea that organisms may "select" or "direct" the nature of gene mutation to adapt to a particular environmental pressure. In 1943, Salvador Luria and Max Delbruck presented the first direct evidence that mutations are not adaptive but occur spontaneously. This experiment marked the beginning of modern bacterial genetic study. Their experiment, known as the Luria-Delbruck fluctuation test, is an example of exquisite analytical and theoretical work.

8.17 Classification Based on the Location of the Mutation

Mutations are classified according to the locations at which they occur. Somatic mutations can occur in any cell in the body except germ cells. Germline mutations occur only in gametes. Autosomal mutations occur within genes located on the autosomes, whereas X-linked mutations occur within genes located on the X chromosome.

Mutations arising in the somatic cells are not transmitted to future generations. When a recessive autosomal mutation occurs in a somatic cell of a diploid organism, it is unlikely to result in a detectable phenotype. The wild-type allele will likely mask the expression of most such mutations. Somatic mutations will have a greater impact if they are dominant or, in males, if they are X-linked since such mutations are most likely to be expressed immediately. Similarly, the effect of dominant or X-linked somatic mutations will be more noticeable if they occur early during development when the undifferentiated cells are replicating to give rise to several differentiated tissues or organs.

Mutations in germline cells are of greater significance because they are transmitted to offspring as part of the germline. They have the potential of being expressed in all cells of an offspring. Inherited dominant autosomal mutations will be expressed phenotypically in the first generation itself. X-linked recessive mutations arising in the gametes of a homogametic female may be expressed in hemizygous male offspring. This will start only when the male offspring receives the affected X chromosome. Because of heterozygosity, the occurrence of an autosomal recessive mutation in the gametes of either males or females (even one resulting in a lethal allele) may go unnoticed for many generations until the resultant allele has become widespread in the population.

8.18 Molecular Change and Classification of Mutation

Changing one base pair to another in a DNA molecule is known as a point mutation or base substitution. A change of one nucleotide of a triplet within a protein-coding portion of a gene may cause the creation of a new triplet that codes for a different amino acid in the protein product. These are missense mutations. Sometimes, the triplet may change into a stop codon, resulting in the termination of translation of the protein, therefore called a nonsense mutation. If the point mutation alters a codon but does not change the amino acid at that position in the protein, it is called a silent mutation.

8.19 Tautomeric Shifts

Changes in the nucleotide base pair generally cause base substitutions. There are different ways to bring a tautomeric shift. The chemical nature of the bases of DNA is such that rare but natural, spontaneous fluctuations in the bonds of the bases can occur. These changes may affect how a base forms a hydrogen bond, *e.g.*, when it undergoes a tautomeric shift, adenine pairs with cytosine. Therefore, if a tautomeric shift occurs during replication, the wrong nucleotide can be inserted into the newly synthesized DNA. Base substitutions can also be caused by chemical modification of the bases. The alkylating agents, such as ethyl methane sulfonate and methyl methane sulfonate, can also result in base substitution. These agents donate alkyl groups (such as methyl and ethyl groups) to bases, affecting their base pairing. When guanine, which is an alkylated agent producing 7-ethylguanine, it will base pair with thymine. If it occurs during replication, the wrong nucleotide may be inserted in the synthesized molecule, leading to a mutation that causes mutation. The most stable tautomers of nitrogenous bases take part in the standard base pairings that serve as the basis of the double-helix model of DNA. The less frequently occurring transient tautomers can hydrogen bond with non-complementary bases.

However, the pairing is always between pyrimidine and a purine (**Figure 8.12**). This results in anomalous T ≡ G and C = A pairs

(a) Compared with anomalous base-pairing that occurs as a result of tautomeric shifts

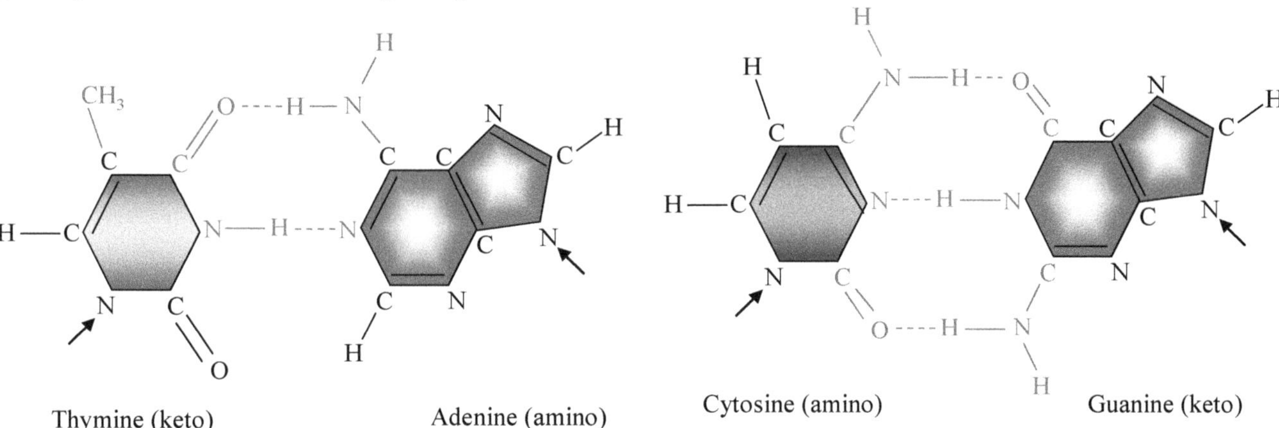

(b) The long triangle indicates the point at which the base bonds to the pentose sugar.

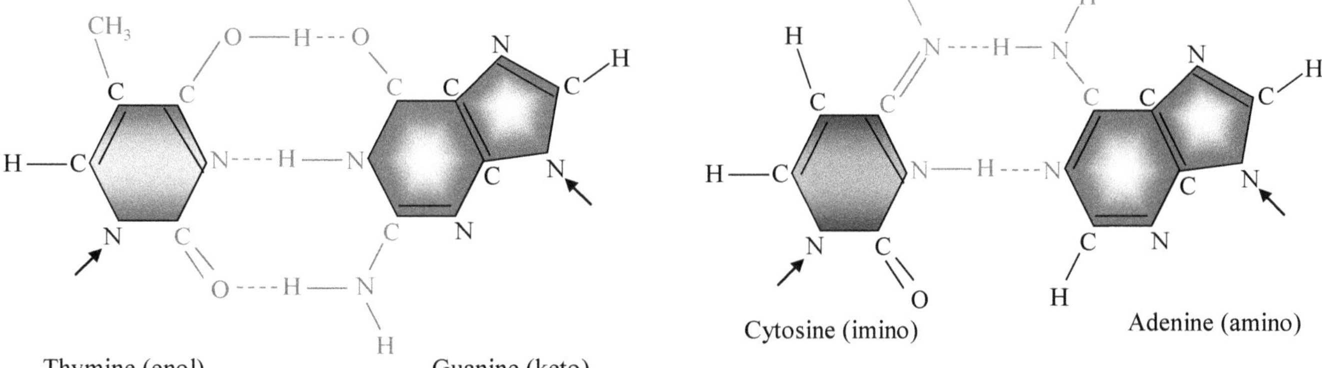

Figure 8.12: Standard Base-Pairing Relationships.

A mutation occurs during DNA replication when a rare tautomer in the template strand pairs with a noncomplementary base. In the next round of replication, the "mismatched" members of the base pair get separated and become templates for its normal complementary base. The result is a point mutation (**Figure 8.13**).

Another type of change is the insertion or deletion of one or more nucleotides at any point within the gene. As illustrated in **Figure 8.14**, the loss or addition of a single letter causes the entire subsequent three-letter words to be changed. These are the frameshift mutations because the triplet reading frame during translation is altered. A frameshift mutation will occur when any number of bases are added or deleted, except multiples of three, which would re-establish the initial reading frame. The analogy in **Figure 8.14** demonstrates that insertions and deletions can potentially change a gene's subsequent triplets. One of the many altered triplets may be the termination codons of UAA, UAG, or UGA. Polypeptide synthesis is terminated when one of these triplets is encountered during translation. A frameshift mutation is generally lethal.

8.20 DNA Repair Mechanisms

Sometimes, the action of a single enzyme can reverse the action. The first known example of direct

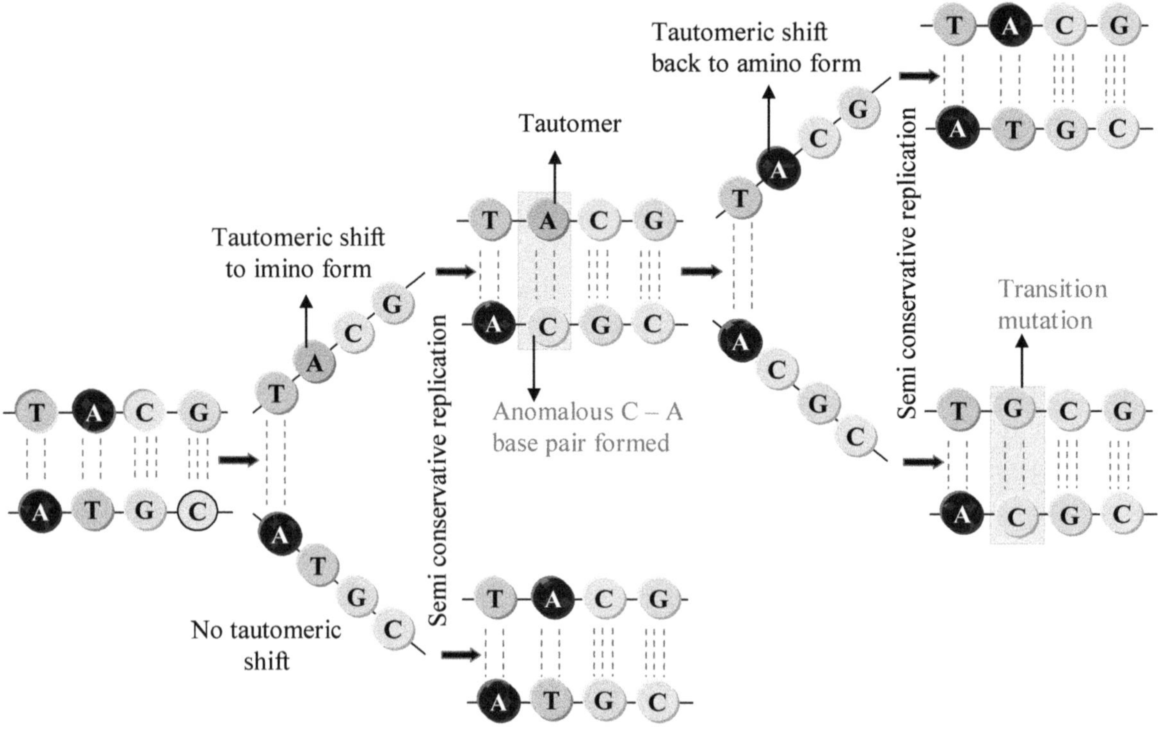

Figure 8.13: Formation of a T = A to C ≡ G Transition Mutation due to a Tautomeric Shift in Adenine.

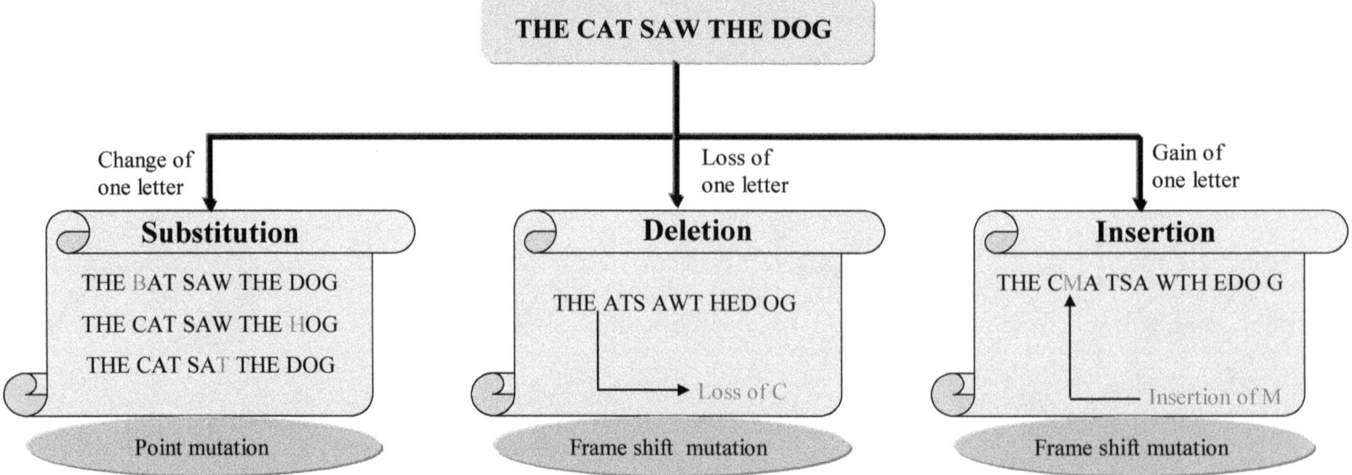

Figure 8.14: Effects of Substitution, Deletion, and Insertion of One Letter in a Sentence Composed of Three – Letter Words, Showing Point and Frame Shift Mutations.

repair comes from the discovery of photoreactivation, a phenomenon in which bacterial cells reposed to UV irradiation exhibited better survival rates if they were kept in the light after UV exposure rather than placed in the dark. This was because a single enzyme, photolyase, uses the energy of an absorbed photon to directly cleave the cyclobutane ring of a CPD, restoring the original pyrimidines, as shown in **Figure 8.15**. Photolyase catalyzed a similar repair of (6-4) PPs. Photolyases are ubiquitous but lost in placental mammals (such as humans) sometime after the split between marsupial and placental mammals. Placental mammals utilize a more complex system to repair UV damage (Nucleotide excision repair removes bulky DNA lesions); however, now humans have started using the bacterial photolyases in clinical applications.

Various cells have developed various mechanisms to avoid DNA damage. In most of the repair pathways, the damaged base or the damaged region is removed and replaced by the correct complementary strand.

8.21 Direct Reversal of Damage

The easiest and most direct way to repair a lesion once it occurs is to reverse it directly and regenerate the normal base. However, it is only sometimes possible as sometimes the reversal is irreversible.

The alkyl groups are removed by a group of enzymes known as alkyl transferase. Many of these enzymes exist, with specificities for different lesions. A human enzyme, alkyl adenine DNA glycosylase (AAG), recognizes and removes a variety of alkylated substrates, including 3-methyladenine, 7-methylguanine, and hypoxanthine. The structure of AAG is bound to a methylated adenine, in which the adenine is removed and bound in the glycosylase's active site. This base-flipping mechanism is used by numerous repair enzymes (including photolyase) to both recognize and repair DNA damage. A surprising feature of the alkyl transferase is that they are otherwise suicide enzymes that are inactivated by their catalysis. The alkyl transferase acts by transferring the alkyl group from the damaged base to a cysteine in the alkyl transferase itself. This restores the base, but the alkyl group is now irreversibly attached to the alkyl transferase, which is rendered permanently inactive. Sometimes, the mutagenesis is caused by a mutagenic photo dimer caused by UV light (**Figure 8.16**). The photo dimer produced is Cyclobutane pyrimidine, which can be repaired by photolyzing, which is found in bacteria and lower eukaryotes but not in humans. The enzyme binds to the photo dimer and splits it. This must take place in the light. Hence, other pathways are required to remove the UV damage.

8.22 Mismatch Repair Corrects Replication Errors

During the replication of DNA, DNA polymerase is used. DNA replication is a very accurate process due to sensitive proofreading mechanisms; still,

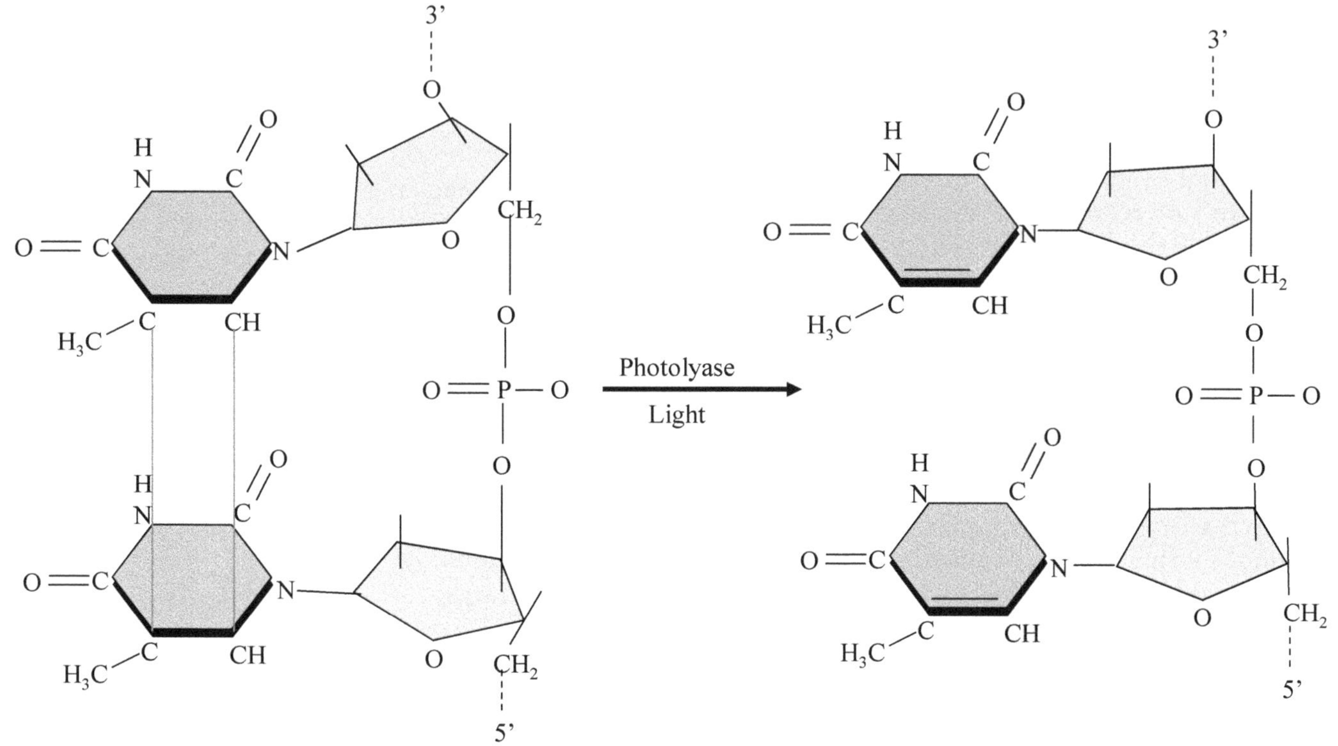

Figure 8.15: Photolyase Reverses UV-induced Pyrimidine Dimmers, using a Light-driven Reaction to Disrupt the Cyclobutane Ring.

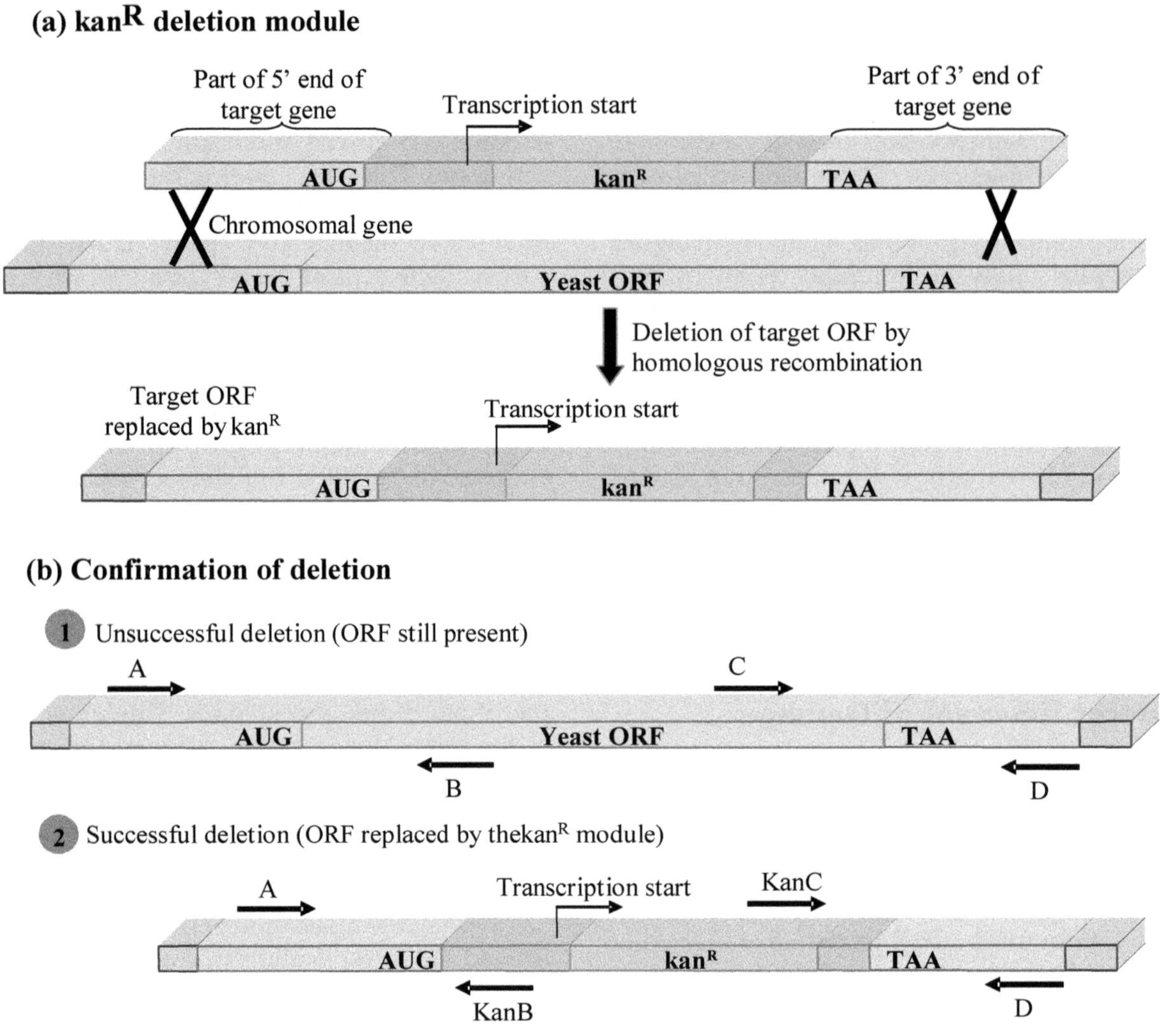

Figure 8.16: Creating and Verifying a Gene Knockout in Yeast.

a significant number of base mispairing escapes proofreading, and additional loops are created by misalignment of the parental and daughter stands, which are not recognized by proofreading. At this stage, the cells use the repair system, which is responsible for surveilling newly replicated DNA, known as the mismatch repair (MMR) pathway. There are various genes in prokaryotes; these are muts, mut H and mut L. The prokaryotic MMR pathway is shown in **Figure 8.17**. In this pathway, the gene MutS binds to the mismatch (as either a dimmer or a tetramer) and is joined by MutL.

The challenge of repairing mismatched bases is that, unlike forms of DNA damage in which a base is noticeably aberrant, both bases in a typical mismatch are normal, undamaged bases. Thus, the mismatch machinery has no intrinsic means of determining which base in a mismatch represents an error. If the mutated base is excised, the wild–type sequence is restored. If it happens to be the original (wild type) base that is excised, the new (mutant) sequence becomes fixed.

When mismatch errors occur during replication in E. coli, it is possible to distinguish the original strand of DNA. Immediately after the replication of methylated DNA, only the original parental strand carries methyl groups; this provides the basis for a system to correct replication errors accurately. MutS has two DNA–binding sites, one of which recognizes mismatches. The other binding site is not specific for sequence or structure, and it has been suggested it may be used to translocate along with DNA until it finds the GATC sequence driven by the ATPase activity of MutS.

The MutS MutL complex activates the MutH endonuclease, cleaving the unmethylated GATC. This unmethylated strand is excised from the GATC site to the mismatch site. The excision can take place in the 5′ to 3′ direction (using RecJ or exonuclease VII) or in the 3′ to 5′ direction (using exonuclease I)

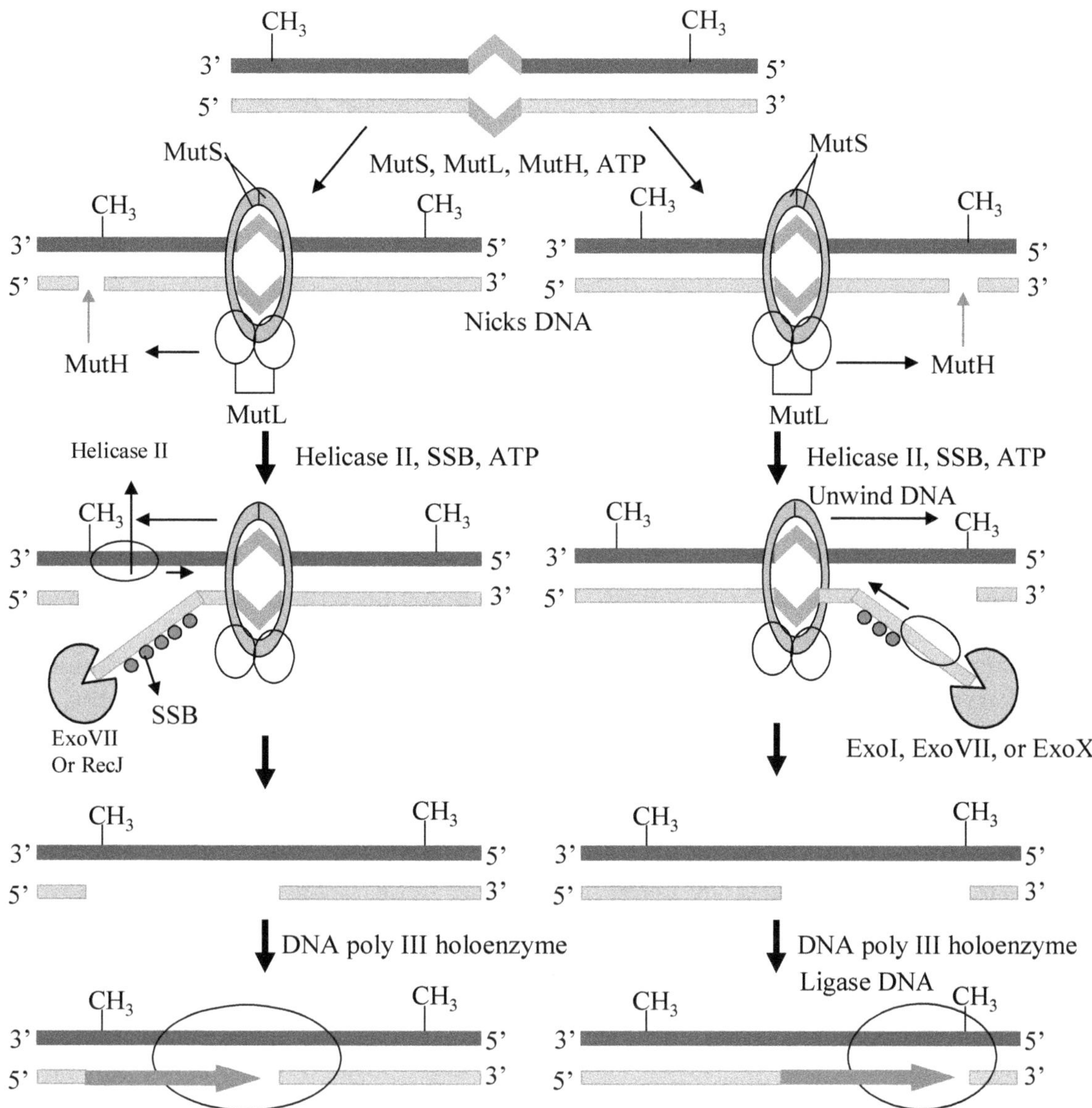

Figure 8.17: Mismatch repair in *E. coli* Involves Recognizing a Mismatch by the MutS and MutL Proteins.

and is assisted by helicase II (also known as UvrD). DNA polymerase III further synthesizes a new DNA strand. The excision can be extensive; mismatches can be repaired preferentially for >1 Kb around a GATC site. The result is that the newly synthesized strand is corrected to the sequence of the parental strand.

Eukaryotic cells have systems that are homologous to the E. coli mut system. In addition to recognizing mismatches, eukaryotic MMR also repairs insertion/deletion loops caused by replication slippage. In a region such as a microsatellite, where a concise sequence is repeated several times, realignment between the newly synthesized daughter strand and its template can lead to stuttering in which the DNA polymerase slips backward and synthesizes extra repeating units. Alternatively, the template strand can form the loop, resulting in a net decrease in the number of repeats synthesized in the daughter strand.

Three MutS homologs, MSH2, -3, and -6, and two MutL homologs, MLH1 and PMS2, are involved in mismatch recognition and repair in eukaryotes. MutSα, an MSH2- MSH6 heterodimer, is involved in recognizing mismatches and small insertion/deletion loops, while MutSβ, an MSH2- MSH3 heterodimer, only recognizes insertion/deletion loops in a range of sizes. MutS α or -β not only recognizes the mismatches/loops but also interacts directly with the components of the replication machinery, including PCNA and the clamp loader RFC. Surprisingly, even though most eukaryotes possess DNA methylation, eukaryotic mismatch repair systems do not use DNA methylation to select the daughter strand for repair.

Instead, MutSα (in the presence of PCNA, RFC, and ATP) activates an endonuclease activity in MutLα (a MLH1 – PMS2 heterodimer), which then nicks the DNA with a strong preference for a DNA strand that already contains a nick. In other words, MutLα cleaves the lagging strand efficiently, thus targeting the correct strand for excision by the MutS-activated exonuclease ExoI. It is not known how specificity for the daughter strand is achieved on the leading strand. **(Figure 8.18)**

Three MutS homologs, MSH2, -3, and -6, and two MutL homologs, MLH1 and PMS2, are involved in mismatch recognition and repair in eukaryotes. MutSα, an MSH2- MSH6 heterodimer, is involved in recognizing mismatches and small insertion/deletion loops, while MutSβ, an MSH2- MSH3 heterodimer, only recognizes insertion/deletion loops in a range of sizes. MutS α or -β not only recognizes the mismatches/loops but also interacts directly with the components of the replication machinery, including PCNA and the clamp loader RFC. Surprisingly, even though most eukaryotes possess DNA methylation, eukaryotic mismatch repair systems do not use DNA methylation to select the daughter strand for repair. Instead, MutSα (in the presence of PCNA, RFC, and ATP) activates an endonuclease activity in MutLα (a MLH1 – PMS2 heterodimer), which then nicks the DNA with a strong preference for a DNA strand that already contains a nick. In other words, MutLα cleaves the lagging strand efficiently, thus targeting the correct strand for excision by the MutSα- activated exonuclease ExoI. It is not known how specificity for the daughter strand is achieved on the leading strand.

8.23 Excision Repair Pathways

The phosphodiester bond is broken on either side of the lesion, causing the excision of an oligonucleotide, resulting in a gap filled by repair synthesis ligase that seals the break.

The base excision repair pathway (BER) pathway repairs an extensive array of DNA single–base lesions, including bases damaged by oxidation and alkylation. Base removal is catalyzed by a class of enzymes known as DNA glycosylases, which (like the alkyl transferases described above) flip the damaged base out of the DNA helix into the active site of the enzyme, where the N-glycosyl bond is cleaved. Different glycosylases recognize different substrates; the reaction of a glycosylase that recognizes uracil (generated by cytosine deamination). Some enzymes, in addition to their glycosylase activity, also contain AP lyase activity that cleaves the bond between the sugar and phosphate 3′ to the AP site. These enzymes are called DNA glycosylases/lyases.

No enzymes exist to replace a base and a basic site, so the next stage of BER is to cleave the phosphodiester backbone so that the DNA containing the basic site can be excised and replaced. Spontaneous AP sites enter the pathway at this stage. AP sites are recognized by AP endonuclease, which cleaves the DNA backbone 5′ to the AP site. This results in a gap with a 5′ deoxyribose phosphate (the basic sugar) and a 3′ –OH. This gap can be repaired in one of two different ways, as shown in **Figure 8.19**.

In long patch repair, DNA pol δ or e (in conjunction with PCNA and RFC) adds 2-8 nucleotides to the 3′ end, simultaneously displacing the DNA containing the 5′ -deoxyribose phosphate to form a flap, which is cleaved by a flap endonuclease. In short patch repair, a conserved repair polymerase, DNA pol β, inserts a single nucleotide. DNA pol β contains an intrinsic 5′ -deoxyribose phosphate lysate activity that removes the 5′ -deoxyribose phosphate. DNA ligase seals the remaining nick in both pathways. When BER is initiated by glycosylase/lyase action, AP endonuclease removes the sugar left at the 3′ end so that when DNA pol β inserts a single nucleotide, the product is a suitable substrate for ligase.

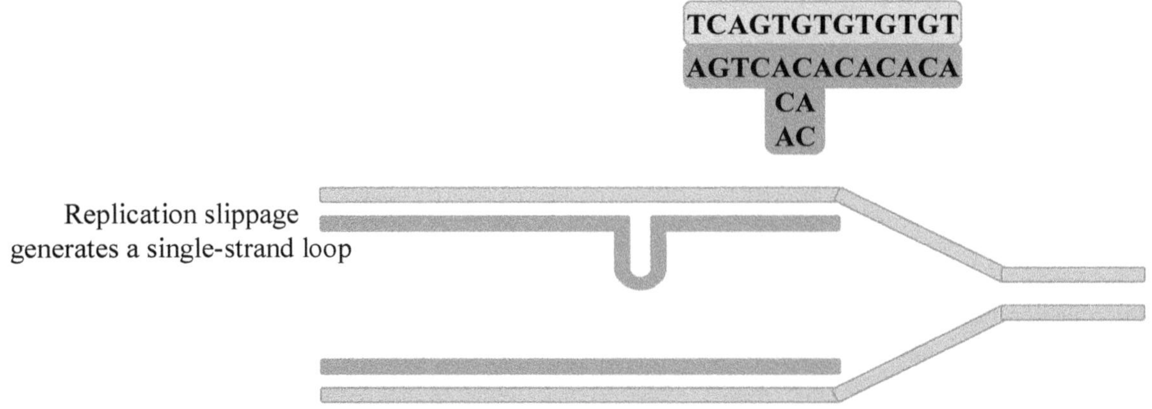

Figure 8.18: Replication Slippage may cause Insertion/Deletion Loops.

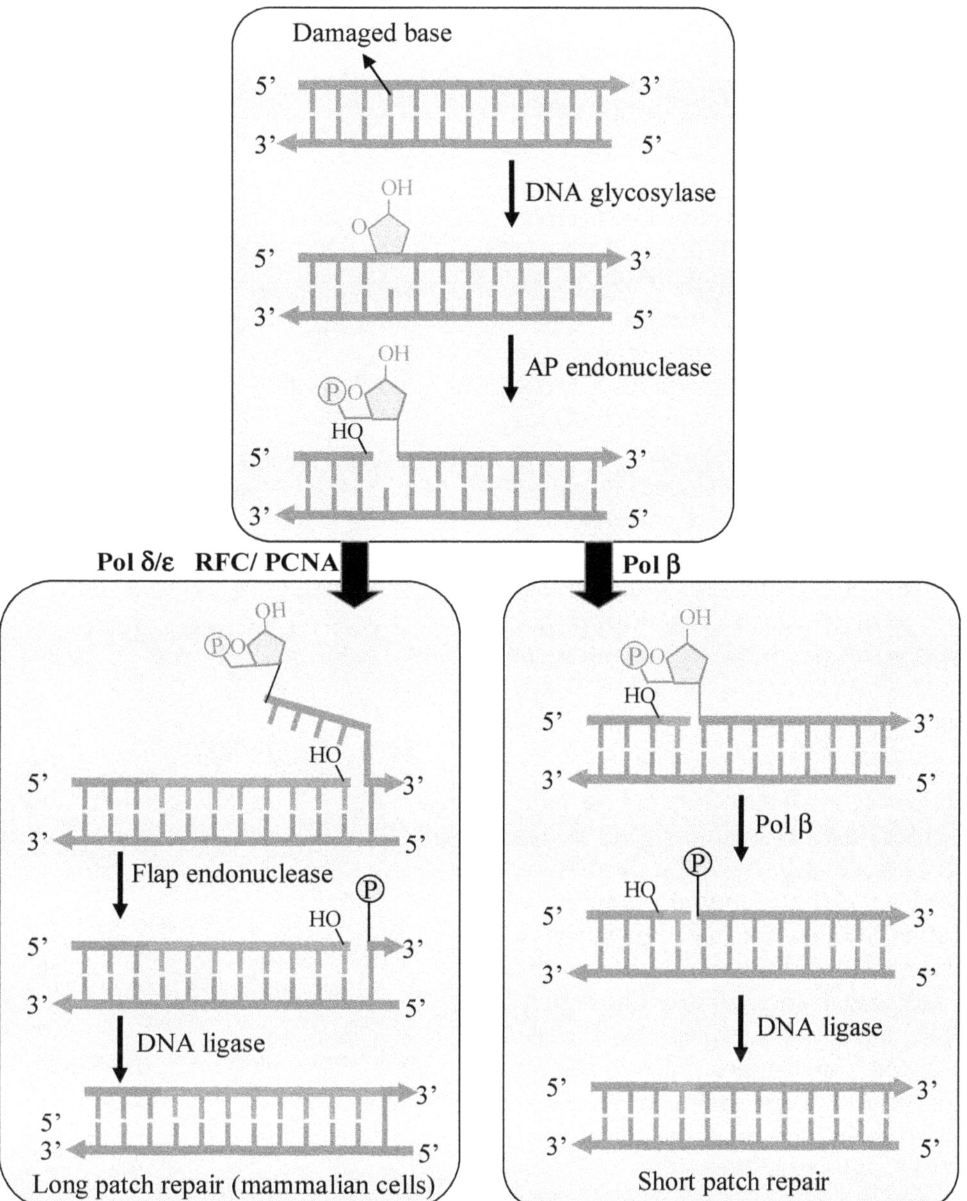

Figure 8.19: The Gap Resulting from Glycosylase/AP Endonuclease Action can be Repaired via one of the two Pathways in Eukaryotes.

(a) Long patch repair causes polymerization by pol δ and e and removal or 2-8 nt. (b) Short patch repair involves pol β and add a single nucleotide and remove the 5'–deoxyribose phosphate. The nick in both pathways is sealed by ligase.

8.24 Nucleotide Excision Repair Removes Bulky DNA Lesions

Like MMR and BER, NER is an excision pathway that removes the damaged base. In this case, a sizeable region containing the damaged base fills in the resultant gap using the intact strand as a template for a DNA polymerase. The NER pathway is used to repair a variety of bulky DNA lesions, including the dimeric photoproducts resulting from UV irradiation, bulky adducts such as aflatoxin-damaged bases, and the same DNA cross-links. In general, results severe into a distortion of the DNA helix, the more readily the lesion is recognized and repaired by NER.

In E. coli, genes responsible for NER were identified by mutants that were unable to repair UV-induced damage in the dark (*i.e.*, when photolyase enzyme was unable to repair the damage) and thus were named uvr for UV resistance. Photolyases are DNA repair enzymes that repair damage caused by exposure to ultraviolet light. This enzyme requires visible light from the violet/blue end of the spectrum and is known as photoreactivation. Therefore, are uvr genes –uvrA, uvrB, uvrC and uvrD- encode the core proteins required for NER. They act essentially in the order of their names, as seen in **Figure 8.20.** First, a heterotrimer consisting of two UvrA and one UVrB subunit scans the DNA using the helicase activity of UvrB. Upon binding to damage, UvrA is released in an ATP-dependent manner, leaving UvrB-DNA complex in which a region of UvrB is inserted into the DNA helix to unwind the region of interest to the lesbian. UvrC, an endonuclease, then binds the UvrB-DNA complex and cleaves the damaged strand in a precise manner: one cut is made 4 nucleotides 3′ to the lesbian, and a second cut is made seven nucleotides 5′ to the lesbian. The UvrD helicase then removes the excised fragment, leaving a gap to be filled in by DNA pol I and ligase seal the final nick. A conserved NER system repairs bulky adducts in eukaryotes-in mammalian cells; this is the only system that can repair UV-induced photoproducts. The importance of NER for repairing UV damage in mammals is revealed by XP, a disease caused by loss of NER in humans, resulting in up to 1000 times increase in the incidence of skin cancer.

NER in eukaryotes is divided into two sub-pathways, as illustrated. The major difference between the two pathways is how the damage is initially recognized. In global genome repair (GG-NER), a protein called XPC, complexes with a protein called hHR23b, detects the damage, and initiates the repair pathway. XPC can recognize any damage in the genome. On the other hand, transcription-coupled repair (TC-NER), as the name suggests, is responsible for repairing lesions that occur in the transcribed strand of active genes. In this case, the damage is recognized by RNA polymerase II itself.

Both pathways then use an overlapping set of proteins to affect the repair. The strands of DNA are unwound for ~20bp around the damaged site. This action is performed using the helicase activity of the transcription factor TFIIH (discussed in 4.5 promoters direct the initiation of transcription), which includes the products of two XP genes (XPB and XPD. Then, cleavages are made on either side of the lesion by endonucleases encoded by the XPF and XPG genes. The single-stranded stretch, including the damaged bases, can then be replaced by a new synthesis using pol δ or e. **(Figure 8.20)**

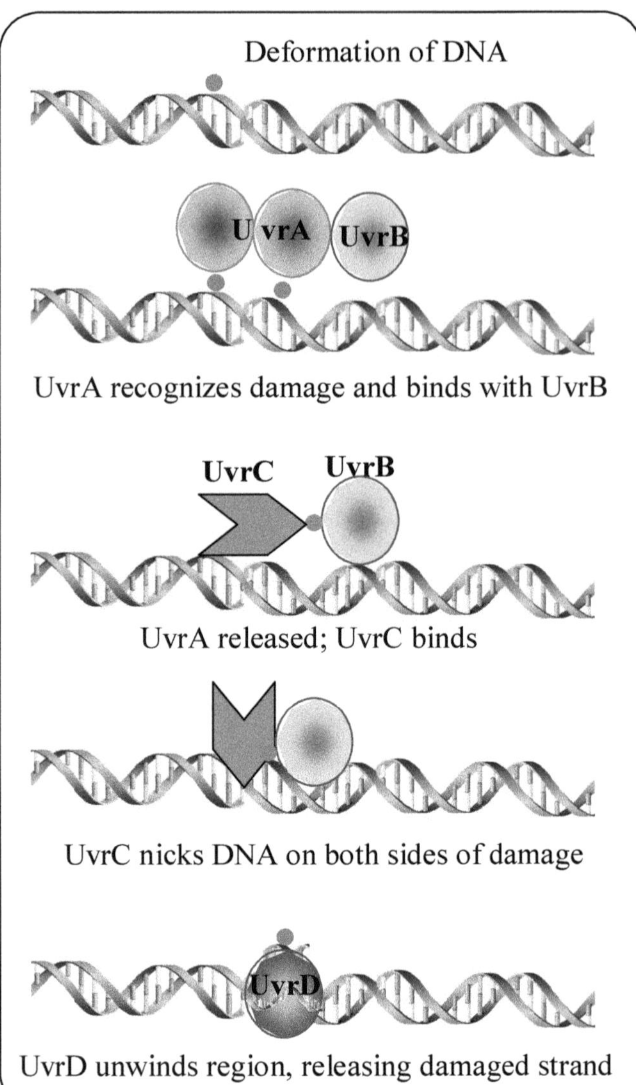

Figure 8.20: In the Uvr Repair Pathway, and UvrA2UvrB Heterotrimer Translocate along DNA and Recognizes Damage, which Triggers the Release of UvrA and the Formation of UvrB-DNA Complex with the Lesion-containing Region Unwound. UvrC binds and nicks on each side of the damage site. The helicase UvrD unwinds the region to release the damaged strand. DNA pol I and ligase complete the repair.

8.25 Two Major Pathways Repair Double-Strand Breaks

DNA double-strand breaks–broken chromosomes are the most severe form of DNA damage that can occur, particularly in eukaryotes. Unrepaired breaks in a linear chromosome result in the loss of any chromosome fragments lacking centromeres, and aberrant repair can result in a chromosome –chromosome fusions, translocations, inversions, and deletions. Even if the two correct broken ends are matched up and rejoined, loss of sequence from the broken ends is a severe risk. Cells have evolved two very different methods to repair double-strand breaks. The first method of double-strand break repair (DSBR) is homologous recombination (HR), which is a collection of related pathways. HR depends on the presence of a homologous donor sequence (a sister chromatid or homologous chromosome) that can be used to accurately replace any sequences that may have been lost from a broken DNA end. In the absence of an available homologous donor, cells instead use a pathway called nonhomologous end joining (NHEJ), which essentially entails the direct ligation of DNA ends, at some risk of loss of sequence prior to repair. Interestingly, both DSBR pathways share most of their components with two normally occurring pathways: HR is at the heart of meiotic recombination during gamete formation and NHEJ is the pathway used during somatic recombination in the immune system.

The basic steps of double-strand break repair via HR are shown in **Figure 8.21**. Many of the factors required for HR are encoded by RAD genes, so named because they were originally identified in yeast mutants sensitive to ionizing radiation (X-rays). The key steps of this repair pathway following recognition of the break are as follows:

1. The 5′–ends of the break are exonucleolytically degraded to produce single-stranded overhanging ends with 3′ –OH termini. This strand resection requires a complex of MRN (Mre11, Rad50, Nbs1) proteins in mammals or MRX (Mre11, Rad50, Xrs2) in yeast. This complex is required for resection but does not have all the enzymatic activities to perform the resection itself and must, therefore, recruit other (unknown) factors. The single-stranded DNA is immediately coated with RPA.
2. With the assistance of Rad52 and other factors, Rad51 replaces RPA to form a Rad51 filament.

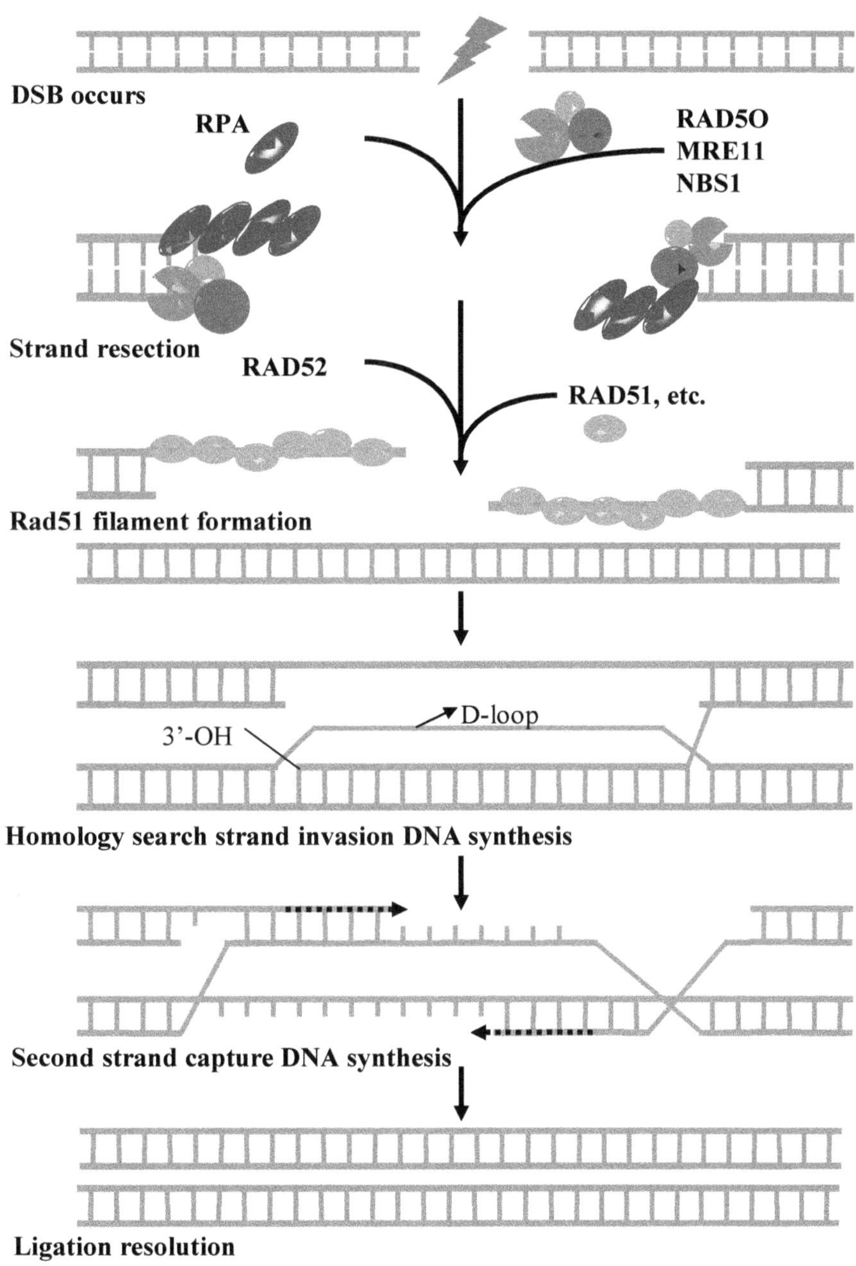

Figure 8.21: Double-strand Break-repairs by Homologous Recombination.

3. The 3′ –end of the single-strand DNA invades a homologous duplex to form a D-loop. Rad51 catalyzes this strand invasion step.
4. DNA polymerase extends from the invading 3′ –end, enlarging the D-loop. The second 3′ –end anneals to the D-loop and primes synthesis by DNA polymerase.
5. Ligase joins the free ends.
6. The resulting crossover structures, known by the name of Holliday junctions, are resolved. The resolution entails either the joint action of helicases and topoisomerases to dissolve the junctions or the action of resolvases that cleave and relegate the junctions.

The net result of HR is that the sequence surrounding a DSB is accurately repaired using intact information from another chromosome. If a sister chromatid is used, identical information is used for repair. If a homologous chromosome is the source of information, any sequences lost from the site of the DSB will be replaced with the sequences present on the homolog which may or may not be identical to the sequences lost during damage. If these sequences differ, the result can be that a previously heterozygous region becomes homozygous during repair, a type of gene conversion.

In the absence of homologous sequences to guide accurate repair, cells must still repair DSBs as efficiently as possible to avoid chromosomal aberrations. This is often necessary during G1 when no sister chromatids are readily available, and homologous chromosomes may be challenging to access (particularly in large mammalian genomes) or absent in typically haploid eukaryotes such as yeast. In this case, the repair is carried out using the NHEJ pathway.

The central player in NHEJ is the Ku heterodimer, a Ku70 and Ku80 protein complex. Ku was first identified as an autoantibody target in patients with the autoimmune disease scleroderma and is highly conserved throughout eukaryotic species. The Ku heterodimer recognizes DNA ends, and two heterodimers form a scaffold that holds the ends together and allows other enzymes to act on them. The MRN/MRX complexes that function in HR also appear to assist Ku in bringing the ends together during NHEJ. DNA–dependent protein kinase (DNA-PK), which DNA activates to phosphorylate protein targets, also participates in end juxtaposition. One of DNA –PK's phosphorylation targets are the protein Artemis, which in its activated form has both exonuclease and endonuclease activities and can trim overhanging ends to provide blunt ends compatible with ligation-specialized DNA polymerases (PolL and PolM in mammals) fill in any remaining single-stranded gaps. These processing steps can result in loss or alteration of sequence near the breakpoint but are essential to creating ligatable ends if the initial DSB does not have blunt ends that terminate properly with 5′ phosphates and 3′ –OH. The actual joining of the double-stranded ends is performed by DNA ligase IV, which functions in conjunction with the protein XRCC4 (LIF1 - Lif1p). Mutations in any of these components render cells more sensitive to radiation.

8.26 Meiotic Recombination is due to Homologous Recombination

We have introduced homologous recombination as a mechanism for DNA repair; it also results in recombination. The frequency of recombination is not constant throughout the genome but is influenced by both global and local effects. The overall frequency may be different in oocytes and sperm: recombination occurs twice as frequently in females as compared to males. Within the genome, its frequency depends upon chromosome structure; for example, crossing over is suppressed in the vicinity of the condensation and inactive regions of heterochromatin, and there are recombination "hotspots" where recombination frequency is very high.

As in the case of double-strand break repair, meiotic recombination is initiated by a double–strand break in DNA. In this case, however, the break is deliberately induced by the action of a meiotic endonuclease, Spo11. Spo11 is related to the type II topoisomerases, and it undergoes a similar reaction cycle in which it induces double-strand breaks and becomes covalently attached to the generated ends, as shown in **Figure 8.22**. Yeast Spo11 makes ~150-200 breaks throughout the genome in each meiosis, resulting in ~1 crossover per chromosome arm. The MRX/N complex, assisted by other factors (Sae2 in yeast), then releases Spo11 to respect the DNA ends as occurs during double-strand break repair. The subsequent steps are very similar to repair (**Figure 8.22**), except that meiosis-specific proteins perform the required functions in some steps. Rad51 is active during meiosis, a Rad 51–related recombination enzyme, Dmc1, is expressed exclusively during meiosis, and both proteins are needed to complete meiosis.

The different stages of meiotic recombination correlate with the visible progress of chromosomes through the five stages of the meiotic prophase, as depicted in **Table 8.2**. The beginning of meiosis is marked by the point at which individual chromosomes become visible. Each chromosome

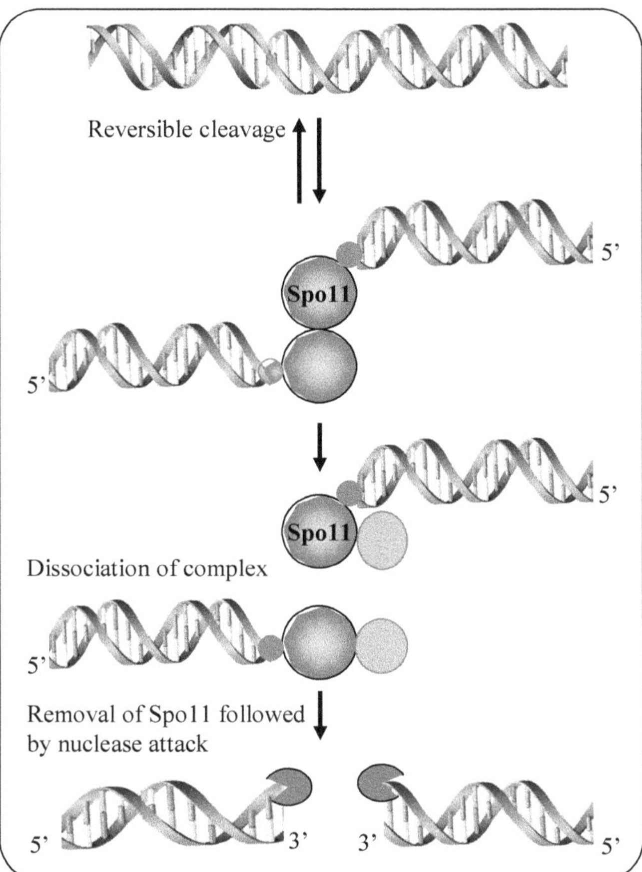

Figure 8.22: The Spo11 Endonuclease Initiates Meiotic Recombination by Creating a Double-strand Break in a Recipient Chromosome. Spo11 is covalently attached to the DNA ends. The MRN/X complex, in combination with other factors, releases Spo11 and respects the ends.

Table 8.2: Recombination Occurs during the First Meiotic Prophase

Progress through Meiosis	*Molecular Interactions*
Leptotene	
Condensed chromosomes become visible, often attached to nuclear envelope	Each chromosome has replicated, and consists of two sister chromatids
Zygotene	**Initiation**
Chromosomes begin pairing in limited region or regions	DNA break is induced by Spo11
Pachytene	**Strand exchange**
Synaptonemal complex extends along entire length of paired chromosomes	Single strands exchange
Diplotene	**Assimilation**
Chromosomes separate, but are held together by chiasmata	Region of exchanged strands is extended
Diakinesis	**Resolution**
Chromosomes condense, detach from envelope; chiasmata remain. All four chromatids become visible	DNA is cleaved and relegated to generate intact products

has completed replication and consists of two sister chromatids containing duplex DNA. The homologous chromosomes approach one another and begin to pair in one or more regions, and pairing extends until the entire length of each chromosome is opposed to its homolog. The process is called synapsis or chromosome pairing. It is thought that initiation of sites of crossovers may be what leads to the initial pairing (as opposed to pairing preceding crossover formation).

When the synapsis process is completed, the chromosomes are laterally associated as synaptonemal complexes with a characteristic structure in each species. However, there is wide variation in the details between species **Figure 8.23**. During synaptonemal complex formation, each chromosome (sister chromatid pair) condenses around a proteinaceous structure called the axial element. The axial elements of corresponding chromosomes then become aligned, and the synaptonemal complex forms a tripartite structure in which the axial elements, now called lateral elements, are separated from each other by a central element. Each chromosome at this stage appears as a mass of chromatin bounded by a lateral element. A fine but dense central element separates the two lateral elements. The triplet of parallel dense strands lies in a single plane that curves and twists along its axis. Two groups of proteins play central roles in the formation of these structures. First, the cohesins form a single linear axis for each pair of sister chromatids from which chromatin loops extend, forming the lateral elements. (Cohesins belong to a general group of proteins involved in connecting sister chromatids so that they segregate properly at mitosis or meiosis.) Second, the Zip proteins form the central element, creating transverse filaments that connect the lateral elements.

The distance between the synapsed homologous chromosomes is greater than 200 nm, which is considerable in molecular terms (the diameter of DNA is 2 nm). Thus, a major problem in understanding the role of the synaptonemal complex is that although it aligns homologous chromosomes, it is far from bringing homologous DNA molecules into contact.

As described for repair via homologous recombination, an intermediate joint molecule is formed in which there is a connection between the two DNA duplexes. The recombinant joint is the point at which an individual strand of DNA crosses from one duplex to the other. An essential feature of a recombinant joint is its ability to move along the duplex. Such mobility is called branch migration. **Figure 8.24** illustrates the migration of a single strand in a duplex. The branching point can migrate in

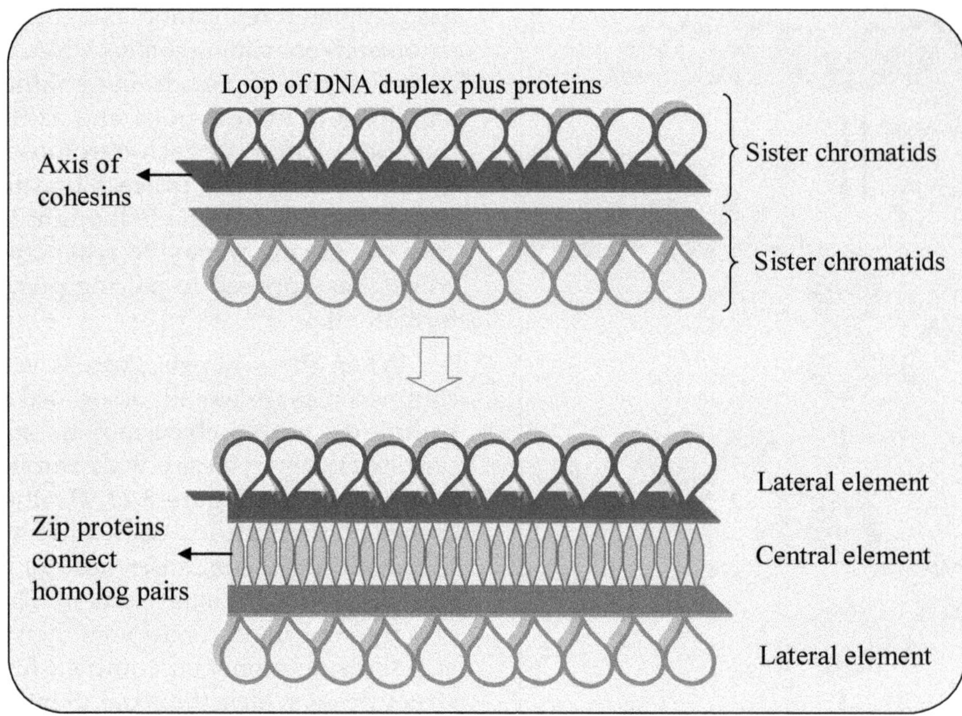

Figure 8.23: Juxtaposition of Synaptonemal Complex.

either direction as the other strand invasion displaces one strand, and branch migration thus results in the structure in **Figure 8.24**, a molecule with two recombinant joints, which can be at varying distances from the original breakpoint.

These crossovers formed during recombination must be resolved to restore two separate duplex molecules. The resolution requires nicking of the DNA backbones. We must easily visualize this process by viewing the joint molecule in one plane as a Holliday junction. This is illustrated in **Figure 8.25**, which depicts a molecule with a single recombinant joint (for simplicity) with one duplex rotated relative to the other. The outcome of the reaction depends on which pair of strands is nicked.

Nicking one pair of strands results in splice recombinant DNA molecules. The duplex of one DNA parent is covalently linked to the duplex of the other DNA parent via a stretch of heteroduplex DNA. There has been a conventional recombination event between markers on either side of the heteroduplex region. In contrast, nicking the other pair of strand's results in patch recombinants. These nicking releases the original parental duplexes, which remain intact except that each contains a length of heteroduplex DNA. For a molecule with two recombinant joints, if both joints are resolved in the same way, the original noncrossover molecules will be released, each with a region of altered genetic information that is a footprint of the exchange event. A genetic crossover

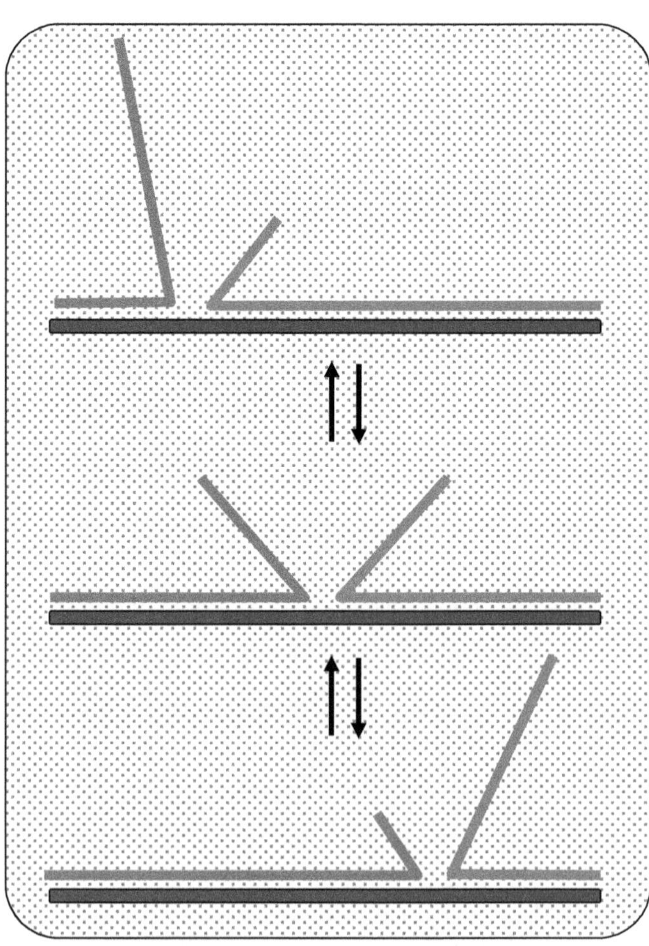

Figure 8.24: Branch Migration can occur in either Direction when an Unpaired Single Strand Displaces a Paired Strand.

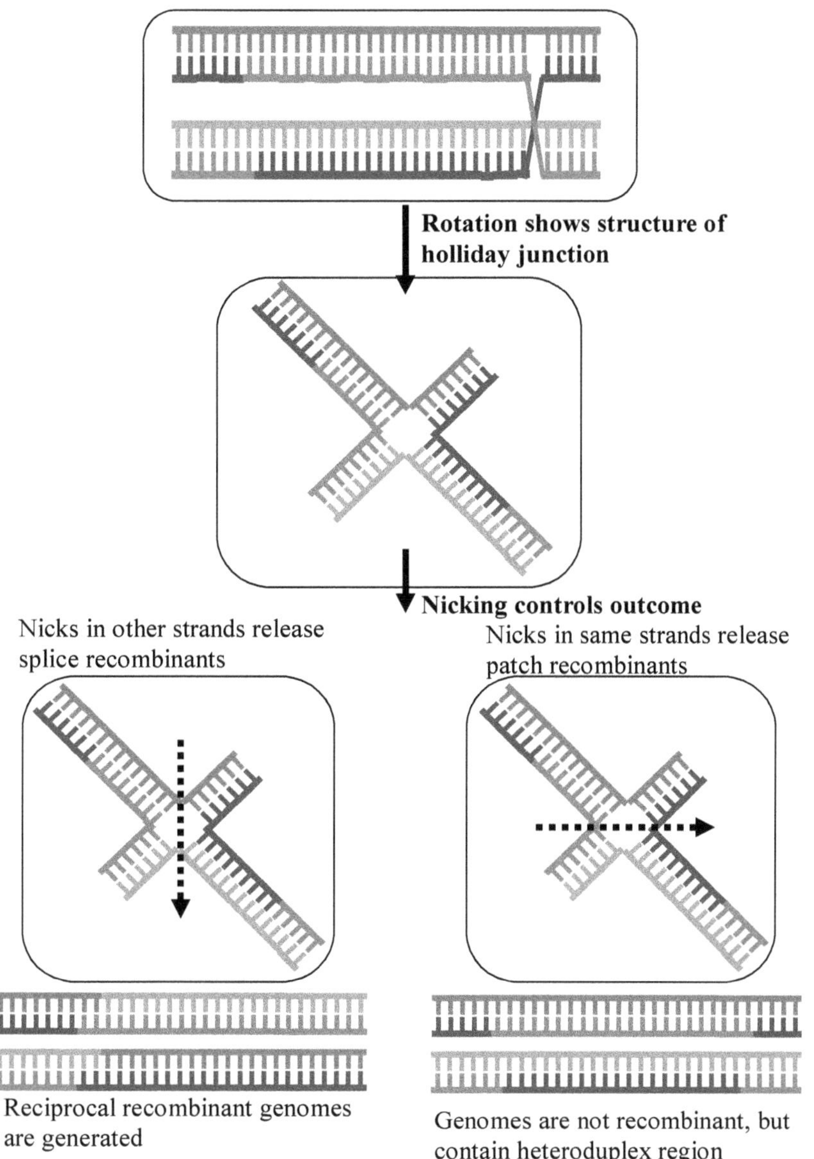

Figure 8.25: The Resolution of the Holliday Junction can Generate Parental or Recombinant Duplexes, Depending on which Strands are Nicked. Both types of products have a region of heteroduplex DNA.

is produced if the two joints are resolved in opposite ways.

Different repair mechanisms are used by the cell (**Table 8.3**). Enzymes are quite helpful in repair systems. Sometimes, it is impossible; hence, an excision repair system removes any damaged base. Depending upon the nature of the lesions, a replication mismatch repair system operates. Post replication mismatch repair system eliminates gaps across from blocking lesions that are not repaired by the other systems.

8.27 Mutators

Repair mechanisms are ingenious; still certain strains in bacteria have been observed, which increases the spontaneous mutation. These strains are called mutators. In humans, these repair defects often lead to severe diseases.

Table 8.3: Comparison of the Genomes of M Leprae, and M Tuberculosis

Characteristics	*M. tuberculosis*	*M. leprae*
Genome size (bp)	4,411,532	3,268,203
Percentage of genome that encodes proteins	90.8 per cent	49.5 per cent
Protein-encoding genes (bp)	3959	1604
Pseudogenes (bp)	6	1116
Gene density (bp/gene)	1114	2037
Average length of gene (bp)	1012	1011

8.28 Repair Defects and Human Diseases

Repair defects may cause various human diseases. Mostly, these are autosomal recessive disorders. Two examples of such diseases are given below.

8.29 Xeroderma Pigmentosum (XP)

It is caused by the defect in a complementation group gene affecting nucleotide excision repair. Individuals suffering from this disorder are susceptible to UV light, develop skin cancers, and are also susceptible to developing neurological abnormalities.

8.30 Hereditary Non-polyposis Colorectal Cancers (HNPCC)

It is a common cancer, and the incidence is approximately 1/200 people worldwide who may develop HOVPCC. The defect is in the genes that encode the human counterparts of the bacterial MutS and MutL proteins. It is dominantly inherited. Cells behave normally if one functional copy of the mismatch repair gene is present, but individuals develop HNPCC if cells have lost this gene, hence mismatch repair deficient.

8.31 Transposons

Transposons are mutagens. They can cause mutations in several ways. They are found in both eukaryotes and prokaryotes.

If a transposon inserts itself into a functional gene, it will probably damage it. Insertion into exons, introns, and even into DNA flanking the genes (which may contain promoters and enhancers) can destroy or alter the gene's activity.

8.32 Transposons

They have been called "junk" DNA and "selfish" DNA. "Selfish" because their only function seems to make more copies of them, and "junk" because there is no obvious benefit to their host. Because of the sequence similarities of all the LINEs and SINEs, they also make up a large portion of the "repetitive DNA" of the cell.

It has been proposed that this DNA may confer some benefit. Retrotransposons often carry some additional sequences at their 3′ end as they insert into a new location. Perhaps these occasionally create new combinations of exons, promoters, and enhancers that benefit the host. For example, thousands of our Alu elements occur in the introns of structural genes. Some of these contain sequences that, when transcribed into the primary transcript, are recognized by the spliceosome; these can then be spliced into the mature mRNA, creating a new exon, which will be transcribed into a new protein product. Alternative splicing can provide not only the new mRNA (and thus protein) but also the old; in this way, nature can try out new proteins without the risk of abandoning the old one. L1 elements inserted into the introns of functional genes reduce the transcription of those genes without harming the gene product. The longer the L1 element, the lower the level of gene expression. Some 79 per cent of our genes contain L1 elements, and perhaps there is a mechanism for establishing the baseline level of gene activity; telomerase, the enzyme essential for maintaining chromosome length, is closely related to the reverse transcriptase of LINEs and may have evolved from it, RAG-1 and RAG-2. The proteins encoded by these genes are required to assemble the repertoire of antibodies and T-cell receptors (TCRs) used by the adaptive immune system. The mechanism resembles that of the cut-and-paste method of Class II transposons, and the RAG genes may have evolved from them. If so, the event occurred some 450 million years ago when the jawed vertebrates evolved from jawless ancestors. Only jawed vertebrates have an adaptive immune system and the RAG-1 and RAG-2 genes that make it possible. In Drosophila, the insertion of transposons into genes has been linked to the development of resistance to DDT and organophosphate insecticides.

8.33 Transposable Elements move within the Genome and may Disrupt the Genetic Function

Transposons are also known as Mobile DNA. These are segments of DNA that can move around to different positions in the genome of a single cell. In the process, they may cause mutations to increase (or decrease) the amount of DNA in the genome. These mobile segments of DNA are sometimes called "jumping genes."

There are three distinct types: Class II Transposons consisting only of DNA that moves directly from place to place. Class III Transposons are also known as Miniature Inverted-repeats Transposable Elements or MITEs. Retrotransposons (Class I) first transcribe the DNA into RNA and then use reverse transcriptase to make a DNA copy of the RNA to insert a new location.

Recent work on transposable elements in plants has led us back to Gregor Mendel and his observation of the inheritance of round and wrinkled peas. Alleles of a single gene control the two phenotypes. It is now known that the wrinkled phenotype is caused by the absence of an enzyme, starch–branching enzyme (SBEI), that controls the formation of branch points in starch molecules. The lack of starch synthesis in wrinkled peas leads to sucrose accumulation higher water content, and osmotic pressure in the developing seeds. As the seeds mature, wrinkled (genotype rr) lose more water than smooth seeds (RR or Rr), producing a wrinkled phenotype.

The structural gene for SBEI has been cloned and characterized in both wild–type and mutant genotypes. In the rr genotype, the SBEI protein is nonfunctional because the SBEI gene is interrupted by a 0.8-kb insertion, producing an abnormal RNA transcript. The inserted DNA has 12 –bp inverted repeats at each end that are highly homologous to the ITRs of the transposable element Ac from maize and to other Ac-like elements from snapdragons and parsley. (**Figure 8.26**)

8.34 Copia Elements in Drosophila

Copia Insertion depends on the presence of the ITR sequences and seems to occur preferentially at specific target sites in the genome. The Copia–like elements demonstrate regulatory effects at the point of their insertion in the chromosome. Certain mutations affecting eye color and segment formation are due to Copia insertions within genes. For example, the eye color mutation white–apricot (Wa), an allele of the white (w) gene, contains a Copia element within the gene. Transposition of the Copia element out of the Wa allele can restore the allele to wild–type.

Copia elements are only one of the approximately 30 families of transposable elements in Drosophila, each of which is present in up to 20 to 50 copies of the genome. Together, these families constitute about 5 per cent of the Drosophila genome and over half of the middle repetitive DNA of this organism. One study suggests that 50 per cent of all visible mutations in Drosophila result from transposons being inserted into otherwise wild-type genes.

Each copia element consists of approximately 5000 to 8000 bp of DNA, including a long direct terminal repeat (DTR) sequence of 276 bp at each end. Within each DTR is an inverted terminal repeat (ITR) of 17 bp (**Figure 8.27**). The short ITR sequences are

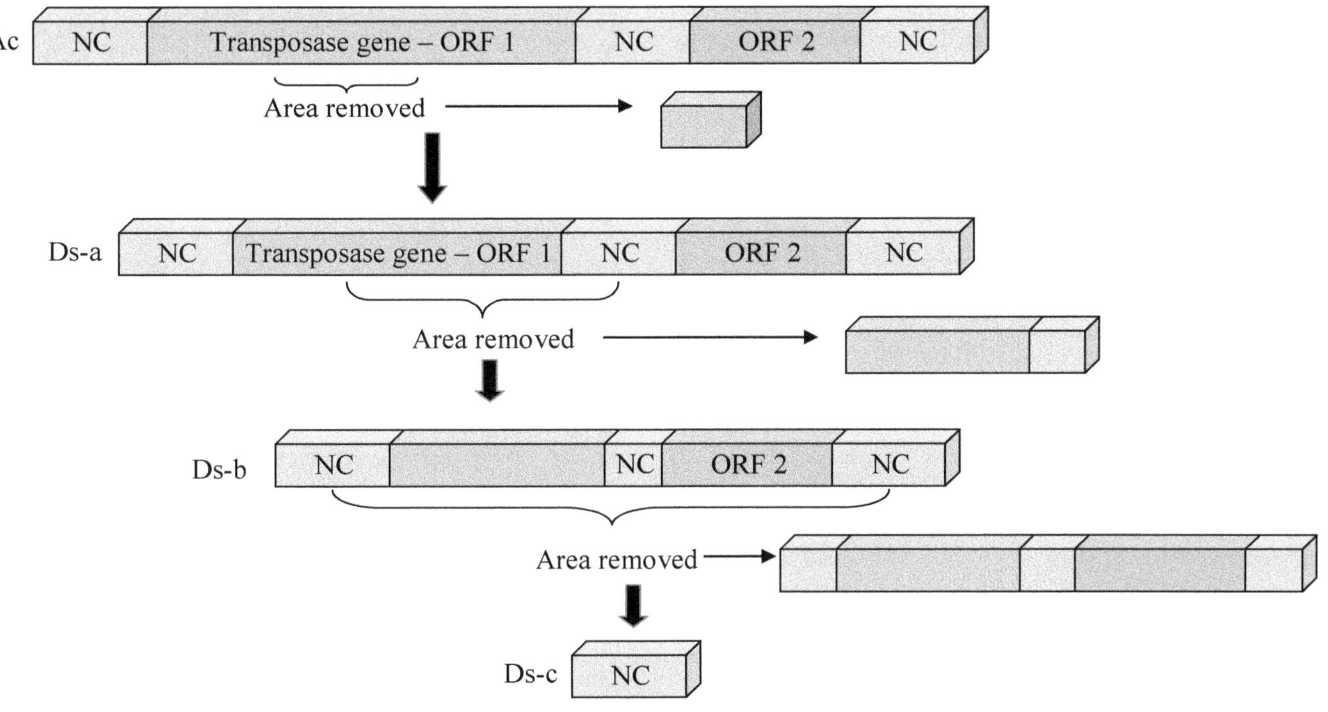

Figure 8.26: Comparison of Ac and Ds Elements.

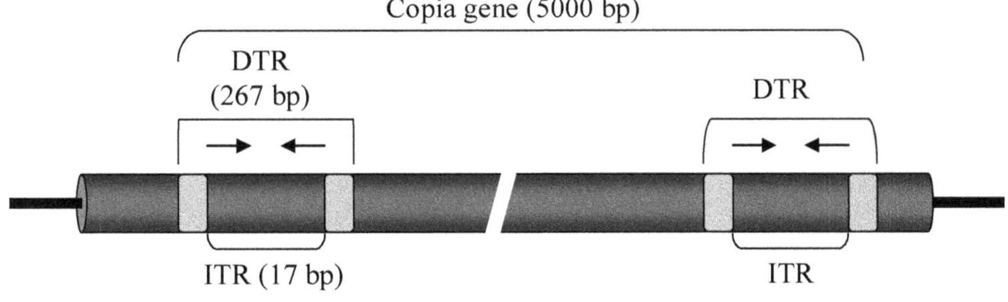

Figure 8.27: Structural Organization of a Copia Transposable Element in Drosophila melanogaster with Terminal Repeats.

characteristic of copia elements. The DTR sequences are found in other transposons in other organisms, but they are not universal.

8.35 Transposable Elements in Humans

Like other eukaryotes, the human genome is riddled with DNA derived from transposons. Recent genomic sequencing data reveal that approximately half of the human genome comprises transposable DNA elements.

The major families of human transposons are the long-interspersed elements and short interspersed elements (LINES and SINES). LINES consist of DNA elements of about 6-kilo base pairs in length, present in up to 850,000 copies. In all, LINES account for 21 per cent of human genomic DNA. SINES are about 100 to 500bp long with about 1.5 million copies present in human cells. SINES comprise about 13 per cent of human genomic DNA. Other families of transposable elements account for a further 11 per cent of the human genome. As coding sequences comprise only about 5 per cent of the human genome, there is about tenfold more transposable element DNA in the human genome than the DNA in functional genes.

Although most human transposons appear inactive, the potential mobility and mutagenic effects of transposable elements have far-reaching implications for human genetics, as seen in a recent example of a transposon "caught in the act." The case involves a male child with hemophilia. One cause of hemophilia is a defect in blood–clotting factor VIII, the product of an X-linked gene. It has been found LINES inserted at two points within the gene. Researchers were interested in determining if one of the mother's X chromosomes also contained this specific LINE. If so, the unaffected mother would be heterozygous and pass the LINE-containing chromosome to her son. The surprising finding was that the LINE sequence was not present on either of her X chromosomes but was detected on chromosome 22 of both parents. This suggests that this mobile element may have transposed from one chromosome to another in the gamete–forming cells of the mother before being transmitted to the son. LINE insertion into the human dystrophin gene has resulted in at least two separate cases of Duchenne muscular dystrophy. In one case, a transposon was inserted into exon 48. In another case, a transposon was inserted into exon 44, leading to a frameshift mutation and premature termination of transportation of the dystrophin protein. There are also reports that LINES have been inserted into the APC and c-myc genes, leading to mutations that may have contributed to the development of some colon and breast cancers. In the latter cases, the transposition had occurred within one or a few somatic cells.

SINE insertions are also responsible for several human disease cases. In one case, an Alu element integrated into the BRCA2 gene, inactivating this tumor suppressor gene, and leading to a familial case of breast cancer. Other genes that Alu integrations have mutated are the factor IX gene (leading to hemophilia B), the ChE gene (leading to cholinesterasemia), and the NF1 gene (leading to neurofibromatosis).

8.36 Class II Transposons

Many transposons move by a "cut and paste" process: the transposon is cut out of its location and inserted into a new location.

This process requires an enzyme – a transposase –encoded within some of these transposons. Transposase binds to both ends of the transposon, which consist of inverted repeats; that is, identical sequences reading in opposite directions a sequence of DNA that make up the target site. Some transposases require a specific sequence as their target site; others can insert the transposon anywhere in the genome.

The DNA at the target site is cut in an offset manner (like the "sticky ends" produced by some restriction enzymes). After the transposon is ligated to the host DNA, the gaps are filled in by base pairing. This creates identical direct repeats at each end of the transposon. Often, transposons lose their gene for transposase, but if somewhere in the cell there is a transposon that can synthesize the enzyme, their inverted repeats are recognized, and they, too, can be moved to a new location (**Figure 8.28**).

8.37 Miniature Inverted-Repeat transposable Elements (MITEs)

The recent completion of the genome sequence of rice and *C. elegans* has revealed that their genomes contain thousands of copies of a recurring motif consisting of almost identical sequences of about 400 base pairs flanked by

Characteristic inverted repeats of about 15 base pairs, such as

5′ GGCCAGTCACAATGG.~400 nt.CCATTGTGACTGGCC 3′

3′ CCGGTCAGTGTTACC.~400 nt.GGTAACACTGACCGG 5′

MITEs are too small to encode any protein. Just how they are copied and moved to new locations is still uncertain. Probably larger transposons that do encode the necessary enzyme and recognize the same inverted repeats which are responsible.

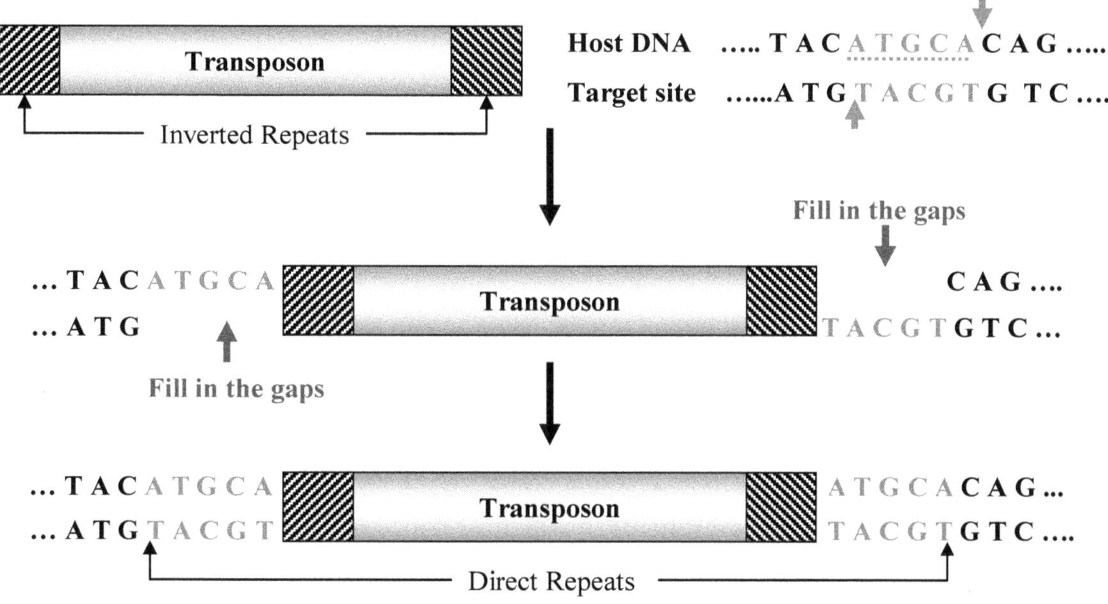

Figure 8.28: Tansposons in Humans.

There are over 100,000 MITEs in the rice genome (representing some 6 per cent of the total genome). Some of the mutations found in certain strains of rice are caused by insertion of a MITE in the gene. MITEs have also been found in the genome of humans, Xenopus, and apples.

8.38 Retrotransposons

The RNA copies are transcribed back into DNA using a reverse transcriptase and inserted into new locations in the genome called retrotraspsomes. Many retrotransposons have long terminal repeats (LTRs) at their ends that may contain over 1000 base pairs in each. Like DNA transposons, retrotransposons generate direct repeats at their new insertion sites. In fact, these direct repeats are often a clue that the intervening stretch of DNA arrived there by retro transposition. Approximately 40 per cent of the entire human genome consists of retrotransposons.

HIV-1 causes AIDS, and other human retroviruses (*e.g.*, HTLV-1, the human T-cell leukemia virus) behave like retrotransposons. The RNA genome of HIV-1 contains a gene for reverse transcriptase and one for integrase. The integrase serves the same function as the transposases of DNA transposons. The DNA copies can be inserted anywhere in the genome.

The human genome contains some 850,000 LINEs (representing some 21 per cent of the genome). Most of these belong to a family called LINE-1 (L1). These L1 elements are DNA sequences that range in length from a few hundred to as many as 9,000 base pairs. Only about 50 L1 elements are functional "genes"; that is, can be transcribed and translated. The functional L1 elements are about 6,500 bp long and encode three proteins, including an endonuclease that cuts DNA and a reverse transcriptase that makes a DNA copy of an RNA transcript. L1 activity proceeds as follows: RNA polymerase II transcribes the L1 DNA into RNA. RNA is translated by ribosomes in the cytoplasm into proteins. The proteins and RNA join and re-enter the nucleus. The endonuclease cuts a strand of "target" DNA, often in the intron of a gene. The reverse transcriptase copies the L1 RNA into L1 DNA and inserts it into the target DNA, forming a new L1 element. This copy-paste mechanism can increase the number of LINEs in the genome. The diversity of LINEs between individual human genomes makes them valuable markers for DNA "fingerprinting." Variation in the length of L1 elements is seen; for example, transcription of active L1 elements sometimes continues downstream into additional DNA, producing a longer transposed element. Reverse transcription of L1 RNA often concludes prematurely and produces a shortened transposed element.

While these elements are not functional, they may play a role in regulating the efficiency of transcription of the gene in which they reside. Occasionally, L1 activity makes and inserts a copy of a cellular mRNA (thus a natural cDNA). Lacking introns as well as the necessary control elements like promoters, these genes are not expressed. They represent one category of pseudogene.

8.39 Short Interspersed Elements (SINEs)

SINEs are short DNA sequences (100–400 base pairs) that represent reverse-transcribed RNA molecules initially transcribed by RNA polymerase

III; that is, molecules of tRNA, 5S rRNA, and some other small nuclear RNAs. The most abundant SINEs are the Alu elements. There are over one million copies in the human genome (representing about 11 per cent of the total DNA). Alu elements consist of a sequence of 300 base pairs containing a site recognized by the restriction enzyme AluI. They appear to be reverse transcripts of 7S RNA, part of the signal recognition particle. Most SINEs do not encode any functional molecules and depend on the machinery of active L1 elements to be transposed, copied, and pasted in new locations.

8.40 Role of Mutation in Human Health

8.40.1 ABO Blood Types

The ABO system is based on a series of antigenic determinants found in erythrocytes and other cells, particularly epithelial cells. There are three alleles of a single gene that encodes the glycosyltransferase enzyme. The H substance is modified to either the A or B antigen due to the product of the I^A or I^B allele, respectively–failure to alter the substance results from the null allele.

The glycosyltransferase gene has been sequenced in 14 people with different ABO status. When the DNAs of the I^A and I^B alleles are compared, four consistent nucleotide substitutions are found. It is assumed that the resulting changes in the amino acid sequence of the glycosyltransferase gene product led to the different modifications of the H substance.

The I^O allele situation is unique and exciting. Individuals who are homozygous for this allele have type O blood, lack glycosyltransferase activity, and fail to modify the H substance. Analysis of the DNA of this allele shows one consistent change that is unique compared with the sequences of the other alleles-the deletion of a single nucleotide early in the coding sequence, causing a frameshift mutation. A complete messenger RNA is transcribed, but at translation, the reading frame shifts at the point of the deletion and continues out of frame for about 100 nucleotides before a stop codon is encountered. At this point, the polypeptide chain terminates prematurely, resulting in a nonfunctional product.

These findings provide a direct molecular explanation of the ABO allele system and the basis for the biosynthesis of the corresponding antigens. The molecular basis for the antigenic phenotypes is clearly the result of mutations within the gene encoding the glycosyltransferase enzyme.

8.40.2 Dystrophin: The Protein Product of the DMD Gene

A 13-Kb mRNA would encode a substantial protein. The predicted amino acid sequence was deduced from the triplet codons and was compared with other known proteins. A major region of homology was found with a protein called spectrin, which is an actin-binding protein located in the cytoskeleton of the red blood cell. The actin-binding domain is the N terminal region. The bulk of the molecule (**Figure 8.29**) comprises a large rod domain consisting of 24 repeats of similar sequences of nearly 109 amino acids. A 150–amino acid cysteine-rich region is found after the rod domain. At the X terminus, there is a 420 amino acid region that is believed to interact with other membrane proteins.

The protein itself was identified by immunologic staining. Portions of the cDNA of the mouse muscular dystrophy gene have been subcloned into plasmids that juxtaposed the cDNA with an inducible gene (trypE) (**Figure 8.30**), stimulation of transcription in the bacteria resulted in the production of a fusion protein, which then was purified and used to immunize rabbits. The resulting antisera was used to stain muscle protein samples. The antibody recognized a protein of molecular weight over 400 kDa. This protein was expressed in muscle cells, including skeletal muscle, smooth muscle, and cardiac muscle. It was also found in brain tissue but nowhere else in appreciable abundance. Even in muscle, the protein was of a minor species, comprising less than 0.002 per cent of the total muscle protein.

Homology with spectrin suggested that the new muscle protein might also be associated with cell membranes. Antibodies were used to affect immunofluorescent staining of tissue sections for the protein. Fluorescent staining was found to outline

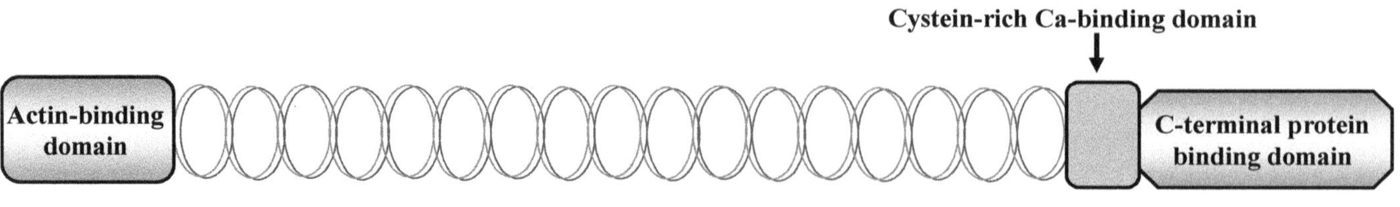

Figure 8.29: Actin Binding Domain of Dystrophin Gene.

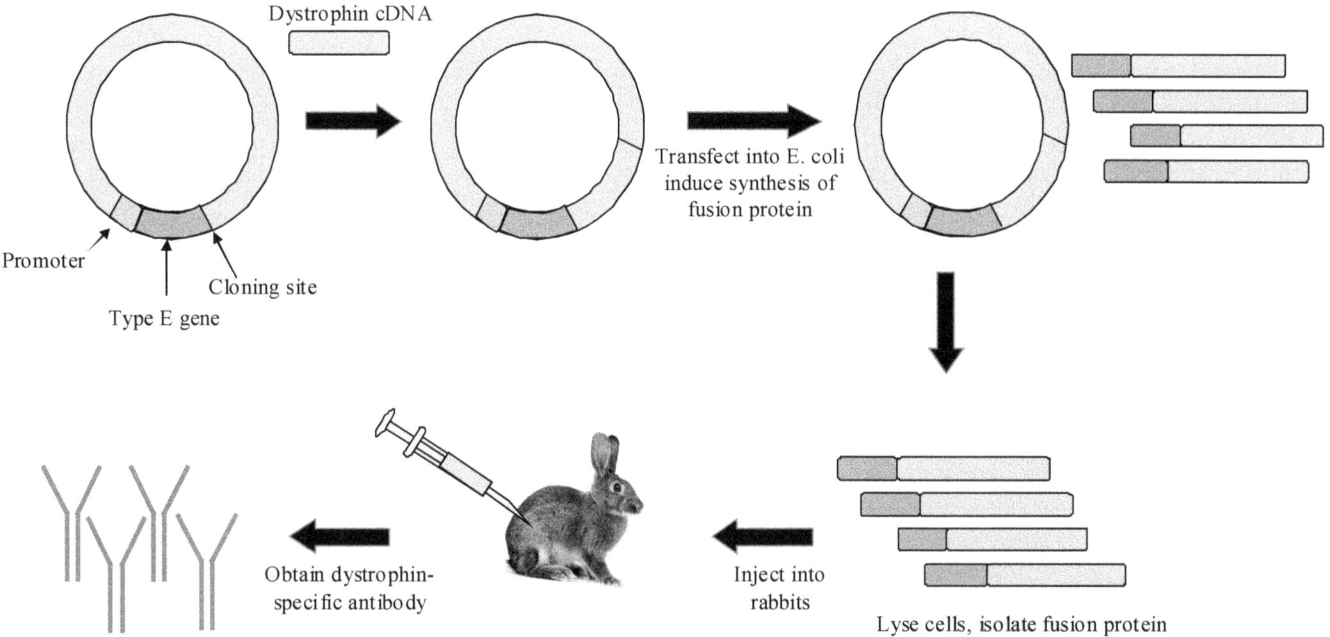

Figure 8.30: Preparation of Antibodies to the Portions of Dystrophin.

each muscle cell in cross-section, confirming cell membrane localization.

This new protein had not been known to exist prior to its discovery through positional cloning of the muscular dystrophy locus. It was therefore dubbed dystrophin, though its function was still unknown. The protein was named after the name of the disease.

8.41 Trinucleotide Repeats Huntington Disease

Within the closing decade of the twentieth century, 14 neurological disorders were shown to result from the expansion of unstable trinucleotide repeats, establishing this once-unique mutational mechanism as the basis of an expanding class of diseases. Trinucleotide repeat diseases can be categorized into two subclasses based on the location of the trinucleotide repeats: diseases involving non-coding repeats (untranslated sequences) and diseases involving repeats within coding sequences (exonic). The large body of knowledge accumulating in this fast-moving field has provided exciting clues and inspired many unresolved questions about the pathogenesis of diseases caused by expanded trinucleotide repeats. For example, the genes responsible for fragile X syndrome, myotonic dystrophy, and Huntington's disease contain a specific trinucleotide DNA sequence repeated many times. While these types of repeated sequences are also present in the non-mutant (normal) allele of each gene, the mutations found in individuals with these disorders significantly increase the number of times the trinucleotide is repeated.

FMR-I gene is responsible for Fragile X syndrome this gene may have several hundred to several thousand copies of the trinucleotide sequence CGG, located in the 5′ un-translated region of the gene. Individuals with up to 54 copies are normal and do not display mental retardation associated with this syndrome. Individuals with 54 to 230 copies are considered as carriers. Although they are normal, their offspring may contain even more copies and express the syndrome. The large regions of CGG repeats in the gene's regulatory region result in the loss of expression of the FMRP protein, thought to be RNA binding protein affecting brain cell function.

Myotonic dystrophy, or DM, is the most common form of adult muscular dystrophy, found in 1 in 8000 individuals. It is a dominantly inherited disorder and not as severe as DMD and it is highly variable in symptoms and age of onset. Mild myotonia (atrophy and weakness) of the musculature of the face and extremities is most common. Cataracts, reduced cognitive ability, and cutaneous and intestinal tumors are also part of the syndrome.

The affected gene, MDPK, is located on the long arm of chromosome 19. It encodes a serine-threonine protein kinase, MDPK. This protein is the product of 15 exons of the gene, the last of which encodes the 3′–untranslated RNA sequence of the mRNA. It is this sequence that houses the multiple copies of the trinucleotide CTG. Individuals with 5 to 37 copies of the CTG repeat are normal, and the number of copies is stable from generation to generation. Individuals with more than 37 copies exhibit symptoms ranging from mild to severe, with onset occurring anywhere between birth and age 60. Both the severity and onset are directly related to the size of the repeated sequence. Minimally affected patients have up to 150 repeats, while severely affected patients have up to 1500 copies of the CTG triplet. Genetic anticipation is exhibited in the offspring. The mechanism by which these repeated regions cause DM is still uncertain.

Huntington's disease (HD) is inherited as an autosomal dominant. It is a neurodegenerative disease. Its gene is located on chromosome 4. The gene contains the trinucleotide CAG sequence, repeated 10 to 35 times in normal individuals. These sequences are located within the coding region of the gene and code a polyglutamine tract. The CAG repeat sequence exists in significantly increased numbers (up to 120) in diseased individuals. A much earlier onset occurs when the number of copies is closer to the upper range. Interestingly, in another disorder, spinobulbar muscular atrophy (Kennedy disease), the involved gene (different from that in Huntington's disease) also contains repeated copies of the CAG triplet. However, only 35 to 60 copies of the sequence cause individuals to be affected.

In Huntington disease, the repeats lie within the coding portion of the gene. In the case of Huntington's disease, this causes the mutant protein known as the huntingtin. This contains an excess of glutamine residues. The gene responsible for fragile X syndrome, the repeat is upstream (the 5′ end) of the gene's coding region, in an area that is most often involved in regulating gene expression. In the case of myotonic dystrophy, the repeat is downstream of the coding region (the 3′ end). These two examples reveal that the location of the gene can vary in different diseases.

Repeated sequences expand from generation to generation. However, the exact mechanisms have yet to be discovered. It is thought that errors during replication are associated with repairing the damaged DNA. The instability of short repeats is more prevalent in humans than in other organisms.

9

Chromosome Mutations

The DNA is packed in the chromosomes; hence, any chromosome change will cause mutation. These changes can be deletion, duplication, inversion, *etc.*, as shown in **Figure 9.1**. Most chromosomal anomalies cause deleterious mutations; change in chromosomal number results in lethal mutations. Sometimes chromosomal breakage also leads to mutations that occur in the middle of the gene, hence will disrupt the whole function. If an organism gains or loses one or more chromosomes, it is called aneuploidy, which mainly occurs during reduction division, *i.e.*, meiosis. The loss of a single chromosome of a diploid organism is called monosomy. The gain of one chromosome is called trisomy. If more than 2n chromosomes are present, it is called polyploidy (**Table 9.1**).

Different terms are being used if the change in chromosomal number occurs.

Table 9.1: Variation in Chromosome Numbers and Terms Used

Term	*Explanation*
Aneuploidy	2n ± x chromosomes
Monosomy	2n – 1
Trisomy	2n + 1
Tetrasomy, pentasomy, *etc.*	2n + 2, 2n + 3, *etc.*
Euploidy	Multiples of n
Diploidy	2n
Polyploidy	3n, 4n, 5n ....
Triploidy	3n
Tetraploidy, pentaploidy, *etc.*	4n, 5n, *etc.*
Autopolyploidy	Multiples of the same genome
Allopolyploidy (Amphidiploidy)	Multiples of different genomes

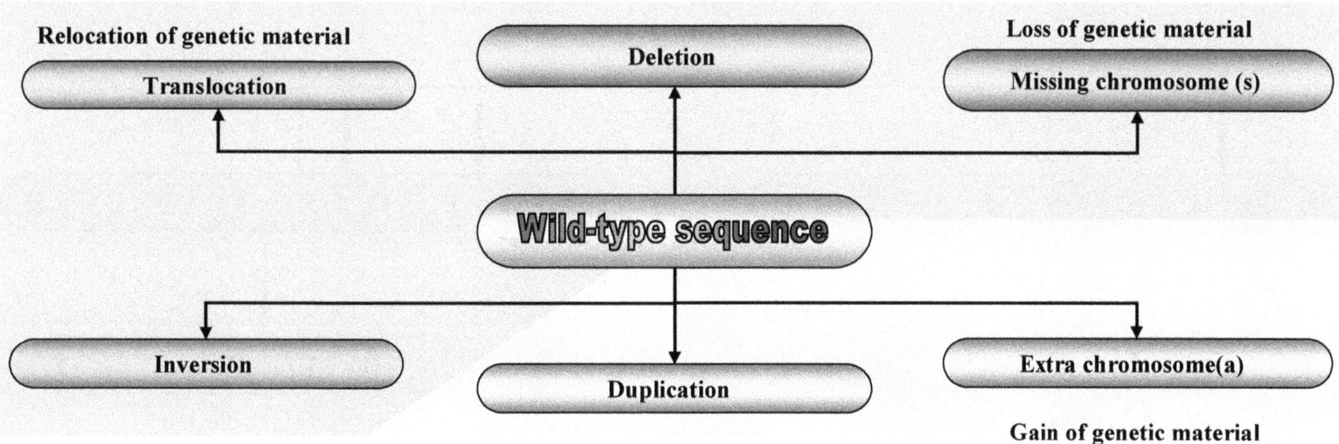

Figure 9.1: Deletion, Duplication, and Inversion of the Chromosome.

9.1 Mechanism of Change

One of the mechanisms of change is the breakage and then rejoining of chromosomes. This may result in a change in the DNA which may cause mutations. There are two mechanisms through which the mutation may be generated (1) breakage and rejoining (ii) illegitimate crossing over between repetitive segments **Figure 9.2**. This is called chromosomal rearrangement.

9.2 Deletion

The chromosomal breakage may occur, and the two ends may join again. If one bears the centromere due to breakage, the resultant chromosome is reduced in size. The deleted segment will be missing, hence acentric and immobile. Chromosomal breakage may be due to X-rays or gamma rays, which have a high energy source. Two breaks result in interstitial deletion (**Figure 9.3**), and a single break can cause terminal deletion, but two breaks are involved when it is near telomeres. Intragenic deletions inactivate the gene. Intragenic deletions cannot be reverted. Even small omissions can cause phenotypic abnormalities in humans. For example, cri-du chat syndrome is due to heterozygous deletion of the tip of the short arm of chromosome 5 (**Figure 9.4**).

Most of the time, these deletions in humans arise spontaneously in the germline of the normal parents of an affected person. However, some human deletions are produced by meiotic irregularities in a parent

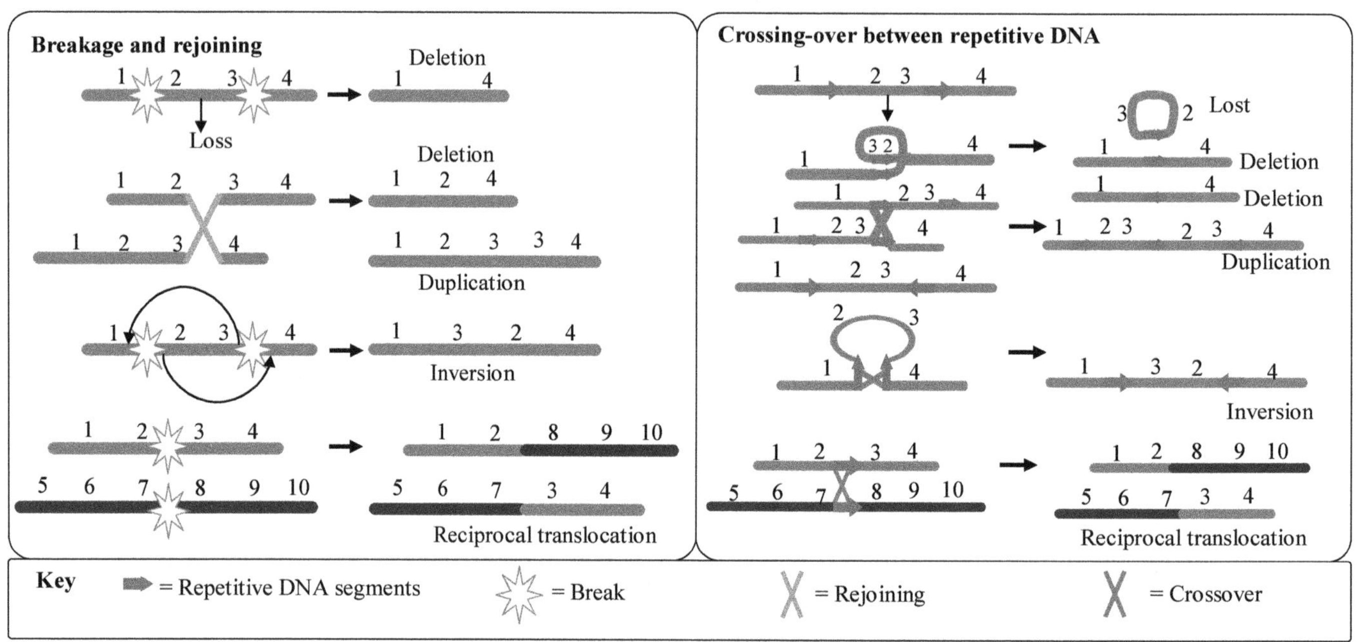

Figure 9.2: Chromosomal Rearrangements.

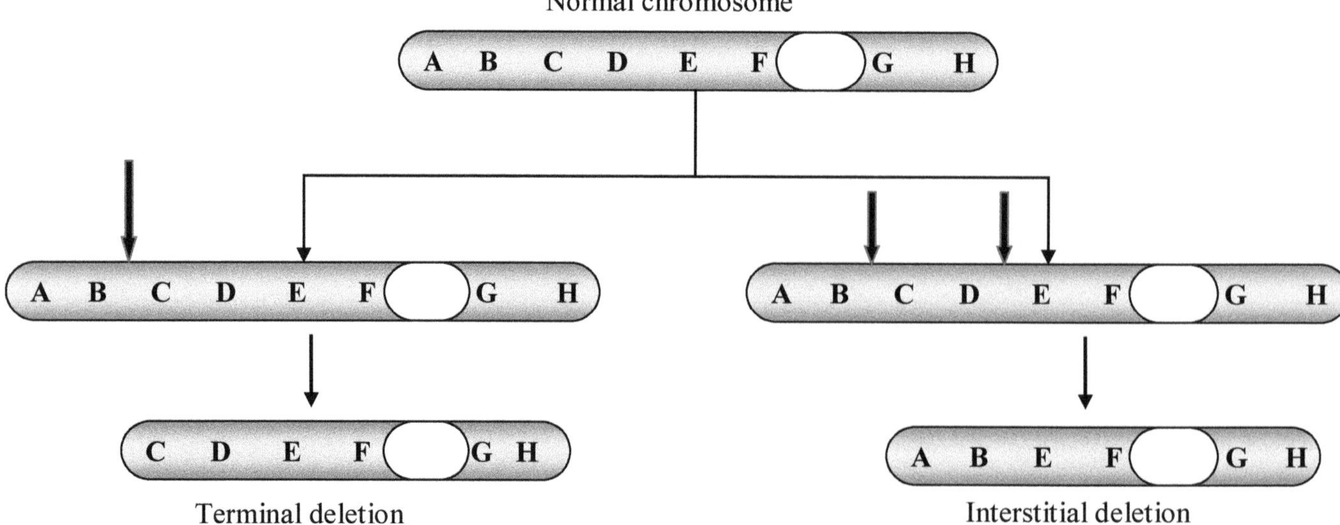

Figure 9.3: Terminal and Interstitial Deletions.

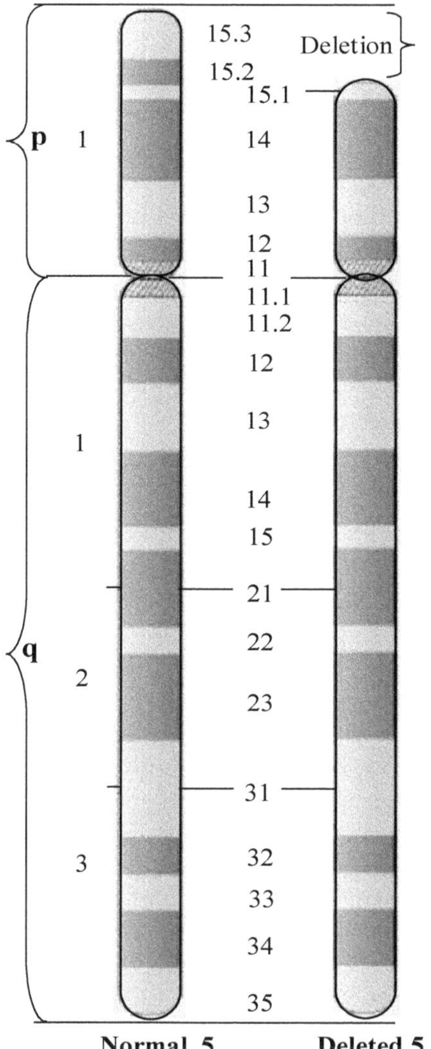

Figure 9.4: The Cause of the cri du Chat Syndrome is the Loss of the Tip of the Short Arm of One of the Homologs of Chromosome 5.

heterozygous for another type of rearrangement. Cri-du-chat syndrome can result from one of the parents being heterozygous for a translocation. Deletions can be mapped by using a technique called in-situ-hybridization. Various tumors also show deletions (**Figure 9.5**).

9.3 Duplication

Extra copies of some chromosomal regions are produced, known as duplication. The duplication can be seen adjacent, or one region could be at a new position on the same or different chromosomes. There may be three copies of the chromosome region in diploid organisms. These heterozygotes are generally called duplications because they contain the product of duplication events. An exciting structure occurs, paired at the meiosis time (**Figure 9.6**). The adjacent duplication is tandem. Now if inbreeding occurs in tandem duplication, the resultant progeny may be homozygous, carrying four copies of the duplicated chromosomal region. Such individuals may show asymmetrical pairing (**Figure 9.7**). Crossing over at meiosis will result in tandem triplication of the chromosomal region. Duplication in the Drosophila X chromosome produces a dominant mutation where the Drosophila produces a slit-like eye instead of the normal oval-shaped eye.

Some of the best evidence that tandem duplication and their reciprocal deletions, which are driven by the unequal crossovers, come from studies of the genes which determine the structure of human hemoglobin. There are α and β subunits; however, they differ in different ages. The fetus has two α subunits and two

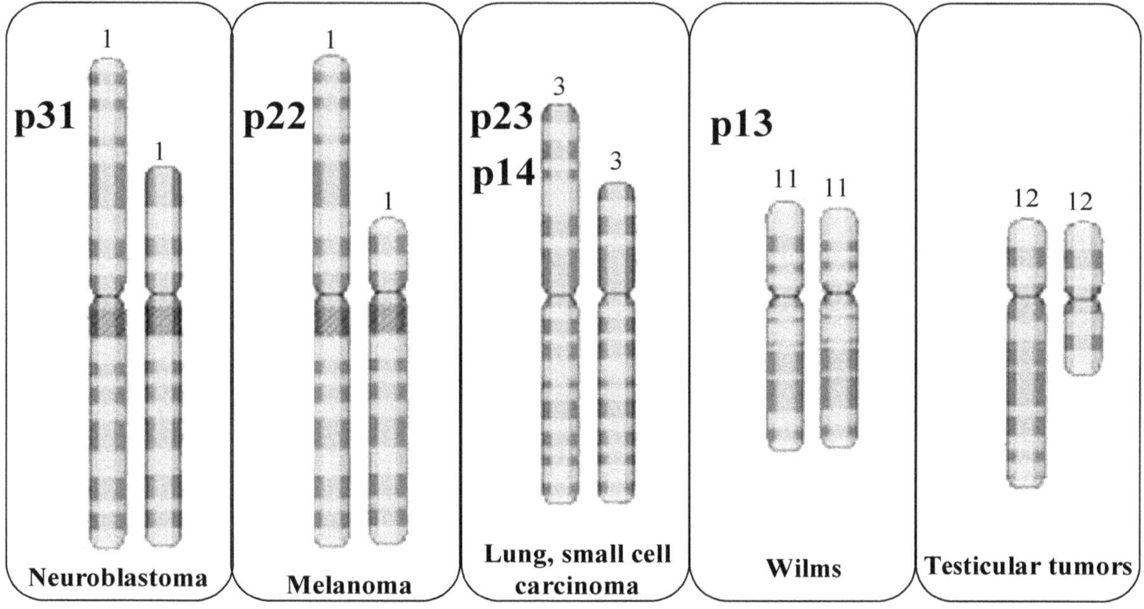

Figure 9.5: Deletions were found Consistently in Several Types of Human Solid Tumors. (Band numbers indicate recurrent breakpoints).

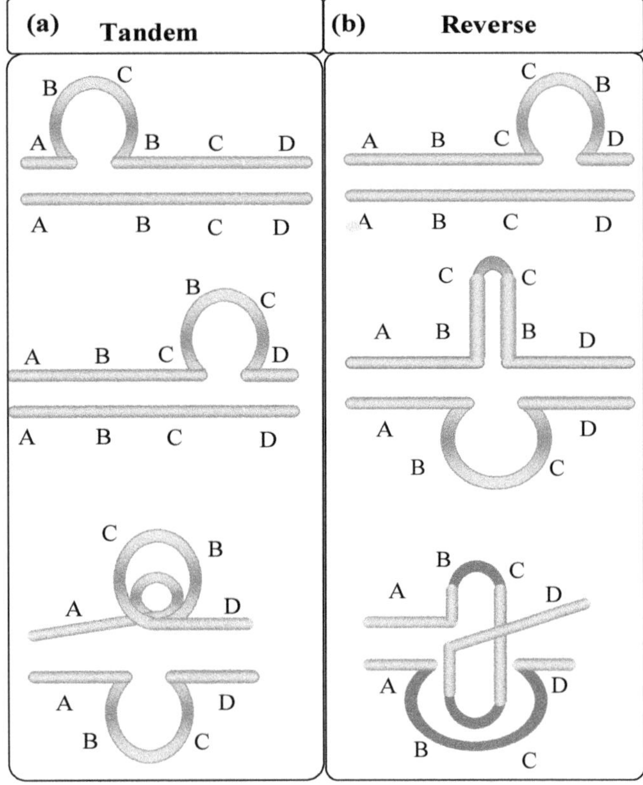

Figure 9.6: Possible Pairing Configurations in heterozygotes of a Standard Chromosome and a side-by-side Duplication.

g subunits (α2 g2) adult has 2α and β subunits (α2/β2). Different gene combinations determine these. Some genes are linked, and some are not. The linked genes are g-δ-β group. Some people have part of δ and part of β (Lepore hemoglobin) and some have part of g and part of β (Kenya hemoglobin). These rare hemoglobin's are examples of unequal crossing over (**Figure 9.8**).

In **Figure 9.8** the origins of the reciprocal cross-over products called anti-Lepore and anti-Kenya are depicted.

9.4 Inversion

Inversions involve two chromosomal breaks and rejoining, with the broken piece reincorporated in the opposite orientation from which it naturally occurs. When they include the centromere, they are called pericentric inversions. When they do not include the centromere, they are called paracentric inversions. Both types of inversions arise in mitotic cells. If they arise in precursors of the gametes, they may produce abnormal genomes as they progress through meiosis. Recombination between homologous chromosomes is an essential part of each normal meiosis. The probability of non-disjunction is significantly increased if there is no recombination.

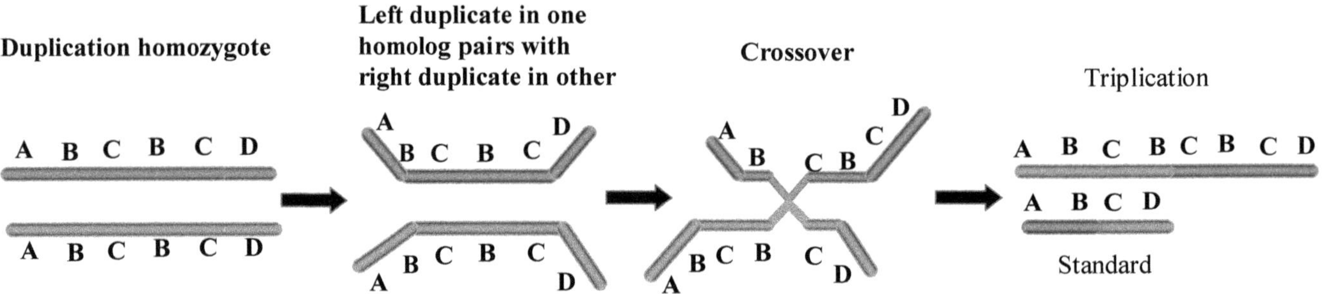

Figure 9.7: Generation of Higher Orders of Duplications by Asymmetric Pairing followed by Crossing-over in a Duplication Homozygote.

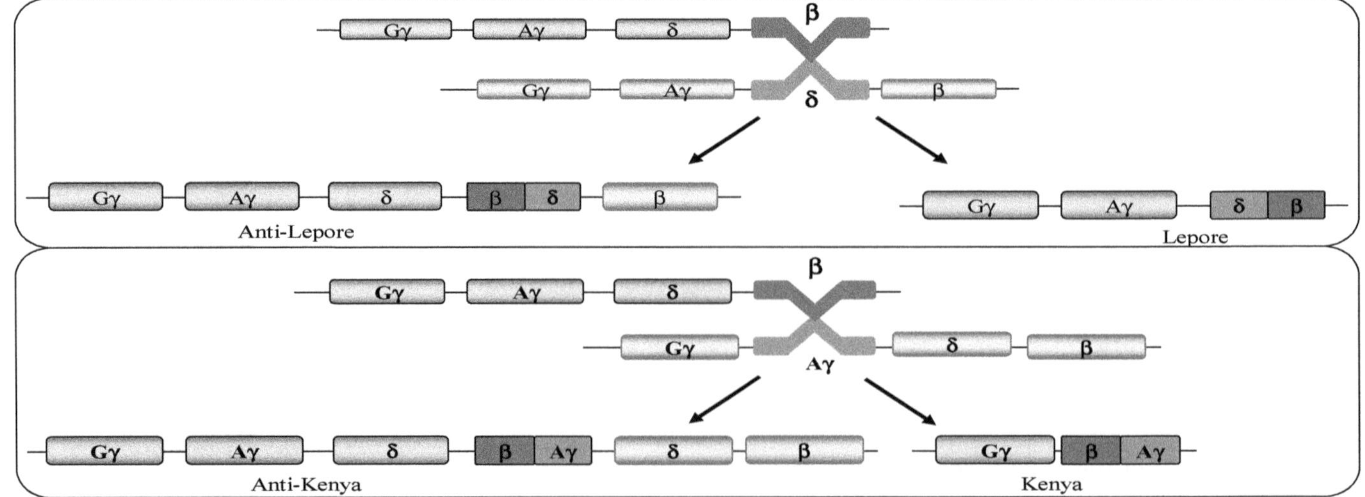

Figure 9.8: Variant Human Hemoglobin Subunits by Unequal Crossing-over in the γ-δ-β Genetic Region. Lepore and Kenya subtypes of hemoglobin are deletions.

However, recombining a chromosome with an inversion and its normal homolog may produce two abnormal chromosomes. A paracentric inversion results through meiosis with recombination within the inversion. Of the four gametic products, one is normal, one has the inversion, one has an acentric chromosome, and one has a dicentric chromosome. The acentric chromosome cannot survive. The dicentric chromosome may be pulled apart during mitosis, with a random loss or gain of genetic material. The possible outcomes of inversions at the DNA level are shown in **Figure 9.9**. Sometimes paired homologs may form an inversion loop (**Figure 9.10**). Inversion can be paracentric when the centromere is outside inversion; spanning the centromere is pericentric.

Paracentric
A B C D E F
A B E D C F
Pericentric
A B C D E F
A D C B E F

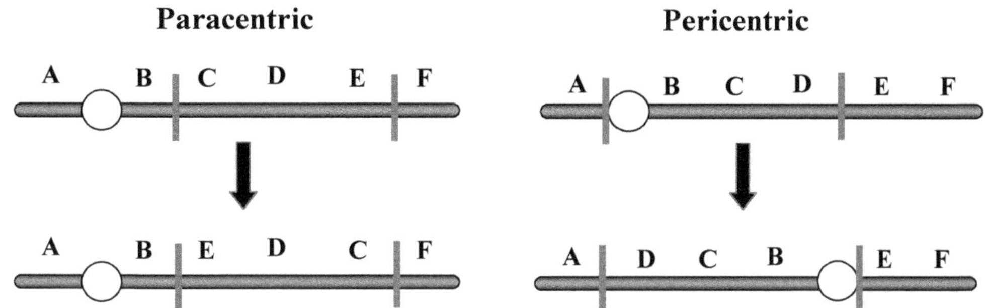

The net genetic effect of pericentric and paracentric is almost the same.

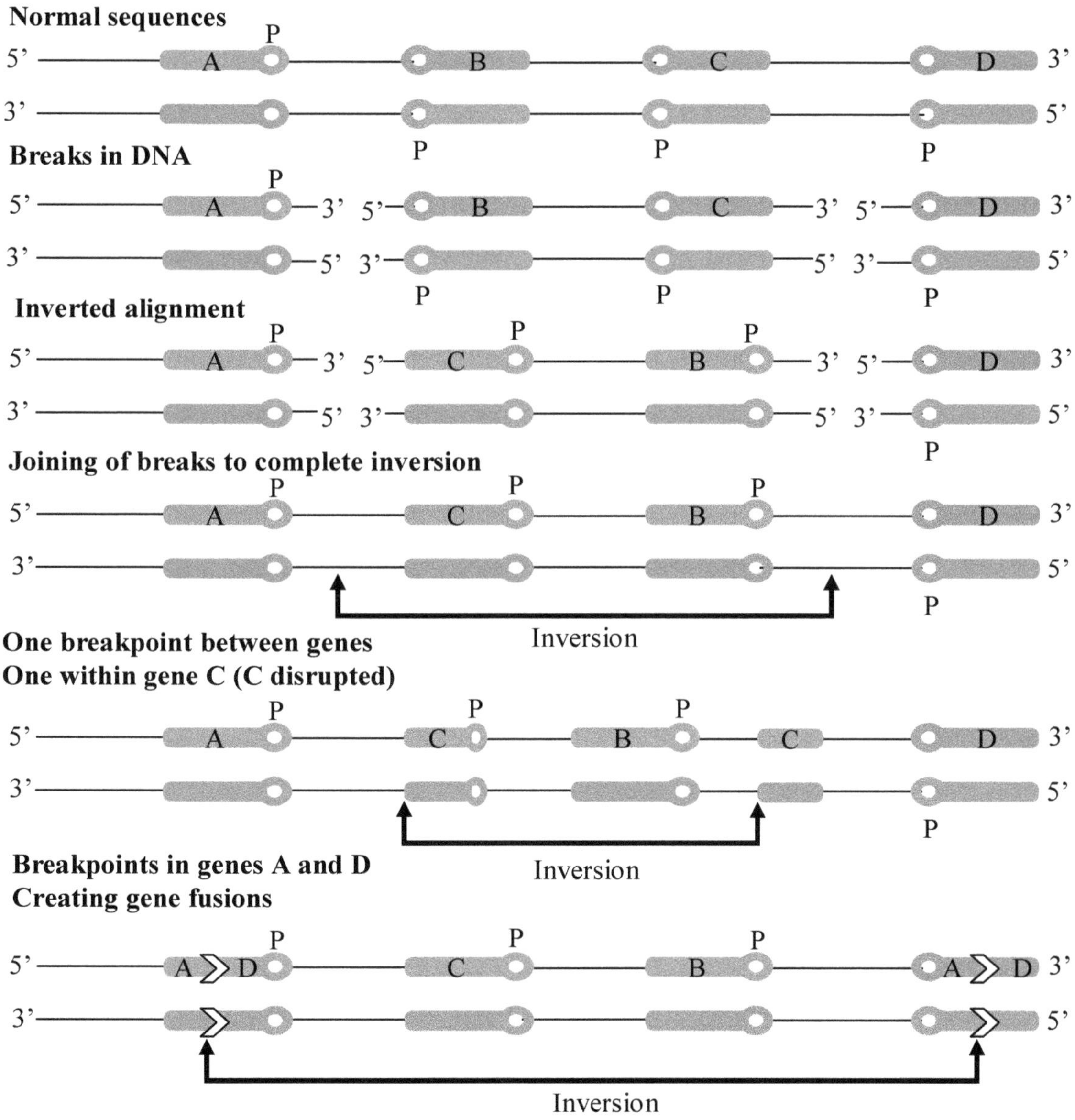

Normal sequences
Breaks in DNA
Inverted alignment
Joining of breaks to complete inversion
Inversion
One breakpoint between genes
One within gene C (C disrupted)
Inversion
Breakpoints in genes A and D
Creating gene fusions
Inversion

Figure 9.9: Effects of Inversions at the DNA Level. Genes are represented by A, B, C, and D.

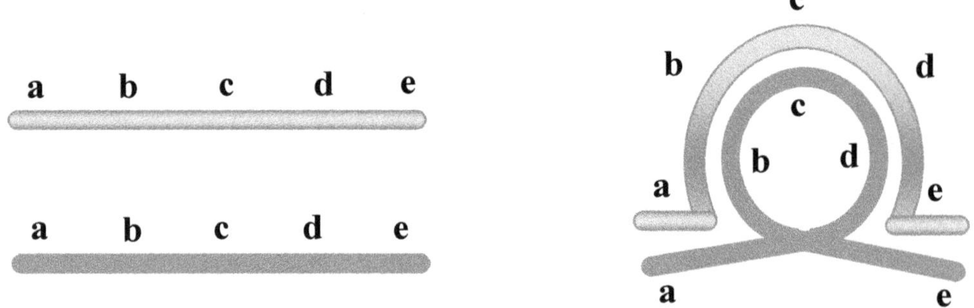

Figure 9.10: The Chromosomes of Inversion Heterozygotes Pair in a Loop at Meiosis.

9.5 Translocation

When chromosomal parts exchange with one another, it results in chromosomal rearrangements. Translocations can be centric fusion reciprocal or merely the loss of material from one chromosome attached to a different chromosome. This type of translocation is called reciprocal. Translocations, too, can be inherited or de novo. Normal individuals can have balanced translocations. However, when they form gametes, they may not include the correct amount of chromosomal material. They can give too much or too little resulting in trisomy or monosomy of chromosomes or portions of chromosomes. If the male is a carrier for translocation, there is a lower risk of recurrence than if the female is the carrier. Many carriers go undetected and are recognized only after the birth of affected children, followed by karyotyping of the child and then for the parents. Couples detected as carriers can be offered amniocentesis and elective termination of chromosomally unbalanced fetuses in future pregnancies.

The D and G group chromosomes see Robertsonian translocations or centric fusions. The p arms of both are usually lost in the fusion. If a dicentric chromosome is formed, it is unstable with two centromeres. It is thought that if D and G group translocations frequently occur, it may be because of the homology between the p arm of the chromosome. On the other hand, they are the only chromosomes that can afford to lose p arms because of the redundancy of genes in this arm. When translocation occurs between two 21 chromosomes, the balanced translocation carriers can never have a normal offspring. They will either contribute no 21 chromosomes, which results in monosomy 21 and is lethal, or they contribute the 21/21 translocation, which results in a Down syndrome child. Robertsonian translocation can result in (i) Carriers with translocations involving chromosome 21, (ii) having normal children, (iii) Down syndrome children, and (iv) more than the usual number of miscarriages. When a carrier parent gives the translocation chromosome and one of the normal homologs, the child will be trisomic and have uniparental disomy. Robertsonian translocations involving the same chromosome have a higher incidence of uniparental disomy. This reflects that both arms have come from the same parent chromosome.

Reciprocal translocations can take place between any two chromosomes. A piece of one chromosome is translocated to another chromosome. When this takes place, the result is having a balanced translocation. At the time of meiosis, the balanced translocation carriers produce a variety of gametes, some of which carry the normal homolog, some carry the balanced reciprocal translocation, and some of which result in unbalanced gametes with duplications or lack of the pieces of chromosomes involved in the translocation. These conditions are partial trisomies and partial monosomies, depending upon which combination the fetus receives. When a child has multiple congenital anomalies (MCA) chromosome analysis is carried out. When a duplication or deficiency of a portion of a chromosome is found, parents should be tested for translocation. Amniocentesis should be counseled for future pregnancies. If the translocations are de novo, there is a negligible risk of recurrence.

Balanced reciprocal translocations differ from Robertsonian translocations involving the D and G group chromosomes, where the p arms are all composed of repetitive DNA. Gametes of balanced reciprocal translocation carriers can contain unbalanced gametes with deletions and duplications but do not result in trisomies or monosomies; the results are partial or trisomy.

9.6 Reciprocal Translocations

In a reciprocal translocation, two non-homologous chromosomes break and exchange fragments. Individuals carrying such abnormalities still have a balanced complement of chromosomes and generally have a normal phenotype but with varying degrees of subnormal fertility. The subfertility is caused by problems in chromosome pairing and segregation during meiosis.

Instead of pairing homologous chromosomes as bivalents, the translocation chromosomes and their homologs must form quadrivalents. Segregation in such a strange situation leads to the formation of several genetically unbalanced gametes and hence, offspring with unbalanced genomes that are often lethal.

This is apparent when one considers the types of sperm that could be produced in a hypothetical male with a reciprocal translocation between chromosomes 1 and 2, as illustrated below.

In the testis or ovary, the fraction of gametes that are unbalanced would depend upon how crossing over and segregation has occurred, which is influenced by factors such as the site of the translocated segments. The net result is considerable difficulty in predicting the fertility of a translocation carrier. In general, however, they show a substantial (often greater than 50 per cent) reduction in infertility.

Some of the offspring of translocation carriers are cytogenetically normal, while others carry the translocation of their parent. Translocations are thus heritable and can be perpetuated in populations.

9.7 Centric Fusions

Acentric fusion is a translocation in which the centromeres of two acrocentric chromosomes fuse to generate one large metacentric chromosome. They are also often called Robertsonian translocations, although that term is used by purists to designate a very similar but distinct translocation in which one of the two centromeres is lost. The karyotype of an individual carrying a centric fusion has one less than the normal diploid number of chromosomes.

Meiosis in animals carrying a centric fusion chromosome involves the formation of a trivalent, which is certainly an abnormal structure. Considerable effort has gone into characterizing the effect of this type of translocation on fertility, particularly in cattle and sheep. In general, centric fusions appear to cause a mild reduction in fertility (5-15 per cent), much less severe than in the case of reciprocal translocations.

9.8 Polyploidy, in which more than two Haploid Sets of Chromosomes are Present, is Prevalent in Plants

Polyploidy describes instances where more than two multiples of the haploid chromosome set are found. The naming of polyploids is based on the number of sets of chromosomes found. A triploid has 3n chromosomes; a tetraploid has 4n; a pentaploid, 5n; and so forth. Polyploidy is uncommon in many animal species but is common in lizards, amphibians, and fishes. It is much more frequent in plant species. Odd numbers of chromosome sets are not usually maintained reliably from generation to generation because a polyploid organism with an uneven number of homologs usually does not produce genetically balanced gametes. For this reason, triploids, pentaploids, and so on are not usually found in species depending solely upon sexual reproduction for propagation.

Polyploidy can originate in two ways: (1) The addition of one or more extra sets of chromosomes, identical to the usual haploid complement of the same species, results in autopolyploidy; and (2) the combination of chromosome sets from different species may occur as a consequence of interspecific mating resulting in allopolyploidy (from the Greek word allo, meaning other or different). The distinction between auto- and allopolyploidy is based on the genetic origin of the extra chromosome sets, as illustrated in **Figure 9.11**.

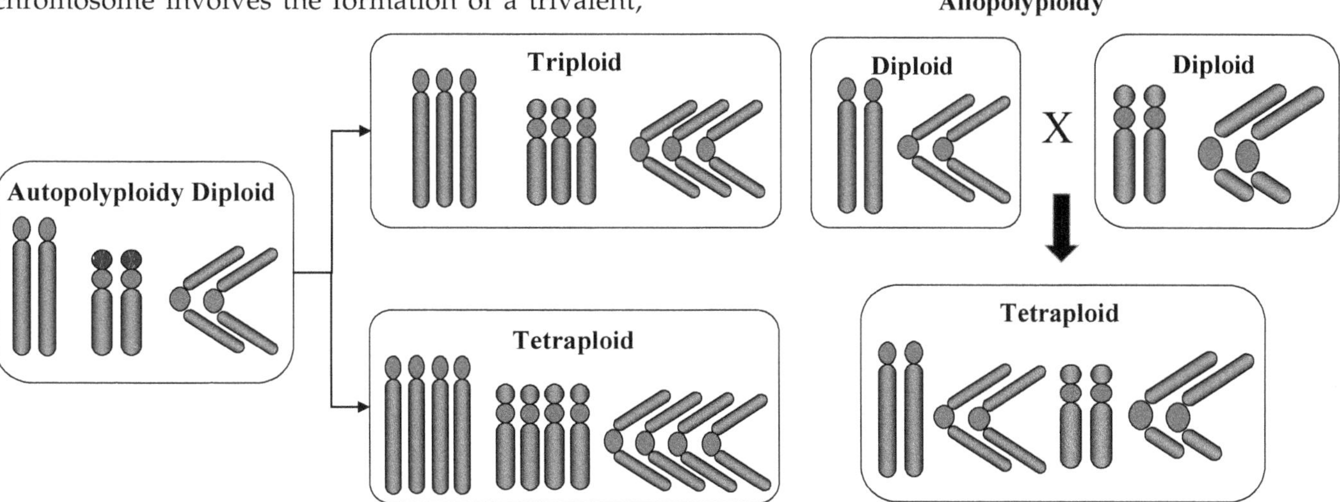

Figure 9.11: Origin of Autopolyploid with an Allopolyploid Karyotype.

Polyploidy can be represented as

A= a1 + a2 + a3 + a4 + ... + an

Where a1, a2, and so on are the individual chromosomes, and where n is the haploid number. Using this nomenclature, a normal diploid organism would be represented simply as AA.

9.9 Autopolyploidy

In autopolyploidy, an additional set of chromosomes are present which are identical to the parent species. Triploids are represented as AAA; tetraploids are AAAA, and so forth.

If the chromosomes do not segregate during meiotic divisions (first–division or second–division non-disjunction), it can produce a diploid gamete. If such a gamete survives and is fertilized by a haploid gamete, a zygote with three sets of chromosomes is produced, or occasionally, two sperm may fertilize with an ovum, resulting in a triploid zygote. Triploids can also be produced under experimental conditions by crossing diploids with tetraploids. Diploid organisms produce gametes with n chromosomes, whereas tetraploids produce 2n gametes. Upon fertilization, the desired triploid is produced.

Because they have an even number of chromosomes, auto tetraploids (4n) are theoretically more likely to be found in nature than are autotriploids; unlike triploids, which often produce genetically unbalanced gametes with odd numbers of chromosomes, tetraploids are more likely to produce balanced gametes when involved in the sexual reproduction.

How polyploidy arises naturally is interesting. Theoretically, if chromosomes have replicated, but the parent cell never divides and re-enters interphase, the chromosome number may be doubled. It is very likely to occur because tetraploid cells can be produced experimentally from diploid cells by applying cold or heat shock to meiotic cells or using colchicines to somatic cells undergoing mitosis.

Colchicine is a natural product. It is an alkaloid derived from the autumn crocus; if added at the time of spindle formation, the replicated chromosomes that cannot be separated at anaphase do not migrate to the poles. When colchicine is removed, the cell can reenter interphase. When the paired sister chromatids separate and uncoil, the nucleus will contain twice the diploid number of chromosomes, therefore 4n. This process is shown in **Figure 9.12**.

The autopolyploid, in general, are larger than their diploid relatives. This is due to a larger cell size rather than a greater number of cells. Although autopolyploids do not contain new or unique information compared with the diploid relative, the flower and fruit of plants are often increased in size, making such varieties of greater horticultural or commercial value. Economically important triploid plants include several potato species of the genus Solanum, Winesap apples, seedless watermelons, and the cultivated tiger lily Lilium tigrinum. These plants are propagated asexually. Diploid bananas contain hard seeds, but the commercial, triploid, "seedless" variety has edible seeds. Tetraploid coffee, peanuts, and McIntosh apples are also of economic value because they are either larger or grow more vigorously than their diploid or triploid counterparts. The commercial strawberry is an octoploid. Chromosomal aberrations can result in serious complications among humans some of the examples are illustrated below:

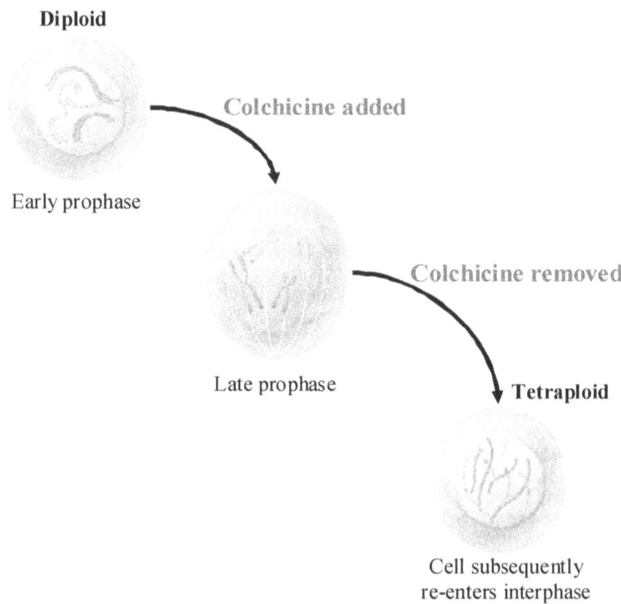

Figure 9.12: Effect of Colchicines on Chromosome.

9.10 Human Aneuploidy

Most of the time, affected fetuses are not able to survive. Trisomy of the autosomal chromosomes is the most common. Monosomies are rarely found even though non-disjunction should fabricate n –7 gametes with a frequency equal to n +7 gametes. This observation made scientists believe that gametes not having a single chromosome are functionally nonfunctional and do not contribute to fertilization. The monosomic embryo is not viable even in the early part of development.

Approximately 15 to 20 per cent of all human conceptions are terminated due to recurrent miscarriage, and about 30 per cent of all resultant abortuses show one or the other chromosomal anomalies. A calculation using these figures (0.20 X

0.30 = 0.06) hypothesizes that approximately 6 per cent of all pregnancies indicate an abnormal number of chromosomes. It is estimated that 10 to 30 per cent of all fertilized eggs in humans contain some error in chromosome number.

The most significant percentages of chromosomal abnormalities are aneuploids. Surprisingly, an aneuploid with one of the highest incidence rates among abortuses is the 45, X condition, which produces an infant with Turner syndrome if the fetus survives. About 70 to 80 per cent of the aborted and live born 45, X conditions include the maternal X chromosome. Thus, the meiotic error leading to this syndrome occurs during spermatogenesis.

9.11 Structural Aberrations

The structural aberrations are illustrated in **Table 9.2**. These include balanced reciprocal translocations, centric fusions, short arm deletions, partial duplications, isochromosomes of xq, ring chromosomes, and pericentric inversions.

Table 9.2: The Structural Aberrations

Karyotype	*Comment*
46,XY,t(5;10)(p13;q25)	A balanced reciprocal translocation involving chromosomes 5 and 10 (breakpoints indicated)
45,XX,t(13;14)(p11;q11)	Centric fusion translocation of chromosomes 13 and 14. A Robertsonian translocation standard carrier
46,XY,del(5)(p25)	Short arm deletion of 5, Cri du chat syndrome
46,XX,dup(2)(p13p22)	Partial duplication of the short arm of chromosome 2 (p13p22)
46,X,i(Xq)	Isochromosome of Xq; Turner female
46,XY,r(3)(p26q29)	Ring chromosome 3 (p26q29)
46,XY,inv(11)(p15q14)	Pericentric inversion of chromosome 11

10

Recombinant DNA Technology

Recombinant DNA technology deals with an essential aspect of all genetic research. The creation of dolly, the discovery of recombinant medicine, and the development of diagnostic kits result from recombinant technology. Kathleen Danna and Daniel Nathans, in 1971, made recombinant technology possible, for which they received a Nobel Prize in Physiology and Medicine. The human genome contains more than 3 billion nucleotides and approximately 21,000 genes. DNA can be cut into smaller fragments using restriction enzymes, which can be manipulated, separated, copied, and investigated individually. This has made researchers see many aspects of gene organization and regulation.

Genetic engineering requires DNA cloning. This is a powerful tool, with the help of which we can study many aspects. Various steps involved in DNA cloning using a plasmid as a vector are mentioned in (**Figure 10.1**).

10.1 Polymerase Chain Reaction (PCR)

10.1.1 History

Short DNA templates, with the help of primers, can be replicated in vitro with an enzyme. This was first described by Kleppe and Co-workers in 1971. At that time, not much attention was given to this discovery. Later, in 1993, Kary Mullis got the Nobel Prize. He has worked in California for a biotechnology company called Cetus Corporation. He was thinking about how to create DNA mutations when he learned that DNA can be amplified in the presence of an enzyme called DNA polymerase.

This enzyme is found in nature in living organisms. Further, this enzyme is used in the processing of cells. This binds with the complementary single DNA strand. The double-strand can be converted into a

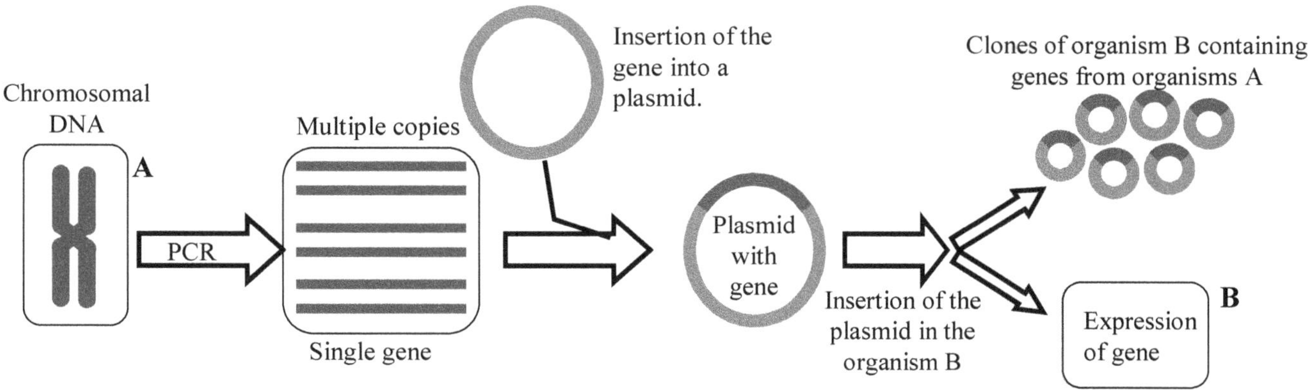

Figure 10.1: Cloning a Gene using a Plasmid as a Vector.

single-strand upon heating. The DNA polymerase may get destroyed while heating, hence requiring DNA polymerase to be added at each cycle.

To solve this problem, people used the DNA from thermophilic bacteria. This bacterium is grown in geysers at more than 110°C (230°F). This DNA polymerase was stable and was used in PCR reactions. This made scientists make the procedure quite simple and automatic.

The first thermostable DNA polymerase was acquired from Thermus aquaticus and known as the "Taq". This is used currently. The shortcoming of taq is that some correct DNA is not synthesized, causing problems and changes (mutations) in the DNA sequences. This happens because Taq polymerase does not have a 3′ → 5′ proofreading activity. In case the DNA polymerase is obtained from the Pwo or Pfu where the proofreading activity is present which decreases the number of mutations significantly. The action of these enzymes is low as compared to Taq. Now there are kits available that have the combinations of both Taq and Pfu due to which polymerization is faster with high fidelity.

PCR can be carried out on DNA that is larger than 10 kilobases, however, in practice, several hundred even thousand bases of DNA are processed. To solve this problem there is a need to keep the balance between accuracy and processivity of the enzyme. In case there are small fragments the probability of errors is decreased. PCR is the technique that increases the quantity of a particular DNA segment. If this amplification is carried out for the sequencing reaction it is called a cycling reaction. Most standard PCR reactions require a range of reagents which include specific DNA primers, buffers, dNTPs (dinucleotide triphosphates, A, G, T, C), Magnesium ions, and template DNA for DNA amplification. There are different steps performed during PCR which are mentioned below:

Denaturation: Single-stranded DNA is created by melting the double-strand. Annealing: Once the primer fits into the single-stranded DNA the attachment between the primer and DNA strand takes place in the presence of DNA polymerase the synthesis starts along with the extension. The extension needs the presence of dNTPs during this process many copies of a gene is generated. The increase in number is on an exponential scale, *i.e.*, 2 to 4 to 8, and so on. In every cycle there is a doubling of PCR amplicons (2n) and on average 30 cycles are run on PCR over a million copies of a single DNA sequence are produced.

10.1.2 Variations on the basic PCR Technique

There are different types of variation in PCR technique which are discussed below: Allele-specific PCR: In this technique, even the single nucleotide can be amplified by designing the primers in such a way so that the primers end overlap the SNP. If primers are mismatched, the PCR products are either absent or incorrect. To overcome this problem, one can use different sets of primers. Assembly PCR - Long gene product can be synthesized by designing long oligonucleotides with short overlapping segments. These primers detect alternations between sense and antisense directions and the overlapping segments. The PCR fragments selectively produce their final product. Asymmetric PCR - One strand is the original DNA, useful for sequencing, hybridization probing where having only one of the two complementary strands is required. PCR is carried out as usual, but with a great excess of primers for the chosen strand. Amplification is slow after the limiting primer has been used up; extra cycles of PCR are required. Recently a modification has been done in this which is called process, Linear-After-The-Exponential-PCR (LATE-PCR), uses a limiting primer with a higher melting temperature (Tm) than the excess primer to maintain reaction efficiency as the limiting primer concentration decreases mid-reaction. Colony PCR - Whole colonies are amplified, the colonies are picked up from the agarose, primers, and the master mixture is added. The extension is 95°C. The denaturation time is at 100°C, and the product of PCR is chimeric. Helicase-dependent amplification is like standard PCR; however, it keeps the constant temperature rather than cycling through denaturation and annealing/extension cycles. There is another enzyme called Helicase, it helps in unwinding the DNA. It is used as an alternative to thermal denaturation. Helicase is used in hot-start PCR. This reduces the nonspecific amplification. The reaction mixture is heated manually at 95°C, and then the polymerase is added. In the case of inter-sequence-specific (ISSR), PCR-only sequences between some repeat sequences are amplified to get the specific fragments. The ISSR can be used only when one of the internal sequences is known. The flanking sequence is identified from the inserts. This procedure is a little complicated as it needs multiple digestions and self-ligations, creating known sequences on either end of the unknown sequences. Another PCR is ligation-mediated [PCR (LM)], which detects vector insertion sites into the genome. Methylation-specific PCR (PCR-MSP) is used where it detects CPG islands. DNA is treated with bisulfate which converts unmethylated cytosine to uracil. As DNA is stranded, one strand has uracil, and another has cytosine. Primers are designed in such a way that they recognize DNA strands with cytosine and another stand with uracil thymine. Like this, unmethylated DNA is amplified. Quantitative PCR is used to measure the quantity of a PCR product. Quantitative real-time PCR (QRT-PCR) uses fluorescent dyes, either cyber-green or fluorophore. In this PCR, the quantity of PCR products is measured in real-time. The 5′ end of a gene (corresponding to the transcription start site) is typically identified by an RT-PCR method, RACE-PCR, short for Rapid Amplification of cDNA Ends. TAIL-PCR - thermal asymmetric interlaced PCR isolates an unknown sequence, flanking a known sequence. Within the known sequence, TAIL-PCR uses a nested pair of primers with differing annealing temperatures; a degenerate primer is used to amplify in the other direction from the unknown sequence. Touchdown PCR is a variant of PCR that aims to reduce nonspecific background by gradually lowering the annealing temperature as PCR cycling progresses. The annealing temperature in the beginning cycles is usually a few degrees above the Tm of the primers utilized, while at the later cycles, it is a few degrees less than the primer Tm. The high temperatures give higher specificity for primer binding, and the lower temperatures permit more correct amplification from the specific products formed during the initial cycles.

10.2 Limitations of PCR

PCR is a robust technique, but it has certain limitations.

1. Sequence information is required prior to primer designing.
2. Amplicon (PCR-product) may not be adequate.
3. The size of the DNA should be small as compared to cloning.

A combination of a high level of an exonuclease-free, N-terminal deletion mutant of Taq DNA polymerase has been used to overcome this limitation, Klentaq1, with a very low level of a thermostable DNA polymerase exhibiting a 3′-exonuclease activity (Pfu, Vent, or Deep Vent) to conduct high fidelity long PCR. 35 kb of bacteriophage lambda can be amplified to high yields from 1 ng of lambda DNA template. It provides a high yield of the amplicon with high fidelity, the ability to use PCR products as primers, and the maximum yield of the target fragment. Other conditions have been identified for effective amplification of longer targets, including amplification up to 22 kb of the beta-globin gene cluster from human genomic DNA and up to 42 kb from phage lambda DNA.

10.3 Uses of PCR

The PCR can be used for different purposes for example it can be used for the isolation of genomic material for sequencing this method is used quite extensively for the detection of genetic and infectious diseases. It is extensively used for DNA fingerprinting. The main applications are outlined below.

Analysis of ancient DNA: PCR is helpful in identifying ancient DNA both from animals and humans. Detection of viral DNA: Viral diseases can be detected by PCR. The viral DNA is amplified. The primer used for amplification is against the virus. In case the amplification occurs, it indicates that the virus is present. Sometimes it is seen immediately after infection, or it can be from several days to several months before actual symptoms occur. If it is possible in the early stages of infection the results are useful for the physician. Physicians can also use DNA quantization techniques to assess the amount of virus ("viral load") in a patient. Now a day gene expression profiling is quite useful because it can directly correlate with protein concentration. This quantification allows knowing how active the gene is.

10.4 Restriction Enzymes Properties

One of the important discoveries in recombinant technology was the discovery of restriction enzymes. These were discovered in 1960 by Werner Arber. Werner Arber Banial Nathans and Hamilton D. Smith received the Nobel Prize in 1978.

Approximately more than 400 different restriction enzymes have been isolated. They are named after the name of the organisms from which they are isolated. Conventionally, a three-letter system is used. Commonly the first letter is that of the genus, and the second and third letters are from the species name. The letters are italicized or underlined, followed by roman numerals. Letters sometimes are added to signify a particular bacterial strain from which the enzymes are obtained. For example, EcoR1 is from E. Coli strain RY13, and HindIII is from Haemophilus influenzae strain Rd. The names are pronounced in ways that follow no set pattern. For example, BamHI is "bam-H-one," BglII is "bagel-two" EcoRI is "echo-R-one" or "eeko-R-one," HindIII is "hin-Dthree," HhaI is "ha-ha-one," and HpaII is "hepa-two." Many restriction sites have an axis of symmetry through their midpoint. **Figure 10.2** shows this symmetry for one of the restrictions, *i.e.*, for EcoRI. The base sequence from 5′ to 3′ on one DNA strand is the same as that from 5′ to 3′ on the complementary DNA strand. Thus, the sequences show twofold rotational symmetry. Several restriction sites are shown in Table 10.1. The most used restriction enzymes

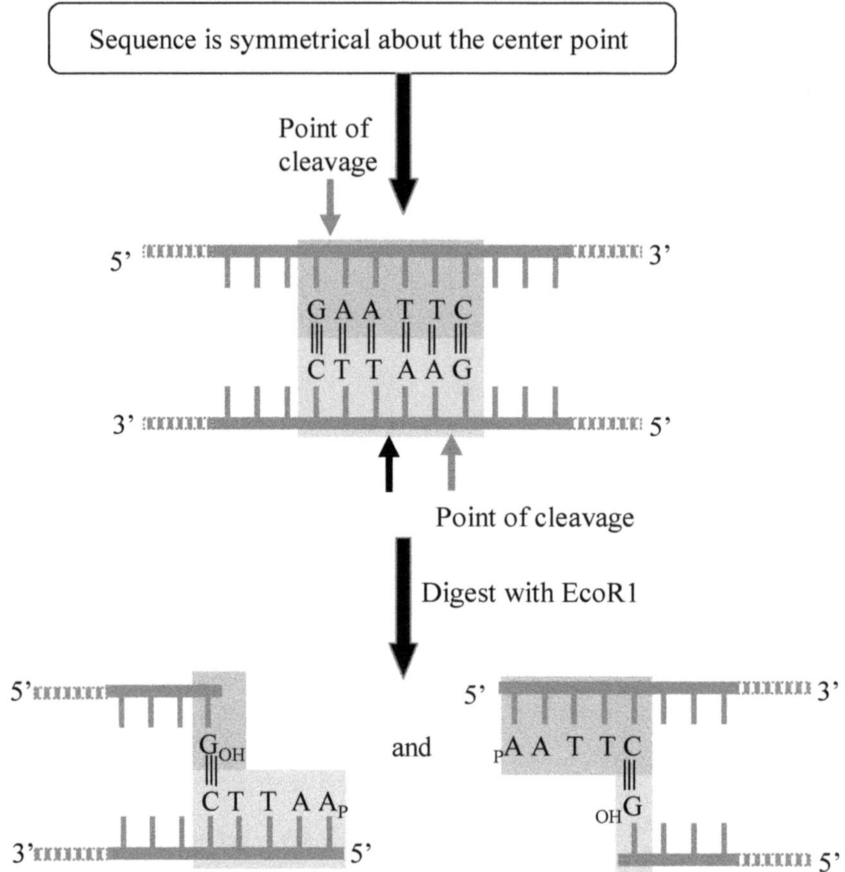

Figure 10.2: Digestion of DNA with the Help of Restriction Enzymes.

recognize four base pairs (for example, HhaI) or six nucleotide pairs (for example, BamHI or EcoRI). Some enzymes recognize eight-nucleotide pair sequences (for instance, NotI ["notone"]). Hinfl ("hin-f-one"), recognizes a five-nucleotide pair sequence in which there is symmetry in the two base pairs on either side of the central base pair, but the central base pair is obviously asymmetrical within the sequence. BstXI ("b-s-t-x-one") is representative of several restriction enzymes with a nonspecific spacer region between symmetrical sequences (**Table 10.1.**).

Table 10.1 Restriction Enzymes

	Enzyme Name	*Pronunciation*	*Organism in which Enzyme is found*	*Recognition Sequence and Position of Cut*
Enzymes with 6-bp recognition sequences	BamHI	"bam-H-one"	Bacillus amyloliquefacies H	5'-G↓GATCC-3' 3'-CCTAG↑G-5'
	BglII	"bagel-two"	Bacillus globigi	A↓GATCT TCTAG↑A
	EcoRI	"echo-R-one"	*E. Coli* RY13	G↓AATTC CTTAA↑G
	HaeII	"hay-two"	Haemphilus aegypticus	RGCGC↓Y Y↑CGCGR
	HindIII	Hin-D-three"	Haemopilus influenzae R_d	A↓AGCTT TTCGA↑A
	PstI	"P-S-one"	Providencia stuartii	CTGGA↓G G↑ACGTC
	SalI	"sal-one"	Streptomyces albus	G↓TCGAC CAGCT↑G

	Enzyme Name	Pronunciation	Organism in which Enzyme is found	Recognition Sequence and Position of Cut
	SmaI	"sma-one"	Serratia marcescens	CCC↓GGG GGG↑CCC
Enzymes with 4-bp recognition sequences	HaeIII	"hay-three"	Haemphilus aesypticus	GG↓CC CC↑GG
	HhaI	"ha-ha-one"	Haemophilus haemolyticus	GCG↓C C↑GCG
	HpaII	Hepa-two"	Haemophilus parainfluenzae	C↓CGC GGC↑C
	Sau3A	"sow-three-A"	Staphylococcus aureus 3A	↓GATC CTAG↑
Enzyme with 8-bp recognition sequences	NotI	"not-one"	Nocardia otitidis-caviarum	GC↓GGCCGC CGCCGG↑CG
Enzyme with Recognition sequence that is not symmetrical	BstXI	"b-s-t-x-one"	Bacillus stearothermophilus	CCANNNNN↓NTGG GGTN↑NNNNNACC

When we digest the DNA with the help of restriction enzymes shorter sequences are more than the longer nucleotide pair sequences. Restriction enzymes cut the enzyme at specific sites. The enzyme will cut the 4 nucleotide sequences more easily as compared to 6 or 8 nucleotide long sequences. Some of the restriction enzymes are shown in **Table 10.1**.

Now, consider DNA with 50 per cent GC and a random distribution of nucleotide pairs. For that DNA, there is an equal chance of finding one of the four possible nucleotide pairs G C A, and T at any one position.

C′ G′ T A

The restriction enzyme Hpall recognizes the sequence 5′-GGCC-3′.

3′-CCGG-5′

The probability of this sequence occurring in DNA is computed as shown below:

1st nucleotide pair: G probability =

C′

2nd nucleotide pair G probability =

C′

3rd nucleotide pair: C probability =

G′

4th nucleotide pair C probability =

G′

The probability of finding any one of the nucleotide pairs is independent of the likelihood of finding one of the other nucleotide pairs. Therefore, the probability of finding the HpaII restriction site in DNA with a random distribution of nucleotide pairs is x x x = 1/256 in short, the recognition sequence for HpaII occurs, on average, once every 256 base pairs in such a piece of DNA.

Blunt ends the ends produced by a straight cut sticky or cohesive ends – the ends produced by a staggering cut.

For example, a restriction enzyme from the bacterium Proteus Vulgaris called PvuII gives a blunt cut at this sequence:

↓

5′ –C-A-G-C-T-G-3′ 5′ – C-A-G C-T-G-3′

3′-G-T-C-G-A-C-5′ 3′-G-T-C G-A-C-5′

↑

Many other restriction enzymes make staggered cuts called sticky ends because of their ability to connect with sections of DNA that have a complementary ending. For example, enzyme EcoR1 performs the following cut:

↓

5′ –G-A-A-T-T-C-3′ 5′-G A-A-T-T-C-3′

3′-C-T-T-A-A-G-5′ 3′-C-T-T-A-A G-5′

↑

More than 500 different restriction enzymes have been discovered; every restriction enzyme has a specific site and produces blunt end or sticky ends.

Restriction enzymes are powerful tools for analyzing DNA because they locate specific sequences and cut the DNA in specific ways.

Another critical component in recombinant biotechnology is DNA. Different techniques can isolate DNA.

After isolating the DNA and then cutting it into manageable lengths, in the next stage various fragments are sorted out, which enables the detailed analysis of each fragment.

10.5 Gel Electrophoresis

Gel electrophoresis is used to separate fragments of DNA (or RNA) according to their size. It is commonly used to identify specific cloned genes. Methods for sorting a sample of DNA into its constituent groups of identical sections make use of two features:

1. Within a sample, fragments of DNA will vary in length.
2. Phosphate groups in DNA give all fragments a net negative charge.

The vast array of fragments produced by the action of restriction enzymes can be separated using a process called gel electrophoresis. In the agarose gel, DNA is applied in the slots and a voltage is applied. The DNA carries a charge hence it will cause it to move across the gel, towards the anode (positive electrode). The smaller pieces of DNA will move more rapidly than the larger fragments as they move through the sieve-like matrix of the gel. There are some identical pieces that get accumulated in the same place. The movement is in a log manner.

10.6 Southern Blotting

A more complex method of displaying the location of DNA fragments is to use short synthetic sequences of DNA called gene probes (radio-labeled oligonucleotides). These can be synthesized so that they identify specific pieces of DNA and are used to mark the presence of specific sequences. Together with the information gained from measuring how fast the fragment moved through the gel, we can discover not only how big each fragment of DNA is, but also start to discover some of the information it is carrying.

10.7 Gene Probes

To search for a particular gene hidden away inside a large section of DNA, we must first construct a piece of DNA with a sequence of bases that complement the gene. This will bond (anneal) the specific DNA sequences (gene) but will not stick to any other part of the DNA. In 1970, reverse transcriptase (RNA-dependent DNA –polymerase) was isolated from certain RNA tumor viruses. This enzyme builds DNA strands from any messenger RNA (mRNA) molecule. Some mRNAs are easily purified because tissues are produced in vast numbers at specific times. For example, the mRNA that encodes silk protein (fibroin) is present in large quantities in the glands of silkworms, and the mRNA molecules that code for milk proteins are numerous in cells in lactating mammary glands. For the probe to get attached to the specific mRNA, it is essential. To create the probe, one can follow the following procedure.

1. Isolate all mRNA.
2. Incubate mRNA with reverse transcriptase in a solution containing adenine, guanine, thymine, and cytosine nucleotides as a source of the four bases.
3. The enzyme produces a copy of complementary DNA (cDNA).
4 Detect cDNA using oligonucleotide gene probes.

10.8 Oligonucleotides

While gene probes are relatively large sections of DNA, oligonucleotides are small, often consisting of not more than 20 bases being small. They will only attach to a DNA strand if there is a perfect match in their sequence; gene probes will attach if most of the bases are complementary.

The task is to build a chain of four different nucleotides arranged in a suitable sequence. Oligonucleotides are made by adding nucleotides one at a time, with chain growth occurring in the 3′ to 5′ direction.

10.9 Building Oligonucleotides

Oligonucleotides are of great use in looking for mutations within genes, whereas little as a single base may have been altered. This is becoming extremely important in genetic screening, as a single base change, resulting in the synthesis of an abnormal protein, may be the root cause of a specific disease.

10.10 Reading the Base Sequence

A great deal of information can be gained about a gene's operation once the sequence of bases is known. Gene probes can locate a particular gene on a chromosome, but they cannot determine the order of the bases within the entire gene. Restriction mapping, where the genome is cut with restriction enzymes, starts revealing small code sections as each enzyme cuts only at specific sequences. Digesting a section of DNA with many restricted enzymes increases the number of known areas. However, as the location at which a restriction enzyme cuts is dictated by a six–or–eight-letter code out of the many thousand bases in a section of DNA, their use can, at best, reveal the location of only a hundred or so bases.

One possible method is to mark with a radioactive label the end nucleotide on a sequence, clip it off and analyze it. Working one base at a time along a sequence is incredibly time–consuming, and when this was the only method, work progressed at a very slow pace. However, new systems were devised with the advent of gene cloning, which provides a way of making large amounts of specific DNA fragments, and gel electrophoresis, which can separate fragments of different lengths.

10.11 Cloning Vector and Gene Cloning

Having isolated a specific gene or created a new piece of DNA de novo so that novel proteins can be constructed, the next stage is to incorporate it into some carrying unit. This vector will transport it into a host cell. There are several vectors (*e.g.*, λ and specific single-stranded DNA species), cosmids (vectors with features of both plasmids and bacteriophage vectors), and artificial chromosomes. The characteristic features of these vectors are given below. Many copies of the DNA can be multiplied and allowed to be expressed, producing a protein. This process is called gene cloning.

10.12 Plasmids

Plasmid DNA is either circular or double-stranded; it contains one gene for antibiotic resistance, a site where replication always starts, and sites where specific restriction enzymes can make cuts and open the loop. This site is known as ori (origin). An E Coli plasmid should have (i) An ori sequence required for E. coli to replicate and (ii) It should have a dominant selectable marker. These are ampicillin or tetracycline resistance (iii) One or more unique restriction enzyme cleavage sites (**Figure 10.3**). The DNA fragment to be inserted is prepared to have sticky ends that will match those created by the restriction enzyme.

Opened plasmids and DNA fragments are mixed and joined into extended loops as their sticky ends come in contact. This joining is made permanent by adding an enzyme called DNA ligase. Some of the plasmids rejoin without including the additional DNA fragment. In a process called transformation, plasmids are introduced into a bacterial cell. Bacteria and plasmids are mixed in a medium containing calcium chloride, which makes the bacterial cell wall permeable. Plasmids then pass through the cell wall into the cytoplasm. Plasmid uptake is not uniform, and only a small proportion of the bacteria will take up plasmids, while few will receive more than one copy.

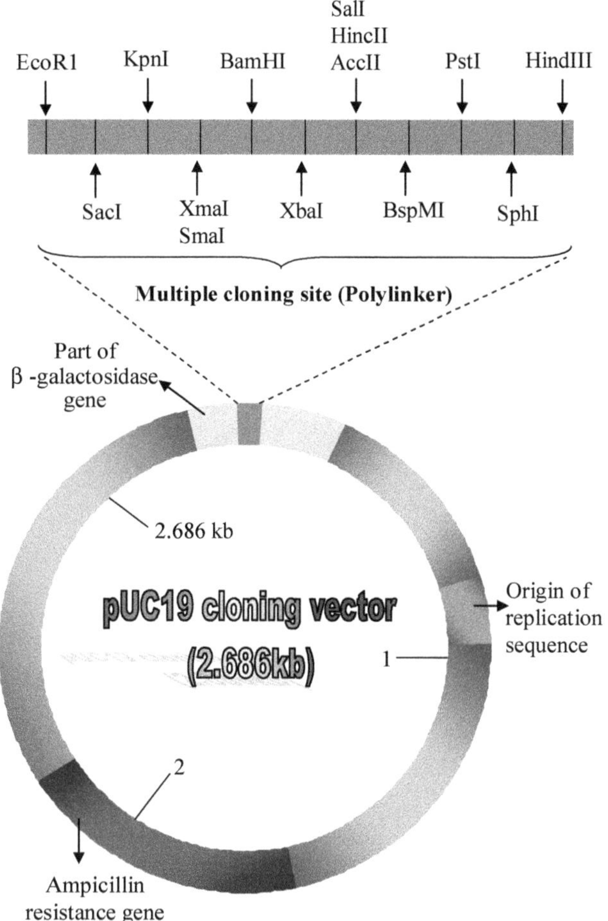

Figure 10.3: The Plasmid Cloning Vector pUC19.

As plasmids contain a gene that confers antibiotic resistance, adding an antibiotic to the culture medium will kill any bacteria with the modified plasmid.

Once inside the host cell, the bacterial DNA synthesizing machinery replicates the plasmids autonomously. This means that the number of copies of cloned DNA can be amplified (**Figure 10.4**).

10.13 Lambda (λ) Phage Vectors

Approximately 10kb of DNA can be inserted into a plasmid vector. Plasmid vectors mostly carry 10kb of inserted DNA, but sometimes, more significant pieces of DNA are needed to be inserted. The genetically modified strains of λ phage vectors can be used with foreign DNA without affecting the phage's ability to infect cells and form plaques (**Figure 10.5**).

As they reproduce, they lyse their bacterial host cells, forming clear spots –plaques – on Petri plates, where the cloned DNA can be recovered.

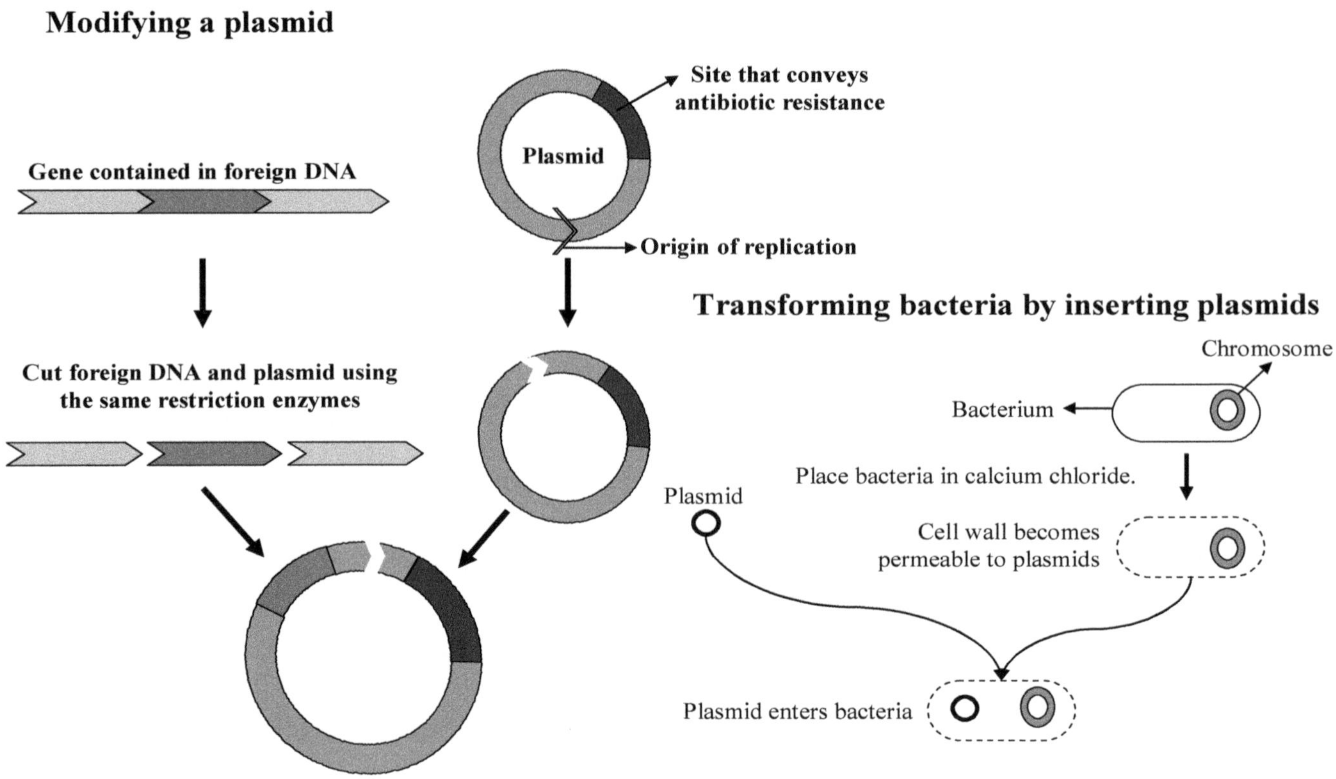

Modifying a plasmid

Site that conveys antibiotic resistance

Gene contained in foreign DNA

Plasmid

Origin of replication

Cut foreign DNA and plasmid using the same restriction enzymes

Transforming bacteria by inserting plasmids

Chromosome

Bacterium

Place bacteria in calcium chloride.

Plasmid

Cell wall becomes permeable to plasmids

Plasmid enters bacteria

Figure 10.4: Modification of a Plasmid.

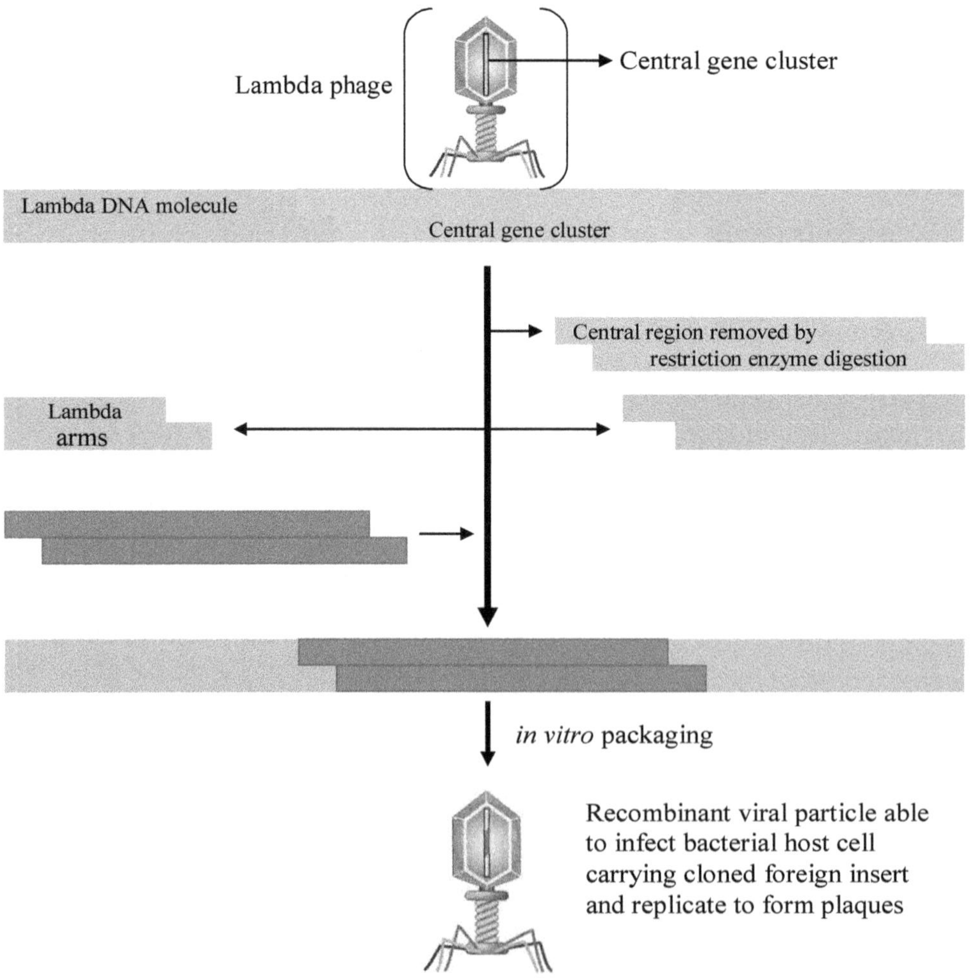

Lambda phage

Central gene cluster

Lambda DNA molecule

Central gene cluster

Central region removed by restriction enzyme digestion

Lambda arms

in vitro packaging

Recombinant viral particle able to infect bacterial host cell carrying cloned foreign insert and replicate to form plaques

Figure 10.5: Lambda (λ) Phage as a Vector.

Phage vectors can carry bigger inserts up to 20kb, more than twice as long as DNA inserts in plasmid vectors. This is an important advantage when cloning large genes or small genomes. In addition, some phage vectors accept only inserts of a minimum size; they do not carry relatively useless small inserts only a few dozen or a few hundred nucleotides in length.

10.14 Cosmid Vectors

These are hybrids in nature and can be created by combining parts of the lambda chromosome with parts of plasmids. Cosmids constitute the cos sequence of phage lambda, essential for packing phage DNA into phage protein coats, and have plasmid sequences needed for replication. The antibiotic resistance gene is used (**Figure 10.6**) to identify host cells carrying recombinant cosmids.

Cosmids become recombinant once the DNA fragment is inserted. These are packaged into lambda protein heads, resulting in infective phage particles. Inside the bacterial host cell, cosmid has the property to replicate as a plasmid. A small portion of the lambda genome is retained; hence, the cosmids can have DNA inserts that are much larger than those of lambda vectors. It has about 50 kb of inserted DNA; however, phage vectors can only keep DNA inserts of 10-15kb length.

Because the vector is small (5.4 kb long), it can accept foreign DNA segments between 33 and 46 kb. The cos sequence allows cosmids carrying large inserts to be packaged into lambda viral coat proteins as though they were viral chromosomes. The viral coats carrying the cosmid can infect a suitable bacterial host, and the vector, carrying a DNA insert, will be transferred into the host cell. Once inside, the ori sequence replicates the cosmid as a bacterial plasmid.

Some hybrid vectors have the origins of replication derived from other sources (*e.g.*, animal viruses such as SV40). These can replicate in different cell types known as the shuttle vectors. They have selectable genetic markers in both types of host cells and are used to shuttle DNA inserts between E. Coli and another kind of host cell, such as yeast. The gene expression studies can be carried out in these vectors.

10.15 Bacterial Artificial Chromosomes

The complex and large eukaryotic genomes require cloning vectors that can carry huge DNA fragments. For more prominent human genes, it is necessary to carry 2000kb long vectors with a large cloning capacity.

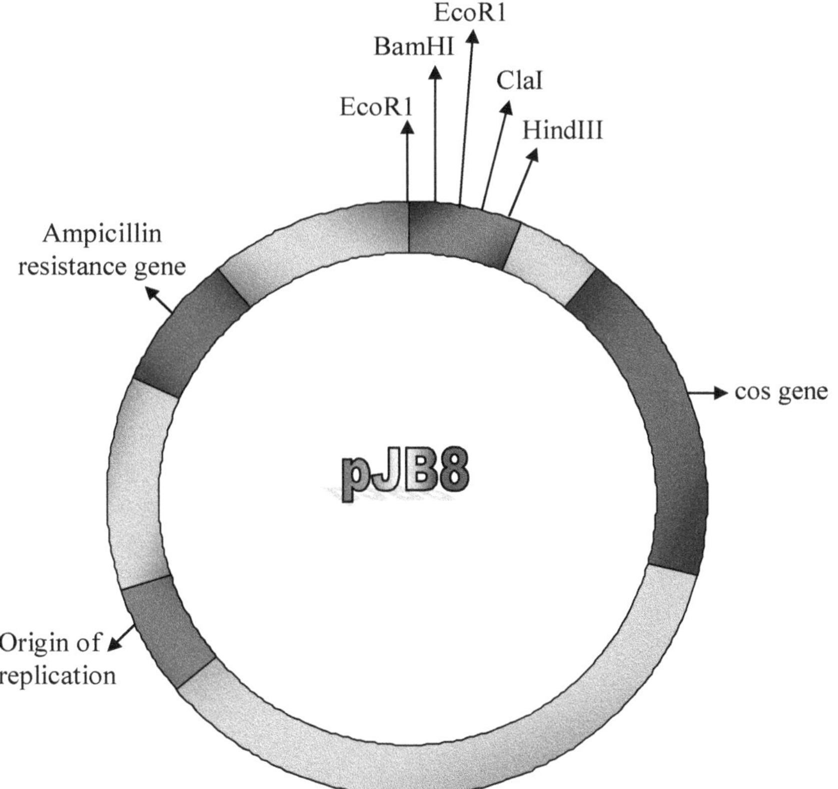

Figure 10.6: The Cosmid pJB8 Contains a Bacterial Origin of Replication (ori), a Single cos Sequence (cos), an Ampicillin Resistance Gene (amp, for selection of Colonies that have taken up the cosmid), and a Region Containing Four Sites for Cloning (BamH1, EcoR1, Cla1, and Hind III).

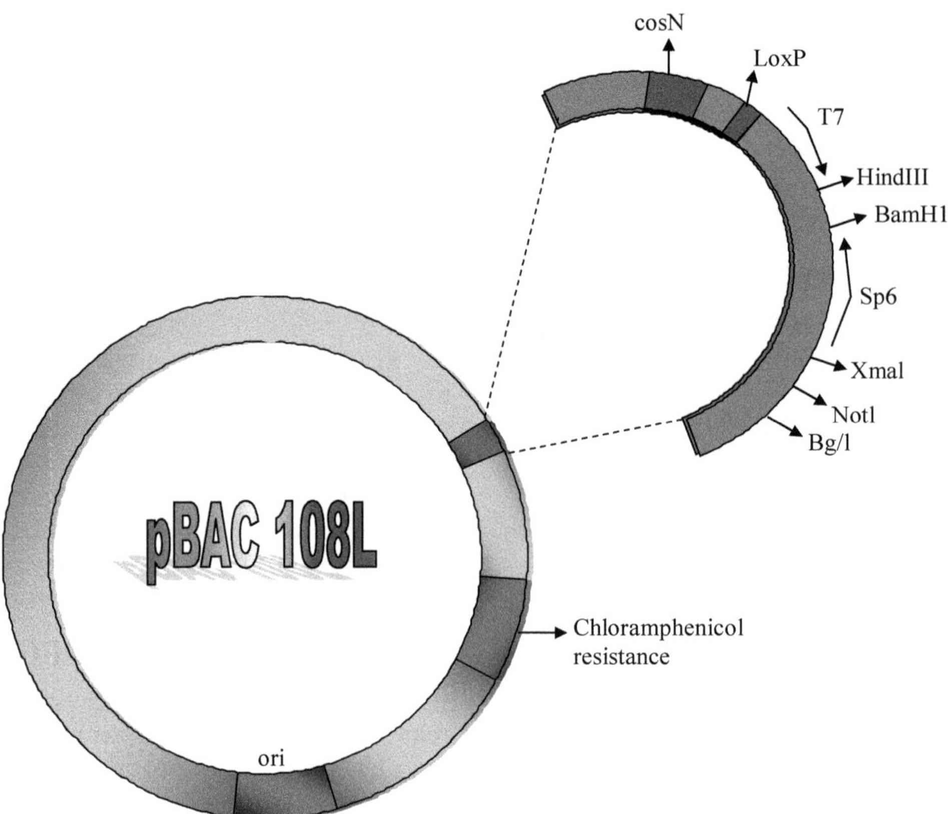

Figure 10.7: A Bacterial Artificial Chromosome (BAC).

The bacterial artificial chromosome has high cloning capability dependent upon the fertility plasmid called the F factor. F factor is used at the time of bacterial conjugation. They carry a bacterial chromosome of 1Mb in length. They are a vector for eukaryotic DNA, which can carry an inset of 300kb (**Figure 10.7**). F factors have at least one antibiotic-resistant marker restriction enzyme recognition site and polylinkers. Polylinkers contain promoter sequences that help in generating RNA molecules helpful in studying the expression of the cloned genes. These are also used as probes in chromosomal walking and sequencing of the cloned insert.

There are hybrid vectors that have the origins of replication derived from other sources (*e.g.*, animal viruses such as SV40). These can replicate in different cell types known as the shuttle vectors. They have selectable genetic markers in both types of host cells and are used to shuttle DNA inserts between E. Coli and a different kind of host cell, such as yeast. The gene expression studies can be carried out in these vectors.

10.16 Plasmid Shuttle Vectors

There are shuttle vectors that can be used to transform mammalian cells in culture and vectors to transform other animal cells, plant cells, and yeast cells. These are called shuttle vectors, as they can be introduced into two or more different host organisms. For example, some shuttle vectors can be transformed and replicated in *E. Coli* detection, which is based upon antibiotic resistance and can also be transformed into yeast, selected by a nutritional maker, such as the URA3 gene conferring uracil-independent growth on a Ura3 mutant yeast cell. Different types of yeast – *E. Coli* shuttle vectors have been developed, some of which replicate freely as single copies in the nucleus, and some of these integrate into a yeast nuclear chromosome, and they replicate when the chromosome replicates.

10.17 Expression Vectors

Expression vectors are present in both prokaryotic and eukaryotic host cells. An expression vector to be used in an E. Coli host cell is pET (**Figure 10.8**). The gene to be expressed is cloned in the polylinker, which has recognition sequences. It is kept near the T7 viral promoter and the bacterial lac operator. The host cell genome is modified so that the gene for T7 viral polymerase can be carried, the lac promoter, and the lac operator. After recombinant plasmids are inserted into host cells, an expression is induced by adding the lactose analog IPTG to the medium. IPTG displaces the repressor from the lac operator, activating the T7 polymerase gene on the bacterial chromosome, and the gene in the polylinker T7 polymerase binds

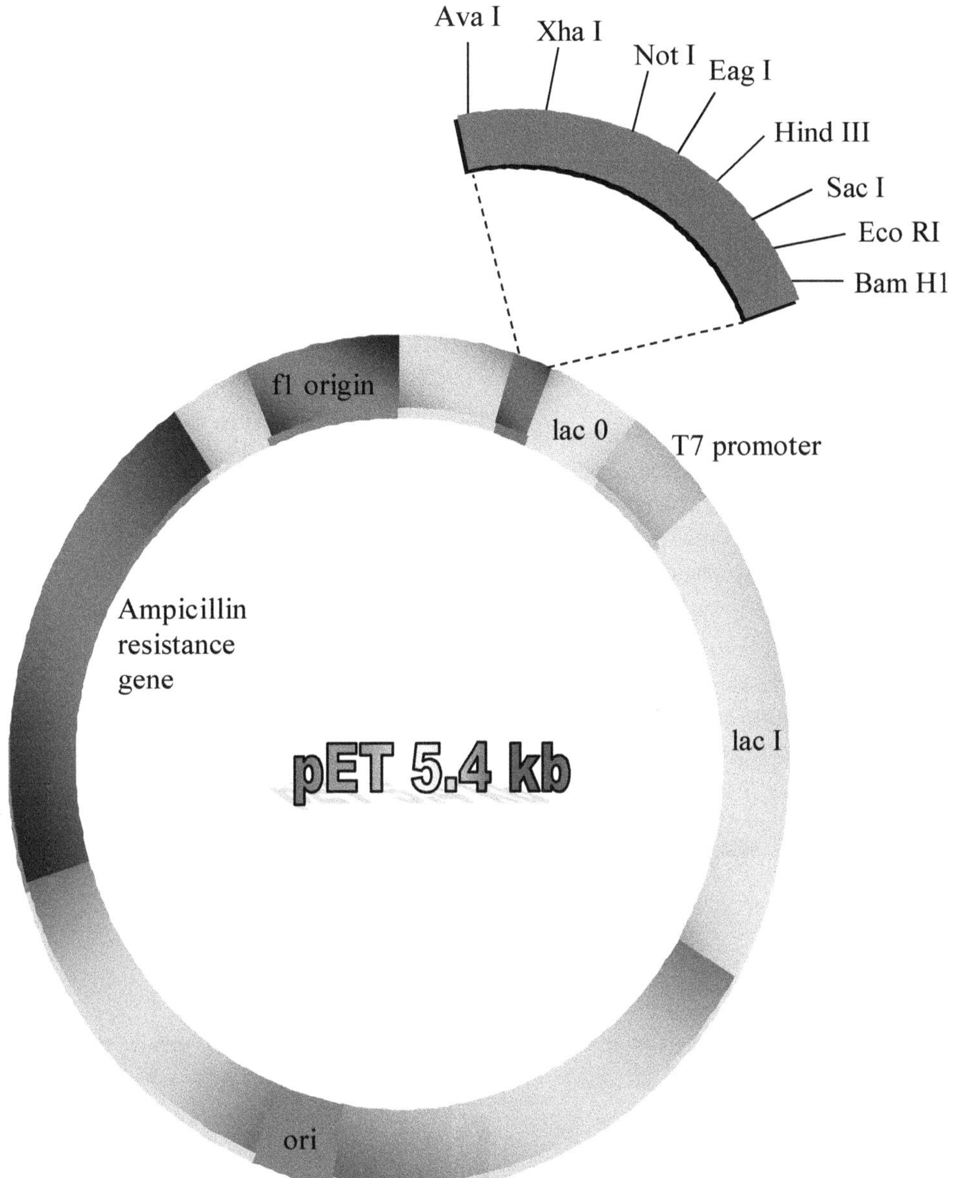

Figure 10.8: A pET Expression Vector. A genetically engineered host is used in pET.

to the T7 promoter and transcribes the gene in the polylinker, producing large quantities of the encoded protein.

10.18 Artificial Chromosomes

Sometimes, large cloning vectors are required to carry substantial pieces of DNA, forming recombinant DNA molecules resembling small chromosomes.

10.19 Yeast Cells are Used as Eukaryotic Hosts for Cloning

For cloning, the yeast Saccharomyces cerevisiae is used widely. To study the function of some eukaryotic proteins, it is necessary to use a host cell that can post-translationally modify the protein, so it folds into a functional form. Bacterial host cells cannot carry out these modifications and rapidly degrade due to the abrupt folding of the proteins. Yeast is a wise organism and can be used for producing proteins for vaccines and therapeutic agents. Some of the recombinant proteins have been synthesized in the yeast cell these are:

1. Clotting factor XIIIA
2. A1 – antitrypsin
3. Platelet-derived growth factor
4. Epidermal growth factor
5. Malaria parasite protein
6. Hepatitis B virus surface protein

The YAC artificial chromosome has telomeres at each end, an origin of replication (which initiates DNA synthesis), and a centromere. These components are joined to selectable marker genes (TRP J and URA3) and a cluster of restriction enzyme recognition sequences for inserting foreign DNA. Yeast chromosomes range in size from 230 kb to over 1900 kb, which allows cloning DNA inserts from 100 to 1000 kb in YACs. These large inserts have been helpful in the human genome project.

10.20 Bacterial Artificial Chromosomes

As mentioned above, YACs can accommodate large DNA inserts, and BACs (Bacterial artificial chromosomes) have the advantage that they can be manipulated like regular bacterial plasmids. Once transformed into E. Coli, the F factor origin of replication, keeps the copy number of the plasmid at one per cell. BAC does not undergo rearrangements in the host; therefore, they have superseded YACs in physical mapping studies of genomes.

Because of their ability to accommodate large DNA inserts, BACs form the basis of vectors used to study gene regulation in vertebrates such as mice and zebrafish. The promoter and regulatory genes are known to span a large section of DNA. Therefore, a gene and a large segment of DNA upstream of the gene can be cloned in a BAC, and the clone can be transformed into an organism.

10.21 DNA was First Cloned in Prokaryotic Host Cells

Prokaryotic hosts are the most common laboratory strain of the bacterium E. Coli known as K12. E. Coli strains such as K12 are genetically well-characterized and can create many vectors. In **Figure 10.9,** the different steps involved are shown.

The genes can be transferred even into the eukaryotic cells. The vector used for this purpose is YAC. When the vector used is a plasmid, it is called transformation. When the virus is used, it is called transfection.

10.22 Plant Cell Hosts

Bacterial plasmid vectors are used for gene transfer in plants. The tumors are produced in plants due to a soil bacterium called Agrobacterium

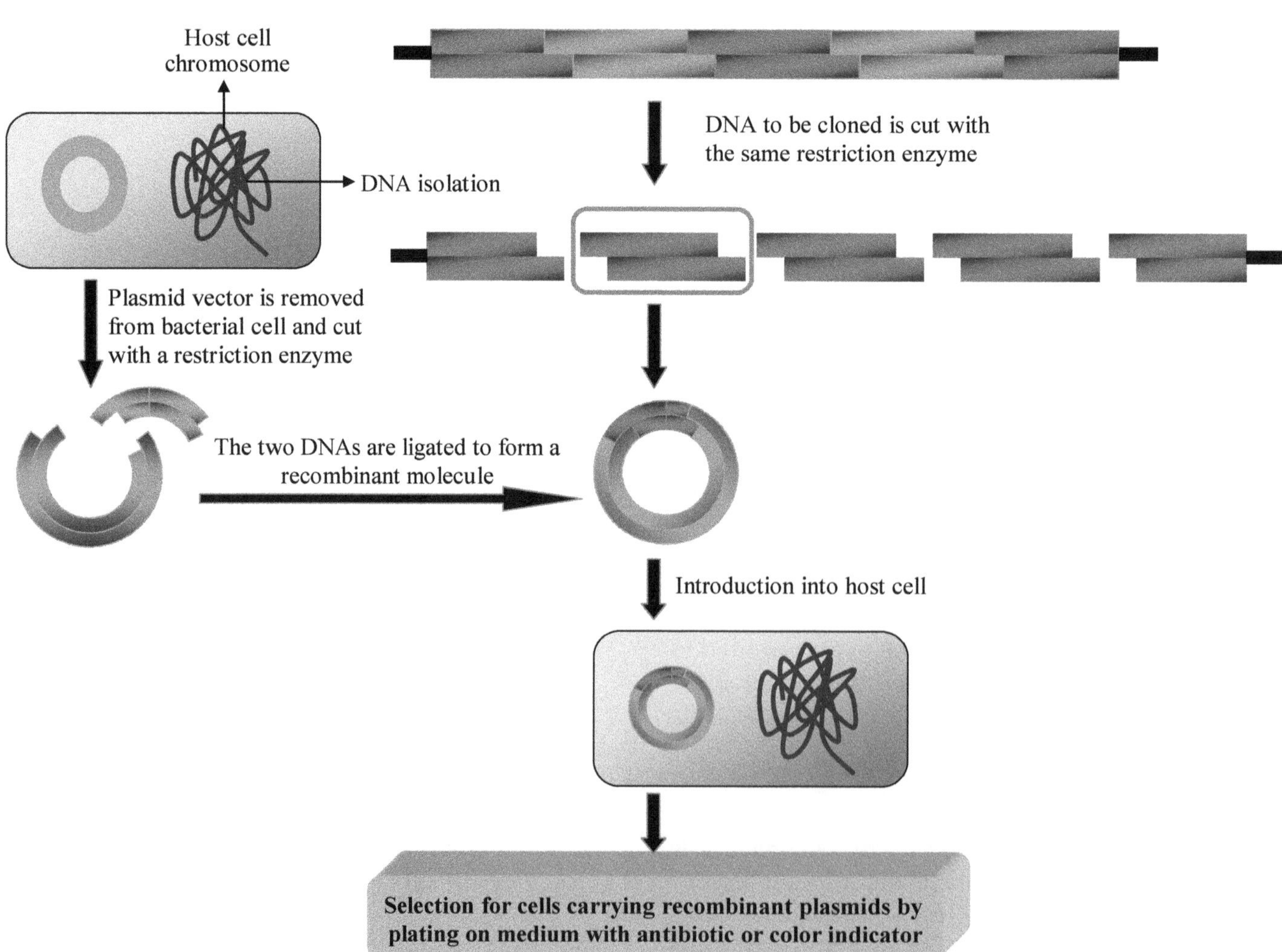

Figure 10.9: Plasmid and DNA to be Cloned are Cut by the same Restriction Enzyme.

tumifaciens which infects the plant's tumor-inducing plasmid (Ti).

Gene transfer into higher plants uses bacterial plasmid vectors. The soil bacterium Agrobacterium tumifaciens infects plant cells and produces tumors in many species of plants. Tumor formation is associated with the presence of a tumor-inducing (Ti) plasmid carried in the bacteria (**Figure 10.10**).

When such a kind of infection occurs, the Ti plasmid T-DNA is produced and transferred into the genome of the host plant cell. Many genes in the T-DNA segment are responsible for controlling tumor growth. Outsider genes can be introduced into the T-DNA segment, and the recombinant plasmid is transferred into plant cells due to infection with A. tumifaciens. Foreign DNA is inserted into the plant genome, and the T-DNA integrates into a host cell chromosome. Plant cells carrying a recombinant Ti plasmid can grow in tissue culture to form a cell mass (callus). If experimentally, the culture medium is changed, the cells get induced and result in and shoots, the formation of root and shoots containing outsider genes. These plants are called transgenic plants.

10.23 Mammalian Cell Hosts

Several methods can be used to transfer DNA into mammalian cells. These methods could be endocytosis or encapsulation of DNA into artificial membranes, liposomes. The fusion takes place between the cell membranes. Transgenic animals are produced if the genes are transferred into fertilized eggs.

The YAC vectors are used for this purpose. YAC vectors increase the efficiency of gene transfer into the germ line of mice. Recombinant YAC is introduced into the nucleus of mouse-fertilized cells or embryonic stem cells. This is done through microinjection in fertilized eggs or embryonic stem cells.

Once the zygotes are formed, which are transgenic, these are then implanted into the foster mothers for further development. YAC is used, then fused with the yeast cell carrying a YAC with a mouse stem cell, transferring the YAC and all or most of the yeast genome into the stem cell. These transgenic ES cells are injected into the early stages of the mouse embryo. This technique has enormous applications.

After infection of the host cell, the RNA is transcribed by reverse transcriptase into a double-stranded DNA (dsDNA) molecule. The dsDNA integrates into the host genome and is passed to daughter cells during cell division. The retroviral genome can be engineered to remove several viral genes, creating vectors that can take the foreign DNA, including human genes. Many other vectors can be used, for example, avian genetically modified cells, retroviruses, and single-stranded RNA molecules.

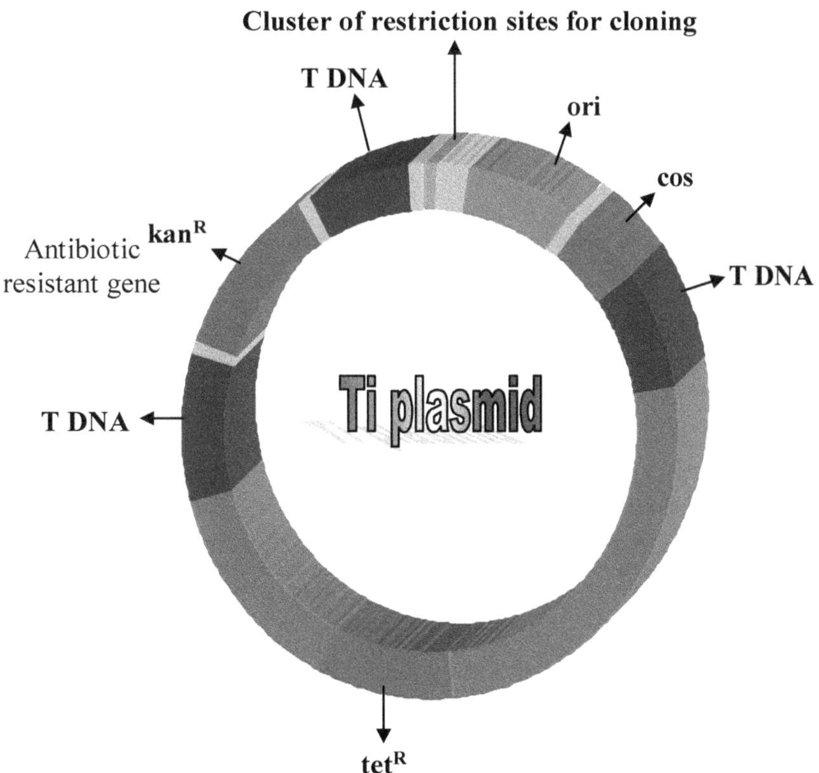

Figure 10.10: A T$_i$ Plasmid Designed for Cloning in Plants.

10.24 Recombinant DNA Libraries

Genomic libraries are constructed from many organisms, including humans and many viruses. There are chromosome libraries, which are collections of clones of fragments of individual chromosomes, and complementary DNA (cDNA) libraries, which are collections of clones of DNA copies of mRNAs isolated from cells.

10.24.1 Genomic Libraries

The following procedure constructs the genomic libraries.

1. Restriction digestion of genomic DNA clone the DNA fragment is carried out, followed by inserting into the cloning vector. The problem with this technique is that a particular enzyme may split the gene into different fragments. The gene may be cloned into many pieces. Another drawback is that the average size of the fragment produced by the digestion of eukaryotic DNA with restriction enzymes is small (approximately 4 kb for restriction enzymes with six base-pair recognition sequences; (**Table 10.2**). There are only a few genes that are larger than 4 kb, but also an entire genomic library would have to contain a vast number of recombinant DNA molecules, hence screening for the specific gene would be highly difficult.

Table 10.2: Restriction Sites for Restriction Enzymes in DNA with Randomly Distributed Nucleotide Pairs

Nucleotide Pairs in Restriction Site	*Probability of Occurrence*
4	$(1/4)^4$ = 1 in 256bp
5	$(1/4)^5$ = 1 in 1,024bp
6	$(1/4)^6$ = 1 in 4,096 bp
8	$(1/4)^8$ = 1 in 65,476bp
N	$(1/4)^n$

2. The problems of genes split into multiple fragments and the large number of recombinant DNA molecules can be minimized by cloning longer DNA fragments in an appropriate vector. Longer DNA fragments can be generated by shearing the high-molecular-weight (usually 100- to 150-kb) DNA mechanically. For this purpose, DNA is passed through a syringe along with the needle, which allows the production of DNA fragments of considerable size that overlap. However, because the ends of the resulting pieces will not be generated by cutting with restriction enzymes, additional enzymatic manipulations are required to add appropriate ends to the molecules for insertion into vector cloning sites.

3. Partial digestion of DNA can produce fragments with six base pair recognition sequences (**Figure 10.11**). Partial digestion can be carried out by using less amount of enzyme, or less time may be used. The fragments can be visualized on the simple agarose gel. If the DNA is digested with the enzyme Sau3A, which has the recognition sequence 5′-GATC-3′, the ends are complementary to the ends produced by the digestion of a 3′-CTAG-5′

cloning vector with BamHI, which has the recognition sequence 5′-GGATCC-3′.

3′-CCTAGG-5′

That is, in

5′-↓GATC-3′

3′-CTAG-5′

Sau3A cuts to the left of the upper G and the right of the lower G, giving an overhang with the sequence 5′ -GATC. ….3′ as follows:

5′- and 5′ GATC-3′

3′-CTAG 5′ -5′

Similarly, in the sequence

5′- G↓GATCC-3′

3′- CCTAG°G-5′

BamHI cuts between the two G nucleotides, also giving a 5′ overhang with the sequence 5′-GATC.3′ as follows:

5′-G and 5′ GATCC-3′

3′-CCTAG 5′ G-5′

The Sau3A and BamHI "sticky" ends can pair to produce a hybrid recognition site.

The recombinant DNA molecules produced by ligating the Sau3A-cut fragments and the BamHI-cut vectors are then introduced into *E–Coli*, where the molecules are cloned.

The use of the above method produces a library of recombinant molecules. However, not all eukaryotic genome sequences are equally represented in such a library.

The number of clones needed to be included are all sequences in the genome, which depends on the size of the genome being cloned and the average size of the DNA fragments inserted into the vector. The probability of having at least one copy of any DNA sequence in the genomic library can be calculated from the formula.

$$N = \frac{\ln(1-P)}{\ln(1-f)}$$

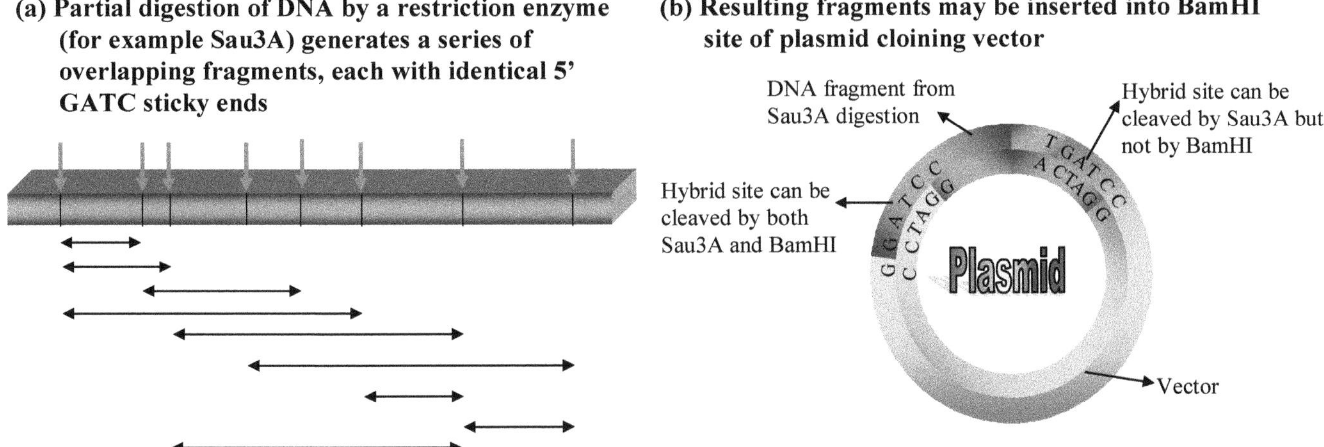

Figure 10.11: Partial Digestion with a Restriction Enzyme to Form DNA Fragments of Appropriate Size.

where N is the necessary number of recombinant DNA molecules, P is the probability desired, f is the fractional proportion of the genome in a single recombinant DNA molecule (that is, f is the average size, in kilobase pairs, of the fragments used to make the library, divided by the size of

the genome, in kilobase pairs), and it is the natural logarithm. For example, for a 99 per cent chance that a particular yeast DNA fragment is represented in a genomic library of 15-kb fragments, where the yeast genome size is about 12,000 kb, 3,682 recombinant DNA molecules would be required. For the approximately 3,000,000-kb human genome, more than 920,000 clones would be needed-hence the use of YAC or BAC vectors for making libraries of larger genomes. Whatever the genome or vector, to have confidence that all genomic sequences are represented, one must make a library with several times more than the calculated minimum number of clones.

10.24.2 Chromosome Libraries

Genomic library screening is very time-consuming for large genomes. One can prepare individual chromosome libraries in the genome; this can reduce the screening time. This gives humans 24 different libraries, one each for the 22 autosomes, the X, and the Y. Then, if a gene has been localized to a chromosome by genetic means, researchers can focus their attention on the library of that chromosome when they search for its DNA sequence.

Chromosomes can be separated using flow cytometry; chromosomes from cells in mitosis are stained with a fluorescent dye and passed through a laser beam connected to a light detector. This system sorts the chromosomes based on differences in dye binding and the resulting light scattering. Once the chromosomes are sorted and collected from several cells, a library of each chromosome type can be made.

10.24.3 cDNA Libraries

DNA copies, called complementary DNA (cDNA), can be made from all mRNA molecules in a population of eukaryotic cells at a particular time or tissue. These cDNA molecules can then be cloned to produce a cDNA library. Since a cDNA library reflects the gene activity of the cell type when the mRNAs are isolated, cDNA libraries help compare gene activities in different cell types of the same organism or of the same cell type at different times, as in cell differentiation during development.

The clones in the cDNA library represent the mature mRNAs found in the cell. In eukaryotes, mature mRNAs are processed molecules, so the sequences obtained are not equivalent to genomic clones. Intron sequences are present in genomic clones but not cDNA clones; hence, cDNA clones are smaller than the equivalent gene clone. For any mRNA, cDNA clones can be helpful for subsequently isolating the gene that codes for a particular mRNA. The gene clone can provide more information than cDNA clone.

cDNA libraries are readily made from mRNAs, because, unequally among RNAs, eukaryotic mRNAs contain a poly(A) tail. These polys (A)+ mRNAs can be purified from a mixture of cellular RNAs bypassing the RNA molecules over a column to which short chains of deoxythymidylic acid, called oligo(dT) chains, have been attached. As the RNA molecules pass through the column, the poly(A) tails on the mRNA molecules base pair to the oligo(dT) chains. As a result, the mRNAs are captured on the column while the other RNAs pass. The captured mRNAs are then released and collected by decreasing the ionic strength of the buffer passing through the column so that the hydrogen bonds are disrupted. This method causes a significant enrichment of poly(A) + mRNAs in the mixed RNA population, to about 50 per cent versus approximately 3 per cent in the cell.

Figure 10.12 shows how a cDNA molecule can be made from the mRNA molecules. Key to this synthesis is the presence of the 3′ poly(A) tails on the mRNAs. After the mRNA has been isolated, the first step in cDNA synthesis is annealing a short oligo(dT) primer to the poly(A) tail. The primer is extended by reverse transcriptase (RNA-dependent DNA polymerase) to make a DNA copy of the mRNA strand. The result is a DNA-mRNA double stranded molecule. Next, RNase H ("R-N-aze H," a type of ribonuclease), DNA polymerase I, and DNA ligase are used to synthesize the second DNA strand. RNase H partially degrades the RNA strand in the hybrid DNA-mRNA; DNA polymerase I makes new DNA fragments, using the partially degraded RNA fragments on the single-stranded DNA as primers; and, finally, DNA ligase ligates the new DNA fragments together to make a complete chain. The result is a double-stranded cDNA molecule, a faithful DNA copy of the starting mRNA.

In **Figure 10.13,** the cloning of cDNA is shown. DNA cloning is done by using a restriction site linker, or linker, which is a short, double-stranded piece of DNA (oligo deoxy ribonucleotide) about 8 to 12 nucleotide pairs long that includes a restriction site, in this case, the site for *BamHI.* Both the cDNA molecules and the linkers have blunt ends, and they can be ligated together at high concentrations of T4 DNA ligase. Sticky ends are produced in the cDNA

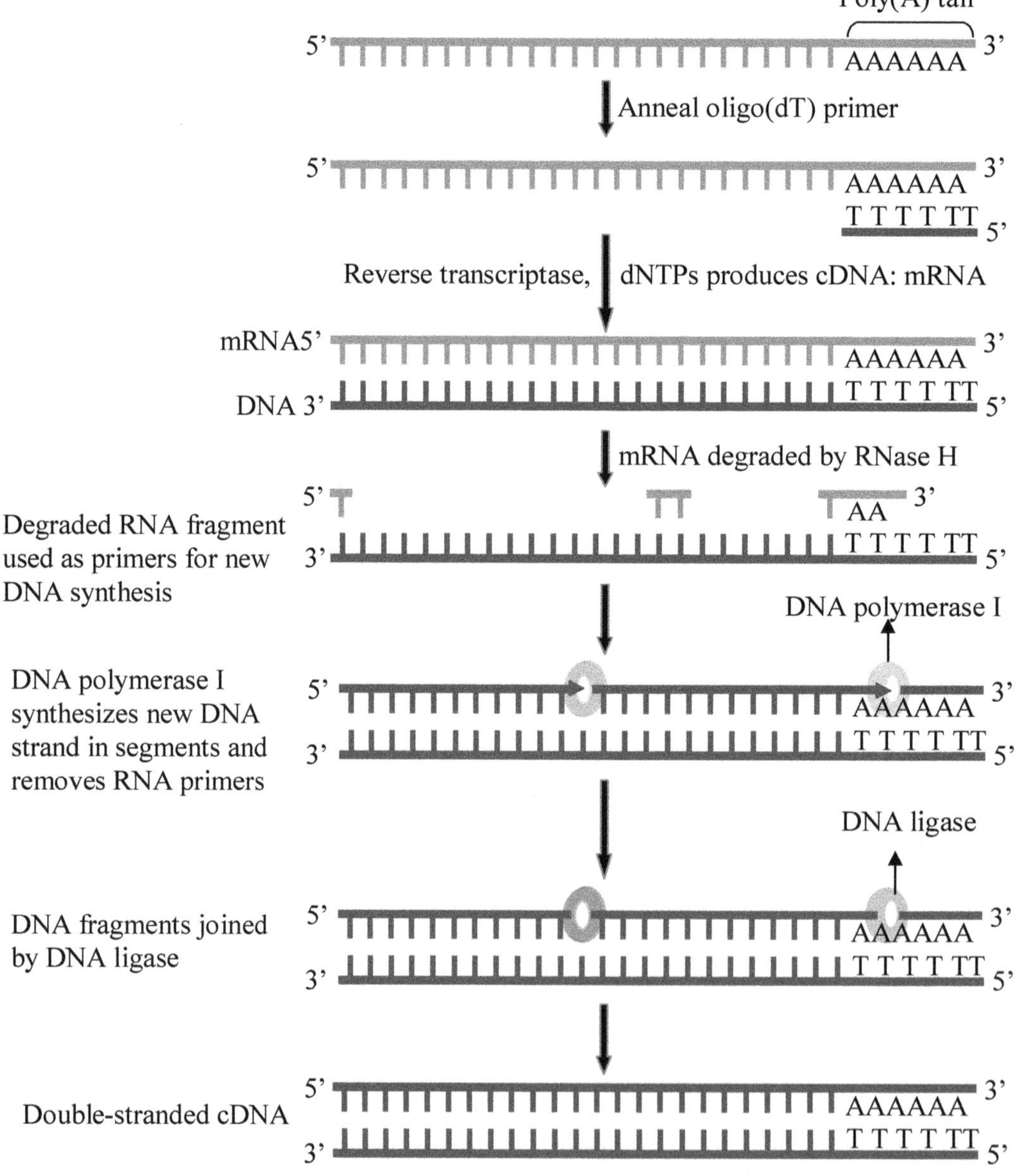

Figure 10.12: Formation of Double-Stranded Complementary DNA (cDNA) from a Polyadenylated mRNA by using Reverse Transcriptase, RNase H, DNA Polymerase I, and DNA Ligase.

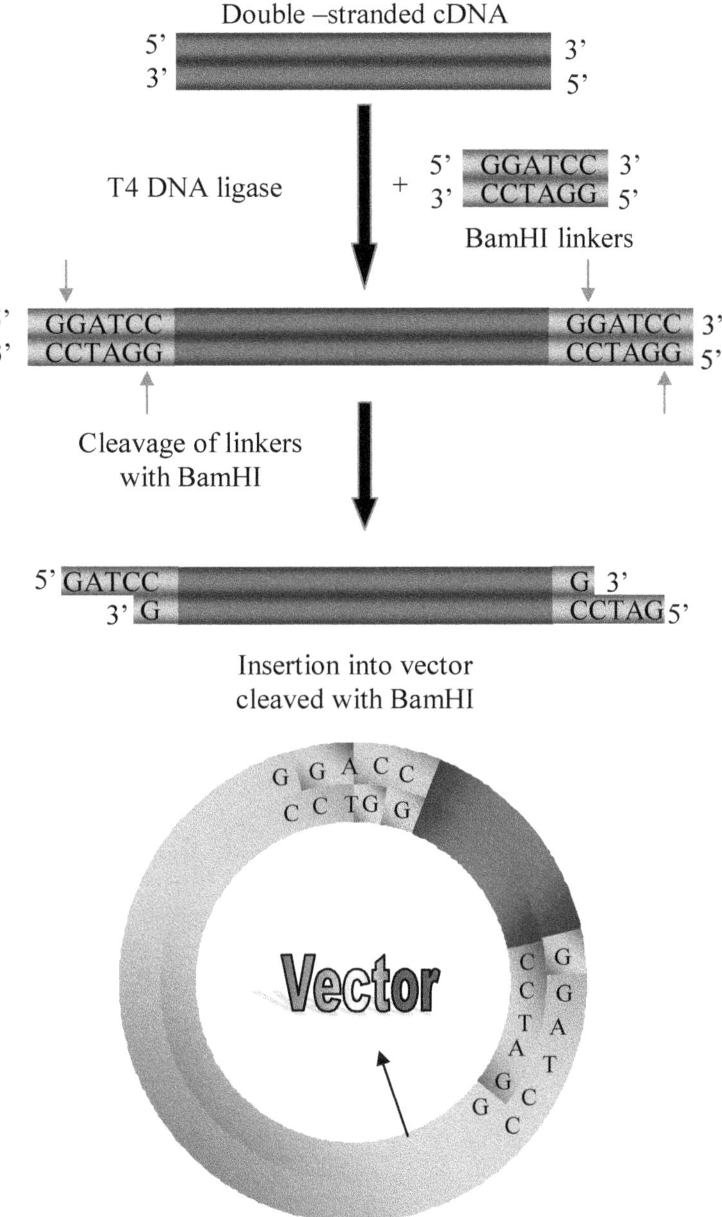

Figure 10.13: Use of BamH1 Linkers for Cloning cDNA.

molecule by cleaving the cDNA (with linkers now at each end) with *BamHI*. The resulting DNA is inserted into a cloning vector that has also been cleaved with *BamHI*, and the recombinant DNA molecule produced is transformed into an *E. Coli* host cell for cloning.

A problem when the linkers are used for cloning cDNAs is that there may be a restriction site within the cDNA for the enzyme used to cleave the linkers, which would mean that the cDNA would also cut when the linkers are cut, resulting in cloning the cDNA in pieces. To solve this potential problem, an adapter can be added to the cDNA, which has a sticky end that is suitable for cloning, so that the cDNA is never digested with a restriction enzyme. For example, if we anneal together 5′-GATCCAGAC-3′ with 5′ -GTCTG-3′ to form the adapter.

5′-GATCCAGAC-3′

GTCTG-5′

ligate it to a cDNA, the blunt end of the adapter will get covalently attached to the blunt end of the cDNA, leaving the 5′ overhang GATC at each end. The overhang will base pair with a vector digested with BamH1 (**Figure 10.13**), and the cDNA will be cloned in one piece.

cDNA molecules can also be cloned by blunt-end cloning. The cDNA molecules have blunt ends, so they can be inserted into the vector cut with a restriction enzyme, such as Sma1, that generates blunt ends.

10.24.4 Finding a Specific Clone in a DNA Library

Unlike libraries of books, clone libraries have no catalog, so they must be searched through screening to find a clone of interest. Fortunately, several screening procedures have been developed.

10.24.5 Screening a cDNA Library

cDNA is searched by attempting to search a cDNA clone that encodes a specific protein (**Figure 10.14**). For this purpose, antibodies binding to specific proteins are used. Further, the cDNAs must be cloned in an expression vector (**Figure 10.14**), in which the cDNA is inserted between a promoter and a transcription termination signal. In the host, an mRNA corresponding to the cDNA is transcribed, and the mRNA is translated to produce the encoded protein.

For screening, purpose the first step is to transform the E. coli with cDNA clones constructed agar in an expression vector (**Figure 10.14**), and then the cells are grown on a plate so that each bacterium gives rise to a colony. These clones are preserved, for example, by picking each colony off the plate and placing it into the medium in a well of a 96-well microtiter dish (**Figure 10.14**). Replicas of the set of clones are dotted onto a membrane filter placed on a Petri plate of a selective medium appropriate for the recombinant molecules, for example, ampicillin for plasmids carrying the ampicillin resistance gene. Colonies grow on the filter in the same pattern as the clones grow in the microtiter dish. The filter is peeled from the dish, and the cells are lysed in situ. The proteins that were within the cell, including those expressed from the cDNA, become stuck to the filter, which is then incubated with an antibody to the protein of interest (**Figure 10.14**).

If the antibody is radioactively labeled, any clones that show the protein of interest can be identified by placing the dried filter against X-ray film, leaving it in the dark for a period (from 1 hour to overnight) to produce an *autoradiogram* (**Figure 10.14**). The process is called *autoradiography*. When the film is developed, dark spots are seen wherever the radioactive probe is bound to the filter in the antibody reaction. (The dark spots result from the decay of the radioactive atoms, which changes silver grains in the film.) These spots correspond to the correct cDNA clones. Once a cDNA clone for a protein of interest is identified, it can be used, for example, to analyze the genome of the same or another organism for homologous sequences, to isolate the nuclear gene for the mRNA from a genomic library, or to quantify mRNA synthesized from the gene.

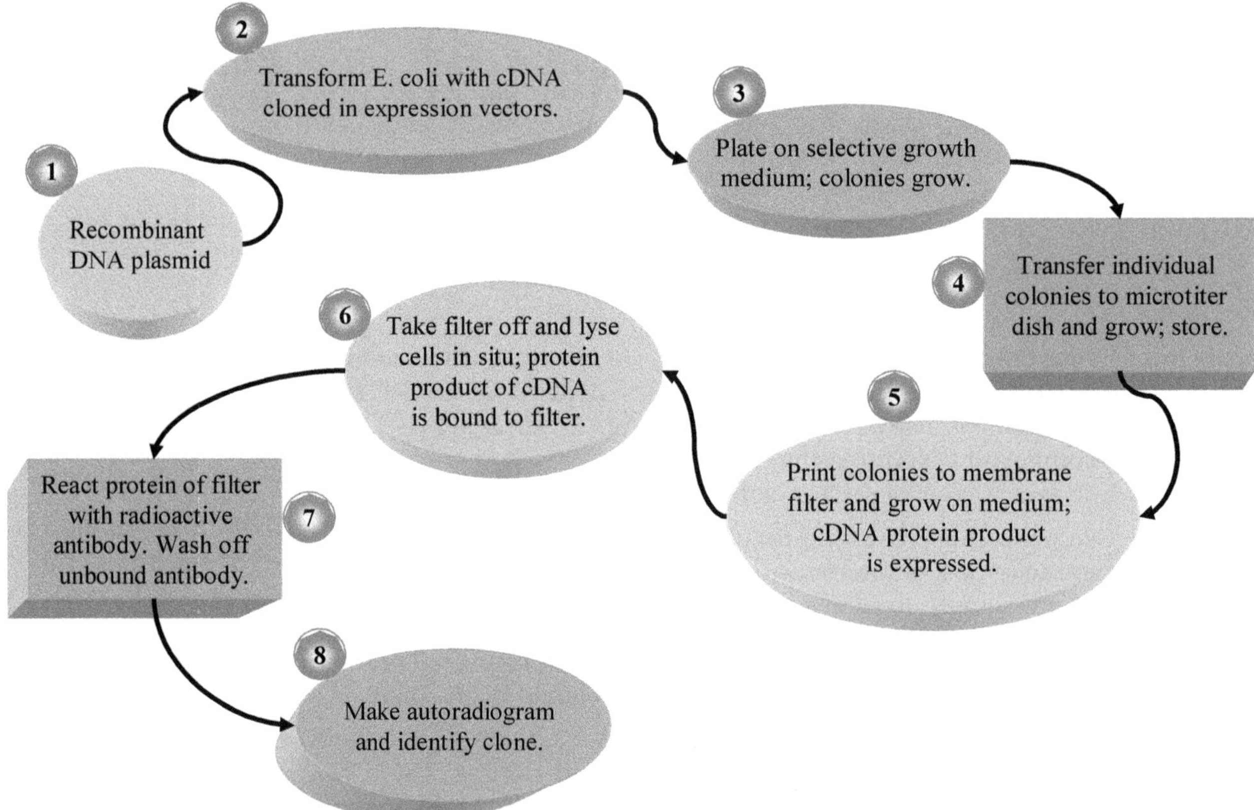

Figure 10.14: Screening for Specific cDNA Plasmids in a cDNA Library by Using an Antibody Probe.

10.24.6 Screening a Genomic Library

Given the existence of a probe, such as a cloned cDNA, it is possible to identify in a genomic library the cloned gene that codes for the mRNA molecule from which the cDNA was made and then isolate that gene for characterization. The procedure is given below.

Screening a genomic library made in a plasmid vector is like the screening just described for a cDNA library. First, *E. Coli* cells are transformed with the genomic library (**Figure 10.15**), and the cells are plated onto a plate of selective medium, where colonies are formed. After this, the colonies are replicated and plated onto another plate with a selective medium with a membrane filter on its surface (**Figure 10.15**). Colonies grow on the membrane filter, which is then lifted off the plate and processed to lyse the bacterial cells; DNA is denatured so that single-stranded DNA is obtained. This DNA binds firmly to the filter (**Figure 10.15**).

Next, the filter is placed in a heat-sealable plastic bag and incubated with the cDNA probe (**Figure 10.15**), labeled radioactively or non-radioactively. To prepare the labeled DNA for use as a probe, the DNA is denatured by boiling and then quickly cooled on ice to produce single-stranded DNA molecules. These labeled molecules are added to the membrane filters to which the denatured (single-stranded) DNA from each colony has been bound. The labeled molecules diffuse over the filter, and, with time, they find the filter-bound DNA with which they can pair by complementary base pairing. The hydrogen bonding DNA-DNA hybrids form between the probe and the colony DNA. If the cDNA probe is derived from the mRNA for β-globin, for example, that probe will hybridize with the filter-bound DNA that encodes the β-globin mRNA (that is, the genomic β-globin gene). After the hybridization step, the filters are washed to remove the unbound probe. They are subjected to the appropriate detection procedure, depending upon whether the probe is radioactive or non-radioactive: autoradiography for a radioactive probe, chemiluminescence, or colorimetric detection for a non-radioactive probe (**Figure 10.15**). From the positions of the spots on the film or filter, the locations

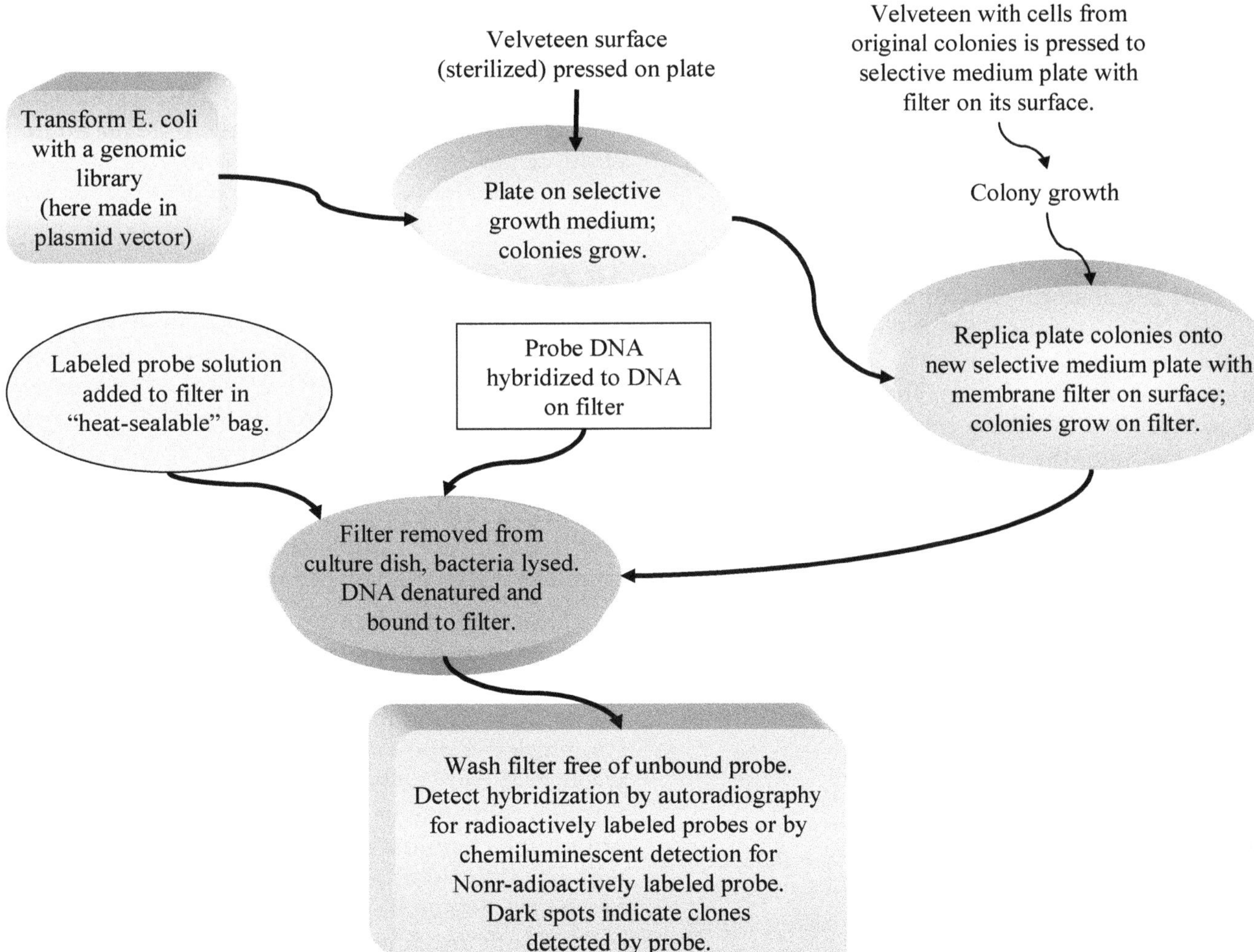

Figure 10.15: Using DNA Probes to Screen Plasmid Genomic Libraries for Specific DNA Sequences.

of the bacterial colony or colonies on the original plate can be determined, and the clones of interest are isolated for further characterization.

10.25 Identifying Genes in Libraries by Complementation of Mutations

For organisms in which genetic systems of analysis have been well developed with well defined mutations, it is possible to clone genes by complementation of those mutations. In brief, this approach depends on the expression of the wild-type gene introduced into the cell by transformation, overcoming the defect of a mutant form of the gene in the genome. This can be done with the yeast Saccharomyces cerevisiae, as it is easy to manipulate genetically and for which efficient integrative and replicative transformation systems using E. Coli -yeast shuttle vectors are available.

To clone a yeast gene by complementation, first, a genomic library is made of DNA fragments from the wild-type yeast strain in a yeast- E. Coli shuttle vector. The library transforms a host yeast strain carrying two mutations: one to allow transformants to be selected (uraJ, for example) and another in the gene for which the wild-type gene clone is sought. Consider the cloning of the ARGl gene, the wild-type gene for an enzyme needed for arginine biosynthesis, by the complementation of an argl mutation. A yeast strain carrying the argl mutation has an inactive enzyme for arginine biosynthesis and therefore needs arginine to grow. A genomic library uses DNA from a wild-type (ARGl) yeast strain. When a population of uraJ argl yeast cells is transformed with the genomic library prepared in the shuttle vector, some cells receive plasmids containing the normal (ARG1) gene for the arginine biosynthesis enzyme. The plasmid's ARG1 gene is expressed, enabling the cell to grow on the minimal medium, that is, without arginine, despite a defective argl gene in the cell's genome. The ARGl gene is said to overcome the functional defect of the argl mutation by complementation of that mutation. The plasmid is then isolated from the cells, and the cloned gene is characterized.

10.26 Using Heterologous Probes to Identify Specific DNA Sequences in Libraries

cDNA probes can be used to identify and isolate specific genes, and a huge number of genes have been cloned from both prokaryotes and eukaryotes with those probes. It is also possible to identify specific genes in a genomic library by using clones of equivalent genes from other organisms as probes. For example, a mouse probe could probe a human genomic library. Such probes generally are called heterologous probes, and their effectiveness depends on a good degree of homology between the probes and the genes.

10.26.1 Chromosome-Specific Libraries

Chromosome-specific libraries can be constructed. However, the process is very difficult. Primarily these are prepared by cell sorting. The individual chromosomes can be isolated for this purpose. The mitotic cells are collected, and the metaphase chromosomes are stained with two fluorescent dyes, one binding to AT pairs and the other to GC pairs. The laser beam is used to cause stimulation which allows the cells to fluoresce, and the photometer sorts and fractionates the chromosomes by differences in dye binding and light scattering. After the chromosomal isolation DNA is extracted, a restriction enzyme digests the DNA the fragmented DNA is cloned into a vector.

For the library construction, individual chromosomes have been isolated. Various methods have been used for this purpose. These are pulsed-field gel electrophoresis which is used to isolate yeast chromosomes for the construction of chromosome-specific libraries. A cloned library of yeast chromosome III (315 kb) was the starting point for the Yeast Genome Project. Sequencing of chromosome III revealed that about 50 per cent of all the genes on this chromosome were unknown. Now the entire yeast genome was published in 1996.

10.26.2 Screening a Library

Plasmid library can be screened. For this purpose, the library is first cloned. To start with clones, the literary is grown on nutrient agar plates where thousands of colonies would be formed. These can be transferred to a filter paper by gently pressing the filter paper against the agar on the plate. Now the filter paper can be processed to lyse the bacterial cells and denature the double-stranded DNA to single-stranded which bind on the filter (**Figure 10.16**).

These are further processed by incubating the filter with the labeled probe. If the filter paper contains any of the complementary DNA sequences, the probe will bind to these sequences. After binding is completed, the extra single-stranded DNA has washed away. The bound DNA can be seen either through radioactivity if the probe was radiolabeled or through chemiluminescence. The autoradiograph will reveal the location of colonies carrying the gene of interest. Phage libraries can also be analyzed by using a little modified method. In the first step, the phage containing DNA insert is spread over a lawn of bacteria growing on a plate. Bacterial cells will get

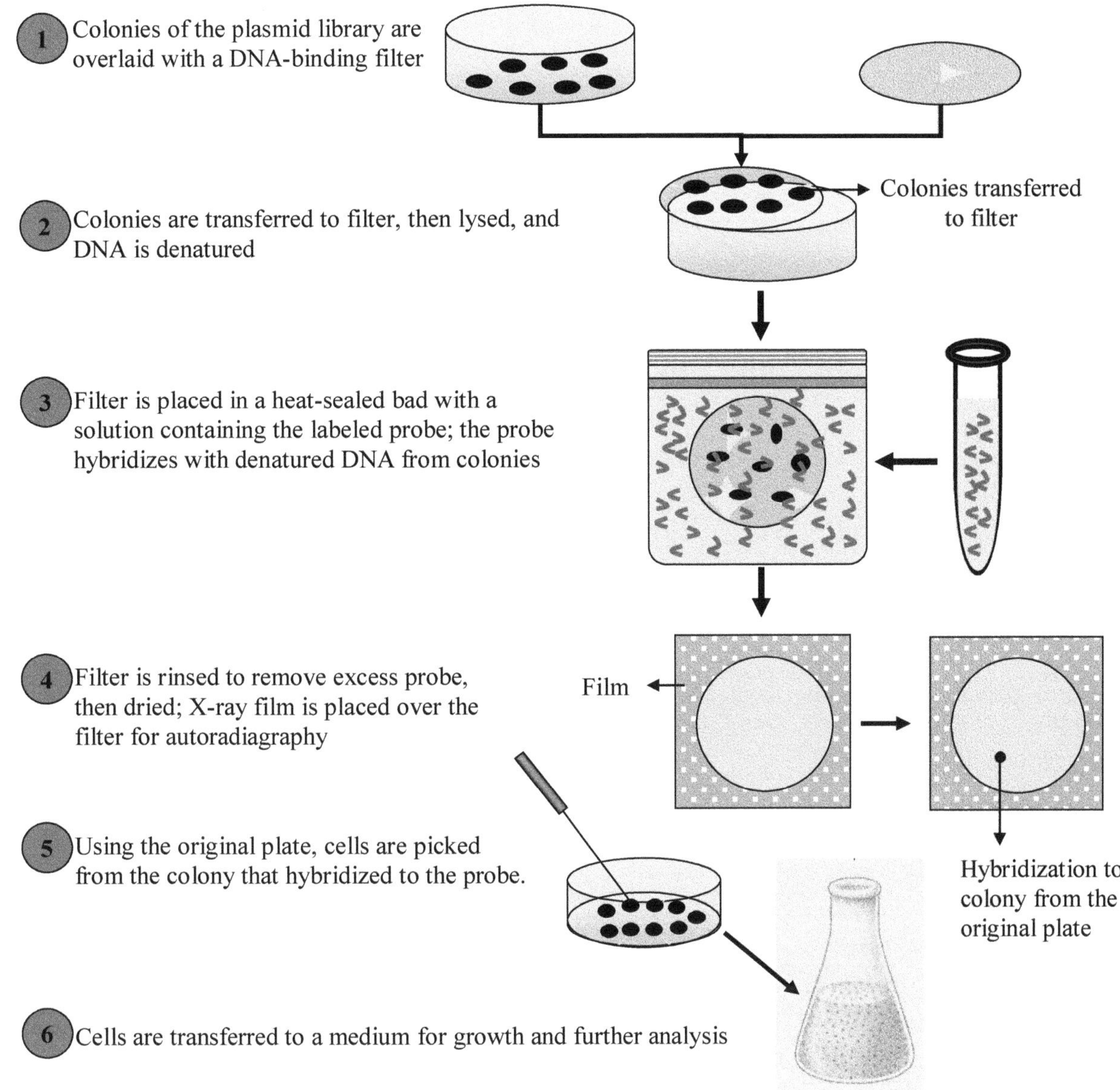

Figure 10.16: Screening a Plasmid Library to Recover a Cloned Gene.

infected by the phage and form plaques as they grow. The plaques are transferred onto a nylon membrane which is further denatured to convert the double-stranded DNA into single-stranded and then these are screened with the help of probes.

10.27 Cloned Sequences can be Characterized in Several Ways

The recovery and identification of genes and other DNA sequences by cloning or PCR is a powerful tool for analyzing genomic structure and function. Much of the Human Genome Project is based on such techniques. The technique used is RFLP, as shown below (**Figure 10.17**). The clones carrying RNA can be characterized by northern blotting and DNA by Southern blotting.

10.28 Northern Blot Analysis of RNA

It is like southern blot analysis, called northern blot analysis or Northern blotting. In this technique, RNA is used instead of DNA. In northern blot analysis, RNA is extracted from cells or a tissue and then separated by size with gel electrophoresis. The RNA molecules are transferred and bound to a filter in a procedure identical to Southern blot analysis. A labeled probe is used for hybridization, which helps detect the proper bands and location of RNA fragments complementary to the probe. Each time the RNA size is detected using the marker.

Northern blot analysis is used for mRNA, which encodes the gene. In some cases, several different mRNA species encoded by the same gene have been identified in this way, suggesting that different

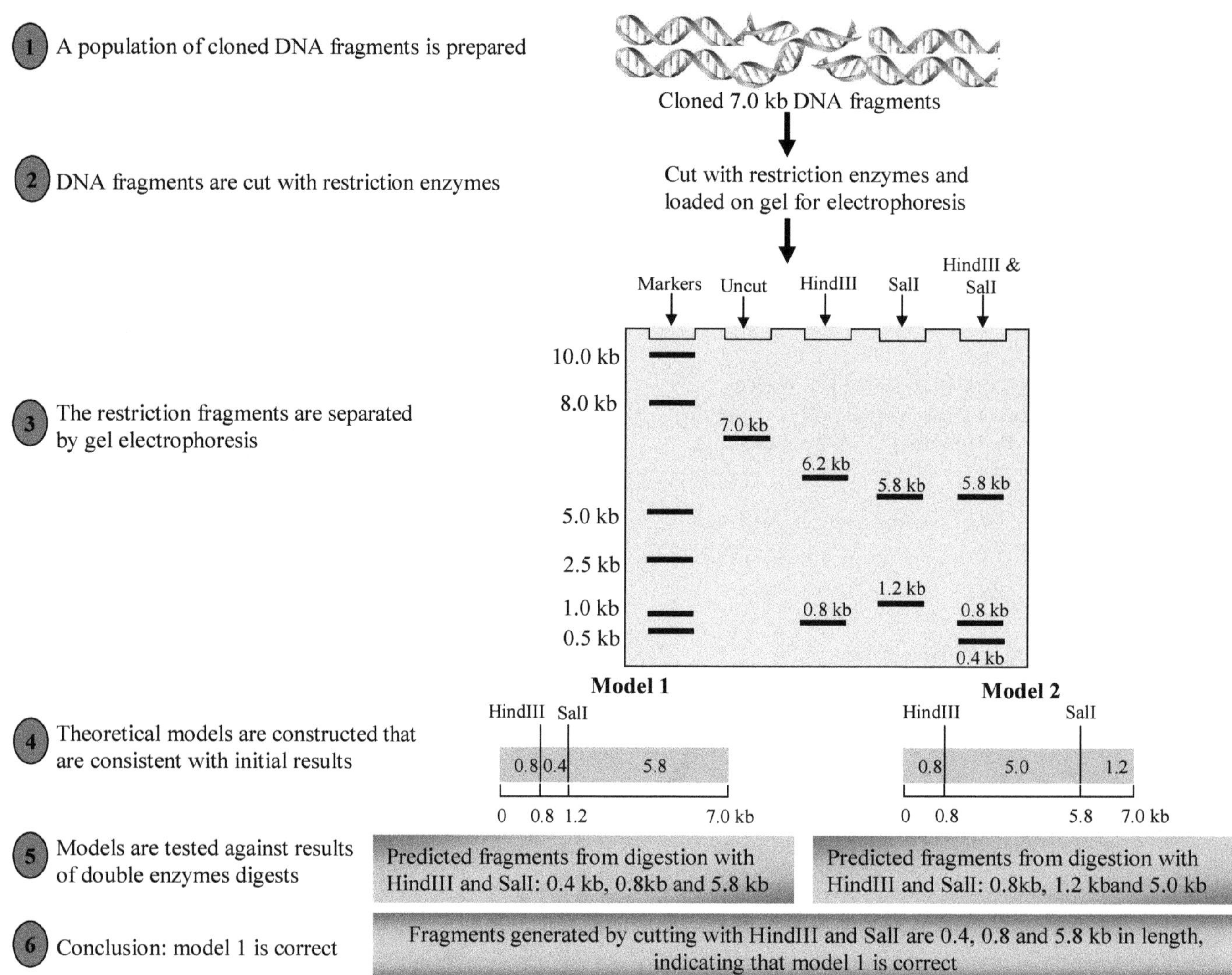

Figure 10.17: Constructing a Restriction Map.

promoter sites or different terminator sites are used or that alternative mRNA processing can occur. This type of experiment helps determine levels of gene activity, for instance, during development, in different cell types of an organism, or in cells before and after they are subjected to various physiological stimuli.

10.29Applications of Recombinant DNA Technology and the Ethical Issues

Applications of recombinant DNA technology are many, especially in the genomic era. These are in human health, which includes gene therapy, somatic therapy, and germline therapy.

10.30 Human Health

10.30.1 Gene Therapy

It was in 1953 when Carly Todd, became the first patient in the UK to be treated with gene therapy. She had a monogenic disorder called adenosine deaminase (ADA) deficiency. The cells in her body could not produce working copies of the enzyme ADA because of a fault in the gene sequence. While all cells use ADA, the white blood cells are particularly vulnerable if absent. Consequently, Carly had almost no white blood cells. Leukocytes (white blood cells) play an essential role in our immune systems; she could not fight against infections (a condition known as severe combined immunodeficiency).

It was thought that if a correct version of the gene could be put into her bone marrow cells then they would be able to make healthy white blood cells. In a simple operation, doctors collected a sample of her bone marrow. They isolated and cloned the gene for ADA and packaged it in a viral vector. Mixing the bone marrow cells and the modified viruses in a laboratory allowed the gene to be passed into some cells. These cells were then returned to Carly. The hope was that over the next few years, these cells would multiply, populate her bone marrow, and produce sufficient healthy white blood cells to form a natural defense against infections. Many trials were going on.

10.31 Isolated and Cloned

The gene responsible for the disease must be known, isolated, and cloned. This is still the case for only a few conditions. Before inserting the gene, one should know the correct controlling regions. Without these, it could either fail to cause any protein production, which would be a waste of time, or it could cause an over-production of the protein, which may be damaging.

10.32 Location of Effect

Gene therapy will only work if the gene can be placed into the cells that mainly need to use it. Again, this narrows the field as the process by which a faulty enzyme causes a disease, and hence the exact cells that need to be treated are known in a tiny group of genetic diseases. Having discovered which cells need targeting. In the case of ADA deficiency, this was done by removing the appropriate cells from the patient, treating them outside her body, and then returning them.

However, this will not always be possible. For example, cystic fibrosis affects cells throughout the body that produce secretions. For treating the respiratory problems associated with cystic fibrosis, the gene can be placed in an aerosol that can be inhaled in a way the asthma drugs are inhaled, but treating secreting cells, not on the body's surface, may prove more difficult.

10.32.1 Somatic versus Germline Therapies

While treating a disease using gene therapy can be considered an extension of conventional medicine, altering the genetic makeup of sperm and ova has broader implications. At the moment, germ-line therapy is banned in the UK even though it holds the possibility of eradicating diseases (*e.g.* hemophilia) from future generations.

Once a gene has been inserted into the germline, it has the potential to be inherited by all future generations. The consequences of this are unknown. There is not yet enough information about what happens to genes once they have been placed in cells and whether they affect the function of other genes.

10.32.2 Transgenic Organisms

Transgenic organisms: an organism that has had its genetic make-up altered by transferring into it a gene from another species. As a result, it manufactures a protein that it does not usually produce.

One of the most significant areas of potential commercialization, and not surprisingly of controversy, is recombinant DNA technology to move a gene deliberately from one organism to another. This procedure is often carried out to produce commercially valuable proteins like hormones or drugs.

10.32.3 Micro-organisms

As we have seen, once a gene has been isolated, it is relatively easy to move it into a bacterium. Once in place, the bacterium can manufacture the protein the gene codes. Placing the human insulin gene in bacteria has been successfully used commercially to produce human insulin, a vital drug for people with diabetes. Another example is the yeast cells that had DNA incorporated so that they manufacture a hepatitis B vaccine.

10.32.4 Animals

If a gene is inserted into the nucleus of a fertilized egg cell before it starts to divide then every subsequent cell will contain the new gene. This technique has been used to place a gene that codes for alpha-1 antitrypsin into a sheep embryo. The gene was surrounded by controlling sequences that allow it to be expressed only in mammary glands. As a result, this transgenic sheep produces milk that contains this protein. Alpha-l antitrypsin is used to treat patients with a particular type of inherited lung disease - congenital emphysema.

More radically, transgenic animals may become a valuable source of transplant organs. When patients receive transplanted organs, one of the biggest problems they face is that their immune systems recognize the proteins on the surface of cells are foreign - they come from a different individual. The immune systems then set about destroying the new organ. To prevent this rejection process, transplant patients are given drugs that suppress their immune systems. However, this leaves them vulnerable to disease. Using recombinant DNA technology, it may be possible to transfer to a developing pig embryo the parts of a person's genetic sequence that produce proteins that allow recognition of their organs. When transplanted, the pig's organs would then carry proteins that would fool the patient's immune system into thinking that the organ is not foreign. The organ would not be rejected, and no immunosuppressants would be required.

10.32.5 Agriculture

Recombinant DNA technology is being applied to increase yields of both plants and animals, to increase resistance to disease or pollution, and to create new crops that can utilize previously wasted resources.

A popular method for introducing DNA into plants uses a strain of bacteria found in soil (Agrabacterium tumefaciens). In their natural state, the bacteria infect plant cells, inserting part of their DNA and causing cancer–like growth. New genes can be inserted into a section of the bacterium's DNA to transfer the genes to plant cells. As tobacco plants are particularly susceptible to infection by this bacterium, much of the fundamental research has been conducted on them to use the knowledge gained to develop food crops.

10.32.6 Improving Yields

In cereal crops like wheat or rice, much attention has been paid to increasing the number of grains that grow in each head while causing the plant to grow with a shorter stem, thus wasting less energy on this inedible part of the plant. In the past, such development was carried out in prolonged breeding programs, but now the rate of development can be accelerated.

Recombinant DNA technology has also been used to manufacture bovine somatotrophin (BST), a hormone that increases by 10 to 15 per cent the amount of milk that a cow can produce. The gene has been placed inside bacteria which makes them produce the hormone. However, this example highlights some of the potential problems of implementing technology. In Europe, there is surplus milk. Making cows produce more milk means we will need less of them.

10.32.7 Nitrogen Fixation

A major area of research is dedicated to finding ways of moving nitrogen-fixing genes (NIF genes) into crops. Nitrates are vital nutrients for most plants, and some bacteria are particularly good at creating them by biochemically reducing nitrogen. Plants of the legume family (*e.g.* peas, clover, *etc.*) foster these bacteria in highly specialized root nodules. Using energy stores from the plant, the bacteria fuel nitrogenous enzymes to reduce atmospheric nitrogen to ammonia; this is then taken up by the plants and converted into several nitrogenous compounds. When the plant dies, it is left in the soil. If cereal crops are grown on this enriched soil in the following year, the extra nitrates lead to a better yield. Alternatively, farmers spread expensive nitrate fertilizers on their land to boost crops.

However, all of this might change. The genes which produce the necessary NIF enzymes have been isolated, sequenced, and cloned in E. coli. Now scientists are looking at ways of placing them into cells in crops such as wheat or rice. This would enable these crops effectively to fertilize themselves, saving much money and increasing yields. Placing the gene in crops grown in developing countries would have an enormous impact on their ability to grow food. Currently, they cannot afford the nitrogenous fertilizers used in more affluent countries.

10.33Resistance to Disease, Pests, or Herbicides

Many crops are particularly vulnerable to attack from specific organisms, *e.g.*, tomatoes are attacked by aphids, potatoes by viruses, and wheat by fungi. Farmers fight back by spraying their crops with chemicals that destroy the invading organisms.

Recombinant technology helps to produce proteins that can kill the undesired worms *e.g.*, a tomato plant has a gene that produces a protein that kills tomato fruit worms.

10.34 Risks

Such genes may get into weeds, giving them resistance to attack and allowing them to flourish; when the crop is eaten, one will consume the genetically produced chemical, which may be harmful. The attacking organisms may become resistant to the chemical, so the process gives no protection.

10.35 Novel Crops

One of the problems of modern agriculture is that a crop is frequently grown hundreds or thousands of miles from the customer. It can be challenging to get it to shop before it rots. Fruit and vegetables are often harvested before they are ripened and then ripened artificially just before they are sold. This leads to a loss of flavor.

In the case of tomatoes, the problem may have been solved. Researchers have now produced a tomato resistant to becoming soft and rotting once it has ripened. Fruit normally softens because it produces an enzyme called polygalacturonate (PG). This enzyme breaks down pectin in cell walls, so the fruit becomes soft. By inserting a gene that has a sequence that is precisely the opposite of the gene coding for PG (an antisense gene), the cells produce an anti-sense strand of mRNA as well as a sense strand from the normal gene. These two mRNA strands are so perfectly matched that they stick together, thus rendering the sense strand useless. Without the mRNA no enzyme forms, and without the enzyme, the fruit stays firm.

Some people are allergic to some of the proteins found in milk. In the same way that sheep can have genes inserted causing them to produce valuable pharmaceutical proteins in their milk they could have genes inserted that modify the milk removing these allergenic proteins. In the future, there could be many new niche markets for supplying tailor-made foods.

10.36 Food and Drink

We have already seen how recombinant DNA technology affects food produced in plants in standard agricultural situations. Now, we turn to the other areas of the food industry that use the techniques.

10.37 Microbial Protein

Bacteria can be designed to grow on virtually any energy-rich source. Some have been adapted to use methane gas as a nutrient, and others grow successfully on paper pulp. Provide a new source of protein to the food industry. The growing bacteria can be harvested, and their proteins purified. These proteins may be of particular value as the number of people eating a vegetarian diet increases.

10.38 Benefits

Growing farm animals to supply us with protein is a very inefficient use of energy as it can take between 10 and 20 kg of protein in feeds to produce 1 kg of meat protein. However, bacteria are much more efficient. Waste materials such as pulped newspapers could form the essential nutrient supply for new bacteria.

10.39 Risks

However, these bacteria can contain the proteins to which people are allergic. If a microbe is designed to digest cellulose efficiently, care is required in how it is contained.

10.40 Microbes in Food Production

Industrial cheese production uses a lot of enzymes called chymosin (commonly called rennin) to coagulate the protein casein found in milk. Traditionally this enzyme is obtained from the abomasum (fourth stomach) of suckling calves when slaughtered for meat. However, the number of calves being slaughtered is decreasing, but the quantity of chymosin required for cheese making is increasing.

Bacteria are modified by transforming them by including a gene that causes chymosin production. There are seven basic steps.

1. mRNA for all the sequences needed to make chymosin is isolated from calf stomach cells.
2. This mRNA is transcribed into cDNA.
3. cDNA is cloned into a vector.
4. The vector transforms E. Coli.
5. Standard fermentation techniques are used to grow E. Coli.
6. E. Coli is harvested, broken down, and the enzyme is purified.
7. Tests show that no bacteria are present in the extracted enzyme, and feeding trials with rats showed that they were not harmed by consuming a dose of the enzyme that was 100,000 –fold more significant than expected for the average human consumption. It is, therefore, safe to use in the cheese industry.

Such bacterial chymosin is used in the production of vegetarian cheese.

10.41 Detection of Foodborne Pathogens and Spoilage Agents

Recombinant DNA technology has also come to the help of food manufacturers who need to know whether a product is safe to eat or is contaminated with pathogenic bacteria. A series of gene probes have been built that carry sequences that can specifically identify the presence of a wide variety of different food pathogens.

Some bacteria, such as Listeria monocytogenes, are only dangerous if alive. While conventional tests take one of three days to give results, gene probes can do the job in a few hours. Gene probes for sections of DNA in L. monocytogenes have been built, and these can be used to see whether the bacteria are present. But as DNA lasts thousands of years once the bacteria are dead, this won't distinguish between dead and live organisms. However, various probes have been built, coding for the mRNA sequences. As mRNA is only present in living cells, this will only detect live bacteria.

10.42 Labeling Genetically Modified Foods

Owing to public concern about the use of recombinant DNA technology to move genes between plants, some people think that genetically modified foods should be labeled. This would also include any foods which have been manufactured using genetically modified crops as ingredients.

10.43 Environment

10.43.1 Releasing Genetically Modified Organisms

Releasing genetically modified plants, animals, or microbes into the environment is regulated by law, and a license must be obtained for each product before any release is allowed. There are few risks associated with the process of adding a new gene to an organism, but there are risks that result from designing organisms that have new combinations of genes.

10.43.2 Knock-on Effect

Recombinant DNA technology can be used to alter bacteria or small animals, such as beetles or worms, so that they selectively destroy pests. This will save using agricultural pesticides, which are harmful and expensive. However, by introducing a new predator into an ecosystem, we deliberately alter its balance. It is impossible to predict fully the outcome of such a change. Such an alteration may not necessarily be harmful, but it needs to be monitored carefully each time a new organism is released.

The danger is that if any organism does cause harmful knock-on effects, there may be significant problems removing it from the environment.

Some fish have been genetically altered to have an extra growth hormone gene. They also have an additional gene, making them produce a type of anti-freeze in their blood. This enables them to survive in freezing water. If they were released, either deliberately or accidentally, these fishes might initially supply an extra source of food. As time passes, they may disturb the native fish populations and deplete food reserves.

10.43.3 Self-control

At the same time as adapting an organism to perform a particular function, it is possible to adapt it so that its ability to survive is limited. This can control its spread once released.

Genes can be added that cause an organism to require nutrients that are generally not found. When you deliberately stop supplying this nutrient, the organism dies. A suicide gene may be inserted. For instance, a bacterium could be designed to destroy a particular pollutant, such as crude oil, and a gene could be inserted that kills the bacterium if there is no eroded oil around. Therefore, the bacteria can be sprayed onto an oil spill, they will destroy the oil, and when all of the oil is gone, they will kill themselves.

10.43.4 Economic Considerations

Many crops can only be grown in countries because those countries have hot climates. The ability to move genes into tropical plants that will then allow them to be grown in temperate regions may, at first sight, seem a good idea. However, such crops would then destroy the economies of many tropical countries. For instance, many Caribbean countries depend on selling bananas to foreign countries.

On the other hand, recombinant DNA technology could be used deliberately to design new crops to grow in arid conditions.

10.43.5 Handling Pollution

As we have seen, microbes can be designed to grow on many waste materials to produce valuable materials such as food. They can also be usefully employed to control pollution.

Bacteria have been designed to break up oil pollutants. Bacteria have been designed to destroy noxious gases released from factories. Fumes are pumped through pipes running under gravel or wood chippings, which supply a large surface area on which the bacteria can grow. As the gases leave the pipes, they pass through the filter bed, and the bacteria ingest and destroy the noxious components. Plants and animals can be designed to grow in polluted environments.

10.44 Forensic Science

When investigating a criminal offense, such as a burglary, serious sexual assault or murder, the police try to find evidence that will indicate who the criminals are. Recombinant DNA technology has now been used in several different ways. These are:

10.45 DNA Fingerprinting

The best-known example is DNA fingerprinting, based on two assumptions: (i) Each person's cell carries an identical set of DNA, and (ii) The DNA code for every person is unique.

This means that if a few drops of blood or the few hair follicle cells attached to pieces of hair (or semen in the case of a sexual assault) are left at the crime scene, the DNA can be extracted, multiplied, and then analyzed. The results can then be matched to those of a similar analysis on samples taken from a suspect. If they match, this indicates that the suspect was at the crime.

11

Linkage and Chromosome Mapping in Eukaryotes

Term linkage was given in 1903 by Walter and Sutton. It is defined as an association between two or more genes on a chromosome that tends to cause the characteristics determined by these genes to be inherited as an inseparable unit. The reciprocal exchange of segments occurs at the time of the first meiotic division; currently, the homolog's chromosomes get paired, and recombination occurs; the frequency with which the exchange occurs is called recombinant frequency. Exchange of genes may occur reciprocally or between non-reciprocal genes resulting in two types of gametes.

Recombinant gametes are produced if the loci of two genes are linked and not far apart from each other. If 50 per cent recombination occurs, the ratio would be 1:1:1:1, resulting in 4 types of gametes meaning two parental and two recombinant gametes, as shown in **Figure 11.1**.

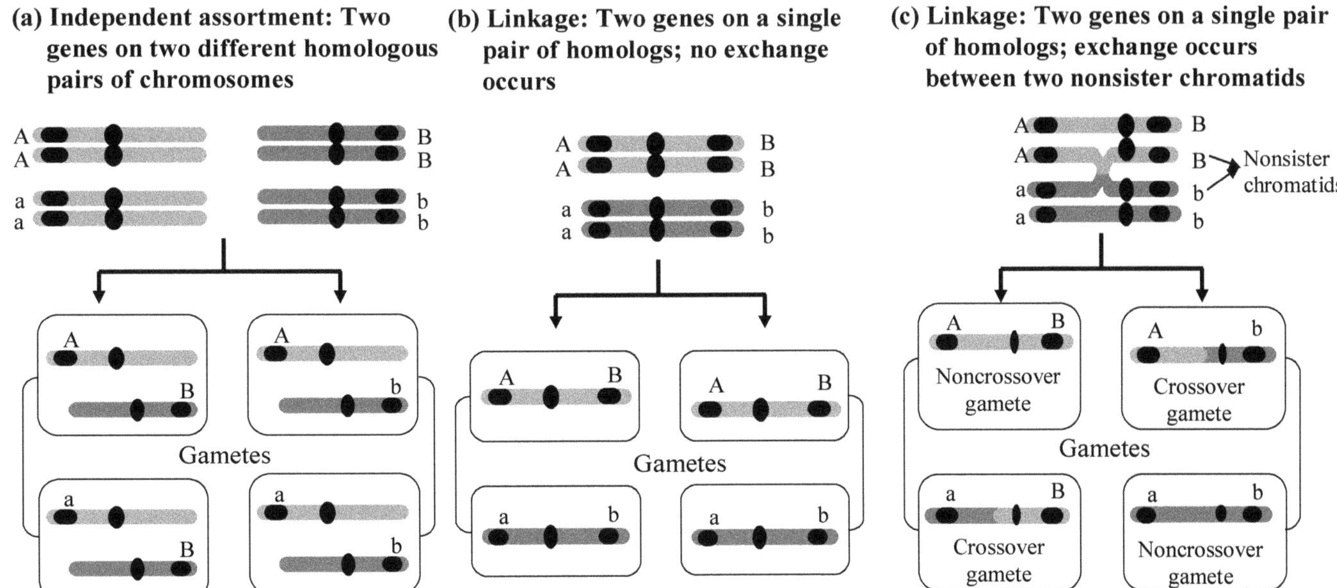

Figure 11.1: Two Non-sister Chromatids where the Exchange takes place followed by Independent Assortment.

11.1 Deviation from the Law of Independent Assortment

The distance between the two genes on the chromosome can be calculated based on recombinant frequency. If the genes are placed far away, recombination will occur; however, if the genes are located very close to each other, no recombination will occur. Based on this, the recombination frequencies can be calculated. The method used is the testcross, in which the individual to be analyzed genetically is crossed to a homozygote carrying the recessive alleles. This is called genetic crossing. For example, the heterozygous F1 sweet peas could be crossed to homozygotes carrying the recessive alleles of the genes for the flower color and pollen length.

Among 1000 progeny, approximately 920 resemble one or the other of the parental strains, and the remaining 80 are recombinant. The frequency of recombinant gametes formed by the heterozygous F1 plants would be 80/1000 = 0.08.

Recombinant gametes generally are not more than 50 percent. This is possible when genes are not located near each other, and their location is at the two different chromosome bands. It may also be possible when genes are located on different chromosomes; 50 per cent of recombination means that genes are assorted independently. Suppose genes A and B are on different chromosomes, and an AA BB individual is crossed to an aa bb individual. The Aa Bb offspring are then test-crossed to the double recessive parent from this cross. Because the A and B genes assort independently, the F1 will consist of two classes (Aa Bb and aa bb) which are phenotypically like the parents in the original cross, and two classes (Aa bb and aa Bb) are phenotypically recombinant. Each F1 class will have a frequency of 25 percent. Thus, the total frequency of recombinant progeny from a testcross involving two genes on different chromosomes will be 50 percent. The frequency of less than 50 per cent of recombination shows that the genes are linked. Linked genes are in the linkage phase and are arranged in heterozygous individuals (**Figure 11.2**). Bateson and Punnett's sweet pea experiment shows that in the F1 generation, plants received two dominant alleles, R and L, from one parent and r and l from another. So, the plant genotype RL/rl separates alleles inherited from two parents. If we write RL/rl that is on one chromosome are dominant alleles while on the other chromosome, there is a recessive allele called coupling linkage phase. If we write Rl/rL, a combination of dominant and recessive alleles on one side, then linkage is in a repulsion phase.

11.2 Physical Basis of Crossing Over

Recombinant gametes mean that crossing over has occurred during the first meiotic division. At the time of duplication, the chromosomes are paired. Four homologous chromatids are present; this is called a tetrad; Only one crossover occurs at any point; once the crossover occurs, the chromatid breaks. After the break, the broken pieces return, and the product is called recombinant. One must note that the other two chromatids are not recombinant. Hence out of the four homologous chromatids, only two are recombinant chromatids.

Most of the evidence related to the crossing over between two chromatids within a tetrad came from the study on fungi in the class Ascomycetes (Saccbmwnyces cerevisiae). The importance of this fungus is that mutation can be created quickly, and recombination can be studied.

The study on S. cerevisiae shows that crossing over occurs once the chromosome is duplicated. If it occurred before duplication, an ascus can never have more than two kinds of ascospores. It has been shown that only two chromatids are involved in exchange at any one point. However, the other two chromatids may cross over at a different point. Thus, there is a possibility for multiple exchanges in a tetrad of chromatids. There are examples showing two, three, or even four separate exchanges. These are called double-triple or quadruple crossovers.

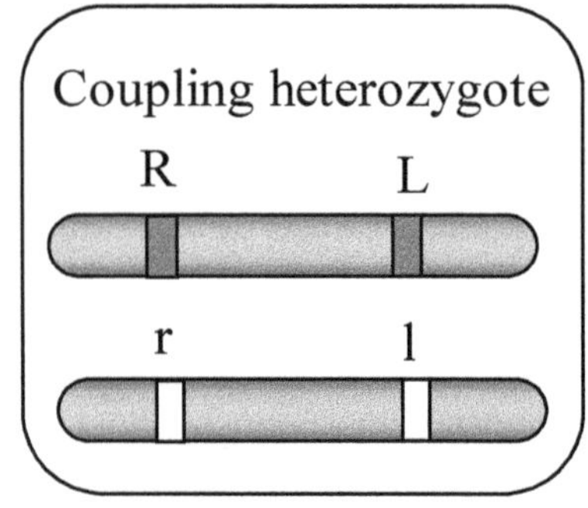

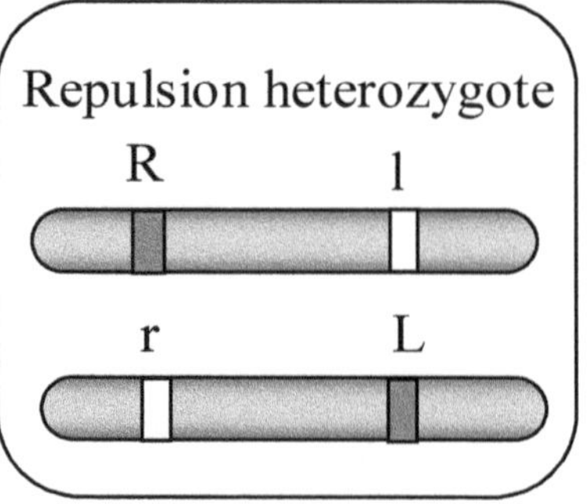

Figure 11.2: Coupling and Repulsion Phases in Linkage.

However, an exchange between sister chromatids does not produce genetic recombinants as the sister chromatids are identical.

Different theories have been projected. One hypothesis is that the chromatid fibers become intertwined with each other, and stress causes them to break. Although such tangles may occur, sometimes the breaks are induced by enzymes acting on the DNA inside the chromatids. En zymes are also accountable for correcting these breaks and for reattaching chromatid fragments to each other.

11.3 Crossing over the Result in Recombination

It was in 1931 when Harriet Cregton and Barbara McClintock revealed that exchange occurs between two chromosomes. The experiments were conducted on maize. Maize chromosomes are morphologically different. They took chromosome nine and considered two genes; one gene was responsible for kernel color (C= colored c= colorless), and the other gene caused the texture (WX = starchy, wx = waxy), out of these two of the 9th chromosomes; one was expected, and other had cytogenetic aberration at each end, *i.e.*, a heterochromatic knob is situated on a different chromosome which is not near to one another. After crossing over, the resultant chromosome carried CWx and fWX, one of the abnormal markers lost in the previous generation.

These findings strongly suggested that the physical exchange between paired chromosomes caused recombination.

11.4 Chiasmata and the Time of Crossing Over

Only in the late prophase the crossing over can be seen, which is the cytological proof of the 1st meiotic division. At this stage, the chiasmata are clear. Once the chiasmata are formed, the chiasmata chromosomes repel each other to some extent; close contact centromere between the centromeres is maintained, and chiasmata can be seen accurately due to this separation. Thus, the number of chiasmata is nearly proportional to the chromosome length. Formation of chiasmata late in the first meiotic prophase is seen in the crossing over. The frequency of the recombinations can be altered by giving heat shocks. There is little effect if the heat shocks are given in the late prophase. However, in the early prophase, the recombination frequency was changed. Thus, the event responsible for recombination, namely, crossing over, occurs early in the meiotic prophase. Additional evidence-based molecular studies show that these phenomena may occur during DNA synthesis. Although almost all the DNA is synthesized during the interphase that precedes the onset of meiosis, a small amount is synthesized during the first meiotic prophase. This limited DNA synthesis has been interpreted as a part of a process to repair broken chromatids, which may be associated with crossing over. DNA synthesis occurs in the early to mid-prophase but not later. The existing proof, therefore, shows that crossing over occurs in the early to mid-prophase, long before the chiasmata can be seen.

11.5 Genetic Distance can be Measured by Crossing Over

The distance between two loci affects the chances of recombination. Recombination is not a selective phenomenon; it occurs randomly in each tetrad [**Figure 11.3 (a) and (b)**]. In these figures, the location of the loci is shown, and have demonstrated that recombination occurs only when the loci are farther away.

Two non-sister chromatids can undergo a single crossover (a), Linkage is not disturbed after the exchange, (b), Exchange produces recombinant gametes which can be detected; interestingly, the other two chromatids present in the tetrad are not involved during this exchange; therefore, the gamete remains unaltered. If a single crossover occurs between two linked genes, only recombination occurs in 50 per cent of the gametes (**Figure 11.4**).

If a single exchange is found, 20 per cent recombinant gametes are formed. Crossing over actually occurs only in 40 per cent of the tetrads. The recombination is seen only in 50 per cent of the crossovers. If the distance between the two linked genes is more than 50mu the crossover is possible in 100 per cent of tetrads **Figure 11.4**. However, the genes are found on different chromosomes and assorted independently.

11.6 Three-point Cross

If there are three dominant loci, ABC, and their counterpart recessive alleles abc are present. Upon linkage, the combinations are ABC, CAB, ACB and are ABC/abc, ABc/abc, Abc/aBc or Abc/aBc. These are created by back cross experiments only when three of the loci are found on the same chromosome.

There will be eight combinations When there is an equal number of phenotypic and genotypic combinations (**Table 11.1**). In case some combinations are found more frequently, this indicates that some genes are linked; the other way to check is to see if the single phenotypic class from the AaCc x aa cc, AaBb x aa bb, and BbCc crosses occur with a frequency of 25 per cent of total offspring or not. If it is 25 per cent,

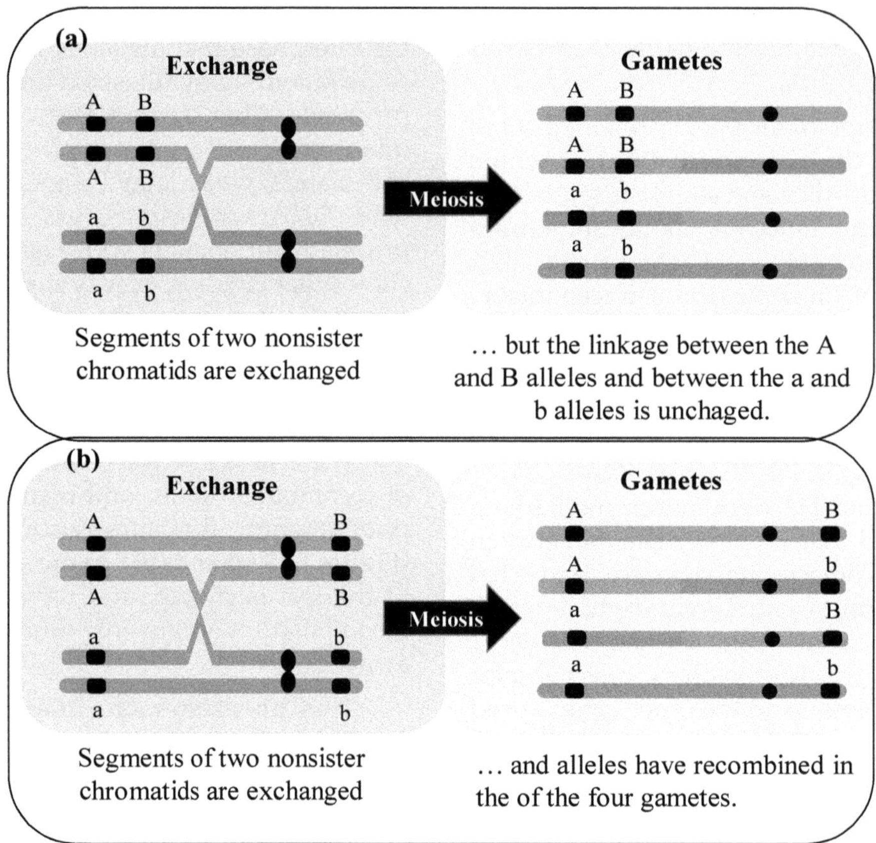

Figure 11.3: Single Crossovers between Two Non-sister Chromatids and Production of Gametes.

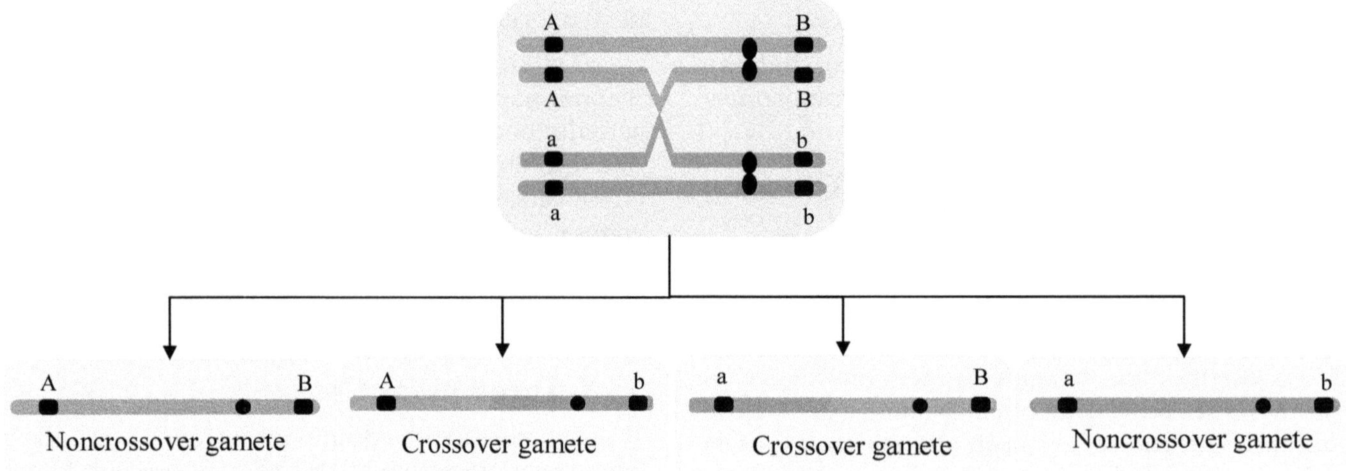

Figure 11.4: The Result of a Single Exchange between Two Non-sister Chromatids Occurring in the Tetrad Stage. This results in two parental noncrossovers and two recombinant crossovers.

Table 11.1: Results from a Cross between AaBbCc x aabbcc

Phenotypes	*ABC*	*ABC*	*AbC*	*Abc*	*aBC*	*aBc*	*abC*	*Abc*
Genotypes	ABC/abc	ABC/abc	AbC/abc	Abc/abc	aBC/abc	aBc/abc	abC/abc	Abc/abc
Numbers of progeny	21	77	395	4	3	397	73	30

it means independent assortment has occurred. If independent assortment is not present, these loci are linked to one another.

The locus in the middle can be found by looking at the common progeny with less frequent ones. The data generated can be classified into four classes. The criteria of classification are that they should be numerically similar. It is seen that 395 are ABC and 397 are aBc phenotypes. There are aBC (3) and Abc (4) phenotypes. If no cross-over occurs, recombination is less frequent; in this case, two classes are formed noncrossovers (Nco) and triple heterozygous parents. The phase of the genes in the triple heterozygous parent is AbC/aBc, ACb/acB, or Cab/caB. It isn't easy to find out which locus is in the center. It is considered that the chromosomes represented in the least frequent class (Abc, aBC) are due to single crossover (DCO); this is quite rare as compared to single crossover (SCO).

When a crossover is found on each side of the middle of the locus, the middle alleles are lost. These are about the two bordering loci. If the B locus is in the center, depending on the parental genotype AbC/aBc, the DCO class would be ABC and abc. However, the data show that the ABC and abc chromosomes do not belong to the standard class. This causes the B locus not to lie between the A and C loci. Thus, the phase and gene order of the triple heterozygous parent is ACb/acB.

Map distances are measured by considering all the crossovers between two loci and then dividing them by the sum of the recombinant and nonrecombinant chromosomes. The frequency of males and females may be different; recombination is seen more frequently in females than in males. A female genetic map is 1.5 times greater than a male genetic map. Genetic maps in males/females are calculated depending upon the frequency of crossovers in the chromosomes that originate from males and females respectively.

A cross between ABCD/abcd x abcd/abcd can be created for genetic mapping. This results in phenotypic and genotypic types. In total, there are 16 in number. If all offspring have all the types, the map distances between loci can be evaluated from the appropriate single double and triple crossovers. An increase in the number of genes complicates the time-consuming process. Multipoint data is formed, which is not perfect; hence, it is difficult to create map distances. If we know the gene order, it is easy to constitute gene rearrangement on the chromosomes, which could be rare or of a common type.

The genetic map represents the linear arrangement of the genes. Crossing over can also be calculated if we know the distances of the linkage groups. This helps in mapping the genes. This whole information helps in constructing evolutionary history.

A single exchange will occur if two genes are located near one another. In case of a double crossover, two independent events would occur side by side. The probability of two independent events co-occurring is as per product law.

If the crossing-over results due to exchange is 20 per cent of the time ($p = 0.20$) between A and B, and 30 per cent of the time ($p = 0.30$) between B and C. The probability of recovering a double-crossover gamete arising from two exchanges (between A and B and between B and C) is predicted to be $(0.20)(0.30) = 0.06$, or 6 per cent. The double-crossover gametes are always less as compared to single crossover.

If a chromosome has three genes and these are narrowly placed. The double-crossover gametes are seen in frequency provided the distance between A-B is only three mu, and the B-C distance is 2mu, the double-crossover frequency will be $(0.03)(0.02) = 0.0006$, or 0.06 per cent equivalent to 6 events.

Three or more linked genes can be mapped. If crossover over occurs, then the effect would be the creation of gametes that will be heterozygous at all loci; gametes can be assessed accurately by observing the phenotypes of the resulting offspring. This is necessary because the gametes and their genotypes cannot be seen in a straight line; many offspring must show significant mapping results for all the crossovers.

In Drosophila, there are linked recessive mutant genes; let us suppose for yellow body color (y), white eye color (w), and echinus eye shape (ec). The sequences of the three genes would be y-w-ec.

All males are hemizygous in F1 generation for all the three wild-type alleles if crossed with females which are homozygous for the mutant alleles. The resultant males will be the wild type for the body color, eye color, and eye shape. They show a wild-type phenotype. The females show three mutant phenotypes yellow body color, white eyes, and echinus eye shape.

F1 generation consists of a cross-product consisting of heterozygous females at all three loci and males because the Y chromosome is hemizygous for the three mutant alleles. F1 female genotype fulfills the first criterion of mapping three linked genes; it is heterozygous at three loci and hence can serve as the source of recombinant gametes generated due to crossing over. In F1 parents, all three mutant alleles are on one homolog, and all three wild-type alleles are on the other. The heterozygous F1 female may have y and ec mutant alleles on one homolog and the w allele on the other. In the P1 cross, one parent was yellow, echinus, and the other was white.

All the gametes have an X chromosome, and suppose all three mutant alleles are on this chromosome. In contrast, the Y chromosome is not genetically active for the three loci under consideration. The genotype of the gamete produced by the F1 female will be expressed phenotypically in the F2 male and the females.

The chromosomal map can be created, as shown in **Figure 11.5**. The F2 phenotype shows both non-crossover and crossovers. This reveals the need for the F2 phenotypes to be evaluated because F1 females form parental gametes. The X chromosome remains unchanged even after crossing over. Further, the segregation results in equal proportions of the two gametes only if one is wild type and the other is mutant. These are called reciprocal gametes or phenotypes.

The two noncrossover phenotypes are seen because they are in higher proportions. **Figure 11.5** shows that gametes 1 and 2 are found in more significant numbers. Therefore, flies that show yellow, white, and echinus phenotypes and flies that are normal (or wild type) for all three characters have a noncrossover category and represent 94.44 per cent of the F2 offspring.

The second category that can be seen is due to double-crossover phenotypes. They should be present in fewer numbers because of their low probability of occurrence. It contains two independent but also single-crossover events. Two reciprocal phenotypes can be identified, which result in 7 gametes, which reveal the mutant trait yellow, echinus but standard eye color; and gamete 8, which shows the mutant trait white but standard body color and eye shape. Together these double-crossover phenotypes constitute only 0.06 per cent of the F2 offspring.

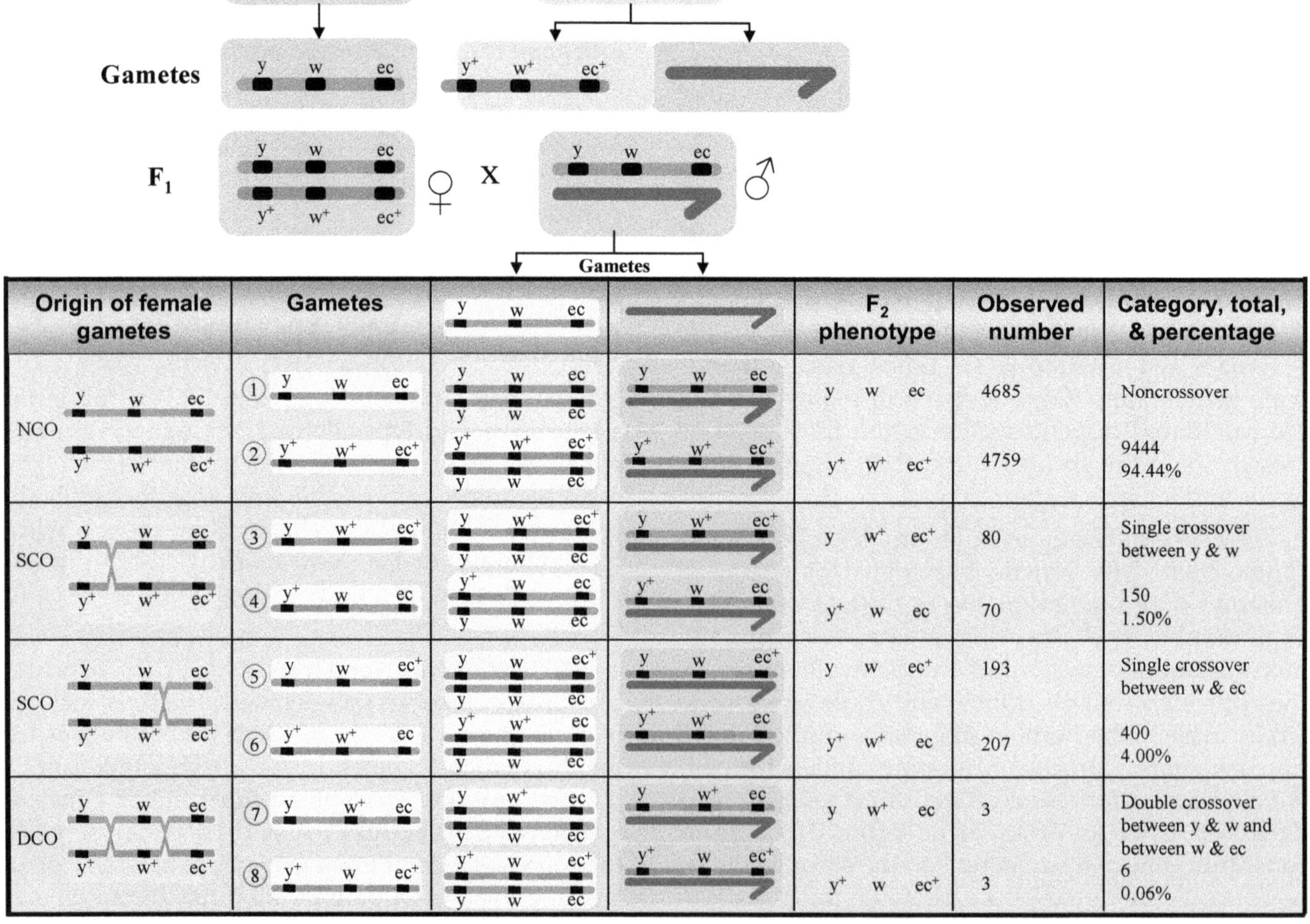

Origin of female gametes	Gametes	y w ec	(Y)	F_2 phenotype	Observed number	Category, total, & percentage
NCO	① y w ec	y w ec / y w ec	y w ec	y w ec	4685	Noncrossover
	② y^+ w^+ ec^+	y^+ w^+ ec^+ / y w ec	y^+ w^+ ec^+	y^+ w^+ ec^+	4759	9444 94.44%
SCO	③ y w^+ ec^+	y w^+ ec^+ / y w ec	y w^+ ec^+	y w^+ ec^+	80	Single crossover between y & w
	④ y^+ w ec	y^+ w ec / y w ec	y^+ w ec	y^+ w ec	70	150 1.50%
SCO	⑤ y w ec^+	y w ec^+ / y w ec	y w ec^+	y w ec^+	193	Single crossover between w & ec
	⑥ y^+ w^+ ec	y^+ w^+ ec / y w ec	y^+ w^+ ec	y^+ w^+ ec	207	400 4.00%
DCO	⑦ y w^+ ec	y w^+ ec / y w ec	y w^+ ec	y w^+ ec	3	Double crossover between y & w and between w & ec
	⑧ y^+ w ec^+	y^+ w ec^+ / y w ec	y^+ w ec^+	y^+ w ec^+	3	6 0.06%

Figure 11.5: A Three-point Mapping Cross Involving the *Yellow* (*y* or *y* +), *White* (*w* or *w* +), and *Echinus* (*ec* or *ec* +) Genes in *D. melanogaster*.

The rest of the four phenotypes have only two categories resulting in single crossovers. Gametes 3 and 4 are reciprocal phenotypes due to single-crossover events occurring between the yellow and white loci and are equal to 1.50 per cent of the F2 offspring. Gametes 5 and 6 constitute 4.00 per cent of the F2 offspring, having reciprocal phenotypes because of a single-crossover event between the white and echinus loci.

The distance between y and w or between w and ec equals the percentage of all observed exchanges. Any two genes under consideration include all the appropriate single crossovers and all double crossovers. The y and w genes include gametes 3, 4, 7, and 8, amounting to 1.50 per cent + 0.06 per cent, or 1.56mu. Similarly, the distance between w and ec equals the percentage of offspring resulting from an exchange between these two loci: gametes 5, 6, 7, and 8, totaling 4.00 per cent + 0.06 per cent, or 4.06mu.

11.7 Determining the Gene Sequence

Let us consider three genes on a chromosome which may be y-w-ec. There are three combinations possible, as shown below:

(I) w-y-ec (y in the middle)
(II) y-ec-w (ec in the middle)
(III) y-w-ec (w in the middle)

The following steps can be taken.

1. The first step is to find out how the alleles are placed or arranged on each homolog of the heterozygous parent giving rise to noncrossover and crossover gametes among female gametes. One should investigate the double-crossover event occurring inside the arrangements, which will produce the observed double-crossover phenotypes that are less in number. Various steps involved are shown in **Figure 11.6**; first, use a y-w-ec cross. This results in arrangements that can be I, II, and III. Let us suppose that y is located between w and ec; the arrangement of alleles in the homologs of the F1 heterozygote is.

   $$\frac{w}{w+}\ \frac{y}{y+}\ \frac{ec}{ec+}$$

   If P1 generation is crossed: all the females of the P1 generation will contain the X chromosome having the w, y, and ec, alleles; on the other hand, PI males will contribute an X chromosome bearing the w+, y+, and ec+ alleles.

2. A double crossover within this arrangement may show the following gametes:

   w *y + ec* and *w + y ec +*

   Let us further assume that y is in the center; the F2 double-crosser phenotypes will correspond to the above gametic genotypes, revealing offspring with white, echinus phenotype, and offspring will show the yellow phenotype. However, the determination of the double-crossover phenotypes reveals yellow, echinus flies, and whiteflies. Hence this order is not correct.

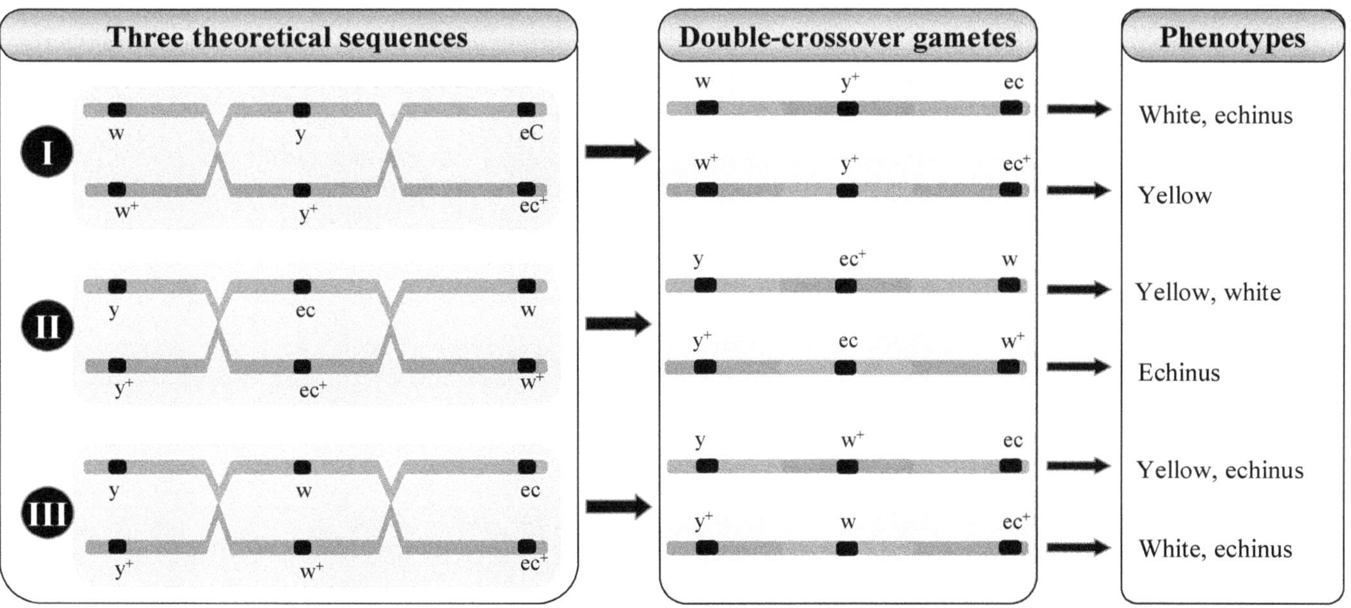

Figure 11.6: The Possible Sequences of the *White, Yellow,* and *Echinus* Genes in F2 Generation.

3. In the case of type II arrangement with the ec/ ec + alleles in the middle or arrangement III with the w/w+ alleles in the middle:

   (II) $\frac{y}{y+}\ \frac{ec}{ec+}\ \frac{w}{w+}$ or (III) $\frac{y}{y+}\ \frac{w}{w+}\ \frac{ec}{ec+}$

   Arrangement II shows expected double-crossover phenotypes; however, these are not like the experimental double-crossover phenotypes. The predictable phenotypes are yellow, white, and echinus flies in the second filial generation. Arrangement III shows the observed phenotypes-yellow, echinus flies, and whiteflies. This assumption with the w gene in the center is in the correct order.

11.8 A Mapping Problem in Maize

Sometimes inter-locus distance is not known in maize mapping. It can be done by using the following criteria. (1) One of the parents must be heterozygous for all traits considered for the study; (2) knowing the phenotype-genotype should be apparent (3) the sample size should be large enough to get correct results.

There are three types of recessive mutant genes in maize plants. These are bm (Brown midrib), v (virescent seedling), and pr (purple aleurone). All these alleles are located on chromosome number 5; however, the arrangement and map distance are unknown, but these are linked genes. Let us suppose these are heterozygous in female maize plants. Hence for mapping purposes, the alleles should be homozygous in male plants. If it is not so, then mapping becomes difficult or impossible (**Figure 11.7**).

This is possible by grouping the offspring into two categories for each pair of reciprocal phenotypic classes. The reciprocal classes are taken where no crossing over has occurred (NCO). There are two possibilities for these events, *i.e.*, SCO and DCO.

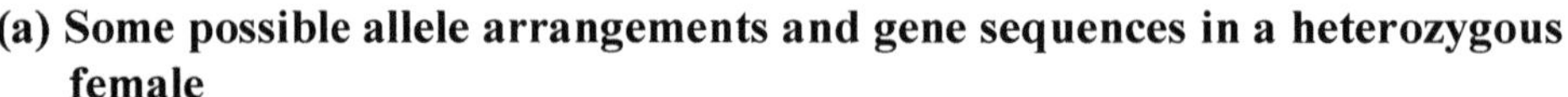

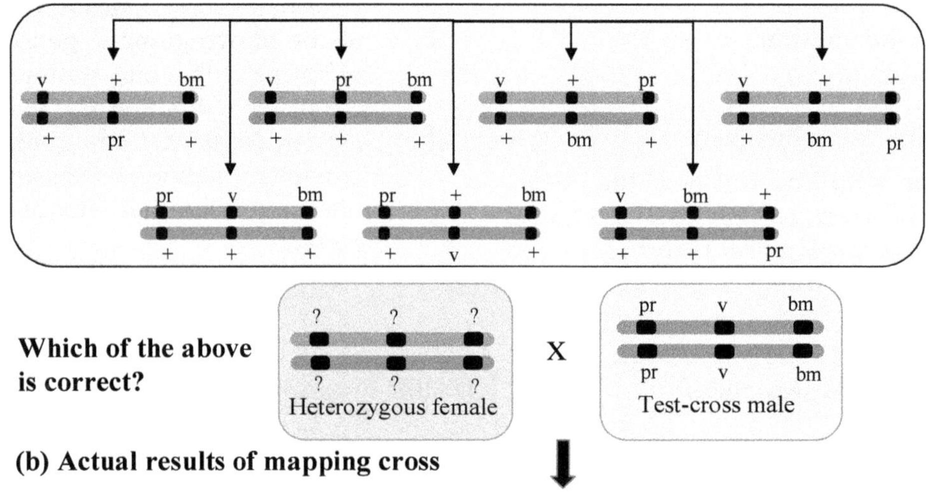

(b) Actual results of mapping cross

Phenotypes of offspring			Number	Total and percentage	Exchange classification
+	v	bm	230	467	Noncrossover
pr	+	+	237	42.1%	(NCO)
+	+	bm	82	161	Single crossover
pr	v	+	79	14.5%	(SCO)
+	v	+	200	395	Single crossover
pr	+	bm	195	35.6%	(SCO)
pr	v	bm	44	86	Double crossover
+	+	+	42	7.8%	(DCO)

Figure 11.7: Possible Allele Arrangements and Gene Sequences in a Heterozygous Female.

1. Arrangement of alleles in the female parent

Let us first take two crossovers with the highest frequency. All + *v bm* and *pr* + + should be arranged as shown in **Figure 11.8(a)**. All the homologs segregate into gametes, not affected by any recombination event. No other arrangement of alleles will be observed in the non-crossover classes. (Remember that + *v bm* is equivalent to *pr* + *v bm,* and that pr + + is equivalent to *pr v* + *bm* +).

2. To evaluate the correct sequence of genes

We recognize that the arrangement of alleles is.

+	v	bm
pr	+	+

The first step is to determine the two non-crossover classes seen with a higher frequency.

Let it be +v bm and pr ++, the alleles on the homologous chromosomes must be arranged as shown in **Figure 11.8(a)** in the female parent. These chromosomes (homologous) gamete segregate. One must remember that +vpm = pr + vbm and pr ++ = prv + bm+.

Another question arises What is the correct sequence of the genes considering the following combinations?

	+	v	bm
(1)	pr	+	+
	+	bm	v
(2)	pr	+	+
	v	+	bm
(3)	+	pr	+

Upon double-crossing over, the pr gene is in the middle.

Still, another question remains: to determine the distance between each pair of genes based on the recombination events occurring between them, one must consider both single and double cross-over events.

The other two orders (**Figures 11.8 c and d**) maintain the same arrangement:

+	bm	v		v	+	bm
			or			
pr +		+		+	pr	+

	Allele arrangement and sequence	Test-cross phenotypes			Explanation
(a)	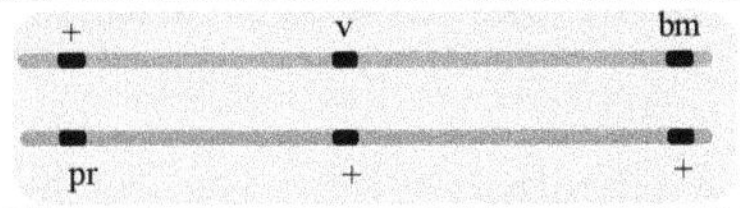 + v bm / pr + +	+ / pr	v and +	bm / +	Noncrossover phenotypes provide the basis of determining the correct arrangement of alleles on homologs
(b)	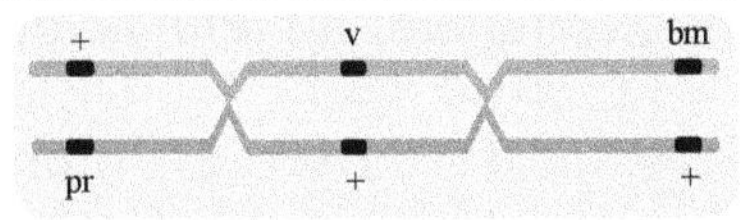 + v bm / pr + +	+ / pr	+ and v	bm / +	Expected double-crossover phenotypes if v is in the middle
(c)	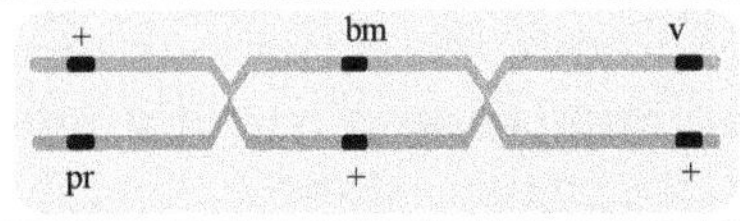 + bm v / pr + +	+ / pr	+ and bm	bm / +	Expected double-crossover phenotypes if bm is in the middle
(d)	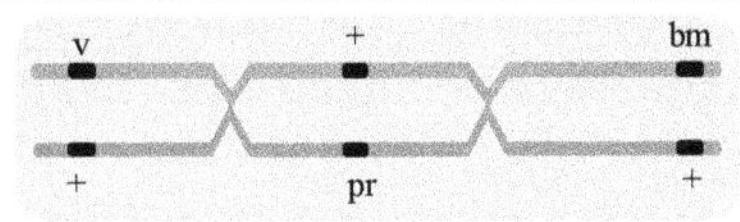 v + bm / + pr +	v / +	pr and +	bm / +	Expected double-crossover phenotype if pr is in the middle (this is the actual situation)
(e)	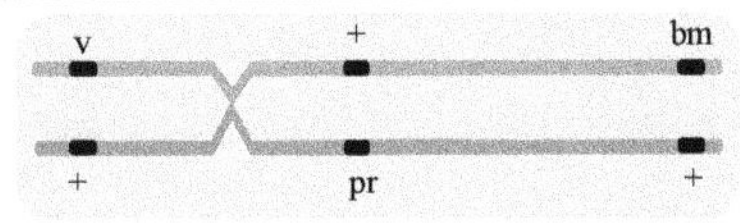 v + bm / + pr +	v / +	pr and +	+ / bm	Given that (a) and (d) are correct, single, crossover phenotypes when exchange occurs between v and pr
(f)	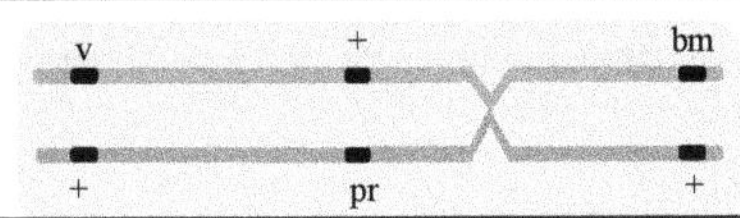 v + bm / + pr +	v / +	+ and pr	+ / bm	Given that (a) and (d) are correct, single-crossover phenotypes when exchange occurs between pr and bm.
(g)	Final map: 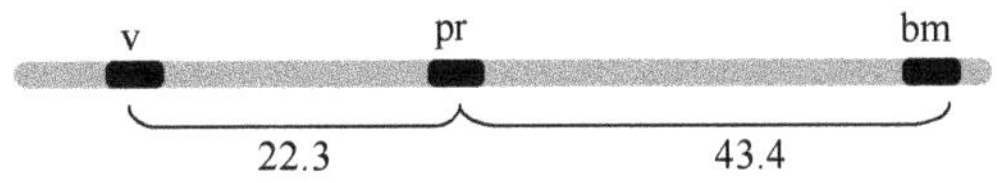 v — pr — bm (22.3, 43.4)				

Figure 11.8: A Map of the Three Genes in the Test Cross, the Arrangement of Alleles, and the Sequence of Genes in the Heterozygous Female Parent is Unknown.

The right yield order is the observed double-crossover gametes (**Figure 11.8d**). Therefore, the *pr* gene is in the middle. From now on, work on the problem using this arrangement and sequence, with the *pr* locus in the middle.

3. **Determination of distance between each pair of genes?**

Once the loci v-pr-bm sequence is known, the next step is determining the distance between *v* and *pr* and between *pr* and *bm*. The map distance between two genes is calculated. This depends upon the recombination events. During recombination, both single and double crossovers are taken into consideration.

From **Figure 11.8 (e),** phenotypes *v pr* + and + + *bm* are due to SCO between the *v* and *pr* loci, which occurs in 14.5 per cent of the offspring. The single cross-over is 7.8 per cent. If double crossover and single crossovers are added, the distance between the *v* and *pr* loci is calculated to be 22.3 mu.

Figure 11.8 (f) demonstrates that the phenotypes *v* + + and + *pr bm are* due to single crossovers between the *pr* and *bm* loci and are found in 35.6 per cent of offspring. The distance between *pr* and *bm* is calculated to be 43.4 mu. The result is shown in **Figure 11.8(g)**.

11.9 Distance between the Genes Determines the Mapping

When the double exchange takes place, it is not likely to find out the complete crossovers. If a double exchange occurs (**Figure 11.9**), the initial arrangement of the alleles is maintained.

The multiple-strand exchanges are complex; hence mapping is not easy. If the two genes are placed apart, higher are the chances that hidden crossovers can be found. If genes are near each other (**Figure 11.9**), it isn't easy to find the correct information about the recombination frequency and the map distance.

11.10 Coefficient of Coincidence

If the distance between genes is established, we know that the distance between *v* and *pr* is 22.3mu, and the distance between *pr* and *bm* is 43.4 mu. In case two single crossovers make up a double crossover occur independently of one another; the expected frequency of double crossovers (DCO exp) can be calculated:

$$DCO_{exp} = (0.223) \text{ X } (0.434) = 0.097 = 9.7 \text{ per cent}$$

In the maize cross, let us assume double crossovers (DCO example) only 7.8 per cent. These are the observed crossovers, while the expected crossovers would be 9.7 per cent.

To solve this difference, the coefficient of coincidence (C) is calculated:

$$C = \frac{\text{Observed DCO}}{\text{Expected DCO}}$$

In the maize cross, as per the above data

$$C= \frac{0.078}{0.097} = 0.804 \text{ C}$$

The simple equation would suggest.

$$I = I - C$$

In the maize cross, it would be.

$$I = 1.000 - 0.804 = 0.196$$

In a situation where we get a correct inference that is also complete and with no double crossovers, then I = 1.0. If we find some DCOs more than the expected value: I is the positive number, and positive interference can be deduced. If more DCOs are found that are higher than the expected value, I is the negative number, and harmful interference may be seen. In the maize example, I is the positive number (0.196), indicating that 19.6 per cent fewer double crossovers have occurred than the expected number.

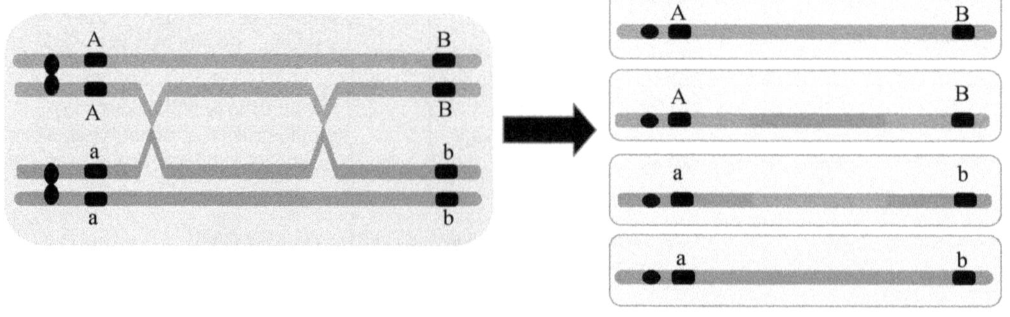

Figure 11.9: A Double Crossover is Undetected Because No Allele Rearrangement Occurs.

11.11 Extensively Mapping Genes: Drosophila, maze, and mouse

Many mutations have been found in Drosophila, maize, and mice; mapping is also possible, due to which extensive chromosome maps have been constructed. Genetic mapping can be carried out in humans using polymorphic genetic markers. These markers can be RFLPs, short tandem repeat polymorphisms, *etc.*

11.12 Genetic Mapping of Human Chromosome

11.12.1 Mapping of a Genetic Disease Locus to a Chromosome Location in Humans

It has been said that pedigree analysis can help in genetic mapping in humans, but except for X-chromosome inherited diseases, it is difficult to map the diseased gene. As we now have various markers like STR, SNPs, VNTRs, *etc.*, it is easy to map the gene. How to map the disease gene onto the chromosome is explained below.

1. Collect blood samples from the affected individual and family members. If multiple generations are available, linkage becomes easier.
2. Lymphocytes can be separated from the blood. Lymphocytes are treated with EBV and lymphoblastoid cell lines are maintained.
3. DNA is extracted from these cell lines.
4. Polymorphic markers from all autosomes can be studied.
5. LOD score is calculated between the markers and the disease.

Two–local LOD score is estimated for each locus, and its relationship with the genetic disease using informative parent and their offspring is established. After this information is gathered, the linkage can be calculated.

A large multi-generation pedigree with nonmalignant familial neonatal convulsions (BFNC) has been seen to study the chromosome location, and the disease genes are found. The characteristic features of this disease are episodes of violent and uncontrolled jerking of the face, body, arms, and legs within the first six months of life. In about 90 per cent of the cases, these symptoms lighten away in the first year. Seizures caused in the early months have no noticeable effect on subsequent neurological or intellectual functioning. This is a rare disorder autosomal dominant trait with high penetrance and can be diagnosed clinically. Two markers D20S20 and D20S19 showed linkage with the BFNC locus (**Table 11.2**)

Further larger kindred were studied, and it was seen that D20S19 marker probe can differentiate ten different alleles. RFLP identified these, and the D20S20 probe identifies two alleles of its locus.

Table 11.2: Pair-wise LOD Scores between a Benign Familial Neonatal Convulsion Locus and Two Chromosome 20 Polymorphic Loci

	Recombination fraction (q)					
Locus	0.00	0.05	0.10	0.20	0.30	0.40
D20S20	3.12	3.05	2.88	2.30	1.57	0.69
D20S19	2.87	2.92	2.83	2.36	1.63	0.76

The haplotype (8, 2) co-segregated with the disease in many individuals. It was expected that individuals III-18, IV-4, and IV-14, who have received an (8, 2) chromosome from the affected parent, would also be affected because the BFNC locus is located on this chromosome in this family. It was thought this could be because of incomplete penetrance, but it has not been proved. This could be because of misdiagnosis or a difference in the onset of the disease.

When the offspring of an affected parent show (8, 2) chromosomes without any BFNC (*e.g.*, IV -7 and IV-11), it is probable to find out the origin of the (8, 2) chromosome. For instance, individuals IV-7 inherited a (14, 1) chromosome from the affected parent and did not inherit the (8, 2) chromosome with the defective BFNC gene. By contrast, the genotypes of the individuals IV-9, however, V-1 are not readily explained because they may have inherited an (8,2) chromosome from outside the family that carries the normal BFNC gene, or they may have cases of incomplete penetrance. In other families with BFNC, polymorphic alleles other than (8, 2) of the D20S19 and D20S20 sites will link with the BFNC locus. However, it shows no biological relevance.

Two-point LOD scores were calculated for the D20S19 and D20S20 loci, and it was seen that approximately 4 and possibly less than 1cM, respectively, from the BFNC locus. These are approximate values still; they locate the polymorphic loci from 1 to 4 million base pairs from the BFNC locus. Generally, the resolution of a two–point LOD score is 1 to 2 cM for very closely linked loci (q = 0.00). Because the D20S19 and D20S20 loci map to 20q13.2 – 13.3, the BFNC locus must be near or within this chromosome region.

If large markers are used, but no linkage is established, such cases should be excluded. For example, pairwise LOD scores among 20 families with alcoholic individuals and several polymorphic DNA markers for chromosome 4q failed to detect any linkage (**Table 11.3**). These results show that the risk allele for alcoholism is not studied on 4q131 to 4q32.

Failure of linkage may be due to genetic heterogeneity; more families taken for linkage analysis may increase the LOD score values. Families selected for such studies should form a homogeneous population because LOD scores may be wrong if families of different ethnicities are taken. Also, linkage studies from a single large pedigree do not, of course, preclude genetic heterogeneity. In some families, BFNC locus was found at 8q24 in other families also.

One crucial point is avoiding false linkage for correctly mapping the disease gene.

11.12.2 Mapping of Human Chromosomes using many Loci

The Center d etude du Polymorphism Human Penetrance (CEPH) in Paris takes up the genetic mapping project. In this project, only large families have been taken.

The mapping of human chromosomes is quite tricky. The number of loci is quite a lot on a single chromosome. For n number of loci, there are N/2 possible orders. Just for 10 loci, there would be a large number (1, 814, 4000) of possible combinations. Out of these, some may be invalid several computer programs have been developed, distances have been determined, and many loci have been examined. Genetic mapping takes quite a lot of time. To solve this problem, finding the most similar order for some of the linked loci at a time is best, and then combining the best order. Statistical tools can be used to find out the best order. To calculate the linked loci, the LOD score is used.

The following steps are used for mapping. These markers should be highly polymorphic to find many markers on a single chromosome. The families selected for this purpose are CEPH, which show many polymorphic markers on a particular chromosome. Approximately 10,000 tests are required to genotype 40 CEPH families. Twenty markers are used, the data is generated, and the database is created and added to the existing database.

The computer program CRIMAP is used for data analysis. For example, the pair of functional loci for 16p and 16q has LOD scores of 29.71 (q = 0.18) and 83.89 (q=0.07), respectively. Like this, a map can be created where the different loci are 20cM apart that may be used as a framework for introducing more loci. Loci remain part of a particular map; in many studies, the possibility of a particular placement is 1000:1 or higher. DNA typing should be repeated to avoid genotyping errors.

CEPH data has shown that there are approximately 6000 polymorphic markers. Multi locus mapping has revealed approximately 1000 loci, which were 6cm apart over the entire genome.

11.13 Disease Gene into a Linkage Map

A high LOD score shows that genes are closely placed on the chromosomes.

For mapping purposes, we can use LOD score linkage mapping. A LOD score is determined for each possible order with specific map distances. For example, the likelihood ratio for locus x, positioned midway between loci 1 and 2, is.

Table 11.3 Pairwise LOD.

Genetic Marker	*Value of q (recombination frequency)*						
	0.00	*0.001*	*0.01*	*0.05*	*0.10*	*0.20*	*0.30*
INP10	-5.56	-5.44	-4.63	-2.75	-1.61	-0.56	-0.16
ADH	-4.01	-3.93	-3.36	-2.00	-1.14	-0.36	-0.06
EGF	-5.16	-5.07	-4.33	-2.58	-1.55	-0.58	-0.18
D4S191	-3.90	-3.82	-3.14	-1.62	-0.83	-.018	0.00
D4S194	-10.02	-9.37	-6.80	-3.54	-1.88	-0.57	-0.12
D4S175	-6.62	-6.44	-5.27	-2.69	-1.25	-0.11	0.13
D4S192	-5.62	-5.48	-4.52	-2.42	-1.24	-0.25	0.02
MNS	-6.12	-5.95	-4.85	-2.60	-1.36	-0.35	-0.06
MLR	-6.38	-6.12	-4.90	-2.84	-1.70	-0.65	-0.22
FG	-3.78	-3.57	-2.71	-1.34	-0.57	-0.06	0.17
A101	-9.05	-8.75	-7.19	-4.29	-2.57	-0.94	-0.29
D4S243	-6.37	-6.08	-4.68	-2.31	-1.01	-0.01	0.20

$$\text{Log10} \frac{L(1-(0.025) - x - (0.025) - 2 - (0.10) - 3)}{L(x - (0.5) - 1 - (0.05) - 2 - (0.10) - 3)}$$

L is the likelihood for a specific order, and the distance 0.5 in the denominator provides for the case of non-linkage between locus x and other loci. The computer program uses the observed data with four positions where locus x may be situated in each possible region. It determines 16 LOD scores and the map distances between loci.

When the disease is mapped on a particular chromosome, it shows a high LOD score value of more than +3. These results can be represented graphically by plotting the LOD scores for each position in each region. After this, it becomes easier to calculate the map distances. The first position is named position zero, where the highest LOD falls may be the position of the disease gene.

11.14 Homozygosity Mapping

If only a two–point linkage analysis is carried out, it isn't easy to map the recessive loci. There is a problem if we carry only two–a point linkage analysis. In such cases, it isn't easy to map the recessive loci. In the case of inbreeding, it is easier to find the position by carrying out the homozygosity mapping. In this case, we have a common recent ancestor by identity by descent (IBD) (**Figure 11.10**).

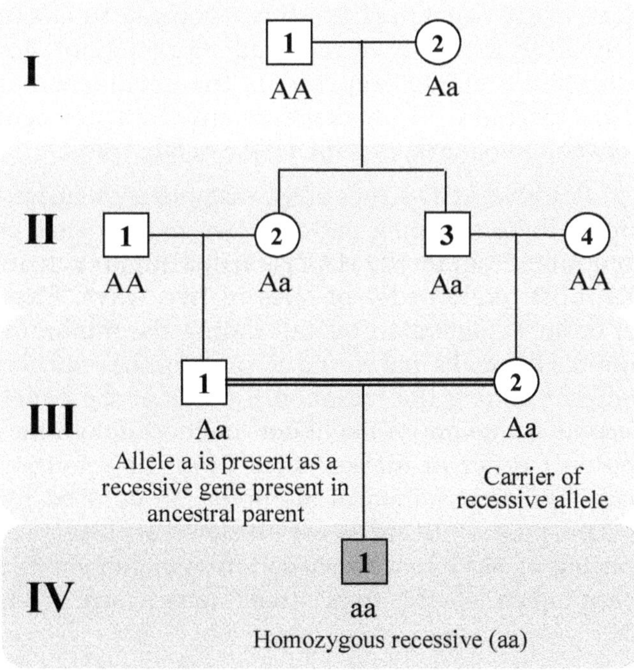

Figure 11.10: Homozygosity and Consanguinity.

The homozygosity could be because of homozygosity–by–descent (HBD) or allelic identity-by-descent (IBD). It is unlikely that a set of linked loci could be detected in nonconsanguineous marriages. Consanguineous families with the same homozygosity-by-descent recessive trait are genotyped with a large set of polymorphic markers to calculate the homozygosity-by-descent. The likelihood is that these markers may stay close to the disease gene. Even two-point linkage analysis can reveal that the closely associated loci are linked to the disease locus. However, the past approach is multipoint linkage analysis.

11.15 Linkage Disequilibrium Mapping

Linkage disequilibrium (LD) is the nonrandom association between two loci on the same chromosome. The distance between the two loci is less than 1 to 2 cm. LD is generally calculated from the population-based data. The genetic distance (q) is calculated between the marker locus and the disease gene is by the following formula.

This is because they belong to the same ancestry, which can be traced up to approximately 250 or more years (≥ 20 generations). This phenomenon is called the founder-father effect. Once the genetic linkage between a polymorphic locus and the disease gene is determined, the linkage disequilibrium mapping can be done. Suppose individuals with the genetic disease also belong to the founder population and are haplotyped with several additional polymorphic markers on the same chromosome. In that case, statistical analysis is carried out to find out which loci are in linkage disequilibrium with the disease gene. The genetic distance (q) between a marker locus and the disease gene that is in linkage disequilibrium is calculated using the following equation:

$$P_{excess} = \frac{P_{affected} - P_{normal}}{(1 - P_{normal})} = (1 - \mu g q^{-l})(1 - \theta)^{g}$$

where θ is the recombination fraction between the marker and disease loci; μ is the mutation rate for the disease gene; g is the number of generations since the common ancestor; q is the worldwide frequency of the disease allele:

θ = recombination fraction

P_{normal} is the proportion of the marker allele in normal chromosomes.

$P_{affected}$ is the proportion of the marker allele in chromosomes with the disease gene.

P_{excess}, a measure of disequilibrium, is the fraction of the excess occurrence of a chromosome with the disease gene and a marker allele compared to the chromosome with the no disease gene and the marker allele.

The value of μ for a specific disease gene is unknown; the typical human gene mutation rate (1 x 10-6) is used. For many rare recessive genetic disorders, q is 0.001.

In one linkage disequilibrium study of Finnish families with an autosomal recessive disorder producing short stature and abnormalities of the joints (diastrophic dysplasia, DTD), the distance between the disease gene and the closest allele was ~0.065 cM or ~65 kb. This distance was confirmed when a single cloned piece of DNA contained both sites 60 kb apart.

11.16 Radiation Hybrid Mapping

In the radiation hybrid mapping, we get the fragmented DNA, and these fragments are used further. The non-irradiated cells rodent cells are fused with radiated cells. Some cells do not fuse. The fused cells have selected markers. These cells are cultured, and cell lines are maintained.

The unique DNA sequences are amplified; these are not polymorphic and are analyzed using a radiation hybrid panel. Monomorphic PCR -identifiable chromosome-specific sites are called sequence-tagged sites (STSs). The PCR-identified site of each probe for each member of a radiation panel is recorded. Any radiation hybrid that does not show amplification of a sequence for any of the probes is eliminated from the panel.

RH mapping is like meiotic mapping. The closer the two sites are to one another on a chromosome, the more likely they will be retained on a radiation-induced DNA fragment. Further, they will not be separated by recombination upon meiosis. RH mapping shows a pattern of either the presence (+) or absence (-) of each marker for each RH panel member. The order of sites and the retention pattern are noted.

The frequency of breakage based upon the radiation-induced breakage between two sites is independent of markers, and the retention of the fragment with one of the markers is independent of any other fragment. The frequency is calculated based on the following formula.

$$\frac{(A^+ B^-) + (A^- B^+)}{T(R_A + R_B - 2R_A R_B)}$$

(A^+B^-)= number of radiation hybrids in the panel with the retained marker B but not the marker A.

T= Total number of radiation hybrids in the panel.

RA= fraction of all the radiation hybrids in the panel that have marker A.

RB = fraction of all the radiation hybrids in the panel with markers B.

This symbol is like the recombination frequency. Unlike meiotic recombination, the range of θ values for radiation breakage frequency extends from 0 to 1. Z shows that a particular radiation dose never separates two marker sites and therefore is closely linked. When θ equals 1, the markers are always separated by a specific dose of radiation; this represents that they are not linked.

The likelihood (Lθ) of obtaining the observed retention pattern for a pair of markers is

$$[(1-\theta) R_{AB} + R_A R_B]^{A+B+} [\theta R_A (1 - R_B)]^{A+B} + [\theta(1 - R_A) - R_B]^{A-B+} \times [(1 - \theta)(1 - R_{AB}) + \theta(1 - R_A)(1 - R_B)]^{A-B-}$$

RAB is the fraction of all radiation hybrids that retain both markers and (A- B-) this represents an observed number of radiation hybrids lacking both markers A and B. A LOD score can be calculated from the relationship log [Lθ/L θ(θ - 1)], where Lθ represents the likelihood for a particular retention pattern, and Lθ (θ = 1) denotes the likelihood that this is expected if the two markers are always separated by radiation-induced breakage. A LOD score of +3.00 or more shows that two markers are linked.

The distance (D) between two markers on an RH map is 1n (1 - θ). The distance units are called centiRays (cR), and 100cR equals 1 Ray. When D values are cited, the radiation dose used to create a specific RH panel must be shown because fragment size varies inversely to the irradiation amount. For example, 1 cR8000 represents the occurrence of radiation-induced breakage of around 1 per cent between two markers with a dose of 8000 rads.

Pairwise LOD scores or the distance is calculated for multiple mapping between two marker sites. A computer program RHMAP is carried out to evaluate the most likely order of sites in two ways. First, an order is generated by calculating the minimum number of breaks that would account for the retention patterns among the radiation hybrids of the panel. Second, a maximum likelihood method determines the best order of marker sites from the retention data. Different statistical methods can be used for this purpose: 33,627 STSs were ordered at an average spacing of 94 kb to a 50,000-rad human genome RH mapping panel with an average fragment size of 800 kb.

11.17 Genotyping Single-Nucleotide Polymorphisms

SNP's can be used as a marker for linkage studies. It has been reported that sequence-tagged sites of different individuals have different SNPs, which could be candidate SNP's sites. Currently, it has been seen that there are public and private databases that contain millions of possible SNP sites. Sets of evenly spaced, good-quality SNPs that can be used for genotyping have been assembled from these resources. There are many techniques for identifying SNPs; many of these have diagnostic value, and some of the SNPs are associated with disease-causing genes.

11.18 Physical Mapping of the Human Genome

Physical mapping of the human genome depends upon the frequency of recombination. A physical map is constructed by creating retrieving, a genomic DNA library, and clones with overlapping segments. Based on the overlaps and other positional information, a continuum of clones can be seen for a chromosome region, a whole chromosome, or the entire genome. Various procedures generate sets of contiguously ordered clones (contigs) from human YAC, BAC, PAC, Pl, and cosmid libraries. BAC human DNA libraries are the source of cloned DNA for assembling the human physical map.

11.19 Assembling Contigs from BAC Libraries

DNA fingerprinting and STS-content mapping are commonly used to construct large-insert contigs. The human physical mapping was initiated using DNA fingerprinting of more than 415,000 BAC clones. This was used to sequence the human genome. The DNA of each clone was digested with a restriction endonuclease, followed by separation using agarose gel electrophoresis. The bands from each clone are scanned, recorded, and compared with the other banding patterns for matches. The distinctive pattern of DNA fragments from each clone is equivalent to a fingerprint; hence, the term DNA fingerprint is used. Most of the matching of the bands is computerized. However, manual inspection is required to identify redundant and non-overlapping clones and resolve any discrepancies that are not detected using the computer program. At this stage, because the chromosome origins of the members of the partially assembled contigs are unknown, STSs and fluorescent in situ hybridization (FISH) with inserts are used to tie the clones to specific chromosome locations.

Sequence-tagged site content mapping has become essential for constructing large-insert physical maps. As noted previously, an STS is a short single-copy segment of DNA (~100 – 300 bp) and can be detected specifically with a unique pair of PCR primers. Large numbers of STSs are required to confirm that a contig is continuous over a long span of a chromosome. Generally, assembling a high-resolution, continuous whole chromosome contig requires an STS site every 50 to 100 kb. This helps physically map a chromosome comprising about 160Mb DNA and requires 1500 to 3,000 STSs. At least 30,000 STSs are required for high-resolution physical mapping of the complete human genome. Various strategies are used to generate STSs. For example, DNA from a flow-sorted human chromosome preparation is treated with a restriction endonuclease and cloned into a small insertion (1000bp) vector. Clones are selected at random, and inserts are sequenced. Clones with inserts less than 100bp and those that contain sequences from human repetitive DNA elements are discarded. Computer nucleotide determines these by matching the insert sequence with all known human repetitive DNA sequences.

The criteria for choosing PCR primers for each insert include the length, which should be 20 nucleotides with no internal secondary structure, and complementarily between the two primers of one set should not be there. After the primers are synthesized, a potential STS is tested for PCR amplification, with human DNA as the template. The PCR product is run on agarose gel, and bands are detected either directly or by using hybridization, a calorimetric assay. The chromosome specificity of human genome-specific STSs can be verified using a DNA sample from the appropriate mono-chromosomal hybrid cell line. Region specificity is determined by PCR amplification assays with a chromosome deletion hybrid cell panel for the target chromosome. In addition, DNA from mono-chromosomal cell hybrid lines carrying other chromosomes can be assayed for PCR amplification to ensure each STS is chromosome-specific. The order of STSs can be determined with an RH mapping panel. Thus, by screening the STS content of each BAC clone, overlapping clones can be ordered, and the chromosome region spanned by each contig is identified. Meter a prodigious effort, a whole-genome BAC map comprising 1246 contigs and about 26,000 clones was assembled.

12

Genetics of Complex Traits

Understanding genetic disorders can be a difficult task, particularly when it comes to assessing the influence of a single gene on a phenotype. While single-gene disorders have limited phenotypes, it is worth noting that gene-gene interactions can still affect phenotype expression. Therefore, the principle of "one gene-one phenotype" does not always hold true. On the other hand, polygenic characters are even more complex and involve multifactorial inheritance, which means that various genetic and environmental factors contribute to their development. Interestingly, there is a differential expression threshold between genders for some multifactorial disorders. This means that the number of males and females affected by a disorder may differ female genders.

There is familial segregation of these disorders. This depends upon how family members are related to one another. For example, the risk is higher if a brother or sister has the trait or disease than if the first cousin has the trait or disease. These traits are also known as threshold traits. Let us take the example of height; in the case of height, there exist differences between men and women. Most of the time, women are 3 inches shorter than men. There is no selection bias in mating for tallness in humans. However, men tend to marry women who are shorter than themselves. In this case, children will receive genes for both tall and short height. To calculate the expected height of the child, mid-parental height is considered.

12.1 To Determine a Trait, Interaction of Two Genes may be Required

Let us suppose that there are two genes, A and B, with alternative alleles as a and b, in the pure breeding of lentils, all the hybrids produced in F1 are brown in color. However, in the F2 generation, four different seed colors were seen. The ratio of these 4 different seed colors was 9:3:3:1 *i.e.*, 9 brown, 3 tan, 3 grey 1 green. This showed two independently assorting genes determine the seed coat color in a dihybrid cross.

Multifactorial inheritance is relatively complex and does not follow simple Mendelian inheritance. Let us first take the example of a single trait, *i.e.*, the seed color of lentils if different colored seeds are self-crossed with the different types of F2 lentil plants, then the self-pollinated F2 green individuals show that they are pure breeding in the F3 generation, and all were having green seeds. But across generations, Tan, or tan and green, similarly grey produced grey or grey plus green. Brown produces all brown or assorted tan plus brown or grey plus brown or produces all four colors. Two gene hypotheses can explain the observations mentioned above.

There are four coat color phenotypes: brown (A-B-), tan (A-bb), grey (aa B-), and green (aa, bb). When hybridization occurs, it is by the Mendelian ratio, which would be 9:3:3:1. These results show two genes that determine the coat color. Sometimes we find that crossing two lines of pure breeding white-flowered peas, it is expected that all F1 would be white, but we get purple flowers also. This will produce a ratio of 1 purple: 7 white in the F2 generation. The explanation is that two genes work in tandem (**Figure 12.1**).

There is a biochemical explanation that determines the color of the flower, there are different enzymes involved in producing the flower color. There is a precursor in the presence of these enzymes the colorless precursor can be converted into colorful pigment. Let there be three genotypic classes (A- bb, aa B-, and aa bb), and they may group according to phenotype since they do not recognize proficient forms of essential enzymes, giving rise to no color. It is easy to interpret the "7" part of the 9:7 ratios from 3:3:1 of the 9:3:3:1 ratio of two genes. The 9:7 ratio is the phenotype of the complementary genes that interact with the dominant alleles (A-) and produce the required trait. The other genotypic categories (A- bb, aa B-, and aa bb) are not capable of producing the trait.

Sometimes one gene may mask the other gene; this phenomenon is called epistasis. The Labrador retrievers have black, chocolate brown, or golden yellow colors. The appearance of the coat color depends on the allelic combination of two independently assorting coat color genes. The B dominant allele determines the black color, while the recessive bb homozygote is brown. With the second gene, the dominant E allele has no visible effect on black or brown alleles to yield golden color. The genotypes mask entirely the influence of other genes; hence it isn't easy to find out the genotype for black and brown color.

12.2 ABO Blood Group

The ABO blood group was discovered at the University of Vienna by Karl Landsteiner in 1900 and 1901. This was possible when people observed that non-matched blood transfusions resulted in transfusion reactions. Efforts were made to discover the reason and ABO blood groups were discovered. In 1930, Karl Landsteiner won the Nobel Prize for discovering blood types. The ABO blood group exists both in humans and other primates. There are four major types: A, B, AB, and O. To test the A and B

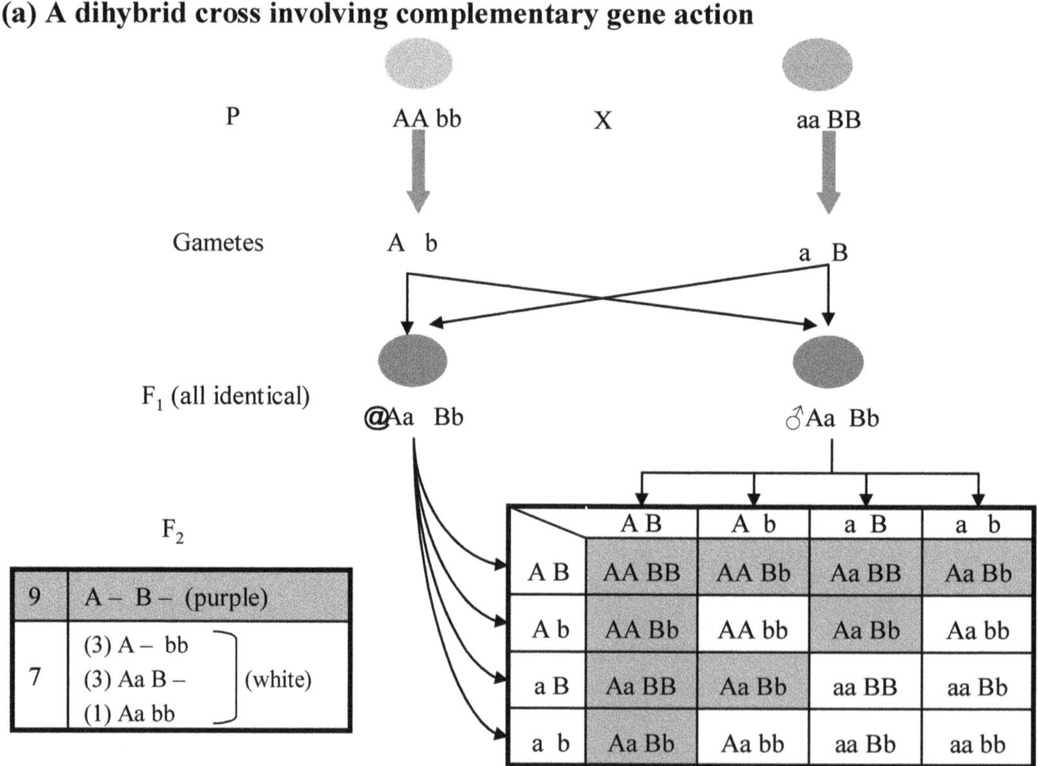

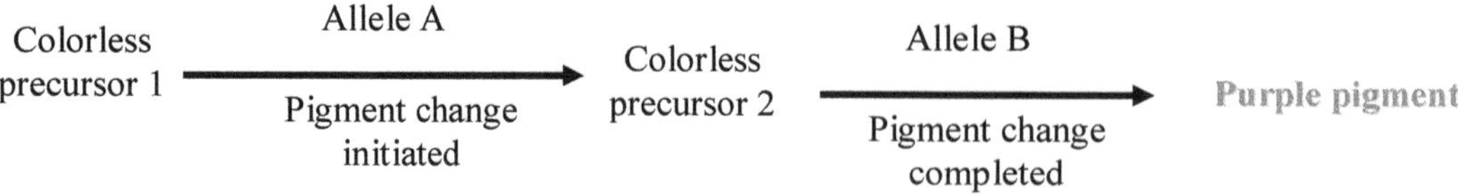

Figure 12.1: Complementary Actions of Genes.

antigens, specific antibodies are required; these are shown in **Table 12.1**

Table 12.1: ABO Antigens and Corresponding Antibodies

ABO Blood Type	*Antigen A*	*Antigen B*	*Antibody Anti-A*	*Antibody Anti-B*
A	+ve	-ve	-ve	+ve
B	-ve	+ve	+ve	-ve
O	-ve	-ve	+ve	+ve
AB	+ve	+ve	-ve	-ve

+ = presence of antigen/antibodies, -ve = absence of antigen or antibodies.

A person in the A blood group will show A antigen on the surface of red blood cells. In case blood is transfused into this person it will produce anti-B antibodies in serum which will act as an alien causing agglutination of RBCs and resulting in a transfusion reaction.

Individuals with blood group O do not have antigens A or B; hence upon transfusion, no rejection episode occurs even when transfusion takes place between blood group 'O' to A or B antigen-positive individuals. These individuals are called universal donors. However, they should receive blood group 'O' only. Interestingly people in the AB group do not have any antibodies and any blood type can be transfused hence are known as the universal recipient. The typing ABO blood group is relatively easy, only the RBCs and antibodies have to be mixed and results are interpreted in the form of agglutination.

12.3 Inheritance Patterns of ABO Blood Groups

They are inherited in a co-dominant manner. Genes of the ABO blood group are located on chromosome 9. Blood group O is recessive over the A and B blood types. AO genotype will show A phenotype. People who are type O show OO genotypes revealing that none of the parents had either blood group A or B. The A and B alleles behave in a co-dominant manner. In this case, if an individual has inherited A from one parent and B from the other the resulting phenotype will be AB.

12.4 Bombay Blood Group

All the individuals having blood groups A, B, O, or AB contain antigen H. In approximately 1/10,000 individuals, the H antigen is not found. This antigen was discovered in Bombay, India hence called the Bombay blood group. Bombay phenotype is (h/h, also called Oh). In Europe, it occurs in 1 in million people. H-deficient individuals have no side effects however, if a blood transfusion is given severe transfusion reaction is caused. The sequence of oligosaccharides determines the specificity of the H antigen. The antigenicity of the H antigen depends upon the terminal disaccharide fucose-galactose. Fucose has an alpha-(1-2)-linkage. The H antigen is attached to oligosaccharide chains that appear on the surface of the RBC's oligosaccharide chains attached to the proteins and lipids. FUT1 encodes a fructosyltransferase that catalyzes the synthesis of the H antigen. The FUT2 gene encodes a soluble form of the H antigen found in body secretions. Individuals with blood group A make an enzyme that adds polysaccharide A, which is a sugar polymer and identified as substance H. Blood group B individuals formulate an altered form of the enzyme that adds polysaccharide B onto the base; and type O individuals make neither A-adding nor B-adding enzyme and thus have only substrate H. All people of A, B, or O phenotype carry the dominant wild-type H allele as the second gene and thus produce some H substance. Only in the rare Bombay–phenotype individuals have the genotype hh for the second gene; in this case, neither a nor B antigen gets attached to the H substance. Individual looks like a blood group without an H substance. The offspring receiving the IA allele for the ABO gene and recessive h allele for the H-substance from his mother plus i allele and a dominant H allele from its father would be blood group A (genotype IAi, Hh), even though neither of its parents is phenotypically A or AB.

12.5 Environmental Factors

Environmental factors influence/modify the genetic factors. Let us take the example of summer squash influenced by environmental factors; in other words, show dominant phenotypic ratios of 12:3:1 or 13:3 in summer squash, the dominant B allele results in white color and is enough to cover the effects of any combination of A and alleles. As a result, yellow (A-) or green (aa) color is articulated only in bb individuals.

A double dominant AA BB genotype exists for feather color in white leghorns; white wy *wyandottes* chickens are homozygous recessive for both genes (aa bb). A cross between these two pure-breeding white strains produces an all-white dihybrid (Aa Bb) in the F1 generation, but birds with color in their feathers emerge in the F2, and the ratio of white to colored is 13:3 (**Figure 12.2**).

This may be explained because dominant epistasis occurs in which B is epistatic to A; the A allele (in the absence of B) produces color, and the a, B, and b alleles produce no color. The interaction is characterized by a 13:3 ratio because there are 9 A- B-, 3 aa B-, 1 aa bb

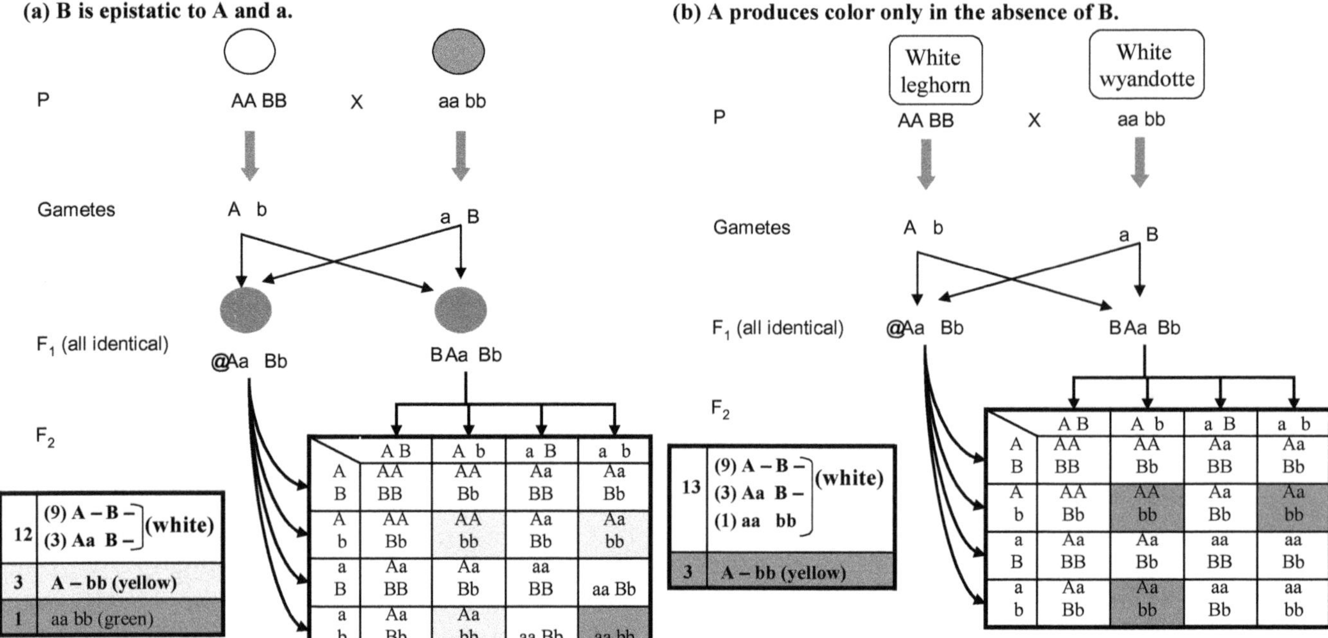

Dominant B allele causes white color which mask the effects of combination of A and a alleles. As a result, yellow (A-) or green (aa) color is expressed only in bb individuals.

In F2 generation resulting from a dihybrid cross between white leghorn and white Wyandotte chickens, the ratio of white birds to birds with color is 13:3. one copy of A gene is absent but B is needed to produce color.

Figure 12.2: Phenotypic Ratios of 12:3:1 or 13:3 due to Domination Epistasis in Summer Squash Fruit.

genotypic classes, which combine to produce only one phenotype: that is white. In white leghorns, it is seen that there exists a double dominant AA/BB genotype for feather color. While in *wyandottes* chickens, there is a recessive genotype, *i.e.*, aa/bb. When pure bread white leghorns and white *wyandottes* chickens are crossbred, they produce all-white dihybrid (Aa/Bb) in the F1 generation. In the F2 generation, the birds showed both white and colored feathers in a 13:3 ratio; this can be explained based on epistasis B is epistatic to A; the A allele (in the absence of B) produces colored feathers while a, B, and b allele produces no color; therefore, we find 13:3 ratio. There are 9A-B-, 3aa, B-1aa bb genotypic combinations resulting in only one phenotype, *i.e.*, white color.

If there exist assorting genes that result in a 9:3:3:1 ratio of the four Mendelian genotypic classes in the F2 generation, it can produce variable phenotypic ratios, depending upon the gene interaction. There can be four, three, or two phenotypes composed of different types of the four genotypic classes. **Table 12.2** shows the correlation of the phenotypic ratios with the genetic phenomena.

Deafness is a complex trait, having approximately 50 different genes that arose as a result of mutation

Table 12.2: Summary of I and II of Complementary, Recessive Dominant Epistasis and Recessive Epitasis

Gene interaction	*F_2 Genotypic Ratios from an F_1 di hybrid Cross*					*F_2 Phenotypic Ration*
	Example	*A-B-*	*A- bb*	*aa B-*	*a bb*	
Dominant epistasis I: Dominant allele does not show the effect of other gene	Summer squash: color	9	3	3	1	12:3:1
Dominant epistasis II: The dominant allele of one gene shows only its effect and not the other gene.	Chicken: feather color	9	3	3	1	13:3
Recessive epistasis: One gene affects both genes	Retriever: coat color	9	3	3	1	9:3:4
Complementary: A dominant allele of each of the two genes is required to show the phenotype	Sweet pea: flower color	9	3	3	1	9:7
None: four distinct F_2 phenotypes	Lentil: seed coat color	9	3	3	1	9:3:3:1

and resulted in deafness. These genes are required at the time of development in case there is a loss of function in any of these genes, it will result in deafness.

For normal hearing, a dominant wild-type allele is required many times; a deaf couple produces an average child. This can be explained because both father and mother are heterozygous, *i.e.*, having one normal gene and the child receiving the normal genes with normal genes for a particular locus. On the other hand, in deaf children, both parents are homozygous for a mutation for the same gene, and all their children are homozygous for the same mutation (**Figure 12.3**).

12.6 Complementation Test

Original mutations affect two different genes, known as complementation showing one normal and the other mutant gene may result in different effects. If an offspring receive two recessive mutant alleles, one from each parent will show the mutant phenotype; complementation is not found because the two mutations have independently altered the same gene alternatively. Complementation shows genetic heterogeneity. The complementation test cannot be used if either of the mutations is dominant over the wild type.

One of the heterogeneous traits is severe mental retardation, where genetic and environmental factors play an essential role. If there is a homozygous mutation for the phenylalanine hydroxylase enzyme gene, it results in severe mental retardation. The presence of this enzyme converts phenylalanine to tyrosine into other amino acids. In case this enzyme is absent, it results in the accumulation of phenylalanine immediately after birth; this builds up phenylalanine resulting in the disturbance of the normal functioning of the brain development causing mental retardation.

The genetic condition due to the above phenomenon is known as phenylketonuria or PKU. In case it is not treated, it results in severe mental retardation. It is recommended that phenylalanine should be eliminated from the infant's diet, which can prevent mental retardation.

12.7 Role of Breeding Studies to Understand Inheritance

The phenotypic ratios in the first and second generations can be diagnostic for a particular mode of inheritance (9:7 or 13:3 ratios reveal that two genes are interacting). They can provide the first line of evidence to build up the hypotheses. Breeding studies

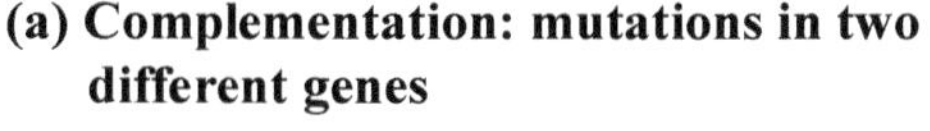

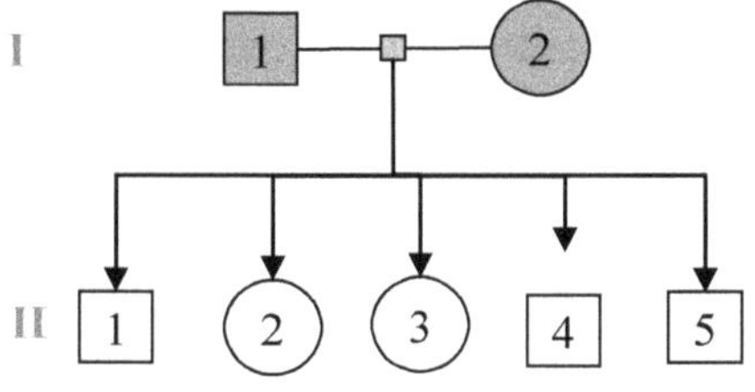

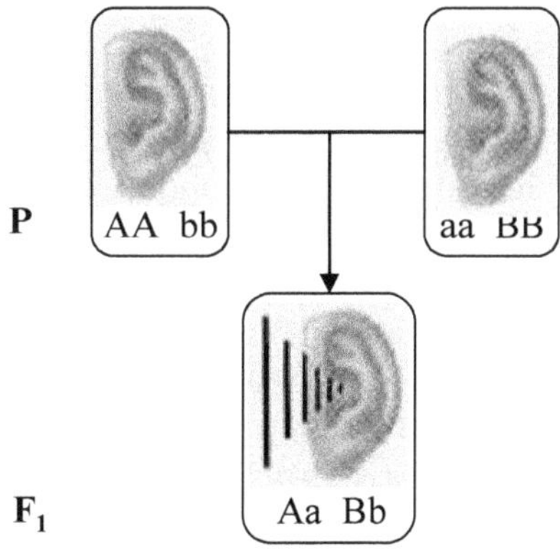

Genetic mechanism of complementation

(b) Noncomplemention: mutations in the same gene

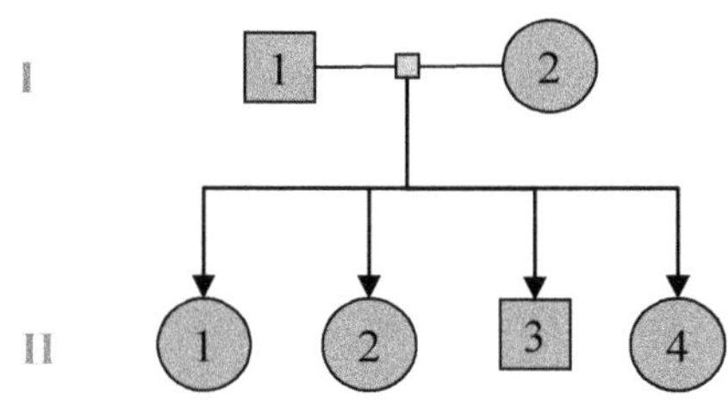

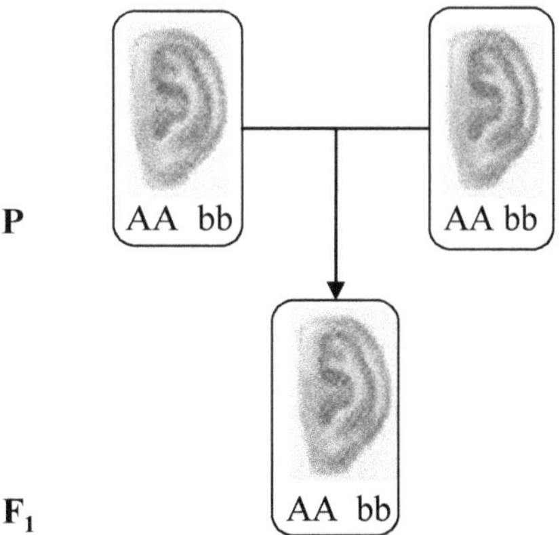

Genetic mechanism of noncomplementation

Figure 12.3: Genetic Heterogeneity in Humans: Deafness as an Example.

can help determine if the hypothesis is correct or not while the recessive gene of Agouti's mice.

If pure-breeding occurs between white albino mice and pure-breeding brown mice, it will form the black hybrids, and a cross between the black F1 hybrids produces 90 black, 30 brown, and 40 albino mice. The overall ratio is 9:3:4 of recessive epistasis; hence it may be hypothesized that two genes, one epistatic to the other, interact to produce three phenotypes of the mouse; This has been proved while taking 160 progeny in a ratio of 90:30:40, taking into consideration only gene albinos would be homozygotes for one allele (bb), brown mice would be homozygotes for a second allele (BB), and black mice would be heterozygotes (Bb) that have their own "intermediate" phenotype because B shows incomplete dominance. In case there is a cross between Bb (black), it is expected that there would be 1 BB brown: 2 Bb black: 1 bb albino, or 40 brown: 80 black: 40 albinos. Is it possible that the 30 brown, 90 black, and 40 albino mice counted were due to the inheritance of a single gene? If we get the ratios, 40:80:40 and 30:90:40 do not seem to be significantly different.

Wilhelm Johannsen (1903-1909), a Danish biologist, observed for the first time that environmental factors also influence a portion of the variation in the quantitative trait. His experiments were on the broad bean Phaseolus vulgaris. Under normal conditions where intercrossing occurs, bean size varies considerably; the smallest beans weigh about 150 mg, and the largest weigh about 750 mg. Johannsen selected either the largest or the smallest beans from the population and carried out inbreeding for several generations. However, he observed variation in the pure line beans. He proposed that this may be due to other factors. Although the small and large bean strains were phenotypically distinct from each other, within each strain there was still significant phenotypic variation. Johannsen recognized that inbred strains should be homozygous for most of the genes. Thus, the residual variation could not be due to segregating alleles; instead, it must have been due to uncontrolled environmental factors.

Scandinavian biologist Herman Nilsson–Ehle provided evidence to show that the genetic component of this variation may contribute to several different genes. Nilsson–Ehle studied color variation in wheat grains. When he crossed a white–grain variety with a dark red-grain variety, he found an F1 with an intermediate red phenotype. The self-fertilization of the F1 produced an F2 with seven distinct classes, ranging from white to dark red. In F2, the phenotypic types were of three independently assorted genes of grain color.

Nilsson–Ehle hypothesized that each gene had two alleles, one causing red grain color and the other white grain color, and that the allele for red grain color was semi-dominant over the allele for white grain color. Based on this hypothesis, the genotype of the white-grained parent could be represented as aa bb cc, and the genotype of the red-grained parent could be represented as AA BB CC. The F1 genotype would be Aa Bb Cc, and the F2 would contain the genotypes that would differ in presenting number for the pigment color.

The American geneticist Edward M. East extended Nilsson- Ehle's study and studied a trait that did not show simple Mendelian ratios in the F2 generation. He investigated the length of the corolla I of the tobacco flowers. The length of the corolla was approximately 41mm; this was a pure line in another, 93mm. These two phenotype variations are presumably due to environmental influences.

To know how many genes are involved, a crude guess can be created by comparing the F2 plants with each inbred parental strain. One strain where corollas were homozygous for one set of alleles, and the strain with the longer corollas was homozygous for another set of alleles. Furthermore, let's suppose that the long–corolla alleles are semi-dominant, that all length-controlling genes assort independently, and that each gene contributes equally to the phenotype. If corolla length were determined by one gene, with alleles a (for short corolla) and A (for long corolla), then of the F2 plants will have short corollas, and will have long corollas. If two genes determined corolla length, we would expect 1/16 of the F2 plants to resemble the short–corolla parent and 1/16 to resemble the long – corolla parent. If three genes were involved, the frequency of each parental type in the F2 would be 1/64, and if four genes were involved, it would be 1/256. With five genes, the parental frequencies in the F2 would be 1/1024 for each class. East studied 444 F2 plants and failed to find even one with either parental phenotype. This failure would seem to rule out the hypothesis of four or fewer genes controlling corolla length. At least five genes are responsible for the differences in corolla length between East's two inbred strains.

12.8 Oculocutaneous Albinism Type 1

Oculocutaneous albinism type I (OCA1) is a recessive disorder that is due to mutations. It is a heterogeneous condition. Five genes are responsible for this condition resulting in the reduced pigmentation of hair, skin, and eyes.

It is found in 1 in 40,000 individuals throughout the globe. There is zero pigmentation in the skin, eyes, and hair. There occurs a stereoscopic vision as the eyes are affected. Tyrosinase is absent in oculocutaneous albinism type I, essential for melanin production. This gives color to hair, skin, and eyes- type B is a milder category and shows a partial absence of tyrosinase. It is due to mutations in the tyrosine gene.

The difference between types A and B depends on the pigmentation variation in affected individuals. In OCA1A, albinism is seen throughout life, whereas with OCA1B, the pigmentation is higher with age. Figure 12.4 shows the biosynthesis of melanin Tyrosinase (TYR) gene mutations located at 11q14-11q21 OCA1. TYR mutations with aberrant activity are called hypomorphic mutations linked to the OCA1B phenotype. Tyrosinase catalyzes under three separate consecutive steps in the pathway that precedes the synthesis of eumelanin (black pigment) and pheomelanin (red pigment) (**Figure 12.4**). A TYR mutation in the homozygous form affects the pigment production pathway, and the precursors for either eumelanin or pheomelanin do not accumulate significantly.

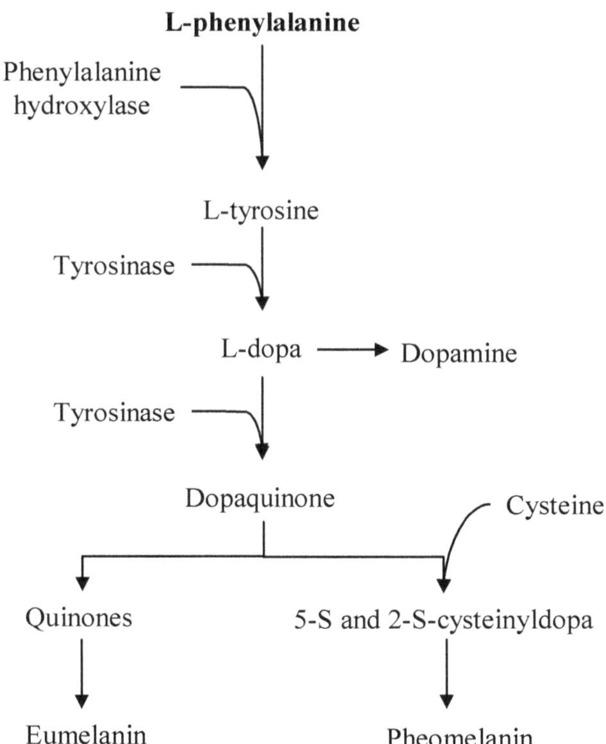

Figure 12.4: Biosynthesis of Melanin.

All four types of OCA are inherited in an autosomal recessive manner. Four genes responsible for the different types of disease are TYR, OCA2, TYRP1, and MATP). A molecular diagnosis is necessary to know the gene defect and the OCA subtype. Molecular testing of TYR and OCA2 is carried out depending on the clinical symptoms. Differential diagnosis includes ocular albinism, Hermansky-Pudlak syndrome, Chediak-Higashi syndrome, Griscelli syndrome, and Waardenburg syndrome type II. Parental diagnosis is probable if the disease-causing mutation is documented in the family. Glasses (possibly bifocals) and dark glasses or photochromic lenses may help decrease visual activity and photophobia. Correction of strabismus and nystagmus is necessary, and sunscreens are optional. OCA has normal life expectancy, intelligence, and fertility; the skin of the OCA patient should be undertaken routinely to keep away skin cancer.

12.9 Glucose-6-phosphate Dehydrogenase Deficiency

Approximately 400 million individuals worldwide are affected by G6PD deficiency which is an X-linked recessive trait. There are approximately 300 reported variants. As it confers protection against malaria, there could be a high degree of polymorphism. There are various reasons why hemolysis is caused among individuals with G6PD deficiency. Their causes are shown in **Table 12.3.**

Table 12.3: Causes of Hemolysis among ndividuals with G6PD Deficiency

Analgesics/antipyretics
Acetanilide, acetophenetidin (phenacetin), probenicid, pyramidone
Antimalarials
Hydroxychloroquine, mepacrine (quinacrine), pamaquine, pentaquine, primaquine, quinocide
Cytotoxic/antibacterial
Chloramphenicol, co-trimoxazole, furazolidone, furmethonol, nalidixic acid, neoarsphenamine, nitrofurantoin, nitrofurazone, ρ-aminosalicylic acid
Sulfonamides/Sulfones
Dapsone, sulfacetamide, sulfamethoxypyrimidine, sulfanilamide, sulfapyridine, sulfasalazine, sulfisozazole
Miscellaneous
α-methyl-DOPA, dimercaprol (BAL), hydralazine, mestranol, methylene blue, nalidixic acid, naphthalene, niridazole, phenylhydrazine, toluidine blue, trinitrotoluene, urate oxidase, pyridium

12.10 Pathophysiology

G6PD enzyme causes oxidation of glucose-6-phosphate to 6-phosphogluconate. There occurs a reduction in the oxidized form of nicotinamide adenine dinucleotide phosphate (NADP+) to nicotinamide adenine dinucleotide phosphate

(NADPH). NADPH is an essential cofactor in many reactions, maintaining the glutathione in its reduced form.

The enzyme glutathione peroxidase, reduced glutathione, also converts harmful hydrogen peroxide to water. RBC requires G6PD activity as it is the only source of NADPH protecting the oxidative stresses; hence people who lack this enzyme are advised not to take oxidative drugs because they cause high stress. G6PD can be below average or increased levels. Thus, when cells are under oxidative stress, the red blood cells of G6PD–deficient individuals are not protected and disintegrate, leading to hemolysis followed by anemia.

G6PD deficiency results in prolonged neonatal jaundice along with hemolytic anemia. Neonatal jaundice is treated with bile lights, a special light system. If red blood cell loss is high, blood transfusions are given regularly. Newborn babies are free of hemolytic anemia, but when they experience infection and ingest certain oxidative compounds or eat broad (fava) beans, hemolytic anemia is caused in G6PD deficient cases. Infection is the common factor that causes hemolysis, especially infectious hepatitis, pneumonia, and typhoid fever, eliciting the severest responses. The susceptibility to drugs was initially discovered when individuals were given primaquine to treat malaria and those who were G6PD–deficient; these studies established the severe hemolytic episodes. More drugs have been discovered that may cause hemolytic anemia in G6PD deficient cases (**Table 12.3**)–the broad bean constituent's divicine, convincing, and isoamyl initiates hemolysis. After the diagnosis of G6PD deficiency is established, the affected individuals should avoid broad beans.

G6PD deficiency is found in tropical and subtropical regions. The high frequency of G6PD is associated with a high rate of malaria. Due to selective pressures, there is an increased frequency of G6PD in endemic areas. Most G6PD is caused by single missense mutations found in different exons. The severe form is chronic non-spherocytic hemolytic anemia (CNSHA); the genes are located on the exon ten, which is involved in G6PD dimer formation. It is an essential monogenic trait affected by environmental factors.

12.11 Cystic Fibrosis

Cystic fibrosis (CF) is an autosomal recessive disorder seen in individuals of Northern European ancestry. Its occurrence is 1 in 3200 live births in Northern European ancestry. The affected sites are mucus and sweat glands, secretory glands in the lungs, pancreas, liver, intestines, sinuses, and sex organs. Blockage of the intestine (meconium ileus) is a severe problem in approximately 15 per cent to 20 per cent of individuals, both at birth and later in life. The diagnosis can be made based on augmented chloride concentration in sweat. CF gene is located on chromosome 7. The protein involved is the cystic fibrosis transmembrane regulator (CFTR). CFTR protein is abnormal in patients with CF affecting the chloride channels in the cells. As CF is an autosomal recessive trait, children suffering from CF should inherit the defective genes from each parent. The common mutation is a 3-base pair deletion of the phenylalanine codon (F508del; ΔF508) that codes for the amino acid at position 508 of the CFTR protein. The frequency of the ΔF508 mutation is approximately 30 per cent to 80 per cent. There is variation in the occurrence of this gene.

CFTR protein is multidimensional; it regulates intracellular chloride ion concentration by conducting chloride ions out of the cell and regulates other channel and transporter proteins. CFTR also helps in the regulation of multiple proteins (**Figure 12.5**).

CFTR gene mutations are responsible for the production and other functions of CFTR, which can be classified into six classes. (i) It is the category dealing with the mutations that are truncated, and hence no active CFTR is produced. (ii) Mutations that are non-truncating and produce less activity. (iii) This mutation is involved in the channel gating and affects the nucleotide-binding domains (NBD1, NBD2) and decreases chloride transport. The NBD and R domains are essential as they play roles in opening and closing the CFTR channel. (iv) In this category, mutations alter amino acids of the transmembrane domains that create pores in the channel. Some mutations are related to the decrease in the amount of functional CFTR production. Class VI (regulatory) mutations affect the regulation of CFTR. Generally, classes I, II, and III account for >85 per cent of the cases of CF. The homozygous ΔF508 genotype is more severe than other CFTR mutations. In the case of ΔF508 homozygotes, the lung dysfunction case varies from less to more severe forms of CFTR. CFTR genotypes and CF phenotypes suggest that modifier genes or environmental factors may be responsible for determining the phenotypic outcome (**Table 12.4**).

12.12 Oligogenic Disorders

All the genetic disorders are supposed to be inherited in a Mendelian fashion. Monogenic and complex diseases do not follow this rule. Some disorders do not follow Mendelian inheritance as many other factors can modulate a disease phenotype. A few genes modulate oligogenic disorders and can provide a conceptual bridge between diseases

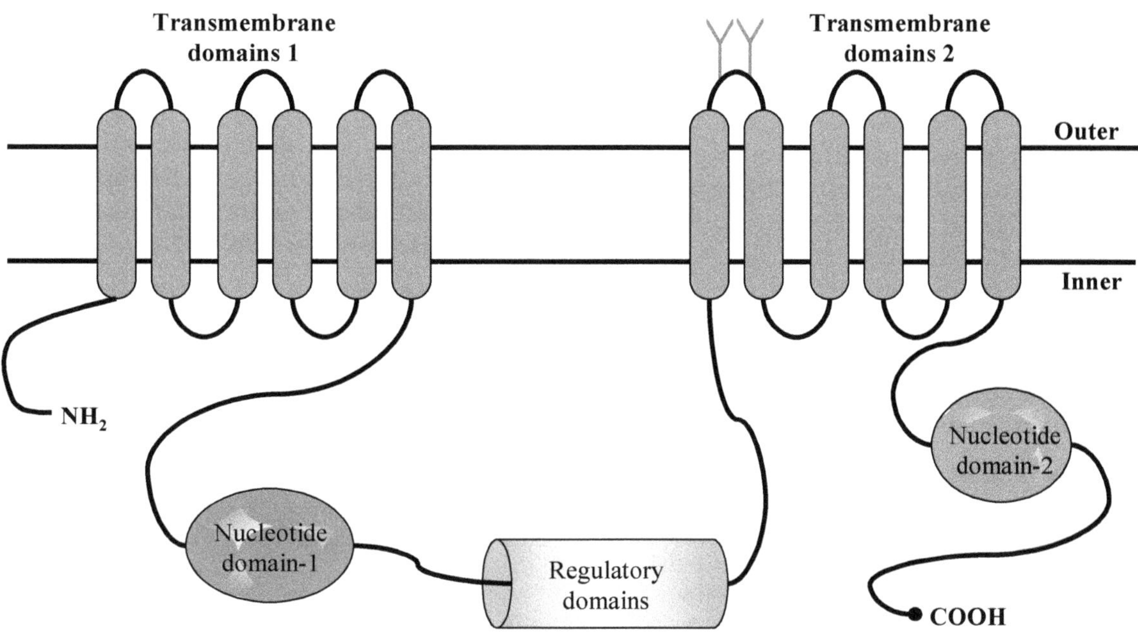

Figure 12.5: Schematic Representation of CFTR Chloride Channel. The Y-shaped structure represents the glycosylation site.

Table 12.4: Modifier Genes Involved in Cystic Fibrosis.

Protein	*Function*	*CF – Associated Allele*	*Amount of Modifier Gene Product*	*Pulmonary Disorder*
β-2 adrenergic receptor	Stimulation of adenylyl cyclase activity	Down-regulating ADRB2 allele	Low	Decreased Severity
Glutathione-S-transferase	Antioxidant; detoxification f oxidative stress	GSTM1-0 (null allele)	None	Severe
Nitric oxide synthase	Nitric oxide homeostasis	Lowered nitric oxide production NOS1 allele	Low	Increased severity due to enhanced risk of early colonization by Pseudomonas aeruginosa
Major histocompatibility complex, class II antigens	It helps in presenting antigens to T cells	HLA-DRB1*DR4; HLA-DRB1*DR7	Normal	Increased severity due to enhanced risk of early colonization by Pseudomonas aeruginosa
Mannose-binding lectin	Phagocytosis of bacteria	MBL2-0 (null allele)	None	Severe
Transforming growth factors-β	Modulates fibroblast proliferation and collagen production in airway epithelium	1. Low –producing TGFB1alleles. 2. High-producing TGFB1 allele	1. Low 2. High	1. Decreased severity 2. Increased severity
TNF-α	Proinflammatory cytokine	Increased production TNF2 allele	High	Increased severity

classically characterized as monogenic and poorly understood polygenic or complex disorders.

Most of the alleles follow Mendelian laws of segregation and independent assortment. However, disease traits do not follow this. A single locus is responsible for the causation of monogenic disorder; however, even in the most classic monogenic disorders, the 1:1 or 3:1 Mendelian ratio of dominant to recessive phenotypes is found, and it is difficult to explain this kind of deviation. It has been reported now that environmental factors alter this ratio. Along with the environmental interaction of many genes, it also causes modulation. One of the examples is cystic fibrosis; there is a single locus with additional genes and environmental factors that may affect the severity of the disease.

The combined activity of alleles at some loci to produce a phenotype is responsible for some conditions. Oligogenic traits can be two-locus (digenic) systems (**Table 12.5**). It is necessary to distinguish between digenic inheritance and a modifier gene effect. Digenic mutations occur at two different loci, which produce a clinical phenotype, while modifier genes may reduce or enhance the phenotypic impact of alleles at another locus.

Table 12.5: Digenic Inheritance

Gene 1	Gene 2	Condition
BBS2	BBS1, BBS4, or BBS6	Bardet –Beidel syndrome
BBS1	BBS4, BBS6, or BBS7	
RET	EDNRB	Hirschsprung disorder
NPHS1	NPHS2	Nephrotic syndrome
GJB2	GJB6	Nonsyndromic recessive deafness
ROM1	RDS	Retinitis pigmentosum
PRARG	PPP1R3A	Severe insulin resistance
MIFT	TYR	Waardenburg syndrome type 2 and ocular albinism

12.13 Non-syndromic Deafness an Autosomal Recessive in Nature

One of the most common sensory disorders is hearing impairment (HI) or deafness in humans. It can be syndromic or non-syndromic, prelingual or postlingual, and conductive or sensorineural. In the case of non–syndromic deafness sensorineural hearing loss, approximately 40 genes have been found with the autosomal recessive condition. About 120 genes have been cloned for deafness. There exist approximately 95 autosomal recessive non-syndromic deafness loci (DFNB). Nearly 28 loci and 15 genes have been involved in deafness mapped in Pakistan, and approximately 1.6 are involved in deafness. However, there is a large phenotypic variation, possibly due to the involvement of different mutations of the same gene in 1,000 cases. Most cases of recessive HHI show single-locus mutations. However, mutation occurs in two different genes (GJB2/GJB6) or two different mutations in a single allele (GJB2). GJB2 plays an important role in 50 per cent of the cases of prelingual, non-syndromic autosomal recessive hearing loss. Others are SLC26A4, MYO15A, OTOF, CDH23, and TMC1. All these genes are the most frequent cause of autosomal recessive non-syndromic hearing loss. Cochlear implants can solve this problem; cochlear implants can cure congenital sensor neural deaf people. Experimental gene therapy is in process. Monogenic and digenic inheritance of non–syndromic deafness is shown in **Table 12.6**

Table 12.6: Monogenic and Digenic Inheritance of Non-syndromic Deafness

Genotype	Phenotype	Inheritance
$GJB2^{+}/GJB2^{+}$; $GJB6^{+}/GJB6^{+}$	Normal	
$GJB2^{-}/GJB2^{-}$; $GJB6^{+}/GJB6^{+}$	Deafness	Monogenic
$GJB2^{+}/GJB2^{+}$; $GJB6^{-}/GJB6^{-}$	Deafness	Monogenic
$GJB2^{+}/GJB2^{-}$; $GJB6^{+}/GJB6^{-}$	Deafness	Digenic

12.14 Bardet-Biedl Syndrome

Bardet–Biedl syndrome is one of the uncommon syndromes; it affects almost all body parts. It is quite a heterogeneous syndrome with different signs and symptoms in affected individuals, even among members of the same family. The symptoms are loss of vision (affecting the retina), and night vision is not good when the individual reaches middle childhood. Blind spots appear in the peripheral vision. As time passes, these black spots merge and produce tunnel vision. Individuals with this syndrome primarily develop obesity with an abnormal weight gain during childhood which remains throughout life. Type 2 diabetes develops with hypertension and hypercholesterolemia. Sometimes, polydactyly is seen. There are learning problems in these individuals. Male individuals produce less sex hormone.

Additional features are impaired speech, less developed motor skills, and behavioral problems. Its prevalence is 1/140,000 to 1/160,000 individuals from North America.

It is reported that there are mutations in the 14 genes (BBS genes). These genes play a role in cell structure and cilia formation. Approximately one-quarter of the individuals have the BBS1 gene, and 20 per cent have the BBS10 mutation. In approximately 25 per cent of the individuals, the cause is not known. It shows Tri allelic inheritance. People with BBS6 mutation are mostly heterozygous and hence do not fit into the normal mode of inheritance; the Tri allelic inheritance is between BBS2 and BBS6 loci.

12.15 Polygenic

If many loci control a phenotype of a particular trait with an additive effect, the trait is known as the polygenic trait. These are also known as threshold traits. These traits are closely related to fitness but show high heritability. Directional selection may be responsible for large heritability. Direction selection occurs in the presence of evolutionary forces like mutation–selection results in a change in the frequency of one type towards fixation. The selection intensity necessarily declines to permit mutation to restore genetic variation.

A simulation model has been used to test this hypothesis. A hypothetical model is being used if no drift has taken place. It is shown that over 80 per cent of the original genetic variance can be maintained at equilibrium provided the population (N) and the number of loci (n) is reasonably large (N>5000, n=50). However, unless the selection coefficient is minimal (<0.001) the equilibrium frequency of the phenotypes (<2 percent) is considerably below what is otherwise observed. Mutation can play a significant role in maintaining genetic variation in threshold traits. Still, some form of selection, such as frequency-dependent selection, is required to maintain the phenotypic variation (**Figure 12.6**).

12.16 The Distribution of a Trait in a Population Implies Nothing about its Inheritance

The data on the continuous traits can always be shown in a table or graphical form, as in **Figure 12.7**. The distribution of different traits can be shown in the form of mean and variance. To discuss the mean and the variance in quantitative terms, we shall use the data in Table 12.7. The height can be 1 (53-55 inches) to 11 (73-75 inches). The symbol Xi designates the midpoint of the height interval numbered I; for example, X1 = 54 inches, X2 = 56 inches, and so on. The number of women in height interval i is designated fi; for example, f1 = 5 women, f2 = 33 women, and so on. The total size of the sample, in this case, 4995, is denoted as N. The mean and variance characterize the height distribution among these women and the distribution of many other quantitative traits.

The mean, or average, is the peak of the distribution. The mean of a population is estimated from a sample of individuals from the population, as follows:

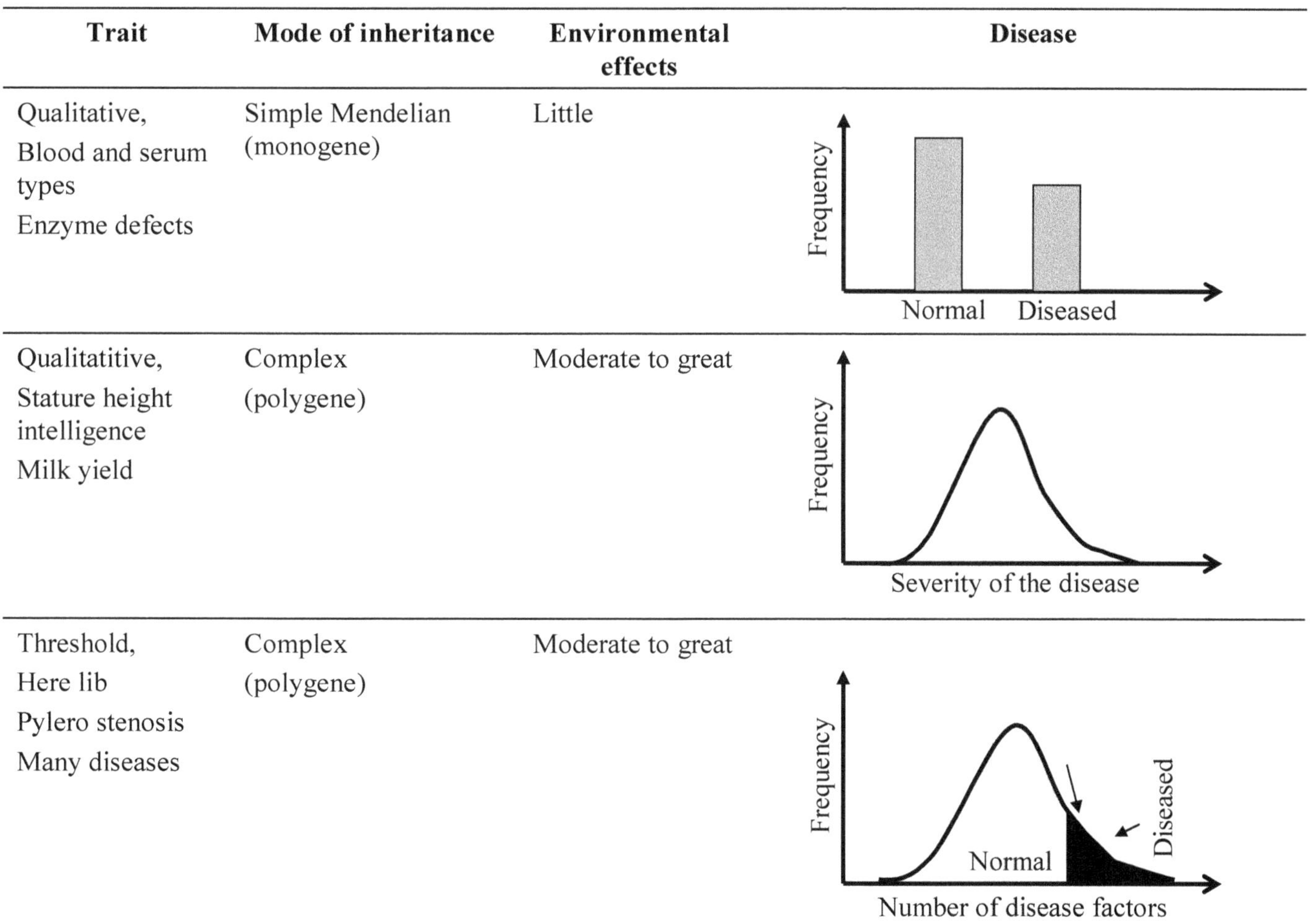

Trait	Mode of inheritance	Environmental effects	Disease
Qualitative, Blood and serum types Enzyme defects	Simple Mendelian (monogene)	Little	
Qualitatitive, Stature height intelligence Milk yield	Complex (polygene)	Moderate to great	
Threshold, Here lib Pylero stenosis Many diseases	Complex (polygene)	Moderate to great	

Figure 12.6: Intelligence as a Threshold Trait.

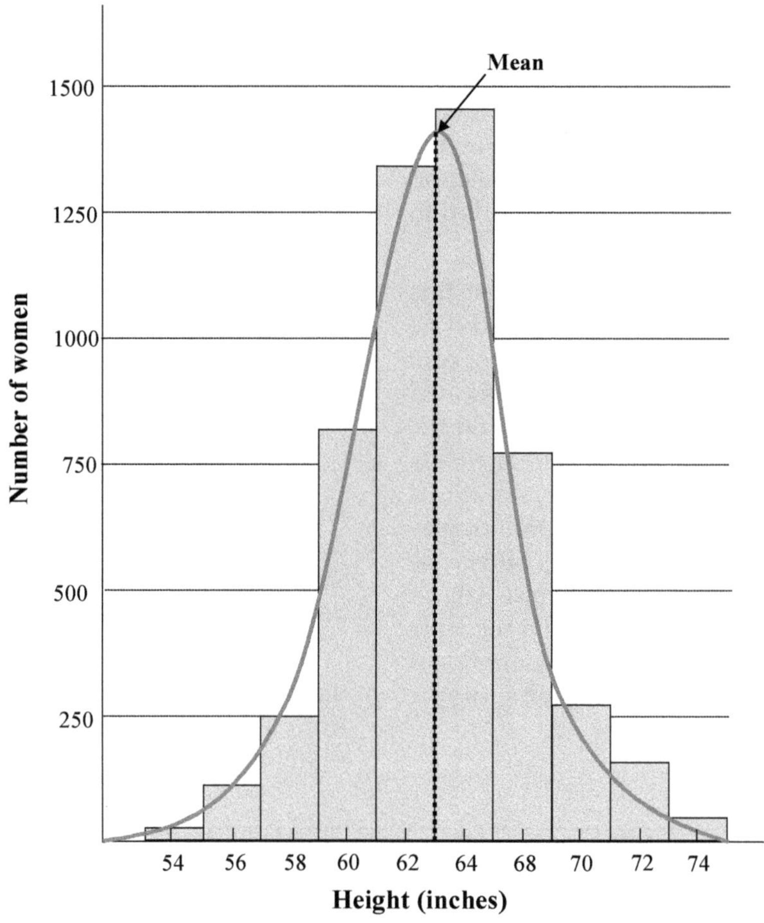

Figure 12.7: Normal Distribution of Height among 4995 British Women.

$$X = \frac{\Sigma fi\ Xi}{N}$$

in which *x* is the estimate of the mean and *k* symbolizes summation over all classes of data (in this example, summation over all 11 height intervals). In **Table 12.7**, the mean height in the sample of women is 63.1 inches.

Table 12.7: Height Distribution in Inches

Interval Number (i)	*Height interval (inches)*	*Midpoint (Xi)*	*Number of Women (fi)*
1	53-55	54	5
2	55-57	56	33
3	57-59	58	254
4	59-61	60	813
5	61-63	62	1340
6	63-65	64	1454
7	65-67	66	750
8	67-69	68	275
9	69-71	70	56
10	71-73	72	11
11	73-75	74	4
			Total N= 4995

The variance is a measure of the spread of the distribution and is estimated in terms of the squared *deviation* (difference) of each observation from the mean. The variance is estimated from a sample of individuals as follows:

$$S^2 = \frac{\Sigma f_i\ (Xi - 'X)^2}{N - 1}$$

S^2 is the estimated variance, and Xi, fi, and N are as in **Table 12.7**. Note that (X_i - 'X) is the difference from the mean of each height category and that the denominator is the total number of individuals minus 1. The variance describes the extent to which the phenotypes are clustered around the mean, as shown in **Figure 12.8**. A significant value implies that the distribution is spread out, and a small value implies that it is clustered near the mean. From the data in **Table 12.7**, the variance of the population of British women is estimated as $S^2 = 7.24\ in^2$.

A quantity is closely related to the variance standard deviation of the distribution defined as the square root of the variance. For the data in Table 12.7, the estimated standard deviation s is obtained from equation 14.2 as S = (S2) ½ = (7.24 in2) ½ = 2.69

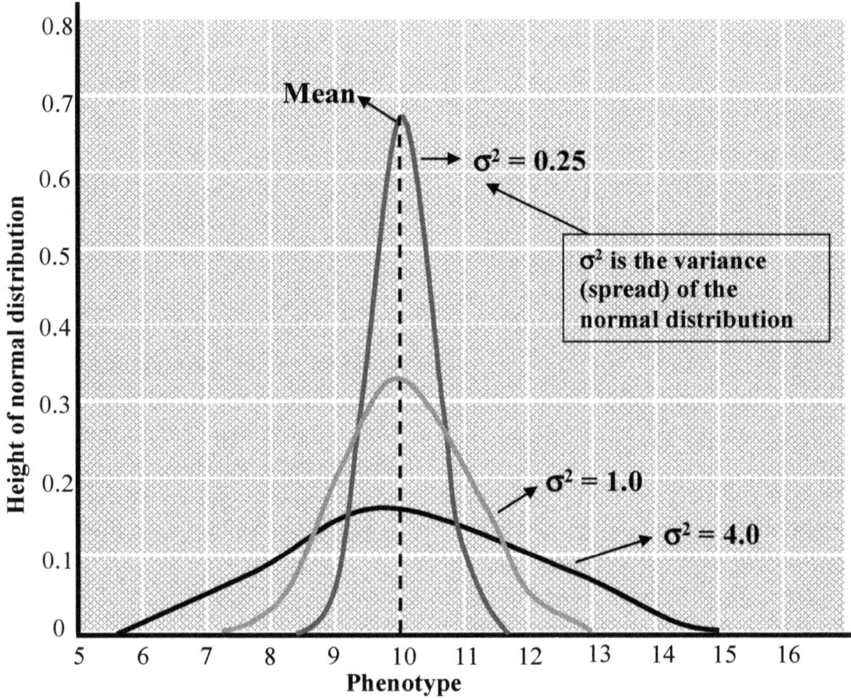

Figure 12.8: Variance of Distribution around the Mean.

inches. The standard deviation has the helpful feature of having the same dimension units as the mean-in this example, inches.

The distribution is depicted as a smooth arching curve in the case of symmetrical data (**Figure 12.8**). This is called the normal distribution. Because the normal curve is symmetrical, half of its area is determined by points with values greater than the mean and half by points with values less than the mean, and thus the proportion of phenotypes that exceed the mean is 1/2. The normal distribution is completely determined by the value of the mean and the variance.

A normal distribution's mean and standard deviation (square root of the variance) provide much information about the distribution of phenotypes in a population, as illustrated in **Figure 12.9**. Approximately 68 per cent of the population have

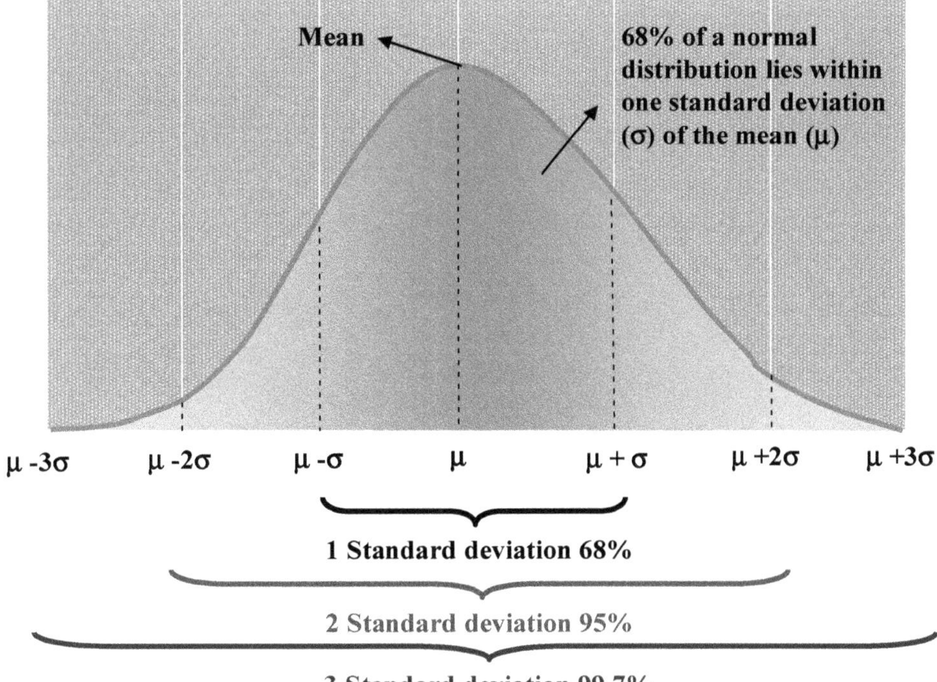

Figure 12.9: Normal Distribution.

a phenotype within one standard deviation of the mean (in the symbols of **Figure 12.10**, between $\mu - \sigma$ and $\mu + \sigma$).

The proportions of individuals lying within one, two, and three standard deviations from the mean are approximately 68 percent, 95 percent, and 99.7 percent, respectively. In this normal distribution, the mean is symbolized μ and the standard deviation σ.

1. Approximately 95 per cent lie within two standard deviations of the mean (between $\mu - 2\sigma$ and $\mu + 2\sigma$).
2. Approximately 99.7 per cent lie within three standard deviations of the mean (between $\mu - 3\sigma$ and $\mu + 3\sigma$).

Applying these rules to the data in **Figure 12.10**, in which the mean and standard deviation are 63.1 and 2.69 inches, approximately 68 per cent of the women are expected to have heights in the range from 63.1-2.69 inches to 63.1+2.69 inches (that is, 60.4–65.8), and approximately 95 per cent are expected to have heights in the range from 63.1–2 X 2.69 inches to 63.1+2 X 2.69 inches (that is, 57.7–68.5).

Accurate data frequently falls under normal distribution. Normal distributions are usually the rule when the phenotype is determined by the cumulative effect of many individually small independent factors. This is the case for many multifactorial traits.

Genetics of multifactorial traits is required to assess the relative importance of genotype versus environment. Sometimes, it is possible to separate genotype and environment for their effects on the mean in experimental organisms. For example, inbred lines were grown in environments that differ in planting density or amount of fertilizer; the possible conditions are shown below.

1. To compare yields to the same genotype grown in different environments and thereby rank the environments relative to their effects on yield, or

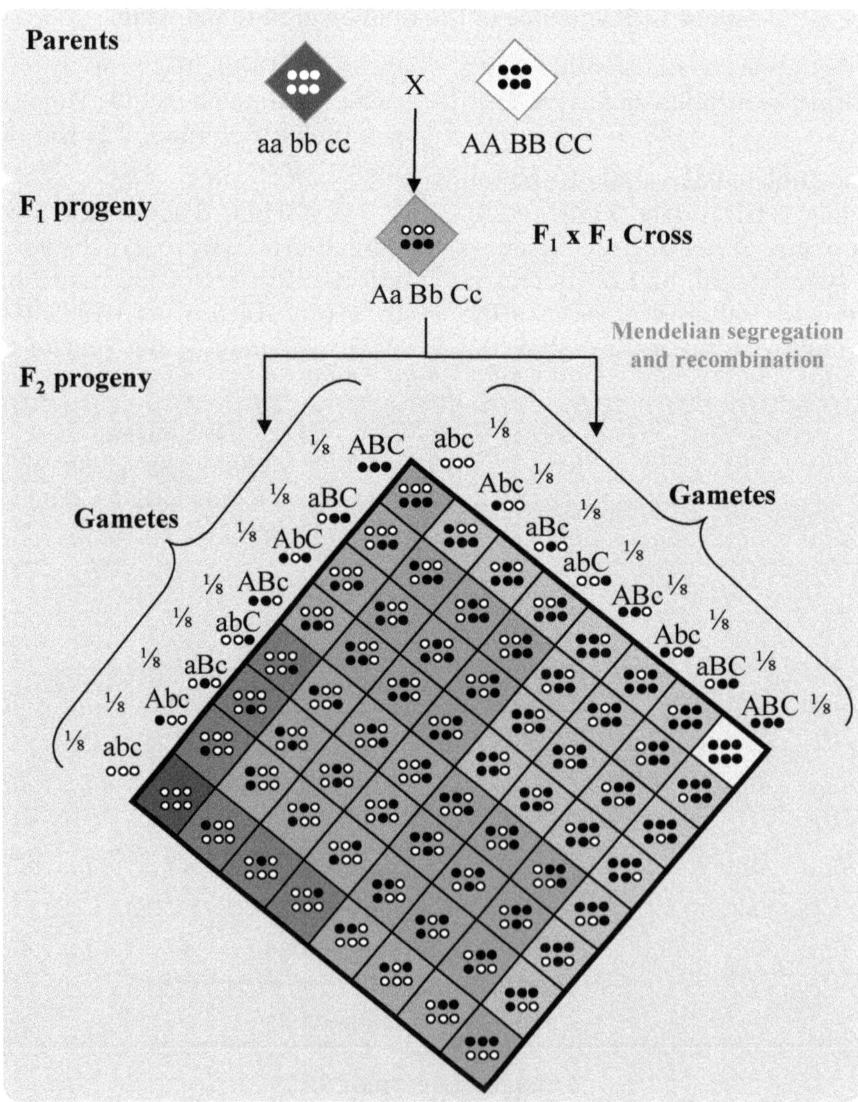

Figure 12.10: Three Independent Gene Segregation Affecting the Quantitative Trait.

2. To compare yields of different genotypes grown in the same environment, rank the genotypes relative to their effects on yield.

Such fine discrimination between genetic and environmental effects is only sometimes possible, particularly in human quantitative genetics. For example, about the height of the women in **Figure 12.10**, the environment could be considered favorable or unfavorable for tall stature only in comparison with the mean height of a genetically identical population reared in a different environment.

Unfortunately, there is no reference population. Likewise, the population's genetic composition could be judged as favorable or unfavorable for tall stature only in comparison with the mean of a genetically different population reared in an identical environment. This reference population does not exist, either. By comparing genetic and environmental effects, it becomes possible to calculate the mean and see the interaction of one another. However, it is still possible to find genetic versus environmental contributions to the variance because instead of comparing the means of two or more populations, we can compare the phenotypes of individuals within the same population. Differences in phenotype are due to differences in genotype and others from differences in environmental conditions, and it is often possible to dissect these effects.

The genotypic variance is due to differences in the genotypes. The genotypic variation results in phenotype variation. The genotypic variation in statistical terms is known as a genotypic variance. A cross of two inbred lines differing in genotype for three unlinked genes A/a, B/b, and C/c, and the genetic variation in the F2 generation caused by segregation and recombination is evident in the color differences. Relative to a meristic trait (one whose phenotype is determined by counting, such as ears per stalk in corn), if we assume that each uppercase allele is favorable for the trait and adds one unit to the phenotype, whereas each lowercase allele is without effect, then the aa bb cc genotype has a phenotype of 0 and the AA BB CC genotype has a phenotype of 6. There are seven possible phenotypes (0 through 6) in the F2 generation. The distribution of phenotypes in the F2 generation is shown in **Figure 12.11**. The normal distribution approximating the data has a mean of 3 and a variance of 1.5. In this case, we assume that all the variation in phenotype in the population results from differences in genotype among the individuals.

Figure 12.11 also includes a bar graph with diagonal lines representing the theoretical distribution when the trait is determined by 30 unlinked genes segregating in a randomly mating population, grouped into the same number of phenotypic classes as the three–gene case. Suppose 15 genes are nearly

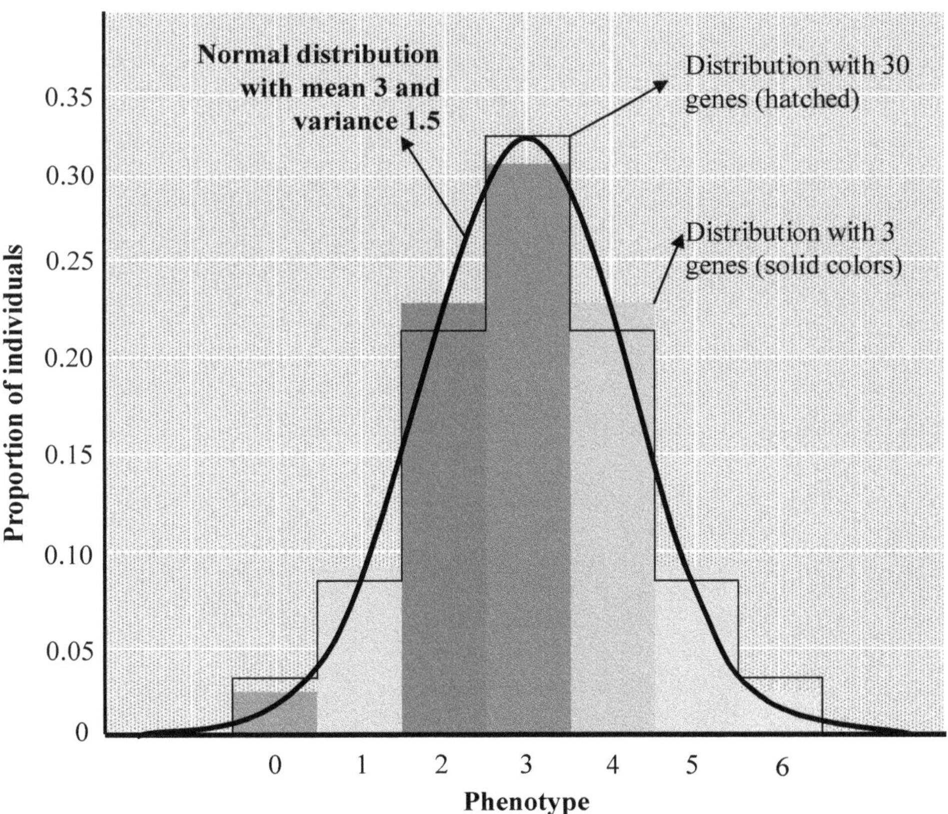

Figure 12.11: The Distribution of Phenotypes Determined by the Segregation of the Three Genes.

fixed for the favorable allele, and 15 are nearly fixed for the unfavorable allele due to selective pressures. The contribution of each favorable allele to the phenotype has been chosen to make the mean of the distribution equal to 3 and the variance equal to 1.5. Note that the distribution with 30 genes is virtually identical to that with three genes and that the same normal curve approximates both.

If such distributions were encountered in actual research, the researcher wouldn't be able to distinguish between them. The critical point is that, even without environmental variation, the distribution of phenotypes provides no information about the number of genes influencing a trait and no information about the dominance relations of the alleles. However, the number of genes influencing a quantitative trait is vital in determining the potential for long-term genetic improvement of a population employing artificial selection. When there is a combination of three genes **Figure 12.11**, the best possible genotype would have a phenotype of 6. Still, in the 30–gene case, the best possible genotype (homozygous for the favorable allele of all 30 genes) would have a phenotype of 30.

12.17 The Environmental Variance Results from differences in the Environment

The distribution of seed weight in edible beans reveals that the mean distribution is 500mg and the standard deviation is 95mg; all the beans in this population are genetically identical and homozygous because they are highly inbred. Such data has no importance of environmental effect, as shown in **Figure 12.12**.

It is always difficult to separate the genotypic and environmental variation. As depicted, the trait can have one of the three distinct and nonoverlapping phenotypes determined by the effects of two additive alleles. The genotypes are in random-mating proportions for an allele frequency of ½, and the distribution of phenotypes has a mean of 5 and variance of 2.

If the variance is genotypic, then this variance is genotypic variance, which is symbolized as σ2g. The three panels at the upper right (**Figure 12.13**) depict the distribution of phenotypes in environmental variation of each of the three genotypes. In each case, the variance in phenotype, due to environment alone,

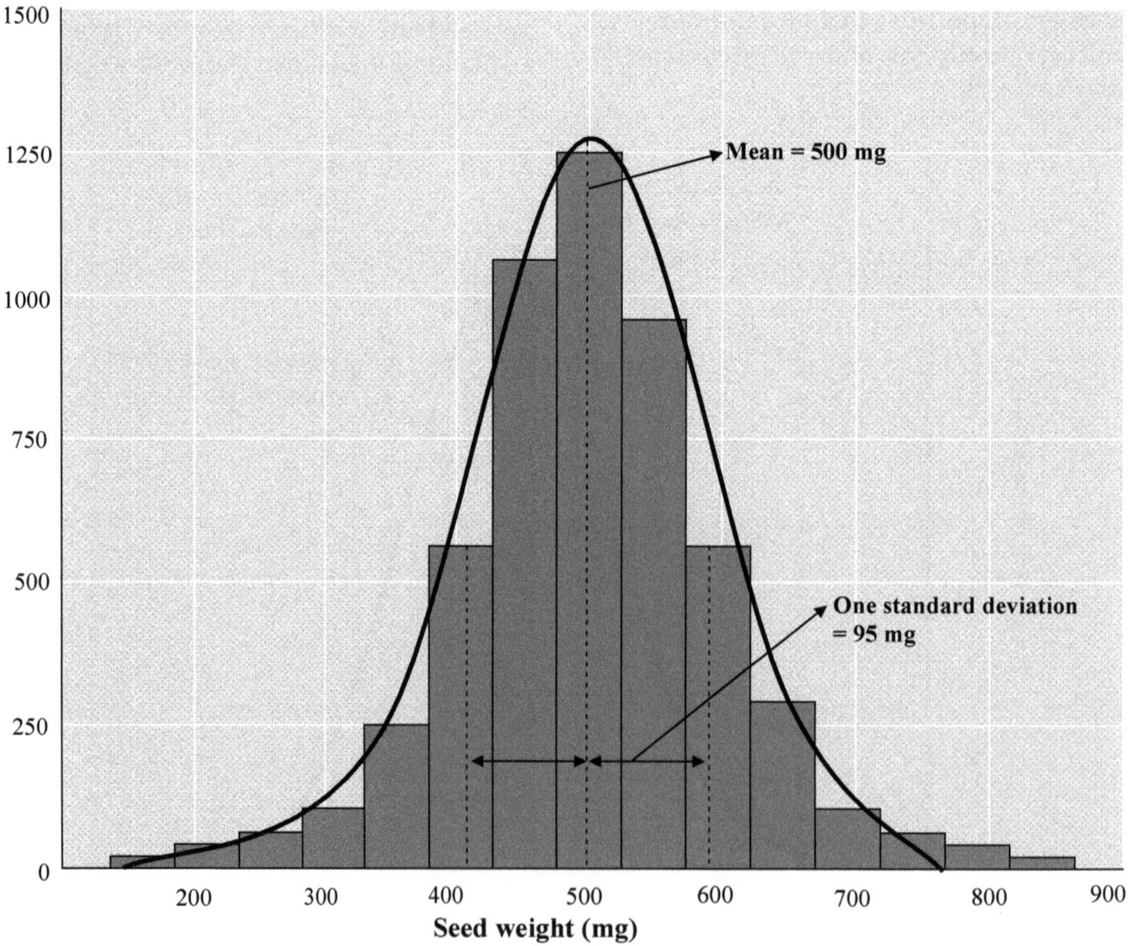

Figure 12.12: Distribution of Seed Weight in Homozygous Beans for Weight.

is 1. Because this variance is due to environmental differences, it is environmental variance, which can be written as σ2e. The total variance in phenotype is the addition of environmental and genetic factors. Because we assume that genotype and environment have separate and independent effects on phenotype, we expect a; to be greater than σ2g or σ2e alone. In fact,

$$\sigma 2t = \sigma 2g + \sigma 2e$$

When genetic and environmental effects contribute independently to phenotype, the total variance equals the sum of the genotypic and environmental variance.

12.18 Genotype and Environment can Interact, or they can be Associated

The genotype and environmental factors interact with one another. Studies have been conducted on alcohol dependence. Alcohol dependency has been done on adopted individuals in 1970; the risk is almost the same as in the case of biological parents. In another study, two genes involved in the metabolism of toxic substances associated with smoking–CYP1A1 and GSTT1–predict the birth weight of the offspring only in mothers who smoked during pregnancy and not otherwise. These two examples explain the effect of environmental and genetic factors, *i.e.*, GxE interaction (**Table 12.8**).

12.19 Heritability

Heritability is the proportion of a population's phenotypic variation due to genetic factors. As we have seen, continuous traits are influenced by multiple genes and by environmental factors. Many essential traits, such as variation in body size, fecundity, and developmental rate, are polygenic, and the genetic contribution to this variation is essential for understanding how natural populations evolve. Heritability is nonetheless often misunderstood, and the term is frequently misused. To assess heritability, we must first measure the variation in the trait, and then we must partition that variance into components attributable to different causes like the environment and genetics.

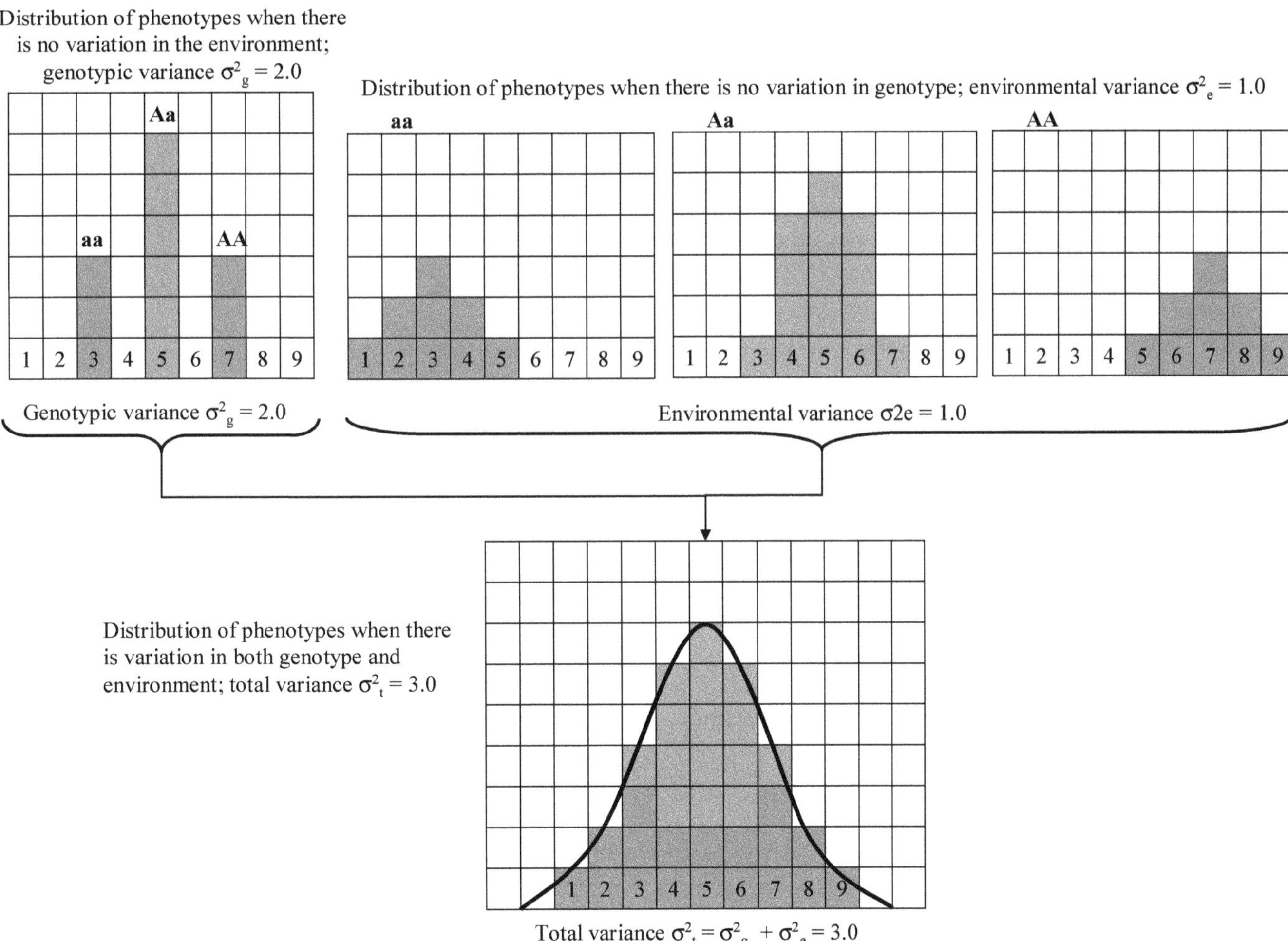

Figure 12.13: The Combined Effects of Genotypic and Environmental Variance.

Table 12.8: Contributions of Genetic and Environmental Effects, the Interaction of Genotype and Environment

	Mean Genetic Sharing	*Genetic Effects (G)*	*Shared Environment Effects (SE)*	*Non-shared Environment Effects (NSE)*	*Genetic x Shared Environment Effects (GxSE)*	*Genetic x non-shared Environment Effects (GxNSE)*
Monozygotic twins reared together	100 per cent	1.0	1.0	0.0	1.0 x 1.0 = 1.0	1.0 x 0.0 = 0.0
Fraternal twins and full sibling pairs reared together	50 per cent	0.5	1.0	0.0	0.5 x 1.0 = 0.5	0.5 x 0.0 = 0.0
Genetically unrelated siblings reared together	0 per cent	0.0	1.0	0.0	0.0 x 1.0 = 0.0	0.0 x 0.0 = 0.0
Identical twins reared apart	100 per cent	1.0	0.0	0.0	1.0 x 0.0 = 0.0	1.0 x 0.0 = 0.0
Fraternal twins and full sibling pairs reared apart	50 per cent	0.5	0.0	0.0	0.5 x 0.0 = 0.0	0.5 x 0.0 =0.0
Total population variance	–	1.0	1.0	1.0	1.0 x 1.0 = 1.0	1.0 x 1.0 = 1.0

Note: A coefficient of 1.0 indicates a high contribution, a coefficient of 0.5 indicates an intermediate contribution and a coefficient of 0 indicates no contribution.

12.20 Components of the Phenotypic Variance

There are multiple phenotypic variations within any given population due to different factors. It is represented as a measure of all variability for a trait (*i.e.*, the whole stick). Some variations are due to genetic differences between individuals (different genotypes within the group); alternatively, some differences from the mean may be due to different environments that the individuals experience–genetic variance, represented by the symbol VG.

As noted, additional variation often results from environmental differences experienced by the individuals. The environmental variance is symbolized by VE and, by definition, includes any non-genetic source of variation.

There are many conditions where both genetic and environmental factors contribute. The following formula can calculate phenotypic variance.

$$VP = VG + VE$$

However, both VP, VG, and VE are complicated components.

Genetic variance, VG, can be subdivided into components arising from different types of gene action and interactions between genes. Some genetic variances are due to the additive variance of the different alleles on the phenotype (VA). For example, an allele G may contribute 2 cm in height to a plant, while the allele G contributes an average of 4 cm. In this case, the gg homozygote would contribute 2 +2 = 4 cm in height, the Gg heterozygote would contribute 2 + 4 = 6 cm in height, and the GG homozygote would contribute 4 + 4 = 8 cm in height. To determine the genetic contribution to height, we would then add the effects of alleles at this locus to the effects of alleles at other loci that might influence the phenotype. Such genes cause additive effects, and variation resulting from this sort of gene action is called additive genetic variance, symbolized by VA. Some alleles contribute to the pigment of the kernel, and others do not. The added effects of all the individual contribution alleles determine the phenotype of the kernel. Thus, the genotypes AA bb, aa BB, and Aa Bb all produce the same phenotype because each genotype has two contributing alleles–the phenotypic variance arising from the additive effects of genes and the additive genetic variance.

The genetic source of variation can be divided into many subcategories, *i.e.*, additive variance (VA), dominance variance (VD), and epistatic variance (VI). Hence, the total genetic variance for a particular phenotypic trait would be.

$$VG = VA + VD + VI$$

To understand this equation, we should know what each category deals with. The additive genetic variance refers to the deviation from the mean phenotype due to the inheritance of a particular allele, which affects the phenotype. Under dominance, genetic variance deviation is due to the interaction between alternative alleles at a specific locus. Let us take the example of a plant that produces white flowers. Suppose its genotype is A1A1, and red flowers are due to the expression of both alleles. However, suppose there is an interaction between the alleles such that their expression is unequal. In that case, the phenotype of the heterozygous (A1A2) would more closely resemble one of the homozygotes (A1A1/A2A2). If A2 is dominant over A1, all flowers would be red. The third category is epistatic variance. This involves interaction between alleles. Alleles are associated with different loci.

12.21 Broad-sense and Narrow-sense Heritability

One of the most important questions for quantitative geneticists is the extent to which variation occurs between individuals results from genetic differences. Hence one is interested to know how much of the phenotypic variance, Vp, can be attributed to genetic variance, V G' This quantity is called the broad-sense heritability and can be thought of as how much of the stick of variation is made up of genetic variance. Broad-sense heritability is calculated as a proportion:

Broad-sense heritability = H2B = VG/VP

The heritability of a trait can range from 0 to 1. A broad-sense heritability of 0 indicates that no variation in phenotype among individuals results from genetic differences. In contrast, the broad-sense heritability of 1 suggests that all the phenotypic variance is genetically determined. Broad-sense heritability ignores partitioning the genetic variance into additive, dominant, or interactive components and assumes that genotype-by-environment interaction (VGxE) is not important.

More frequently, we are interested in the proportion of the phenotypic variation resulting from additive genetic effects. This is because of the critical relationship between additive genetic variation and artificial and natural selection. This is because only in the case of additive interactions between alleles can we unambiguously determine an individual's genotype from their phenotype. With either dominance or epistasis, we must know either the genotypes of the individual's parents or conduct test crosses in the hope of knowing an individual's genotype. Only the additive portion of genetic variation allows accurate predictions of the offspring's average phenotype from an individual's phenotype.

12.22 Understanding Heritability

Heritability can be understood under the following lines.

1. Broad-sense heritability does not define the complete genetic basis of a trait.
2. Heritability does not indicate the genetic proportion of an individual's phenotype.
3. Heritability is not fixed for a trait.

Genes are not the only factors that influence height in humans. Diet, an environmental effect, is also a significant determinant of height. Since most individuals in small New England towns probably receive an adequate diet, at least in terms of calories, this part of the environmental variance for height would not be significant. In a developing nation, however, some individuals might receive adequate nutrition, whereas the diet of others might be severely deficient. Since more significant differences in diet exist, the environmental variance for height would be more prominent, and as a result, the heritability of height would be less. Thus, heritability calculated for human height might differ substantially for residents of the small New England town, residents of the USA, and residents of India.

These examples illustrate that heritability can be applied only to a specific group of individuals in a specific environment. If the genetic composition of the group is different or the environment is different, heritability estimates cannot be transferred. Changing groups or environments does not alter how genes affect the trait. Still, it may change the genetic and environmental variance for the trait, which would alter the heritability.

12.23 Calculations of Heritability

12.23.1 Parent-offspring Regression

Heritability is a measure of the proportion of phenotypic variance. Many families are collected to determine a particular trait's genetic component. Data is analyzed statistically, and the relationship between phenotypes and genotypes is determined using correlation and regression, so the mid-parental value is calculated. If the variation between parents is due to additive genetic variation, then the mid-parental value predicts the mean phenotype of the offspring. Other combinations of relatives can also be used.

When the slope of the parent-offspring regression is the mean offspring phenotype is intermediate to the phenotype of the parents and genes with additive effects, determine all the phenotypic differences. If the slope is less than 1 but greater than 0, additive genes, genes with dominance or epistasis, and environmental factors will all affect the phenotype. If points are scattered, the likelihood of heritability is zero.

12.23.2 Selection Results in the Evolution

Both natural selection and artificial selection are dependent upon genetic variation.

12.23.3 Estimating the Response to Selection

In response to selection, the mean value of the phenotype changes in the populations. This is called selection response (R).

The selection response depends on the narrow-sense heritability and the selection differential. The selection differential differs between the selected parents' mean phenotype and the population's before selection. In the case of body size in fruit flies, the original population had a mean weight of 1.3 mg, and the mean weight of the selected parents was 3.0 mg, so the selection differential is 3.0 mg - 1.3 mg = 1. 7 mg. The selection response is related to the selection differential and the narrow-sense heritability by the following formula, known as the breeder's equation:

$$R = H^2_N S$$

With values for two of the three parameters in the preceding equation, the selection response (0.7 mg) and the selection differential (1.7 mg), we can solve for the narrow sense heritability:

Narrow-sense heritability = H2N = Selection response/selection differential

$$H^2_N = 0.7 \text{ mg}/1.7 \text{ mg} = 0.41$$

Selection experiments such as this provide another means for estimating the narrow-sense heritability.

12.24 Twin Studies

The dizygotic twins are like two sibs as two ova are fertilized against two sperms. The twins can be both males, both females or male and female. They share the same intrauterine environment.

MZ twins are found with a frequency of 3.5 per 1000 live births, suggesting that this is a random event and not under genetic control. The frequency of D2 twins in East Asia, sub-Saharan Africa, and both in the Middle East and Europe is about 1, 12.5, and 4.5 per 1000 live births, respectively; DZ twinning has a genetic component; however, the environmental factors, such as a diet, that induce double ovulation cannot be ignored. The exact incidence of twinning in the first week is unknown in the first week of fertilization; however, it may be found in 10 per cent to 15 per cent of the conceptions. An ultrasound can detect twins 6 to 8 weeks after gestation. Ultrasound may detect false twin pregnancy as a single child is born later. This is called "vanishing twin syndrome" in early development.

The twins can be concordant for the traits that mostly happen in monozygotic twins, or they can be discordant, found in dizygotic twins. A concordance can be written as c and dis-concordance as D.

The calculation of concordance value is:

$$C_{Pairwise} = C/C + D$$

Proband-wise concordance can also be calculated.

$$C_{Probandwise} = 2C_2 + C_1/2C_2 + C_1 + D$$

12.25 D Represents the Discordant Twins

Monozygotic twins show more similarity as compared to dizygotic twins. The concordance has a genetic basis. Similarity is assessed for all disorders, like behavior, cognitive ability, and physical features. MZ concordance is mortally greater than 75 per cent, revealing that approximately 25 per cent of the contribution is to the environment.

Concordances can be calculated only for meristic and continuous traits. A correlation coefficient is a standard statistical analysis used to determine the relationship between two sets of observations. The formula used is shown below.

$$r = \frac{2\,\Sigma(x - x)(x' - x)}{\Sigma\,(x - x)^2 + \Sigma(x' - x)^2}$$

where x and x′ are twin 1 and twin 2 of each twin pair, respectively, and 'x is the mean of all observations. For numerous conditions, the intraclass correlation coefficients for MZ (rMZ) and DZ (rDZ) twins can be seen. If any trait shows a greater value for rMZ and a high difference between the observed rMZ and rDZ values, a genetic component might be involved. A better representation would be to calculate heritability estimates which calculate the proportions of the total variation from the mean of a normally distributed trait that could be attributed to genetic and environmental factors. A commonly used broad-sense heritability estimate for twin studies is $h^2B = 2(r_{MZ} - r_{DZ})$, where r_{MZ} and r_{DZ} are the intraclass correlation coefficients for MZ and DZ twins, respectively.

Mostly variance is calculated to calculate heritability. The deviation from the variance (S^2) represents the mean of a normally distributed sample and is the sum of the square of each observation minus the mean divided by the sample size minus 1, that is,

$$S^2 = \frac{\Sigma(x - x)^2}{N - 1}$$

The phenotypic variance, which is written as (Vp) is the total variance for the quantitative trait V_P

$= (V_G + V_E)$ V_G is the genotypic variance while V_E is environmental variance Heritability is equivalent to $V_G/V_P = V_G/V_G + V_E$

Another term used to calculate heritability is additive genetic variance, which considers all the alleles that act independently but contribute to the final phenotype.

Another heritability estimate is the narrow sense heritability (h^2_N). This is calculated as V_A/V_P. All these terms are shown in **Table 12.9**.

Table 12.9: Intra-class Correlation Coefficient among Monozygotic and Dizygotic Twins

Relationship	*Variance*			
	V_A	V_C	V_E	V_D
$r_{MZ} = r_{DZ} = 0$			+	
$r_{MZ} = r_{DZ} > 0$		+	+	
$r_{MZ} = r_{DZ} = 1$		+		
$r_{MZ} = 2r_{DZ}$ with $r_{MZ} < 1.0$	+		+	
$r_{MZ} > ½\ r_{DZ}$	+	+	+	
$r_{MZ} < ½\ r_{DZ}$	+		+	+

Note: V_A, additive genetic effect; V_C, common environmental effect; V_E, individual–specific environmental effect; V_D, nonadditive genetic effect.

12.26 When Twins are Parents?

It is essential to see what happens when twins are parents. The most cited example is alcohol dependence, and other psychiatric disorders are off; however, for these disorders, not many genes have been identified that may be responsible for these diseases. Some genetic and environmental interactions have been studied in Asian populations. ALDH2–2 allele is one of the important genetic factors along with the environmental risk factors; not much data is available from other populations. In **Table 12.10,** some examples have been shown.

12.27 Artificial Selection

Artificial selection can be preceded by selecting the best parental organism and the following inbreeding. The best method to estimate how good is the artificial selection is to calculate heritability. Without sexual reproduction, each offspring will have the same genotype as its parent.

In sexually reproducing populations that are genetically heterogeneous, broad-sense heritability is not relevant in predicting progress resulting from artificial selection because segregation and recombination must necessarily break up superior genotypes. For example, if the best genotype is heterozygous for each of the two unlinked loci, A/a; B/b, then because of segregation and independent assortment, among the progeny of a cross between parents with the best genotypes-A/a; B/b X A/a; B/b- only 1/4 will have the same favorable A/a; B/b genotype as the parents. The rest of the progeny will be genetically inferior to the parents. For this reason, to the extent that high genetic merit may depend on combinations of alleles, each generation of artificial selection results in a slight setback in that the offspring of superior parents are generally not quite as good as the parents themselves.

12.28 The Quantitative Traits in Humans can be Used for Mapping the Gene

A gene that affects a quantitative trait can be represented as a quantitative –trait locus (QTL). Locating QTLs in the genome is essential to manipulating genes in breeding programs, cloning, and studying genes to identify their functions.

For mapping, the gene copy number variation SNP and short tandem repeat markers are used in humans. More than 300 highly polymorphic genetic markers have been mapped in the tomato genome, with an average spacing between markers of 5 map

Table 12.10: Twins as Parents can Distinguish the Genetic and Environmental Influences on Risk of Psychopathology in the Children of Alcohol-Dependent Parents, Including other Shared Environmental Effects Associated with Parental Alcohol Dependence

Parent's History of Alcohol Dependence	*Alcohol Dependence History of the Parent's Twin*	*Level of Risk to Children Due to*		
		Genetic Effects	*Familial Environmental Effects*[a]	*Genotype x Environmental Interaction Effects*
Alcohol dependent	Any	High	High	High
Nondependent	Alcohol dependent, identical twin	High	Low	Low
Nondependent	Alcohol dependent, fraternal twin	Intermediate	Low	Low
Nondependent	Nondependent	Low	Low	Very Low

units. The chromosome maps have a subset of 67 markers segregating in crosses between the domestic tomato and a wild South American tomato. Although additional QTLs of more minor effects undoubtedly remained undetected in these types of experiments, the effects of the mapped QTLs are substantial: The mapped QTLs account for 58 per cent of the total phenotypic variance in fruit weight, 44 per cent of the phenotypic variance insoluble solids, and 48 per cent of the phenotypic variance in acidity.

Because the locations of the QTLs can be specified only within 20 – 30 map units, it is unclear whether the coincidences result from pleiotropy, in which a single gene affects several traits simultaneously, or from the independent effects of multiple, tightly linked genes. However, the locations of QTLs for different traits coincide frequently enough that, in most cases, pleiotropy is likely to be the explanation.

12.29 Location of Quantitative Trait Loci

The genes can be localized using multi-generation families and twin studies. The LOD score method is not possible for polygenic traits because recombinants cannot be detected, and the mode of inheritance of a particular QTL is not easily modeled.

12.30 Case-control Association Studies

Case-control studies are essential for studying multifactorial disorders. For this purpose, an adequate number of cases (patients) and normal healthy controls are selected. Various polymorphic markers are selected. Once data for both cases and controls are accumulated, they are compared by using various statistical tests. Once the significance is determined, these are studies for various functional studies. Functional studies can be conducted on human or animal models. The guidelines used for case-control studies are summarized in **Table 12.11**.

12.31 Genome-wide Studies

A genome-wide association study is conducted on a larger sample size and also DNA markers, which are many; hence the power of the study is higher. If genome-wide association shows a gene or a marker to be associated, it is used to formulate better strategies to prevent the disease. These studies have been used for asthma, cancer, diabetes, heart disease, and mental retardation. In genome-wide association, we must collect cases and controls. GWAS is performed using single nucleotide polymorphism (SNPs). Sometimes genetic differences occur more in individuals suffering from the disease than in healthy individuals. This may be the disease associated with a particular genetic variation. It should be noted that associated variants may not cause the disease but may be tagged along with the actual causal variants. Some additional information is required for this purpose, like DNA sequencing, *etc.*, to pinpoint the region where the variation is located. Most STRs are used for genome scanning purposes.

Linkage disequilibrium can be easily detected between marker sites and quantitative trait loci. Linkage disequilibrium requires a mode of inheritance, penetrance, and other genetic features; all these parametric tests are nonparametric and study the significant associations between DNA markers and a chromosome site in affected individuals. Genome scans are quantified as LOD scores and plotted to determine the function and location in cM. A LOD score of more than 3.6 is considered significant of association, > 2.2 suggestive, and > 1.5 interesting.

Many times there is a lack of consistency in many studies, and it is not readily explained. These discrepancies may be due to technical errors. More powerful analytical methods must be developed to overcome this larger sample size. The human genome

Table 12.11: Case-control Association Studies

Issue	*Key Questions*	*Possible Solutions*
Selection of candidate	Is candidate gene biologically reasonable?	Demonstration of biologically functional effect
Gene polymorphism	Is the candidate gene a positional candidate?	Within linked region in man or syntenic from animal model
Population stratification	Are cases and controls matched?	Matching on ethnicity
		Family-based association designs
		Negative results with multiple unlinked markers
Hardy-Weinberg (H-W) equilibrium	Is the control group in H-W equilibrium?	Calculation of H-W equilibrium with goodness-of-fit test (2 alleles) or simulation (multiple alleles)
Multiple comparisons	How many alleles were tested?	Bonferroni correction
	How many genetic loci were tested?	Estimation of empirical *P* values

comprises blocks with a high degree of LD up to 10 to > 100kb; these regions are separated by segments with little or no linkage disequilibrium (LD). Several haplotypes is found in a block with high LD. In these blocks, we can find that the (tag SNPs) will be enough to detect linkage with any gene in the block.

12.32 Transmission/Disequilibrium Test (TDT)

TDT (transmission/disequilibrium test) was carried out to avoid the confounding effect of population stratification. In the TDT, we require familial data; the parents should be informative, *i.e.*, heterozygous. In the case of the disease study, the affected allele would be transmitted more frequently from parents to the offspring than the normal alleles.

The recombination frequency (q) is less than 0.5, and linkage disequilibrium (δ) is not zero. The possible combinations of transmission and non-transmission of alleles for A1A1, A1A2, and A2A2 genotypes can be depicted with a four–cell table (**Figure 12.14**). Cells a and d represent the results for homozygotes. Cells b and c denote the two possibilities with a heterozygous genotype. However, it isn't easy to distinguish which allele is transmitted to an offspring. If the parents are homozygotes, it is easier in the case of heterozygotes. Therefore, only the number of transmissions of alleles recorded in cells b and c are used for the TDT. The significance of TDT can be calculated using $\chi 2$. Both the parents should be genotyped, and one of the parents should be heterozygous and have an affected child.

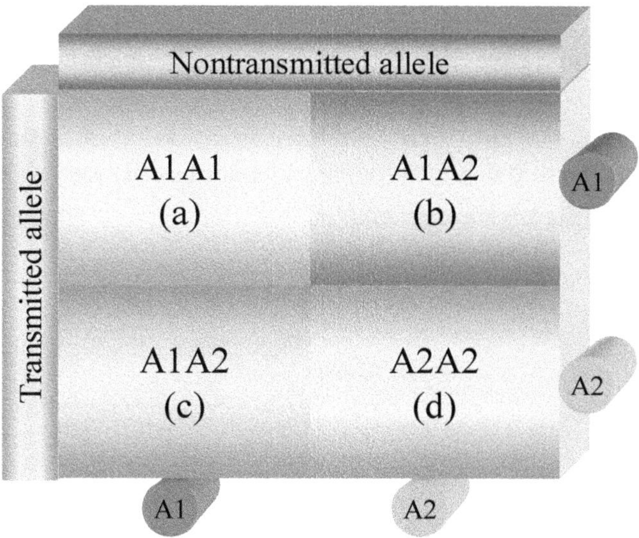

Figure 12.14: Transmitted and Non-transmitted Alleles for A1A1, A1A2, and A2A2 Genotypes.

In such studies, the sample size, *i.e.*, the number of informative families, should be significantly large.

The numbers of transmitted alleles from heterozygous parents are determined and the significance test is run (**Figure 12.15**).

(A)

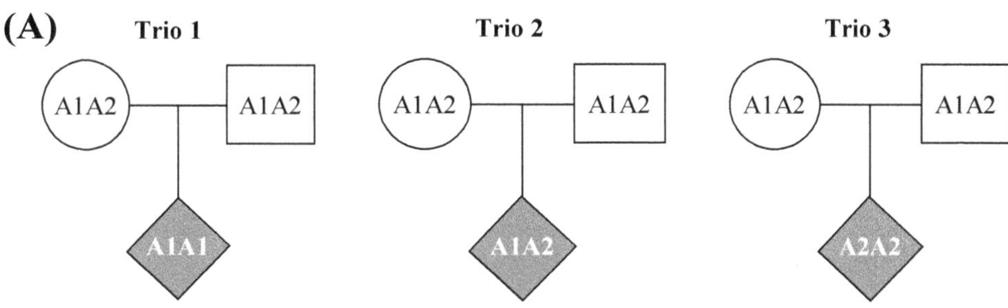

(B)

Trio	Number of Trios	Transmitted alleles	
		A1	A2
1	20	40	
2	10	10	10
3	5		10
		50 (b)	20 (c)

(C)

$$\chi^2 = \frac{(b - C)^2}{(b + C)^2} = \frac{(50 - 20)^2}{(50 + 20)^2} = 12.86;\ dt;\ p<0.001.\ \text{The difference is significant}$$

Figure 12.15: Transmission/Disequilibrium Test.
(A) In this fabricated example, only trios with heterozygous parents were scored for autosome markers.

12.33 Affected sib-pair Linkage Analysis

This analysis involves linkage analysis of large numbers of sib pairs that share the same trait with some polymorphic markers the polymorphic alleles in the sib pairs must be identical by descent (identity–by state; IBS). The state of IBD and IBS is shown in **Figure 12.16**.

As mentioned, for an autosomal locus, the proportion of sib pairs with 2, 1, and 0 IBD alleles is 1/4: ½: . This ratio can be shown by considering two parents with genotypes A1A2 and A3A4. This type of result in mating will have four kinds of offspring (A1A3, A1A4, A2A3, A2A4), with the same frequency. The probabilities of pairs of sibs with 2, 1, and 0 IBD alleles can be deduced using a 4 x 4 table (**Figure 12.17**). For example, in the top leftmost cell in **Figure 12.17**, both sibs received the A1 allele from the same parent and the A3 allele from the other parent; thus,

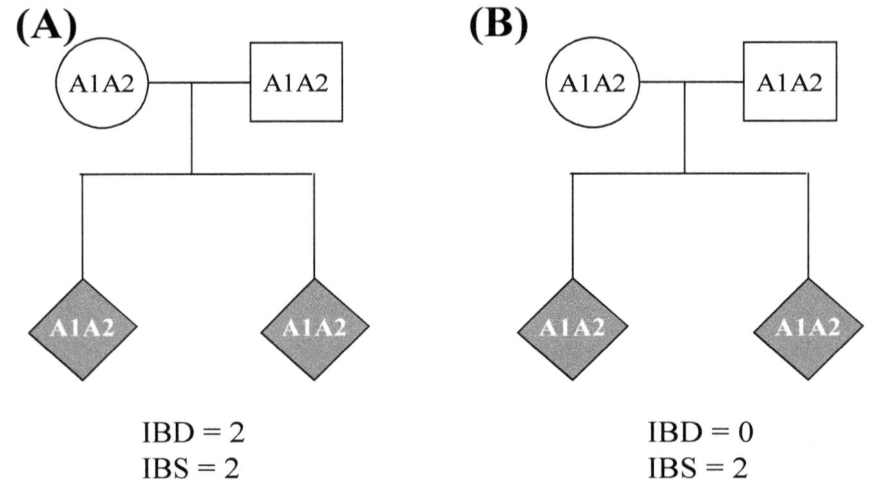

Figure 12.16: Identity by Descent (A) and Identity by State (B). The alleles are color-coded (black, blue) to track the parental source in the sibs.

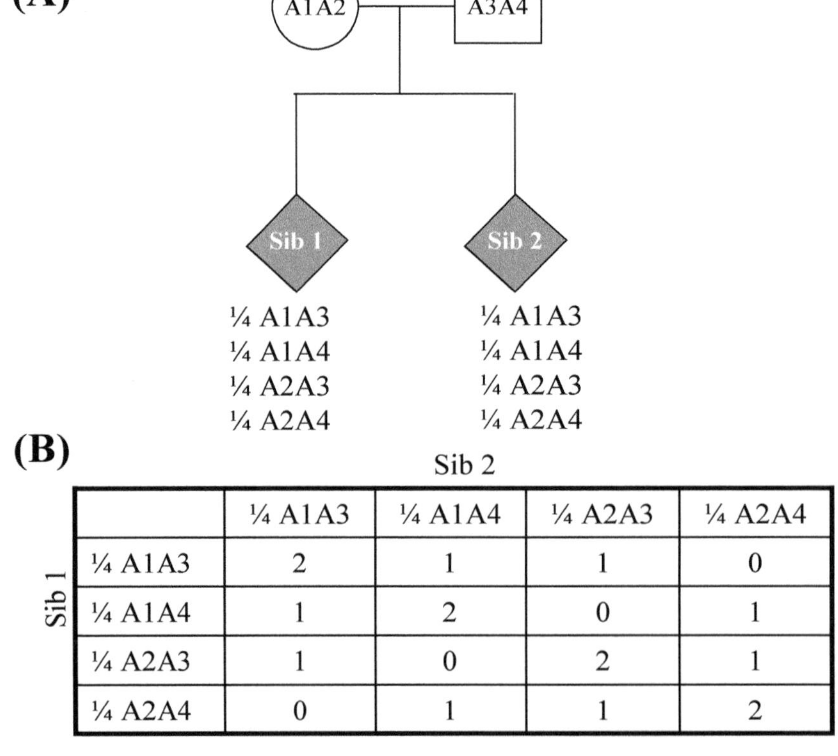

	¼ A1A3	¼ A1A4	¼ A2A3	¼ A2A4
¼ A1A3	2	1	1	0
¼ A1A4	1	2	0	1
¼ A2A3	1	0	2	1
¼ A2A4	0	1	1	2

Tally: 2 IBD = 4/16= 25%; 1 IBD= 8/16 = 50%; 0 IBD = 4/16 = 25%

Figure 12.17: Probabilities of sib Pairs with 2, 1, and 0 IBD alleles. (A) Heterozygous parents with different genotypes can have offspring with four equally likely genotypes. (B) The frequencies of shared alleles that are identical by descent between all possible genotypes for sibs 1 and 2 are noted in A.

the two alleles are identical by descent that is IBD2, and so on. The tally is IBD2, ½ IBD1, and IBD0. ASP linkage analysis detects markers that deviate significantly from the expected ratio.

There is computer software that can be used for such types of calculations. Large pedigrees can be used, taking into consideration a large number of markers.

12.34 Genetic Correlations

The traits do not vary independently when two or more phenotypes are associated. The phenotypic correlation between two quantitative traits can be analyzed by measuring the two phenotypes on several individuals and then calculating a correlation coefficient for the two traits. Pleiotropy is one of the leading causes of genetic correlations for quantitative traits.

Another significant cause of genetic correlations is genetic linkage. It occurs when Mendel's law of independent assortment is not followed due to the occurrence of the closer loci on the same chromosome. These loci are inherited together. When new alleles are first produced by mutation, they are associated with the other alleles that stay on that chromosome. These new alleles will be inherited with the other closely linked alleles, causing genetic correlations, and the persistence of these correlations over time depends in part on the amount of recombination between the loci. An important distinction between linkage and pleiotropy as causes of genetic correlations is that over evolutionary time, even the tightest linkages can be broken, allowing new associations between alleles; with pleiotropy, however, functional constraints of an individual protein might not be able to be dissociated, causing a correlation to persist. Care must be taken when making a correlation, because, beyond pleiotropy and linkage, environmental factors may also influence several traits simultaneously to cause nonrandom associations between phenotypes. For example, adding fertilizer to soil often causes plants both to grow taller and to produce more flowers. If we measured plant height and counted the number of flowers on a group of responsive plants, some of which received fertilizer and some of which did not, we would find that the two traits are correlated; plants receiving fertilizer would be tall and would have many flowers, and those without fertilizer would be short and have few flowers. This phenotype is due to gene action, but the correlation results from the expected effect of an environmental factor, the fertilizer, on both traits.

Genetic correlations may be positive or negative. A positive correlation means that genes causing an increase in the magnitude of one trait bring about a simultaneous increase in the magnitude of the other. A positive correlation exists for many traits related to sexual attractiveness, as predicted by various evolutionary theories, but the actual cause of this correlation is typically ambiguous in general. In chickens, body weight and egg weight have a positive genetic correlation. If breeders select heavier chickens, both the size of the chickens and the mean weight of the eggs produced by these chickens will increase. This increase in egg weight occurs because the genes that produce heavier chickens presumably have a pleiotropic effect on egg weight. In the case of negative genetic correlations, genes that cause an increase in one trait tend to produce a corresponding decrease in another trait. For example, when breeders select chickens that produce more giant eggs, the average egg size increases, but the number of eggs each chicken lays decreases.

Negative correlations between traits represent tradeoffs or genetic constraints that must be balanced under selection pressures. Negative genetic correlations often place practical constraints on the ability of plant and animal breeders to make progress from selection. For example, milk yield and butterfat content have a negative genetic correlation in cattle. The same genes that cause an increase in milk production bring about a decrease in the butterfat content of the milk. Thus, when breeders select for increased milk yield, the milk produced by the cows may go up, but the butterfat content decreases. Knowing the amount and type of genetic correlations before undertaking a breeding program is essential to ensure success.

Genetic correlations among traits strongly influence an organism's ability to adapt to a particular environment; therefore, genetic correlations interest evolutionary biologists. As another illustration, consider two traits in tadpoles: developmental rate and size at metamorphosis. Most tadpoles are found in small ponds and pools, where fish (potential predators) are absent and food is abundant. A significant liability in using this aquatic habitat is that ponds often dry up, frequently before the tadpoles have developed sufficiently to metamorphose into frogs and leave the water. One might expect, then, that natural selection would favor a maximum rate of development in tadpoles so that the tadpoles could quickly metamorphose into frogs. However, many species of tadpoles fail to develop at maximum rates, contrary to this prediction. One reason for a slower rate of development is a negative genetic correlation between developmental rate and body size at metamorphosis. Genes that accelerate development also tend to cause metamorphosis at a smaller size, at least in some populations. Thus, selection for fast metamorphosis also produces smaller frogs, and size is essential in determining the survival of young frogs. Tiny frogs lose water more rapidly in the terrestrial environment, are more likely to be eaten by predators, and have more difficulty finding sufficient food. The negative genetic correlation between developmental rate and body size at metamorphosis constrains the frogs' ability to develop rapidly and attain a large body size at metamorphosis. Knowing such genetic correlations is essential for understanding how animals adapt or fail to adapt to a particular environment.

12.35 Computing Heritability (h^2)

To estimate the heritability of a single trait, you must first compute the trait's covariance matrix, which we will call Ω, where location *i*, *j* in the matrix is filled with the covariance in X between subject *i* and subject *j*

ΩX=Cov[X,X]=[cov(X1,X1) cov(X2,X1):cov(X1,Xn) cov(X1,X2)cov(X2,X2) :cov(X2, Xn) ....,....... cov(X1,Xn) cov(X2,Xn): cov(Xn,Xn)]

Where covariance is defined as:

cov[Xi,Xj]=E[(Xi–E[Xi])(Xj–E[Xj])]

Where, at least in this case, the expectation can be defined as:

E[X]=μ(X)=1NΣi=1:NXi

Similarly, you can compute a similar *kinship matrix* (Φ) where location *i,j* represents i-j's relationship (*rij*) as the probability that any given gene is identical by decent (IBD).

Φ can be computed based on the pedigree, as Φ=12*R*

where R is the matrix of each pair of animals' relationship to each other, with the *r* for a parent and a child.5=(2–1), *r* for siblings.5=(2–2+2–2), shown in **Table 12.12**.

Table 12.12: The Relation Shive and the Degree of the Relationship

r	*Relationship*	*Degree of Relationship*
100 per cent	identical twins; clones	0
50 per cent	parent-offspring	1
50 per cent	full siblings	2
37.5 per cent	3/4 siblings or sibling cousins	2
25 per cent	grandparent-grandchild	2
25 per cent	half-siblings	2
25 per cent	aunt/uncle-nephew/niece	3
25 per cent	double first cousins	4
12.5 per cent	great grandparent-great grandchild	3
12.5 per cent	first cousins	4
12.5 per cent	quadruple second cousins	6
9.38 per cent	triple second cousins	6
6.25 per cent	half-first cousins	4
6.25 per cent	first cousins once removed	5
6.25 per cent	double second cousins	6
3.13 per cent	second cousins	6
0.78 per cent	third cousins	8
0.20 per cent	fourth cousins	10

Once you have these two matrices, you can estimate the putatively genetic and environmental variance of a quantitative phenotypic trait in the form:

$$\Omega \approx 2\ \Phi\ \sigma\ 2g + In\sigma\ 2e$$

12.36 Among Humans' Pedigree Data can be Used for QTL Analysis

The quantitative trait loci can be analyzed by using pedigree analysis. The example and usefulness of pedigree analysis are shown by analyzing ocular cutaneous albinism (OCA), which has little or no pigment in the skin and hair. The horizontal inheritance pattern is seen as shown in **Figure 12.18.** OCA is determined through the recessive allele of one gene, with albino family members being homozygotes for that allele. It is illustrated in **Figure 12.18,** where a family with two albino parents has produced three normal children.

This shows that albinism is an example of heterogeneity: Mutant alleles at any one of several different genes can cause this condition. The reported mating was, in effect, an inadvertent complementation test, which revealed that one parent was homozygous for an OCA–causing mutation in gene A, while the other parent was homozygous for an OCA-causing mutation in a different gene, B. But this is not always true. Sometimes a genotype is not expressed at all; even though the genotype is present, the expected phenotype does not appear. Other times, the trait caused by a genotype is expressed to varying degrees or in various ways in different individuals. Factors that alter the phenotypic expression of genotype include modifier genes, the environment, and chance.

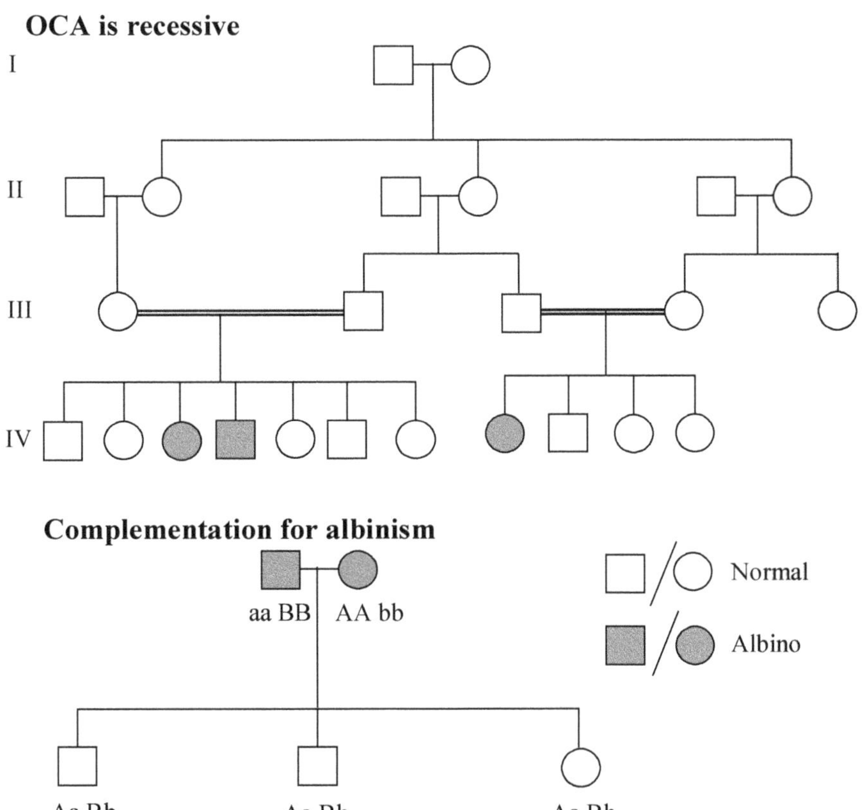

Figure 12.18: Inbred Families Suffering from OCA Indicate that the Trait is Recessive. A family in which two albino parents have non-albino children demonstrates that homozygosity for a recessive allele of either of two genes can cause OCA.

Phenotype often depends on the penetrance and expressivity of the genes. Both expressivity and penetrance can be variable; one of the examples is retinoblastoma, where either one eye would be affected, or both eyes may get affected. Other modifier genes are also responsible for both phenomena.

13

Genomics

Genomics is an important branch of biology in that studies are carried out on genetic material that provides a landscape of the genes related to humans or other species like animals, plants, microbes, *etc.* DNA is studied extensively to understand the organism's function; the role of genes and environmental interaction is studied further. The science of genomics was well studied by taking the example of *Mycobacterium leprae*, a simple organism that has provided an opportunity to understand complex genomes. The Mycobacterium leprea genome is 3.27 Mb and isolated from an Indian isolate of the *leprosy bacillus* constituting 1,605 functional genes and 50 genes for stable RNA species. The genetic diversity within Mycobacterium tuberculosis is quite high. The knowledge about its diversity has provided new insight into the macro-and micro-evolution of the tubercle bacillus. Understanding the differences between M. tuberculosis and related less pathogenic mycobacterium is expected to reveal critical bacterial virulence mechanisms and provide opportunities to understand host resistance to mycobacterium infection. This has provided an opportunity to understand the whole reductive evolution. M tuberculosis contains several pseudogenes; only 50 per cent of the genes are functional and implicated with the metabolic pathways. This genome is under enormous gene decay. The information gathered from sequencing illustrated the power of genomics. These studies are essential to provide information about the organization, function, and evolution of genetic information about the entire genome. The science of genomics has been classified into structural, functional, and comparative genomics. Structural genomics deals with the organization of sequences, functional genomics provides information on how these sequences behave or function, and comparative genomics compares the gene content, function, and organization of genomes of diverse organisms.

13.1 Brief History of Human Genome Project (HGP)

In February 2001 (February 15, 2001), the publically funded HGP consortium published its 1st draft of the human genome in the journal Nature. The human genome was funded commercially simultaneously, and results were published commercially. There are three billion base pairs in the human genome. In 1911 Alfred Sturtevant created the first *Drosophila* gene map 1911. The double-helical structure of the DNA molecule was illustrated in 1953 by Francis Crick and James Watson. The other two researchers shared the Nobel Prize (along with Maurice Wilkins) in the category of "physiology or medicine." In the mid-1970s, Frederick Sanger developed techniques to sequence DNA, for which he was awarded the Nobel Prize in chemistry in 1980. As DNA sequencing is challenging to process, people started thinking about automation as they wanted to sequence the entire human genome. This started in 1980. The United States Department of Energy (DOS) started working in 1986 and established an early genome project in 1987. The memorandum of understanding was signed in 1988 between two government agencies. In a 2001 article in the journal Genome Research,

Francis Collins wrote, "Building detailed genetic and physical maps, developing better, cheaper and faster technologies for handling DNA, and mapping and sequencing the more modest-sized genomes of model organisms were stepping stones towards large-scale sequencing of the human genome." In 1993, the NCHGR established a Division of Intramural Research (DIR), in which genome technology was developed and used to study specific diseases. By 1996, eight NIH institutes and centers had also collaborated to create the Center for Inherited Disease Research (CIDR), for the study of the genetics of complex diseases. In 1997, the NCHGR received full institutional status at NIH, labeled as the National Human Genome Research Institute. Collins remained as the director for the new institute. The list of various events is illustrated in **Table 13.1**.

The HGP aimed to sequence 3 billion DNA base pairs and to find out all the estimated 20,000 to 25,000 human genes, to develop technology for high-throughput sequencing, and to build up the capacity to collect sequences at a rate of 50 Mb per year by the end of 1998. Further, it was decided to develop a model organism to study the conservation across a diverse range of species. It was thought to analyze the data, and for this purpose, different strategies were used to sequence the whole genome.

13.2 Creation of Genetic Maps

The genetic map may be considered as the chromosomal position, which is situated linearly. The location and distance between the loci denote the per cent recombination (map units, centimorgans). This is called the linkage map. The distance between the two genes can be solved through recombination and genome sequencing. The linked genes are positioned on the same chromosome and close to one another. If crossing over occurs between two homologous chromosomes, then the recombination occurs; this enables us to find out the location of the loci. In case the recombination frequency between two loci is 50 per cent, it indicates that the loci are located at a far distance from each other. If it is less than 50 per cent, they are located nearby on the same chromosome.

The recombination rate is relative to the distance between two loci, which are situated nearby and can be clubbed together; these are called linkage groups. The distances are measured as centimorgan map units. The map unit is written as centimorgan (cM). 1cM is when a recombination fraction of 0.01 is seen; 1 in 100 gametes should be recombinant, and the rest of the 99 will be non-recombinant. Physically, 1cM is equivalent to 0.7 and 1Mb DNA sequences; however, there is no invariant correlation between physical and genetic distances. Recombination can be determined

Table 13.1: Human Genome Project Goals and Completion Dates

Area	*HGP Goal*	*Standard Achieved*	*Date Achieved*
Genetic map	2- to 5-cM resolution map (600 – 1,500 markers)	1-cM resolution map (3,000 markers)	September 1994
Physical map	30,000 STSs	52,000 STSs	October 1998
DNA sequence	95 per cent of gene-containing part of human sequence finished to 99.99 per cent accuracy	99 per cent of gene-containing part of human sequence finished to 99.99 per cent accuracy	April 2003
Capacity and cost of finished sequence	Sequence 500 Mb/year at < $0.25 per finished base	Sequence >1,400Mb/year at <$0.09 per finished base	November 2002
Human sequence variation	100,000 mapped human SNPs	3.7 million mapped human SNPs	February 2003
Gene identification	Full-length human cDNAs	15,000 full-length human cDNAs	March 2003
Model organisms	Complete genome sequences of *E. coli*, *S. cerevisiae*, *C. elegans*, *D. melanogaster*	Finished genome sequences of *E. coli*, *S. cerevisiae*, *C. elegans*, *D. melanogaster*, plus whole-genome drafts of several others, including *C. briggsae*, *D. pseudoobscura*, mouse and rat	April 2003
Functional analysis	Develop genomic-scale technologies	High-throughput oligonucleotide synthesis	1994
		DNA microarrays	1996
		Eukaryotic, whole-genome knockouts (yeast)	1999
		Scale-up of two-hybrid system for protein-protein interaction	2002

only when there is informative meiosis leading to informative markers; these markers are heterozygous (**Figure 13.1**).

13.3 Phase

In case crossing over is absent then, the gametes are of parental origin. In the case of crossing over is present the resultant products are recombinant gametes. The allelic composition of parental and recombinant gametes depends upon whether the original crossing over involves genes in a coupling or repulsion phase. In **Figure 13.2** the gamete for linked genes from coupling and repulsion phases are demonstrated. It is known that recombinant gametes are mostly found with a low frequency. The most frequent gametes are the original cross-over gametes coupling or repulsion phase. Coupling gametes cross and give rise to two dominant or recessive alleles. The repulsion phase crosses from gametes containing one dominant and one recessive allele.

It is tough to determine how close two genes are to a chromosome. However, by definition, one map unit (m.u.) equals one per cent recombinant phenotypes measured in centimorgans. A total of 2839 gametes in the coupling phase were studied; 305 (151 pr+ vg+ 154 pr vg+) were recombinant. Linkage distance can be calculated by dividing 10.7cM [(305/2839) *100)] number of recombinant gametes with the total number of gametes.

The same calculations can be done with the results from the repulsion phase cross. For this

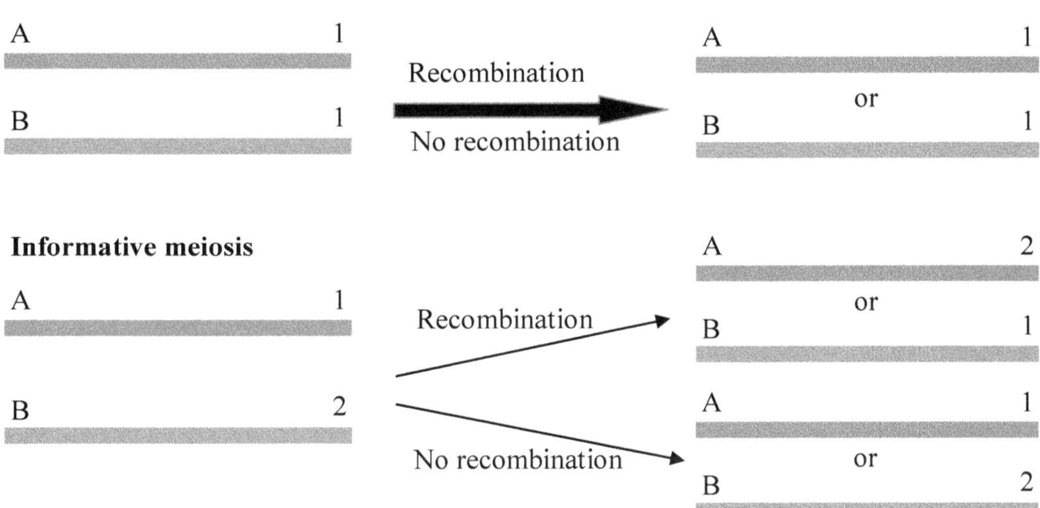

Figure 13.1: Informative and Non-informative Meioses.

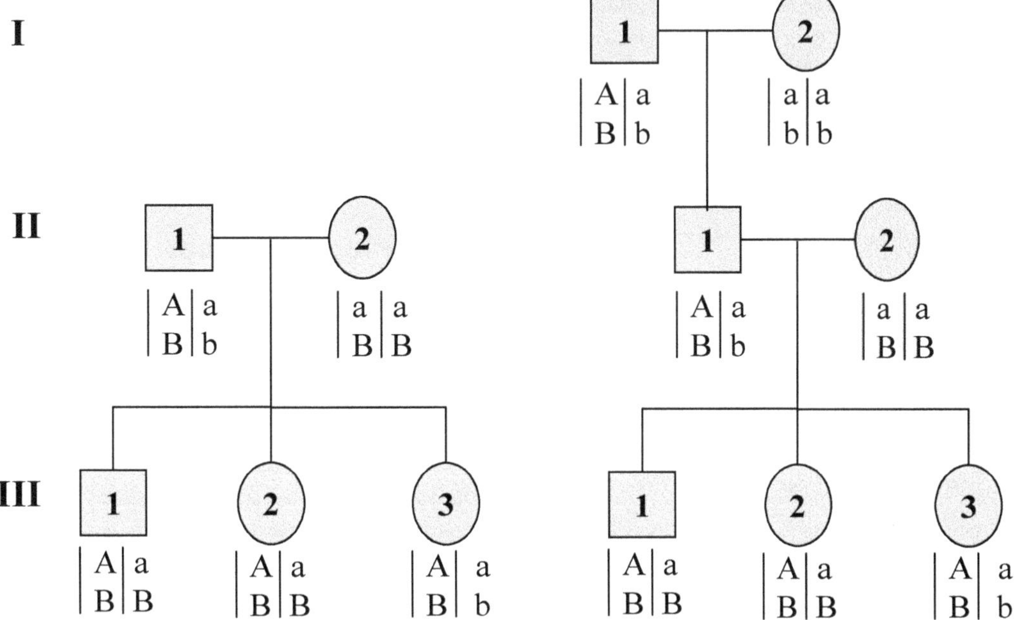

Figure 13.2: Determination of the Phase using a Three-generation Pedigree.

experiment, 2335 gametes were analyzed, and 303 (151 pr^+ vg^{++} 154 *pr vg*) were the result of recombination. The estimated linkage distance between *pr* and *vg* from these experiments is 13.0cM [(303/2335) *100] was calculated. To find out the correct value it is required to repeat the experiment several times. The final point we need to address is the maximum distance we can measure. Because of how the calculations are performed, we can never have more than 50 per cent recombinant gametes. Therefore, the maximum distance those two genes can be apart and still measure that distance is just less than 50cM. If two genes are more significant than 50cM apart, we cannot determine if they reside on the same or different chromosomes. If the experimental error is considered, as distances approach 50cM it is tough to determine if two genes are linked on the same chromosome. In such a situation, other mapping techniques must be used to determine the linkage relationship among distantly associated genes. If we have multi-generation families, it becomes easier to cross-analyze the three points.

13.4 Haplotype

When different loci are close enough and are not separated at the time of meiosis and are inherited as an enblock, they are called a haplotype.

13.5 CEPH Families

A vast project started in 1984 known as Centre d'Etude du Polymorphism Humain (CEPH); families living in Utah were selected randomly under this project. These families were tested for quantitative traits and linkage studies. Heritability, locus variance, *etc.*, were calculated using a simulation study. Statistical significance and the data generated were applied to accurate phenotypic data. Two hundred and sixty-four individuals were selected from 26 Utah CEPH pedigrees. All these individuals were tested for expiratory volume (FEV1) and forced vital capacity (FVC) segregation analysis was performed. No evidence of significant gene inheritance was observed. Heritability was 67 per cent, 46 per cent, and 52 per cent. This was calculated based on FEV1/FVC ratios. The linkage analysis revealed that chromosomes 2 and 5 were involved.

The components revealed no variance, as the LOD score was only 1.5. Together, the simulation study and the pulmonary function measurements analysis have demonstrated that the Utah CEPH pedigrees have sufficient information to detect the linkage of a marker to a quantitative trait locus depending on the underlying genetic model and that the analysis of normal variation can lead to gene localization.

The families used were three or four-generation families. They have been used for constructing genetic maps. A three-generation pedigree is important because it allows for finding out the phase of the markers among parents. These cell lines are available to different investigators and are used as reference pedigrees.

13.6 LOD Score Analysis

In organisms other than humans it is easier to find the big progenies; hence the correct genetic crosses can be obtained, which helps discover the recombination frequencies. This is not possible in humans. Hence, we can use a mathematical analysis called LOD score to solve this issue.

LOD score, or logarithm of odds, is a statistical test used in genetic linkage analysis. The LOD score compares the probability of obtaining the test data if the two loci are linked to the probability of obtaining the test data if the two are not linked. The log ratio of odds is calculated. LOD score analysis gives a numerical value that measures the ratio of odds that two loci are linked at a given recombination fraction q is compared with the chances that they are not linked (q = 0.5). The threshold for declaring linkage is a LOD score of 3 or more. It is improbable that a LOD score 3 will be obtained from one family. To get the correct information, many families must analyze the odds ratios.

Maximum likelihood score (MLS) is the LOD score for the most possible of a series of alternatives. For example, an MLS score can be planned for the most likely recombination fraction (θ) between two markers. This is possible by repeating the LOD score calculation presuming a different value of θ each time. The MLS obtained gives the most likely value of θ. MLS can also be obtained when other parameters of a genetic model are erratic for penetrance. In practice, computer analysis of the observed results calculates LOD scores and MLS values. We can use LIPED software for calculations.

13.7 Physical Maps

Physical mapping can be done by using various techniques like restriction mapping and sequence tag mapping. For mapping purposes, we need different markers. Mapping of the markers helps to identify the location of the genes and the distance between the two genes. Markers used for mapping are enzyme polymorphisms, disease genes, blood groups, and monogenic phenotypic traits. The first map was constructed in 1970.

13.8 Restriction Fragment Length Polymorphism

In restriction fragment length polymorphism, we try to find the variation in homologous DNA sequences. The difference between samples of homologous DNA molecules can be identified in two locations of restriction enzyme sites. In RFLP analysis, the DNA sample is digested with the help of restriction enzymes, and fragments are separated according to their lengths using gel electrophoresis. It was extensively used for genetic fingerprinting; it was an essential method in genome mapping, localization of genes for genetic disorders, and determination of risk for disease.

In the human genome, many sequences are exceedingly polymorphic and still do not have any phenotypic effects; such polymorphisms are called neutral polymorphisms. One such marker is restriction fragment length polymorphism (RFLP). The occurrence or lack of one of the nucleotides makes the DNA vulnerable to digestion with an enzyme. These are called restriction sites. They can be analyzed simply by running on gel or using southern blotting (**Figure 13.3**).

RFLP markers were used for the first time in 1987 to create complete maps and for genetic mapping in monogenic disorders, which were further used for prenatal diagnosis, *etc.*

13.9 Minisatellites

Minisatellites constitute 10-60 nucleotides. They are found in different locations in the human genome. They are helpful in the context of genome mapping. They show a high degree of heterozygosity. The heterozygosity of minisatellites can be used for identifying different genomes. Genetic markers are considered informative if they have a high heterozygosity rate. This value is known as the polymorphism information content or PIC.

The maximum PIC value of a biallelic marker is 0.375 it is less than 0.5 because some meioses are in the non-informative state even if both parents are heterozygous. Short tandem repeat (STRs) has much higher PIC values segregation of VNTR marker in a family.

13.10 Microsatellites

Tandem repeats create microsatellites. The units can be di-, tri-tetra- or pentanucleotides. The simplest repeat unit is found in birds, *i.e.*, ACn, where the two nucleotides A and C are repeated about 8 to 50 times in a bead-like fashion. These are found in the non-coding regions of the DNA; some genetic disorders are caused by trinucleotide microsatellite regions in coding regions. The repeat unit has flanking regions on each side that contain "unordered" DNA. Flanking regions are important as they can be used to develop

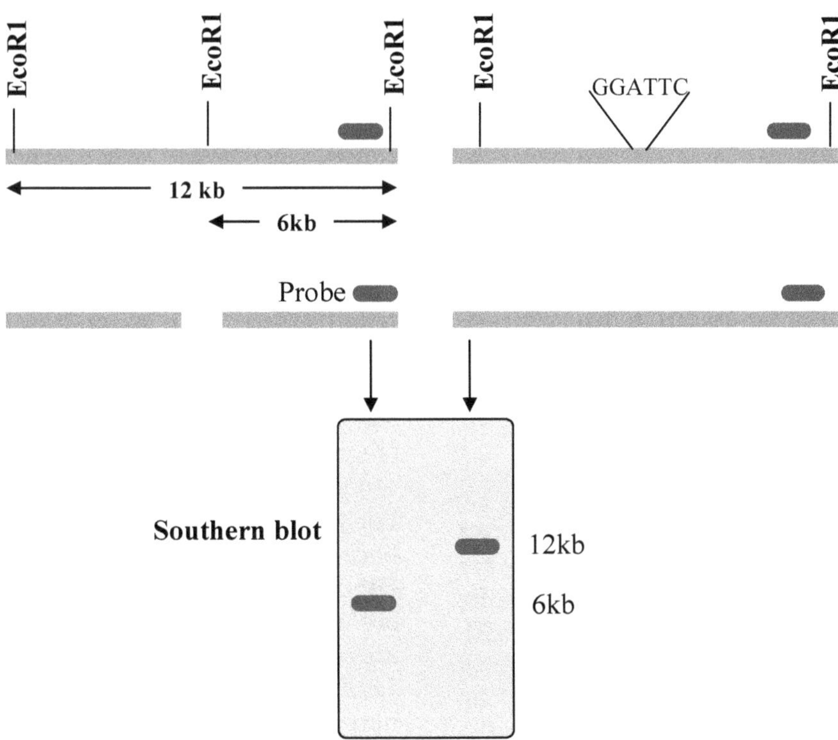

Figure 13.3: A RFLP Marker.

locus-specific primers to amplify the microsatellites with PCR. Mini satellites can be used for genetic mapping studies.

They are primarily located near the telomere. PCR is used to amplify these markers, but sometimes, the large size of the STRs becomes difficult. A more helpful type of marker is the STR's help in studies associated with genetic mapping.

13.11 A Comprehensive Genetic Map of the Human Genome

Creating a map started with the recognition and sequencing of CA repeat loci. The genotype of these loci is known in the members of large CEPH families. DNA banking was done for 1050 individuals in 52 world populations, for all these individuals, lymphoblastoid cell lines (LCLs) were made. The corresponding milligram quantities of DNA are banked at the Foundation Jean Dausset-CEPH in Paris. These LCLs were taken from various laboratories by the Human Genome Diversity Project (HGDP) and CEPH to provide unrestricted supplies of DNA and RNA for studies of sequence diversity and the history of modern human populations. Information for each LCL is limited to the sex of the individual, population, and geographic origin. 134 individuals in eight families were observed to map the autosomes, and 186 meiosis were analyzed. The mapping of the X chromosome was necessary; hence the examination of 20 families, 304 individuals, and 291 meioses were taken.

Once the genotype of the markers was finished, these were analyzed, and the genetic distance between the markers was calculated. This can be attained by using specially developed computer programs. 5264 markers were analyzed and were found to define 2335 positions. The order of 2 032 was estimated with odds of 1000:1 against alternative orders.

In females, the length of the map is 4 396cM but only 2 769cM in males. This variance is due to the high recombination rate among female meiosis compared to male meiosis. The sex-averaged length is 3 699cM, the length of the genetic maps for individual chromosome 1 is 292cM, the smallest, chromosome 21, and its length is 58cM.

The standard distance between markers is 1.6 cm, an essential point as it reveals how evenly placed these markers are. Some regions in the genome are isolated as few markers are found. In 1 per cent of the genome, the distance between markers is over 10cM. This consists of three positions where the interval between adjacent markers is 11cM. This could be due to a substantial physical distance that makes the markers separate. However, it could also be due to higher recombination rates in a particular region of the genome. The physical map showed that at least part of the distance is due to enhanced recombination. In contrast, there were some places where markers revealed no recombination with one another but were physically separated through several Mb of DNA.

DNA markers are known as the sequence-tagged site (STS) and were used for genetic mapping. An STS is composed of hundreds of base pairs or less in length and is generated by PCR, using primers based on DNA sequences and considered to allow the specific amplification of some or all of the sequence. The genomic site for the sequence in question is tagged by its ability to be assayed for a particular sequence. STRs are the best type of DNA markers for generating genetic maps of STSs.

13.12 How to Construct the Clone Contig Map?

Contig is a set of overlapping segments of DNA that can be compared with one another. It is also defined as the length of a contig to be the length of the consensus sequence derived from it.

Contig has high molecular-weight DNA, which is isolated and passed through a syringe with a fine needle to produce a complete set of random, partially overlapping genome fragments; this is how DNA can be sheared mechanically. The fragments produced have blunt ends. For cloning, they are inserted into a vector YAC, with a restriction enzyme such as *SmaI* that generates blunt ends. (DNAs with blunt ends can be ligated together in the standard reaction catalyzed by DNA ligase.) The YAC clone library produced may be of the whole genome if genomic DNA is used as the source or of a particular chromosome if chromosomes are first separated by flow cytometry before DNA is isolated.

YAC clones from the library are roughly situated on a chromosome map by using FISH. Still, the most excellent way to accumulate clone contigs over large regions of the genome is to use DNA fingerprinting techniques to type clones at random and then pull them together with the clones in order based on the overlaps determined between them. The STS (sequence-tagged site) is a DNA sequence marker unique to the genome. Several types of STSs are known. For genetic mapping, STSs were highly polymorphic because of the need to follow genetic recombination. For constructing contig maps, nonpolymorphic STSs are used because unique markers, not variable ones, are required at the chromosomal loci being mapped. This type of STS includes parts of genes and random non-gene sequences. **Figure 13.4** shows a YAC contig map assembled by STS mapping. First, the locations of

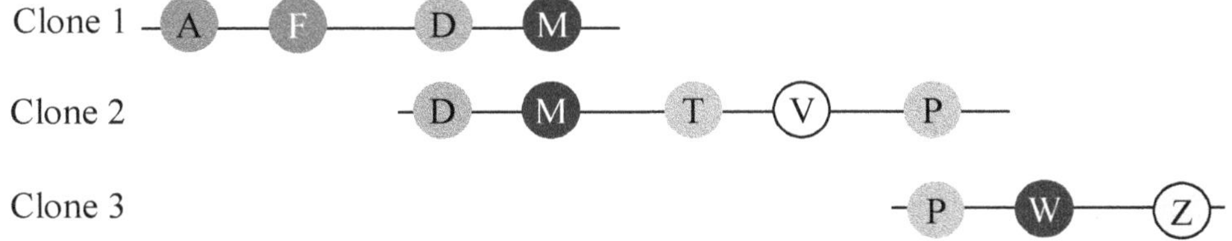

Figure 13.4: Construction of Clone Contig.

STSs on YAC clones are mapped using PCR-derived probes. Then, the STS maps on each YAC clone are compared to find the clones that share some STSs, indicating partial overlap of the cloned DNA.

We can identify the STSs present in each YAC clone through STS mapping. For instance, clone 1 contains STSs A, D, F, and M, clone 2 has STSs D, M, P, T, and V, and clone 3 has STSs P, W, and Z. We can conclude that clones 1 and 2 overlap in a DNA segment that contains STSs D and M, and clones 2 and 3 overlap in a segment that contains STS P. However, the correct order of these segments may not be known.

Radiation hybrid mapping of STS markers was used to overcome the problem of incorrect placement of DNA segments on the physical map. This led to the publication of a map with 15,806 STS markers in 1995, with one marker per 199 kb. Subsequently, the map was improved with 20,104 STS markers, most of which were ESTs (expressed sequence tags). EST markers are formed by PCR using oligonucleotide primers that depend on the sequence of a cDNA. As an EST corresponds to a functional protein-coding gene, it is a unique STS if the gene is unique. The enhanced map had the added benefit of locating many protein-coding genes on the physical map. The marker density on this new map averaged almost one per 100 kb. Combining the genetic and physical maps created a map with markers at the preferred density for sequencing.

BAC (bacterial artificial chromosome) cloning vectors have been used to construct the human genome's more precise, second-generation clone contig maps. A BAC vector contains a part of the E. coli F factor, a circular extrachromosomal element that facilitates mating between E. coli strains. BAC vectors can hold 300 kb or more of DNA; the copy number is one per E. coli cell. Recombinant BAC is produced using electroporation techniques.

Direct DNA sequencing is the most detailed physical map. Frederick Sanger developed it in 1975 and 1977. Sanger and his colleagues developed the dideoxy sequencing method based on the elongation of DNA. In contrast, the second method by Allan Maxam and Walter Gilbert was based on the chemical degradation of DNA. The Sanger method is the more widely accepted method for sequencing. DNA sequencing has revolutionized molecular biology. Human DNA is made up of nucleotides arranged in specific sequences. The human genome has over 3 billion of these genetic letters. DNA sequencing involves synthesizing DNA for a template strand and terminating it with a fluorescently labeled dideoxynucleoside. Four reactions, each with a ddNTp labeled with different dyes, are kept in the same tube. After sequencing, the products are separated using electrophoresis through an automated sequencer. Laser detects the colored DNA bands, and the equipment automatically converts the data into the computer sequence file. PCR is used for DNA synthesis, and the reaction is carried out in the presence of thermostable DNA polymerase. The single primer ensures that only one of the two strands is copied onto the newly synthesized DNA molecule. In DNA sequencing, the double-stranded DNA can be sequenced directly, as both the strands are separated at the time of the melting step of PCR. This allows the single primer to anneal; only a tiny amount of DNA is required per reaction. This is based on the Sanger sequencing method, which is routinely used. Next-generation sequencing is now widely used, and a brief description of various techniques is given below.

13.13 Next-Generation Sequencing

Next-generation sequencing is a high-throughput method of DNA sequencing that does not rely on the Sanger method. This approach allows for the sequencing of many DNA strands simultaneously, reducing the need for fragment-cloning methods often used in Sanger sequencing. The use of sequencing has brought down the cost of sequencing and has resulted in high-quality sequencing results that are particularly useful for whole-genome sequencing. There are various sequencing platforms available, including Roche's (454) Genome Sequencer 20/FLX Genome Analyzer, Illumina's Solexa 1G sequencer, Applied Biosystem's SOLiD system, and the Polonator G.007 (Dover system. ms/Harvard). Below is a brief description of each of these platforms.

13.14 Roche's (454) GSFLX Genome Analyzer

This is based on the principle of pyrosequencing. The DNA is fragmented in the first step. These are attached to the adapter sequences at the ends of DNA fragments and are ready to be attached to the beads, beads are the oligomers. Emulsion PCR is used to amplify the single unique DNA fragments which are lying in the droplets of oil. Each fragment is separated by the other fragment hence there are no chances of contamination there are six important steps involved (1) Attachment of adaptors with the fragments of DNA (2) Denaturation and preparation of emulsion PCR (3) Fragments get bound to the beads (4) the beads are in droplets of PCR reaction mix. Conditions are quite well controlled, usually allowing only one fragment per bead. PCR amplification takes place per bead within each droplet. (5) In the end, the amplicon per bead carries 10 million copies of a unique DNA template. Finally, the emulsion is broken DNA strand is denatured and beads holding DNA fragments are placed in a well of the plate. These plates are picotiter plates and will have a fiberoptic slide. (6) The pyrosequencing is carried out in the wells. After amplification with appropriate constituents, approximately more than 10,000,000 copies are created sequence reaction is carried out. During pyrosequencing, the nucleotides and reagents are delivered sequentially. The nucleotides used during pyrosequencing release are base on a bond with the growing DNA chains the reaction contains specific enzymes and luciferin. As soon as the nucleotide is added the light is emitted which is directly proportional to the number of nucleotides added. Read length is approximately 250bp.

13.15 Illumina's Solexa IG Sequencer

This is also called Soleza and was developed in 2006. The sequencing concept is sequencing by synthesis. It uses bridge PCR the DNA is fragmented each fragment should not be more than 800bp. The nebulizer is used for this purpose. The DNA is modified so that no unwanted unpaired bases are present at the two ends. Unique adapters are ligated to the fragments. Ligated fragments are of 150-200bp and are prepared via gel extraction and amplification with limited cycles of PCR. There exists a flow cell surface that is coated with oligonucleotides which are single-stranded and correspond to the sequences of the adapters. The next step involves extension. The free/distal end of a ligated fragment bridges to the complementary oligo on the surface of the flow cell when extension and denaturation are repeated several times amplification occurs at unique locations across the flow cell surface. This requires an automated flow cell processor.

Illumina's Genome Analyzer is also known as the "Solexa". It was commercially available in 2006. The basic concept behind it was "sequencing by synthesis." This platform uses bridge-PCR rather than emulsion –PCR. It uses high-density molecule arrays consisting of the DNA sample of interest which is randomly sheared to a size which is approximate ~800bp for this purpose compressed air device known as a nebulizer is used. The ends of the DNA are polished by removing any unpaired bases at the ends, and two unique adapters are ligated to the fragments. Ligated fragments of the size range of 150-200bp are isolated via gel extraction and amplified using limited cycles of PCR.

The flow cell surface is coated with single-stranded oligonucleotides corresponding to the adapters' sequences ligated during the sample preparation stage. Single-stranded, adapter-ligated fragments are bound to the surface of the flow cell and exposed to reagents for a polymerase-based extension. Priming occurs as a ligated fragment's free/distal end "bridges" to a complementary oligo on the surface.

Repeated denaturation and extension are carried out, which result in the localized amplification of single molecules in millions of unique locations across the flow cell surface. This takes place in the automated flow cell processor.

13.16 Applied Biosystem's SOLiD System

The applied SOLiD system was developed in 2007. It uses the fundamentals of emulsion PCR; once the amplicons are completed the emulsion gets broken and beads get involved covalently to the surface of the solid planar substrate this creates a disordered array. In each sequencing amplification 8mer fluorescently labeled octamers are added (**Figure 13.5**). When the 8mer oligo matches it can hybridize with the universal primers at the 3′ end. After oligo ligation, a florescent readout involves imaging in four channels and identifies the fixed base. Subsequently, a chemical cleavage removes the sixth through eight bases, leaving a free end for another cycle of ligation. The whole procedure is explained in the form of the flow chart.

The comparison of all the sequencing methods along with the cost involved is shown in Table **13.2.**

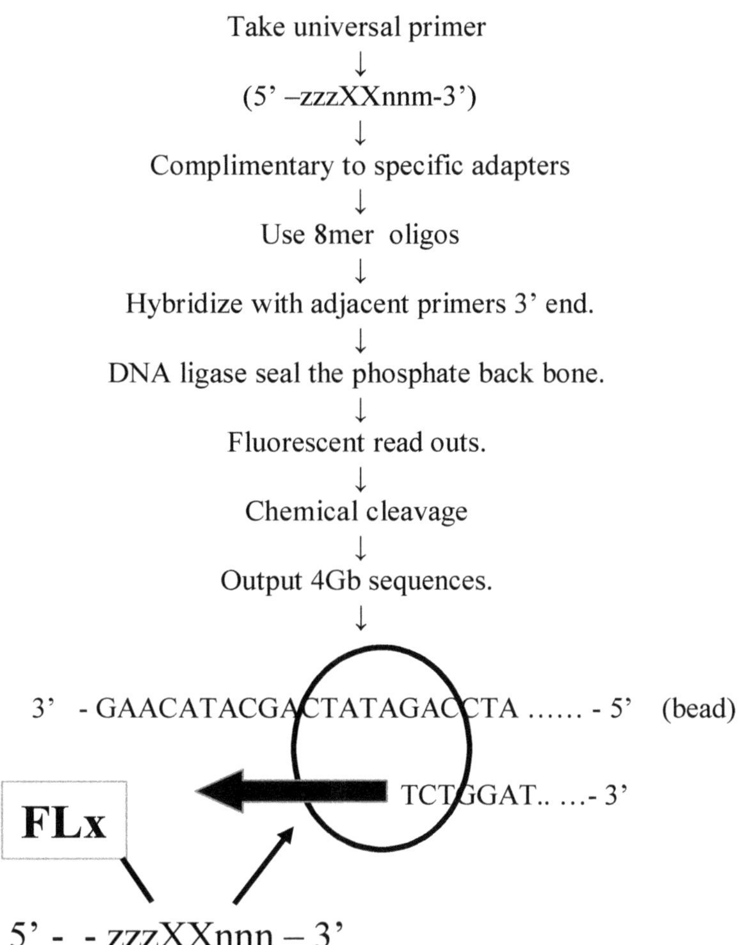

Figure 13.5: Steps Involved in Sequencing with the Abi SOLiD System.

Table 13.2: Comparison of Various Next Generation Sequencing Methods

(a)				
Sequencer	*454 GS FLX*	*HiSeq 2000*	*SOLiDv4*	*Sanger 3730xl*
Sequencing mechanism	Pyrosequencing	Sequencing by synthesis	Ligation and two-base coding	Dideoxy chain termination
Read length	700bp	50SE, 50PE, 101PE	50 + 35bp or50 + 50bp	4 0 0 ~ 9 0 0bp
Accuracy	99.9 per cent *	98 per cent, (100PE)	99.94 per cent *raw data	99.999 per cent
Reads	1M	3G	1200~1400M	–
Output data/run	0.7Gb	600Gb	120Gb	1.9~84Kb
Time/run	24 Hours	3~10 Days	7 Days for SE14 Days for PE	20Mins~3Hours
Advantage	Read length, fast	High throughput	Accuracy	High quality, long read length
Disadvantage	Error rate with polybase more than 6, high cost, low throughput	Short read assembly	Short read assembly	High cost low throughput

(b)				
Sequencers	*454 GS FLX*	*HiSeq 2000*	*SOLiDv4*	*3730xl*
Instrument price	Instrument $500,000, $7000 per run	Instrument $690,000, $6000/(30x) human genome	Instrument $495,000, $15,000/100Gb	Instrument $95,000, about $4 per 800bp reaction
CPU	2* Intel Xeon X5675	2* Intel Xeon X5560	8* processor 2.0GHz	Pentium IV 3.0GHz
Memory	48GB	48GB	16GB	1GB
Hard disk	1.1TB	3TB	10TB	280GB
Automation in library preparation	Yes	Yes	Yes	No
Other required device	REM e system	cBot system	EZ beads system	No
Cost/million bases	$10	$0.07	$0.13	$2400
(c)				
Sequencers	*454 GS FLX*	*HiSeq 2000*	*SOLiDv4*	*3730xl*
Resequencing		Yes	Yes	
De novo	Yes	Yes		Yes
Cancer	Yes	Yes	Yes	
Array	Yes	Yes	Yes	Yes
High GC sample	Yes	Yes	Yes	
Bacterial	Yes	Yes	Yes	
Large genome	Yes	Yes		
Mutation detection	Yes	Yes	Yes	Yes

13.17 Third-Generation Sequencing

Third-generation sequencing is not fully developed. It uses the concept of "single-molecule" sequencing, and it reads the individual DNA fragments do not need amplification hence fewer errors and no need for expensive reagents, such as fluorescent tags. It is faster and cheaper than next-generation sequencing. Nanopore technology is one of the methods used recently.

In this technique, the DNA is passed through a pore by interrupting the flow of ions through the aperture. The pores, made from a ring of seven a-hemolysin membrane proteins, are the same as those pushed into the membranes of other cells by the infectious bacterium *Staphylococcus aureus* to create damaging holes. The identity of each of the four bases traversing the hole might be revealed by distinctive changes in ion flow, which can be read as an electrical signal.

Companies involved are Helicos Bioscience, Complete Genomics, Pacific Biosciences, and Oxford Nanopore.

DNA markers are known as the sequence-tagged site (STS) and were used for genetic mapping. An STS is composed of hundreds of base pairs or less in length and is generated by PCR, using primers based on DNA sequences and considered to allow the specific amplification of some or all of the sequence. The genomic site for the sequence in question is *tagged* by its ability to be assayed for a particular sequence. STRs are the best type of DNA markers for generating genetic maps of STSs.

13.18 Map-based Sequencing

In this approach, short DNA fragments are sequenced and assembled in the light of the whole genome sequence; this requires the initial creation of detailed genetic and physical maps of the genome. Some of these methods are described in the previous pages. Once the DNA is digested, we get short fragments that can be cloned into Cosmid, yeast artificial chromosome (YAC), or bacterial artificial chromosome (BAC).

Overlapping segments of DNA are called contig, which can be compared with one another. It is also defined as the length of a contig to be the length of the consensus sequence derived from it.

Contig has high molecular-weight DNA, which is isolated and passed through a syringe with a fine needle to produce a complete set of random, partially overlapping fragments of a genome. This is the way how DNA can be sheared mechanically. The fragments produced have blunt ends. For cloning, they are inserted into a vector YAC, for example-cut with a restriction enzyme such as *SmaI* that generates blunt ends. (DNAs with blunt ends can be ligated together in the normal reaction catalyzed by DNA ligase.) The YAC clone library produced may be of the whole genome if genomic DNA is used as the source or of a particular chromosome if chromosomes are first separated by flow cytometry before DNA is isolated.

YAC clones from the library are roughly situated on a chromosome map by using FISH. Still, the most excellent way to accumulate clone contigs over large regions of the genome is to use DNA fingerprinting techniques to type clones at random and then pull them together with the clones in order based on the overlaps determined between them. The STS (sequence-tagged site) is a DNA sequence marker unique to the genome. Several types of STSs are known. For genetic mapping, STSs were highly polymorphic because of the need to follow genetic recombination. For constructing contig maps, nonpolymorphic STSs are used because unique markers, not variable ones, are required at the chromosomal loci being mapped. This type of STS includes parts of genes and random non-gene sequences.

Figure 13.6 shows a YAC contig map assembled by STS mapping. First, the locations of STSs on YAC clones are mapped using PCR-derived probes. Then, the STS maps on each YAC clone are compared to find the clones that share some STSs, indicating partial overlap of the cloned DNA.

For example, through STS mapping, we may find that YAC clone 1 has STSs A, D, F, and M; clone 2 has D, M, P, T, and V; and clone 3 has P, W, and Z. We can see that clones 1 and 2 share STSs D and M, and clones 2 and 3 shares STS P. Therefore, clones 1 and 2 must overlap with a DNA segment containing STSs D and M, and clones 2 and 3 must overlap with a DNA segment containing STS P. Direct order may not be known.

Sometimes YAC clones could result in the incorrect placement of the DNA segments on the physical map. To overcome this problem, radiation hybrid mapping of STS markers was used, resulting in the publication in 1995 of a map with 15,806 STS markers with one marker per 199kb. Later, the map was improved by 20,104 STS markers, most of the ESTs (expressed sequence tags). An EST is a marker formed by PCR using oligonucleotide primers that depend upon a cDNA's sequence. Because cDNA is a DNA copy of an mRNA, an EST marker corresponds to a functional protein-coding gene. If the gene is unique, then the EST-derived from it is a unique STS. The enhanced map had the additional benefit of locating many protein-coding genes on the physical map. The marker density on this new map averaged almost one per 100 kb. Combining the genetic and physical maps gives a map with markers at the targeted density preferred for sequencing.

More precise, second-generation clone contig maps of the human genome have been constructed using BAC (bacterial artificial chromosome) cloning vectors. A BAC vector has a part of the *E. coli* F factor, a circular extrachromosomal element that facilitates mating between *E. coli* strains. BAC vectors have a capacity of 300kb of DNA or more, and the copy number is one per *E. coli* cell. Electroporation techniques make recombinant BAC.

13.19 Assembling and Finishing Genome Sequences

The raw sequences should be *assembled* into sequence contigs; in this case, the bases must be placed together in the correct order as they should be. The sequences can be joined together and modified by this method in 1988 by Michael Waterman. They examined the correlation between the oversampling of the genome (also called coverage) and the number of contiguous pieces of DNA (commonly called contigs) that an idealized assembly program can reconstruct. The assembly program should produce one contig for every chromosome that is sequenced.

Single contig assembly has some limitations. The reads originating from different copies of a repeat appear identical to the assembler, causing assembly errors.

13.20 Scaffolding

Contigs and gaps are present in a portion of the genome called a scaffold. Gaps are present where reads from the two sequenced ends, at least one fragment overlap with other reads in two different contigs. As most of the time, the length of the fragments is approximately known, the number of bases between contigs can be known.

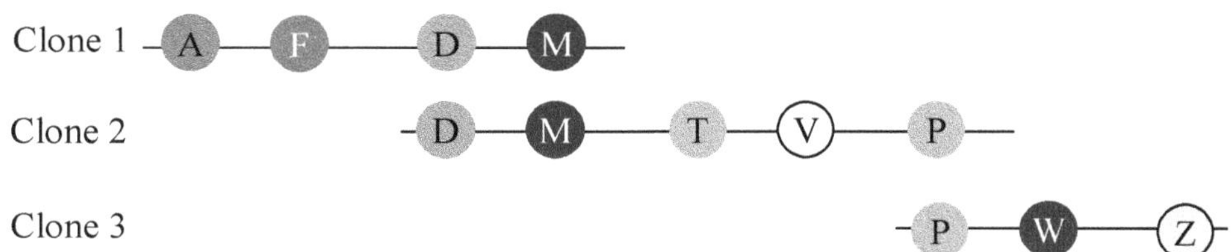

Figure 13.6: Construction of Clone Contig.

A single chromosome may have many scaffolds (*e.g., Chlamydomonas reinhardtii*) or a single scaffold (*e.g.,* Human chromosome 19). This depends upon how the genome is reconstructed or assembled from the available reads. The locations of scaffolds in the genome are unknown. Scaffolds are usually numbered approximately from the largest to the smallest. While assembling, some of the scaffolds may be filtered out, causing skipped scaffold numbers. In some cases, scaffolds may overlap. For example, in the polymorphic genome, regions with a high density of allelic differences between haplotypes may split into separate sets of scaffolds, each representing one allele.

The total sequencing should be known for an original set of chromosomes. However, it is known that sequencing is possible for a small and single piece of DNA for each chromosome. The gaps are filled in through a direct sequencing process known as the finishing or gap closure. Extensive validation is done to get the best results. Finishing helps solve errors in the assembly process; however, it is costly and needs sizeable human intervention and special lab techniques.

13.21 Assembly Algorithms

There are large numbers of assembly programs. Some of these are given below.

13.22 Greedy Assemblers

Under this algorithm, the assembler greedily joins together the reads that are most similar to each other. An example is shown below, where the assembler joins, in order, reads 1 and 2 (overlap = 200bp), then reads 3 and 4 (overlap = 150bp), and then reads 2 and 3 (overlap = 50bp); this creates a single contig from the four reads provided in the input. This is a simple process, but the assembler may be confused due to the presence of complex repeats. This may result in misinterpretation of the assemblies.

13.23 Overlap-Layout-Consensus

The reads are given by the assembler, which can be represented in the form of a graph; nodes represent each of the reads an edge connects only when two nodes corresponding reads overlap.

Then the path is identified through the graph containing all the nodes; known as the Hamiltonian path. This overlap layout consensus can solve many problems. It starts with the overlapping stage, during which all overlaps between the reads are computed, and the graph structure is computed. In the layout stage, the graph is simplified by removing redundant information. In the final consensus stage, the assembler builds an alignment of all genomes reads. It infers that, as a consensus of the aligned reads, the original sequence of the genome is assembled.

13.24 Eulerian Path

This path is used at an early stage. This technique is called sequencing by hybridization. In this technique, scientists find out all strings of length k (k-mers) contained in the original genome instead of generating a set of reads. This approach is based on a graph-theoretic model and breaks up each read into a collection of overlapping k-mers.

13.25 Align-Layout-Consensus

It consists of two types of consensus assembly processes. These are overlap –layout –consensus assembly and De Bruijn graph assembly. Both these processes handle unsolvable repeats by essentially leaving them out. Unsolvable repeats break the assembly into fragments. Fragments are contigs; thus, the overlapping assembly stage (often one of the most computationally intensive assembly tasks) is replaced by an alignment step. The layout stage is also greatly simplified due to the additional constraints provided by the alignment to the reference.

13.26 BAC-by-BAC (Hierarchical) Sequencing

In BAC–by–BAC (hierarchical) sequencing, the genome is broken into large fragments (between 40 and 200kbp) known as Bacterial Artificial Chromosomes or BACs. After that, it is mapped using specialized laboratory experiments. The least tiling path of BACs is selected so that at least one BAC can cover each base in the genome, and the overlap between BACs is reduced. Each BAC is then sequenced through the shotgun method; the resulting assemblies are combined into an assembly for each chromosome using the information provided through tiling paths.

13.27 Representing Assemblies

A description of how assemblies are represented in some of the most common assembly formats is presented here.

13.28 Notable Assembly Programs

TIGR assembler-assembly program was developed at the Institute for Genomic Research (TIGR). This assembler generated the first sequence of the free-living organism Haemophilus influenza.

Phrap - Assembly program was developed at the University of Washington. It is one of the most highly used assembly programs for many bacterial and eukaryotic genomes. This was the most crucial assembly program used later on for human genome assembly.

Celera assembler - It is scientific software developed by Celera Genomics. It uses the shotgun method to assemble a whole eukaryotic genome by successfully assembling the genome of the fruit fly *Drosophila melanogaster*. Before this assembler, the assembly of large genomes was carried out by using BAC-by-BAC sequencing. It is currently used in numerous bacterial and eukaryotic projects.

Arachne - This is developed by MIT and widely used in genome projects at the Broad Institute of MIT and other research organizations. Arachne and Celera Assembler are arguably the best assemblers available to the scientific community to assemble large eukaryotic genomes.

Phusion - It is a sophisticated algorithm to reduce the complexity of the data provided to phrap and to correct mistakes in the resulting assemblies.

Atlas - It was developed at the Baylor College of Medicine. Atlas is specifically designed to optimize the assembly of BAC-by-BAC projects and uses a hybrid approach combining some of the advantages of BAC-by-BAC sequencing and whole-genome shotgun. Like Phusion, Atlas uses phrap as a low-level assembly tool.

Some organisms have been sequenced fully; the details are below.

13.29 Bacterial Genome

The first organism sequenced was the *eubacterium H. influenza*. This organism was taken because its genome size is different than the other bacteria; more importantly, the GC content of the genome is almost identical to humans. The whole-genome shotgun approach sequenced the genome. No genetic or physical map available till the beginning of the project. This was taken up by the Institute for Genomic Research and completed in 1995. *H. influenzae* causes ear and respiratory tract infections in humans. Its size is 1.83 Mb (1,830, 137bp).

Once the sequencing is completed, the next step is *annotation*: identifying and describing putative genes and other important sequences. There are usually no introns in protein-coding genes of bacteria and archaeons, the way to find out the putative protein-coding genes is by using computer to search both DNA strands for open reading frames (ORFs, or potential amino-acid coding regions, *i.e.*, potential protein-coding genes).

The ORF begins with a start codon (AUG; ATG in the DNA) and ends with a stop codon (TAG, TAA, or TGA in the DNA sequence), a multiple of three nucleotides downstream. However, not all ORFs (particularly the short ones) correspond to genes. Thus, somewhat arbitrarily, ORFs larger than 100 codons are considered statistically likely to be protein-coding genes. The identified ORFs for the entire genome are searched against nucleotide and amino-acid databases to attempt to identify genes specifically. With the current state of computer searching algorithms and the amount of defined information in sequence databases, a complete microbial genome sequence can be annotated for all coding regions and other elements, such as repeated sequences, operons, and transposable elements.

The *H. influenzae* genome analysis predicted 1,743 protein-coding genes making up 85 per cent of the genome. Of these predicted genes, 736 did not match any protein in the databases or matched only proteins designated hypothetical. The remaining 1,007 predicted ORFs matched genes with known database functions. This sort of result is typical of genome projects. Many genes have known functions, while a significant fraction has unknown functions, requiring much hypothesis-driven science to determine the functions.

Escherichia coli is an important organism. It is found in the lower intestines of animals, including humans, and survives well when introduced into the environment. Pathogenic *E. coli* strains cause deadly enteric and other infections. In the laboratory, *E. coli* has been a fundamental model system. Hence complete genome sequencing was required. Its sequencing was the 6th in the series.

The circular genome was sequenced using the whole-genome shotgun approach. The genome of *E. coli* is 4.64 Mb (4,639,221 bp). The 4,288 ORFs make up 87.8 per cent of the genome. Thirty-eight per cent of the ORFs had unknown functions. Many bacteria have been sequenced or under the process of sequencing, as shown in **Table 13.3**.

13.30 Archaeon Genome

The *Methanococcus jannaschii* genome was the first genome of an archaeon to be completely sequenced. This organism is neither a bacterium nor a eukaryote. Its genome sequence, therefore, affirmed the existence of a third major branch of life on Earth. Many organisms thrive some time unimaginable, such as in the water heated to over boiling temperatures in the hydrothermal vents and saturated with hydrogen sulfide, or in extreme salinity. Archaea are, in fact, the only detectable life forms which stay in most thermophilic conditions. Archaea coexist with bacteria and eukaryotes, and their ecological importance is being increasingly recognized. They

Table 13.3: Large Number of Bacterial Sequenced

Organism	Size (Mb)	G+C	Funding
Alivibrio salmonicida	3.33 + 1.21	39.0 per cent	Wellcome Trust
Bacteriovorax marinus	3.44	36.7 per cent	Wellcome Trust
Bacteroides fragilis NCTC9343	5.20	43.1 per cent	Beowulf Genomics
Bacteroides fragilis 638R	5.373	43.4 per cent	Beowulf Genomics
Bordetella avium	3.73	61.6 per cent	USDA
Bordetella bronchiseptica RB50	5.34	68.1 per cent	Beowulf Genomics
Bordetella bronchiseptica strain 253	~5	~68 per cent	Wellcome Trust
Bordetella bronchiseptica MO149	~5	~68 per cent	Wellcome Trust
Bordetella parapertussis 12822	4.77	68.1 per cent	Beowulf Genomics
Bordetella parapertussis Bpp5	~5	~68 per cent	Wellcome Trust
Bordetella pertussis Tohama I	4.09	67.7 per cent	Beowulf Genomics
Bordetella pertussis strain 18323	~5	~68 per cent	Wellcome Trust
Burkholderia cenocepacia (formerly *Burkholderia cepacia* genomovar III)	3.87 + 3.22 + 0.88	66.9 per cent	Beowulf Genomics
Burkholderia pseudomallei	4.07 + 3.17	68 per cent	Beowulf Genomics
Campylobacter jejuni	1.64	30.6 per cent	Beowulf Genomics
Chlamydia trachomatis serovar A	1.052	41.3 per cent	Wellcome Trust
Chlamydia trachomatis serovar B	1.052	41.3 per cent	Wellcome Trust
Chlamydia trachomatis Jali	1.051	41.3 per cent	Wellcome Trust
Chlamydia trachomatis L2	1.039	41.3 per cent	Wellcome Trust
Chlamydophila abortus	1.14	39.9 per cent	The Scottish Office
Citrobacter rodentium	5.357	54.7 per cent	Wellcome Trust
Clavibacter michiganensis	3.404	72.38 per cent	USDA
Clostridium botulinum Hall strain A	3.89	28.2 per cent	Beowulf Genomics
Clostridium botulinum 17B	~3.9	~30 per cent	Wellcome Trust
Clostridium difficile 630	4.29	29.06 per cent	Beowulf Genomics
Clostridium difficile 196	~4.3	~30 per cent	Wellcome Trust
Clostridium difficile BI1	~4.3	~30 per cent	Wellcome Trust
Clostridium difficile M120	~4.3	~30 per cent	Wellcome Trust
Clostridium difficile M68	~4.3	~30 per cent	Wellcome Trust
Clostridium difficile SM	~4.3	~30 per cent	Wellcome Trust
Clostridium difficile CF5	~4.3	~30 per cent	Wellcome Trust
Clostridium difficile 855	~4.3	~30 per cent	Wellcome Trust
Corynebacterium diphtheriae	2.49	52 per cent	Beowulf Genomics
Ehrlichia (*Cowdria*) *ruminantium*	1.516	27.5 per cent	NWO
Erwinia amylovora	3.81	53.5 per cent	USDA
Erwinia carotovora	5.06	50.97 per cent	The Scottish Executive
Escherichia coli 042	5.355	50.5 per cent	Beowulf Genomics
Escherichia coli E2348/69	4.966	50.6 per cent	Beowulf Genomics
Escherichia coli H10407	5.15	50.7 per cent	BBSRC
Escherichia coli non-K1 clinical isolate	~5.0	~50 per cent	Beowulf Genomics
Haemophilus influenzae 10810	1.98	38.1 per cent	Wellcome Trust
Haemophilus influenzae F3031	1.99	38.2 per cent	Wellcome Trust
Haemophilus influenzae F3047	~2.1	~40 per cent	Wellcome Trust
Haemophilus parainfluenzae	~2.1	39.6 per cent	Wellcome Trust
Helicobacter mustelae	1.58	42.5	Wellcome Trust
Mycobacterium africanum	4.39	65.6 per cent	Wellcome Trust

Organism	Size (Mb)	G+C	Funding
Mycobacterium bovis AF2122/97	4.4	64 per cent	DEFRA/Beowulf Genomics
Mycobacterium bovis BCG	4.4	65.6 per cent	DEFRA/Beowulf Genomics
Mycobacterium canetti	~4.4	~64 per cent	Wellcome Trust
Mycobacterium leprae	3.27	57.8 per cent	Heiser/Follereau
Mycobacterium marinum	6.637	65.7 per cent	Beowulf Genomics
Mycobacterium microti	~4.4	~64 per cent	DEFRA/Beowulf Genomics
Mycobacterium tuberculosis	4.41	65.6 per cent	Wellcome Trust
Mycoplasma amphoriforme	~1.0	ND	Wellcome Trust
Neisseria lactamica	2.2	52.3 per cent	Wellcome Trust
Neisseria meningitidis Serogroup A strain Z2491	2.18	51.8 per cent	Wellcome Trust
*Neisseria meningitidis*Serogroup C strain FAM18	2.19	51.6 per cent	Beowulf Genomics
Photorhabdus asymbiotica	5.07	42.2 per cent	BBSRC
Proteus mirabilis	4.06	38.88 per cent	Wellcome Trust
Pseudomonas aeruginosa	6.602	66.3 per cent	Wellcome Trust
Pseudomonas fluorescens	6.722	60.5 per cent	BBSRC
Rhizobium leguminosarum	7.751	60.86 per cent	BBSRC
Rhodococcus equi	~4.5	~68 per cent	HBLB
Salmonella bongori	4.46	51.3 per cent	Beowulf Genomics
Salmonella enteritidis PT4	4.686	52.17 per cent	Beowulf Genomics
Salmonella gallinarum 287/91	4.747	52.2 per cent	Beowulf Genomics
Salmonella enterica Hadar	4.786	52.3 per cent	Wellcome Trust
Salmonella enterica Infantis	4.711	~52.3 per cent	Wellcome Trust
Salmonella paratyphi A	4.582	52.2 per cent	Wellcome Trust
Salmonella typhimurium DT104	5.02	52.1 per cent	Beowulf Genomics
Salmonella typhimurium D23580	~5.0	~52 per cent	Wellcome Trust
Salmonella typhimurium DT2	~5	~52 per cent	Wellcome Trust
Salmonella typhimurium SL1344	5.067	52.2 per cent	Beowulf Genomics
Salmonella typhi	4.81	52.1 per cent	Beowulf Genomics
Serratia marcescens	5.11	59.5 per cent	Wellcome Trust/CNRS
Shigella dysenteriae	~4.7	~50 per cent	Beowulf Genomics
Shigella sonnei	4.989	50.7 per cent	Beowulf Genomics
Staphylococcus aureus			
Stenotrophomonas maltophilia	4.851	66.3 per cent	Wellcome Trust
Streptococcus equi	2.25	41.3 per cent	Home of Rest for Horses
Streptococcus pneumoniae			
Streptococcus pyogenes	1.84	38.6 per cent	Beowulf Genomics
Streptococcus suis P1/7	2.01	41.3 per cent	Beowulf Genomics
Streptococcus suis BM407	~2.0	~41 per cent	Wellcome Trust
Streptococcus suis SC84	~2.0	~41 per cent	Wellcome Trust
Streptococcus uberis	1.852	36.6 per cent	Beowulf Genomics
Streptococcus zooepidemicus	~2.3	~40 per cent	HBLB
Streptomyces coelicolor	8.67	72.1 per cent	BBSRC/Beowulf Genomics
Streptomyces scabies	10.148	71.45 per cent	USDA
Tropheryma whipplei	0.93	46.3 per cent	Beowulf Genomics
Wolbachia pipientis endosymbiont of *Culex quinquefasciatus*	~1.6	~34 per cent	Beowulf Genomics
Wolbachia endosymbiont of *Onchocerca volvulus*	~1.1	ND	Beowulf Genomics
Yersinia enterocolitica strain 8081Biotypes	4.62	42.3 per cent	Wellcome Trust
Yersinia pestis	4.65	47.6 per cent	Beowulf Genomics

consist of glycerolipids membranes which are different from that of the bacterial and eukaryotic cells and archaea do not contain peptidoglycan or murein, the predominant component of bacterial cell walls. It is a strict anaerobe, and it derives its energy from the reduction of carbon dioxide to methane. Sequencing has been carried out using the whole-genome shotgun approach. The sequence was reported in 1996. The complete genome has a large, main circular chromosome of 1,664,976bp; a circular, extrachromosomal element (ECE) of 58,407bp; and a smaller, circular ECE 16550bp. The main chromosome has 1,682 ORFs, the larger ECE has 44, and the smaller ECE has 12. Most of the genes are involved in energy production, cell division, and metabolism are like their counterparts in the Bacteria, whereas most of the genes involved in DNA replication, transcription, and translation are like their counterparts in the Eukarya.

13.31 The Flowering Plant Arabidopsis thaliana Genome

Arabidopsis is an important model organism for studying the genetic and molecular aspects of plant development. The 120Mb genome contains about 25,900 genes almost twice that found in the fruit fly *Drosophila melanogaster*. Interestingly, about a hundred Arabidopsis genes are like disease-causing genes in humans, including the genes for breast cancer and cystic fibrosis. The next step is to fill in the gaps in the sequences and explore the structure and function of the genome in detail. Toward that end, an initiative called the Arabidopsis 2010 Project has been set up. It has an ambitious set of goals, including defining the function of every gene in the plant, determining where and when every gene is expressed, showing where the encoded protein ends up in the plant, and defining any proteins with which the protein interacts.

13.32 Eukaryotic Genomes

13.32.1 Yeast *Saccharomyces cerevisiae*

Saccharomyces cerevisiae has been used as a model organism in many ways. It resembles mammals. It is commonly known as the baker's yeast or Brewer's Yeast. It was fully sequenced in 1996. It has 16 chromosomes with 12,067,280bp. Nearly 969,000bp of repeated sequences were estimated not to be included in the published sequence, which revealed 6,183 ORFs, only 233 of which had introns. At the outset of the yeast genome project, only about 1,000 genes have been defined by genetic analysis. About a third of the protein-coding genes have no known function.

13.32.2 The Nematode Worm *Caenorhabditis elegans*

The worm *C. elegans* was the first multicellular eukaryotic genome to be sequenced. Nematodes are smooth, non-segmented worms with long, cylindrical bodies. *C. elegans* is about 1 mm long; it lives in the soil and feeds on microbes. It has a self-fertilizing XX hermaphrodite and an XO male sex. The former has 959 somatic cells, the latter 1,031. The lineage of each adult cell through development is well understood. It has a simple nervous system, exhibits several behaviors, and is capable of simple learning tasks. *C. elegans* has become an important model organism for studying the genetic and molecular aspects of embryogenesis, morphogenesis, development, nerve development and function, aging, and behavior.

The *C. elegans* genome project was initiated by Sydney Brenner and carried out by an international consortium. A nearly complete physical map was constructed in the 1980s; in the 1990s, sequencing began, and a draft was reported in December 1998. The genome size was determined, and it was 100.3Mb, with 20,443 genes, 1,270 of which do not code for protein.

13.32.3 The Fruit Fly *Drosophila melanogaster*

The fruit fly's genome was sequenced with the whole genome shotgun approach, and also supported by a BAC derived physical map.

The sequence of the euchromatic part of the *Drosophila* genome is 118.4Mb in size. Approximately another 60Mb of the genome consists of highly repetitive DNA that essentially cannot be cloned, making the associated sequences not obtainable. There are 14,015 genes, less than that found in the worm, but with a similar diversity of functions. Surprisingly, the number of fruit fly genes is just over twice that are found in yeast, yet the fruit fly seems to be a much more complex organism. We must conclude that higher complexity in animals such as flies and humans does not require a correspondingly larger repertoire of gene products. The fruit fly's value as a model system for studying human biology and disease was affirmed by the finding that *D. melanogaster* has homologs for 177 of 289 genes known to be involved in human disease, including cancer.

13.32.4 Homo-sapiens

It was announced in a joint press conference in June 2000 involving Francis Collins of the National Human Genome Research Institute (NHGRI, representing the HGP) and J. Craig Venter of Celera Genomics that the genome of interest is of Homo sapiens whose DNA was sequenced. The HGP researchers collected female blood and male sperm from many donors but used only some of the samples to extract DNA for sequencing. Hence, neither the scientists nor the donors know whose DNA is being sequenced. They recruited up to 30 donors via self-referral, newspaper advertisement, and outreach activities and chose five males and five females from a variety of ethnic backgrounds from whom they obtained DNA for sequencing. (Latter, Craig Venter admitted that he was one of the donors). In both cases, the human genome sequence being generated is an amalgamation of sequences and will not be an exact match for the genome of any one person in the human population.

The sequencing part of the HGP was carried out by an international human genome sequencing project consortium of scientists at 16 institutions in the United States, Great Britain, France, Germany, and China. This consortium set out to determine the sequence of the euchromatic part of the human genome. Their sequencing approach was built on a foundation of genetic and physical maps as described earlier in the chapter. At the time of the press conference, the consortium centers had generated more than 22.1 billion nucleotides of raw sequence data, representing seven-fold sequence coverage of the genome. Approximately 50 per cent of this genome sequence was in a nearly finished form, and 24 per cent was finished. The achievement of the consortium's final goal of a finished sequence of the euchromatic part of the genome was announced in April 2003.

Celera Genomics used the whole–genome shotgun approach for sequencing and assembled its sequence data with the help of human genome physical map data in public databases that were generated by the Human Genome Sequencing Project Consortium. The Consortium had assembled only part of the genome, and Celera claimed that it alone had achieved the first complete assembly of the human genome sequence. Celera used two independent methods to assemble the sequence, which consisted of 3.12 billion bp (3.12 Gb, where Gb= 1 gigabase = 1 billion base pairs; current figures indicate that the genome is 2.9 Gb). One method used 26.4 million sequences of 550 bp, for a total of 14.5 billion bp, with 4.6 –fold sequence coverage. More than 99 per cent of the genome was covered by this assembled sequence. According to Celera, the calculation to perform the sequence assembly involved 500 million trillion base–to–base comparisons taking more than 20,000 CPD hours on a supercomputer. The second assembly method was used to validate the results from the whole genome direct shotgun sequence assembly and involved relating the Celera sequence data to BAC clone sequence data in the GenBank database. Recall that BAC clones contain large inserts, so this method helps to resolve any ambiguities arising from assembling the sequences of the short fragments involved in the direct shotgun sequencing approach. Unquestionably, Celera's accomplishment is a highly significant event in science. With an assembled sequence in hand, the next step is annotation: determining the genes it contains and analyzing other features of the genome.

The draft genome sequences and initial interpretations of assembled sequences were published by the Human Genome Project Sequencing Consortium in the February 15, 2001, issue of Nature, and by Celera Genomics on February 16, 2001, issue of Science. In the next two years, the human genome sequence was finished and as mentioned at the beginning of this chapter, was announced to the public on April 14, 2003. Analysis of the sequence continues with the goal of determining the number of genes and what each gene encodes. So, how many genes does a human have? As of October 2004, the best estimate is 20,000–25,000 protein-coding genes, far fewer than the 50,000 to 100,000 often predicted before sequencing began. This low number is making scientists drastically change their thinking about the complexity and development of organisms. Interestingly, the human genome shares 223 genes with bacteria, but those genes are not found in yeast, the worm, or the fruit fly. All in all, the two human genome sequences are proving a great resource for scientists to learn about our species, and data mining searching through genome sequences for information will continue for many years. There will be a strong focus on human disease genes, with an eye toward treatment and therapy.

Three mammalian genomes have been sequenced completely, these are humans, the mouse (*Mus musculus*), and most recently, the rat (*Rattus norvegicus*). The human genome is the largest, followed by that of the rat, and then that of the mouse. All three mammals have approximately the same number of genes. Importantly, both the mouse and the rat have been model organisms for studies of mammalian physiology, including those dealing with diseases. The mouse has been a model for mammalian genetics due to its genetic tractability, including its susceptibility to gene knockout mutations. Sequence analysis reveals that approximately 99 per cent of the genes of the mouse and the rat have direct counterparts in the human, including, therefore, genes associated with the disease. Studies of the mouse and rat genomes will undoubtedly provide valuable knowledge about human disease and other areas of human biology, as well as about mammalian evolution.

13.33 Comparative Genomics

Comparative genomics involves comparing entire genomes of different species, to enhance our understanding of the functions and evolutionary relationships of each genome. Comparative genomics can follow two routes. (1) Homology search determines if a new sequence is like any known gene, in other words, looking for evolutionarily related genes by using BLAST or PSI BLAST (2) comparative genomics to search a homology to related species. Since direct experimentation with humans is unethical, comparative genomics provides a valuable way to determine the functions of human genes by studying homologous genes in nonhuman organisms. Identifying and studying homologs to human disease genes in another organism is potentially valuable for developing an understanding of the biochemical function and malfunction of the human gene.

13.34 Functional Genomics

Once the whole genome sequencing is completed it is important to know the function of the genes. One key approach for assigning gene function experimentally is to knock out *i.e.*, deleting the gene and the function of a gene and to determine what the changes in the phenotype are **Figure 13.7a** shows how to do so in yeast, using a PCR-based strategy. Using PCR primers designed based on the known genome sequence, an artificial linear DNA deletion module is made and amplified. This module consists of a part of the gene sequence upstream and has the start codon and part of the gene sequence downstream including the stop codon, flanking a DNA fragment containing the kanamycin selectable marker (kanR) that confers resistance to the inhibitory chemical G418. This linear DNA is transformed into yeast, and G418-resistant colonies are selected. The desired transformants are those generated when the fragment replaces the target ORF by homologous recombination. The strategy completely inactivates the knock out genes because most of it is replaced. In genetic terms, a *loss-of-function mutation* or a *null allele* is produced.

A molecular screening must be used to confirm that the transformant has resulted in the deletion of

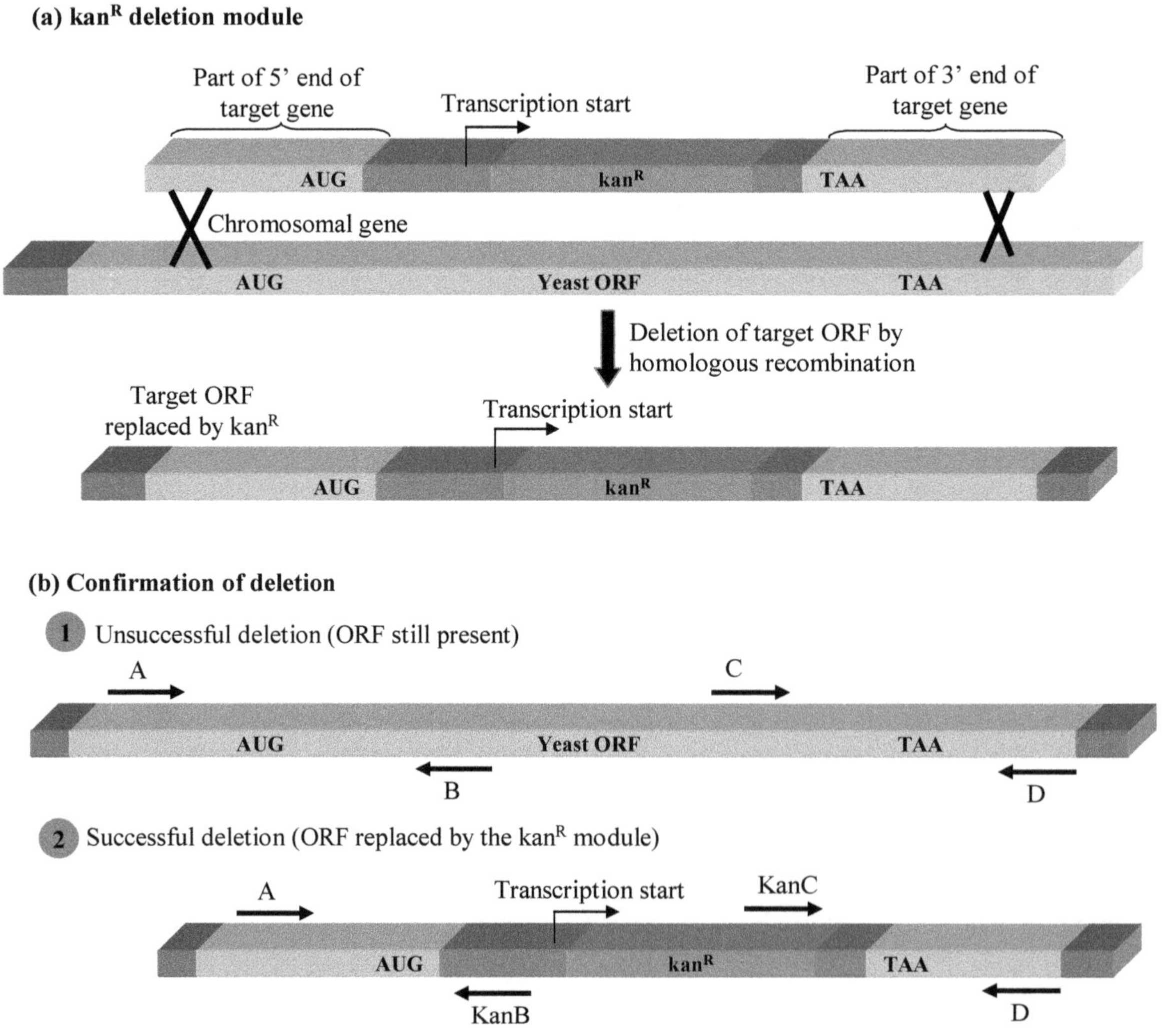

Figure 13.7: Creating and Verifying a Gene Knockout in Yeast.

the ORF of interest. PCR is used for this screen, as illustrated in **Figure 13.7b**. First, let us consider the condition of an unsuccessful deletion in which the ORF is still present (**Figure 13.7b**). Four different PCR primers, A-D, are used. Primers A and D are 200 to 400 bases upstream and downstream, respectively, of the ORE Primers B and C are from within the ORF itself. DNA is isolated from transformants, and separate PCRs are done with primers A and B, on one hand, and primers C and D, on the other. If the ORF is still present, these reactions produce DNA fragments of predictable sizes. If the ORF is deleted, no PCR products will be seen. However, it is still necessary to show definitively that the deletion has been made (**Figure 13.7b.2).** Primers A and D are shown in **Figure 13.7b.l,** and there are two other primers-K and B and KanC-that are specific to the *kanR* DNA fragment. If the deletion has been successful, the *kanR* module has replaced the ORF and PCR, using primers A and KanB, and primers KanC and D generate fragments of predictable sizes.

Using the gene deletion approach, a yeast knockout (YKO) project was completed in which each yeast gene was systematically deleted. Because some genes have essential functions, deleting them gave a lethal phenotype. However, about 4,200 of the approximately 6,200 genes are nonessential, since knocking each of them out individually results in the change of phenotype. This set of 4,200 strains in the yeast deletion collection is a genomic resource for investigating the functions of nonessential genes in the organism. For example, the deletion strains are being studied under various conditions for changes in phenotype to assign function.

Bioinformatics plays a very important role in finding the genes in DNA sequences. For this purpose, the following information is required. (i) textual information about gene functions by matching keywords of the gene functions; (ii) gene function information from the known functions of genes with sequences like a target gene; and (iii) the prior probabilities of gene functions to be associated with an arbitrary gene by mining the known gene functions from curated databases. A supervised learning method is utilized to obtain the weights for combining the three types of evidence to assign appropriate Gene Ontology terms for target genes.

As gene annotation databases continue to evolve and improve, it has become feasible to incorporate the functional and pathway information about genes, available in these databases into the analysis of gene expression data, for a better understanding of the underlying mechanisms. A few methods have been proposed in the literature to formally convert individual gene results into gene function results.

The function of an ORF identified in genome scans may be assigned by searching databases for a sequence match with a gene whose function has been defined. Such searches are called sequence similarity searches and involve computer-based comparisons of an input sequence with all sequences in the database. The searches can be done using an Internet browser to access the computer programs. For example, the BLAST program at the National Center for Biotechnology Information (http://www.ncbLnlm.nih.gov/) enables a user to paste the sequence to be studied into a window, either in nucleotide form or in amino-acid form, and to get results indicating the degree to which the sequence of interest is like sequences in the database.

Similarity searching is an effective way to assign gene function because homology-descent from a common ancestor-is a reflection of evolutionary relationships. That is, a pair of homologous genes in different organisms has a common evolutionary ancestor, so the two genes' nucleotide sequences are similar; their differences have resulted from mutational changes over evolutionary time. Thus, if a newly sequenced gene (*e.g.*, from a genome sequence project) is like a previously sequenced gene, the two genes are related in an evolutionary sense, so the function of the new gene probably is the same as, or at least like, the function of the previously sequenced gene.

Sequence similarity searching can be done with either a nucleotide (DNA) sequence of an amino-acid sequence, but the latter is preferred because, with 20 different amino acids and only four different nucleotides, unrelated genes appear more different from one another at the amino-acid level than at the nucleotide level. Given the information in current databases, there is less than a 50 per cent chance that a new gene sequence will match a gene sequence in the database with a high enough degree of similarity to infer the new sequence's function.

A sequence similarity search can indicate a match either for the whole protein sequence or for parts of it. In the latter case, this means that a domain of the new gene product matches a domain of a previously identified gene product, so at least part of the new protein's function can be inferred. Evolutionarily speaking, such a result implies that the domains have a common ancestor, but the genes may not.

Sequence similarity searching plays an important part in assigning gene functions. About 30 per cent of the genes were known because of standard genetic analysis before genome sequencing, including direct assays for function.

In genes, approximately 70 per cent are open reading frames where the functions have not been determined. However, thirty percent-encode a protein that is related to functionally characterized proteins. Another 10 per cent have homologs in databases, but the functions of those homologs are still not known. Such yeast ORFs are called FUN (function unknown) genes, and those genes and their homologs are called *orphan families.* Genes without homologs are called single orphans.

This concept applies to other organisms also *i.e.,* prokaryotic and eukaryotic. Certain modifications have been adopted for the ORF scanning of the Eukaryotic genome. Certain codon bias options are added, for example leucine is coded by CTG, rarely by TTA or CTA, and then search for the exon-intron boundaries consensus (AGGTA AGT).

There are methods of identifying DNA responsible for conferring a particular phenotype in a cell which comprises a) constructing a cDNA or genomic library of the DNA of the cell in a suitable vector in an orientation relative to a promoter(s) capable of initiating transcription of the cDNA or DNA to double-stranded (ds) RNA upon binding of an appropriate transcription factor to the promoter(s), b) introducing said library into one or more of the cells comprising the transcription factor, and c) identifying and isolating a particular phenotype of the cell comprising the library and identifying the DNA or cDNA fragment from the library responsible for conferring the phenotype. Using this technique it is also possible to assign a function to a known DNA sequence by a) identifying a homolog (s) of the DNA sequence in a cell, b) isolating the relevant DNA homologs (s) or a fragment thereof from the cell, c) cloning the homolog or fragment thereof into an appropriate vector in an orientation relative to a suitable promoter(s) capable of initiating transcription of dsRNA from the DNA homolog or fragment upon binding of an appropriate transcription factor to the promoter(s), and d) introducing the vector into the cell from step a) comprising the transcription factor.

13.35 Gene Expression Analysis to Study Functional Genomics

Once the gene is found and isolated, experiments can be done to study the expression of the gene in normal and mutant organisms to understand the role of the gene in determining the phenotype. When the complete genome sequence is obtained for an organism, a new line of research is possible, including the analysis of the expression of all genes in a cell at the transcriptional and translational levels and the analysis of all protein–protein interactions. Measuring the levels of mRNA transcripts gives an insight into the global gene expression profile of the cell. A new term has been coined for the set of mRNA transcripts in a cell this is known as transcriptome. The transcriptome is a major indicator of cellular phenotype and function. The study of the complete set of proteins in a cell is called the proteome.

Probe arrays are among the most powerful tools for global gene expression. For example, several researchers are using probe arrays to study gene expression in yeast. Yeast sporulation has been studied by Pal Brown and Ira Heriskowitz. Meiosis produces haploid spores in yeast. Yeast sporulation involves four major stages: DNA replication and recombination, meiosis I, meiosis II, and spore maturation. The sequential transcription of at least four classes of genes-early middle mid-late and late-correlates with these stages. Brown and Herskowitz's probe array experiments showed approximately 150 genes are differentially expressed during sporulation. In new research, the researchers induced diploid yeast cells to sporulate, and at seven timed intervals, they took cell samples and used DNA microarrays containing 97 per cent of the known or predicted yeast genes to analyze the temporal program of gene expression during meiosis and spore formation. Both light and electron microscopy were utilized to correlate the sampling time with the exact stage of sporulation. To quantify gene expression, Brown and Herskowitz isolated mRNAs from the cell samples and synthesized fluorescently labeled cDNAs by reverse transcription in the presence of Cy5 (red)-labeled dUTP. For a nonsporulating cell control, they isolated mRNAs from cells at a time point immediately before inducing sporulation, and they synthesized fluorescently labeled cDNAs at this time, using Cy3 (green)-labeled dUTP. For each time point, they hybridized a mixture of reference green-labeled cDNAs and experimental red-labeled cDNAs to DNA microarrays, by using PCR to amplify each ORF and printing the sequences onto a glass slide using a robotic device. After hybridization, they scanned the microarrays with a laser detector to quantify the red and green fluorescence locations and intensities. The relative abundance of transcripts from each gene in sporulating versus non-sporulating yeast cells is seen by the ratio of red to green fluorescence. If an mRNA is more abundant in sporulating cells than in non-sporulating cells, as is the case for the *TEPl* gene, the result is a higher ratio of red-labelled to green-labeled cDNAs prepared from the two types of cells and, therefore, in the same higher ratio of red to green fluorescence detected on the array. In general, a gene whose expression is induced by sporulation is seen as a red spot, and a gene whose expression is repressed by sporulation is seen as a green spot. Genes that are expressed at approximately equal levels in non-sporulating cells and during sporulation are seen as yellow spots.

With this approach, the researchers found that more than 1,000 yeast genes showed significant changes in mRNA levels during sporulation. About one-half of the genes are repressed during sporulation, and one-half are not repressed. At least seven distinct temporal patterns of gene induction are seen, and this observation has provided some insights into the functions of many orphan genes.

Other methods used to study the function of the genome are illustrated below.

13.36 DNA Microarrays and Microchips

Presently we can spot small volumes of DNA on a chip, so the dots of the dot blot are very small. This allows many different DNAs to be spotted on one chip, called a DNA microarray. This was developed by Vivian Cheung and colleagues. They used a robot with 12 parallel pens, each of which can squirt out a tiny volume of DNA solution: 0.25–1.0nL (billionth of a liter). The spots are exquisitely small, only 100–150μm in diameter and the centers of the spots are only 200–250μm apart. The result looks like the schematic diagram in which represents a DNA microarray with 5808 DNA spots on a common microscope slide. After spotting, the DNAs are air-dried and covalently attached by ultraviolet radiation to a thin silane layer on top of the glass.

Another strategy for reducing the size of a blot has been to synthesize many oligonucleotides simultaneously, right on the surface of a chip. In this case, short DNAs (oligonucleotides) are spotted closely spaced spots on a small glass microchip. A 1999 version of this technique was started with a small glass slide coated with a synthetic linker that was blocked with a photo-reactive group that can be removed by light. Steven Fodor and his colleagues were founder people in pioneering this method. They masked some of the areas of the slide and illuminated it, so the blocking agent was removed only from the unmasked areas. Then they added a nucleotide (also blocked with a photo-reactive group) and chemically coupled it the previous step. The result: A nucleotide was attached to a subset of the tiny spots on the chip. Next, masked a different subset of spots illuminated the others to remove the blocking groups and attached another nucleotide. On the spots that were unmasked in both steps, dinucleotides were formed. By repeating this process, they could build up different oligonucleotides on each spot.

The resulting chip is known as a DNA microchip or oligonucleotide array. The technology is so miniaturized that about 300,000 oligonucleotides can be attached to a chip only 1.28 x 1.28 cm (about 1.2" square). And the process is so efficient that a set of 4" different oligonucleotide can be built in only 4 x n cycles. So, if our goal is to generate all the possible 9 – mers (4^9, or about 250,000 different oligonucleotides), we can do it in only 4 x 9 =36 cycles. How long must an oligonucleotide take to uniquely identify one human gene product in a mixture of all the others? Knowing the sequence of the human genome will help us answer this question with great accuracy. However, we can already do a calculation to give us a minimum estimate. A given sequence of n bases will occur in DNA about every 4^n bases. In other words, a DNA sequence needs to be n base long to occur about once in a DNA 4^n bases long. Thus, we need to solve the following equation for n to find the minimum size of an oligonucleotide we would expect to find only once in the whole human genome, which may be as much as 3.5×10^9 bases long: $4^n = 3.5 \times 10^9$

The answer is that if n = 16, 4n > 3.5×10^9. So, our oligonucleotides need to be at least 16 bases long. Again, however, this is a minimum estimate, so it would be a good idea to start with longer oligonucleotides need to be at least 16 bases long to be reasonably sure that they occur only once the longer oligonucleotide can be used to identify human genes.

Even before the publication of the sequence of the first human chromosome, scientists at Affymetrix, Inc. were already producing microchips containing 25– mers designed to recognize single genes. They based their design on the available sequence, including many ESTs already in the database. To enhance the reliability of their chips, they included multiple oligonucleotides designed to hybridize to single transcripts, so the results obtained with each of these oligonucleotides could be checked against one another.

The oligonucleotides on a microchip or the cDNAs on a microarray can be hybridized to labeled RNA isolated from cells (or to corresponding cDNAs) to see which genes in the cell were being transcribed.

Once we know the sequence of all the genes in a human cell, we will be able to design a comprehensive DNA chip to assay transcription from every gene, with no guesswork about the oligonucleotides we need to put on the chip. The power of this kind of functional genomics increases our knowledge related to the sequence of the organism's genome.

13.37 Serial Analysis of Gene Expression

In 1995, Victor Velculescu, working with Kenneth Kinzler and colleagues, developed a novel method of analyzing the range of genes expressed in a given cell. They called this method serial analysis of gene expression (SAGE). The underlying strategy of SAGE is to synthesize short cDNAs, or tags, from all the mRNAs in a cell, and then link these tags together in clones that can be sequenced to learn the nature of the tags, and therefore the nature of the genes expressed in the cell, and the extent of expression of each gene.

Figure 13.8 shows how Velculescu and colleagues carried out this strategy. First, they used a biotinylated oligo (dT) primer to prime reverse transcription of the mRNAs present in human pancreatic tissue, yielding double–stranded cDNAs. The goal was to reduce the size of the cDNAs to short tags that could be ligated together and sequenced readily. Because of the shortness of the tags 9bp in the example given here), it is important to confine them to a small region of the cDNAs to increase the chance that they will uniquely identify one cDNA. To begin the shortening process, Velculescu and colleagues cleaved the biotinylated cDNAs with an anchoring enzyme (AE) to chop off a short 3′– terminal fragment. They chose as their anchoring enzyme NlaIII, which recognizes 4- base restriction sites and therefore yields fragments averaging 250 bp long. They bound these biotinylated 3′– fragments to streptavidin beads, which bind biotin.

Next, they divided the bead-bound cDNA fragments into two pools and ligated one pool to a linker (A) and the other pool to a second linker (B) Both linkers contained the recognition site for a Type IIS restriction endonuclease with the tagging enzyme [TE] and cuts 20bp downstream of this recognition site. The result of cleavage of the cDNA fragments with the tagging enzyme FokI was a set of short fragments, each containing the linker (A or B) followed by the 4bp anchoring enzyme site, followed by 9bp from the cDNA. The 9bp piece of cDNA is the tag. If the tagging enzyme leaves overhangs, these can be filled into yield blunt ends.

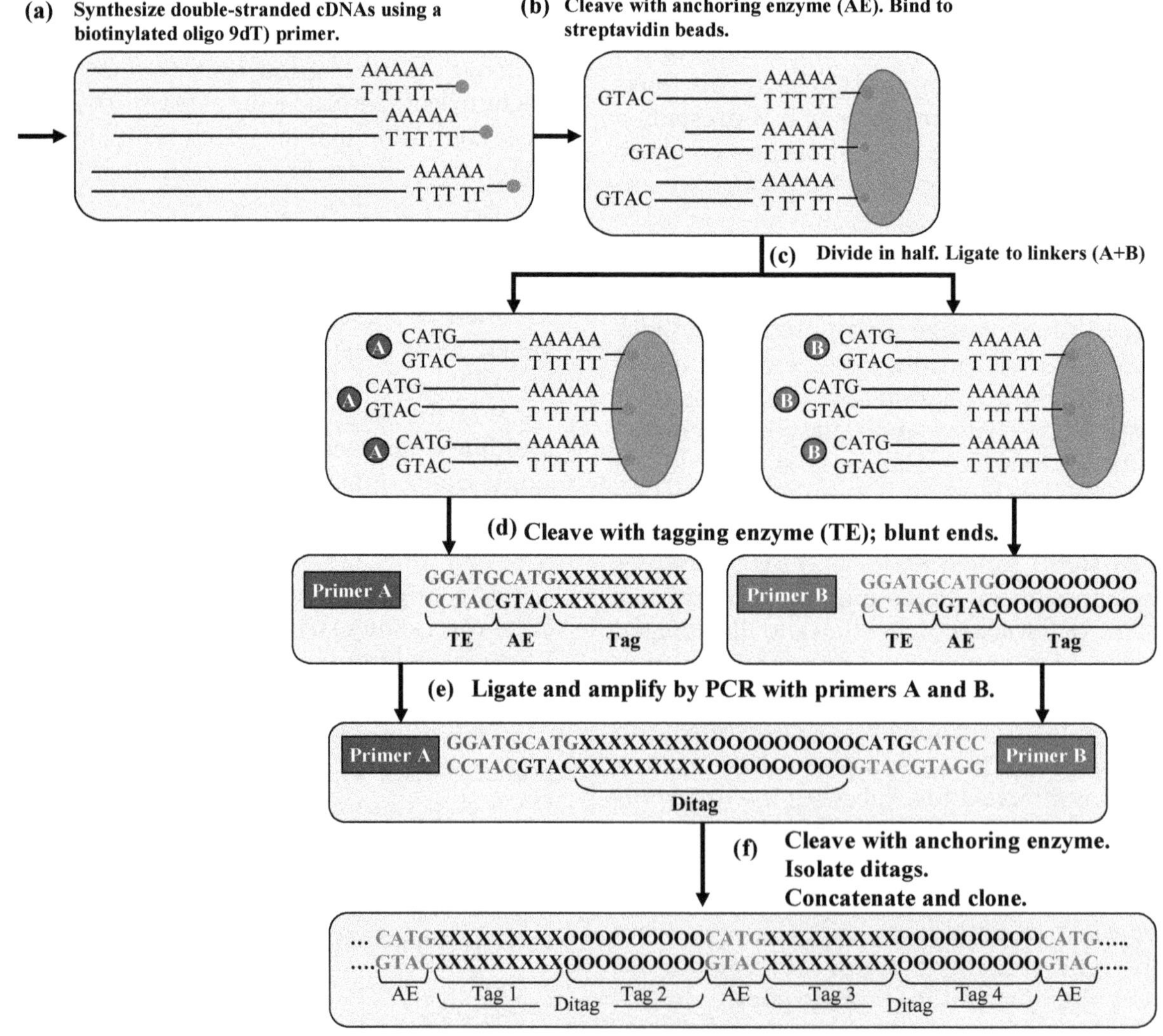

Figure 13.8: Serial Analysis of Gene Expression (SAGE).

Velculescu and colleagues' next task were to ligate the tags together, along with defined DNA so they could tell where one tag left off and another began. To do this, they ligated the blunt end of the tagged fragments together to form fragments with two tags abutting each other in the middle (forming a ditag) and linkers on each end. The linkers contain sites that are complementary to a pair of primers that can be used to amplify the whole fragment by PCR. After the PCR amplification, Velculescu and colleagues cleaved the products with the anchoring enzyme, ligated these restriction fragments together, and cloned the products. Now the ditags can be easily identified because each one is flanked by the 4bp anchoring enzyme recognition sites. Half of each ditag belongs to one tag, and a half to the other. Clones with at least 10 tags (some had more than 50) can be identified by PCR analysis and sequenced. If enough clones are sequenced, we can get an idea of the range of genes expressed, and tags that show up repeatedly indicate genes that are very actively expressed.

In the SAGE analysis for the expression of the human pancreas, the most common tags GAGCACACC and TTCTGTGTG can be used which correspond to the genes for procarboxypeptidase A1 and pancreatic trypsionogen 2, respectively. These are two abundantly expressed pancreatic proenzymes, which, after cleavage to the mature enzyme form, digest proteins in the small intestine. Many other familiar pancreatic genes were identified among the plentiful tags, but many of the tags did not match any gene sequences in the database, so their identities were unknown. As the database expands to include all human genes, all tags should at least be correlated to genes, even if the functions of some of those genes remain obscure.

13.38 Positional Cloning

Before the genomics era, geneticists seeking the genes responsible for human genetic disorders frequently faced a problem: They did not know the identity of the defective protein, so they were looking for a gene without knowing its function. Thus, they had to identify the gene by finding its position on the human genetic map, and this process, therefore, is known as positional cloning.

The strategy of positional cloning begins with the study of a family or families afflicted with the disorder, to find one or more markers that are tightly linked to the gene causing the disease. Because the position of the marker is known, the disease gene can be pinned down to a relatively small region of the genome. However, that "relatively small" region usually contains about a million base pairs, so the job is not over. The next step is to search through the million or so base pairs to find a gene that is likely to be the culprit. Some tools have traditionally been used. These are: (1) finding exons with exon traps; and (2) locating the CpG islands that tend to be associated with genes. We will see how these tools have been used as we discuss two classical positional cloning experiments finding the genes responsible for Huntington's disease and cystic fibrosis in the next section of this chapter.

13.39 Exon Traps

Once we have a contig stretching over hundreds of kilobases, how do we sort out the genes from the other DNA? If that DNA region has not yet been sequenced, we can sequence it and look for ORFs, but that is very laborious. Several more efficient methods are available; include a procedure invented by Alan Buckler called exon amplification or exon trapping. **Figure 13.9** shows how an exon trap works. We begin with a plasmid vector such as pSPL1, which Buckler designed for this purpose. This vector contains a chimeric gene under the control of the SV40 early promoter.

The gene was derived from the rabbit β – globin gene by removing its second intron and substituting a foreign intron from the human immunodeficiency virus (HIV), with its own 5′ – 3′ – splice sites. They spliced human genomic DNA fragments and placed them into the intron are complete exons, with their own 5′ – and 3′ – splice sites, this exon will become part of the processed transcript in the COS cells. We purify the RNA made by the COS cells, reverse transcribe it to make cDNA, then subject this cDNA to amplification by PCR, using primers designed to amplify any new exon. Finally, we clone the PCR products, which should represent only exons. Any other piece of DNA inserted into the intron will not have splicing signals; thus, after being transcribed, it will be spliced out along with the surrounding intron and will be lost.

13.40 CpG Islands

Another gene-finding technique takes advantage of the fact that active human genes tend to be associated with unmethylated CpG sequences, whereas the CpGs in inactive regions are almost always methylated. Furthermore, the restriction enzyme *HpaII* cuts at the sequence CCGG, but only if the second C is unmethylated. In other words, it will cut active genes that have unmethylated CpGs within CCGG sites, but it will leave inactive genes (with methylated CCGGs) alone. Thus, geneticists can scan large regions of DNA for the sites that can be cut with *HpaII* in a "sea" of

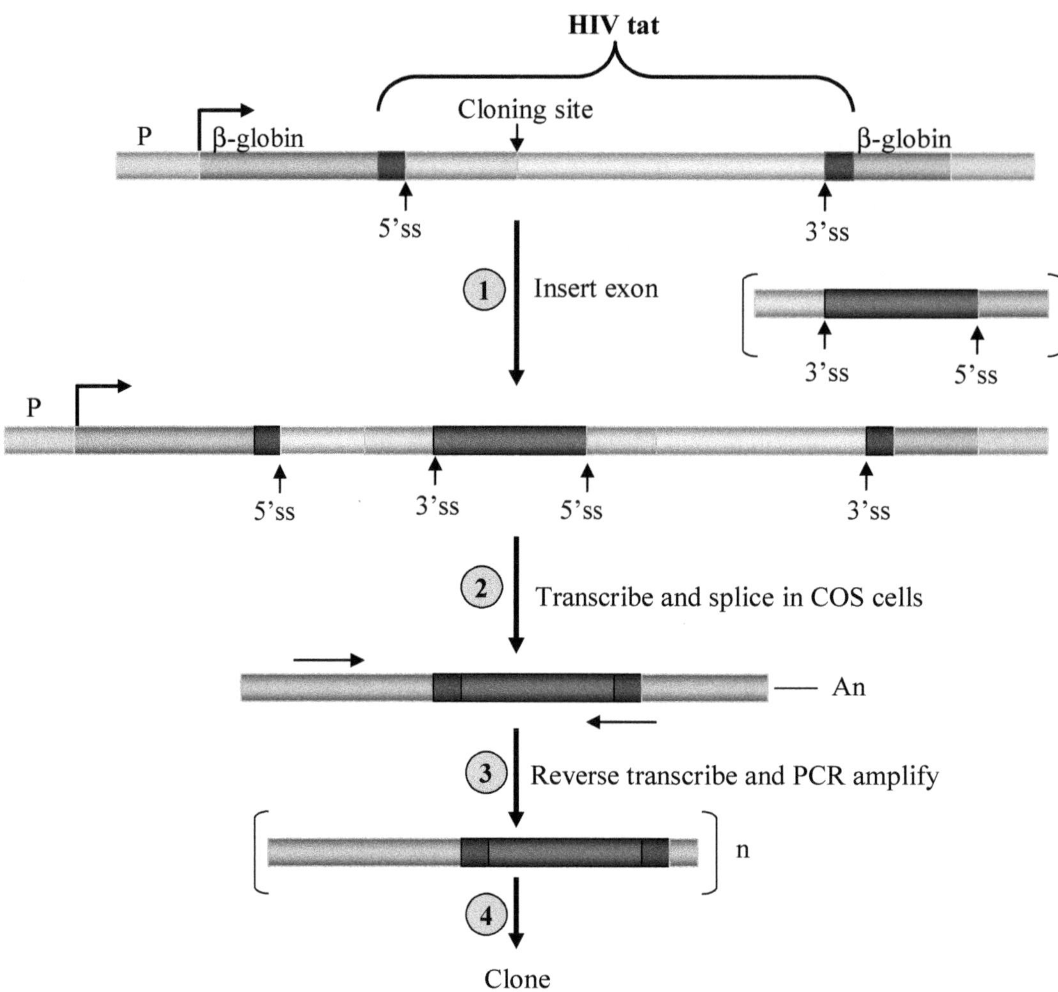

Figure 13.9: Exon Trapping.

other DNA sequences that could not be cut. Such a site is called a CpG island or an HTF island because it yields *HpaII* tiny fragments.

13.41 Gene Rich and Desert Regions in the Human Genome

It is seen that some of the chromosomal regions have many genes. On chromosome 6 the complement genes are located which have approximately 60 genes with diverse functions. Approximately 70 per cent of the DNA in this region is transcribed. This region has high GC content. The exact region why some of the areas on chromosomes have high gene-rich regions is not exactly known however, this could be because of gene functions, or it could be because of the functional importance of these genes. Many sub-telomeric regions appear to be gene-rich, matching both known and unknown expressed genes. This indicates that human sub-telomeric regions are not simply buffers of nonfunctional 'junk DNA' next to the molecular telomere but are instead functional parts of the expressed genome. Opposite to this is that several regions are not DNA rich these are called gene desert regions. Gene deserts are improbably large regions of the human genome that contain no genes. The evolutionary origin of these regions is still unknown, but experimental and computational evidence suggests there are multiple classes of gene deserts in the human genome, each with different evolutionary origins and functional associations. In the human genome's largest gene deserts are found on chromosomes 2 and 4. These are large regions of DNA that do not have protein-coding genes. Human chromosome 2 arose from the fusion of two ancestral ape chromosomes.

13.42 The Proteome

Genome analysis has revealed that there are approximately 25,000 genes in humans and these genes are translated into proteins. The analysis of these genes is known as the proteome. Many times, some of the proteins are conserved hence show evolutionary descent. Approximately 1200 gene families contain about two or more members of which 77 per cent are vertebrate-specific and are related to immune functions. The proteome is much more complex as compared to the genome. The complexity increases from yeast genome to human genome. This may be due to more genes in humans as compared to yeast. The complexity has further increased due to alternate splicing and chemical modifications of proteins. The proteome is the complete set of expressed proteins in a cell at a particular time. Various goals of proteomics are (1) to identify every protein in the proteome, (2) to determine the sequences of each protein and to enter the data into databases, and (3) to globally analyze protein levels in different cell types and at different stages in development. The Human Proteome organization is involved in this project (HUPO).

Proteomics is an extremely important field because it focuses on the functional products of genes, which determine the phenotypes of a cell. Of human interest are diseases, and proteins and peptides are closer to the actual disease process than are the genes that encode them. However, the challenges for proteomics are much greater than those for genomics. The complexity of the human proteome greatly exceeds that of the genome. Whereas there are an estimated 25,000 genes in the human genome, there may be about 500,000 different proteins. This disparity is the result of variations in gene expression, such as alternative RNA splicing producing different translatable mRNAs, as well as posttranslational modifications of proteins that affect their functions. There are various methods used for protein analysis. A short description of these methods is given below.

13.43 Mass Spectrometry

Mass spectrometry (MS) is an emerging technology it detects and quantifies proteins in a very complicated way and generates a complex biological matrix this technique detects proteins in a very precise manner the concept behind it that there is a difference in the composition by a single hydrogen atom which is the smallest atom. Despite its mass spectrometry is difficult for biospecimens. For this purpose, additional technologies such as organelle or protein fractionation or affinity capture have been developed to reduce the complexity of proteins in biospecimens by enriching the proteins of interest, in addition to improving the sensitivity of instrumentation for detection and quantification of proteins.

13.44 Protein Microarrays

Similar in concept to DNA microarrays, protein arrays are rapidly becoming the best way to detect proteins, measure their levels in cells, and characterize their functions and interactions, all on a very large scale. Consequently, they are central proteomics technology, valuable both for basic research and for biotechnology applications. As with DNA microarrays, the use of protein arrays is becoming highly automated, and this makes it possible to do large numbers of measurements in parallel.

Protein arrays also called *protein microarrays* and *protein* chips-involve proteins immobilized on solid substrates, such as glass, membranes, or microtiter wells. At the moment, the density of proteins on the arrays is much lower than that for DNA-on-DNA microarrays. However, with technological advances, we can expect the density of proteins in the arrays to increase.

Protein expression profiling can be done by using protein microarrays. As with DNA microarrays, target proteins are labeled fluorescently (*e.g.*, with Cy3 and Cy5, as is used for DNA), and binding to spots on the arrays is measured by automated laser detection. The resulting complex data are analyzed by the computer. Because of the similarities with DNA microarray technology, the same instrumentation used to analyze DNA microarrays can be used to analyze protein arrays.

Let us consider two types of arrays to illustrate how protein arrays can be used. One type is the capture array, in which a set of antibodies (usually) bound to the array surface is used to detect target molecules-for example, in cell or tissue extracts. The antibodies are made either by conventional immunization procedures or by using recombinant DNA techniques to make clones from which antibody fragments are made. A capture array can be used as a diagnostic device-for instance, to screen for infections (detecting specific proteins made by the infectious agent) or for the presence of tumors (detecting tumor-specific markers in extracts of biopsied material). In proteomics studies, capture arrays are used for protein expression profiling that is, defining the proteome qualitatively and quantitatively. For example, one can quantify proteins in different cell types and different tissues, as well as compare proteins under different conditions, such as during differentiation, with and without a drug treatment, and with and without a disease.

Another type of array is the *large-scale protein array*. In these arrays, large numbers of purified proteins are spotted onto an array substrate and are used to assay one of a wide range of biochemical functions, including protein-protein interactions (an alternative to the cell-based yeast two hybrid systems) and drug target interactions. The proteins for immobilization on the array substrate are produced from an expression library transformed into a host such as *E. coli* and yeast, from which the expressed proteins are purified.

13.45 Nanotechnologies

Nanotechnology is the technique that creates particles of 1 to 100 nanometers. These are very small in size *i.e.*, one billionth of a meter or 1/80,000 the width of a human hair. Nanotechnology is used in proteomics. These small proteins can reach *in vivo* through physical and biological barriers, detecting low abundance targets, and provide a "toolbox" to translate the discovery of protein biomarkers to novel therapeutic and diagnostic tests. Some nanodevices are used to target cancers. Nanowires and nano cantilever arrays can be used in biosensors that measure minute quantities of biomarkers in biological fluids.

13.46 Pharmacogenomics

Pharmacogenomics is a blend of "pharmacology" and "genomics"; it denotes the study of how an individual's genome affects the body's response. That is, medicine operates mostly on the assumption that all humans are the same and all drugs should behave similarly but in practice, it is not seen.

Pharmacogenomics is a very young area of research now, so mostly there is a lot of promise, but very few demonstrated successes. One success story is the cytochrome p450 (CYP) family of liver enzymes. These enzymes break down more than 30 different classes of drugs. However, variations in the genes that encode the enzymes result in enzymes with different abilities to metabolize drugs. A significant concern here is patients who have inactive or only partially active CYP enzymes because they are susceptible to a drug overdose. Genetic tests for variations in cytochrome p450 genes are currently in clinical trials. Such tests would be valuable for adapting drug treatments to patients exhibiting this particular set of genome variations.

13.47 Applications of Genomics

There are many applications of genome sequencing but two of the most important are: (1) probing the pattern of gene expression in each cell type at a given time (functional genomics); and (2) finding genes involved in genetic traits, especially genetic diseases (positional cloning). It is quite possible to do both functional genomics and positional cloning studies without knowing the complete sequence of an organism's genome, but the sequence greatly facilitates both processes. However, these two issues provide different applications along with different social and ethical issues.

13.48 Applications of Functional Genomics

Let us conclude with two classic examples of the use of functional genomics: Pinpointing the genes for Huntington's disease and cystic fibrosis.

13.49 Huntington's Disease

Huntington's disease (HD) is a progressive nerve disorder. It begins almost imperceptibly with small tics and clumsiness. Over the years, these symptoms intensify and are accompanied by emotional disturbances.

Huntington's disease is controlled by a single dominant gene. Therefore, a child of an HD patient has a 50:50 chance of being affected. People who have the disease could avoid passing it on by not having children, except that the first symptoms usually do not appear until after the childbearing years. It is a disease of triplet repeats.

In all unaffected individuals, the number of CAG repeats ranged from 11 to 34, and 98 per cent of these unaffected people have 24 or fewer CAG repeats. In all affected individuals, the number of CAG repeats has expanded to at least 42, up to a high of about 100. Thus, we can predict whether an individual will be affected by the disease by looking at the number of CAG repeats in this gene. Furthermore, the severity or age of onset of the disease correlates at least roughly with the number of CAG repeats. People with several repeats at the low end of the affected range (now known to be 38-40) generally survive well into adulthood before symptoms appear, whereas people with several repeats at the high end of the range tend to show symptoms in childhood. In one extreme example, an individual with the highest number of repeats detected (about 100) started showing disease symptoms at the extraordinarily early age of 2. Finally, two people were affected, even though their parents were not. In both cases, the affected individuals had expanded CAG repeats, whereas their parents did not. New mutations (expanded CAG repeats), although a rare occurrence in HD, apparently caused both these cases of the disease.

Another way of demonstrating that this gene is HD would be to deliberately mutate it and show that the mutation has neurological effects. One cannot perform such an experiment in humans, but it would be feasible in mice if the gene corresponding to HD is known. Fortunately, HD is conserved in many species, including mice, where the gene is known as Hdh. In 1995, a team of geneticists led by Michael Hayden created knockout mice with a targeted disruption in exon 5 of Hdh. Mice that are homozygous for this mutation die in utero. Heterozygotes are viable, but they show loss of neurons with the corresponding lowering of intelligence. This reinforces the notion that Hdh, and therefore HD, plays an important role in the brain exactly what we would expect of the gene that causes HD.

How can we put this new knowledge to work? One obvious way is to perform accurate genetic screening to detect people who will be affected by the disease. In fact, by counting the CAG repeats, we may even be able to predict the age of onset of the disease. However, that kind of information is a mixed blessing, as it can be psychologically devastating. What we need, of course, is a cure, but that may be a long way off.

To find a cure for HD, we need to answer at least two questions. First, what is the normal function of the HD product (now called huntingtin? Geneticists are constructing transgenic mice carrying the mutant form of HD. This should shed light on the function of the gene in the brain and in other tissues, where it is also active. Second, we need to know how the expansion of the CAG repeat causes disease. Gusella speculated that the extra glutamines in the protein give it a new function that somehow disrupts brain activity. We now know that abnormal huntingtin binds to another protein with a string of glutamines: CREB binding protein (CBP). By binding and inactivating this mediator, abnormal huntingtin may disturb nerve function.

13.50 Cystic Fibrosis

Cystic fibrosis (CF) is the most common lethal genetic disease affecting whites. It is caused by an autosomal recessive mutation carried by 1 in 20 people of European descent. More than 12 million people in the USA are affected. This means that 1 in 400 white couples will both be carriers, and 1 in 4 of their children will have the disease. In the United States, over one thousand children are born with cystic fibrosis every year.

It affects tissues called secretary epithelia, which are responsible for transporting water and salt at the interface between the bloodstream and the external environment (*e.g.*, in the lungs, intestine, and sweat glands). The abnormal secretory epithelia in CF patients fail to carry out this transport properly, which causes the buildup of thick mucus in the affected organs. This in turn causes the clinical symptoms of the disease: failure of the pancreas to secrete digestive enzymes into the intestine; bacteria that find a fertile environment for growth in the thick mucus in the lungs; and high levels of salt in the sweat. Which is a convenient marker for testing and diagnosing the disease? Lung infections are very serious. It is primarily because of these infections many CF patients do not survive past their twenties.

Because this disease is so prevalent and so devastating, it has been the subject of intense research. Geneticists knew that if they could identify the defective gene and its product, this knowledge might suggest new avenues to a treatment for the disease. At least it would provide a test for the mutation so couples would know in advance if they were at risk of having affected children. The test could also be used parentally to determine whether a fetus was homozygous for the mutation.

The first step in the search for the gene was to establish its linkage to known markers, especially those whose chromosomal location was known. One of the first markers to be found linked to CF was variability in serum activity of an enzyme called paraoxonase; unfortunately, this marker had not been mapped, and it was not even known what chromosome it resided on.

This RFLP marker is located on chromosome 7. To do this mapping, to map it on chromosome somatic cell hybridization techniques were used. A panel of hamster human hybrid cells was obtained with known human chromosome content, from another laboratory these are then screened for the ability to hybridize to a radioactive probe for DOCRU- 917. Whenever the cell did not contain any part of human chromosome 7, its DNA would not hybridize to the probe. Thus, the RFLP is located on chromosome 7, and because CF is linked to the RFLP, we know that it is on chromosome 7 as well.

This RFLP is tightly linked to a gene called met. This was valuable information because met had been mapped to the middle third of the long arm of chromosome 7. This placed CF in the same vicinity.

Linkage analysis on several different markers culminated in the assignment of CF to band q31 of chromosome 7, between two closely linked markers: met and D7S8 (defined by the pJ3.11 RFLP). The recombination frequency between met and CF in males was 0.013, whereas the recombination frequency between CF and D7S8 was only 0.009. This place met and D7S8 only about 1 – 2 million bp apart, thus confining the search to a relatively narrow region of the human genome.

The first step in searching for a gene in a megabase pair region of DNA is to clone at least part of the DNA in question. This task has been simplified by techniques such as YAC and BAC cloning. In this case, the investigators used another technique called chromosome walking, which generates a collection of clones with overlapping DNAs that cover the whole region of interest.

To perform a chromosome, walk from left to right in the example illustrated here, we start with a genomic library and a clone at the left end of the sequence we wish to examine. We cut a piece from the right end of this starting clone and label it; then we use this radioactive fragment to probe the library by plaque hybridization. Any clone that hybridizes will overlap the right end of the starting clone and is very likely to contain additional DNA to the right. This process is repeated with a probe from the right end of the second clone to find an overlapping clone farther to the right. We continue this process until we have a contig containing overlapping clones representing the whole region. Finally, we can sequence all our clones to obtain the base sequence of the whole region.

Chromosome walking works very well for examining relatively short stretches of a chromosome (100 kb or so) but runs into trouble when we need to canvass hundreds of kilobases of DNA, as in the present case. The biggest problem is that certain regions of DNA are unclonable, as we saw in our discussion of the Human Genome Project. Still, other regions can be cloned but are unstable and are lost.

However, a modification of chromosome walking called chromosome jumping was used to map the CF gene. This technique allows the investigator to jump over unclonable regions and start the chromosome walk a new. The key to the procedure is to prepare a library of clones containing about 100kb of DNA, then to form circles of these DNAs by ligating in a short piece of DNA containing a selectable marker, for example, the supergene, which suppresses amber mutations. This brings together two DNA regions that had been separated by about 100 kb. Next, we cut the circular DNA with a restriction enzyme and subclone the fragments into an l -phage with an amber mutation in a critical gene. This ensures that only clones with the supF gene will survive, and this gene will mark the place where circle formation took place. We select a clone by plaque hybridization to a probe for a given site in our search (say the met locus). Because this cloned DNA contains met, and a circle joining site marked by supF, the chances are high that the DNA lying on the other side of the supF gene originally lay about 100kb away from met. This can then be used as a new start site for chromosome walking. We have just performed a chromosome jump of 100kb.

One important clue in this sort of mystery is to find a stretch of DNA that is conserved in several different species. If the DNA is conserved in several different species, it probably codes for something and is, therefore, a gene. To find conserved DNA quickly, researchers blotted fragmented genomic DNA from humans, cows, mice, and chickens and hybridized this "zoo blot" to radioactive probe DNAs from different parts of the cloned human DNA region. Four radioactive probes cross-hybridize to DNA from other species besides humans. These probes identified four candidate regions in which to focus the search.

The first candidate region mapped was very close to met, and far from D7S8, so it could not contain CF, the other did not contain any ORFs, so it was also eliminated from consideration. The third region was used as a probe to look for mRNAs with the same sequence, but none was found. Thus, it appeared that this region is not transcribed and therefore, could not contain the CF gene.

Fortunately, the fourth candidate region did contain CF. When the research team determined the sequence of this region, they found encouraging signs: the presence of a CpG island. It took exhaustive screening of many cDNA libraries with a probe for this region, but finally, the researchers were rewarded with a cDNA (clone 10-1) containing a 920–bp piece of DNA encoding part of an mRNA from the sweat gland of a normal (non-CF) individual. This meant that the CG–rich region was transcribed in the normal sweat gland, as CF ought to be. Clone 10 -1 also detected a 6.5–kb transcript in a Northern blot of human RNAs, again demonstrating the expression of this region in human cells. Finally, using cDNAs to hybridize genomic clones, this group was able to show that CF spans approximately 250 kb of DNA and includes at least 24 exons.

How do we know this gene is CF? Several lines of evidence supported this conclusion. First, Northern blotting experiments showed that the gene is transcribed in all tissues affected by CF. Second, the base sequence of the gene showed that it encodes a protein with a so-called membrane-spanning domain and so is very likely to be a membrane protein, as the CF product is predicted to be because it is involved in membrane transport. Further sequence analysis suggested at first that the product of the CF gene was probably not itself a channel for chloride ions, but somehow regulated the transport of these ions across the membrane. This led to renaming the protein product of CF as the cystic fibrosis transmembrane conductance regulator (CFTR). However, we now know that CFTR is a chloride channel because it can create such channels when it is the only protein added to an artificial membrane.

A third finding linking this gene with CF is that most CF patients have a 3 –bp deletion in this gene, which would result in a mutant protein with one amino acid (phenylalanine) missing. 70 per cent of the patients have this mutation.

Now that we know the identity of the CFTR gene, how can we use this knowledge to find treatments for the disease? One way is to develop an animal model so we can examine in detail the effects of the disease on lungs and other tissues as the disease progresses. The supply of lung tissue from CF patients who die in infancy is small; by the time most patients die, infections have damaged their lungs so badly that their condition in the early stages of the disease is impossible to determine. Two research groups have bred CFTR knockout mice that should help to circumvent these problems. Now we can see precisely what is happening in diseased tissues from birth (or before) until death. And we can try out therapies on these mice before risking them on human patients. One problem, however, is that the knockout mice do not show the same phenotype as human CF patients, so we will have to be careful in interpreting experiments.

Another way is the knowledge about the CFTR gene which can help in designing therapies for the disease, so far, the genetic findings have suggested several new therapies. The most obvious idea is to supply a wild-type gene to CF patients. We could use any of several viruses or liposomes to carry the CFTR gene to the place where it would do the best for the lungs perhaps as simply as by using an aerosol inhaler.

Instead of providing a new gene, we may be able to ameliorate the disease symptoms simply by giving regular doses of the CFTR protein by an inhaler. One danger would be that the patients could develop an allergy to the protein, but it appears that they have nonfunctional protein in their lungs already, and the difference between mutant and wild –type proteins (only one amino acid) is so slight that the body may not recognize the wild –type protein as foreign, and so might tolerate it. Another strategy is to stimulate a latent activity of the mutant strategy is to stimulate a latent activity of the mutant protein with drugs and CF specialists are already considering some drugs that may be able to do this.

13.51 Other Applications

The SNPs present in the human genome can be linked to human diseases; we could then screen individuals for the tendency to develop those diseases. We might also be able to find sets of SNPs that associate with polygenic traits, such as intelligence, and thus pin down the genes responsible for these traits.

We may also be able to identify SNPs that correlate with good or poor responses to certain drugs. Using this information, physicians should be able to screen a patient for key SNPs, then custom design a drug treatment program for that patient based on his or her predicted responses to a range of drugs. This field of study is called pharmacogenomics.

However, these tasks will not be easy. Already, geneticists are discovering that most SNPs are not in genes at all but intergenic regions of DNA. Even when they are found within genes, they tend to be silent mutations that do not alter the structure of the protein product, and thus do not cause any malfunction that could lead to disease. The reason for this situation is clear: polymorphisms caused by mutations that change the products of genes are generally deleterious and are therefore selected against. That is, the individuals with these damaging mutations generally die before they can reproduce and thus the mutations are lost.

We can detect SNPs correlating with the disease or other traits in any given individual by hybridizing that person's DNA to DNA microarrays containing oligonucleotides with the wild–type and mutated sequences. Such knowledge can be useful in helping to prevent or treat disease.

How do SNPs differ from RFLPs? RFLPs are identical to SNPs if the single–nucleotide difference between two individuals lies in a restriction site, as we observed in the RFLPs involving HindIII sites in Huntington's disease patients. In such a case, a single – nucleotide difference makes a difference in the pattern of restriction fragments. However, RLFPs can also result from the insertion of the chunk of DNA between two restriction sites in one individual, but not another – VNTRs, for example. That would not be an SNP because it involves more than just a single nucleotide difference.

The complete sequences of genomes of lower organisms can also be important in understanding and treating human diseases. For example, as soon as the complete yeast genome had been sequenced, molecular biologists began systematically mutating every one of the 6000 yeast genes to see what effects those mutations would have. They also began systematically screening all 18 million possible protein–protein interactions using a yeast two-hybrid screen. The results of such experiments can tell us much about the activities of gene products that are still uncharacterized. And knowing the activities of all the proteins in an organism, and the other proteins with which they interact should lead to a greater understanding of biochemical pathways, such as the ones that metabolize drugs. This understanding, in turn, should give us important clues about how these pathways work in humans. Moreover, yeast cells can be used as human surrogates to test the effects of knocking out the yeast ortholog of a known human disease gene.

13.52 Bioinformatics

As our databases swell with billions of bases of sequence from the human and other genomes, one crucial problem will be to access and manipulate all that data. Accordingly, a new specialty has arisen, known as bioinformatics. Practitioners of bioinformatics must understand both biology and computerized data processing, so they can manage the data collecting during genome sequencing and then provide user access to the data.

Two types of databases are already established. First, we have generalized databases that include DNA and protein sequences from all organisms. Two generalized databases for DNA sequences are GenBank and EMBL. Swissprot is a generalized protein sequence database. Second, we have specialized databases that deal with a particular organism. For example, Fly Base is a database of the genome of the fruit fly *Drosophila melanogaster.* You can access it online at http://flybase. bio.indiana. edu:82, and search it for genetic maps, genes, DNA sequences, and other information. The Genome Database (GDB) is a website *(http://gdbwww. gdb. org)* that allows you to search the human genome for any gene you are interested in. You must select a database to search, and two of the most popular are EMBL and GenBank. The Institute for Genomics Research (TIGR) has a website *(http://www.tigr.org/ tdbl)* that links to the TIGR databases, including the complete genomes of many microbes.

13.53 Ethics and the Human Genome Project

The Human Genome Project is raising ethical issues. Knowledge of the genome will provide us the knowledge of defective genes. Causing diseases and further developing a test to identify these diseases. There are some diseases where cure is rare *e.g.* cancer. This leads to a large number of ethical questions. Should a patient be told if a test for an incurable genetic disease is positive? (This issue applies now to Huntington disease, a dominant lethal disease in which the symptoms typically do not appear until later in life.) Should employers be able to ask for the results of a genetic test if the employee does not want to know? Should health insurance companies or employers have access to genetic testing data, and, if so, how can the patient protect his or her insurability and employability? Should states be able to collect genetic data? The last two questions raise fundamental privacy issues.

Fortunately, these issues are not being ignored. The federal agencies funding the HGP are devoting 3 to 5 per cent of their annual budgets to study the ethical, legal, and social issues (ELSIs) related to the availability of genetic information. (http://www. ornl.gov/sci/techresources/Human_Genome/ elsi/elsi.shtml) This amounts to the world's largest bioethics program. Four areas are being emphasized by the ELSI program: (1) the privacy of genetic information; (2) the safe and effective introduction of genetic information in the clinical setting; (3) fairness in the use of genetic information; and (4) professional and public education. Appropriate laws and regulations are expected to be developed because of the activities of the ELSI program and continuing dialogues among scientists, physicians, lawmakers, and members of the public. Hence the following issues are taken into consideration in the human genome project. (i) Privacy (ii) The safety (iii) Right genetic information (iv) Education to both professional and public domains.

14

Genetics of Bacteria and Viruses

Among prokaryotes, sexual reproduction is not possible. The bacteria and bacteria phages can grow on a semisolid agar surface in a Petri dish and are helpful in DNA recombinant technology. If we allow bacteria or viruses to grow in a solid agar, it will need continuous monitoring. The media used for this purpose requires an essential medium with a unique composition and pH. Exposure to air is vital for their growth. The essential minimal medium consists of glucose/lactose Na+, K+, Mg2+, Ca2+, and NH4+. To grow on a minimal medium, a bacterium must synthesize all essential organic compounds like amino acids, purines, pyrimidines, sugars, vitamins, and fatty acids. Bacteria that can grow on minimal media are called prototrophs. If any organic substances other than carbon sources are required, it is termed an auxotroph. For example, if histidine is added to the resultant bacterium, it is called his negative or his -ve auxotroph, which contrasts with its prototrophic counterpart.

Growth is sluggish originally, and this stage is known as the lag phase of bacterial growth; after this, it enters the log phase. Once the cell density reaches about 10^9 cells/ml, the nutrients, and oxygen become limiting, and the cells stop growing; at this point, it enters the stationary phase. The culture's log phase depends upon the temperature and the composition of the growth medium. The temperature for the growth of different bacterial species is different.

After numerous cell divisions, a visible colony is formed on the surface of the medium. Once the colonies attain a high count, succeeding serial dilutions are required. The original liquid culture is prepared and plated till the colony number is decreased to the point where it can be counted.

14.1 Genetic Recombination in Bacteria

Genetic recombination is more straightforward in bacteria as bacteria reproduce asexually; hence the genetic material of one bacterium can be mixed with the other. It was discovered in 1946 by Joshua Lederberg and Edward Tatum. They observed that bacteria can conjugate; this is the para-sexual process in which the genetic information from one bacterium is transferred to the other resulting in recombination. This differs from the genetic recombination in eukaryotes, where the term describes crossing over due to reciprocal exchange events; however, the overall effect is similar. The transfer of genetic information from one bacterium to another also occurs through transformation and transduction. Recombination and transduction help to find out the arrangement of genes (**Figure 14.1**).

E. coli K12 was chosen, and its first strain was grown on methionine and biotin, while strain B required threonine, leucine, and thiamine (**Figure 14.1**). The supplemented media was taken in two strains which were first grown separately and then mixed and added into the supplemented media. Several generations took to grow them fully. Once grown fully, these were plated on a minimal medium. Any bacterial cells that grew on a minimal medium are known to be *prototrophs*. It is implausible that any cells with two or three mutant genes can undergo

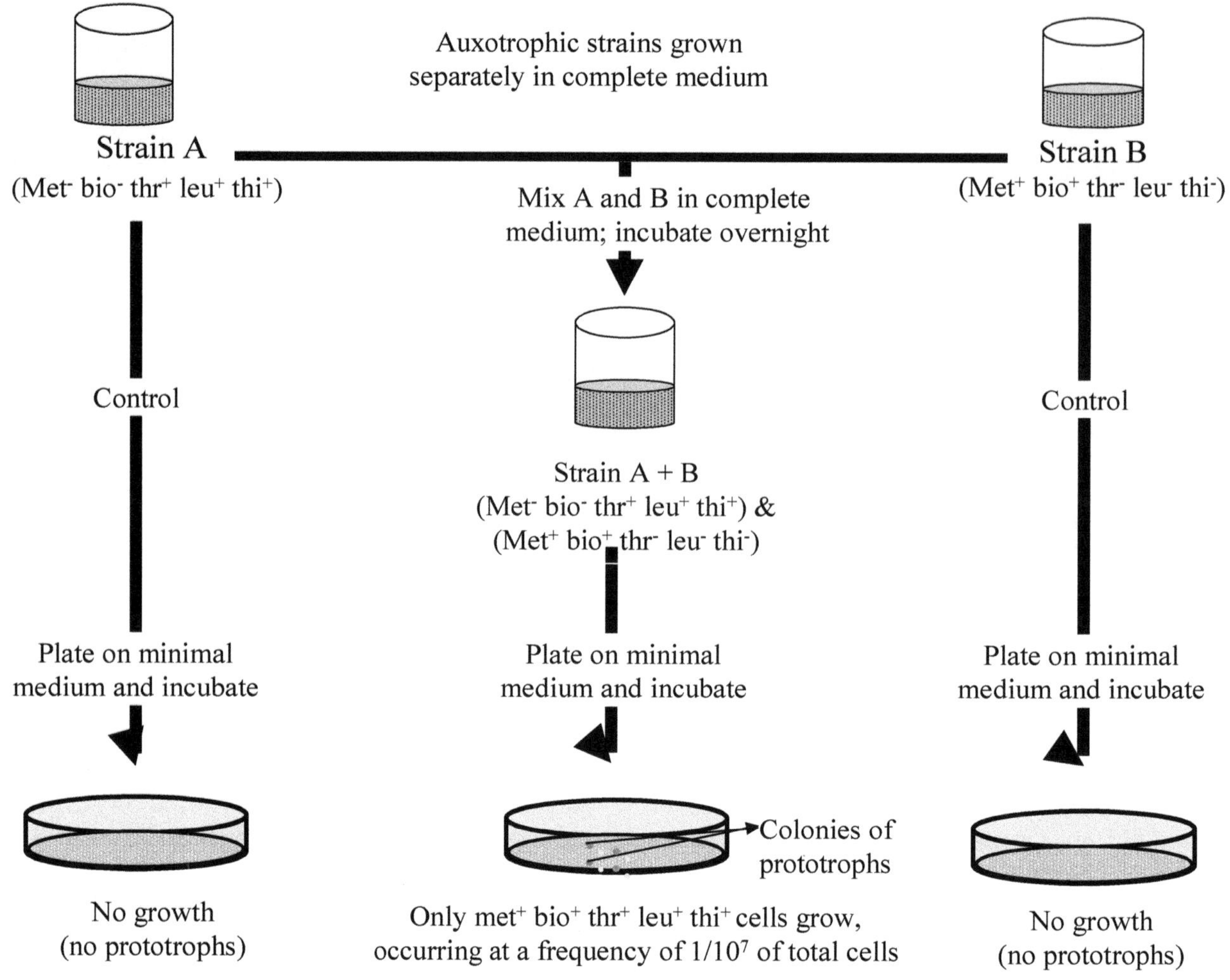

Figure 14.1: Recombination of Two Auxotrophic Strains Producing Prototrophs.

spontaneous mutation simultaneously at two or three locations independently, leading to wild-type cells. Therefore, researchers understood that any *prototrophs* arose due to some form of genetic exchange and recombination between the two mutant strains. The recovered prototrophs were $1/10^7$ or 10^{-7} cells.

14.2 F⁺ and F⁻ Bacteria

Donor cells for their chromosomes are selected; these are called F+ cells (F for "fertility"). These are male cell's recipient bacteria, and they receive the donor chromosome material; these are recombined with their chromosome. They are nominated as F cells or female cells.

For chromosomal relocation from cell to cell, contact is essential. Bernard Devis designed a tube to grow F+ and F- cells (**Figure 14.2**). On one side of the filter F, positive cells are placed. It is sintered glass and F- F-cells on the other side. The medium is flown back and forth; this allows cells to filter the common medium after the incubation is started.

These are called Hfr cells. Conjugation and transfer of part of the bacterial chromosome from an Hfr donor to an F– F-recipient results in recombination.

The physical interaction is the first step in the process of conjugation established by a structure called the F pilus (or sex pilus; pl, pili). Bacteria have many pili, which are tubular extensions of the cell. After the contact is started between mating pairs, chromosome transfer starts. The F+ cells have the fertility factor (F factor). If fertile cells are grown together with infertile cells, the infertile cells regain fertility.

At the time of conjugation, chromosomal genes are transferred. Now,

The F factor is a mobile element; after conjugation and genetic recombination, recipient cells always become F+. Thus, in addition to the rare cases of gene transfer from the bacterial chromosome (genetic recombination), the F factor is passed to all recipient cells.

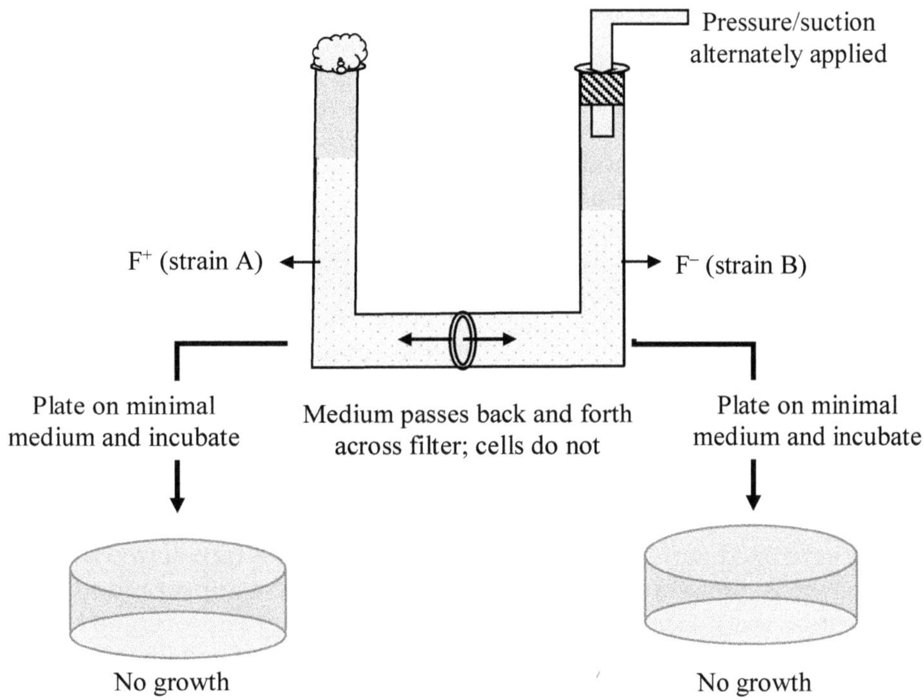

Figure 14.2: The Davis U-tube Apparatus where Auxotroph Strains A and B are grown in a Standard Medium but Separated by a Filter.

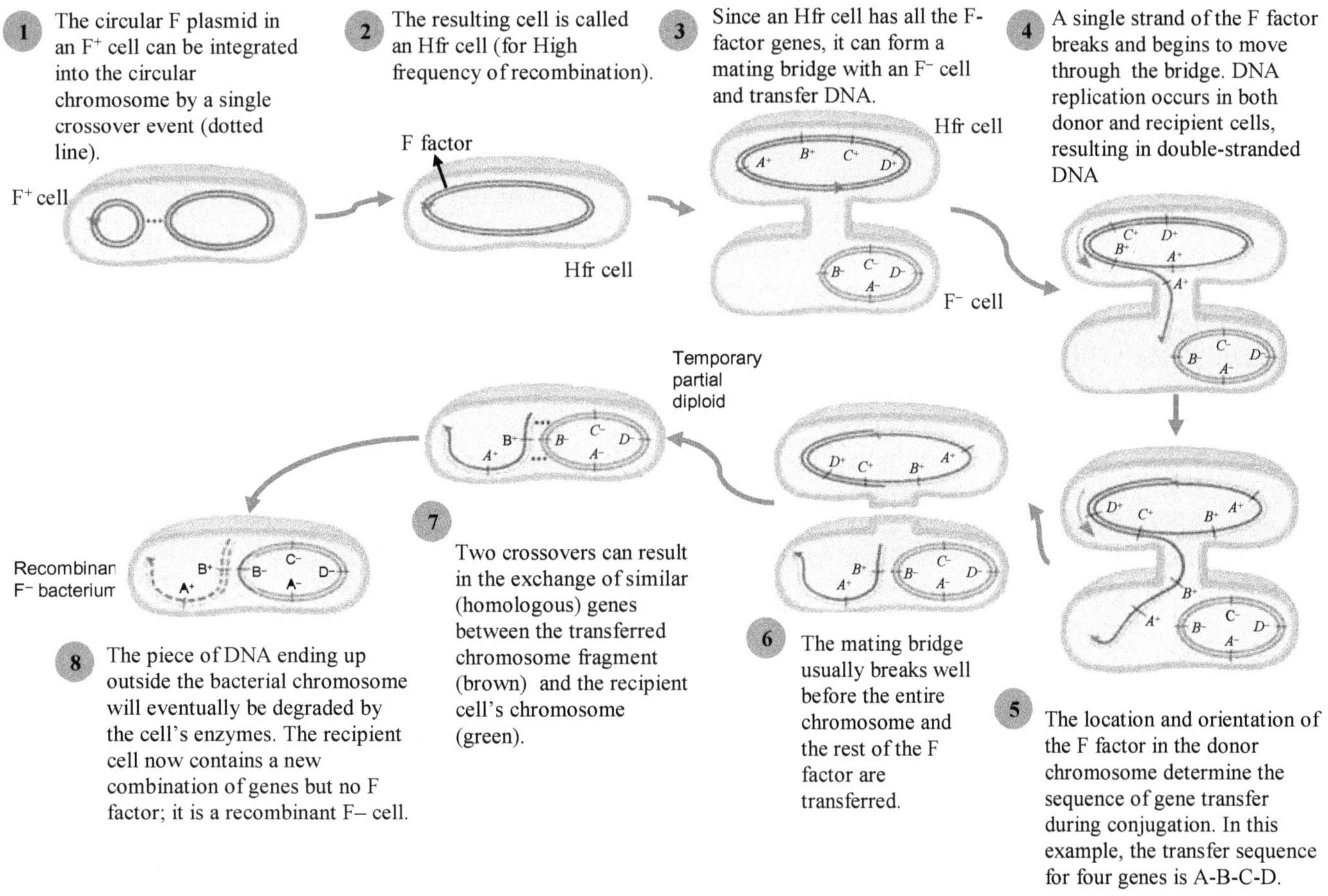

1. The circular F plasmid in an F^+ cell can be integrated into the circular chromosome by a single crossover event (dotted line).

F+ cell

2. The resulting cell is called an Hfr cell (for High frequency of recombination).

F factor

Hfr cell

3. Since an Hfr cell has all the F-factor genes, it can form a mating bridge with an F^- cell and transfer DNA.

Hfr cell

F^- cell

4. A single strand of the F factor breaks and begins to move through the bridge. DNA replication occurs in both donor and recipient cells, resulting in double-stranded DNA

5. The location and orientation of the F factor in the donor chromosome determine the sequence of gene transfer during conjugation. In this example, the transfer sequence for four genes is A-B-C-D.

6. The mating bridge usually breaks well before the entire chromosome and the rest of the F factor are transferred.

Temporary partial diploid

7. Two crossovers can result in the exchange of similar (homologous) genes between the transferred chromosome fragment (brown) and the recipient cell's chromosome (green).

Recombinant F^- bacterium

8. The piece of DNA ending up outside the bacterial chromosome will eventually be degraded by the cell's enzymes. The recipient cell now contains a new combination of genes but no F factor; it is a recombinant F– cell.

Figure 14.3: Conjugation and Transfer of Part of the Bacterial Chromosome from an Hfr Donor to an F^- Recipient, causing Recombination.

14.3 Hfr Bacteria and Chromosome Mapping

Bacterial cells are mostly haploid mapping techniques in prokaryotes that are different than eukaryotes. Hfr bacterial conjugation method is used for mapping. The F- cell contains mutant alleles of two genes *a* and *b* (the F- would be *a*-, *b*-). In case this cell is conjugated with an Hfr cell which is *a*+, *b*+ (in other words, wild type), the F- cell will undergo gene conversion to *a*+, *b*+; during this period, both genes are transferred through conjugation.

By determining how long it takes the *b* gene to transfer after it has been transferred, it is possible to get a rough idea of how far away two genes are on the chromosome.

The *a*+, *b*+ Hfr cells get mixed with *a*-, *b*- F- cells. The time of mixing is known as 'time zero'. At regular intervals, a little amount of the mixture would be removed, and conjugation may be disrupted using a blender. Gene conversion can be tested. If a gene was converted to wild type 8 minutes after time zero, and then the *b* gene was converted to wild type 19 minutes after time zero, the distance between the two genes would be 11 minutes. Bacterial map distances are always expressed in minutes (**Figure 14.3**).

14.4 Recombination in F+ x F- Mattings

F plasmids are of two types F+ and F-. Notably, F plasmids have almost 100 genes that add different properties to the plasmids. These plasmids can divide and, after division, will maintain the original properties. Both F+ and F-ve cells can conjugate F+ and act as a donor, and after conjugation F- cells become F+. This process is relatively quick. Recombination may occur between homologous chromosomes of F+ and F- plasmids. This helps in mapping the bacterial chromosome. The donor male has the fertility factor called F+; recipient females do not have the fertility factor. F+ plasmids have piles, which are required for conjugation, and this needs pair formation. The sex pilus joins the recipient and forms a bridge through which the exchange takes place; the transfer of donor chromosomal genes occurs at a low frequency.

F plasmid is present in the same *E. coli* strains, which are integrated permanently into the chromosome of the cell. These strains are called high-frequency recombination (Hfr); they produce sex pili and can attach to F- F-cells, which allows DNA from the donor cell chromosome (of which the F plasmid is now an integral part) to be transferred. Mating pairs often break apart before there has been time to complete the transfer of donor cell DNA, so recipient cells rarely receive the complete F plasmid and remaining F-. DNA from the donor cell may undergo homologous recombination with DNA in the recipient cell, forming the recombinant cells with genetic properties, unlike either parent cell. This process is called conjugation. This occurs between F+ x F- mating. This is unidirectional and incomplete.

When F+ and F- cells are mixed, conjugation occurs readily, and each F- cell involved in conjugation with an F+ cell receives a copy of the F factor. F+ cells, the F factor spontaneously integrates from the cytoplasm to any random point in the bacterial chromosome, converting the F+ cell to the Hfr state. The F+ x F- matting occurs at a relatively low frequency for genetic recombination (10-7); these newly formed Hfr cells undergo conjugation with F- cells. As the integration of the F factor is random, the genes or genes transferred by any newly formed Hfr donor will also appear random within the larger F+/ F- population. The recipient bacterium will appear as recombinant but will remain F-. If it further undergoes conjugation with an F+ cell, it will then be converted to F+. This procedure involves pair formation.

1. Pair formation - The tip of the sex pilus is associated with the recipient and forms a conjugation bridge between two cells. The DNA passes through this bridge. During this episode, the exchange between the donor and recipient takes place. The DNA is protected from nucleases due to environmental factors. The mating pairs can be removed due to external forces; as a result, the conjugation can be disturbed. The mating pairs remain associated for a very short time. DNA transfer The plasmid DNA is nicked at a specific site called the origin of transfer and is replicated using a rolling circle mechanism. A single strand of DNA passes through the conjugation bridge and enters the recipient, where the second strand is replicated.
2. This process shows the features of F+ X F- crosses. The recipient becomes F+, the donor remains F+, and the transfer of donor chromosomal genes occurs at a low frequency. In practice, however, there is a low transfer of donor chromosomal genes in such crosses.

14.5 The F State and Merozygotes

The F factor mainly carries many other adjacent bacterial genes. These may be labeled condition F to differentiate it from F+ and Hfr. F, like Hfr, is thus an additional case of F+, but this exchange is from Hfr to F.

The presence of bacterial genes within a cytoplasmic F factor forms an exciting situation. An

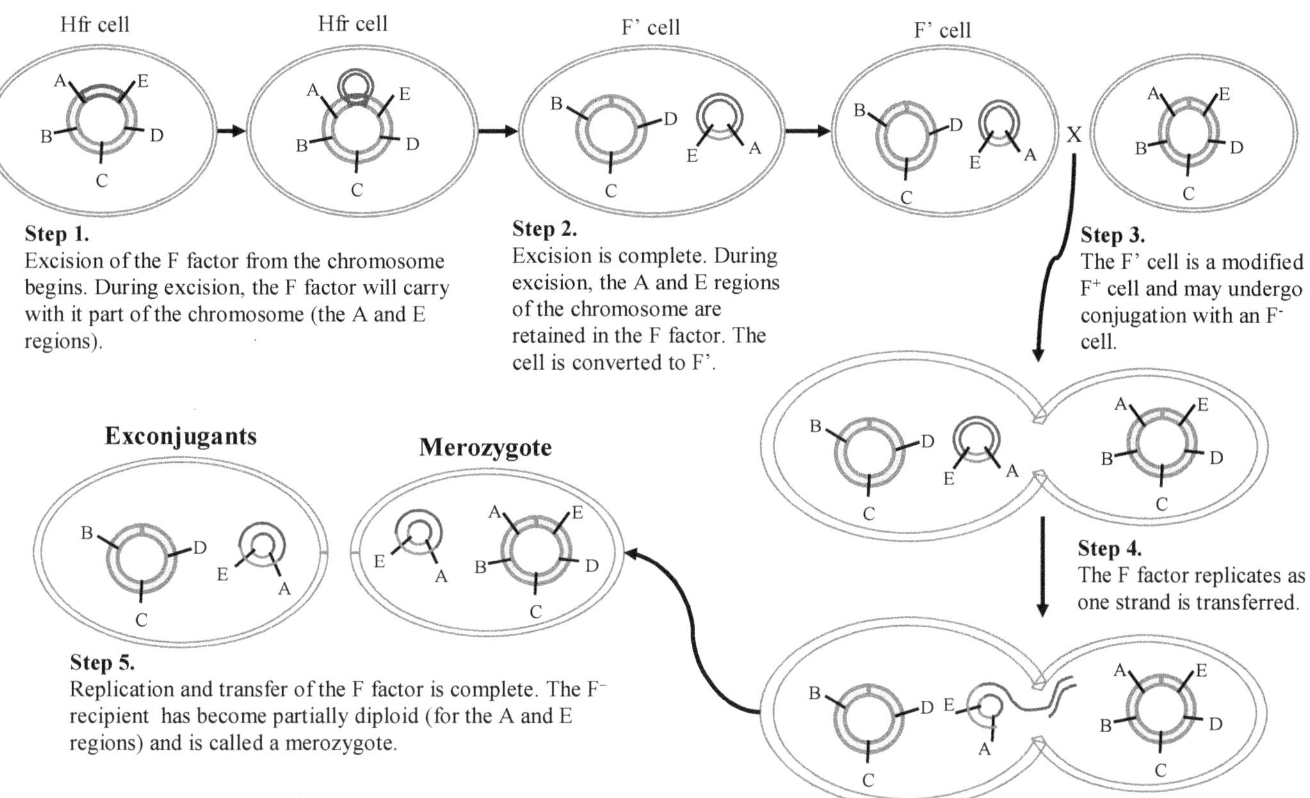

Figure 14.4: Conversion of an Hfr Bacterium to P and Subsequent Mating with a P- -cell.

F' bacterium is like an F+ cell due to conjugation with F- cells (**Figure 14.4**). This is observed in the F factor, which contains chromosomal genes and is transferred to the F cell (step 4). The chromosomal genes are part of the F factor and are now present as duplicates in the recipient cell (step 5) because the recipient still has a complete chromosome. A diploid cell is known as the merozygote.

Merozygote is an organism that, in addition to its original genome (endogenote), contains a fragment (exogenote) of a genome from another organism; the relatively small size of the exogenote permits a diploid condition for only a limited region of the endogenote. Pure cultures of F' merozygotes can be recognized. These can be used to study the genetic regulation of bacteria. It has an opportunity for recombination. Bacterial recombination requires specific proteins. The first protein implicated is the RecA protein. The other proteins are complicated.

14.6 Transformation Helps in Genetic Recombination in Bacteria

Bacterial transformation is how a recipient cell receives genes from free DNA. Entry is at a limited number of receptor sites (**Figure 14.5**). It needs energy and specific transporter molecules. All those substances that slow down energy production or protein synthesis in the recipient cell also inhibit the transformation process.

Once the DNA has entered the bacterial cell, one of the two double-helix strands is digested by nucleases, leaving only a single strand to participate in the transformation procedure (**Figure 14.5**). DNA strand then aligns with its complementary region of the bacterial chromosome. This process engages several enzymes; the segment replaces its counterpart in the chromosome, which is removed and degraded.

The recombination can be detected by removing the transforming DNA from a diverse strain of bacteria that show some genetic variation, such as a mutation. Once integrated into the chromosome, the recombinant region comprises one host and one mutant strand.

14.7 Transformation and Linked Genes

For DNA to be effective during transformation, it is required that it must contain about 10,000 and 20,000 nucleotide pairs, a length equal to about 1/200 of the *E. coli* chromosome; this size encodes many genes that are near to one another. This type of arrangement makes co-transformation of several genes simultaneously. Genes very close to each other to be co-transformed are called linked genes. In contrast to linkage in eukaryotes, which show

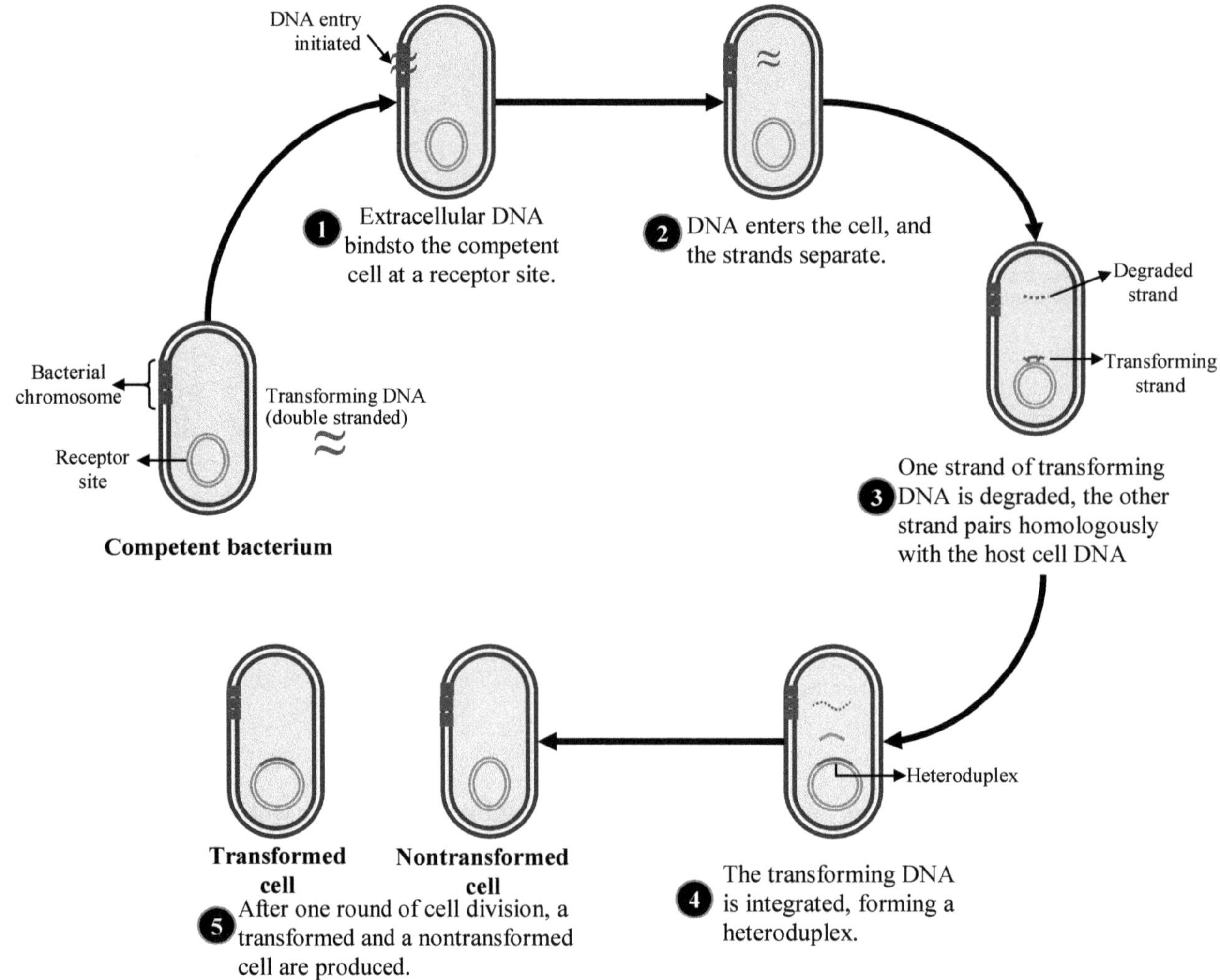

Figure 14.5: Transformation of a Bacterial Cell through Exogenous DNA.

all genes on a single chromosome, linkage here refers to the closeness of the genes. Under certain circumstances, the relative distances between linked genes can be generated from the transformation data. While analysis is more complicated, such data are evaluated analogous to chromosome mapping in eukaryotes.

14.8 Bacteriophages are Bacterial Viruses

Bacterial phage is a parasite. Phages have a rapid replication cycle. A phage can persist by itself but not use bacterial cells for replication. Phage genes encode proteins.

Phages can perform the following functions.

1. Protection of its nucleic acid from environmental chemicals that could alter the molecule (for example, break the molecule or result in a mutation).
2. Release of nucleic acid inside the bacterium.
3. Conversion of an infected bacterium to a phage-producing system yields higher numbers of progeny.
4. Delivery of progeny phage from the infected bacterium.

Some phages have less than ten genes and depend on cellular functions, whereas others have 30 – 100 genes and are more dependent on proteins encoded in their genetic material. Few phage genes may duplicate the host genes.

14.9 Structures of Phages

The phages can be either DNA or RNA phages but never both. A simple phage has only 3-4 genes, while a complex phage may contain 100 genes. Sometimes the phages that contain DNA may have single-stranded DNA. The size of the phage is 24-200 nm in length. The giant phage is T4; its length is 200 nm, and its width is 80 -100 nm. Few phages are *icosahedral*, while others are filamentous.

Nucleic acids are in the head, and the head has a protective covering. Sometimes the tail is attached to the head. The T4 tail has a tail which is a hollow tube. Tail fibers are attached to the base plate.

Most of the time, the phage particles contain a single nucleic acid molecule which may be single or double–stranded, linear, or circular DNA, or single–stranded, linear RNA and one or more proteins. (The one known exception is phage f6, which contains three linear double–stranded RNA molecules whose base sequences differ from one another). The proteins form a shell, called the coat or the capsid, around the nucleic acid; the nucleic acid is thereby protected from nucleases and harmful substances. Phages containing double–stranded DNA are typically 50 per cent DNA by weight and hence are a valuable source of DNA for physical studies **Figure 14.6**.

14.10 Life Cycle of Phage

Each phage has two life cycles; the lytic or virulent phages multiply in bacteria and kill the cell by lysis at the end of the life cycle. The basic lytic cycle is shown in **Figure 14.7**. The lysogenic cycle, which has been observed only with phages containing double–

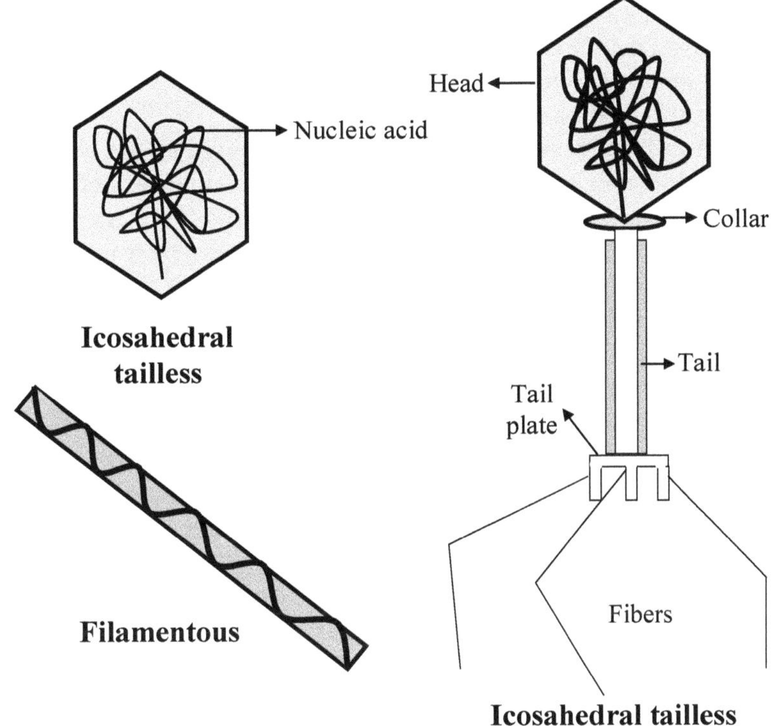

Figure 14.6: Three Basic Phage Structures.

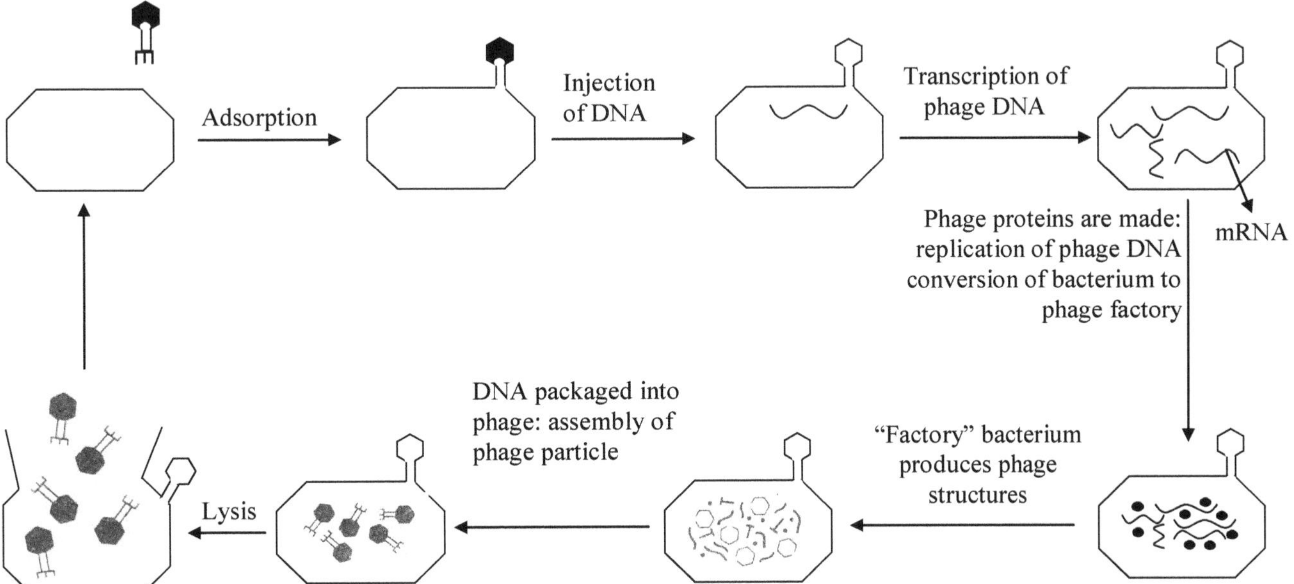

Figure 14.7: Life Cycle of Typical Phage.

stranded DNA, is one in which progeny particles are produced; the phage DNA usually becomes part of the bacterial chromosome. The phage nucleic acid takes over the host biosynthetic machinery, resulting in the phage-specified mRNA and proteins. Sometimes the host chromosome is degraded by the host chromosome. Structural proteins (head, tail) are present in the phage, and the proteins requiring lysis of the bacterial cell are synthesized separately. Nucleic acid is backed into the head after that, into the tail. The assembly of phage components into mature infective phage particles is called maturation. In the lysis and release phase, the bacteria begin to lyse due to the accumulation of the phage lysis protein, and intracellular phage is released into the medium. It is hypothesized that phage enzymes weaken the cell wall of bacteria. A large number of particles are released per infected bacteria may be as high as 1000. The average yield of phages per infected bacterial cell is known as burst size.

14.11 Lysogenic Cycle

Some phages multiply via the lytic cycle, or they can become dormant. In both conditions, phage DNA enters the host chromosome, replicates with the host chromosome, and is passed on to the daughter cells. This stage is known as the prophage. The procedure is called lysogeny, and the bacteria-harboring prophage is called lysogenic bacteria. Prophage has a gene that provides new properties to the bacteria. When a cell becomes lysogenic, extra genes carried by the phage are occasionally expressed on the cell's surface. Genes can bring alterations in the properties of the bacterial cell. The process is known as phage conversion.

Lysogenic phages carry genes that can modify the Salmonella O antigen. Toxin production by *Corynebacterium diphtheriae* Beta phage carries a gene that mediates Corynebacterium diphtheriae strains undergoing lysogeny; these are pathogenic.

Food poisoning is caused by clostridium botulinum, which makes several different toxins encoded by prophage genomes. Lysogenized bacteria are resistant to superinfections by the same or related phages. This is called superinfection immunity.

The lysogenic state of a bacterium can be terminated at any time when exposed to unsuitable conditions. This is called induction. Termination of lysogenic conditions is desiccation, exposure to UV or ionizing radiation, exposure to mutagenic chemicals, *etc.* The separated phage DNA then initiates the lytic cycle resulting in cell lysis and release of phages. These phages can infect new cells at risk and make them lysogenic.

Most temperate phages also undergo a lytic cycle in certain circumstances, as illustrated below: Adsorption of the phage to the precise receptors on the bacterial surface.

1. Passage of the DNA from the phage through the bacterial cell wall. Some types of tailed phages use an injection sequence. In this process, the nucleic acid is transferred into the cell and is not exposed to the medium near the recipient cell; however, little is known about the procedure (**Figure 14.8**).

   Following infection by most phages, a bacterium loses the ability either to divide or to transcribe its DNA; sometimes both functions are lost. This shutdown of host DNA or RNA synthesis is achieved in many ways (for example, degradation of host DNA) depending on the phage species. Shutdown is less common with phages containing single-stranded DNA or RNA.
2. Production of phage nucleic acid and proteins. It takes place using several ways the phage directs the synthesis of a replicative system that specifically makes copies of phage nucleic acid. This programming is accomplished either by synthesizing phage-specific DNA and RNA polymerases or by adding specificity elements to bacterial polymerases.

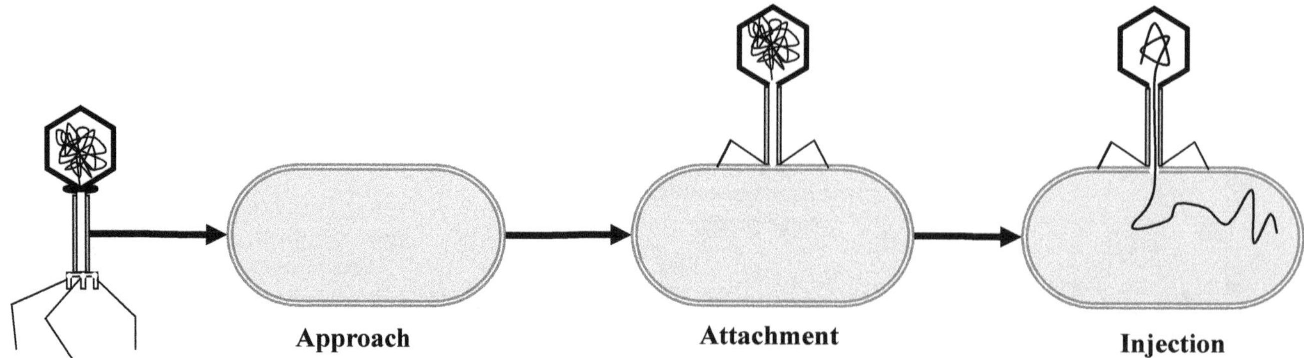

Figure 14.8: Injection Sequence of a Tailed Phage.

RNA-containing phages differ in the use of host replication enzymes-they encode their replication enzymes as bacteria do not have enzymes that replicate RNA. Furthermore, the RNA molecules of phages containing single-stranded RNA serve as their mRNA. Phages also contain single-stranded DNA and differ from the prototype as they use unmodified RNA polymerase throughout the life cycle.

Two types of proteins are compulsory for the assembly process: structural proteins, present in the phage particle, and catalytic proteins, which contribute to the assembly process but do not become part of the phage particle. It consists of the maturation of proteins, which convert intracellular phage DNA to make it appropriate for packaging in the phage particle. The icosahedral phage assembly occurs in numerous stages: (1) Aggregation of phage structural proteins to form a phage head and, when required, to form a phage tail; at this point, the tail is not attached to the head. (2) Condensation of the nucleic acid and entry into a preformed head. (3) Attachment of the tail to a filled head. With filamentous phages, the nucleic acid and the protein form a phage particle in a single step. The mechanism of nucleic acid condensation is not entirely known. Usually, 50-1000 phage particles are produced; the number depends upon the particular phage species. Most phages synthesize an enzyme called a lysozyme or an endolysin late in the infection cycle.

14.12 Production of a Phage Lysate

Phage multiplies much more rapidly than bacteria. That is, bacteria double in one generation time whereas, in one life cycle, the number of phages is augmented by a factor equal to the burst size (**Table 14.1**). A single phage whose average burst size is 100 and whose life cycle lasts 25 minutes infects a one-milliliter bacterial culture with a 25-minute doubling time. In this calculation, it is assumed that adsorption is always instantaneous and complete. Note that, in four generations, the number of bacteria has increased 14-fold, whereas the number of phages has increased 108-fold. There is about an eightfold higher number of phages as bacteria; hence, all bacteria are infected. Thus, after 125 minutes, the bacteria are gone, and the original phage particle has produced 1.4×10^9 progeny.

The second and less common type, for which *E. coli* phage PI is the prototype, differs from the preceding one in that there is no DNA-insertion system, and the phage DNA becomes a plasmid (an independently replicating circular DNA molecule) rather than a segment of the host chromosome.

14.13 General Properties of *Lysogenic* and Lysogenization

The following terms describe various aspects of lysogeny.

1. A phage capable of entering either a lytic or a lysogenic life cycle is a temperate phage.
2. A bacterium containing a complete set of phage genes is called a lysogen.
3. Lysogenization forms lysogen by infecting a bacterial culture with a temperate phage.
4. If the phage DNA is contained within the bacterial DNA, the phage DNA is said to be integrated. The process by which this state of the DNA is achieved is called integration or insertion. Phage DNA in plasmid form is nonintegrated. Both integrated and the one–step growth curve.

Specific kinetic parameters of the phage disease can be investigated by examining an infected culture. A classic experiment is the one–step growth curve. In this experiment, a culture is infected with an MOI of about 0.1 (so no cell is infected with more than one phage). Phage antiserum is then added to inactivate any unabsorbed phage. The infected cells are diluted

Table 14.1: Calculation of the Phage and Bacterial Concentrations after Various Numbers of Bacterial Generations*

Number of Generations	*Phage/ml*	*Bacteria/ml*	
		Approximate	*Precise*
0	1	10^6	10^6
1	10^2	2×10^6	$2 \times (10^6 - 1)$
2	10^4	4×10^6	$(4 \times 10^6) - (2 \times 10^2) - 4$
3	10^6	7.98×10^6	$(8 \times 10^6) - (2 \times 10^4) - (4 \times 10^2) - 4$
4	10^8	1.4×10^7	$(1.6 \times 10^7) - (2 \times 10^6) - (4 \times 10^4) - (8 \times 10^2) - 16$
5	$1.4 \times 10^9 = 100 \times (1.4 \times 10^7)$	0	0

* Initially, a bacterial culture at 10^6 cells/ml concentration is infected with one phage. The doubling time of the bacteria and the life cycle of the phage are equal.

approximately 1000-fold in the fresh, warm medium (to prevent inactivation of progeny phage by the antibody), and aliquots are taken at various times for plating for plaques. Initially, the number of plaques is constant because plaques are formed only by infected unlysed cells; that is, each infected cell produces one plaque. This period is called the latent period. The number of plaques increases at a particular time after infection (a time characteristic of each phage). The infected cells are lysing during this short interval (the rise period). When all infected cells have lysed, the phage concentration remains constant. The ratio of phage produced to the initial number of infective centers is the burst size, and the number of minutes before the increase in plaque number occurs is the lyses time.

14.14 The Single-Burst Technique

In a single-burst experiment, a bacterial culture is infected (at an MOI appropriate to the phenomenon being studied 0.1 to study burst size). Once the phages are adsorbed, the infected cells are diluted in growth medium to an exceedingly low concentration-usually, about 0.05 infected cells/ml. Then, 1-ml aliquots are dispensed into hundreds (or thousands) of test tubes. At this concentration, 95.1 per cent of the tubes will not contain infected cells, 4.8 per cent will contain one infected cell, and 0.1 per cent will contain more than one infected cell. Thus, 4.8/4.9, or 98 percent, of the tubes containing infected cells will contain one infected cell. The single cell in each tube is allowed to lyse, and the contents of each tube are plated in a single petri dish with indicator bacteria so that plaques will form. Thus, the plaques on one plate are formed by the phage progeny of a single infected cell. The number of plaques observed on various plates yields the burst size distribution, ranging from less than 10 to several hundred. The wastefulness of this technique should be noted: to study 100 infected cells requires using about 2000 Petri dishes, of which roughly 1900 will contain no plaques. The single-burst technique is mainly used to study phage cross results in a single cell.

14.15 Specificity in Phage Infection

The phages have been isolated, and the specificity of phage is generalized; for example, it can infect particular genera but not subsets of strains or species within the genera. This may be due to phage adaptation followed by evolution. This may result in resistant varieties.

For example, populations of Candidatus isolated from two separate sludge bioreactors were found to differ primarily in genomic regions encoding phage defense mechanisms; despite global dispersal of the strains among the two sites, no phage species that infects the genus *Pseudomonas* can also infect *E. coli*; furthermore, phages that grow in *Ps.* fluorescence generally fail to grow in *Ps.* aeruginosa. Among the *E. coliphages*, the most extensively studied phages, extraordinary specificity has been observed; for example, the phage fX 174 grows well on *E. coli* strain C but fails to grow on most other laboratory strains. There are exceptions, though; for instance, *E. coli* phage T4 can grow on many strains of *E. coli* and certain species of the genus *Shigella*. The above-cited examples show that the highly resistant bacterial strains/species are only infected by broad host range phages. In contrast, phages with narrow to broad host ranges infect the highly susceptible bacteria. The observed lack of modularity may suggest a true continuum of phage host range. Alternatively, it might reflect that most studies included in the analysis examine interactions within a single bacterial species (*i.e.*, across multiple strains/genotypes) or between phages from one.

An interesting phenomenon occurs if a bacterial culture is infected with both wild-type (*T6h+*) and *T6h* phages at an MOI such that all

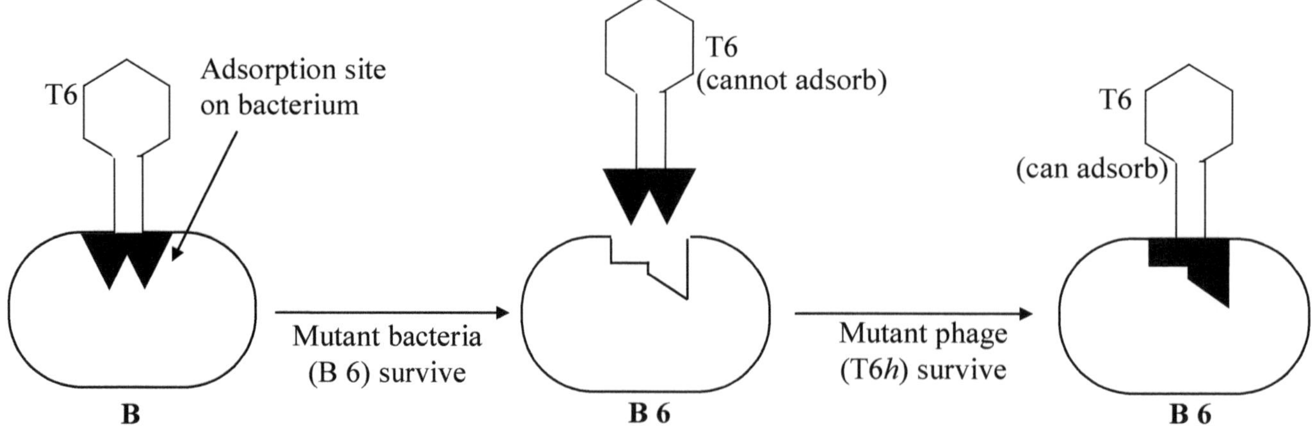

Figure 14.9: A Phage-Resistant Bacterium and a Phage h Mutant.

bacteria are infected with both phage types. One would expect all progeny phage to form plaques on strain B (which plates both phage types), and half of the progeny (the *h* mutants) would plate on B/6. However, only one-fourth form plaques on B/6. The reason is that during the phage-assembly process, no mechanism exists to match a DNA molecule with the *h* genotype with an *h-type* tail fiber. Thus, half of the phage with *h* DNA is in *h* + particles and cannot adsorb to the B/6 bacteria. Similarly, there are *h*+ DNA molecules in *h* particles; they can absorb B/6 but produce *h* + progeny, which cannot go through further cycles of infection. This phenomenon is called phenotypic mixing. If the initial lysate resulting from the mixed infection is allowed to infect a B culture at an MOI well below 1, no phenotypically mixed progeny results (**Figure 14.9**).

14.16 Host Restriction and Modification

Another specificity factor in phage infection is the inability of a "foreign" bacterial RNA polymerase to recognize phage DNA sequences to initiate RNA synthesis. However, even when adsorption and mRNA synthesis are possible, with most bacteria, there is usually another barrier called host restriction and host modification. This is a phenomenon in which a bacterium of type X can distinguish a phage grown in type X bacterium from one grown in a different type, such as Y, and can prevent the phage grown in Y from carrying out a successful infection. The notation used to discuss this phenomenon is that a phage P grown in a bacterium X is denoted p. X. Host modification and the data in Table 14.2 illustrates restriction. Note that A' K, grown in *E. coli* strain K, forms plaques at low efficiency in strain B. Thus, 'l' K is restricted by strain B. The phage population in these rare plaques ('l,' B) has been modified by strain B, so the phage grows efficiently in strain B; however, A' B now fails to grow in strain K-that is, it is restricted. The molecular explanation for this is the following: *E. coli* B contains an enzyme called a restriction endonuclease; specifically, it is the *EcoB* nuclease, a site-specific nuclease that cuts DNA strands only near a specific base sequence (most restriction enzymes cut within a target sequence, but EcoB cuts near the sequence). Phage A' K contains this sequence; when its DNA is injected into *E. coli* B, the phage DNA is broken. *E. coli* B also contains this sequence and would destroy its DNA were the sequence not modified. A site-specific methylating enzyme (*EcoB* methylase) methylates an adenine in the sequence, thereby rendering the sequence resistant to the EcoB nuclease. When K infects strain B, a few parental phage-DNA molecules in the large, infected cells are methylated before they are restricted. All progeny DNA molecules are already methylated on one strand, and the newly synthesized strands are also methylated rapidly, and restriction is avoided in this rare phage. Thus, a small phage population with the B modification (A' B) is produced. *E. coli* K also contains a restriction enzyme (*EcoK*). It attacks a base sequence that is different from the sequence recognized by *EcoB*. An *EcoK* methylase protects *E. coli* K from self-destruction, producing the K modification. A phage that has consistently been grown in strain K-namely, A.K-is methylated in the *EcoK*-specific sequence and is resistant to *EcoK* nuclease. However, A.B. has an un-methylated *EcoK* sequence, so A.B. DNA is usually broken when a strain K cell is infected. Occasionally, a DNA molecule escapes restriction and replicates, and its replicas have a methylated K-specific sequence. Thus, the rare progeny phage that results when A' B successfully infects *E. coli* K is A' K; they lack the B modification and are restricted when infecting strain B.

Table 14.2: The Restriction and Modification Pattern of E. coli Phage λ

Bacterial Strain	*Phage*		
	λ. K	*λ. B*	*λ. C*
K	1	10^{-4}	10^{-4}
B	10^{-4}	1	10^{-4}
C	1	1	1

Note: Numbers indicate relative plating efficiency.

Note in **Table 14.2** that A grows on strain C-i. e., l. C- also fails to grow well in strains B and K, but neither l.B nor l.K is restricted by strain C. The lack of restriction is because strain C has no restriction nuclease active against any base sequence in l DNA. The l, C phage is restricted by both strains B and K because strain C does not have the *EcoB* and *EcoK* methylases.

Host restriction and modification is a widespread process, probably serving to destroy foreign DNA.

14.17 Bacteriophage Mutations

The morphology of the plaques is affected by the mutations of the phage, due to which bacterial cells undergo lysis. When the viruses are isolated from these plaques and re-plated on *E. coli* B cells, the resulting plaque appearance is identical, revealing that plaque phenotype is inherited and caused by the reproduction of mutant phages. The mutant *rapid lysis* can be named as (*r*) because the plaques were larger, apparently caused by the phage's more rapid or more efficient life cycle. We now know that in wild-type phages, reproduction is inhibited once a particular-sized plaque is seen. The *r* mutant T2 phages overcome this inhibition, producing larger plaques.

Salvador Luria discovered another bacteriophage mutation, *host range* (*h*). This mutation shows the range of bacterial hosts that may infect the phage. Although wild-type T2 phages can infect *E. coli* B (a unique strain), they normally cannot attach or be adsorbed to the surface of *E. coli* B-2 (a different strain). The *h* mutation shows the basis for adsorption and subsequent infection of *E. coli* B-2. When grown on a mixture of *E. coli* Band B-2, the center of the *h* plaque shows much darker than the *h* + plaque.

In **Table 14.3**. Some of the mutations shown are helpful in studying bacteriophage genetics.

Table 14.3: Example of some Mutant Types of Phages

Name	*Description*
minute	Small plaques
Turbid	Turbid plaques on *E. coli* B
Star	Irregular plaques
UV-sensitive	Alters UV sensitivity
Acriflavine-resistant	Forms plaques on *acriflavine* agar
Osmotic shock	Withstands rapid dilution into distilled water
Lysozyme	It does not produce lysozyme.
Amber	Grows in *E. coli* K12, but not B
Temperature –sensitive	Grows at 25°C, but not at 42°C

14.18 Intergenic Mapping in Phage

Genetic recombination occurs in the bacteriophages, and it was discovered when mixed infection experiments were undergoing this experiment two distinct mutant strains were allowed to infect the same bacterial culture simultaneously. These studies were designed so that viral particles sufficiently exceeded the number of bacterial cells to ensure simultaneous infection of most cells by both viral strains. As two loci are involved, hence recombination is intergenic.

In one of the studies, the T2/E. *coli* system was used, and the parental viruses were of either the *h* + *r* (wild-type host range, rapid lysis) or *hr*+ (extended host range, standard lysis) genotype. The two parental genotypes would be the only expected phage progeny if no recombination occurs. However, the recombinants *h* + *r* + and *hr* have been detected in addition to the parental genotypes. As with eukaryotes, the percentage of recombinant plaques divided by the total number of plaques reflects the relative distance between the genes.

recombinational frequency = *h* + *r* + + *h r* total plaques X 100

In the case of the *h* + head *hr* + example discussed here, recombinant *h* + *r* + and *hr* chromosomes are formed. These chromosomes can replicate, where new replicates exchange with each other and the parental chromosomes. Furthermore, recombination is not limited to exchanges between two chromosomes- three or more can be involved concurrently. As phage development progresses, chromosomes are randomly removed from the pool and packed into the phage head, resulting in mature phage particles.

15

The Genetic Control of Development

A single-cell zygote is produced by fertilization between egg and sperm during development. Once the organism is fully developed, it contains different cell types with variable functions; for example, they carry oxygen, muscle cells contract, fat cells store nutrients, and nerve cells transmit information. During pattern formation, the interaction between cells of a developing embryo is significant so that each cell will find its position within the emerging body plan. However, all the cells of an early embryo become visible as identical embryos. The central question remains unanswered: What is accountable for deciding the fate of these cells? The cells undergo different functions like differentiation, morphogenesis, and pattern formation. The fertilized egg undergoes several changes before an adult organism is formed.

Developmental genetics have been extensively studied in nematode *Caenorhabditis elegans* and the fruit flies *Drosophila melanogaster*. These studies have demonstrated that the genome of these organisms contains a developmental program that unfolds and results in the expression of different sets of genes in different types of cells. The presence or absence of a particular gene expression depends on the existence of the transcription factor chromatin structure or the synthesis of a particular receptor molecule. Further, gene interaction is another important factor. These regulatory interactions determine the growth, identity, and patterning of the structures they affect. Both genes and the environment are essential for the development of organisms. The genetic control of the developing embryo determines the fate of the embryo will depend on whether the developing embryo will become a nematode, a fruit fly, a chicken, a mouse, or a human.

During development, genes selectively get turned on or off by regulatory proteins. Let us suppose that the developmentally regulated genes are (r) for red color, (g) for green, (b) for blue, and (p) for purple color. The r gene product divides into one region of the egg and sets the polarity of the egg due to the presence of the r gene product. When division occurs, the polarized cell can produce daughter cells with or without the r gene product.

If the transmembrane receptor is absent, all four cells will express the b gene. Expression of the *b* gene in one cell is inhibited in the neighboring cell as the transmembrane receptor is present. The product of the *b* gene causes stimulation of the receptor of nearby cells, and stimulation of the receptor down-regulates the transcription of the b gene. The *b* gene expression is regulated by lateral inhibition; the initial activation of the g gene in cell L is retained in daughter cells 1 and 2 due to the presence of the product of the g gene having a positive auto-regulatory activity and stimulating its transcription. Positive autoregulation is how cells can be amplified. If signals are weak, it results in permanent changes in gene expression. The result of lateral inhibition of *b* gene expression and g gene autoregulation causes cells 1 through 4 in **Figure 15.1**; these be genetically similar but developmentally different as they express different combinations of the g and *b* genes. Specifically, cell 1 has g and b activity,

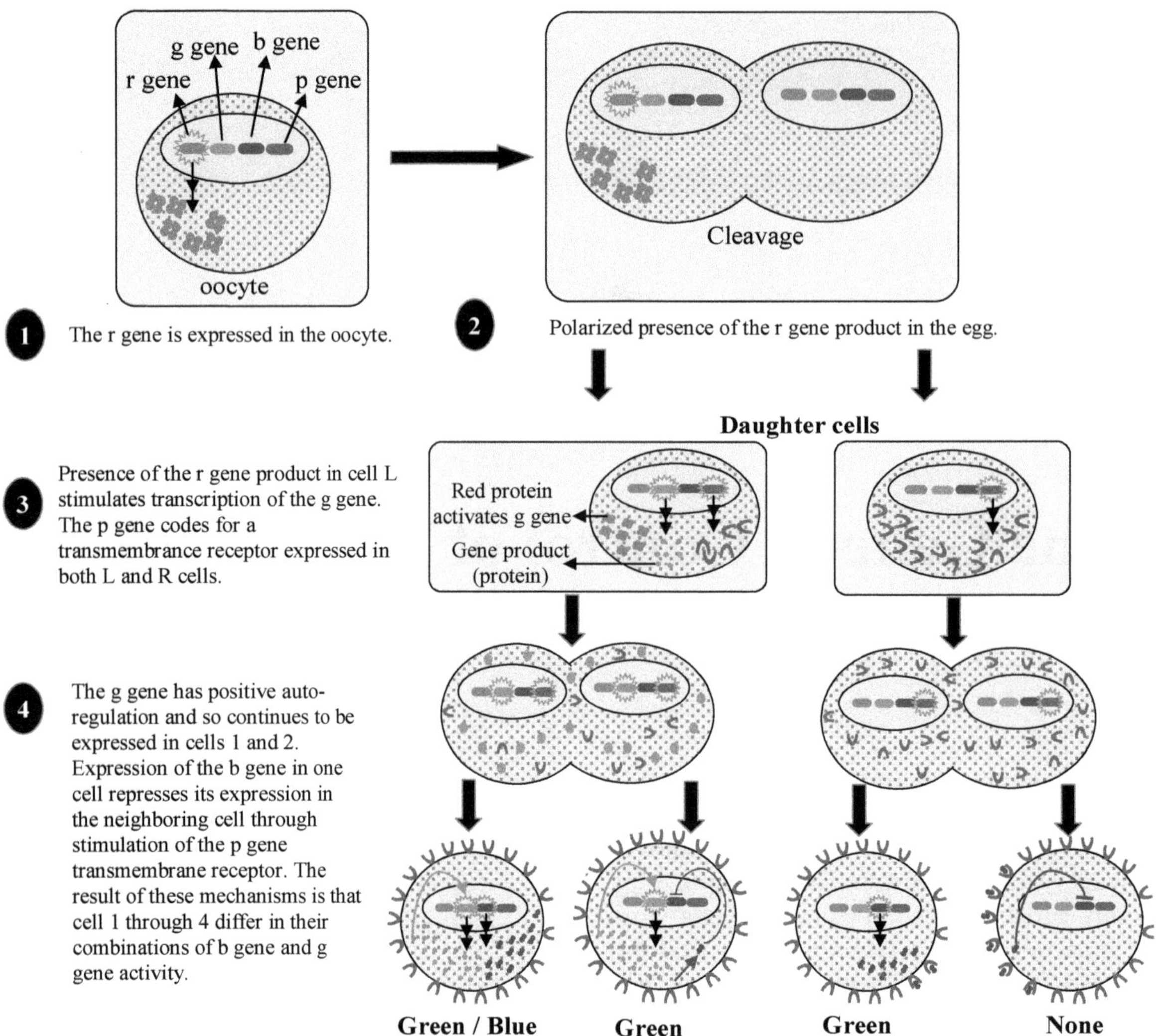

Figure 15.1: Regulatory Mechanisms at the Time of Development.

cell 2 has g activity only, cell 3 has b activity only, and cell 4 has no activity for genes g and b.

15.1 Developmental Mechanisms and Role of Model Organisms

It is well known that both vertebrates and invertebrates are highly evolved. This evolution has occurred due to complex interacting systems controlling the development physiology and behaviors of the organism.

The eye formation in *Drosophila,* a fruit fly, and eye formation in humans is controlled by the same set of genes. *Drosophila* eyes are compound eyes. In flies homozygous for the recessive allele eyeless, eye development is abnormal, and no eyes are formed in the adult fly. The protein responsible for eye formation in wild flies is expressed in all eye cells. In the case of mutant flies, there is a loss of function, and no eye formation occurs. Seven genes are involved in eye formation. The *eyeless gene* is one of Drosophila's seven genes controlling eye formation. The *eyeless gene* is the first gene to be expressed in the process of eye development it acts as a master regulator gene. As it acts at the beginning of eye development, it activates the expression of other genes; hence, a large set of genes is activated, which affects eye formation. This shows a complex network of genes; some of the genes from this network are required to express other genes. Interestingly this gene set is highly conserved and is used by other animals, including humans. The *eyeless* (*ey*) gene is found in drosophila and is the homolog of human *PAX6*. *PAX6* is highly conserved, and its *Drosophila* homology is quite complex and plays a vital role in developmental pathways, which helps in the formation of organs. These pathways help develop enhancer elements in the *PAX6* gene which show importance as sites for upstream regulation of *PAX6*, and possibly even downstream products of *eya*, *so*, or *dac* may play a regulatory role.

The *PAX6* gene comprises a family of genes that help develop tissues and organs during the embryonic phase. This gene has been sequenced and can be diagnosed in the prenatal stage. This family of genes is essential in maintaining the regular role in some cells after birth. PAX6 gene helps in protein production; hence PAX proteins are called transcription factors. It is interesting to note that two highly diverse species reveal differences in size and shape: zebras and zebrafish. They show differential expression in one gene set, which may not be evident in other genes. Conserved DNA sequences are found due to purifying selection. They reveal conserved functions; the conserved coding exons can be studied using comparative genomics. Which otherwise is not seen in the transcriptome profiling. Many conserved non-genetic elements (CNEs) have been found in different genomes. The evolutionarily conserved DNA sequences do not fall in the protein-coding region. They show *cis*-regulatory activities involved during embryonic development. The CNEs are associated with developmental genes and may control the complex and highly regulated spatiotemporal expression patterns of these genes. CNEs are also involved in human diseases, like Van Buchem disease, X-linked deafness, aniridia, and preaxial polydactyly. Orthologous sequences can be detected by comparing the multiple sequences of genome pairs. Model organisms used are yeast (*Saccharomyces cerevisiae*), the fruit fly (*Drosophila melanogaster*), the nematode (*Caenorhabditis elegans*), zebrafish (*Danio rerio*), the mouse (*Mus musculus*), and a small flowering plant *Arabdopsis thaliana.* Most of these organisms have been sequenced fully. Mutations have taken place and got selected at various levels. Selection has played a role in selecting the mutant variants.

15.2 Developmental Mechanisms

Three mechanisms play a role in the developmental process. Three model systems: *D. melanogaster, C. elegans,* and *A. thalania* have been used to show the three developmental processes. There is a specific developmental pathway gene that works as a binary switch.

Fertilization is the process of mitotic cleavage, and the embryo becomes multicellular. Until fertilization, there is no increase in the overall size of the egg, as the cleavage divisions are associated with very little growth. The cleavage divisions at the first step form the blastula, which constitutes ball of about 104 cells containing a cavity. The blastula is followed by folding the part of the blastula wall, and extensive cellular migration takes place. This is followed by a reorganization of cells in the gastrula; the cells become arranged in several distinct layers. These layers are essential to establish the basic body plan of an animal. Developmental processes are different in plants and animals. The development process is under strict genetic control.

There are various signaling interactions between neighboring cells or gradients in the concentration of morphogens. These are directly involved in controlling growth and are formed during development. The four-cell stage human embryo is formed approximately after 40 hours post fertilization. This embryo undergoes two cell divisions, keeps on dividing, and finally gets transformed into a human fetus, which comprises millions of cells. The embryo is not yet implanted in the womb (At the four-cell stage **Figure 15.2**).

There is a lineage relationship between the cells. The transmission of cytoplasm particles illustrates cell-autonomous mechanisms called polar granules from the cells PO to PI to P2 to P3. Polar granules segregate and function as microfilaments in the cytoskeleton. The effects of P2 on EMS and ABp demonstrate cell-signaling mechanisms. The EMS fate is determined by the activity of the *mom-2* gene in P2. The P2 cell produces a signaling molecule called APX-I, responsible for determining the fate of ABp through the cell-surface receptor GLP-1 **Figure 15.2**. In *C. elegans*, many developmental procedures are cell-autonomous, in *Drosophila* and *Mus* (the mouse), regulation by cell-to-cell signaling is more the rule than the exception.

15.3 Activation of Zygotic Genome

There is a complex interaction between maternal and zygotic activities during embryonic development. Maternal mRNAs and proteins enter the egg during oogenesis and undergo the first mitotic divisions. At this stage, only the cell's fate and patterning is defined. Let us take the example of the mouse, where zygotic transcription begins at the two-cell stage, and many maternal mRNA keeps persisting beyond this stage. Further, the activation of the mouse genome is characterized by discrete gene expression patterns. *Drosophila* genome is activated patterning starts with only 59 transcripts induced from cycles 10-11 to cycle 14. Notably, at this stage, 70 per cent of the genes do not have introns; they encode small proteins. The genes not having introns at this stage reflect the requirement of expressing regulatory genes needing a minimal response time. The *Drosophila* embryo also contains the expression of intron-less genes that are still engaged in rapid DNA duplication and mitosis phases. A nuclear membrane is required to assemble the splicing machinery; the selection of intron-less genes might show the production of functional transcripts along with nuclear divisions.

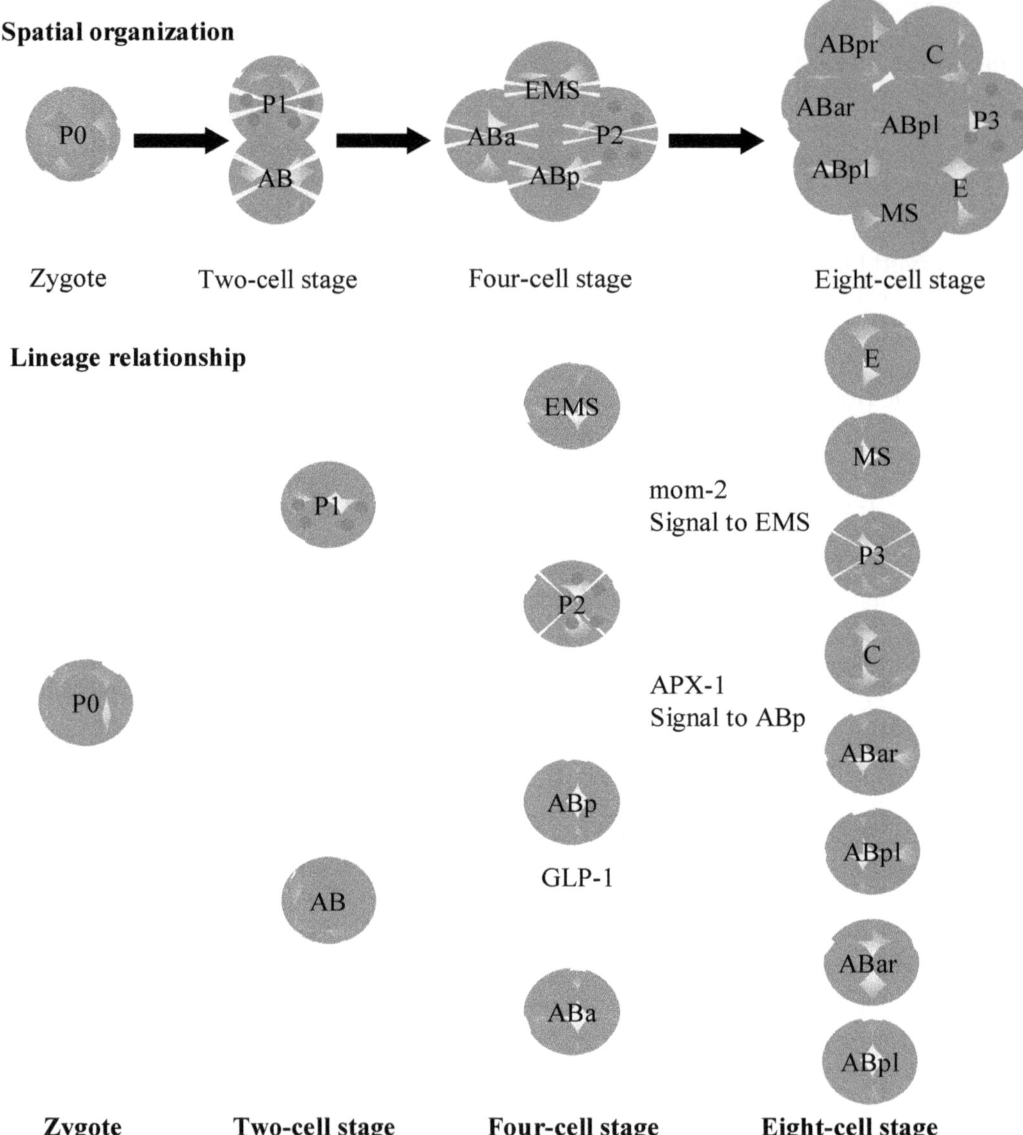

Figure 15.2: Cell Divisions in *C. elegans* at the Time of Early Development.

Genome-wide searches have been made to create a catalog for maternal and zygotic transcripts for many animals. It is surprising to note that many post-transcriptional events take place during embryogenesis.

An embryonic event known as the maternal zygotic transition (MZT) overlaps and helps in the germ layer specification in vertebrates. Maternally supplied mRNAs are degraded during the maternal zygotic transition, while zygotic transcripts are synthesized to reinforce the already specified cell fate or trigger a new cell identity. MZT has been studied in echinoderms, nematodes, insects, fish, amphibians, and mammals. MZT has two interrelated processes (i) maternal mRNAs and proteins are eliminated; (ii) zygotic transcription is initiated. These two significant events, which differ across the species, RNAi-mediated knockdown have been created, which have helped to assess the role of pleiotropic or redundant players because this method can be used to find out the temporal and spatial control while removing the RNAs encoding one or more proteins.

Not only are zygotic genes required for the development, but gene products control the initial events of the development in the oocyte.

1. Storage of the molecules needs to support the cleavage divisions and the rapid RNA and protein synthesis that take place during early embryogenesis.
2. Organization of the molecules in the cytoplasm provides positional information that results from differences between cells in the early embryo.

15.4 Role of Genes in Development

The role of genetics in development can be studied through developmental abnormalities. It is difficult to trace the cell lineage. A hypothetical model is shown in **Figure 15.3**. The cells A.a and A.p are the anterior and posterior daughter cells of parent A cell, and A.aa and A.ap are the anterior and posterior daughter cells of A. a. Let us take a hypothetical cell called A. The anterior and posterior regions of nematodes are designated as a and p. Cells A aa and A ap are the anterior and posterior daughter cells of A a the w, x, and y are terminally differentiated cells.

15.5 Nematode follows a Fixed Program during Lineage Diversification

A soil nematode (*Caenorhabditis elegans*) is a worm that can be cultured easily, it has a short generation time, but large numbers of offspring are produced. The worms can be grown on agar surfaces in Petri dishes, and they get their food from bacterial cells such as *E. coli.* They are a few microns in size, and as many as 105 animals can be grown in a single Petri dish. Sexually mature adults of *C. elegans* produce more than 300 eggs in only a few days. Soil nematode has two sexes: male nematode and hermaphrodite nematode. Hermaphrodite constitutes two XX chromosomes and produces eggs and sperm, which are functional. They can self-fertilize the sperm produced by the male nematode, which fertilizes with the hermaphrodites. The sex-chromosome constitution of *C. elegans* consists of a single X chromosome; there is no Y chromosome, and the male karyotype is XO. Cell division and differentiation are highly stereotyped. The hermaphrodite contains 959 somatic cells, and the male contains 1031 cells. Nematode follows a fixed program during lineage diversification cell fate is determined by intracellular signaling and autonomous development. Cell interaction occurs during development in drosophila, Xenopus, mice, and humans. However, developmental processes are very complicated.

15.6 Signaling Systems in Development

Five signaling systems are used during the early development of metazoans; once the organogenesis starts fresh, five signal systems are added to existing ones. These systems can act independently or coordinate to send and receive developmental signals that release specific transcriptional responses. The signal networks contain anterior-posterior polarity and body axes, coordinate pattern formation, and direct the differentiation of tissues and organs. Signaling systems used during early development are the Notch signaling pathway, Wnt pathway Hedgehog pathway, Receptor tyrosine kinase pathway, and TGF-b pathway, and their role in the development of the vulva in the nematode is shown in **Table 15.1**.

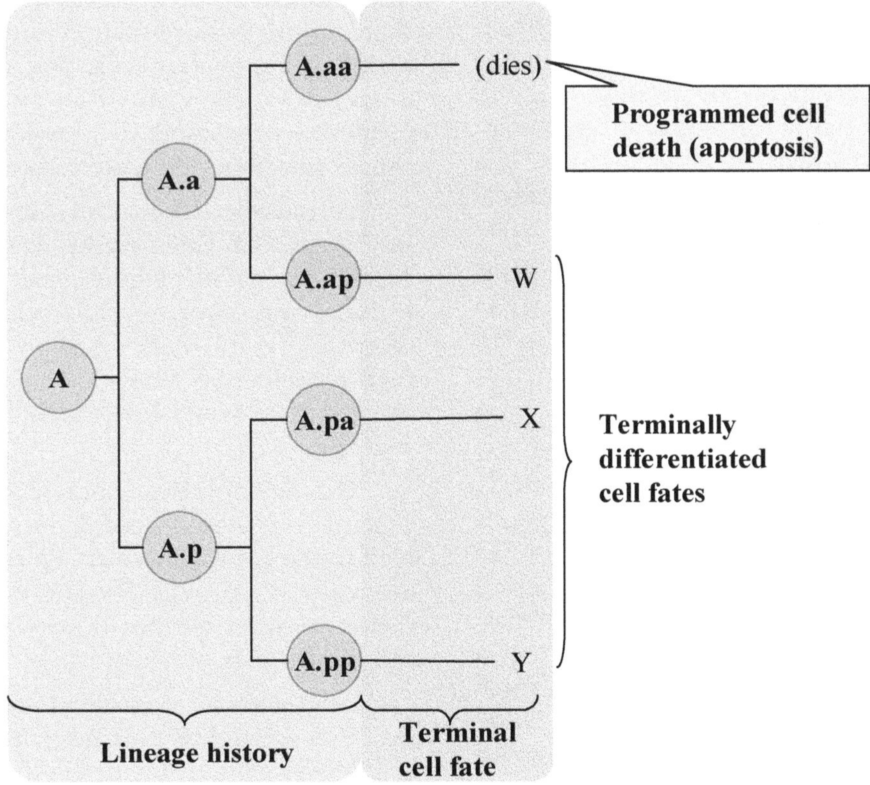

Figure 15.3: Hypothetical Cell Lineages.

Table 15.1: Early Embryonic Development and Signaling Pathways

Nothch/delta pathway	Blood cell development Neurogenesis Retina development
Wnt pathway	Dorsalization of body Female reproductive development Dorsal-ventral differences
Hedgehog pathway	Notochord induction Somitogenesis Gut/visceral mesoderm
Receptor tyrosine kinase pathway	Mesoderm maintenance
TGF-β pathway	Mesoderm induction Left-right asymmetry Bone development

15.7 The Notch Signaling Pathway

During development, the cells interact, and notch signaling works directly with the interacting cells. Various receptors are present in the notch genes. The signals and receptors are membrane-bound. A delta protein binds to the notch receptor; the notch protein gets cleaved off and binds to the cytoplasmic protein, encoded by the SuCH gene. This gene is also called the suppressor of the Hairy gene. Many different developmental processes are controlled in Drosophila, and they help establish the dorsal-ventral boundaries in the wing disc and regulate the fate of cells in the nervous system, muscle, and gut. The complex moves to the nucleus and binds with the transcriptional co-factors, which activate the transcription gene controlling the developmental pathway.

Let us take the example of the Drosophila gene, which is involved in the dorsal-ventral boundary in the wing disc and regulates the cells involved in the nervous system muscle and gut.

There are four notch genes in the development of the mouse. These are involved in the intracellular signaling pathway involving inner cell mass trophectoderm (and blastocyst. Various genes involved are Notch -1-4, Jagged -1, Jag-1, Delta-like (DII-1), DII-3, DII-4, RbsuL, Deltex (Dtx-1), and Dtx-2 genes which are expressed during pre-implantation development since non-fertilization till blastocyst development some genes during implantation Notch 1-4, Jagged –I, Jag-2, Delta like (DII), DII-3, DII-4, Rbsuch, Deltex-1 (Dtx-1) and Stx-2) are expressed in all stages Notch-1, 2, Jag-2, DII-3, RbsuL, and Dtx-2. The expression pattern of these genes can be checked using RT-PCR. If a mutation in these genes occurs, several disorders may be caused, like Alagille syndrome (AGS) and spondylocostal dysotosis (SD).

15.8 *C. elegans* Development

Development is a highly contracted process and is extensively evaluated in C elegance. The reason for studying C elegance is that the organism is well known, the genome sequence is recognized, and fewer cells are necessary for complete development. These factors make the organism less complicated. Full development can take place within two days.

C. elegans is a hermaphrodite organism, and reproduction is determined by cell-cell interaction, which gives the development outcome.

15.9 Genetic Analysis of *Vulva* Formation

C. elegans vulva is situated in the middle of the *vulva* where the eggs are laid. It is developed during larval development, where several rounds of cell-cell interaction occur (**Figure 15.4**). several rounds of cell–cell interaction takes place (**Figure 15.4**).

Two neighboring cells that are developmentally equivalent take part during development. These are Zl.ppp and Z4.aaa which interact with each other, and during this process, one becomes the gonadal anchor cell, and the other becomes a precursor to the ventral uterus. This takes place during the second larval stage (L2) and is under the control of the Notch receptor gene/*in-l2*. In the recessive *lin-l2(O)* mutants, there occurs loss of function mutants, and both cells develop into anchor cells. One mutation/*in-l2(d)* is a gain-of-function mutation causing the formation of uterine precursors. Expression of the *lin-12* gene causes selection of the uterine pathway.

Various developmental stages are demonstrated in **Figure 15.5**. Cells secrete LAG-2 (Delta) signals containing a notch signal gene. LAG (delta) signal induces neighboring cells which produce more receptors ventral uterine precursor cells, and another high producer becomes notch cells. There is always a balance between LAG-2 (delta) and the L1N-12 receptor.

The anchor cells are located in the gonad, and six precursor cells are placed in the skin, or hypodermis, next to the gonad; these are named P3.p to P8.p and are called Pn.p cells. The destiny of each Pn.p cell depends upon the position relative to the anchor cell **Figure 15.6**.

Lin-3 gene is expressed in the anchor cells. The protein secreted by this gene involves vertebrates' epidermal growth factor (EGF). All six Pn.p cells show a receptor encoded by *let-23*, a gene equivalent

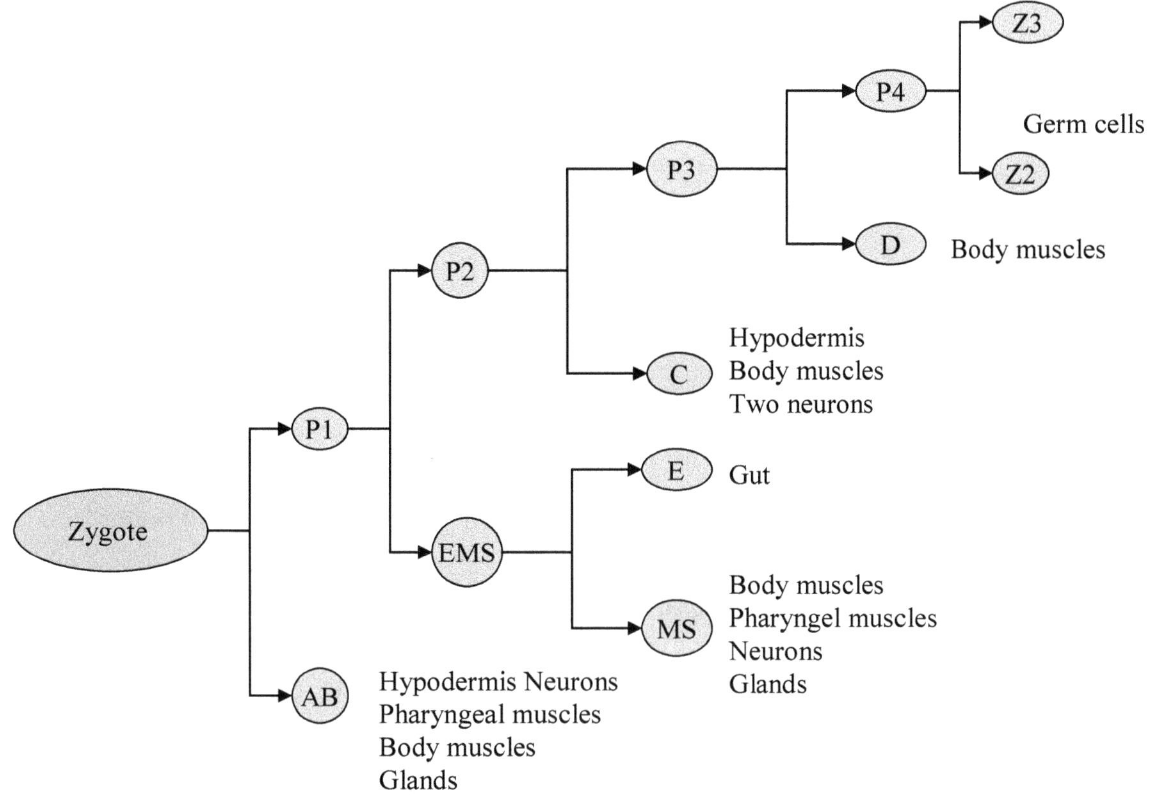

Figure 15.4: *C. elegans*, Demonstrating Early Divisions in the Tissues and Organs.

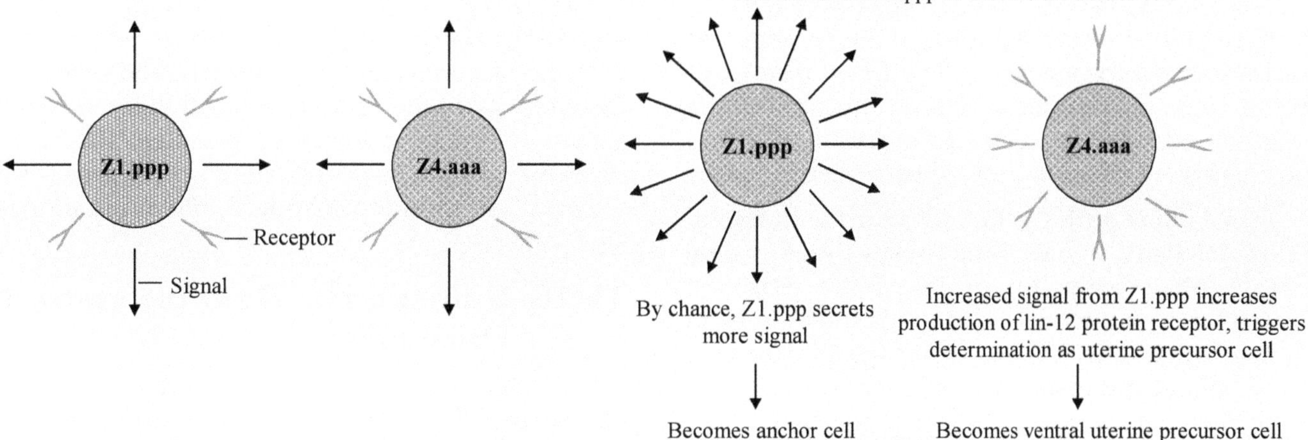

Figure 15.5: Cell–Anchor Cell Determination and Interactions.

to the vertebrate EGF receptor. LIN 3 to the LET-23 receptor binds and induces the intracellular cascade of events and tries to discover which precursor cells will give rise to the primary vulval precursor cell or secondary vulval cells. Sometimes, loss-of-function *let-23* involves recessive mutations as all Pn.p cells show no signal; hence no vulva is formed.

Anchor cells are significant as they convey the signals to the Pn.p cells involved with the *let-60* gene. *The let-60* gene will cause precursor cells to expand as if they have not received a signal from the anchor cell. This happens only when there is a recessive mutation if dominant mutations exist, it will allow all precursor cells to react and form numerous vulvas.

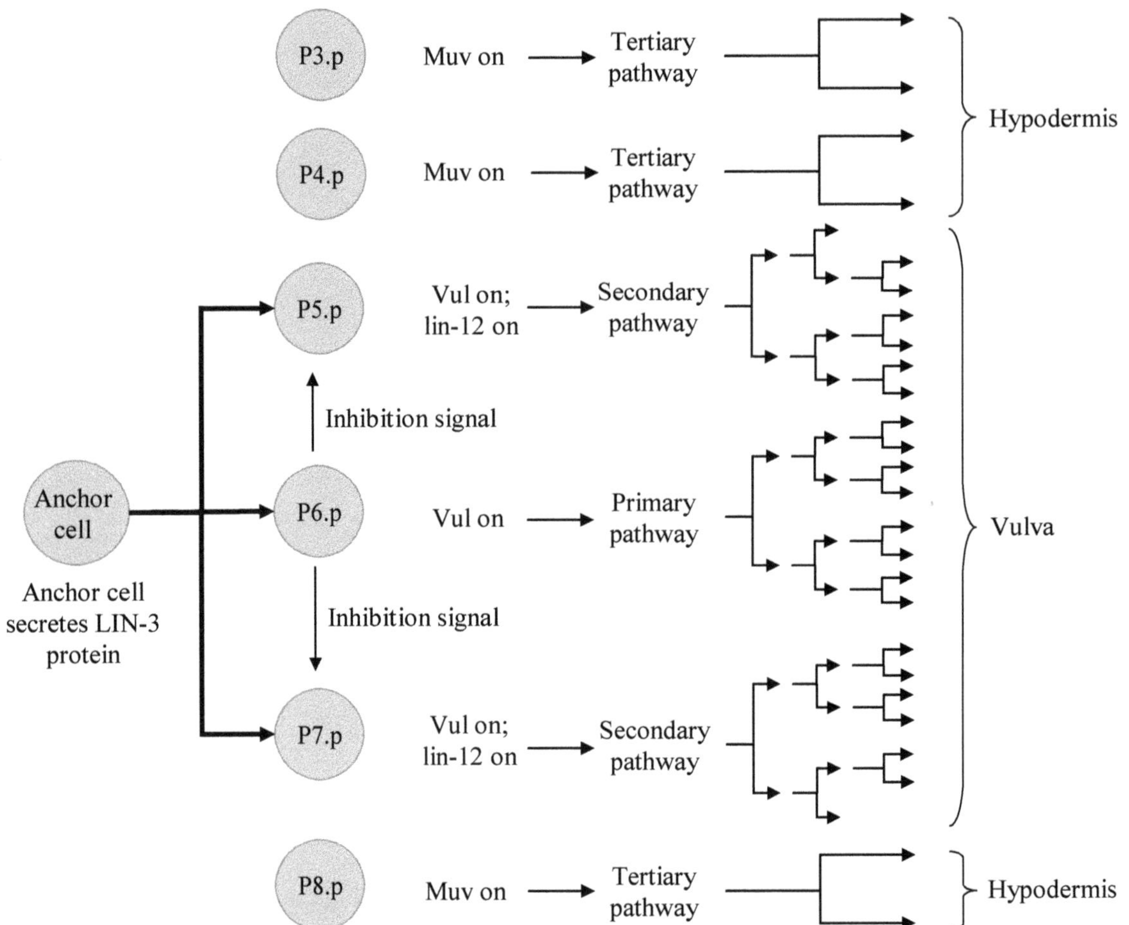

Figure 15.6: Cell Lineage Determinations in *C. elegans* Vulva Formation.

The *C. elegans let-60* is a homolog of the *ras* gene. ras is a human proto-oncogene. The *let-60* dominant gain-of-function mutant results in multiple vulva formation, which has a Gly-Glu mutation at amino acid 13; this is similar to *c-ras* which is an oncogene.

LIN-3 binds with the LET-23 after the initiation of high-intensity signals. This causes the Vulvaless (Vul) gene expression in P6.p, the primary differentiation pathway, and divides three times to produce vulva cells. The cells P5.p and P7.p receive a lower signal and result in secondary fate.

lin-12 in the two neighboring cells (P5.p and P7.p) get activated by P6p cells. This signal prevents P5.p and P7.p primary cells. When *Vul* and *lin-12 are* both active primary vulva cells. The three remaining precursor cells (p3.p, P4.p, and P8.p) receive no signal from the anchor cell. *Multivulva* (*Muv*) gene is expressed in these cells *Muv* represses *Vul,* and the three cells finally develop into skin cells.

Cell-cell instruction takes place in three phases. (i) Two cells interact and establish the identity of the anchor cell; the anchor cell interacts with three vulval precursor cells to discover the identity of the primary vulval precursor cell (usually P6.p) and two secondary cells. The primary vulval cell interacts with the secondary cells to inhibit the ability to adapt the pathway of the primary cell. All these molecular signals are for reception and processing. The human immune system requires four notch genes that generate an immune response and bone marrow stem cells.

15.10 Embryonic Development in *Drosophila*

The specific genes should be turned on or turned off during differentiation. The turn-off and on mechanisms have been well evaluated in the model organisms.

15.11 Overview of *Drosophila* Development

The adult development period occurs immediately after fertilization and shows the following phases. In the embryo, there are three larval stages and the pupal stage. The adult fly is out from the pupal about ten days after fertilization. The Drosophila egg has anterior, posterior, and ventral regions. The anterior region of the egg has a micropyle, a specialized conical

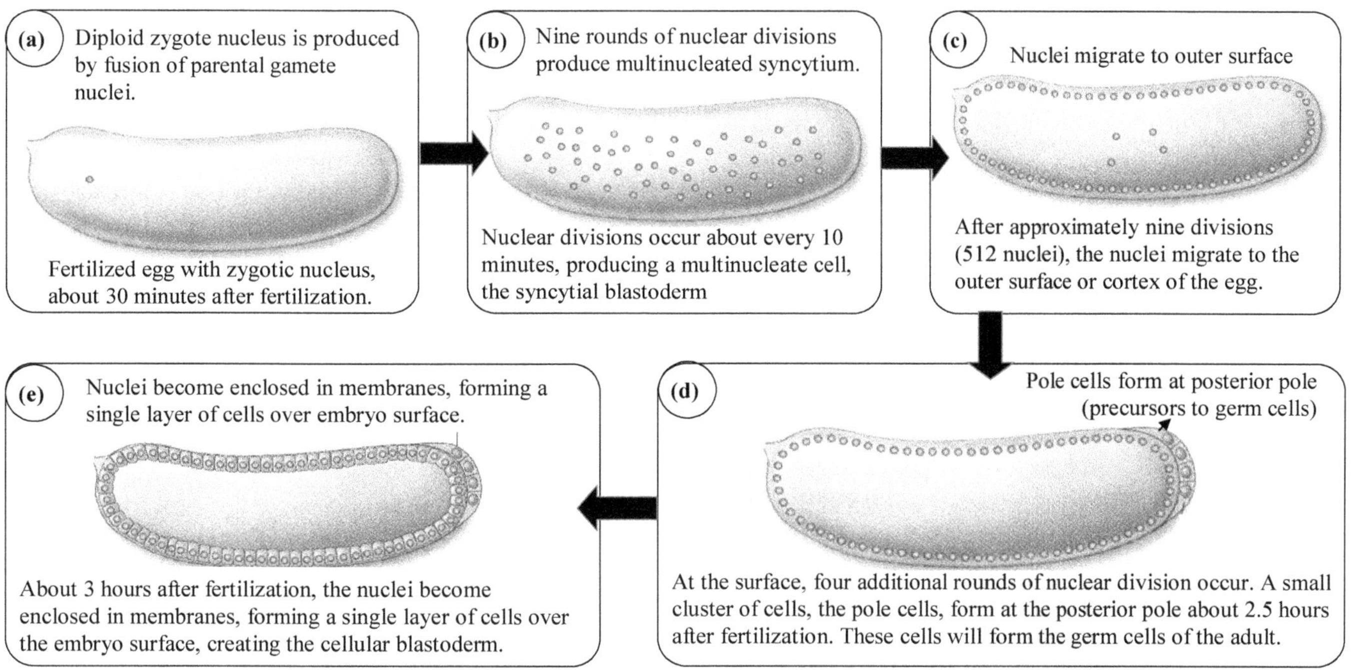

Figure 15.7: Embryonic Development of *Drosophila*.

structure from where sperm enters into the egg. The posterior end is round and contains many openings known as aeropyles. The dorsal side is flat with many appendages, and the ventral side is curved **Figure 15.7**.

There are 9 rounds of cell divisions after fertilization. There exist 512 nuclei; the nuclei move toward the outer surface of the egg or toward the cortex, where further cell division occurs (**Figure 15.7 (c) and (d)**). All 512 nuclei have cytoplasm maternally derived transcripts and proteins around them. At the time of development, the gene expression appears.

The transcriptional programs are triggered in the rest of the migrating nuclei; the embryo shows anterior-posterior and dorsal-ventral axes of symmetry. The posterior cytoplasm contains maternal components from the germ cells.

15.12 Regulation of Anterior-Posterior Body Axis

The blastoderm contains the plasma membrane around individual nuclei and forms the cellular blastoderm (**Figure 15.7(e)**). Segmental rearrangement takes place, which reveals the expression patterns of various embryonic genes. The adult segments get shifted slightly in the position. The polarity of each segment is determined at the beginning of the self.

Drosophila derives its overall organization from the larval body plan and the head, thoracic, and abdominal segments. The larval body gets segmented during the pupal stage, and many adult body structures are constructed from small cell clusters called imaginable discs created during larval stages. There are 12 bilaterally paired discs: eye-antennal discs, leg discs, wing discs, and so forth-and one genital disc.

There are different are more than 20 segmental loci (**Table 15.2**). This form the gap genes formed due to the deletion of a group of adjacent segments, pair-rule genes affect every other segment and remove a specific part of each affected segment. Segment polarity genes cause deformity in homologous portions of each segment.

Table 15.2: Segmentation Genes in *Drosophila*

Gap Genes	*Pair Rule Genes*	*Segment Polarity*
Krüppel	hairy	engrailed
knirps	even shaped	wingless
hunchback	runt	cubitus interrupted
giant	fushi-tarazu	Hedgehog
tailless	odd-paired	fused
Bucke be	odd-skipped	armadillo
	sloppy paired	patched
		gooseberry
		paired
		naked
		disheveled

15.13 Gap Genes

Gap genes are activated after fertilization and are the first set of zygotic genes. Many genes and gene products are expressed along the anterior-posterior axis, and many others are expressed in the maternal gradient system. A significant gap is created if the mutation occurs in any of these genes, affecting the segmental patterns. A mutation known as the *hunchback* causes loss of head and thorax structures, Krüppel mutants lose thoracic and abdominal structures, and *knirps* mutants lose most of the abdominal structures. Transcription occurs in the gap genes, which divide the embryo into broad regions of the head, thorax, and abdomen **(Figure 15.8)**.

15.14 Pair-rule Genes

The pair rules genes, which divide the gap genes according to segment. Gap genes are expressed as narrow bands or strips around the embryo. Eight pair-rule genes help the embryo to divide into a series of strips. There exists an overlapping of the strips at their boundaries. This involves cells in the strips showing different combinations of pair-rule genes in an overlapping fashion. Some of the pair-rule genes encode transcription factors. The gap gene product's expression and action induce the pair-rule gene transcription.

15.15 Segment Polarity Genes

The term 'segmentation gene' is a classification given to a broad class of genes subdivided into three smaller classes of genes. The segmentation gene group has gap genes, pair-rule genes, and segment polarity genes.

The segment polarity genes show expression controlled by the transcription factors encoded by pair-rule genes. Segment polarity genes in each segment become active even in a single cell band extending around the embryo's circumference. The embryo is segmented into 14 in number, and of the segment, polarity genes control the cellular identity within each segment. Some segment polarity genes are *engrailed and* encode transcription factors. Hemoglobin protein is inhibited by engrailed protein.

Segmental genes are essential because they help establish the segmental boundaries, the homeotic (from the Greek word for "same") genes, which are activated as a target of the zygotic genes. Expression of homeotic genes shows that each body segment forms adult structures. *Drosophila* has antennae, mouthparts, legs, wings, thorax, and abdomen. For example, the wild-type allele of *Antennapedia* (*Antp*) specifies a leg's formation on the thorax's second segment. Dominant gain-of-function *Antp* mutations cause this gene to be expressed in the head, and mutant flies have a leg on their head instead of an antenna.

15.16 Hox Genes in *Drosophila*

Hox genes are located on chromosome 3 in *Drosophila* (**Table 15.3**), a cluster of genes. One cluster is known as the *Antennapedia* (*Antp-C*) complex. It contains five genes that specify the structures of the head and the first two thoracic segments.

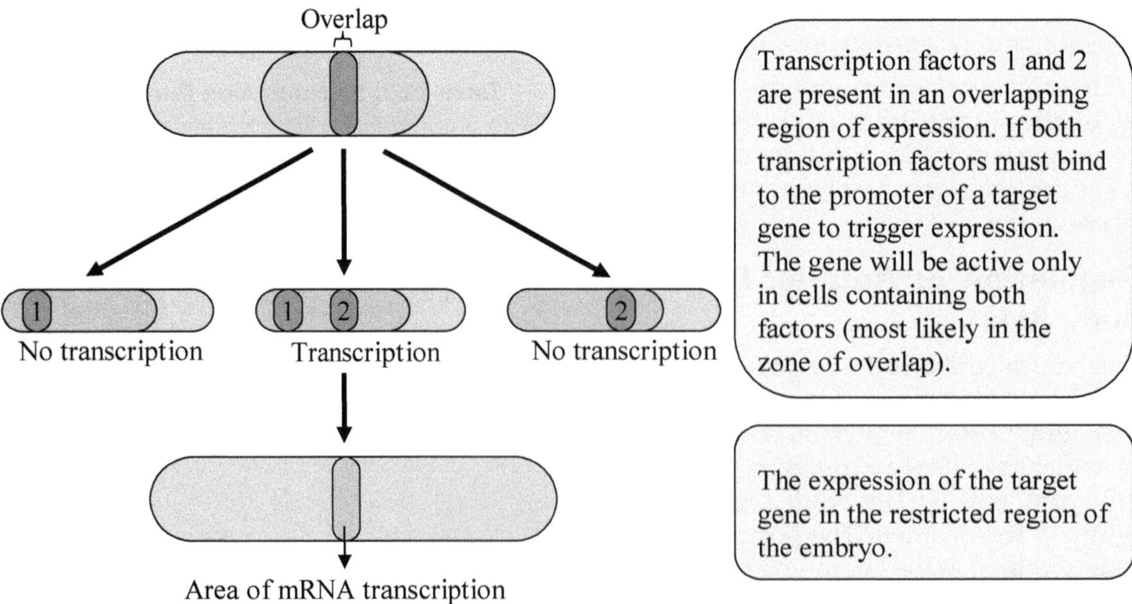

Figure 15.8: Overlapping Regions Result in a New Pattern of Gene Expression.
(a) Transcription factors 1 and 2 are present in an overlapping region of expression. Both transcription factors must bind to the promoter of a target gene to trigger expression. The gene will be active only in cells containing both factors (most likely in the overlap zone). (b) The expression of the target gene in the restricted region of the embryo.

Table 15.3: Hox Genes of *Drosophila*

Antennapedia Complex	*Bithorax Complex*
Labial	Ultrabithorax
Antennapedia	Abdominal A
Sex combs reduced	Abdominal B
Proboscipedia	

The second cluster is called *bithorax* (*BX-C*) complex; it is made up of three genes, and it specifies structures in the posterior portion of the second thoracic segment, the entire third thoracic segment, and abdominal segments.

The homeobox genes require a transcription factor of 180bp. It can bind hence also known as the binding domain of the homeobox gene. The homeobox encodes a sequence of 60 amino acids known as a homeodomain. Second, the expression of the genes is collinear. Genes at the 3′ -end of a cluster are expressed at the anterior end of the embryo, those in the middle are expressed in the middle of the embryo, and genes at the 5′ -end of a cluster are expressed towards the posterior region of the embryo. These genes were identified for the first time in *Drosophila; they* are found in the genomes of most eukaryotes that have segmented body plans, including zebrafish, *Xenopus,* chickens, mice, and humans.

15.17 Developmental Genes in *Drosophila* Encounter Mutation

If the mutation takes place in the blastoderm of Drosophila, there would be abnormal development. This will result in abnormal growth in case cells belonging to one region are placed in different regions. Once there is abnormal growth, the cells will not be integrated in the proper place.

15.18 Mutations in a Maternal-Effect Gene Result in Defective Oocytes

If there is a proper developmental process, translocation of maternal mRNA molecules will be present in the oocyte. If the protein synthesis is blocked, the early cleavage would be arrested. The Zygote genome is also required for the transcription of the zygote.

The blastoderm formation is arrested. Genes that function in the mother that are required for the embryo's development are called maternal-effect genes, and developmental genes that function in the embryo are called zygotic genes.

The zygotic genes interpret and respond to the positional information laid out in the egg by the maternal-effect genes.

Mutations in maternal-effect genes are easy to identify because homozygous females produce eggs unable to support normal embryonic development, whereas homozygous males produce normal sperm. It is shown below:

m/m ♀ x +/+ ♂ → +/m progeny

(Abnormal development)

+/+ ♀ x m/m ♂ +/m progeny

(Normal development)

The +/m progeny of the reciprocal crosses are genetically identical, but development is upset when the mother is homozygous m/m.

The role of maternal-effect genes is to establish the Drosophila oocyte's polarity even before fertilization occurs. Maternal-effect mutations demonstrate the genetic control of pattern formation and identify the molecule - important in morphogenesis.

The group of genes required is coordinate genes, including the *anterior genes,* which affect the head and thorax. The key gene in this class is *the bicoid.* Mutations in *bicoid* produce embryos lacking the head and thorax that occasionally have abdominal segments in reverse polarity duplicated at the anterior end. The *bicoid* gene product is a transcription factor for genes determining anterior structures. Because the *bicoid* mRNA is in the anterior part of the early-cleavage embryo, these genes are activated primarily in the anterior region. The *bicoid* mRNA is produced in nurse cells (the cells surrounding the oocyte) and exported to a localized region at the anterior pole of the oocyte (**Figure 15.9**).

At the time of syncytial cleavages, the protein product is less localized and the cleavage results in an anterior-posterior concentration gradient with the maximum at the anterior tip of the embryo. The Bicoid protein is a principal morphogen that is responsible for.

1. Mapping blastoderm fate, the protein is a transcriptional activator having a helix-turn-helix motif for DNA binding. Genes affected by the Bicoid protein have multiple upstream binding domains consisting of nine nucleotides with consensus sequence S′-TCTAATCCC-3′. The binding sites differ by as many as two base pairs from the consensus sequence bind the Bicoid protein with high affinity; on the contrary, the sites that contain

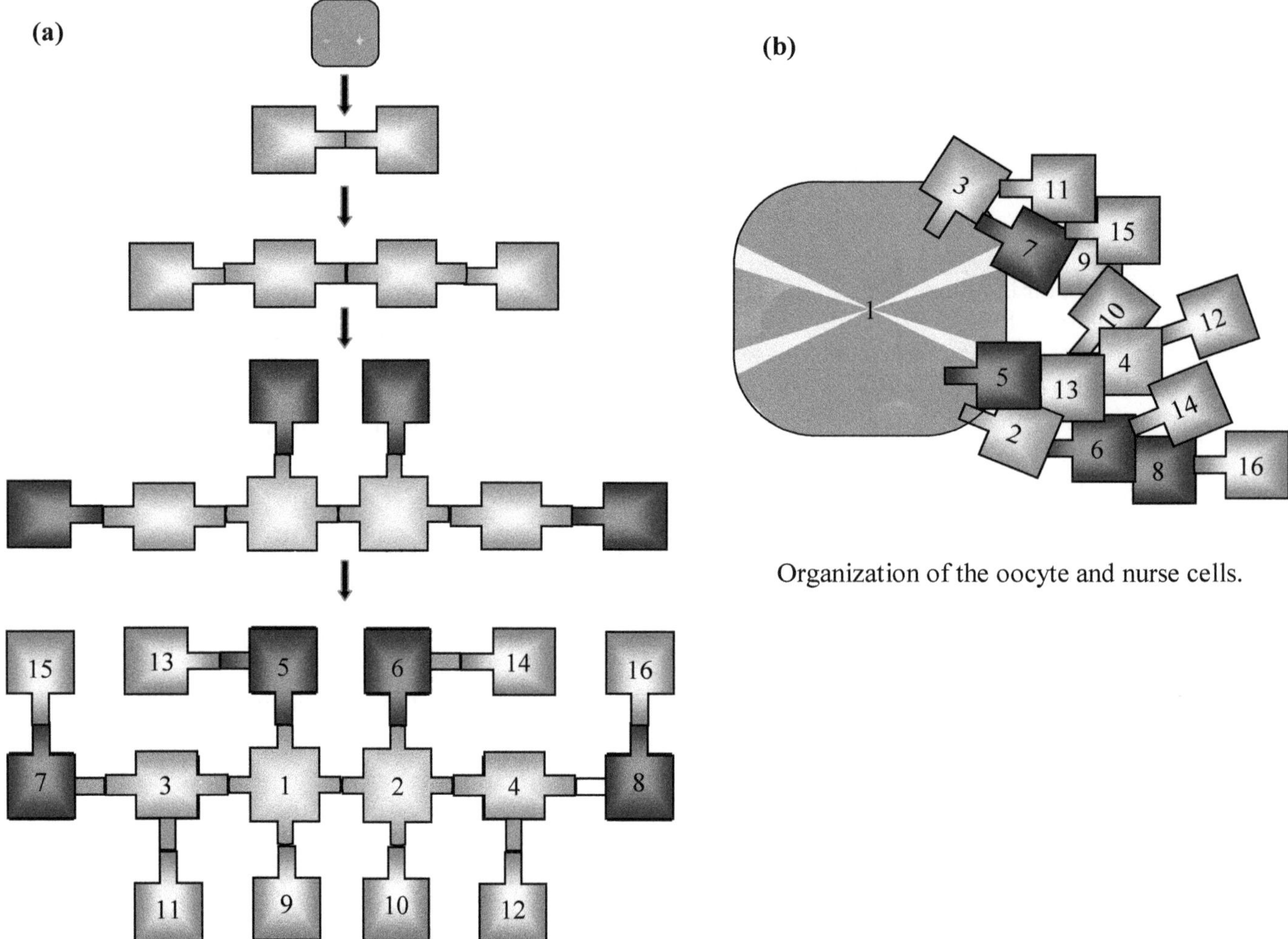

Figure 15.9: Pattern of Cell Divisions in the Development of the *Drosophila* Oocyte.

(A) The cells surrounding the oocyte are connected to it and each other by cytoplasmic bridges; these cells synthesize products transported into the oocyte and contribute to its regional organization. (B) Geometric organization of the cells.

four mismatches bind with low affinity. The combination of high- and low-affinity binding sites determines the concentration of Bicoid protein required for gene activation; genes with many high-affinity binding sites can be activated at low concentrations, but those with many low-affinity binding sites need higher concentrations. Such differences in binding affinity mean that the gene expression level can differ from one regulated gene to the next along the Bicoid concentration gradient. It is the local concentration of the Bicoid protein that regulates the expression of critical gap genes along with the embryo, for example, *hunchback.*

2. The other group of coordinate genes constitutes the *posterior genes,* which interact with the abdominal segments. Some of the mutants also lack pole cells. One of the posterior mutations, *Nanos,* yields embryos with defective abdominal segmentation but normal pole cells. In this case, the *Nanos* mRNA is localized tightly to the posterior pole of the oocyte, and the gene product is a translation repressor. Among the genes whose mRNA is not translated in the presence of Nanos protein is the gene *hunchback.* Hence *hunchback* expression is controlled jointly by the Bicoid and Nanos proteins; the Bicoid protein activates transcription in an anterior-posterior gradient, and the Nanos protein repress translation in the posterior region.
3. The third group of coordinate genes is the *terminal genes,* which simultaneously affect the most anterior structure (the acron) and the most posterior structure (the telson). The key gene in this class is *the torso,* which codes for a transmembrane receptor that is uniformly

distributed throughout the embryo in the early developmental stages. The torso receptor is activated by a signal released only at the poles of the egg by the nurse cells in that location.

Apart from the three sets of genes that determine the anterior-posterior axis of the embryo, the fourth set of genes determines the dorsal-ventral axis. The morphogen for dorsal-ventral determination is the product of the gene *dorsal,* which is present in a pronounced ventral-to-dorsal gradient in the late syncytial blastoderm.

Two small sets of homeotic, or HOX, genes are among the genes that transform the periodicity of the Drosophila embryo into a body plan with linear differentiation that includes the development of wings from segment T2 and halteres from segment T3. Homeotic mutations cause the transformation of one body segment into another, which is recognized by the misplaced development of structures usually present elsewhere in the embryo. One class of homeotic mutation is illustrated by *bithorax,* which transforms the anterior part of the third thoracic segment into the anterior part of the second thoracic segment, with the result that halters get transformed into an extra pair of wings. The other class of homeotic mutation is shown by the *Antennapedia,* which results in the transformation of the antennae into legs. The *HOX* genes are represented by *bithorax,* and *Antennapedia* are, in fact, gene clusters. The cluster containing *bithorax* is designated BX-C (stands for *bithorax-complex*), and that containing *Antennapedia* is called ANT-C (stands for *Antennapedia-complex*). Both the gene clusters were initially found by their homeotic effects in adults. Later they were shown to affect the identity of larval segments. The BX-C primarily concerns developing larval segments T3 through A8, affecting T3 and AI. The ANT-C is primarily involved in developing the head (H) and thoracic segments Tl and T2.

Homeotic genes are transcriptional activators of other genes. Most *HOX* genes contain one or more copies of a characteristic sequence of about 180 nucleotides called homeobox, which are the critical genes concerned with the development of embryonic segmentation in organisms as diverse as segmented worms, frogs, chickens, mice, and human beings. Homeobox sequences are present in exons and code for a protein-folding domain with a helix-turn-helix DNA-binding motif.

15.19 Regulatory Hierarchy and *HOX* Genes

In *Drosophila,* a *HOX* gene is called *Ultrabithorax* (*Ubx*) *which* is usually active in segment T3, which gives rise to the halteres in segment T3. Reduced *Ubx* expression makes the haltere more wing-like, and in the absence of *Ubx* expression, the haltere disk forms a structurally normal wing. Ubx controls the haltere development. However, this interpretation is incorrect as it implies that *Ubx* expression is necessary and sufficient for developing the haltere (the pair of knb-like structures protruding posterior to the wings). The fact is that *Ubx* homologs are expressed in T3 in probably all insects, but segment T3 carries halteres only in the dipteran insects, of which *Drosophila* is an example (**Figure 15.10**).

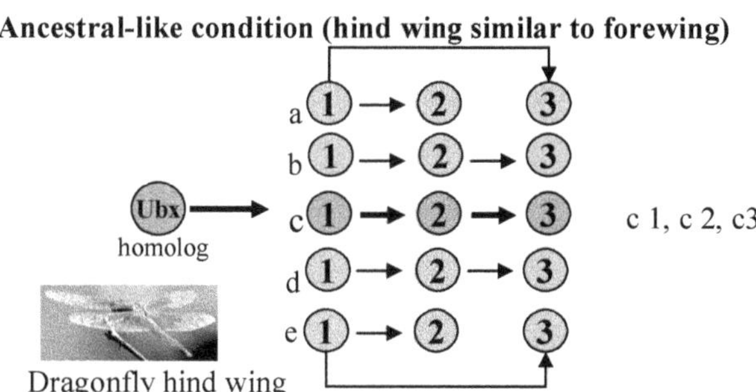

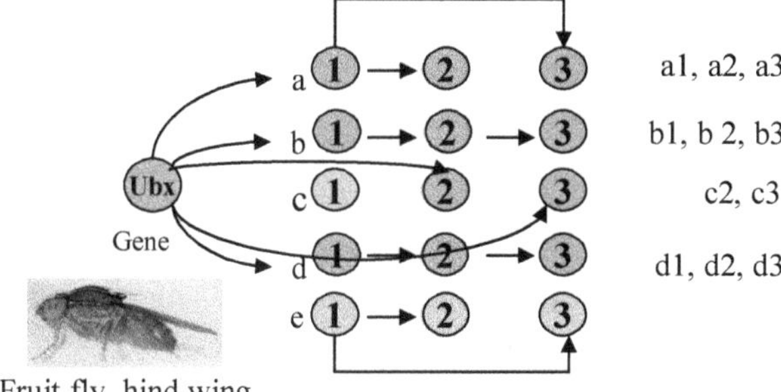

Lepidopteran lineage (somewhat modified hind wing)

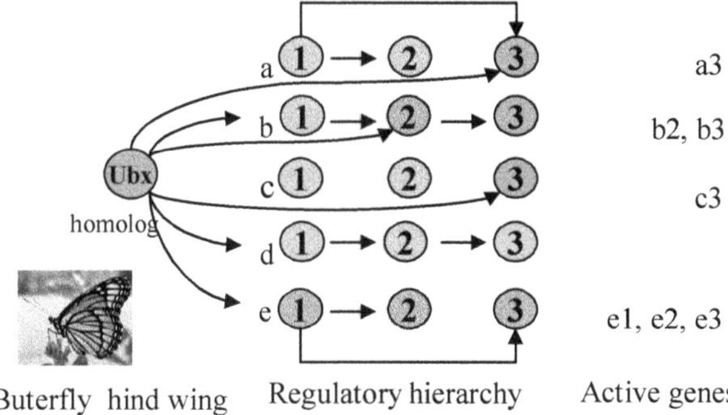

Figure 15.10: Evolutionary Scheme by which the *Ubx* Homolog Occurs.

It is proposed that Ubx controls the haltere development in dipterans due to the gradual evolution of this gene. **Figure 15.10** shows the small circles that reveal the genes, and the arrows show gene activation in segment T3.

In **Figure 15.10,** the ancestral state is shown in part A in which the gene *cl* has evolved so that the Ubx homolog activates it. The active genes in T3 consist of the genes activated by the *Ubx* homolog (*cl*), plus genes further along the regulatory hierarchy activated by these genes (*c2* and *c3*), plus other genes (not shown) whose expression in T3 is independent of the *Ubx* pathway. These genes account for the hind wing's slight modifications compared to the forewing. Without the *Ubx* homolog, the hind wing would lose these slight modifications and become even more like a forewing.

Mutations occur as evolution proceeds. These mutations include the loss or gain of Ubx-responsive enhancers. A newly evolved *Ubx* responsiveness, or loss of *Ubx* responsiveness, may not affect the expression of the same genes in other tissues because *Ubx* is explicitly expressed in the number of genes. If the novel pattern of n gene expression is favored by natural selection, then the mutation will become incorporated into the species. After a long evolution, numerous genes will evolve this way, progressively modifying the hind-wing pattern in a manner favored by natural selection. In dipterans, for example, the selection was for extreme modification of size and structure to form the haltere. The result of the evolution is shown diagrammatically in **Figure 15.10** by depicting *c1* as having lost *Ubx* activation and *a1* and *a3, b1,* e2 and *e3,* and *d1* as having gained it. (Note that the *Ubx* gene can control genes anywhere in the regulatory hierarchy, not only at the beginning). Similarly, in the lepidopteran lineage (part C), there was a selection for less extreme modification, which is shown as a loss of *c1* activation, but the gain of *a3, b2, e3,* and *el* and *e3* activation. In each lineage, the hind wing was modified gradually but independently while a sequence of genes was brought under *Ubx* control or released from *Ubx* control. The result is that in dipterans, the *Ubx* gene "controls" haltere development, whereas in lepidopterans, it "controls" a less extreme modification of the forewings. In each lineage, the *Ubx* homolog is critical in developmental control but regulates different genes. Each lineage mutation in *Ubx* will make the hind wing more like the forewing.

15.20 Normal Development Needs Program Cell Death

Programmed cell death is also known as apoptosis, a genetically controlled developmental program modifying the organism. Programmed cell death is the formation of digits noted in the vertebrate limb. It was the first time seen in *C. elegans*. In *C. elegans,* normal development relies on programmed cell death. The number of cells that die during the worm's development is always the same: 131 of 1090 in hermaphrodites and 147 of 1178 in males.

In *C. elegans,* 15 genes are involved in apoptosis. Which involve four processes: There are decisions about cell death, implementation, engulfment of dying cells, and degradation of cell debris within the engulfing cells.

Expression of *ced-3* and *ced-4* are required for cell death. If mutations occur, these will inactivate either of these genes resulting in the survival of cells that normally die. *ced-3* and *ced-4* expressions are controlled by *ced-9.* Gain-of-function mutations that cause constitutive expression or over-expression of *ced-9* prevent cell death. The loss-of-function mutation inactivates *ced-9* causing embryonic lethality (**Figure 15.11**) *ced-9* is a *binary switch gene* for apoptosis. Cells that express *ced-9* survive, and the ones that do not express ced-9 will die.

The *ced-9* gene of *C. elegans* has a human homolog, *bcl-2,* a proto-oncogene that controls apoptosis. Mutations cause over-expression of *bcl-2,* preventing death in cells that otherwise usually die. In humans, overexpression of *bcl-2* is found in

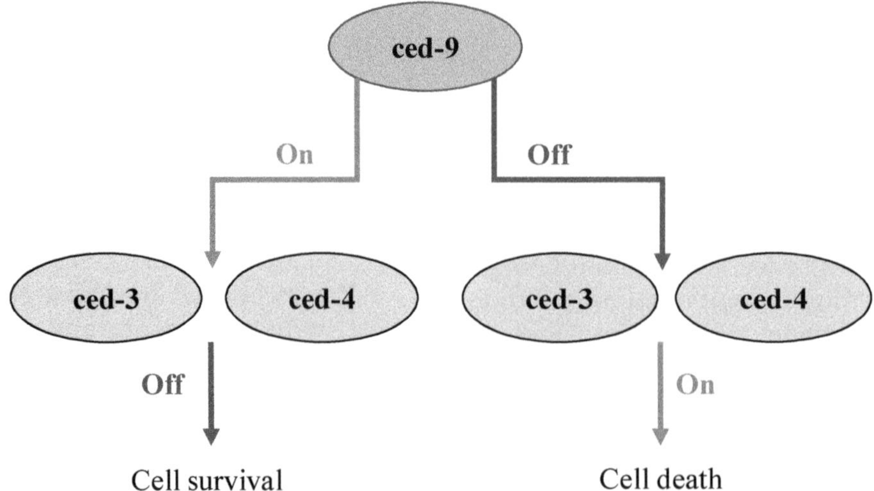

Figure 15.11: The Genetic Pathway Controls Cell Death; the gene *ced-9* Acts as a Binary Switch.

follicular lymphoma, a form of cancer. Transfer of a cloned human *bcl-2* gene into *ced-9* null mutant *C. elegans* embryos prevents apoptosis, indicating that nematodes and mammals share a common pathway for this process.

15.21 Molecular Mechanism of Development among Vertebrates

These genes involved in the development among vertebrates can be organized into four clusters (a,b,c, and d), each 120 kb long; in mice and humans, each cluster is located on a different chromosome. The four Hox gene clusters were created **(Figure 15.12)** by at least 4-time duplication approximately 500 to 600 million years ago.

The molecular analysis of the vertebrate Hox genes has shown striking structural and functional similarities to the homeotic genes of the bithorax and Antennapedia complexes in *Drosophila*. These genes may be members of a larger cluster of genes known as HOM-C, which was divided by a chromosome rearrangement during the evolution of the flies. In more primitive insects such as the flour beetle Tribolium castaneum, the bithorax, and Antennapedia complexes are found in a single cluster. Vertebrates and invertebrates' comparisons reveal that the essential organization of the HOM/Hox genes is preserved during evolution. The structural homologs of the ANT-C genes are at one end of each vertebrate Hox gene cluster, and the structural homologs of the BX –C genes are at the other end. Moreover, within each cluster, the physical order of the Hox genes corresponds to the position of their expression along the anterior-posterior axis of the embryo, just as it does for the homeotic genes in *Drosophila*. With one exception (the Deformed gene in *Drosophila*), all the HOM and Hox genes are transcribed in the same direction, with expression proceeding from one end of a cluster to the other end, both spatially (anterior to posterior in the embryo) and temporally (early to late in development). This phenomenon of collinearity suggests that the HOM and Hox genes provide a common molecular mechanism for establishing the identities of specific regions in many different types of embryos.

15.22 Knockout Mutation, Insertional Mutagenesis Transgenic Mouse Models

In humans, there are many diseases with various phenotypes; however, humans cannot be used as a model system for different diseases. Hence, model systems need to create artificial conditions that mimic human diseases like cardiovascular diseases, cancer, diabetes, *etc.* The mouse is one of the favorite model organisms for human diseases because we can get many generations quickly, and maintenance cost is also minimal.

Now we can make custom-made mice using gene-targeting stem cells, knock out mouse models, and create mutations using site-directed mutagenesis. This allows researchers to study gene functions related to the entire organism. Certain diseases that afflict humans but usually do not strike mice, such as cystic fibrosis and Alzheimer's, can be induced by manipulating the mouse genome and environment.

More than 500 loci are responsible for genetic diseases in mice and are also involved in the developmental processes. Many mice are required for the accuracy of results; these are examined for phenotypic differences. In case differences are

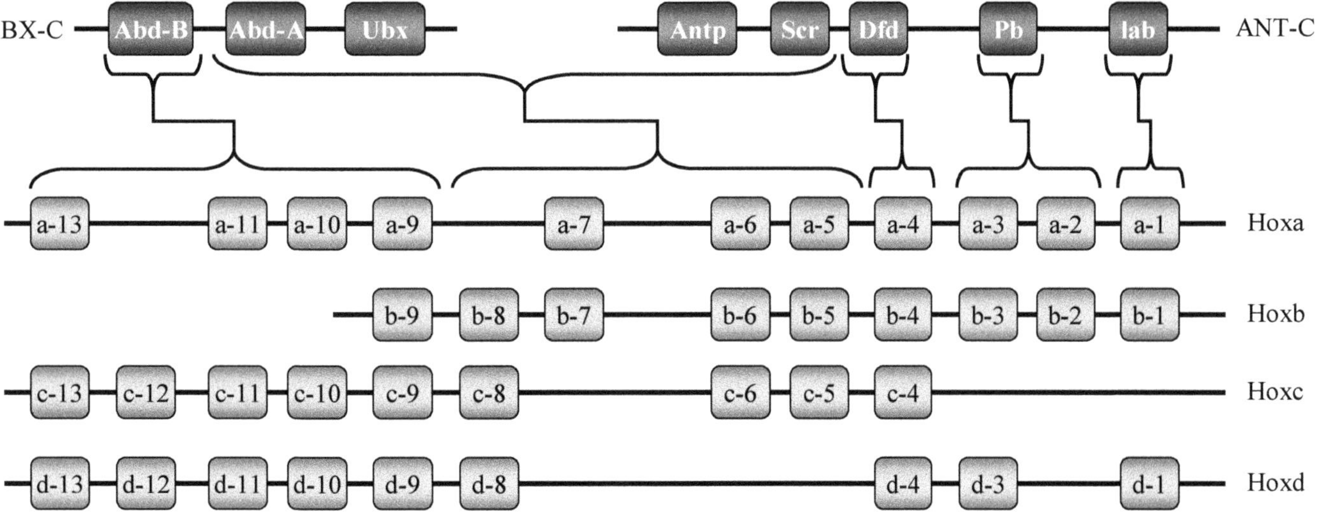

Figure 15.12: Hox Genes Homologous to the *Drosophila* Genes in the BX-C and ANT-C.

found, these are tested for genetic transmission. Mutation can be mapped and tested for allelism and epistatic interactions with other mutations. All these techniques are quite costly.

Presently specific DNA molecules are introduced into the mouse genome. This DNA may create mutations when we insert the genes in the mouse chromosomes. One can clone these mutations in these mutated genes, and they are easy to clone because the inserted DNA has tagged them. The mutated gene can be isolated, and a recombinant DNA library can be constructed from the mutant animal. Therefore, insertional mutagenesis in the mouse is analogous to transposon tagging in *Drosophila* (**Figure 15.13**).

Trans genes can be created, which can be introduced. One relies on the injection into the fertilized eggs or embryos, and the other is based on the injection or transfection of DNA into large populations of cultured cells derived from very young mouse embryos.

Pluripotent cells are derived from embryos in the pre-implantation stage and develop into non-committed cells, which are homogenous. These cells can stay longer and do not lose their pluripotency. Their karyotype is also quite normal. Sometimes these cells undergo massive/genetic manipulation, but they can regenerate. Healthy embryonic stem cells can be induced and offer great therapeutic potential. These cells are essential in studying how different diseases develop pluripotent stem cells that can undergo genetic reprogramming.

These methods have resulted in the creation of many transgenic mice with many mutations.

A "knockout" mutation can help the researcher determine the normal gene's role during development. To create a knockout mutation, the sequence of the cloned gene must be altered *in vitro* and then introduced into cultured embryonic stem cells. The mutated transgene can replace its normal allele on the chromosome through homologous recombination,

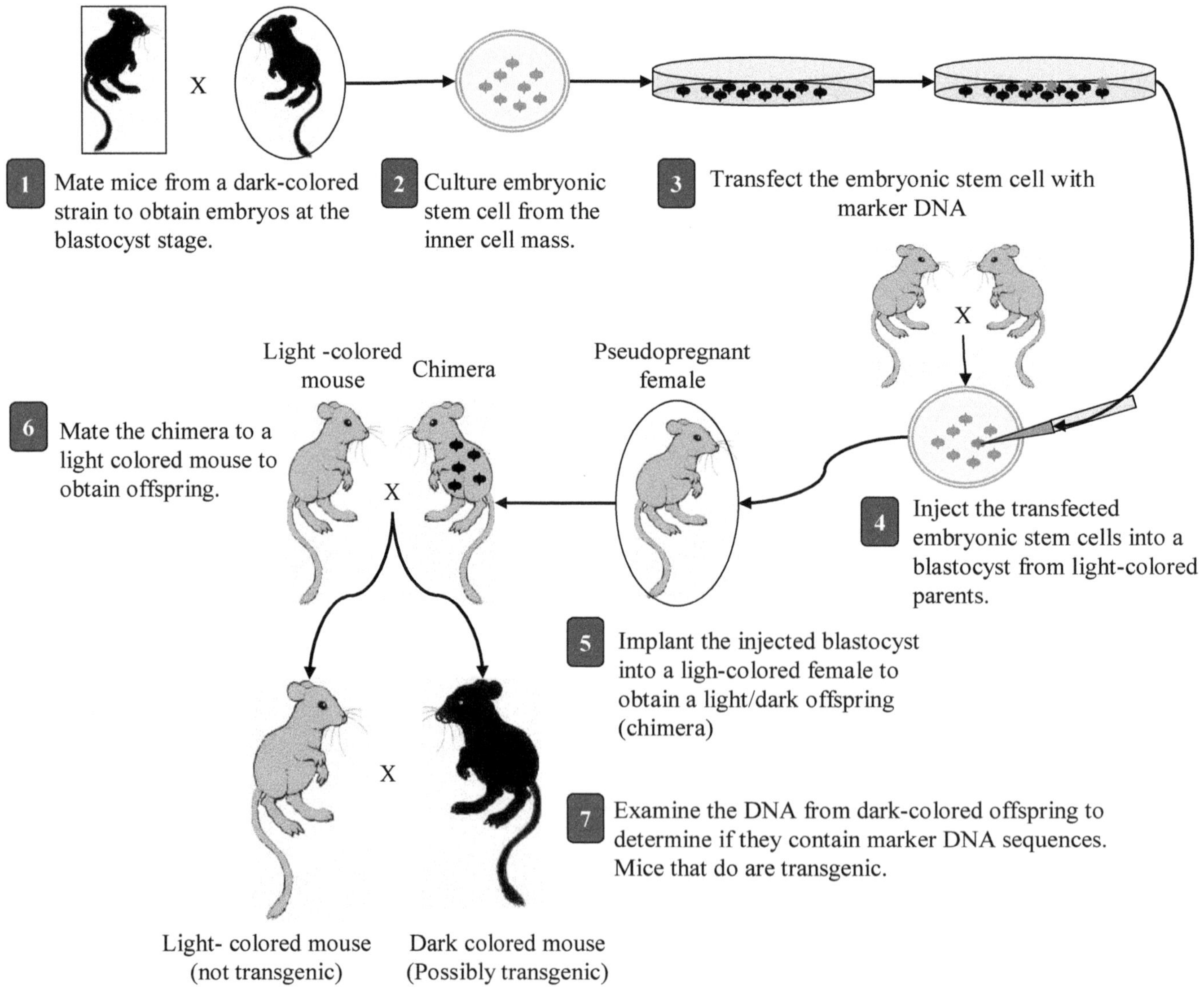

Figure 15.13: Transgenic Mice by Embryonic Stem (ES) Cell Transfer.

known as targeting. In embryonic stem cells, chimeras can be developed using targeted knockout mutations. Transgenic strains can be increased in number by breeding experiments. These strains will carry the knockout mutation. Mice homozygous for the Hoxc 8 mutation develop an extra pair of ribs posterior to the standard set of ribs; they also have clenched toes on their forepaws. The extra-rib phenotypes in these mutant mice are reminiscent of the segmental transformations seen with homeotic mutations in *Drosophila*. The genetic analysis of development in mice has provided clues about human development. Mutations that affect the cell lineages during development have been studied in nematodes; they have different general features shown below: Most genes that affect development are active in many cells.

Complex cell lineages often include simpler, genetically determined lineages within them; these components are called *sub-lineages* because they are expressed as an integrated pattern of cell division and terminal differentiation.

The lineage of a cell may be triggered autonomously within the cell itself or by signaling interactions with other cells.

Regulation of the development is controlled by genes that determine the different sub-lineages that cells can undergo and the individual steps within each sub-lineage.

15.23 Developmental Mutations can be Classified According to the Lineages

Inherent defect results in malformations. Both environmental and genetic factors are responsible for what can be transmitted, as these may occur in the ovum or the sperm. Mutations may bring changes in the developmental process.

Polygenic malformations are cleft lip, cleft palate isolated cleft palate, neural tube defect, club foot congenital heart defect, *pyloric stenosis, etc.* The risk for the defects can be calculated by considering first-degree relatives, second-degree relatives, the number of affected individuals in family consanguinity, nature of severity of the malformation. The environment influences neural tube defects and can be prevented by folic acid supplementation if given in the early phase of pregnancy.

A cleft lip is found in 1/700 individuals. Transforming growth factor (TGFA) is a growth factor expressed during palatogenesis in the mouse, but mice with inactivation of TGFA have abnormal skin, hair, and eyes; however, lack of TGFA does not result in cleft palate in the mouse. However, in humans, it is found that cleft is associated with TGFA; it can act as a modifier in cleft lip formation which is consistent with an oligogenic model for clefting in humans. Malformations generally affect the other sites, giving rise to a malformation sequence. For example, *innervation* problems of the tongue result in reduced pressure on the lower jaw and reduced growth of the mandible, also called micrognathia. Another congenital malformation is arthrogryposis, where congenital contracture of the fingers with partial joint fusions occurs. It is mostly due to the secondary effect caused by reduced fetal movement, either because of reduced muscle mass or muscle innervations. Both examples illustrate that the most apparent abnormality is not necessarily the most informative. Most primary defects lead to secondary effects, and sometimes it isn't easy to separate cause from effect.

15.24 Disruptions

It occurs in an average fetus; every fetus tries to develop normally but sometimes fails to result in disruptions. This could be because of vascular, infectious, or teratogenic origin. These can affect any tissue; one of the examples is the presence of an amniotic band which can destroy any body organ. Other disruptions can be caused by intrauterine infections: rubella, which destroys specific organ systems (in this case, the inner ear) that would have developed regularly if the insult had not occurred. Teratogenic substances such as thalidomide or valproate may cause disruptions because they interfere with normal signaling and may cause proliferation.

15.25 Deformations

Mechanical forces may create deformations. Generally, they occur in the latter part of the pregnancy and have a marked control of fetal development. Deformations may be inherent, *i.e.* due to asymmetrical development of the fetus itself, or extrinsic, due to constraint in *utero* of an otherwise normal fetus. The deformities can be of the foot, or they may affect the other parts of the body. Mechanical forces may result in potter sequence causing complex effects. Amniotic fluid leakage may result in serious oligohydramnios, causing reduced production of fluid (aplasia of kidneys, urethral valves, *etc.*); thoracic growth is controlled, and the entire growth and maturation of the lungs are upset, making them unable to aerobic expansion and oxygen exchange the legs are usually hyper flexed in front of the fetus. Any gradation of hip dislocation may occur.

15.26 Dysplasias

Dysplasia is caused by gene defects that affect the formation and growth of tissues. In compression to malformations that are due to abnormal events during early embryonic patterning, dysplasia affects the embryo at a later point when the patterning phase is finished. The gene defects can cause dysplasia which remains even after birth, and cause deformities of the skeleton, causing skeletal dysplasia, premature osteoarthritis, myopia, *etc*. Mutation in collagen causes brittle bone disease. Mutations in COLA1 result in skeletal dysplasia, which may be in the form of achondrogenesis, which is lethal, resulting in severe dwarfism or average growth with osteoarthritis and myopia. COL1A1 COL1A2 are expressed in bone. These are responsible for the mineralization of the bone; if the mutation occurs, it results in brittle bone disease (COL1A1 COL1A2).

16

Molecular Evolution

Due to genetic and environmental factors, genes undergo permanent changes causing molecular evolution. There are two types of mutations: synonymous and non-synonymous mutations. Synonymous mutations are silent mutations that do not result in an amino acid change in the protein and are hence non-functional. The frequency of synonymous mutations is roughly five times higher than non-synonymous alterations. Defects in DNA replication or repair processes are the main processes for genetic variation. DNA repair mechanisms cannot distinguish between synonymous and non-synonymous mutations; therefore, both synonymous and non-synonymous mutations take place with almost equal frequency. Nonsynonymous changes are removed due to natural selection as they are disadvantageous.

16.1 Synonymous and Non-synonymous Sites

The rate of evolutionary changes in DNA sequences of mammalian genes is known, and some regions of the genome of various species have high or low mutation rates. However, as per the norms of evolution, these genes should be removed by selection. At least 5 per cent of the human genome is under substantial constraint, most non-coding. Coding DNA is often correlated with complexities.

Some pseudogenes function as sources to determine the causes responsible for genetic diversity. Recombination sometimes results in the creation of pseudogenes to make variants of the functional gene. This occurs in immunoglobulins found in birds and mice, and the genes involved are historic in mouse horse globin and beta-globin genes, eta genes globin genes, and humans contain at least one globin gene for each of the different globin chains: α, β, γ, δ, ε, and ζ. Evidence from the study of hemoglobin variants and the biochemical heterogeneity of the chains in fetal hemoglobin (HbF) showed that the α- and γ-globin genes were duplicated. Persons were identified whose red cells contained more than two structurally different α-globin chains that could be best be explained by duplication of the α-globin gene locus, and the characterization of the structurally different Gγ and Aγ globin chains of HbF required duplication of the γ-globin gene locus showing the importance of class switching.

Genomic DNA has been divided into two major classes: short interspersed repetitive elements (SINEs) and long interspersed repetitive elements (LINEs). The Alu (a SINE – Short Interspersed Nuclear Element) repeat sequences occur in gene regulation, which is required in enhancing and silencing gene activity, possibly due to the receptor-binding site. Mobile elements are approximately 45 per cent of the human genome. These elements amplify and, because of the adverse effects of their transposition, are observed and may be seen in a notable number of human diseases. The mobile elements are seen in all eukaryotes; however, the activity of the classes of elements differs widely between genomes. They play an essential role in insertional mutagenesis and unequal homologous recombination events. These events have evolutionary significance.

The significant effect for phylogenetic analyses may be due to insertions in the SINE, showing that there is a need for careful examination of the presence of a SINE at a precise locus in many individuals showing common ancestry. However, sovereign insertions at the identical locus may be exceptional SINE insertions are not homoplasy-free phylogenetic markers.

Pseudogenes are almost as profuse as functional genes. It is considered that, although pseudogenes result in evolution paths and may interfere with genomic analysis, the procedure through which these imprints come up is required to generate an entire family of genes engaged in different characteristics. The biological function of pseudogenes is unclear, but it is known that they are involved in evolution. Over time, these genes have become fixed in the species. It is assumed that pseudogenes are involved in processes like gene conversion or recombination with functional genes, resulting in pseudogenes that show evolutionary conservation of gene sequence, reduced nucleotide variability, excess synonymous over nonsynonymous nucleotide polymorphism, and have functional roles due to the presence of specific DNA sequences.

16.2 Some of the Human and Mouse Junk DNA are Similar

Approximately 500 stretches of DNA in the human genome have not been changed in the last many years. It is known that mice and humans share a common ancestry because they separated from mice and rats millions and millions of years back. These stretches of DNA are "ultra-conserved" regions, which do not participate in coding for the protein; hence they can be treated as junk DNA. In these regions, the adverse selection is three times more vigorous than nonsynonymous changes in coding regions. How the adverse selection acts on every base in a DNA stretch of approximately one kilobase (1000bp) is unknown. It has been reported that approximately 500 regions of DNA have 1000bp, which is similar in rats and humans. This may constitute almost 500,000 similar genetic sites in DNA. Astonishingly, similar regions are primarily similar to chicken, dog, and fish sequences, but are not present in fruit flies. The last universal ancestor for these creatures lived nearly 400 million years ago. Of course, it is only reasonable to presume that if nature has gone under so much of a problem to conserve these ultra-conserved regions over all these years, then they must be essential and valuable and not junk; they must be involved in the management of the crucial genes and embryo progress.

Different pseudogenes in different species have shared specific mutations; it is assumed that they might have common ancestors. The nucleotide changes may not be wholly random in specific gene locations. Mutational "hotspots" have been identified in many genes and pseudogenes. Mutations in the hot spot regions are point mutations. There is a gene called GULOP which is the site for pseudogenes. In mammals, this codes for an enzyme L-glucose-γ-lactone oxidase. This is involved in the last step in the synthesis of vitamin C. In humans, it is defective; hence we cannot synthesize vitamin C; therefore, scurvy is caused. Many such mutations are also found in chimpanzees, orangutans, and macaques, indicating that the inactive GULOP gene in all primates descends from a common ancestor with an inactive gene. GULOP locus in the human genome shows that the pseudogene is missing in the 5′ exons in the intact rat gene. Exons I to VI are not present in the human genome. An attempt has been made to see if there was original inactivation of the GULOP gene in ancestral primates, but nothing conclusive has been reported. The human locus is also missing exon XI. In this case, the deletion is probably due to recombination between *Alu* sequences that flank the deletion site. This deletion probably took place after the original inactivation event.

It is proposed that the GLO gene copies in the human and guinea pig lineages get inactivated due to mutations. The mutations have also been noticed in guinea pigs and primate ancestors whose diets were rich in ascorbic acid, and the absence of GLO enzyme activity was not a disadvantage.

It is assumed that the many functional genes found in primates are found as pseudogenes in humans. GLO gene contains pseudogene sequence was compared recently between humans, chimpanzees, macaques, and orangutans; all four pseudogenes were established to share a common crippling single nucleotide deletion that results in the remainder of the protein to be translated in the wrong triplet reading frame.

GLO pseudogene sequences are thought to have evolved in parallel. Most of the mutations are in the form of deletion or substitution found in humans and guinea pigs. Humans and guinea pigs were thought to be separated at the time of the common ancestor with rodents. Hence mutational dissimilarity between a guinea pig and a rat is not common in humans with better than random odds. Most of the differences were shared by humans, including the one at position 97. Some non-random bias towards evolutionary ancestry has been noted. The likelihood of the same substitutions in both humans and guinea pigs occurring at the observed number of positions is approximately 1.84x10-12, comparable with the mutational hotspots.

Interestingly mutational hot spots seen in guinea pigs and humans match entirely with the mutations that set humans and primates apart from the rat. The researchers clustered substitutions that are precise to the rat lineage with separate substitutions common in guinea pigs and humans.

When one looks at this in a more extensive data set, only one position of the 10, 81, stands out as a probable cause of a shared derivative trait; one position is questionable, *i.e.*, 97, and the rest of the eight positions share the ancestral sites.

The difference in many loci is not restricted to the rat/mouse gene pools; hence, it may show mutational hot spots beyond the general overall "hotspots" or inclination for mutations in this particular genetic sequence.

What is fascinating about many of these mutational losses is that they often share similar mutational changes. The GULO mutation may result from the same genetic instability that is the same between the like creatures such as humans and the great apes. This may be one of the causes of similar mutations which are shared between humans and guinea pigs.

They may be due to hot spots; mutational differences are seen between humans and rats/guinea pigs compared to apes. This shows that humans and apes are more closely related than humans, rats/guinea pigs.

The fundamental question remains to be answered why do guinea pigs and humans have a significant share of deletions of I, V, and VI number exon along with four stop codons Danish pigs underwent mutations and showed a loss of GULO functionality. During this mutation, a sizable part of exon VIII was lost. This loss also matches the loss of primate exon VIII. Further, frameshift occurs in intron 8, which causes a loss of correct coding for exons 9-12. In primates, the same loss is also seen. That's quite a few key similarities that did not result from common ancestry for the GULO region.

Several genetic mutations resulting in functional losses, which are known to usually affect the same genetic loci similarly outside the same descent, have been reported. Achondroplasia is the most common type of short-limbed dwarfism. The condition occurs in 1 in 15,000 to 40,000 newborns. *FGFR3* gene mutations cause achondroplasia. The *FGFR3* gene instructs the formation of a protein involved in developing and maintaining bone and brain tissue. Two specific mutations in the *FGFR3* gene are responsible for almost all cases of achondroplasia. It is believed that mutations of the FGFR3 protein cause over-activation, which interferes with skeletal development and leads to disturbances in bone growth seen with this disorder. A remarkable observation on the FGFR3 gene is that a significant part of the mutations is introduced at the same two spots (755 C->G and 755-757 CGC-> TCT), independent of common descent. The short legs of the Dachshund are also due to the same mutation(s); the same allelic mutation is found in sheep also.

16.3 Mutational Hotspots

It has been noticed that across phylogeny, mutational hot spots influence genome evolution. The mammals split off from other animals more than 200 years ago. Because of this time, traces of random mutations would have vanished, due to which common genetic errors would have also been demolished. However, one question remains unanswered. Why some un-functional mutations are maintained Certain interspecies pseudogenes of this type might be sharing a common ancestor while the various types of animals themselves that have some of these genetic sequences, which may not be linked through common descent so much as they are incompletely related through common infection. There are no markers to find out the common descent.

The retroposons SINE/LINE are considered the best evolution marker. SINE/LINE can be inserted and duplicated, but the parent SINE is retained at its original location. This results from many copies of hybridization taking place due to convergence. Rearrangements are always complicated. Even simple insertions and deletions within coding regions have been painstakingly unlikely to be homoplastic, but numerous examples of convergence and parallelism of these events are now recognized. Although nucleotides and amino acids are widely acknowledged to exhibit homoplasy, some authors have proposed that widespread concurrent convergence in many nucleotides is virtually impracticable. Nonetheless, evolution studies in experimental models have shown examples of such convergence.

16.4 Introns and Common Ancestry

Introns are sections of DNA that do not code for how proteins are made. As introns do not code for proteins, it is thought that mutations in this region do not have any effect. Introns in chimps and apes are similar to human introns. The similarity of introns in humans, chimps, and apes reveals common ancestry.

The magnitude is less than 20 base pairs to almost 500,000 base pairs. The size of the introns within a given gene may be much larger than the coding regions of the gene itself (**Figure 16.1**).

Introns result from the random insertion of DNA; introns may be harmful or neutral. The neutral introns get fixed into the gene pool of a population.

16.5 Endogenous Retroviruses (ERVs)

Endogenous retroviruses (ERVs) are viral components that can be inserted into the genomes of humans and apes. These elements support the theory of common descent. ERVs are sometimes located in the exact location of humans and apes, explaining the shared common ancestry between the two species. The common ancestor of humans and apes must have been the one who experienced the ERV insertion into its genome. Later, when human and ape ancestors split off from this common ancestral lineage, the identical ERV sequences were maintained in the same places in the genomes of both lineages. However, this theory has many problems.

16.6 Function of EVRs

EVRs play a beneficial role in placental morphogenesis and mammalian reproduction. These elements may control or aid in transcribing over 20 per cent of the human genome and may enhance premature transcriptional termination.

EVRs are not junked DNA as was thought initially. During the process of evolution, the gene density decreased. It is seen that *E. Coli* has a gene density of about 2 Kb per gene, Drosophila 4 Kb per gene, and mammalian about 30 Kb per gene. Much of the decreased density is due to increased accumulation of non-coding or 'parasitic DNA' elements, such as type one and two transposons. However, the current evolutionary theory does not support this; some non-defective genomes are kept in the mammalian genome due to their functionality.

ERV-derived promoter sequences are around 51,197, which helps in transcription within the human genome, including 1743 cases where transcription is initiated from ERV sequences located in gene proximal promoter or 5′ untranslated regions (UTRs).

It is seen that approximately 1.16 per cent of the human genome sequence and PET tags that capture transcripts initiated from ERVs cover 22.4 per cent of the genome, suggesting that ERVs play an essential role in human transcription at a large scale. ERVs also behave protectively against various infections due to exogenous retroviruses: A possible biological roles has been hypothesized, which suggest that ERVs help the host resist infections of pathogenic exogenous retroviruses, affording a selective advantage to the host bearing them. Few avian and murine ERVs can block infection of exogenous retroviruses at entry by receptor interference; mouse Fv-1 blocks infection at a pre-integration step, also can be viewed as an ERV may also aid in influencing the immune system:

Human teratocarcinoma is reported to make a retrovirus-like particle that expresses *gag-pol* and *env* genes through different vectors. Also, ERV 3 can express the *env* gene in embryonic placental tissues. This could be one of the reasons why viral particles exist in human tissues. Some HERVs are expressed in mammary tumors. The placental tissues express the feline RD114, ERV-3, and HERV K10+.

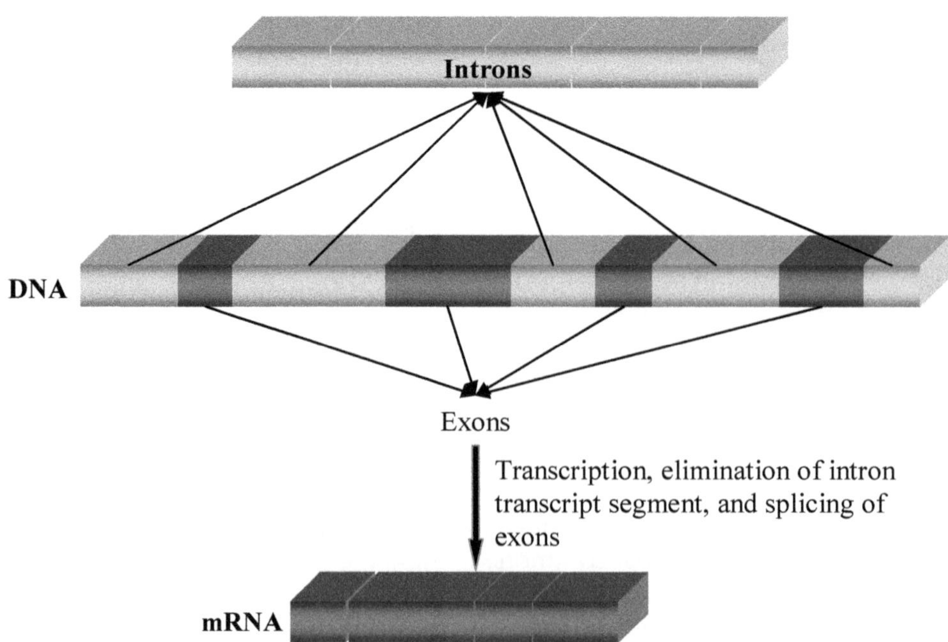

Figure 16.1: Transcription Eliminates Intron Transcript Segment.

Also, the ERV *gag* gene product can act as immunomodulatory. The p70 (*gag*) of mouse IAP has been cloned and expressed and shown to be identical to IgE binding factor (IgE-BF), which regulates B-cell ability to produce IgH. More recently, it has been reported that the endogenous *gag* is Fv-1, an-Herv. L is like an endogenous virus that confers resistance to MLV tumors. However, there are contradictory reports related to the role of p15E, but the immune-suppressing activity in culture assays has been established.

16.7 Origin of ERVs from Exogenous Retroviruses - or visa versa?

ERVs and ERV-Ls may be an intermediate between retrotransposons and exogenous viruses. There is no current knowledge about exogenous retroviral insertions into the modern human genome, and it has been reported that there are no known infectious exogenous counterparts of any human endogenous retroviruses. There exist similarities between ERVs in human and ape genomes. The ERV sequences were derived from exogenous infective retroviruses.

Viruses or viral elements were derived initially from functional genetic sequences that have suffered degenerative changes and loss of genetic controls, resulting in various parasitic features observed in the viruses today.

16.8 Non-random Viral Insertions

ERVs show a preference for certain specific locations in different genomes. Retroviruses can be seen anywhere in the genome; some are local "hot spots" and are present for integration. Statistical methods have provided some answers related to particular sites over the other Leukemia and HIV viruses, which have shown a preference for integration at transcribed regions of the host genome. Preferences have been noticed near the transcriptional start site in murine leukemia virus. There may be the interaction of the pre-integration complex with specific proteins or specific DNA sequences or structures associated with transcription; however, until data, no exact relationship has been established. Many hot spots exist near different integration sites.

16.9 The Odds against Similar ERV Germline Insertions

In different populations, ERV can be found in the same position, possibly due to common descent. The same virus may infect many individuals through the same source; there may be a population where individuals get clustered, infected, and fixed in future generations.

16.10 Inconsistent Phylogenies

The K family of ERVs is also known as the HERV – k provirus. This is present in chimpanzees and gorillas, but interestingly not in humans. Some portion of ERVs is also called CERV-2 and CERV1; these are not present in humans and orangutans, old world monkeys, and new world monkeys but present in chimpanzees, gorillas, bonobo, gorilla, chimpanzee, and gorilla arose from an exogenous source **(Figure 16.2).** It has been stated that there is no overlap

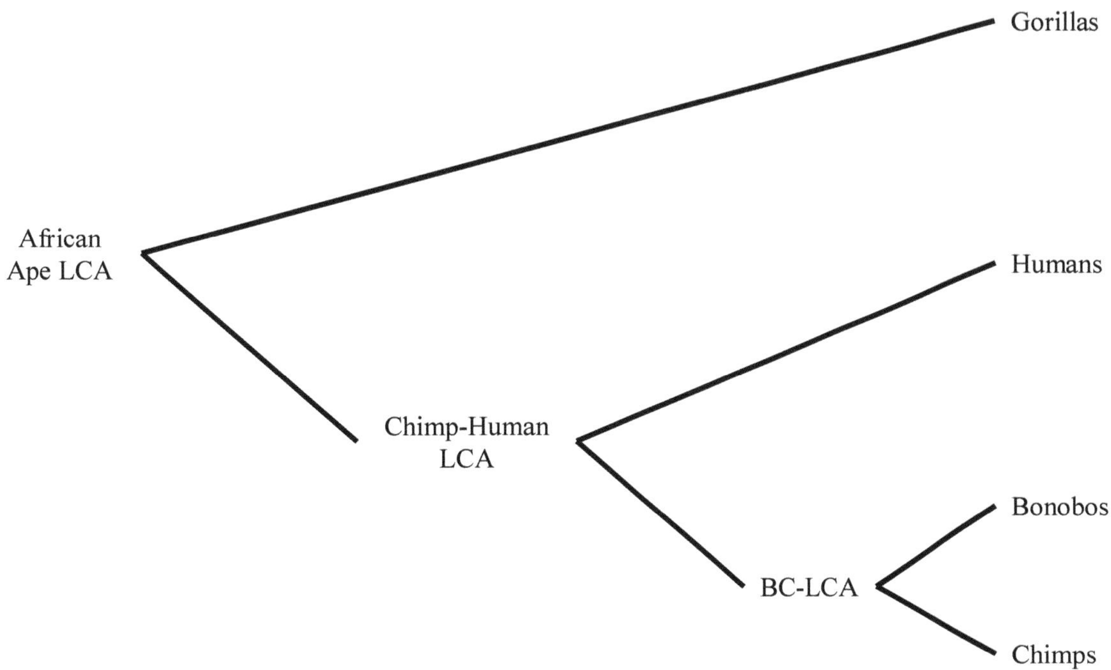

Figure 16.2: Phylogenetic Relationships of Gorillas, Humans and other Primates.

between the location of insertion among chimpanzees, gorillas, macaques, and baboons, revealing that there were no endogenous copies found in the common ancestor. The PTERV1 *phylogenetic* tree is consistent with the most accepted species tree for primates. It shows a horizontal transmission.

The problem in the phylogenetic analysis related to ERV is due to bias and interpretations involved in the EVRs data like wrong alignments *etc.* 1-8 per cent of each in 2005 Chimpanzee Sequencing and Analysis Consortium, have been created. Both chimpanzees and the human genome were compared, and it was found that approximately 200,000 ERVs, or portions of ERVs, exist within the genomes of both humans and apes, amounting to 127 million base pairs (around 4 per cent of the total genomic real estate).

16.11 Fusion of Chromosome 2

The telomeres are found at the end of the chromosome. A telomeric sequence is found in the middle of chromosome 2. There are remnants of an extra centromere and similar banding patterns with the equivalent chromosomes (2p and 2q) of apes. It is assumed that the fusion of chromosomes occurred sometime in the past. In **Figure 16.3** the fusion is demonstrated.

The similarity between humans and apes may be due to common origin which may be due to common descent via slow genetic modifications selected by mindless nature over time from some shared common ancestor.

16.12 Interstitial Telomeric Sequences

There are many interstitial telomeric sequences in humans and apes. One of the ITS sequences found in humans and chimpanzees is the chromosome 2q13IST site. This can be because of evolutionary breakpoints or fusion events other than ITSs. These are involved in the primates. ITS are not DNA scars. When TTAGGG repeats are added to chromosomes by telomerase ITSs sites are created; these are related to many proteins and show an essential role in recombination in the genome.

There is a repeat unit TTAGGGTTAGGGTTAGGG found in human and ape genomes everywhere. It may be a junk sequence that was created during the evolutionary process. However, presently it has been shown it is not the case that ITSs often have functional significance to the genome. The chromatin structures of telomeres can silence genes and are linked to the epigenetic mode of inheritance. These are involved in various evolutionary processes.

There are many ITS sequences within the human and in the ape genomes; only one of them, at the 2q13 ITS site, is shared by humans and chimps. In other words, of the many known ITSs in the genomes of humans and apes (and cows, chickens, rats, *etc.*), only 2q13 ITS can be linked with an evolutionary breakpoint or fusion event; other ITSs are not involved in the primates. Most of the known ITS sequences are not "DNA scars." The ITSs are sites where TTAGGG repeats have been added to chromosomes by telomerase, and many of these ITS sites are related to distinct sets of proteins linked to valuable roles within the genome, such as recombination hotspots, *etc.*

16.13 Pyknons

Some DNA sequences are conserved in nature. These sequences are not transcribed. These are termed pyknons that have extra non-overlapping instances in the untranslated and protein-coding regions of 30,675 transcripts from 20,059 human genes.

They are approximately 1/6th of the human genome of both intergenic and intronic regions. 127,998 Pyknons cover 898,424,004 DNA nucleotide positions in both forward and reverse directions.

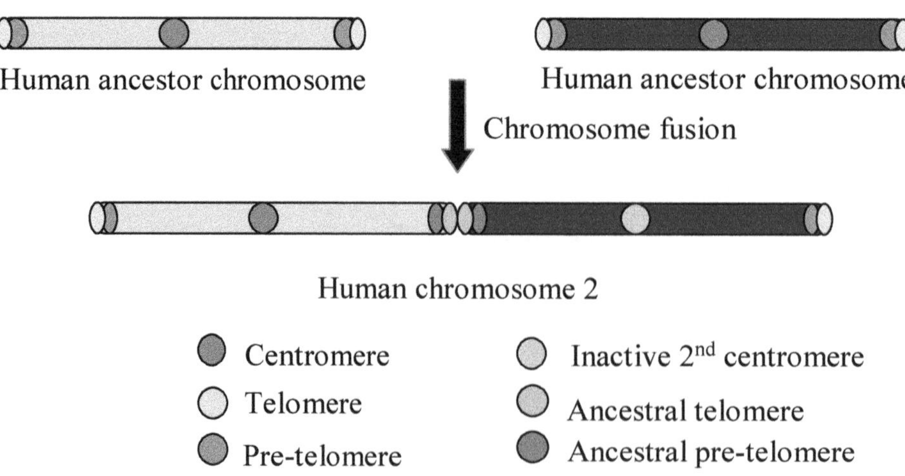

Figure 16.3: Fusion of Chromosome 2 in Humans.

Pyknons are associated with biological processes and can be compared with other genomes. Pyknons have genes in the 3′ UTRs region of other vertebrates and invertebrates, where they are overrepresented in similar biological processes as in the human genome. Pyknons play a role in post-transcriptional gene silencing and RNA interference.

They show similar functions in other genomes. Notably, >600 million nucleotides associated with non-genetic copies of pyknons in the human genome are missing from the mouse and rat genomes.

16.14 Human-Ape Differences

Both humans and apes show that 14 per cent of sequences are conserved at the mRNA level. Of this, 14 per cent, conserved sequence primate-specific is 10 per cent, while human-specific is only 1 per cent. After evolution, new miRNAs show different distributions, which are mentioned below:

1. 50 per cent of the human miRNA are conserved in primates
2. Approximately 30 per cent are conserved in mammals
3. Approximately 9 per cent are conserved in nonmammalian vertebrates or vertebrates
4. Interestingly, 8 per cent are specific for humans.

The expression of miRNA varies in humans and chimpanzees. This may be because miRNAs recently have been concerned with synaptic development and memory formation as the species-specific miRNAs expressed in the brain; The Brain has 10,000 different cell types, and this cell variation might have created differences in brain function.

Pseudogenes are also shown to have similar functionality as miRNAs. Transcripts of processed pseudogenes can have regions with significant antisense homology, which may propose a regulatory role for transcribed pseudogenes through an RNAi-like mechanism.

The pseudogenes can be protein-coding and are involved in the process of transcription. The pseudogene that encodes the nitric oxide syntheses is involved in transcription in Lymnea stag- nails; the gene expression is reduced once the RNA duplex is formed.

FGFR-3 gene is a pseudogene but is involved in the transcription in fetal tissues. It transcribes inside the cell it may be involved in the rapid degradation or inhibition of translation.

The transcribed pseudogenes can control the transcription of homologous protein-coding genes. Transcription of a pseudogene in Lymnea stag-nails, which is homologous to the nitric oxide synthase gene, decreases the expression levels for the gene through the formation of an RNA duplex; this is thought to arise via a reverse-complement sequence found at the 52 end of the pseudogene transcript. The transcription of the *makorin1-p1 TPΨg* gene in a mouse was required to stabilize the mRNA from a homologous gene *makorin1*. This regulation was deduced from an element in the 52 areas of the gene and the pseudogene. More recently, the murine *FGFR-3* pseudogene is transcribed in fetal tissues in an antisense direction. This prompted the following consideration.

FGFR-3 is 60nt long, and its transcript plays a regulatory role in *FGFR-3* expression. If these antisense transcripts could hybridize to sense *FGFR-3* transcripts inside the cells, this may lead to rapid degradation or inhibition of translation.

Human and chimpanzee genomes show 1,418 of 22,000 genes in their complement of genes which is approximately 6 per cent of the difference; there is a 1.5 per cent difference between orthologous nucleotide sequences. The Y-chromosome structures, layouts, genes, and other sequences should be the same in both species to a large extent, given only six million years or so since chimpanzees and humans probably diverged from a common ancestor. There is approximately an overall 70 per cent similarity based on the sequence similarity.

The origin of the Y chromosome is that it is distinct and evolved over a far greater period.

Approximately 6 million years of separation has led to the difference in MSY gene content in chimpanzees and humans, which is more comparable to the differences in the autosomal gene content in chickens and humans at 310 million years of separation.

However, since each respective Y-chromosome appears fully integrated and interdependently stable with its host organism, the most logical inference from the Y-chromosome data, without any prior commitment to the evolutionary story of origins, is that humans and chimpanzees were each specially created as distinct creatures or evolved over a far greater period.

16.15 Codon Usage Bias

Under codon usage bias, specific codons are used more often than other synonymous codons during the process of translation of genes; it varies within and among as it is noticed that 60 per cent of the leucine codons are found in *Escherichia* coli are CUG, and 80 per cent in yeast are UUG. This occurs during translation, *i.e.*, from genes to proteins. This is manifested at the time when a balance between mutation and translational selection takes place. This bias is more in the genes that are expressed more. It varies between different genomes. The factors influencing this are expression level, gene length, composition bias (per cent G+C content and GC skew), recombination rates, and RNA stability. Some synonymous codons pair with different tRNAs with the same amino acid. Therefore, a mutation to a synonymous codon does not change the amino acid, but it may change the tRNA used by a ribosome during translation. The A site binds to an aminoacylated tRNA, and the P site binds to a peptidyl-tRNA (a tRNA bound to the synthesized peptide). Thus, a codon context occupies these two sites of an active ribosome at any given time. Hence, optimization of codon context sequence may profoundly affect the translation selection of genes. The Human influenza virus depends upon the host cell machinery for its replication; codon usage bias could play a role in host adaptation and the virulence of the virus. To investigate this, the 59 codons, which can display bias in their usage, were examined for influenza virus sequences of human, avian, swine, and canine and equine host specificity along with human coding sequences. Relative synonymous codon usage (RSCU) was used to measure codon bias.

Recent pandemic H1N1/2009 viruses have demonstrated the codon bias. Codon usage patterns of the H1N1/2009 viruses in the CA have shown that the locations of the genes that are close to many swine H1N2 and H3N2 triple-reasserting viruses suggest that one of those viruses might be the precursor of H1N1. Similarly, the canine H3N8 virus was in the equine viral cluster, reflecting its *full* equine origin.

Previously, the 1918 H1N1 pandemic was suggested to be caused by the direct zoonotic transmission of an avian virus from birds to humans. However, other phylogenetic analyses of 1918 influenza viral sequences resulted in alternative hypotheses.

Human genes and most viral genes show a negative correlation in codon usage over time of viral isolation. Changes in viral codon usage might not affect or even harm viral gene translation. There are exceptions, *e.g.*, HA (H1N1) and NA (H3N2) genes which showed statistically significant positive correlations.

There exist translational pressures on the codon usage in human influenza viruses. However, this can be seen only in a few viral segments; the exact reason is unknown. The translational section may be weak as compared to other selective pressures. The reduced GC content in human influenza viruses may inhibit the activation of the human immune system, or the structures formed are less stable at low temperatures.

The evolution of protein is quite varied, it may be fast or slow, but nothing is clear when it will occur in conjunction with other proteins. One must keep in mind that regulatory genes evolve faster than structural genes.

Histones are positively charged, essential for DNA binding proteins in all eukaryotes. Almost every amino acid in histone H4 interacts directly with specific chemical residues associated with negatively charged DNA. If there is a change in the amino acid sequence of histone H4, it will affect its ability to interact with DNA. Hence histones are the most slowly evolving protein.

It is known that amino acid substitutions within many genes are mostly deleterious. The process of natural selection selects the favorable genes. One of the examples is genes associated with the major histocompatibility complex. The highest degree of polymorphism is seen differently in different individuals vulnerable to different infections; hence there is a high degree of polymorphism.

Higher concentrations of mutagens such as oxygen free radicals (*i.e.*, O2-) resulting from the metabolic processes in the mitochondria may also play a role in higher substitution rates. Selection pressure may usually eliminate many mutations in nuclear genes relaxed in the mitochondria because most cells contain several dozen–, of containing to a dozen copies of the mtDNA molecule. Regardless, changes in the proteins, tRNAs, and rRNAs encoded by the mitochondrial genome appear to be less detrimental to individual fitness than are similar changes in the proteins, tRNAs, and rRNAs encoded by nuclear genes.

Mammalian mtDNA is inherited from the mother. Mitochondria are in the cytoplasm, and only the mother's egg cell contributes cytoplasm to a zygote. As a result, mtDNA does not undergo reduction division, and all offspring are identical to the maternal genotype for mtDNA sequences. This helps select the ancestry.

Increasing amounts of DNA sequence data from various species are available for testing. The molecular clock for any given gene is constant in all evolutionary lineages. Substitution rates in rats and mice are essentially the same. In contrast, since their divergence, molecular evolution in humans and apes appears to have been only half as rapid as in Old World monkeys. The molecular clock differs in different species.

16.16 Molecular Phylogeny

Drawing the molecular phylogeny of different species has shown similarity revealing that different species have descended from the common ancestor. Molecular phylogeny has been assessed by looking at the DNA and RNA levels. This study was conducted after the advent of protein sequencing, PCR, electrophoresis, and other molecular biology techniques. Over the past 30 years, as computers have become more powerful and more generally accessible, and computer algorithms more sophisticated, researchers have been able to tackle. Now genomic data is available; hence it is easier to construct evolutionary models. After constructing evolutionary events, they are graphically represented as the evolutionary tree. This is quite a complex process. Phylogenetic data sets can consist of hundreds of species, each of which may have varying mutation rates and patterns that influence evolutionary change. Consequently, there are numerous different evolutionary models and stochastic methods available. The optimal methods for phylogenetic analysis depend on the nature of the study and the data used.

16.17 Construction of Phylogenetic Relationship Trees

Approximately 4 billion years ago, all living beings, recent and past, shared the same ancestor. This type of sharing is called phylogenetic and can be graphically shown as the phylogenetic tree, sometimes called an evolutionary tree. This shows evolution related to the species. A phylogenetic tree depends on observations based on morphological changes, molecular characteristics, *etc.* A phylogenetic tree is like any other tree with branches that connect two (occasionally more) adjacent nodes. Terminal nodes indicate the taxa for which molecular information has been obtained for analysis. Internal nodes represent common ancestors before the branching that gave rise to two separate groups of organisms. Branch lengths are often scaled to reflect the divergence between the taxa they connect. Where it is possible to distinguish one internal node as representing a common ancestor to all the other nodes on a tree, it is possible to make a rooted tree. Unrooted trees specify only the relationship between nodes and say nothing about the evolutionary path. Roots for unrooted trees can usually be determined through an outgroup. As in the example of an outgroup used for the relative rate test described earlier, outgroups are taxa that have unambiguously separated the earliest from the other studied ones. In the case of humans and gorillas, when baboons are used as an outgroup, the tree's root can be somewhere along the branch, connecting baboons to the common ancestor of humans and gorillas. When only three taxa are being considered, there are three possible rooted trees but only one unrooted tree (**Figure 16.4**).

16.18 How to Read a Phylogenetic Tree

One has to assume the following facts.

1. All species share a common ancestor, and all species arose from this common ancestor; this is known as the monophyletic grouping.
2. All species share a common ancestor but have not been derived from it. This type of grouping is called paraphyletic grouping.
3. The species do not share an immediate common ancestor; this type of grouping is called polyphyletic grouping.

The main phylogenetic tree types are rooted/unrooted, as shown in Figure 16.4. In the rooted -tree, the last common ancestor is considered. The branching pattern depends upon the events that have taken place over time. In the case of the unrooted tree, the most common ancestor is excluded. Another essential feature is to consider the out-group and in-groups. Very distantly related taxa or relatively

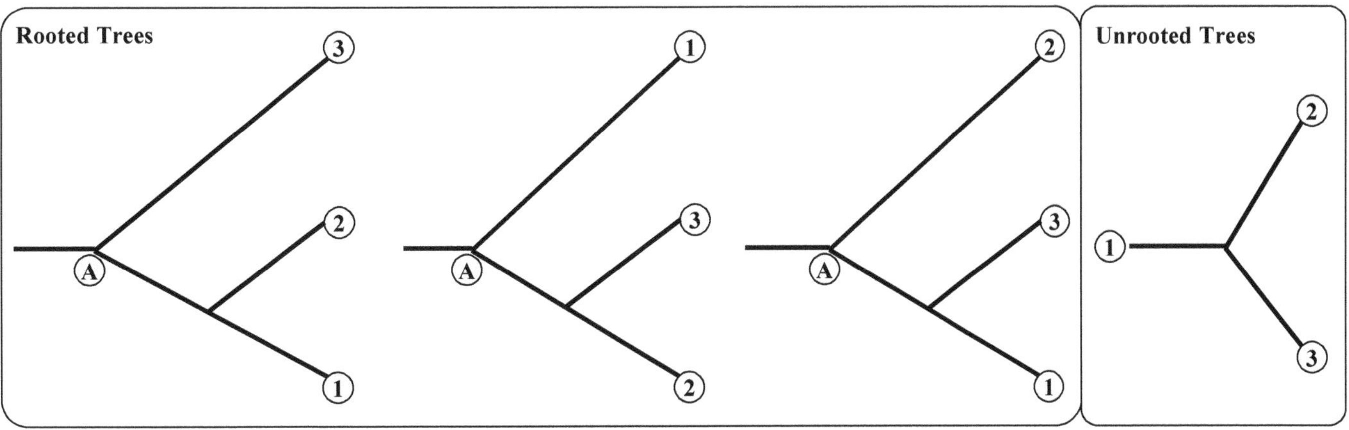

Figure 16.4: Rooted and Unrooted Phylogenetic Tree.

related taxa are considered outgroups. Terminal nodes are called operational taxonomic units (OTU) in a phylogenetic tree. The branches that do not join any terminal/leave OUT are called branches to be in "ancestral states"/hypothetical taxonomic units. These must have arisen during evolution and are not represented now. The speciation events are the internal branch points in a species. Gene families are created due to duplication events. The internal branches may be bifurcated or multi-furcated.

Step 1: For phylogenic analysis, suitable markers are required.

Step 2: Sequence alignment needs to be done multiple times.

Step 3: A proper evolutionary model needs to be selected.

Step 4: Phylogenetic reconstruction is required.

Step 5: Finally, the right phylogenetic tree should be constructed.

16.19 Number of Possible Trees

After the phylogenic tree is constructed, many possible rooted and unrooted trees are reported. The actual number of possible rooted (NR) and unrooted (Nu) trees for any number of taxa *(n)* can be determined with the following equations:

$$NR = (2n - 3)!/[2n-2 (n - 2)!]$$

$$Nu = (2n - 5)!/[2n-3 (n - 3)!]$$

Values for *n* must be greater than or equal to 2 for the first equation and greater than or equal to 3 for the second equation and can be extremely large. In practice, though, the value for *n* is often described in terms of dozens or, at most, hundreds of taxa or individuals where unimaginably large numbers of possible trees can describe the relationship between them.

16.20 Gene versus Species Trees

Phylogenies are also called "species trees" since "tree" is another name for phylogeny. A species tree reveals the overall pattern where species share a standard ancestral population more recently and which share a common ancestral population more distantly in the past. Phylogeny can be described based on how much two species share history and how much they are distinct from each other. If two species share a history for a longer time, the likelihood is that they will be more alike. A common history is seen in humans and chimps; for millions of years after that, the gorillas separated from the (human/chimpanzee) common ancestral population. This is the reason for the similarity between individual genes (and their alleles); this information helps construct the "gene trees."

16.21 Reconstruction Methods

Recently, phylogenetic relationships have been constructed based on molecular data. The overall similarity is measured using statistical tools. Depending upon the similarity index, the distance matrix is constructed, which can be used for molecular evolution.

16.22 Distance Matrix Approaches to Phylogenetic Tree Reconstruction

The oldest distance matrix method is the simplest of all the methods for tree reconstruction. Initially proposed in the early 1960s to help the evolutionary analysis of morphological characters, the un-weighted pair group method with arithmetic average ages(UPGMA) is a statistical method that requires data that can be condensed to measure the genetic distance between all the pairs of taxa being considered. The UPGMA method is used to construct the phylogenetic tree; let us consider four taxa called A, B, C, and D.

Let us construct a matrix to let dAB represent the distance (perhaps as calculated by the Jukes-Cantor model) between taxa A and B, *dAC* is the distance between taxa A and C, and so on. UPGMA begins by clustering the two taxa with the smallest distance separating them into a single, composite taxon. In this case, assume that the smallest value in the distance matrix corresponds to dAB, in which case taxa A and B are the first to be grouped (AB). After the first clustering, a new distance matrix is computed, with the distance between the new taxon (AB) and taxa C and D being calculated as d(AB)C = ½ (dAC + dBC) and d(AB)D = ½ (dAD + dBD). The taxa separated by the smallest distance in the new matrix are clustered to make other new composite taxa. The process is repeated until all taxa have been grouped. If scaled branch lengths are used on the tree to represent the evolutionary distance between taxa, then branch points are positioned halfway between the grouped taxa (*i.e.*, at dAB/2 for the first clustering).

A strength of distance matrix approaches, in general, is that they work equally well with morphological and molecular data and combinations of the two. They, like maximum likelihood analyses, also take into consideration all the data available for a particular analysis, whereas the alternative parsimony approaches described later discard many "noninformative" sites. A weakness of the UPGMA approach, in particular, is that it presumes a constant rate of evolution across all lineages, which the relative rate tests tell us is not always true. Several distance matrix-based alternatives to UPGMA, such as the transformed distance method and the neighbor-joining method are more complex but capable of incorporating different rates of evolution within different lineages.

16.23 New Distance Matrix Algorithm

Using a dynamic programming algorithm to find the score of two sequences takes O(mn) time. To reduce computational time, a new method to calculate a score of two sequences was proposed. Let X and Y be two sequences with lengths of n and m, respectively. The score of their alignment, namely the length of the longest common subsequences, can be calculated using a dynamic programming algorithm in O (mn) time. To reduce the computation time, a score estimating algorithm to approximately estimate the score of a two-sequence alignment was considered. The algorithm estimates the score of the alignment of two sequences in O (m+n) time. The proposed algorithm consists of 4 steps, each of which scans the two sequences from a different direction which denote X and Y as the upper and lower sequence, respectively. The four steps are left-upper, right-upper, left-lower, and Right-Lower. The step of Left-Upper starts from the first character in X, say X[0], and searches for the first matching character in Y from left to right. If there is no character in Y matching X[0], it restarts the scan to search for the character matching X[1] in Y. After such character, say Y[j], is found, the algorithm searches for the first character matching with Y[j] in the rest part of X from left to right. If there is no character in X matching Y[j], it restarts the scan to search for the character matching Y[j+1] in Y. These are alternately repeated until reaching the end of the sequences. As a result, the number of matching characters can be obtained in the scan. The count is the number of the matching characters in X and Y.

16.24 Parsimony-Based Approaches to Phylogenetic Tree Reconstruction

While distance and maximum likelihood-based tree reconstruction methods are grounded in statistics, parsimony-based approaches rely more heavily on the biological principle that mutations are rare events. The word *parsimony* itself means stinginess or cheapness. It refers to the fact that parsimony approaches attempt to minimize the number of mutations a phylogenetic tree must invoke to account for the sequences of all taxa being considered. These parsimony approaches assume that the tree that invokes the fewest number of mutations is the best and is deemed a tree of maximum parsimony.

As mentioned, the parsimony-based approach does not use all sites when considering molecular data. Instead, it focuses only on positions within multiple alignments that favor one tree over an alternative in terms of the number of substitutions they invoke. Not all positions within multiple alignments favor one tree over an alternative from the parsimony perspective. Consider the following alignment of four nucleotide sequences:

	Site					
	1	2	3	4	*5**	*6**
Sequence	G	C	G	A	T	G
	G	T	G	T	T	G
	G	T	T	G	C	A
	G	T	C	C	C	A

As shown in **Figure 16.5**, only three possible unrooted trees can be drawn that describe the

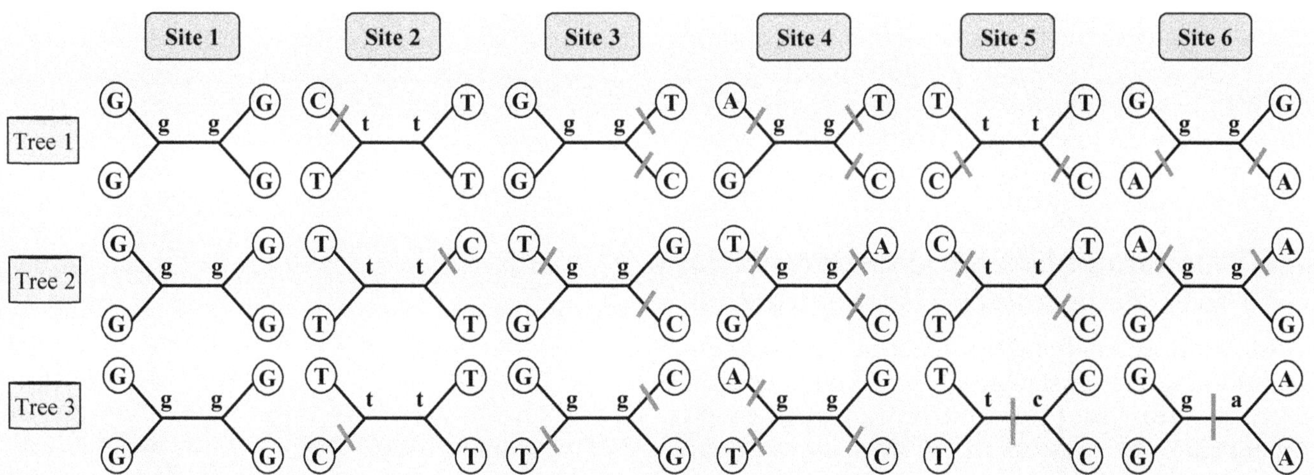

Figure 16.5: Three Different Unrooted Trees Describe all Possible Relationships between Four Taxa. These trees are based on the sequence data set.

relationship between four taxa. The unrooted tree that groups sequences 2 separate from sequences 3 and 4 would require only one mutation to have occurred in the branch that connects both groupings. Either of the two alternative trees that group the taxa differently would require two mutations and therefore do not represent the most parsimonious arrangement of the sequences. In contrast, all three possible unrooted trees for site 1 are indistinguishable from the parsimony perspective because no mutations must be invoked for any of them. Similarly, site 2 is uninformative because one mutation occurs in all three of the possible trees. Likewise, site 3 is uninformative because all three trees require two mutations, and site 4 is uninformative because all three trees require three mutations. In general, for a site to be informative, regardless of how many sequences are aligned, it must have at least two different nucleotides, and each of these nucleotides has to be present at least twice. In such an alignment, only the fifth and sixth sites (marked with asterisks) qualify as *informative sites* from a parsimony perspective.

Maximum parsimony trees are determined by first identifying all informative sites within an alignment and then determining which of all possible unrooted trees invokes the fewest number of mutations for each of those sites. The tree or trees that invoke the fewest mutations when all sites within an alignment are considered is the most parsimonious. An advantageous product of the parsimony approach is the generation of inferred ancestral sequences at each node of a tree (**Figure 16.5**). These inferred ancestral sequences go a long way toward making a non-issue of the infamous "missing links" of the fossil record and, when analyzed carefully, can give remarkably clear insights into the nature of long-dead organisms and even the environment in which they lived. Of course, the parsimony approach described here assumes that all nucleotides are just as likely to mutate into any of the three alternative nucleotides. More complicated parsimony algorithms take the difference in transition and trans version frequencies into account, although none is exceptionally reliable when rates of substitutions between branches of a tree differ dramatically.

16.25 Maximum Likelihood Approaches to Phylogenetic Tree Reconstruction

Maximum likelihood approaches represent an alternative and purely statistically based phylogenetic reconstruction method. With this approach, probabilities are considered for every individual nucleotide substitution in sequence alignments. For instance, transitions are observed roughly three times as often as transversions. In a three-way alignment where a single column is found to have a C, a T, and an A, it can be reasonably argued that a greater likelihood exists that the sequences with the C and the T are more closely related to each other than they are to the sequences with an A. Calculation of probabilities is complicated because the sequence of the common ancestor to the sequences being considered is generally not known. They are further complicated because multiple substitutions may have occurred at one or more of the sites being considered, and all sites are not necessarily independent or equivalent. Still, objective criteria can be applied in calculating the probability for every site and every possible tree that describes the relationship of the sequences in multiple alignments. The number of possible trees for even a modest number of sequences makes this a very computationally intensive proposition. Yet, the one tree with the single highest aggregate probability is, by definition, the most likely to reflect the actual phylogenetic tree.

The dramatic increase in the raw power of computers has begun to make maximum likelihood approaches feasible, and trees inferred in this way are becoming increasingly common in literature. Note, however, that no one substitution model is as yet close to general acceptance and because different models can very easily lead to different conclusions, the model used must be carefully considered and described when using this approach.

16.26 Bootstrapping and Tree Reliability

Longer sequence alignments require a longer time to analyze than shorter ones when the parsimony approach is used. However, because of the relationship between the number of taxa and the corresponding number of unrooted trees illustrated, adding more sequences has a much more dramatic effect on the time required to find a preferred tree. Once data sets involve 30 or more species, the number of possible trees is so large that it is simply impossible to examine all possible trees and assess the fit of the data to each, even when the fastest computers are used. Alternative trees are not all independent of each other, however, and many parsimony algorithms use shortcuts to avoid having to perform an exhaustive search. However, no tree reconstruction method is certain to yield the correct tree. Numerous variations on each approach have been suggested, and intensive simulation studies have been performed to compare the statistical reliability of almost all tree construction methods. The results of these simulations are easy to summarize. Data sets that allow one method to infer the correct phylogenetic relationship generally work well with all the currently popular methods. However, if any changes in the simulated data sets or rates of change vary among branches, then none of the methods works reliably. As a rule, if a data set yields similar trees when analyzed by two or three of the fundamentally different tree reconstruction methods, that tree can be reasonably reliable.

It is also possible for portions of inferred trees to be determined with varying degrees of confidence. Bootstrap tests allow a rough quantification of those confidence levels. The basic approach of the bootstrap test is straightforward: A subset of the original data is drawn (with replacement) from the original data set, and a tree is inferred from the new data set. In a physical sense, the process is equivalent to taking the printout of multiple alignments; cutting it up into pieces, each of which contains a different column from the alignment; placing all those pieces into a bag; randomly reaching into the bag, and drawing out a piece; copying down the information from that piece before returning it to the bag, and then repeating the drawing step until an artificial data set has been created that is as long as the original alignment. This whole process is repeated to create hundreds or thousands of resembled data sets, and portions of the inferred tree that have the same groupings in many of the repetitions are those that are primarily well supported by the entire original data set. Numbers corresponding to the fraction of bootstrapped trees yielding the same grouping are often placed next to the corresponding nodes in phylogenetic trees to convey relative confidence in each part of the tree. Bootstrapping has become very popular in phylogenetic analyses, even though some methods of tree inference can make it very time-consuming to perform.

Despite their often-casual use in scientific literature, bootstrap results need to be treated with some caution. First, bootstrap results based on fewer than several hundred iterations (rounds of re-sampling and three-generation) are not likely to be, especially when large numbers of sequences are involved. Simulation studies have also shown that bootstrap tests tend to underestimate the confidence level at high values and overestimate it at low values. And, since many trees have very large numbers of branches, there is often a significant risk of succumbing to "the fallacy of multiple tests"-some results may appear to be statistically significant by chance simply because so many groupings are being considered. Still, some studies have suggested that commonly used solutions to these potential problems (*i.e.*, doing thousands of iterations; using a correction method to adjust for estimation biases; collapsing branches to multi bifurcations wherever bootstrap values do not exceed a very stringent threshold) yield trees that are closer representations of the true tree than the single most parsimonious tree.

16.27 Human Origins

Another field in which DNA sequences are used to learn evolutionary associations in human evolution. In disparity with the extensive phenotypic distinction observed in size, body shape, facial features, and skin color, genetic differences among human populations are astonishingly small. For example, analysis of mtDNA sequences shows that the mean difference in sequence between two human populations is about 0.33 percent. Other primates show more significant differences. For example, the two subspecies of orangutan have mtDNA sequences that differ as much as 5 percent. This high genetic resemblance indicates that all human groups are intimately related. Another surprising observation emerges upon careful examination of those genetic differences that do exist between different human groups: The greatest differences are not found among populations located on different continents but instead are found between human populations residing in Africa. All other human populations show fewer differences than we find among African populations alone. Many experts interpret these findings to mean that humans originated and experienced their early evolutionary divergence in Africa. By this theory, a small group of humans migrated out of Africa. It gave rise to all other human populations only after several genetically differentiated populations had evolved in Africa. This hypothesis has been called the out-of-Africa theory. Sequence data from both mitochondrial DNA and the nuclear Y chromosome (the male sex chromosome) are consistent with this hypothesis in that they suggest that all people alive today have mitochondria that came from a "mitochondrial Eve" and that all men have Y chromosomes derived from a "Y chromosome Adam" roughly 200,000 years ago. While the out-of-Africa theory is not universally accepted, DNA sequence data are playing an increasingly important role in the study of human evolution and the study of the evolution of many lineages. Processes like gene conversion, gene duplication, exon shuffling, *etc.* are achieving the new functions in molecular evolution.

17

Population Genetics

The basic concept of population genetics is diversity leading to variation. Variation is observed at the phenotypic level. Phenotypic variation has a genetic basis. A population is known as the Mendelian population when the individuals present in this population constitute a group of interbreeding individuals who share the same gene pool. However, Mendelian populations may change with time. The population's total gene pool is considered to understand the evolutionary process's genetics. Variation can be studied by investigating gene, genotype, and phenotype frequencies.

17.1 Darwin's View of Evolution

Charles Robert Darwin was an English naturalist and geologist renowned for his contributions to the theory of evolution. He embarked on a five-year survey voyage worldwide on the HMS Beagle. His studies of specimens from various locations around the globe led him to formulate the theory of evolution and his views on natural selection. In 1859, he published "The Origin of Species". During his time in Patagonia, Darwin studied fossils that resembled living animals, and he evaluated different fossils to correlate them with the anatomical features of existing species. He also observed 13 finch species similar to birds from the South American mainland. He hypothesized that these finches descended from migrants from the mainland and, over time, adapted to different ecological conditions.

Alfred Wallace, who also studied plants and animals in South America, independently arrived at the same hypothesis as Darwin. In agreement with Darwin, he submitted a manuscript to the Linnean Society in London for publication and proposed the theory of natural selection.

17.2 The Hardy-Weinberg Law

All populations of living organisms exhibit genetic variation. When populations mate randomly with no selection bias or evolutionary forces acting on them, the gene frequencies remain constant even after many generations. This is known as the Hardy-Weinberg law, independently discovered in 1908 by the English mathematician G.H. Hardy and the German physician Wilhelm Weinberg. The equation $p2 + 2pq + q2 = 1$ is provided by the Hardy-Weinberg Law to relate the genotype frequencies and allele frequencies in a randomly mating population consisting of two alleles such as A and a. If a population is in Hardy-Weinberg equilibrium with no evolutionary forces affecting it, the proportion of genotypes will remain the same. However, two conditions must be met for the population to be in equilibrium: it must be truly randomly mating and infinitely large.

If there are two alleles for a particular character, they are designated as A and a, and at a gene locus, three genotypes are possible: AA, Aa, and aa. The alleles A and a frequency may be considered as p and q, respectively. The equilibrium frequencies of the three genotypes will be (p + q)2 = p2 + 2pq + q2 for AA, Aa, and aa, respectively. When there are three alleles, p, q, r, then (p + q + r)2 = p2 + q2 + r2 + 2pq + 2pr.

In the first generation, a parental generation has two alleles, p for A and q for a. Three possible genotypes are possible in the next generation, which are the products of the probabilities of the resultant alleles in the parents. The probability of genotype AA among the progeny is the probability p that allele A will be seen in the paternal gamete multiplied by the probability p allele A will occur in the maternal gamete, or p2. Similarly, the probability of the genotype aa is q2. The genotype Aa can arise when A from the father mates with a from the mother, which will be found with a frequency pq, or when a from the father combines with A from the mother, which also has a probability of pq; the result is a total probability of 2pq for the frequency of the Aa genotype in the progeny.

When the population is in Hardy Weinberg equilibrium, there is no change in the allele frequencies. An AA genotype of the offspring means that all the alleles in the individual have an A allele plus half the frequency of the Aa genotype or p2 + pq = p(p + q) = p (because p + q = 1). Similarly, the allele frequency among the offspring is q2 + pq = q(q + p) = q. These are precisely the frequencies of the alleles in the parents. The population is in a state of equilibrium if they are mating randomly. The equilibrium gets disturbed if there is selective mating or assortative mating. Most of the time, it is due to social reasons. Interracial marriage is an example of assortative mating. Suppose a hypothetical community in which 80 per cent of the population is white and 20 per cent is black. With random mating, 32 per cent (2 x 0.80 x 0.20 = 0.32) of all marriages would be interracial, whereas only 4 per cent (0.20 x 0.20 = 0.04) would be between two blacks.

17.3 Hardy-Weinberg for Multiple Alleles

It is easy to calculate the Hardy-Weinberg (HW) equilibrium when there are only two alleles, p, and q, as explained above for an autosomal locus with two alleles. We can also use the square of the allelic frequencies to calculate the equilibrium frequencies for a locus with multiple alleles. For example, an autosomal locus with three alleles, A1, A2, and A3, has six possible genotypes. According to the HW law, the genotype frequencies are in equilibrium, but it depends on the allelic frequencies. If the frequencies of alleles A1, A2, and A3 are p, q, and r, respectively, then the equilibrium genotypic frequencies will be the square of the allelic frequencies, (p + q + r)2 = p2 + 2pq + q2 + 2pr + 2qr + r2. The total of the allelic frequencies should be 1.

17.4 Hardy-Weinberg Expectations for X-linked Loci

In X-linked loci, females have two copies of the gene, but males are hemizygous. For females, let us suppose XA and Xa alleles; there would be five genotypes. Females have two X-linked alleles, so the expected female genotypes are the square of the allelic frequencies. If the frequencies of XA and Xa are p and q, respectively, then the equilibrium frequencies of the female genotypes are (p + q)2 = p2 (frequency of XAXA) + 2pq (frequency of XAXa) + q2 (frequency of XaXa). Males have only one X-linked allele, so the frequencies of the male genotypes are p (frequency of XAY) and q (frequency of XaY).

The p2 is the expected proportion of females with the genotype XAXA. If females comprise 50 per cent of the population, then the expected proportion of this genotype in the whole population is 0.5 X p2. The frequency of an X-linked recessive trait in males is q, whereas the frequency among females is q2. Let us suppose the X-linked allele is rare; the trait will, therefore, be much more frequent in males than in females. A well-known example is hemophilia A, a clotting disorder caused by an X-linked recessive allele with a frequency (q) of approximately 1 in 10,000, or.0001. At Hardy-Weinberg equilibrium, this frequency will also be the disease frequency among males. The frequency of the disease among females, however, will be q2 = (0.0001)2 = 1 in 10 million.

17.5 Estimating Allelic Frequencies with the Hardy-Weinberg Law

Assuming that a population is in Hardy-Weinberg equilibrium for a recessive disorder such as cystic fibrosis, the frequency of the recessive genotype (aa) can be determined through q2, with allelic frequency as the square root of the genotypic frequency, *i.e.*, q = √f(aa). For instance, if the frequency of cystic fibrosis in a population is 1 in 2000, or 0.0005, q would be √0.0005 = 0.02, meaning about 2 per cent of alleles in the Caucasian population encode cystic fibrosis. By subtracting q from 1, we can calculate the frequency of the normal allele, *i.e.*, p = 1 – q = 1 - 0.02 =.98. The frequency of normal homozygous and carriers of the cystic fibrosis gene would be f(AA) = p2 = (0.98)2 = 0.960 and f(Aa) = 2pq = 2(0.02) (0.98) = 0.0392. Hence, around 4 per cent (1 of 25) of people from this population are heterozygous carriers of the allele that results in cystic fibrosis.

17.6 Nonrandom Mating

One of the assumptions of the Hardy-Weinberg law is that mating should be random in a population. Nonrandom mating can affect how alleles merge from genotypes, thus changing the gene frequency. Positive assortative mating can occur when there is a predisposition for like individuals to mate, such as tall people mating preferentially with other tall people and short people mating preferentially with other short people. Negative assortative mating can also occur when unlike individuals' mates. Inbreeding is yet another form of preferential mating between related individuals. Inbreeding, which is positive assortative mating for relatedness, affects all genes, not just those that resolve the trait for which the mating preferences occur. As a result, inbreeding causes a deviation in the Hardy-Weinberg equilibrium, leading to an increase in homozygotes and a decrease in heterozygotes in a population.

The probability that two alleles are "identical by descent" can be measured by the inbreeding coefficient, designated as F. Inbreeding coefficients can range from 0 to 1, where a value of 0 indicates that mating in a large population is at random. A value of 1 indicates that all alleles are identical by descent. Inbreeding coefficients can be calculated from the analyses of pedigrees or by calculating the decrease in the heterozygosity of a population (**Table 17.1**). With inbreeding, the frequency of the genotypes will be f(AA) = p2 + Fpq, f(Aa) = 2pq – 2Fpq, and f(aa) = q2 + Fpq. The proportion of heterozygotes reduces by 2Fpq with inbreeding, and half of this value (Fpq) is added to the proportion of each homozygote.

Table 17.1: Generational Increase in Frequency of Homozygotes in a Self-fertilizing Population Starting with p = q =.5

Generation	*Genotypic Frequencies*		
	AA	*Aa*	*Aa*
1	¼	½	¼
2	¼ + 1/8 = 3/8	¼	¼ + 1/8 = 3/8
3	3/8 + 1/16 =7/16	1/8	3/8 + 1/16 =7/16
4	7/16 + 1/32 = 15/32	1/16	7/16 + 1/32 = 15/32
n	$1 - (½)^n/2$	$(½)^n$	$1 - (½)^n/2$
∞	½	0	½

In self-fertilization plants, F = 1, increasing the frequency of homozygotes.

Inbreeding is harmful because it results in an increase in the proportion of homozygotes and thereby increases the probability, which can be destructive. Let us suppose that a recessive allele (a) that causes a genetic disease has a frequency (q) with a frequency of.01. If the population mates randomly (F = 0), the frequency of individuals affected with the disease (aa) will be q2 =.012 =.0001; so only 1 in 10,000 individuals will have the disease. However, if F=.25 (the equivalent of brother-sister mating), then the expected frequency of the homozygote genotype is q2 + 2pqF = (.01)2 + 2(.99) (.01) (.25) =.0026; thus, the genetic disease is 26 times frequent at this level of inbreeding.

Homozygosity is increased as the inbreeding increases resulting in the individuals becoming homozygous for the same allele. Suppose a species undergoes inbreeding for a few generations. In that case, many harmful recessive alleles will be removed due to natural or artificial selection, so the population becomes homozygous for valuable alleles.

The allelic frequencies may alter due to mutations, migration, genetic drift, and natural selection. Allele frequency can be defined as allele frequency of that allele is the fraction or percentage of loci that the allele occupies within the population.

If there are ten individuals in a population and at a given locus with two possible alleles, *A* and *a*, and the genotypes of the individuals are:

AA, *Aa*, *AA*, *aa*, *Aa*, *AA*, *AA*, *Aa*, *Aa*, and *AA*

then the allele frequencies of allele *A* and allele *a* are:

pA = (2+1+2+0+1+2+2+1+1+2)/20 = 0.7

pa = (0+1+0+2+1+0+0+1+1+0)/20 = 0.3

The number 20 has emerged because all individuals are diploid.

17.7 Factors Affecting Genomic Variation

17.7.1 Mutation

Mutations are believed responsible for evolution as harmful alleles are deleted through natural selection. Some mutations are beneficial and are selected for. Neutral mutations do not influence an individual's fitness and as such, they accumulate in the species.

Variation within a population is crucial for evolution. While recombination can bring about variation, mutations are more common.

17.7.2 The Effect of Mutation on Allelic Frequencies

Mutations can change gene frequencies, often resulting in a shift in allele frequencies. For example, in a small population of 25 diploid individuals with only one locus and two alleles, G1 and G2 (**Figure 17.1**), with frequencies p and q respectively, if there are 45 copies of G1 and five copies of G2, the gene frequency would be p=.90 and q=.10. If a mutation changes G1 to G2, there will be 44 copies of G1 and six copies of G2, resulting in an increase in the frequency of G2 from.10 to.12. If this mutation continues, it will cause an increase in G2 and a decrease in G1. The change in G2 can be labeled as (Dq) and is dependent on the rate of G1 to G2 mutation (μ) and p, the frequency of G1 in the population. When p is extensive, many copies of G1 are accessible to mutate to G2, and the amount of change will be comparatively large.

As mutations occur more frequently and the mutation probability decreases, fewer copies of G1 are available to mutate to G2. The change in G2 due to mutation is equal to the mutation rate, time, and allelic frequency: $Dq = \mu p$. This results in a change in the gene frequency. G2 can mutate back to G1 with a rate v, unlike the forward mutation m. This increases G1 and decreases G2 frequency. The forward and reverse mutations maintain a balance: $Dq = \mu p - vq$.

17.7.3 State of Equilibrium

In a population, if the frequency of G1 is high, then G2 will be low. G1 can mutate to G2, called the rate of forward mutation. As the frequency of G2 increases, there will be a reduction mutation which is high with the G1 frequency, so the number of forward mutations decreases. If the reverse mutation takes place, G1 will increase. This results in no change in the allelic frequency. The equation $Dq = \mu (1 - q) - vq$ is applicable, where $p = 1 - q$, and μ is the mutation rate. At equilibrium, Dq will be 0, so $q = \mu/\mu + v$ where q represents G2 at equilibrium.

17.8 Population Size

The gene frequency remains the same in large populations, whereas in small populations, the frequency gets affected. For instance, if there are

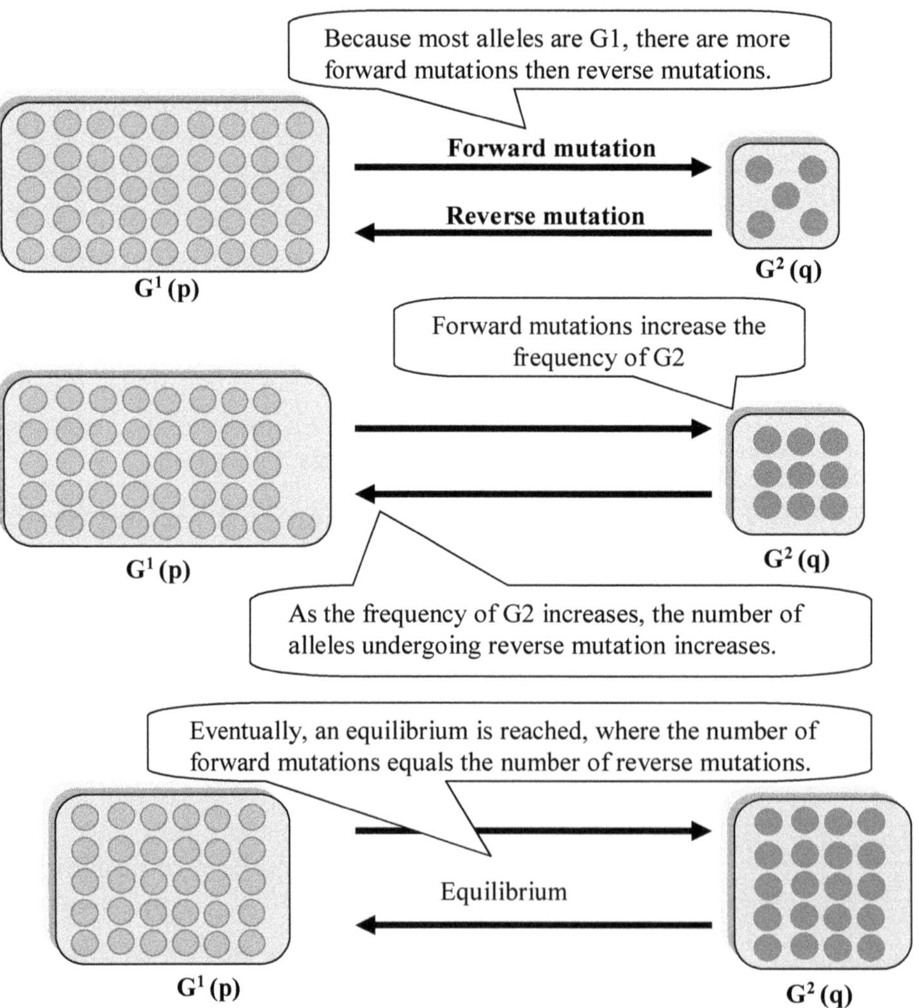

Figure 17.1: Recurrent Mutation Changes Allelic Frequencies.

500 red and 500 white beans in a bag, and 10 beans are drawn from the bag at different intervals, the values of p and q will differ noticeably from one sample to the next. In contrast, when the bag contains 10,000 beans with equal numbers of white and red beans and 2,000 beans are drawn per sample, scored, and replaced, the values of p and q in each sample are very close to the original frequencies. In small populations, the frequencies of alleles, due to sampling, differ remarkably from one generation to the next, resulting in a random genetic drift. Allele frequencies are statistically impossible to remain the same in a small population even if all are in Hardy-Weinberg equilibrium.

The effective population size (Ne) refers to the mean number of individuals who contribute genes to the next generation. When the values for Ne are small, the analyses always show the fluctuation of gene frequencies in successive generations. After several generations, one of the alleles is lost, that is, p or q = 0, while the other allele is the only one at the locus in the population, that is, p or q = 1.0 (**Figure 17.2**). When the frequency of an allele is 1.0, the gene is genetically fixed.

If the sample size increases, the genetic fixation takes many generations. When computer models are run with many independent populations of the same size and with the same initial allele frequencies, the patterns of fluctuation of allele frequency vary from population to population. Eventually, one of the original alleles in each population becomes fixed. Because genetic fixation under these conditions is strictly a random event, there is a 50: 50 chance that one of the alleles will either be lost or fixed (**Figure 17.3**).

17.9 Genetic Structure of Isolated Human Populations

The small human groups are likely to experience nonrandom mating; however, depending on the social behavior of the population, some individuals may transmit most of their genes to the next generation than others. Human groups often expand, but the population may be subdivided into smaller communities due to environmental pressures. In turn, small populations may merge with other groups. Historically, cultural changes have isolated human populations from one another. These examples show there is always a theoretical framework for the change of allele or genotype frequencies. Practically, natural populations can't obey the laws of Hardy Weinberg equilibrium.

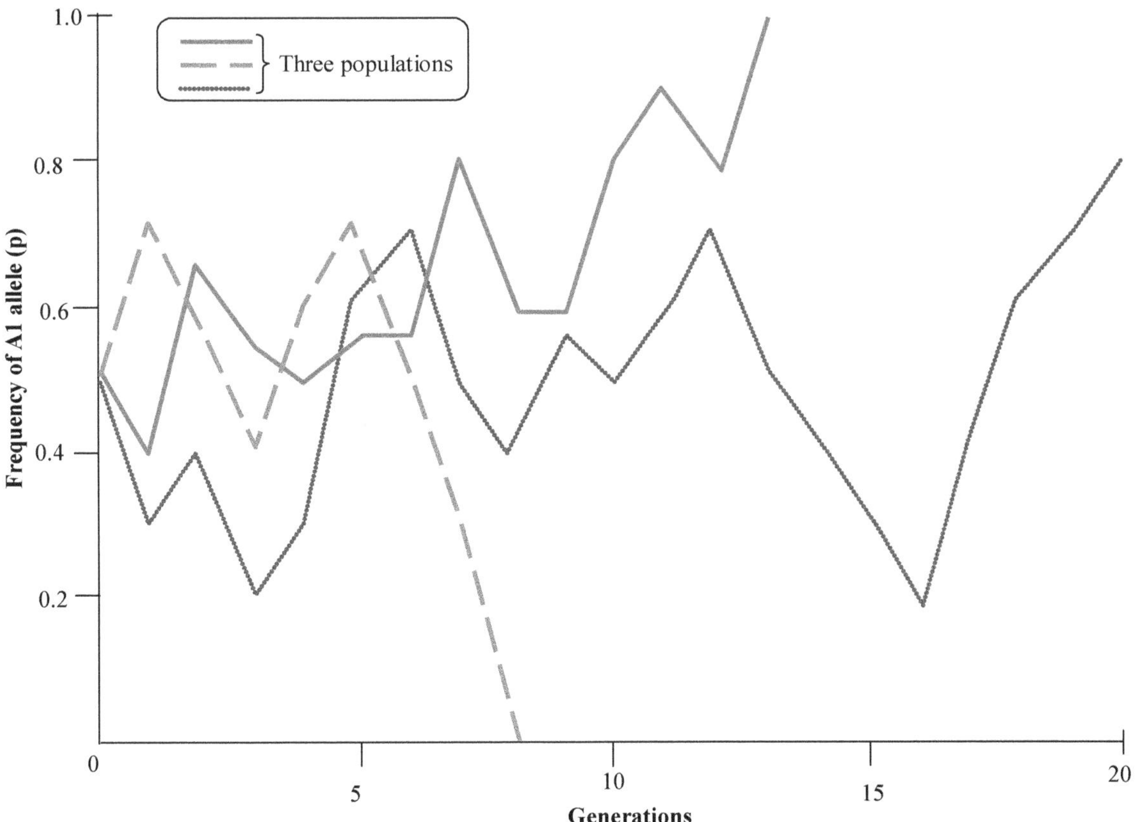

Figure 17.2: Computer Simulations Demonstrating the Result of Random Genetic Drift over Time. The original frequency of the A1 (p) and A2 (q) allele in each population is the same, that is, p = q= 0.5. The effective population size is maintained at ten male and ten female parents per generation.

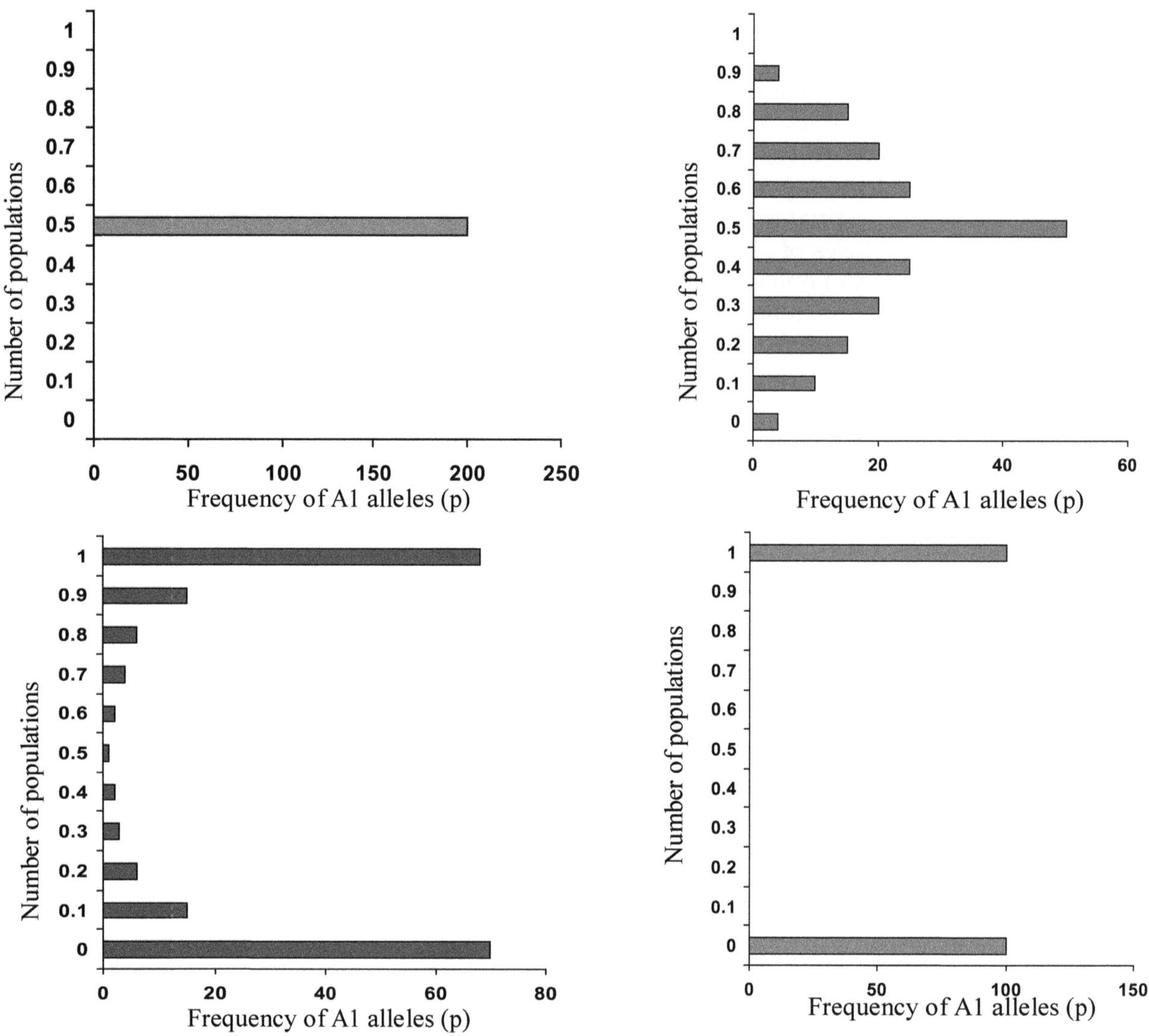

Figure 17.3: Frequency of the A^1 Allele in 200 Small Populations over Time.
The effective population is maintained at 10 male and 10 female parents per generation. Hardy-Weinberg conditions are kept in successive generations. The initial frequency of the A^1 (p) and A^2 allele (q) in each population is the same, that is, $p = q = 0.5$ (A). (B), (C), and (D) likely frequency distribution of the A^1 allele in 200 populations after 5, 30, and 90 generations, respectively.

The Yanomame tribe is a small population that lives in the rainforest. They depend on hunting and farming for their food. Their villages are separated by 15 to 40 miles and have a population of 50 to 250 individuals. Some villages have merged with others, while others have divided into smaller enclaves. Movement of individuals has occurred voluntarily and involuntarily between Yanomame villages and other tribes. The Yanomame population is a great example of how gene frequencies change due to various factors.

A study of 16 alleles from nine gene systems in 48 Yanomame villages revealed a range of allele frequencies for many loci. In some villages, the frequency of an allele varied from 0.2 to 0.8. Some villages did not have an allele that was present in other villages. Genetic fixation was seen in some villages but not in others. This type of change is known as genetic micro differentiation. The factors that contributed to genetic micro differentiation in Yanomame villages were unequal allele frequencies between the sexes, which was attributed to female infanticide, differential fertility due to the headman having more than one wife and therefore having more children on average than other males in the village, and marriages between individuals from different villages.

17.10 Founder Effect

Founder effect is a phenomenon observed in populations with very few members. There are several examples of founder effects in different parts of the world. For instance, in Finland, genetic isolates of small numbers of founders have grown to more than 5 million individuals in approximately 100 generations. As a result, 35 genetic disorders have been seen in the Finnish population, which are collectively referred to as "Finnish disease heritage." Similarly, in the Charlevoix-Saguenay-Lac Saint-Jean region of northeastern Quebec in Canada, a population that originated in 1675 with a small number of founders (N» 600) has now grown to more than 310,000 individuals. During 12 generations, this population has grown and remained relatively isolated. The frequencies of heterozygotes for different disease-causing loci have increased 5 to 40 times compared to the general population. Each of these homozygous recessive disorders occurs infrequently in other populations (**Table 17.2**).

17.11 Migration

The movement of populations from one location to other results in a change in gene flow, known as migration. We can better understand the impact of migration on allele frequencies by considering an example of two populations, I and II. The frequency of a particular allele in population I is qI, and in population II, it is qII. In the next generation, a representative number of individuals move from population I to population II (unidirectional migration). All conditions of Hardy-Weinberg equilibrium are met except for migration. After migration, population II will have two types of individuals - the original population and the migrant individuals (m). The frequency of allele a in the merged population II (q'II) is given by:

$$q'II = qI(m) + qII(1 - m)$$

Here, qI(m) is the contribution made by the copies of allele a in the migrants, and qII(1 – m) is the contribution made by copies of allele a in residents. The change in allelic frequency due to migration (Dq) is equal to the new frequency of allele a (q'II) minus the original frequency of the allele (qII):

$$DqII = q'II - qII$$

Expanding the term qII(1 – m), we get:

$$Dq = qIm + qII - qIIm - qII$$

A simplified equation would be:

$$Dq = qIm - qIIm$$

$$= m(qI - qII)$$

The frequency of q is directly proportional to the migration rate (m); if the rate of migration increases, the allelic frequency of the migrant population also increases. The allelic frequencies of the two populations will be (qI – qII). With each generation of migration, the frequencies of the two populations become more and more similar until, eventually, the allelic frequency of population II equals that of

Table 17.2: Autosomal Recessive Disorders that occur with Increased Frequencies in the Charlevoix-Saguenay-Lac Saint-Jean Region of Quebec, Canada

Disorder	*Clinical Features*	*Frequency of Heterozygotes*
Cystinosis	Lysosomal storage disorder; excess excretion of amino acids and phosphate; softening of bones (osteomalacia); kidney damage	0.0256
Cytochrome C oxidase deficiency	Decreased blood pH and bicarbonate; muscle weakness (hypotonia); abnormal facial shape (dysmorphism); excess lactate in blood and cerebrospinal fluid (metabolic acidosis); liver malfunction	0.0323
Histidinemia	Histidine ammonia-lyase (histidinase) deficiency; usually benign; occasionally mental retardation, kidney damage, and/or neurological defects	0.0313
Lipoprotein lipase deficiency	Increased triglyceride levels in the blood; abdominal pain; damage to the pancreas; enlarged liver and spleen; sudden occurrences of yellow lesions on the body.	0.0233
Mucolipidosis	Lysosomal storage disorder; damage to bone, cartilage, and connective tissue	0.0256
Pseudo-vitamin D-deficient rickets	Calcium deficiency (hypocalcemia); increased secretion of the parathyroids; bone deformities and abnormal bone development (rickets)	0.0385
Pyruvate kinase deficiency	Destruction of red blood cells (hemolytic anemia); lethargy; chills; fever; back and abdominal pains	0.0156
Sarcosinemia	Sarcosine dehydrogenase deficiency; mental retardation	0.0345
Sensorimotor polyneuropathy	Sensory loss; muscle weakness and wasting numbness; decreased tendon reflexes	0.0435
Spastic ataxia	Abnormal gait; involuntary muscle movements; lack of muscle coordination; wasting of distal muscles; loss of control of eye muscles	0.0476
Tyrosinemia type I	ρ-hydroxyphenylpyruvate oxidase deficiency; liver and kidney damage; severe pain in arms and legs	0.0455

population I. When qI – qII = 0, there will be no further change in the allelic frequency of population II, even though migration continues. If migration between two populations takes place for a large number of generations and no other evolutionary forces act, equilibrium is reached, at which the allelic frequency of the recipient population is equivalent to the source population (refer to **Figure 17.4**).

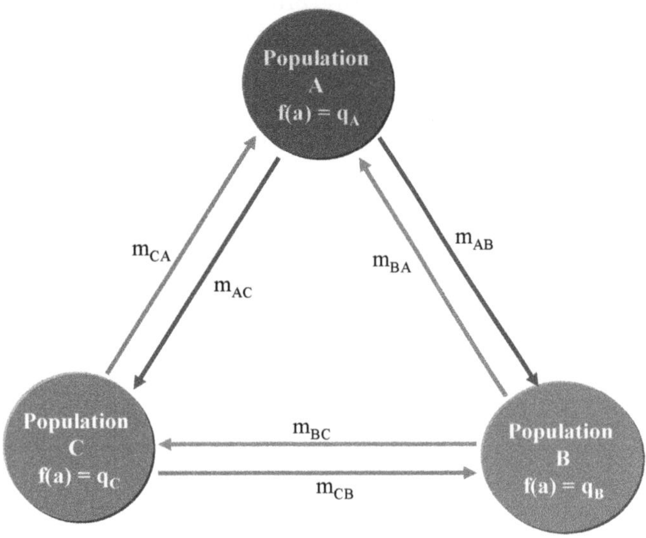

Frequency of a after migration

$$q'_A = q_B m_{BA} + q_C m_{CA} + q_A(1 - m_{BA} - m_{CA})$$

$$q'_B = q_A m_{AB} + q_C m_{CB} + q_B(1 - m_{AB} - m_{CB})$$

$$q'_C = q_A m_{AC} + q_B m_{BC} + q_C(1 - m_{AC} - m_{BC})$$

Figure 17.4: Multidirectional Migrations.

Migration affects the populations in two ways. First, it causes the gene pools of populations to become more similar. Later, we will see how genetic drift and natural selection lead to genetic differences between populations; migration counteracts this tendency and tends to keep populations homogeneous in their allelic frequencies. Migration adds genetic variation in the populations. Different alleles may arise in different populations due to mutational events. Further, these alleles spread to new populations by migration hence increasing the genetic variation within the recipient population.

17.12 Genetic Drift

Genetic drift is the random fluctuation in the allele frequencies caused by a random sampling of gametes. If there are no new mutations, genetic drift can cause all genetic diversity to be lost. This effect is more significant in smaller populations, but it can still affect loci not under-selection in larger populations. Genetic diversity within and among populations is affected by genetic diversity within populations and increases genetic diversity among populations.

A bottleneck is an evolutionary event where a significant percentage of a population or species is killed or not reproducing, hence it is reduced by 50 per cent or more. The event is named after the graph of this change that resembles the neck of a bottle, from comprehensive to narrow. In a population bottleneck, where the population suddenly contracts to a small size, genetic drift causes sudden and dramatic changes in allele frequency that occur independently of selection. In such instances, many beneficial adaptations may also get removed even if the population grows larger later. This is responsible for the founder effect, where only a few individuals are left after drift; they reproduce, and expansion takes place, resulting in larger populations; the allele frequencies are like the original population. The founder effect may be responsible for high frequencies of some genetic diseases generated due to inbreeding. Several genetic diseases are prevalent in Ashkenazi Jewish populations, such as breast cancer caused by mutations at BRCA1 and BRCA2, Tay-Sachs, and Gaucher disease.

17.13 The Magnitude of Genetic Drift

The effect of random genetic drift is inversely proportional to population size. Allele frequencies change because the genes appearing in offspring are not the perfect representative of the sampling of the parental genes. Genetic drift follows the law of probability. It removes genetic variation from the population at a rate inversely proportional to population size. The smaller the population size, the greater the effect of genetic drift. Drift also affects the probability of survival of new mutations. The probability that an allele will reach fixation is equal to its frequency in the population, and an allele with a frequency of 0.2 (20 per cent) has a 20 per cent chance of fixation. New alleles introduced by mutation almost inevitably begin at low frequencies and have a low fixation probability. Drift can lead to the loss of rare alleles and the fixation of only common alleles. Drift has little effect if populations are large.

To calculate the error of the genetic drift, we can calculate the variance in the allele frequencies. Suppose we observe many separate populations, each with N number of individuals with allelic frequencies p and q after one generation of random mating. In that case, genetic drift is expressed in terms of the variance in allelic frequency among the populations (Sp2). The amount of change resulting from genetic drift (the variance in allelic frequency) is determined by two parameters: the allelic frequencies (p and q) and the population size (N). Genetic drift is maximum when p and q are equal (each 0.5) and the population size is small.

For ecological and demographic studies, population size is usually defined as the number of individuals in a group. The evolution of a particular gene pool depends upon the genes that contribute to a particular population, and these genes are transmitted to the next generation. The effective population size is the sum of reproducing males and females, and the male-female ratio affects the other factors in reproductive success.

17.14 Causes of Genetic Drift

Genetic drift occurs due to sampling error, which may happen for various reasons. The sample size may decrease as the number of generations is limited by factors such as space, food, or other critical resources. Genetic drift can significantly impact the composition of the population's gene pool. A third way that genetic drift emerges is through a genetic bottleneck, which occurs when a population experiences a sudden and drastic reduction in size.

17.15 Natural Selection

Natural selection maintains favorable genetic mutations while eliminating harmful ones. When genetic drift and natural selection act together, they may result in the differentiation of populations. Take the example of the FOXP2 gene, where two amino acids differentiate humans and chimpanzees, indicating the role of this gene in the evolution of speech and language.

Other significant examples of genes that show the role of selection are Duffy antigen resistance to HIV infection progression - CCR5, lactase persistence-LCT, drug metabolism - CYP1A2, color vision - OPN1LW, and alcohol metabolism - ADH1B and ALDH2. Most of the functional variants of these genes are geographically or ethnically restricted, like the CCR5 haplotype in Afro-Americans and European-Americans, conferring differential levels of progression of HIV infection.

17.16 Fitness and Selection Coefficient

Any novel mutation in an organism's genome has three different effects on fitness (w= 1+s). A mutation may be lethal (s<0), reducing fitness, fertility, and survival rate. It may be neutral (s » 0), having no effect on fitness, or associated with an advantage (s>0), increasing fitness, fertility, and survival. The fate of mutation depends upon selection, random drift, and population size.

The range of fitness is between 0 to 1, and fitness can be denoted as 'w.' For example, three genotypes can be shown as follows:

Genotypes:	A1A1	A1A2	A2A2
Mean Number of Offspring Produced:	10	5	2

Fitness (W) for each genotype is as follows:

A1A1: W11 = 10/10 = 1.0

A1A2: W12 = 5/10 = 0.5

A2A2: W22 = 2/10 = 0.2

The fitness of each genotype is denoted as W11, W12, and W22, respectively. Another related variable is the selection coefficient (s), which is the relative value of selection against a genotype. The selection coefficient is equal to 1 – W; thus, the selection coefficients for the three genotypes mentioned above are:

A1A1: S11 = 0

A1A2: S12 = 0.5

A2A2: S22 = 0.8

17.17 Selection Model

Different selection models are shown in **Table 17.3.**

Table 17.3: Method for Determining Changes in Allelic Frequency Due to Selection

	A^1A^1	A^1A^2	A^2A^2
Initial genotypic frequencies	p^2	$2pq$	q^2
Fitness	W_{11}	W_{12}	W_{22}
Proportionate contribution of genotypes of the population	p^2W_{11}	$2pqW_{12}$	q^2W_{22}
Relative genotypic frequency after selection	p^2W_{11}/'W	$2pqW_{12}$/'W	q^2W_{22}/'W

Let us take an example of birds that survive during winter and have three genotypes.

$A_1 A_1, A_1 A_2, A_2 A_2$

$p = A_1$

$q = A_2$

Under Hardy Weinberg's equilibrium

$P^2 + 2pq + q^2 = 1$

However, some birds die in winter hence it is suggested that some fitness is applicable. Let us take mean fitness which is v

$\varpi = P^2 w_{11} + 2pqw_{12} + q^2{}_{22}$

The mean fitness 'W is the average fitness of all individuals in the population and allows the frequencies of the genotypes after selection. The frequency of a genotype after selection will be equal to its proportionate contribution divided by the mean fitness of the population (p^2W_{11}/'W for genotype A^1A^1, $2pqW_{12}$/'W for genotype A^1A^2, and q^2W_{22}/'W for genotype A^2A^2), as shown in **Table 17.3**. When the new genotypic frequencies are calculated, the new allelic frequency of A^1 (p') is

$$p' = f(A^1) = f(A^1A^1) + ½ f(A^1A^2)$$

and that of q' can be obtained by subtraction:

$$q' = 1 - p' \text{ (next-generation genotypic frequencies)}$$

If the selection is for recessive, dominant, and co-dominant traits, as well as traits in which the heterozygote has the highest fitness (**Table 17.4**).

17.18 The Results of Selection

Natural selection is a slow process in which biological traits become either more or less common in a population as a function of the effect of inherited traits on the differential reproductive success of organisms interacting with their environment. It is a crucial mechanism of evolution. However, the relative fitness of the genotypes may affect selection. In case there are three genotypes (A1A1, A1A2, and A2A2) with fitness W11, W12, and W22, six different types of natural selection can be noticed (**Figure 17.5**). The dominant allele A1 confers a fitness benefit; in this case, genotypes A1A1 and A1A2 are equal and higher than the fitness of A2A2 (W11 = W12 > W22). Because the heterozygote and the A1A1 homozygote both have copies of the A1 allele and produce more offspring than the A2A2, the frequency of the A1 allele will rise over time, and A2 will reduce. When one is favored over the other, this is the directional selection.

Type 2 selection (**Figure 17.5**) is a directional selection against a dominant allele A1 (W11 = W12 < W22). In this case, the A2 allele increases, and the A1 allele decreases. Type 3 and type 4 selection are also directional selection. Still, in these cases, there is incomplete dominance, and the heterozygote has a fitness that is intermediate between the two homozygotes (W11 >W12 >W22 for type 3; W11 <W12 <W22 for type 4). When A1A1 has the highest fitness (type 3), the A1 alleles are more over time, and the A2 alleles are less. When A2A2 has the highest fitness (type 4), the A2 allele increases over time, and the A1 allele decreases. Finally, directional selection leads to

Table 17.4: Formulas for Calculating Change in Allelic Frequencies with different Types of Selection

Type of Selection	*Fitness Values*			*Change in q*
	A^1A^1	*A^1A^2*	*A^2A^2*	
Selection against a recessive	1	1	1 – s	- spq2/1 – sq2
Selection against a dominant trait	1	1 – s	1 – s	- spq2/1 – s + sq2
Selection against a trait with no dominance	1	1 – ½ s	1 – s	- ½ spq/1 - sq
Selection against both homozygotes (over dominance)	1 – s11	1	1-s22	pq (s11p – s22q)/1 – s11p2 – s22q2

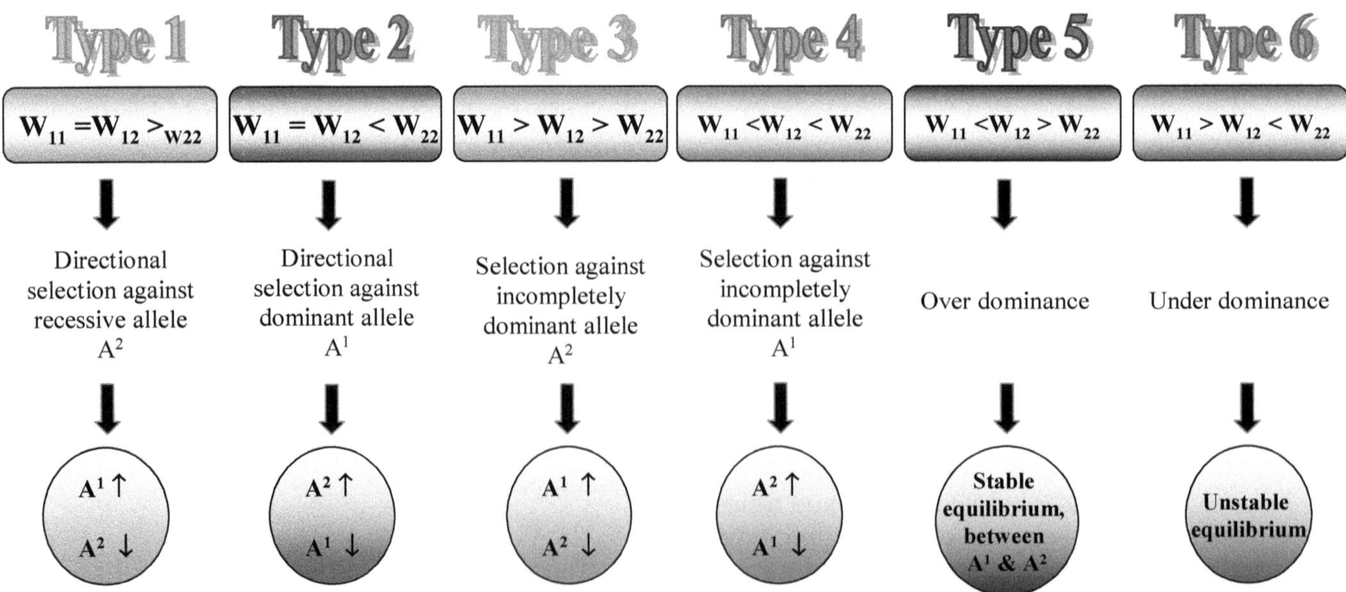

Figure 17.5: Different Types of Selections.

the fixation of the favored allele and elimination of the other allele; so far, no other evolutionary forces have acted on the population.

It has been shown that selection types 5 and 6 lead to the state of equilibrium, which results in no change in allelic frequency. Type 5 selections are referred to as overdominance or heterozygote advantage. Here, the heterozygote has privileged fitness over the fitness of the two homozygotes (W11 <W12 >W22). Due to overdominance, both alleles are favored in the heterozygote state, and no allele is out of the population. The allelic frequencies may initially change because one homozygote has superior fitness to the other; the change direction will depend on the relative fitness values of the two homozygotes. The allelic frequency at equilibrium (^q) depends on the relative fitness (usually expressed as selection coefficients) of the two homozygotes:

q^ = f(A2) = s11/s11 + s22

where s11 represents the selection coefficient of the A1A1 homozygote and s22 represents the selection coefficient of the A2A2 homozygote.

Under dominance, *i.e.*, type 6, the heterozygote has lower fitness than both homozygotes (W11> W12 < W22), leading to an unstable equilibrium.

17.19 The Rate of Change in Allelic Frequency Due to Natural Selection

The allele frequency change is caused due to selection (**Figure 17.6**). In case dominant alleles increase at a greater rate than the recessive alleles, directional selection will take place. This is due to the increase in homozygote and heterozygote selection. When we find incomplete dominance, the heterozygote has a selective advantage, however, at a smaller amount than the homozygote, so incomplete dominant alleles augment the frequency at a lower rate than dominant alleles. Recessive alleles increase at a much lower rate because only the homozygote is preferred due to the procedure of selection.

The selection rate depends upon the allelic frequency in the population. If an allele (A2) is lethal and recessive, W11 = W12 = 1, whereas W22 = 0. The frequency of the A2 allele will decrease due to the nonproduction of offspring as A2. A2 is lethal; hence the decrease will be relative to the frequency of the recessive allele. When allele frequency is high, the change takes a comparatively longer time. Still, as the frequency of the allele decreases, a higher proportion of the alleles are in the heterozygous genotypes, where the heterozygotes have the same phenotype as the fortunate homozygote. The recessive allele gets removed at a prolonged rate.

17.20 Mutation and Natural Selection

If there are harmful mutations, natural selection will eliminate harmful mutations. The population remains in a state of equilibrium.

Table 17.4 recessive allele is - spq2/(1 - sq2). Where q is very low, q2 is almost zero, so 1 - sq2

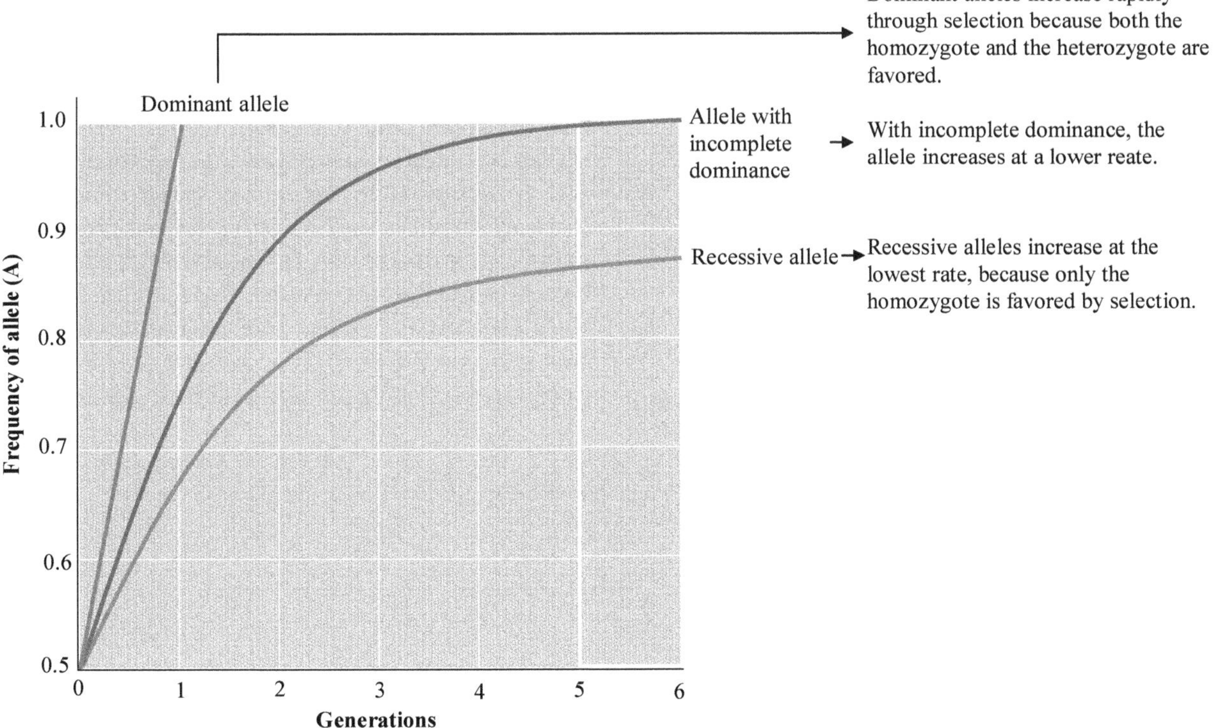

Figure 17.6: Change in Allelic Frequency Depends upon Selection.

will be approximately 1 - 0 = 1. Thus, when q is very low, the decrease in frequency due to selection is around - spq2. The rise in frequency of an allele due to forward mutations is μp. At equilibrium, the effects of mutation and selection are balanced; so

spq2 = μP

or

q2 = μp/sp = μ/s

Taking the square root of each side, we get

q^ = √ μ/s = frequency of the dominant allele at equilibrium

q^ = μ/s

Achondroplasia is due to a defect in bone growth that results in disproportionate dwarfism. At the time of fetal development, most of the skeleton comprises cartilage. Most of the time, the cartilage is converted into bone. When a person is suffering from dwarfism, no conversion of cartilage takes place. This is due to the FGFR3 gene mutation. Dwarfs are fertile and produce approximately 74 per cent of children, like those without achondroplasia. The fitness of people with achondroplasia is.74. The selection coefficient (s) is 1 - W, or.26. If we assume that the mutation rate for achondroplasia is about 3 X 10-5 (a typical mutation rate in humans), then we can predict that the equilibrium frequency for the achondroplasia allele will be 4 = (.00003/.26) =.0001153. This frequency is like the frequency of the disease.

17.21 Effects of Evolutionary Forces

Population variation occurs due to the action of different evolutionary forces on populations. These forces include mutation, selection, genetic drift, and more. As a result, gene frequency changes when populations are under the influence of these forces. Mutations, for instance, can bring about a change in allelic frequency. While lethal mutations are typically removed and have little effect, beneficial ones may be selected and become more prevalent. Consequently, populations remain in a state of equilibrium due to both lethal and beneficial mutations.

Migrations also alter the gene frequency, and equilibrium is achieved only when the allelic frequencies of the founder and recipient populations are the same. Genetic drift, conversely, can cause one allele to become fixed in the population, but this only happens when genetic drift affects a small fraction of the population.

Natural selection can be unidirectional, causing one allele to become fixed or over-dominantly, resulting in an equilibrium state where both alleles are favored. While natural selection can increase deviation in populations due to the survival of identical alleles in different populations, it can also reduce divergence between populations by selecting the same allele in different populations. On the other hand, mutation always causes divergence, which is at a higher rate between populations because different mutations arise in different populations. Additionally, genetic drift increases divergence between populations because changes in allelic frequencies due to drift are random and are likely to change in different directions in separate populations.

Migration, however, decreases divergence between populations because it mixes the genetic composition of the two populations, making them more alike. Migration and genetic drift can sometimes act in opposite directions.

17.22 Measures of Genetic Variation

The amount of polymorphism in the population should be evaluated to determine the genetic variation. Let us take 30 different loci and observe that two or more alleles are present in 15 of these loci, *i.e.*, 15/30 = 0.5 percent. The expected heterozygosity is the proportion of persons expected to be heterozygous at a locus under the Hardy-Weinberg conditions; (*2pq*) p and q are the two alleles in the population. The expected heterozygosity is often preferred over observed heterozygosity as expected heterozygosity is self-governing in the breeding system of an organism. If the species is self-fertilizing, heterozygosity is less expected, and this can be measured by looking into the expected heterozygosity. Expected heterozygosity is calculated for many loci and averaged over all the loci considered. The genetic variation in the population can be seen by considering many markers.

18

Cancer Genetics

It is now widely understood that cancer is an essential disease, although its complexity means that environmental and non-genetic factors also contribute significantly to its development. Nevertheless, substantial progress has been made in comprehending the tumorigenesis process, mainly due to discovering genes that become cancerous when mutated. Essentially, cancer begins when cells start dividing uncontrollably. Typically, cell division is genetically regulated, but the proliferation of cells becomes rapid once a somatic mutation occurs in particular cells that undergo carcinogenesis. This rapid process invades normal cells, requiring a series of somatic mutations for genetic predisposition in some cancers.

The origin of different types of tissue is crucial for cancer classification. For example, carcinomas stem from epithelial cells, sarcomas from bone or connective tissues, and leukemia and lymphomas from blood cells. Solid tumors are considered organs, consisting of a range of cell types and maintained by a small population of cancer stem cells. Pathologists classify tumors based on histology as observed under the microscope, while genetic tests for specific chromosomal rearrangements or genome-wide expression profiling allow for further refinements of the classification. Although they do not explain how cancer develops, classification is essential for prognosis and management.

Cancer genetics aims to understand the multi-step mutational and selective pathway that transforms normal somatic cells into proliferating and invasive cancer cells. Different molecular events contribute to cancer prognosis, with genetic modifications being a multi-step process that results in cell division and growth, leading to malignancy. Both oncogenes and tumor suppressor genes are accountable for cancer, with these genes responsible for DNA repair, the cell cycle, cell-cell contact, and programmed cell death.

18.1 The Evolution of Cancer

18.1.1 How Normal Cell is Developed into a Tumor Cell?

The journey of a normal cell to a tumor cell is a fascinating one, taking place within an individual's lifetime and involving microevolution of a somatic cell to a malignant cell. Certain mechanisms protect us from developing cancer, but approximately 6-7 mutations are required to alter a normal cell into a malignant one. Cancer can be considered a disease of the post-reproductive period (**Figure 18.1**), with tumor precursor cells possessing stem cell-like properties that can be innate or acquired due to mutations. Oncogenes are the genes that undergo mutation, with the non-mutant versions being called proto-oncogenes.

18.2 Tumor Suppressor Genes

Tumor suppressor genes regulate cell growth, programmed cell death, and DNA repair. When these genes malfunction, they can't control cell growth, which can lead to cancer. The Rb gene, for instance, is responsible for childhood retinoblastoma

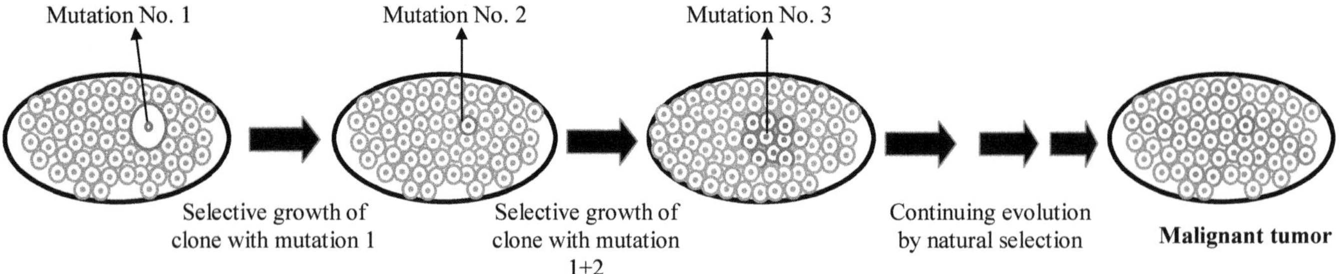

Figure 18.1: Mutation Results in the Malignant Tumor.

and was the first tumor suppressor gene discovered. In contrast to oncogenes promoting cancer growth, tumor suppressor genes are inactivated. Some of these genes may run in families and cause familial cancer, but cancer caused by acquired tumor suppressor gene mutations is not inherited. The TP53 gene is commonly found in over 50 per cent of human cancers and is involved in apoptosis and cell arrest. Other genes like CDKN2A, APC, MLH1, MSH2, MHS6, WT1, WT2, NF1, NF2, and VH2 also play a key role in cell division, transcriptional regulation, and signal transduction.

One such suppressor gene is the BRCA gene, which helps regulate the cell cycle during cell

Table 18.1: Some Proto-oncogenes and Tumor Suppressor Genes

Tumor Suppressor	*Normal Function*	*Abnormal Function Cancers*	*Alteration in Cancer*
p53	Cell cycle checkpoints and Apoptosis,	Brain, lung, colorectal, breast, and many other cancers	Mutation, inactivation by viral oncogene products
RB1	Cell cycle checkpoints, bind to E2F	Retinoblastoma, osteosarcoma, and other cancers	Mutation, deletion, and inactivation by viral oncogene products
APC	Cell-cell interaction	Colorectal cancers, brain, thyroid cancers	Mutation
Bcl2	Apoptoses régulations	Lymphomas and leukemias	Over expression blocks apoptoses
XPA-XPG	Nucleotide excision repaire	Xeroderma pigmentosum, skin cancers	Mutation
BRCA2	DNA repaire	Breast, overian, prostate cancers	Point mutations
Proto-oncogene	Normal function	Associated cancers	Alteration in cancer
Hr-ras	Signal transduction molecules bind to GTP/GDP	Colorectal, bladder, many types of other cancers	Point mutations
c-erbB	Transmembrane growth factor receptors	Glioblastomas, breast cancer, cervix cancer	Gene amplification, point mutations
c-myc	Transcription factor, regulates cell cycle, differentiation, apoptosis	Lymphomas leukemias, lung cancer, many types of other cancers	Translocation, amplification point mutations
c-fos	Transcription factor, responds to growth factors	Osteosarcomas, many types of other cancers	Over expression
c-kit	Tyrosine kinase, signal transduction	Sarcomas	Mutations
c-raf	Cytoplasmic serine-threonine kinase, signal transduction	Stomach cancer	Gene rearrangements
RARα	Hormone-dependent transcription factor, differentiation	Acute promyelocytic leukemia	Chromosomal translocations with PML gene, fusion product
E6	Human papillomavirus encoded oncogene, inactivates p53	Cervical cancer	HPV infection
MDM2	Binds and inactivates p53, abrogates cell cycle checkpoints	Osteosarcomas, liposarcomas	Gene amplification, leading to over-expression
Cyclins	Bind to CDKs, regulate cell cycle	Lung, esophagus, many types of other cancers	Gene amplification, causing over-expression
CDK2, 4	Cyclin-dependent kinases regulate cell cycle phases	Bladder, breast, and many types of other cancers	Overexpression, mutation

division. It prevents cells from growing and dividing too rapidly, thus inhibiting the growth of cells in the breast's milk ducts. The protein produced by the BRCA gene is also responsible for DNA repair and interacts with the RAS51 gene to protect against DNA damage.

A normal epithelial cell turns into an invasive cancer cell requiring specific mutations in one cell. All mutations do not occur in one individual. If a typical mutation rate is 10–7 per gene per cell generation, the chances of this are one of the 1013 cells in a person is 1013 x 10–42, or 1 in 1029. Sometimes two mutations may occur. Some mutations enhance cell proliferation, creating an expanded target population of cells for the next mutation. The combination of two or more mutations may be essential in some cancers. Some mutations influence the whole genome, at the DNA or the chromosomal level, affecting the mutation rate. Malignant tumor cells usually reveal genomic instability, as shown by karyotype abnormalities. Cancer begins with tissue hyperplasia or benign growths followed by successive random mutations, which finally lead to cancer.

18.3 Genomic Instability and Cancer

Genomic instability is a common phenomenon in all types of cancer. There are various types of genomic instabilities, such as chromosomal instability (CIN), which can result in aneuploidy, chromosome loss, DNA amplification, and chromosome deletions. Chromosomal abnormalities can be a diagnostic marker for identifying the cancer stage. For instance, in chronic myelogenous leukemia (CML), a C-ABL gene situated on chromosome 9 is translocated into the BCR gene on chromosome 22, resulting in the Philadelphia chromosome (**Figure 18.2**) and the formation of the BCR-ABL fusion. This fusion codes for a chimeric BCR-ABL protein, stimulating the cells to divide.

Xeroderma pigmentosa is a form of skin cancer caused by ultraviolet exposure. Subjects lacking the nucleotide excision repair gene with mutations in any of the seven genes essential for DNA repair result in XP. When XP cells are faulty, they cannot repair DNA lesions such as thymine dimmers, which can be induced by UV light.

The mutations in the DNA repair genes result in cancer like nonpolyposis (Hereditary) colorectal cancer. This autosomal dominant condition is found in 1/200 individuals (**Figure 18.3**).

Patients with HNPCC show greater genomic instability and also mutation rates. Approximately eight genes are predisposed to HNPCC; four of these genes are involved in mismatch repair. These genes are named MSH2, MSH6, MLH1, and MLH3- if these are inactivated, they result in the rapid growth of genome-wide mutations and develop into colorectal and other cancers. A family of HNPCC is shown in **Figure 18.3**.

The mutator hypothesis states that if mutations take place in the DNA repair genes, they may cause cancer. In sporadic non-hereditary cancers, the molecular basis of genomic instability is not well characterized. Still, high-throughput sequencing results show that mutation in DNA repair genes is

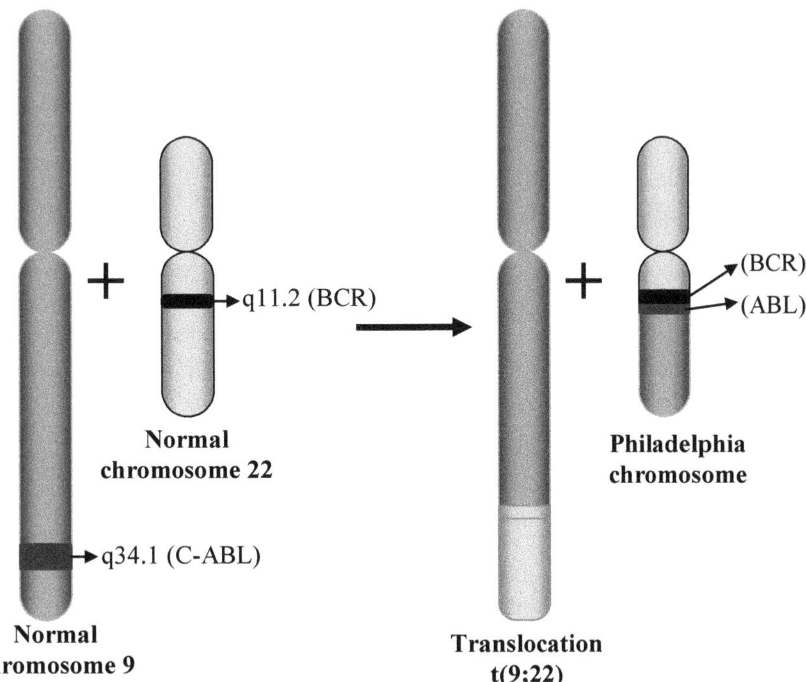

Figure 18.2: A Reciprocal Translocation Involving the Long Arms of Chromosomes 9 and 22.

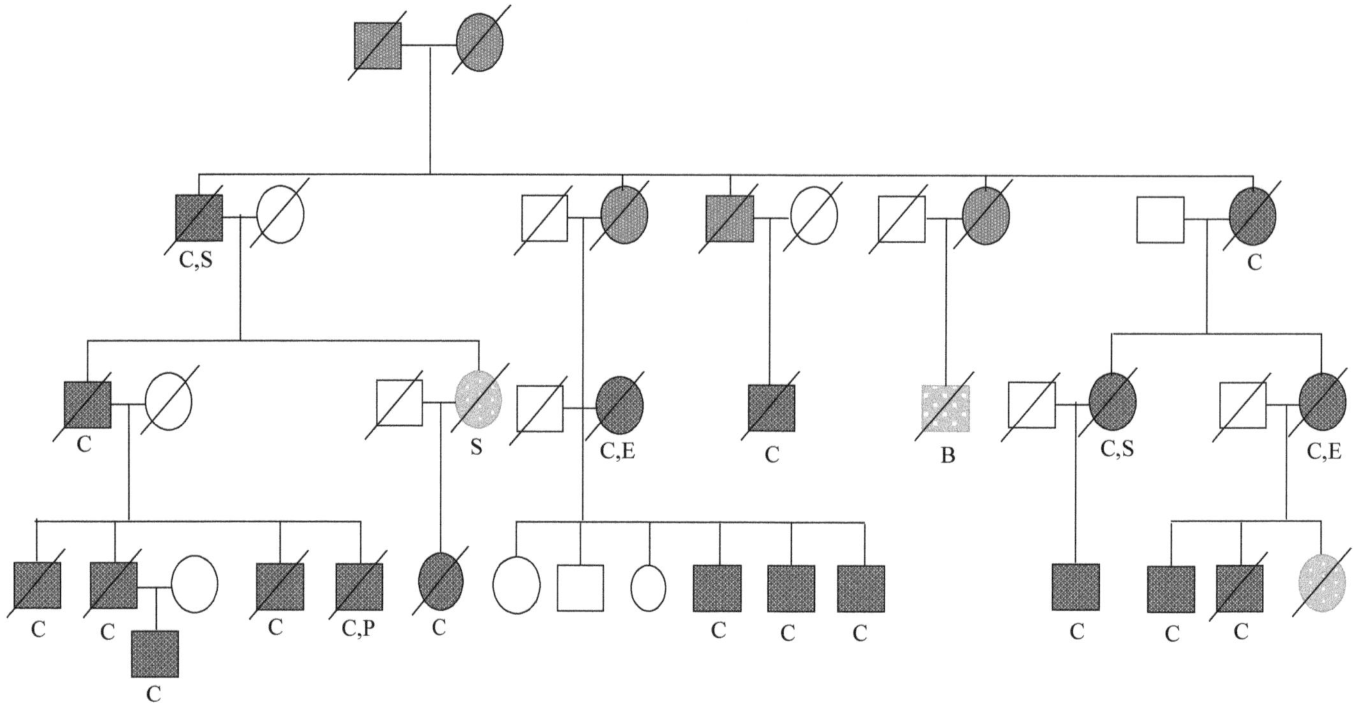

Figure 18.3: Pedigree of a Family with HNPCC.

not frequent before therapy, which does not support the mutator hypothesis. The mutation patterns of the tumor suppressor TP53 (which encodes p53), ataxia telangiectasia mutated (ATM), and cyclin-dependent kinase inhibitor 2A (CDKN2A; which encodes p16INK4A and p14ARF) maintain the oncogene-induced DNA replication pressure; this demonstrates that genomic instability and TP53 and ATM mutations cause DNA damage.

Genomic instability is typical in many cancers, especially in hereditary cancers. DNA repair genes undergo mutation, which may result in cancer and its progression. Genomic instability is unclear in non-hereditary cancers at a molecular level. The mutation in the tumor suppressor TP53 causes ataxia-telangiectasia (ATM), and cyclin-dependent kinase inhibitor 2A (CDKN2A, which encodes p16INK4A and p14ARF) support the oncogene-induced in the DNA replication stress model, causing genomic instability and TP53 and ATM mutations resulting in DNA damage.

18.4 Cell Cycle Regulation in Cancer

The growth and differentiation of cells are under tight genetic control. Cell proliferation at each step in the cell cycle requires apoptosis, and cells respond to external growth signals.

18.5 Signal Transduction Pathway for Cell Proliferation

When a cell that is not dividing receives a signal for cell division, it starts dividing. During cell division, the cell binds to a specific receptor, which triggers some catalytic processes. This signal is transduced, and the overall process is called the signal transduction pathway (**Figure 18.4**). Interestingly, cell proliferation signal transduction pathways exist in many cancers, where a growth factor binds to two receptor molecules. This leads to dimerization and changes in the pattern of each receptor that induces the intracellular kinase domain of one receptor to phosphorylate tyrosine residues in the domain of the other molecule and vice versa. This results in phosphorylation, and cytoplasmic proteins bind to the phosphorylated tyrosine residues. The protein binding of other proteins with the inactive protein (RAS-GDP) activates RAS-GTP. The RAS activation reaction starts when guanosine diphosphate (GDP) binds to the RAS protein, which converts into guanosine triphosphate (GTP). The inactive form of the RAF1 serine/threonine kinase binds to the active RAS–GTP protein and gets phosphorylated by a membrane-associated protein kinase. This phosphorylation activates the serine/threonine kinase activity of RAF1. One of the targets of active RAF1 is an inactive MAPKK (mitogen-activated protein kinase), which gets phosphorylated by RAF1. The phosphorylated MAPK (MAPK-P) enters the nucleus and phosphorylates various transcription factors, including ELK1. Phosphorylated ELK1 activates the transcription of the FOS gene. Then, the FOS protein combines with the JUN protein, which activates the expression of other genes that help in cell proliferation.

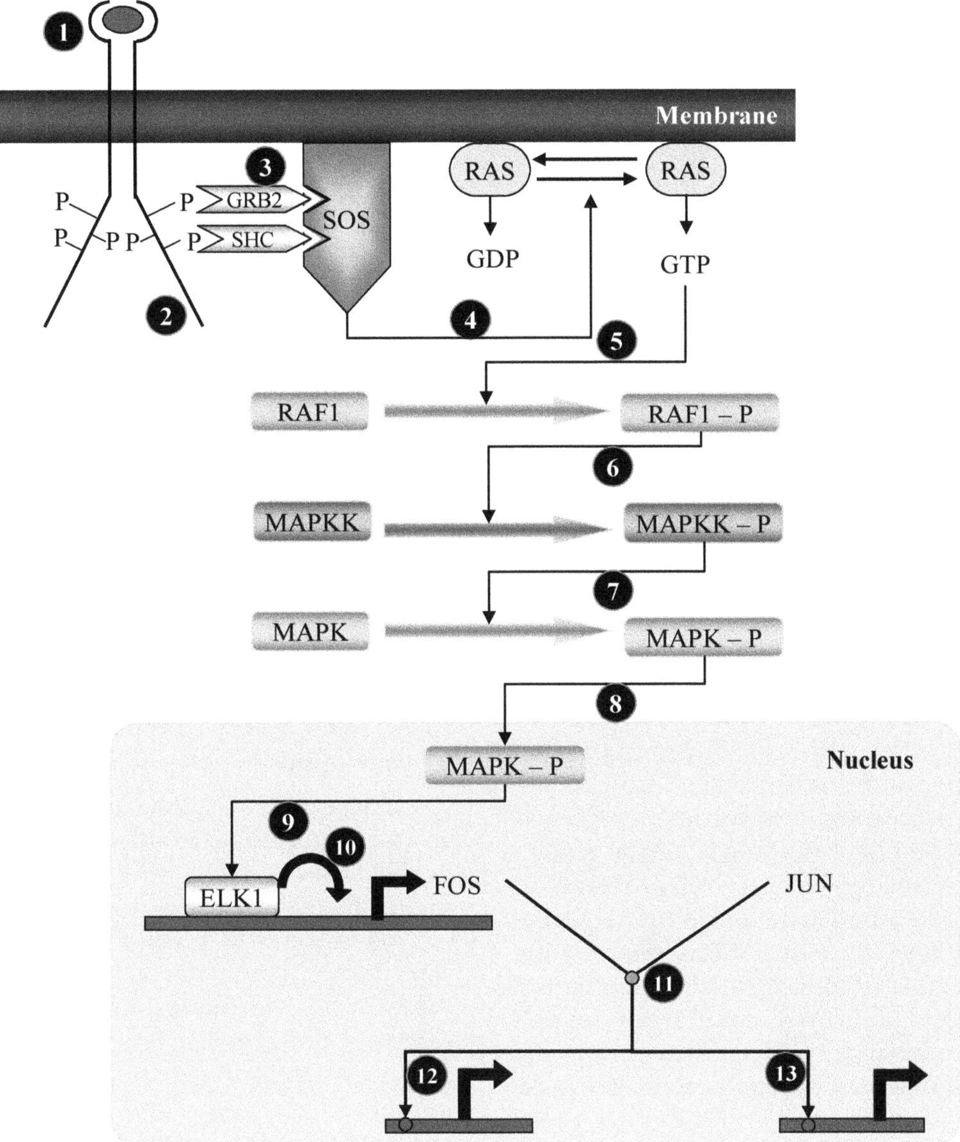

Figure 18.4: A Cell Proliferation Signal Transduction Pathway.

The cell may undergo a switch on and off mechanism. The RAS protein plays a vital role in it. The active RAS form (RAS-GTP) is transformed to the inactive form (RAS-GDP) by the process of hydrolysis of GTP mediated by GTPase activating proteins (GAPs).

18.6 The Ras Proto-oncogenes

The Ras gene family comprises signal transduction molecules that combine with the cell membrane and regulate cell growth and division. A Ras protein usually sends signals from the cell membrane to the nucleus, stimulating the cell to practice cell division in response to external growth factors. Ras proteins cycle between an inactive and an active state. This takes place by binding either to GBP or GTP. When a cell finds the growth factor (such as platelet-derived growth factor or epidermal growth factor), growth factor receptors on the cell membrane bind to the growth factor, resulting in autophosphorylation of the cytoplasmic portion of the growth factor receptor. These nucleotide exchange factors result in Ras releasing GDP binding GTP and activating Ras. The active, GTP-bound form of Ras transfers its signals through cascades of protein phosphorylation in the cytoplasm.

Activation of transcription factors triggers cascades that lead to the expression of genes responsible for driving the cell from a state of rest to a stage of cell division. These signals are transmitted to the nucleus, where they hydrolyze GTP to GDP, rendering them inactive. In the case of an oncogene, mutations in the proto-oncogene ras prevent the Ras protein from hydrolyzing GTP to GDP, thus freezing the protein in its "on" conformation, which constantly stimulates the cell to divide. A comparison of the amino acid sequences of Ras proteins in normal and cancer cells shows that the oncogenic mutation occurs in these proteins. The ras proto-oncogene encodes a protein of 189 amino acids, with glycine at position 12 and glutamine at position 61. Ras proteins in several tumors have a single amino acid substitution at one of these positions, converting the ras proto-oncogene into a tumor-promoting oncogene. K-ras and N-ras are mutant alleles of the ras proto-oncogene. Normal Ras proteins act as molecular switches controlling cell growth and differentiation, turning on when GTP binds to the protein and off when GTP is hydrolyzed to GDP. The conformation changes as the protein switches states. Oncogenic mutant Ras proteins provide a continuous signal for cell growth if they remain in an on state. Different gene families are involved in cell proliferation and signal transduction pathways. Interestingly even a single cell has many receptors responding to different growth factors. The signal transduction pathways interact with different cells; however, in other cells, there is no crosstalk between different pathways. Several genes are involved in these pathways, and mutation in these genes can occur anytime. Mutation in the tyrosine-protein kinase receptor results in repeated phosphorylation. This happens only when the growth factor is absent, because of which the cell proliferation signal transduction pathway permanently gets turned on. The RAS gene mutations prevent the conversion of RAS–GTP to RAS-GDP, which would stimulate nonstop cell growth components of cell proliferation signal transduction pathways in specific cell types. Genes that may probably result in cancer are known as proto-oncogenes. Proto-oncogenes may convert into oncogenes when point mutations or chromosome rearrangements turn off gene expression. This results in cancer; some of these genes have an EGFR function; if a point mutation occurs in this gene, it causes glioma. If the mutation occurs in tyrosine-protein kinase (KIT), it results in gastrointestinal stromal tumor AML, testicular germ cell, RET (tyrosine-protein kinase receptor) gene mutations cause papillary thyroid carcinomas, and multiple endocrine neoplasia types 2 JAK2 functions it proceeds as tyrosine-protein kinase receptor and if a mutation occurs it results into ALL. LCA is also a tyrosine-protein kinase receptor if the mutation result in T cell acute leukemia NRAs binds to GTP, and mutation in this gene causes AML and thyroid cancer. In **Table 18.2,** the detailed list is shown.

18.7 The Cell Cycle and Signal Transduction

At the time of cell division, the cell undergoes different phases (G1 (DNA synthesis) G1 followed by the S phase (cellular and chromosomal DNA is replicated), G2 (Cell continues to grow and prepare for cell division), M phase (duplicated chromosomes are condensed, sister chromosomes are separated and move to opposite poles, followed by cell division.

During mid-G 1 at an early stage, the cell makes an assessment either to enter into the next cell cycle or to leave the cell cycle into quiescence. Continuously dividing cells do not leave the cell cycle but go to G 1, S, G2, and M phases until they get signals to stop the growth. If the cell stops growing, it enters the G0 phase of the cell cycle. During G0, the cell remains metabolically vigorous; however, no growth or division occurs. Most differentiated cells in multicellular organisms remain in the G0 phase. Some cells, like neurons, never re-enter the cell cycle (**Figure 18.5**).

Table 18.2: Association of Proliferation Signal Transduction Pathway Oncogenes with Human Cancers

Oncogene	*Function*	*Cancer*
EGFR	Epidermal growth factor receptor	Glioma
KIT	Tyrosine-protein kinase receptor	Gastrointestinal stromal tumor, acute myelogenous leukemia, testicular germ cell tumor
RET	Tyrosine-protein kinase receptor	Papillary thyroid carcinomas, multiple endocrine neoplasia type 2
JAK2	Tyrosine-protein kinase receptor	Acute lymphocytic leukemia.
LCK	Tyrosine-protein kinase receptor	T cell acute lymphoblastic leukemia
NRAS	GTP-binding protein	Acute myelogenous leukemia, thyroid cancer, melanoma, multiple myeloma
HRAS	GTP-binding protein	Sarcoma
KRAS2	GTP-binding protein	Colon, lung, pancreatic and thyroid cancers, acute myelogenous leukemia
ABL1	non receptor tyrosine kinase	Chronic myeloid leukemia acute lymphocytic leukemia
MYC	Transcription factor	Small cell lung cancer, breast cancer, cervical carcinoma, Burkitt lymphoma
MYCN	Transcription factor	Neuroblastoma, lung cancer
MYCL1	Transcription factor	Small cell lung cancer

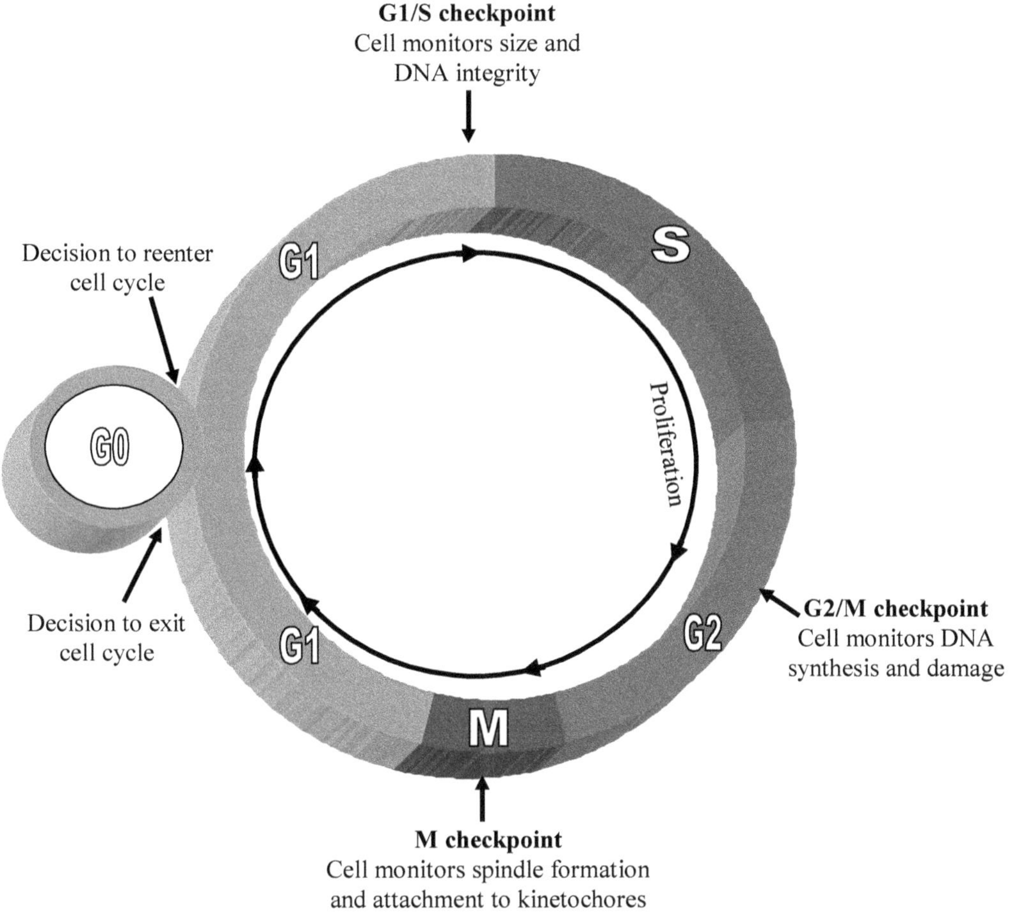

Figure 18.5: Checkpoints and Proliferation of the Cell through different Cell Cycles.

One must remember that in cancer, stop signals are absent. The malignant cells may not react to external signals from surrounding cell signals that normally inhibit cell proliferation within a mature tissue.

18.8 Cell Cycle Control and Checkpoints

The cells in eukaryotic organisms go through the G1 phase to the S phase. During this process, two cell cycle kinase complexes, CDK4/6-Cyclin D and CDK2-Cyclin E, work together to relieve the inhibition of a dynamic transcription complex that contains the retinoblastoma protein (Rb) and E2F. In G1-phase uncommitted cells, Rb binds to the E2F-DP1 transcription factors, forming an inhibitory complex with HDAC to repress key downstream transcription events. Upon entering the S-phase, sequential Rb phosphorylation events by Cyclin D-CDK4/6 and Cyclin E-CDK2 take place, dissociating the HDAC-repressor complex and permitting transcription of genes required for DNA replication.

Akt can phosphorylate FoxO1/3 in the presence of growth factors, which inhibits their function by nuclear export, leading to cell survival and proliferation. Various stimuli, such as TGF-β, DNA damage, replicative senescence, and growth factor withdrawal, cause checkpoint control. These stimuli act through transcription factors to induce specific members of the INK4 or Kip/Cip families of cyclin-dependent kinase inhibitors (CKIs). The oncogenic polycomb protein Bmi1 is a negative regulator of INK4A/B expression in stem cells and human cancer.

TGF-β also inhibits cdc25A transcription, a phosphatase directly required for CDK activation. At a critical convergence point with the DNA-damage checkpoint, cdc25A is ubiquitinated and targeted for degradation via the SCF ubiquitin ligase complex downstream of the ATM/ATR/Chk-pathway. However, timely degradation of cdc25A in mitosis (M-phase) via the APC ubiquitin ligase complex allows progression through mitosis. Growth factor withdrawal activates GSK-3β to phosphorylate Cyclin D, which leads to its rapid ubiquitination and proteasomal degradation.

Ubiquitin/proteasome-dependent degradation and nuclear export are common mechanisms used to reduce the concentration of cell cycle control proteins effectively.

18.9 The Tumor Suppressor Gene p53

The p53 gene is not involved in tumor growth; hence it may act as a tumor suppressor gene. p53 is a multifunctional transcription factor that controls cell cycle progression, DNA integrity, and cell survival. Suppose only one functional copy of the p53 gene is inherited by a person from his parents. In that case, he is liable to develop cancer as an independent tumor is formed in various tissues in early adulthood. This is a rare condition known as Li-Fraumeni syndrome. However, mutations in p53 are found in many other tumor types, leading to the complex network of molecular actions resulting in the development of tumors. p53 is located on chromosome 17. Its protein binds to DNA in a cell, which turns on another gene to make a protein called p21 which interacts with a cell division-stimulating protein (cdk2). When p21 forms a complex with cdk2, the cell cannot undergo cell division. Mutation in p53 can no longer efficiently bind with DNA, which results in the non-availability of the p21 protein; therefore, there is no 'stop signal' during cell division. Because of this, there is uncontrolled cell growth resulting in tumor formation.

18.10 Cell-cell Interaction

Cancer cells have exceptional features in that they leave the original tumor site and go into the lymphatic system; they attack the neighboring tissue and form the secondary tumors resulting in the obliteration of the extra-cellular matrix and basal lamina. The extracellular matrix and basal lamina are constituted of proteins and carbohydrates. They create scaffolding for tissue growth under the normal situation, which reduces the migration of cells.

The genes which are implicated in the process of metastasis are not very well defined. It is considered that genes encoding adhesion molecules and proteolytic enzymes may be occupied during metastasis. The evidence supporting this hypothesis is that epithelial tumors have a lower-than-normal level of the E-cadherin glycoprotein caused by Cell-cell adhesion in normal tissues. The metalloproteinases, which are proteolytic enzymes, survive at a significant level than normal levels in highly malignant tumors and are not susceptible to the normal control conferred by regulatory molecules such as tissue inhibitors of metalloproteinase (TIMPs).

18.11Cancer Predisposition may be Inherited

Many types of cancers have a hereditary or familial component (**Table 18.3**). Many genes are accountable for activating the expansion of cancer. However, for the causation of cancer, one other somatic mutation in the other copy of the gene must occur to cause cancer. The second somatic mutation may be a point mutation, deletion, or chromosomal aberration resulting in gene disturbance. In the case of hereditary cancers, the first gene is in the heterozygous state, and the second, wild type, allele, is mutated in a tumor as loss of heterozygosity.

Table 18.3: Cancers Showing Inherited Predisposition

Tumor Predisposition Syndromes	*Chromosome*
Li-Fraumeni syndrome	17p
Early-onset familial breast cancer	17q
Retinoblastoma	13q
Gorlin syndrome	9q
Neurofibromatosis, type 2	22q
Familial melanoma	9p
Multiple endocrine neoplasia, type 2	22q
Hereditary non-polyposiscolon cancer	2p
Von Hippel-Lindau syndrome	3p
Familial adenomatous polyposis	5q
Neurofibromatosis, type 1	17q
Wilms tumor	11p
Multiple endocrine neoplasia, type 1	11q

Adenomatous polyposis (FAP) shows that there are genetic factors involved in the risk of developing cancer. Individuals who develop colon cancer have one mutant copy of the adenomatous polyposis gene. This gene is situated on the long arm of chromosome 5. Mutations could be deletions, frameshifts, and point mutations. The normal function of this gene is to inhibit the growth of the cells by interacting with the f3 catenin protein; when the heterozygous mutation occurs, the cell cycle control is lost, due to which polyps are formed. Heterozygous individuals for this condition develop hundreds to thousands of polyps. These individuals inherit one mutant copy of the APC (adenomatous polyposis) gene.

The mutation takes place in polyp cells containing the APC gene. The ras proto-oncogene is responsible for adenomatous polyposis. This is the 2nd mutation. The mutation causes growth initiation resulting in the development of intermediate adenomas. In the third step, the loss of function of both alleles of the DCC (deleted in colon cancer) gene is recognized. The DCC gene product may be involved with cell adhesion and differentiation. Mutations in both DCC alleles form late-stage adenomas with several finger-like outgrowths (villi). In the late stage, the progress to cancerous adenomas takes place; it is associated with the loss of function of p53 genes. The function of P53 is to arrest the cell cycle in response to DNA damage. The loss of p53 causes high mutation rates throughout the genome. Many unknown numbers of genes are also associated with metastasis. This shows that cancer is a multistep process. In the first step, one of the alleles of the APC gene is inactivated on the chromosome in FAP cases; one mutant APC allele is inherited. Subsequent mutations involving genes on chromosomes 12, 17, and 18 in cells of benign adenomas can lead to a malignant transformation that causes colon cancer.

18.12 Viruses and Cancer

It has been estimated that 15 per cent of all human cancers worldwide are attributed to viruses; therefore, viruses are involved in the global cancer burden. Epstein-Barr virus, human papillomavirus, hepatitis B virus, and human herpes virus-8 are the four DNA viruses that can cause the development of human cancers. Human T lymphotropic virus type 1 and hepatitis C viruses are the two RNA viruses that contribute to human cancers (**Table 18.4**).

18.13 How the Virus Causes Cancer

The DNA copy enters the infected cell's nucleus, where it integrates randomly into the host cell's genome. The infected copy of DNA is called a provirus. This contains powerful enhancer and promoter elements in its U5 and U3 sequences at the ends of the provirus (**Figure 18.6**). The provirus is integrated into the host genome; the replication occurs along with the host's DNA during the cell's normal cell cycle. Retroviruses do not kill the cell but infect the surrounding cells (**Table 18.5**).

18.14 Environmental Factors Responsible for the Causation of Cancer

As mentioned earlier, cancer is caused by changes in certain genes that disrupt the normal functioning of our cells. Various food contaminants, such as herbicides, and other hazards have been associated with cancer, as well as certain cosmetics. Agents that cause cancer are known as carcinogens. However, it is essential to note that exposure to a carcinogen may not necessarily result in cancer, as other factors such as diet, lifestyle, health, age, and gender can also play a role.

To summarize, both genetic and environmental factors can contribute to the development of cancer. Mutations at the somatic or germline level can cause

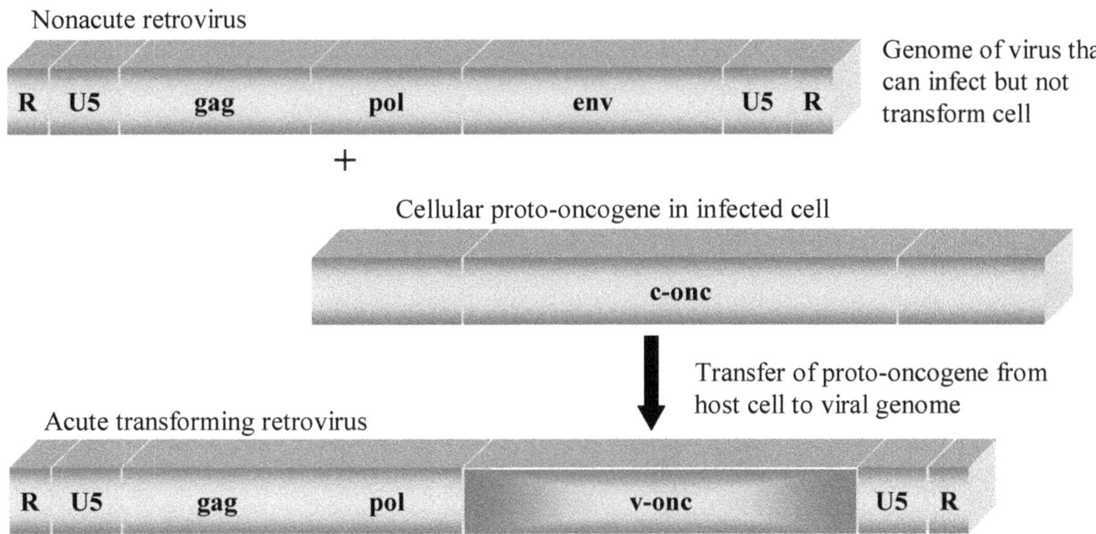

Figure 18.6: Genome of Typical Retrovirus.

Table 18.4: Viral Oncogenes

Oncogene	*Origin*	*Species*	*Cellular Function*
v-abl	Abelson murine leukemia virus	Mouse	Tyrosine kinase
v-erbB	Avian erythroblastosis virus	Chicken	EGF receptor
v-fos	FBJ osteosarcoma virus	Mouse	Transcription factor
v-myc	Avian myelocytomatosis virus	Chicken	Transcription factor
v-ha-ras	Harvey murine sarcoma virus	Mouse	GTP –binding protein
v-sis	Simian sarcoma virus	Monkey	Platelet-derived growth factor
v-src	Rous sarcoma virus	Chicken	Tyrosine kinase signal protein

Table 18.5: Human Viruses Associated with Cancer

Virus	Cancer	Oncogenes	Mechanism
Human T–cell leukemia virus	Adult T-cell leukemia	pX	Stimulates cell cycle
Hepatitis B virus	Liver cancer	HBx	Signal transduction stimulates cell cycle
Human herpesvirus 8	AIDS-related Kaposi's sarcoma	Several possible	Unknown
Epstein –Barr virus	Burkitt's lymphoma, naso-pharyngeal cancer	Unknown	Unknown
Human papillomavirus 16, 18	Cervical cancer	E6, E7	Inhibit p53 and pRB tumor suppressors

cancer, as can chromosomal mutations (karyotypic abnormalities), copy number variations, and loss of heterozygosity. In addition, global hypomethylation - a decrease in the methylation of DNA - is often observed in cancers, but its cause is unknown. Some promoter CpG islands become hypermethylated, leading to gene silencing and ultimately cancer.

19

Genetic Counseling

Genetic counseling includes diagnostic testing and managing genetic disorders. It is a developing branch in clinical genetics. The human genome project has revolutionized our ability to identify specific genes responsible for certain diseases. Genetic testing for these conditions is now widely available and has led to increased screening of newborns and greater access to genetic counseling. It is vital, however, that we tackle the ethical and legal implications of these groundbreaking advancements. When patients with developmental delays, congenital malformations, or progressive mental illnesses with advanced maternal age visit their primary physician, it's essential for the physician to consider genetic diseases. If other family members also have these conditions, the primary physician should refer the patient to a clinical geneticist for further evaluation. Genetic counseling can then be recommended to determine the risk of heritable diseases the patient may pass on to their child. Trained genetic counselors can investigate the family's medical history, and provide information on inheritance patterns, risk of recurrence, and treatment options, making it a valuable resource for families at risk of genetic diseases.

19.1 Steps Involved in Genetic Counseling

The initial stage of genetic counseling involves collecting the disease history of all available family members. The counselor's primary focus is on identifying the disorder, determining the risk of recurrence in future births and generations, and presenting alternatives to assist family members in making informed decisions. It is also recommended to seek support groups to help cope with the situation. Genetic counseling is suggested for those with a family history of known genetic disorders, infertility, miscarriage, stillbirths, or early infant mortality. People with a strong family history of heart disease at an early age should seek advice to change their lifestyle and receive counseling. Advanced maternal age is a risk factor for birth defects, and statistics show a 3 per cent chance of a baby having birth defects in all pregnancies. Additionally, older women over 30 have a 2 per cent chance of having a baby with Down syndrome.

Determining the actual risk for a disorder can be challenging since mutations in different genes can lead to the same symptoms. Retinal degeneration, for instance, has over 400 inherited syndromes with similar features, which makes it difficult to identify the specific gene responsible for the condition. The most critical step in genetic counseling is to create a complete family tree (pedigree) to determine the mode of inheritance of the disorder. After drawing the pedigree, the medical history of both the patient and other family members is taken. Once the exact cause is known, the patient can be referred for genetic testing to confirm the diagnosis.

19.2 Principles of Genetic Testing

Genetic conditions can be diagnosed by considering personal and family history, physical examination, and laboratory tests. Over 10,000 conditions have a recognized genetic basis, and the Online Mendelian Inheritance in Man (OMIM) database (www.ncbi.nlm.nih.gov/OMIM) can search for relevant keywords. The database is updated regularly, and new mutations for different diseases are added as they are discovered. Links to laboratories where tests are available can be found at (www.genetests.org). Gene Reviews is a valuable online resource for clinicians who may encounter situations requiring genetic tests. Mental retardation is among the many conditions that can be caused by genetic factors, such as Fragile X syndrome, which can be diagnosed using genetic tests. Patients with Fragile X syndrome have characteristic features, such as large ears and a protruding chin. Cystic Fibrosis is another condition that can be diagnosed through genetic testing, typically via a sweat chloride test. However, some patients with pulmonary disease who test negative for sweat chloride should be investigated for CFTR gene mutations.

As in CFTR, disease mutations have serious consequences; hence other family members are keen to know the status of the disease. If the family has a person affected by a genetic disease or belongs to some ethnic group where the incidence of a particular genetic disease is high, the carrier screening is used to know the reproductive risks by testing the parents. If a carrier couple is identified, they may like to get the prenatal diagnosis done to determine whether their fetus is affected by this condition. The prenatal testing is performed at 10-11 weeks taking the chorionic villus sample under ultrasound guidance. Amniocentesis is done at 15-18 weeks of pregnancy to obtain cells from the amniotic fluid. These couples might also choose to have a pre-implantation genetic diagnosis by selecting implantation of only those embryos that are not affected.

Genetic testing provides information about a diagnosis, treatment, and prevention; however, it has limitations. If an individual is tested for genetic testing and turns out to be positive but does develop the disease, he will suffer from unnecessary anxiety. On the contrary, an individual with a negative test may develop the disease; both situations are devastating. Genetic testing is used for chromosomal translocations; this test helps test some leukemia; for example, the translocation between chromosomes 1 and 19 in leukemic cells is a diagnostic test for the acute promyelocytic form of this disease, and the translocation between chromosomes 9 and 22 is a diagnostic test for the chronic myelogenous form.

Positive results show that the gene is disease-causing. On the contrary, negative results show no disease is there. Whenever the target gene is present, it means a true positive (TP) result, and every negative result indicates a true negative (TN). Sometimes interpretation is difficult.

False results can cause serious consequences. An average person will suffer from anxiety in case of false-positive results, undergoing unnecessary treatments, or depression. In practice, if a positive test is seen, the person should be reassessed either by repeating the test or finally confirming it by a more advanced test. A false-negative result can have grisly consequences as the treatable disease may sometimes be missed. To overcome these adverse effects, clinical tests should be adjusted to reduce the incidence of false negatives; however, this leads to an increase in the cost.

Various indices, like sensitivity and specificity, have been calculated to know the clinical relevance of diagnostic tests. Various parameters that should be used are shown below.

Parameters that should be calculated in a screening test are shown below.

	Fetus +ve for NTD	Fetus –ve for NTD
+ on test	a	C
-ve on test	b	D

Sensitivity = Proportion of affected picked up = a/a+ b

Specificity = True negative = d/c+d.

False positive = c/(a+b+c+d)

False negative = b/(a+b+c+d)

Positive predictive value = a/(a+c)

Odds ratio: It is defined as the odds of being a case after a positive test (a:c) compared to the odds of being a case after a negative result. b: d) = (a/c)/(b/d) = ad/bc

Relative risk: This is the risk after a positive test is compared with the general population risk or risk after a negative test.

= [(a/(a+c)]/[a+b+c+d]

Or = [a/(a+c)]/[b/(b+d)]

19.3 Screening Programs for Genetic Disorders

Every year, millions of children are tested for various genetic disorders. Many individuals are also screened for inherited disorders through a screening test performed on a large group of people who do not show any symptoms. This screening program can be divided into prenatal screening for newborns and screening for adults. These tests can be performed at different stages, as summarized in **Figure 19.1**.

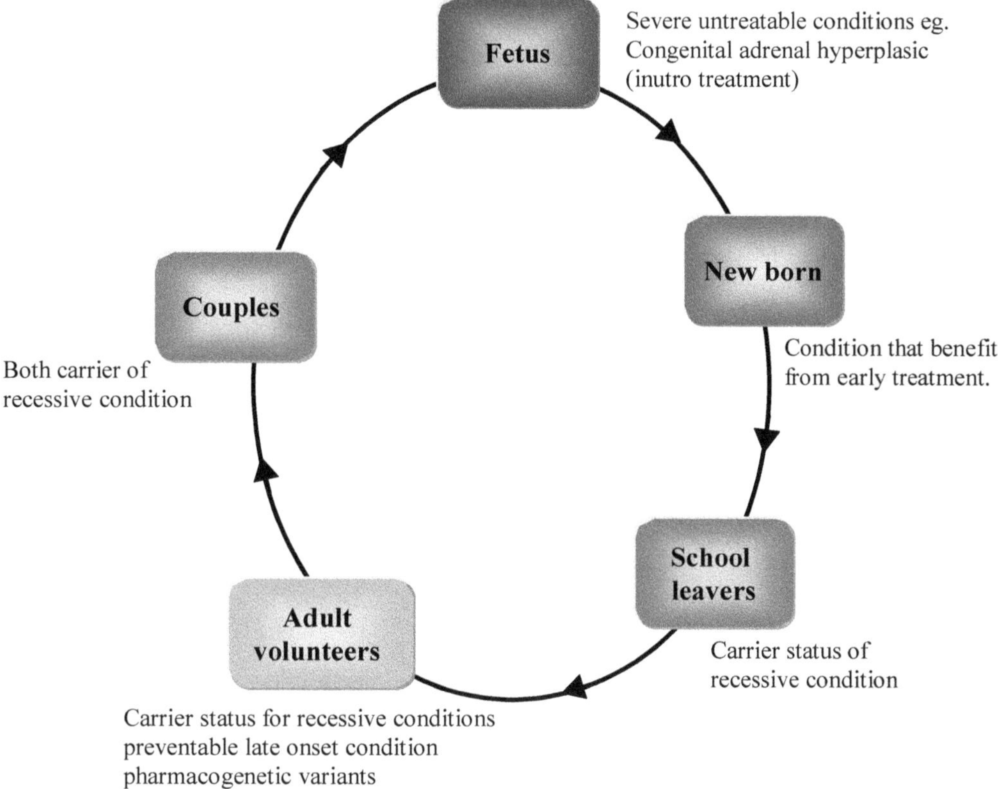

Figure 19.1: Genetic Screening Required at different Time Points.

19.3.1 Prenatal Screening

Prenatal screening is provided to pregnant women. Most common screening target diseases can be tested using ultrasound. Ultrasound detects structural anomalies. Ultrasound may detect NTDs and other abnormalities (**Figure 19.1**).

19.3.2 Newborn Screening

Prenatal screening is necessary for conditions like PKU, CF, galactosemia, sickle cell anemia (SCA), and thalassemia. These are important for the better maintenance of the life of newborns.

Conventionally, the essential requirements are to find out the genetic disorder that may be beneficial if proper treatment is given before the onset of the symptoms. In case the carrier frequency of the disease is high in the population, the individuals getting married should undergo a screening test.

Most genetic diseases cannot be treated through prenatal screening. However, proper counseling can reduce the risk by informing the couple how much would be the risk of occurrence of the disease in their children if two carriers individuals get married and opt for a child. Screening programs are now possible due to the development of newer technology like next-generation sequencing, exome sequencing, and tandem mass spectrometry. In the USA, newborn screening is done for phenylketonuria (PKU) and congenital hypothyroidism; galactosemia and sickle cell anemia; biotinidase deficiency (is performed only when required), maple syrup urine disease congenital adrenal hyperplasia, and homocystinuria; some newborns are for medium-chain acyl-CoA dehydrogenase deficiency, cystic fibrosis, and tyrosinemia type 1. However, the number of screening tests varies from state to state in the USA.

Newborn screening is cost-effective. The best example is PKU early treatment has prevented severe mental retardation in many cases. 29 conditions have been recommended by the American College of Medical Genetics for which the newborn screening should be done. These conditions are shown in **Table 19.1**.

Screening programs primarily target monogenic disorders. With the advancements in advanced techniques such as micro-array whole-exome sequencing, next-generation sequencing, and others, more candidate genes for genetic diseases will be discovered. For example, deafness/hearing loss is quite common, with 1 in 500 births in the US affected by it. There are over 60 genes responsible for deafness, which makes it difficult to differentiate between genetic and non-genetic causes. However, detecting deafness before six months of age and providing

Table 19.1: Newborn Screening for 29 Conditions (American College of Medical Genetics)

Sl.No.	Condition	OMIM Number	Treatment
1	HbS/β Thalassemia	141900	As for sickle cell anemia
2	HbS/C disease	141900	As for sickle cell anemia
3	Medium chain acyl-CoA dehydrogenase deficiency	201450	Nutritional supplements; avoid fasting
4	Very long chain acyl-CoA dehydrogenase deficiency	201475	High carbohydrate, low-fat diet
5	Congential adrenal hyperplasia	201910	Genital surgery; hormone replace
6	Beta-ketothiolase deficiency	203750	Vigilant medical care, low-protein
7	Argininosuccinic aciduria	207900	Low-protein diet, supplements
8	3-methyl crotonyl-CoA carboxylase 1 deficiency	210200	Low protein diet
9	Carnitine-acylcarnitine translocase deficiency	212138	Carnitine
10	Citrullinemia	215700	Low-protein diet, supplements
11	Cystic fibrosis	219700	Vigilant medical care
12	Galactosemia	230400	Avoid milk and dairy products
13	Glutaric academia Type I	231670	Low protein diet with oral carnitine
14	Homocystinuria	236200	Diet, supplements including vitamins
15	Lsovaleric acidemia	243500	Low protein diet,
16	3-hydroxy-3-methylglutaryl-CoA lyase deficiency	246450	High carbohydrate, low fat, low fasting
17	Maple syrup urine disease	248600	Low protein diet
18	Methylmalonic academia (mutase deficiency)	251000	Vitamin B_{12} and another supplement
19	Methylmalonic academia cobalamin disorders A and B (Cbl A, B)	251100	Vitamin B12 low protein diet
20	Biotinidase deficiency	253260	Daily oral biotin
21	Multiple carboxylase deficiency	253270	Oral biotin
22	Phenylketonuria	261600	Low-phenylalanine diet
23	Tyrosinemia Type I	276700	NTBC
24	Sickle cell anemia	603903	Vigilant medical care, avoid infections
25	Propionic academia	606054	Low protein diet, supplements
26	Trifunctional protein deficiency	609015	Low-fat diet, supplements, avoid
27	Long-chain 3-hydroxy acyl-CoA dehydrogenase deficiency	609016	High carbohydrate, low-fat diet.
28	Congenital hypothyroidism	**	Oral thyroid hormone
29	Hearing loss	**	Communication strategy; Hearer cochlear implant

** = Congenital hypothyroidism and hearing loss have many causes, which may or may not be genetic.

the appropriate training can lead to normal speech development and academic achievement, regardless of its genetic or non-genetic origin.

Electrophysiological procedures are non-invasive and take only 5 minutes to determine the level of hearing in infants. One such test involves placing electrodes on the head of a sleeping infant to measure whether the sound is received by the brain (auditory brainstem response, ABR). Another option is placing a tiny receiver in the ear of a sleeping infant to detect sounds produced by the inner ear (otoacoustic emissions, OAE). If the brain doesn't respond to the sound in the ABR test or if the inner ear emissions are not generated in the OAE test, then the hearing is impaired. These options can be considered for early detection of hearing loss and appropriate intervention.

19.3.3 Adult Screening

Screening for genetic diseases in adults is crucial and more prevalent in certain conditions. For instance, Tay-Sachs carrier screening is commonly performed among Ashkenazi Jews, resulting in a 95 per cent reduction in the incidence of Tay-Sachs disease over the last three decades. Hereditary hemochromatosis (HHC) is an autosomal recessive condition that generally appears between the ages of 40 to 60. It involves the excessive accumulation of iron in the liver, heart, pancreas, skin, and joints, leading to symptoms such as weakness, lethargy, and weight loss. The excess iron can cause organ damage, but it can be prevented by removing blood before symptoms appear. Once the disease manifests, it can't be reversed. It is possible to diagnose the risk more accurately by detecting HFE gene mutations. Hemochromatosis is more prevalent in the USA, Ireland, the UK, and some parts of northern Europe. HFE gene mutations are responsible for 60 per cent to 90 per cent of these cases, with most patients being homozygous for a single mutation (C282Y/C282Y) and others homozygous (H63DIH63D) or compound heterozygotes (C282YIH63D). These mutations and their corresponding genotypes can be determined accurately and inexpensively by DNA tests. As the cost of each test is low, a nationwide adult HHC screening program is recommended.

Several carrier screening programs have been initiated worldwide for genetic diseases like phenylketonuria, congenital hypothyroidism, sickle cell anemia, congenital adrenal hyperplasia, maple syrup urine disease, homocystinuria, tyrosinemia, medium-chain acyl-CoA dehydrogenase deficiency, cystic fibrosis, galactosemia, and biotinidase deficiency.

Genetic screening is also necessary in families with a strong history of disease. For instance, in families with a high incidence of breast cancer, screening for BRACA1/2 mutations is necessary. For families with high cholesterol levels, all members should be tested for genetic hypercholesterolemia.

19.4 Treatment of Genetic Disorders

Treatment for metabolic disorders can range from supportive care to more specific interventions. Supportive care focuses on reducing pain and enhancing quality of life. Some conditions require restrictive diets or supplements, while others may benefit from inhibiting enzyme reactions to prevent the accumulation of toxic molecules. In certain cases, defective or absent proteins may need to be substituted, and protein activity may need to be re-established. Organ and bone marrow transplants, as well as nucleic acid-based therapies, may also be necessary.

When a metabolic deficiency is detected, it can be treated with either a supplementary diet or intravenous administration of the required compound. In some cases, enzymatic activity can be restored through ingestion of an essential cofactor, known as substitutive therapy.

Recombinant DNA technology is now being used to provide larger amounts of enzymes and structural proteins for therapeutic use. Gene therapy has also been attempted in some diseases but with limited success. Oligonucleotide-based therapies, such as siRNA, have also been tried in some cases.

Extensive information about genetic counseling is available, but there is still a gap between patients' attitudes toward acceptance of genetic testing and their compliance.

20

Gene Therapy

Sometimes defective genes can be corrected by replacing the faulty genes with a normal copy of the genes. Initial experiments were carried out in 1980. Ashanthi DeSilva, a 4-year girl suffering from SCID (Severe Combined immunodeficiency) was transfused with her T cells which were modified with a retroviral vector carrying a normal ADA gene. This was carried out in 1990. In 2000 the first gene therapy was provided. Alain Fischer from Paris successfully corrected children with SCID-X1, or "bubble boy" syndrome. The 1st death occurred because of gene therapy in October 1999. Gene therapy's success lies in transferring the correct gene at the right place and producing the right amount of protein. If it is done rightfully, a complete cure is possible. Different methods can be used for the replacement of the gene. These are physical, chemical, viral, and non-viral vector-based methods vectors are most used for gene therapy. The new advancement in gene therapy is new organ transplants, implants, the use of human artificial chromosomes, receptor-mediated delivery to find virally directed enzymes, and prodrug therapy.

Dominant-negative proteins are produced due to mutations, and these proteins may block a normal protein. One can replace the mutant protein with a functional copy of the gene; however, this strategy is not very helpful due to the presence of a dominant-negative protein, which may inhibit normal functioning. If antisense RNA is placed in the functional domain of the gene correction, it is possible to reduce the expression of the defective gene. Gene therapy can be used not only for genetic diseases but also for cancers.

Most of the gene therapy experiments have been conducted in developed countries. Out of different trials, the only gene therapy that has been approved is for immune deficiency called adenosine deaminase (ADA) deficiency. Stem cell therapy has been used for clinical purposes. Immune deficiency is caused by a mutation in a gene located on the X chromosome known as Severe Combined Immune Deficiency (SCID). This primarily affects males. A French research group led by M. Cavazzana-Calvo successfully treated this condition (SCID-XI) using gene therapy. This treatment resulted in 7/10 individuals restoring normal immune function. However, in 2002 and 2003, two children developed leukemia after gene therapy. The virus used for gene therapy activated the cancer-causing gene. Clinical trials were stopped but started again for patients with no other options. Gene therapy has been tried for inflammatory joint diseases, such as arthritis, neurological diseases, *etc.*

20.1 Types of Gene Therapy

Both somatic cells and germ cells can be used for gene therapy. If gene therapy is carried out in somatic cells, the change is not passed from generation to generation. If change is brought in the germ cells, it is transferred to the next generation. Many ethical issues are involved when germline gene therapy is used. If the target gene is replaced by a transgene and is replaced in the cell it is called *the Ex Vivo approach*. These cells can be re-implanted inside the human body, *e.g.* lymphocytes, fibroblasts, myoblasts, umbilical cord blood, stem cells, bone marrow cells, hepatocytes, *etc.* For this purpose, we need the following requisites. (1) To find out the gene accountable for the causation of the disease. (2) Gene cloning. (3) To determine the mutations. (4) To understand the pathophysiology of the gene using linkage and gene mutation (5) Selection of the gene and transferring it to the target cell. (6) Gene expression and protein function. (7) Recognition of gene transfers and its safety.

mRNA translation needs all essential components so that therapeutic proteins and RNA can be synthesized. Polyadenylation sequences are required for this function. Gene delivery vehicles should meet the following criteria (i) No or minimal side effects with high safety. (ii) High efficiency and specificity, (iii) high packaging capacity, (iv) large quantity, (v) control of gene expression. For this function, both viral and nonviral vectors can be utilized.

20.2 Methods of Gene Therapy

20.2.1 Viral Vectors

Viral vectors are more useable vectors for transferring genetic information. *RNA viruses (retroviruses)* are the most common vectors which can be used for gene transfer. These are used as vectors because of the following reasons (i) Almost all the cells can be infected and after infection, these cells show expression of the integrated viral genes, (ii) Cells should be infected at the same time and care should be taken to consider the maximum number of cells, (iii) DNA gets incorporated under appropriate conditions single copy at a random single site, the chemical and physical techniques often result into the insertion of multiple copies of the transferred gene (iv) integration is generally accidental concerning the host genome it is accurate with the viral genome which gets integrated into DNA, (v) both long-term harboring and infection of the retroviral vector does not damage cells. Retroviral vectors are under the process of investigation due to their usefulness.

The retrovirus is composed of an RNA-protein core and has a glycoprotein cover that enters the cell, The RNA acts as a template in the cell, and reverse transcription takes place, due to which the double strand of DNA is formed. This DNA can be precisely mixed as a single copy called a provirus. Genetic information gets integrated at a random location in the host's genome. It is important to note that LTR is required for reverse transcription. The host's RNA polymerase creates Retroviral RNA from the provirus DNA. A part of this RNA is used in the cell's translational machinery, which makes the viral proteins that reach the final viral particles along with the genomic RNA. The viral particles can infect other cells. It is now known that almost all the regions coding for viral proteins *(gag, pol, and env)* can be deleted, and some or all these sequences can be replaced with other DNA. If the viral gene is deleted, the retroviral vector becomes defective. The defective provirus is infected with a "helper" virus, which further carries all the vital functions.

Moloney murine leukemia virus (MMLV) or MSV (murine sarcoma virus) is isolated and inserted into a suitable plasmid. The viral genes are then replaced with the required exogenous genes using standard recombinant DNA techniques. The culture cells (for example, NIH 3T3 cells) are transferred through a suitable gene transfer process (for example. calcium phosphate). After infecting the cells with a helper virus (such as intact MMLV), infectious viral particles, having both the retroviral vector and the helper virus, are separated from the cells into the surrounding medium. The bone marrow cells are incubated with virus particles and injected intravenously into the mouse. The original mouse bone marrow cells are killed using lethal doses of X-rays. Once this is over, the expression profile of the gene is investigated to find out whether this procedure was successful or not. The neomycin resistance (neo') functional gene is transferred into mouse hematopoietic progenitor cells. The presence and expression of this gene in granulocytic progenitor cells made these cells opposed to the neomycin-like antibiotic G418, as shown by *in vitro* colony assays. These cells are further introduced into the colony assay. Southern blot analysis and colony assays can show the neo' gene is present and is also active in the spleen of the recipient animals.

20.2.2 Adenovirus

It is a 36kb double-stranded linear DNA virus that can be replicated extra chromosomally in the nucleus. It carries only 5'-3' inverted terminal repeats (ITR'S) packaging signal (Y) and q element. A large packaging capacity of 36kb contains either large or multiple transgene expression cassettes. The drawback of this vector is that it is immunogenic. There are approximately 100 different adenoviruses. Adenoviruses replicate only in human cells. For wild-type adenovirus, DNA replication begins approximately five hours after infection and is completed in Hela cells at 20 to 24 hours. Many adenoviral particles facilitate the preparation of very high titers of adenoviral vectors.

20.2.3 Adeno-associated Virus

The adeno–associated virus (AAV) is 4.7kb single-stranded DNA. It belongs to the family of viruses parvoviral. It can infect but not independently duplicate, hence dependent on other viruses. There are 60 subunits of the virus, and it is 26nm in length. It is a gene therapy vector because it is not lethal in humans. It gets more efficiently transduced so it can divide on its own. It can express for a longer duration due to episomal persistence of AAV genomes, even though site-specific addition at human chromosome 19 AAVS1 site is created, resulting in a new variant known as AAV2 with low efficiency. In recent years a new generation of AAV vector (ScAAV) has been developed, which is like the double-stranded DNA template, which can undergo transcription upon infection. The packaging size of the AAV vector is quite small. There are three potential risks of AAV. These are insertional mutagenesis generation of wild-type AAV and administration of contaminated adenovirus. It is thought that it can activate the proto-oncogene. New modifications like the creation of a transgenetic expression cassette design and split vector 150kb in size have increased its efficacy.

20.2.4 Herpes Simplex Virus (HSV)

HSV is associated with the subfamily of alpha herpes virus. It is 152 kb in size. It has a bigger linear DNA which is double-stranded. It is 1150kb long and enclosed within an icosahedral protein known by the name of the capsid and is surrounded by a lipid bilayer called an envelope. It contains approximately 100-200 genes with various proteins involved in forming the capsid. The HSV-1 virion is enveloped and is about 110nm in diameter.

It has two categories HSV-1 and HSV-2 both of which result in infection in humans. HSV-1 and HSV-2 are complex genomes with two unique parts known as the long unique region (UL) and the short unique regions (UR). There are 74 known ORFs, UL contains 56 viral genes, whereas the UR contains only 12 viral genes. Transcription of HSV genes is catalyzed by RNA polymerase II resulting in the transcription of the HSV gene, which can result in infection. Early gene expression follows DNA replication, and the production of certain envelope glycoproteins permits the synthesis of enzymes. Some genes are expressed late, and they encode proteins that cause the formation of virion particles. The HSV vectors are responsible for delivering genes to neurons and further causing latent infection in these cells. These vectors can treat such neurological conditions as malignant gliomas, Parkinson's disease, known as single-gene disorders, and cerebral ischemia. But all this could be successful only if the technical difficulties are sorted out, which are difficult to sort out. The difficulties involved are efficient delivery of the vector to target the cells, which needs continuous monitoring of foreign gene expression, dependent upon the immune response. HSV is the most promising vector for cancer gene therapy.

In **Table 20.1,** the advantages and disadvantages of various gene therapy vectors are shown.

20.3 Non-viral Vectors

The non-viral vector is a method of gene transfer that does not require viral particles. These could be lipids, nucleic acids, peptides, chemicals, *etc.*

DNA suspension is prepared with calcium phosphate in the chemical method, and a monolayer

Table 20.1: Viral Vectors and their Characteristic Features

Vector	*Advantages*	*Disadvantages*
Retrovirus	(1) 8 kilobases capacity is enough for the majority of gene therapy experiments (2) Continuous integration and lack of immunogenicity.	(1) Titer not very high (2) Only dividing cells can be used (3) Mutagenesis may occur at the insertion sites, due to which the risk of developing malignancy increases.
Adenovirus	(1) Target cell proliferation not essential (2) High titer and efficacy.	(1) Unstable integration and immunogenic. (2) Need not be repeated because of neutralizing antibody formation.
Adeno-associated virus	(1) Stable integration. (2) Integrated into non-dividing cells.	(1) Small capacity of 5 kilobases. (2) Low titer.
Herpes simplex virus	(1) Prolonged expression, however, produces latent infection	(1) Immunogenic. (2) Cytotoxic. (3) Difficult to develop.
Lentivirus	(1) Dividing and non-dividing cells can be used. (2) Stable expression.	(1) Safety concerns. (2) Production is difficult.

of cells grows in the tissue culture dish. Competence can be increased by taking the following steps. (i) Instead of calcium phosphate, **Diethylaminomethyl-dextran hydrochloride** can be used (ii) Glycerol can shock the cells. Two hours of incubation is required for this procedure. However, efficiency varies from cell to cell line. The transfected cells will survive, while others will die. Under most favorable conditions and very receptive cells (for example, mouse L cells), It is imperative to note that obtaining one cell in the 10x3 matrix, which can effectively incorporate and express the exogenous DNA, is highly achievable. The usual efficiency falls within the range of 10 to 50 million, which is quite significant. a process essential to perceive the infrequently transfected cell. Various selective media can protect cells from lethal selective agents added to the incubation medium or that complement a genetic defect (such as HPRT or TK). Cells in suspension are not effective. If a chemical method is used the cell recovery is less.

20.4 Physical Techniques

Physical techniques can be used, and these are microinjection and electroporation. DNA directly crosses a cell membrane under the electric current in this technique. The genes are transferred, which may be used for the number of dissimilar cells. This technique has the advantage of high efficiency (1/5 injected cells can be permanently transfected). Micro-injection is used in transferring genes into fertilized mouse eggs. An experiment has shown that if plasmid DNA is microinjected pronuclear of a recently fertilized mouse egg the ovum is placed into the oviduct of a pseudo-pregnant female, the egg could develop into a normal mouse carrying the plasmid DNA in every cell of its body. These mice are known as transgenic mice. This is carried out by attaching a rat growth hormone gene to an active regulatory sequence (specifically the promoter region which directs the synthesis of metallothionein messenger RNA in mice). A recombinant DNA construct is obtained that actively produces growth hormone in the genetically defective mouse. Although the level of growth hormone production is inappropriately controlled- that is influenced by signals that generally regulate metallothionein synthesis experiments do show that microinjection can be used as a delivery system that can place a gene into every cell of an animal's body. Due to the low success rate, there are several questions about its use in humans. There is a high failure rate; most of the eggs are damaged due to microinjection and also during transfer procedures. This technique can cause mutations due to a lack of control over the DNA integration in the genome. The integration of a rabbit *b*-globin gene exogenously in transgenic mice in a chromosomal location that is not correct may result in the expression of the b-globin gene in inappropriate tissue, namely muscle or testes. There is no control over where exogenous DNA will integrate with any gene transfer procedure due to the damaging effect caused by a harmful insertion site. It could be great when it occurs in the egg but may be insignificant when it is found in one or a few of many bone marrow cells. In a homozygous defect, it becomes more complex as documentation is difficult in an ovum, and the procedure carries a high risk; it is not proper to try any treatment under such circumstances. Some genetic disorders cause infertility in a homozygous state due to lethal effects.

20.5 Expression

Once the gene is corrected, it should provide proper expression, which should be functional; if it is not, then the main problem arises. Exogenous promoters, which are pretty active, are present in transgenic mice. Transgenic mice are created by microinjecting exogenous DNA into the fertilized eggs. Various promoters that are active are metallothionein transferrin, immunoglobulin, and elastase. The promoters get inactivated after microinjection in the mouse oocytes; the reason for inactivation is not very clear. One of the proposed hypotheses of inactivation is that the DNA sequences may get methylated in mouse tissues and hence become inactive but relatively unmethylated in tissue culture cells where they are active therefore, the mouse zygote appears to respond to this foreign DNA by covering it with methyl groups, which remain on the DNA throughout the lifetime of the animal. Attempts to decrease the methylation of the genomic DNA by treating adult mice carrying an exogenous b-globin promoter with the hypomethylating drug 5-azacytidine are not successful.

20.6 Safety

Safety is one of the vital concerns in gene therapy. Gene therapy provides both advantages and disadvantages. The immune response is triggered during gene therapy. To resolve this issue, recombinant vectors are used. There exist mechanisms in mouse retroviral vectors. Mouse sequences are different from known primate retrovirus sequences, and there is almost no homology. The proviral vector should have mouse LTRs along with enhancer and promoter regions deleted and a human gene controlled by the suitable human genomic regulatory signals with the present constructs, following safety measures should be taken.

Table 20.2: Various Diseases Treated using Gene Therapy

Disease	Target Cells	Transfected Gene(s)
Hemophilia A	Liver, muscle,	Factor VIII
Hemophilia B	Bone marrow cells, fibroblasts	Factor IX
Familial hypercholesterolemia	Liver	LDL receptor
Severe combined immunodeficiency	Bone marrow cells, T cells	Adenosine deaminase (ADA)
Hemoglobinopathies	Red blood precursor cells	a-globin, b-globin
Cystic fibrosis	Lung airway cells	CFTR
Gaucher disease	Bone marrow macrophages	Cells glucocerebrosidase
Cancer	tumor cells	p53, Rb, interleukins growth-inhibitory genes apoptosis genes

20.7 Plasmid-based Expression Vectors

The calcium-phosphate process for the transfer of the plasmid-based expression vector into human bone marrow has not yet been shown to be of a competent delivery system. However, sometimes stem cells may be pathogenic. Some diseases related to gene therapy are shown in **Table 20.2**.

20.8 Difficulties in Achieving Successful Gene Therapy

It is challenging to develop a universal vector for all diseases. More so many ethical issues are involved in conducting gene therapy.

20.9 Ethical and Social Concerns in Somatic Cell Gene Therapy

Ethical issues came to light in 1960. These were associated with recombinant DNA research; till then, gene therapy had not started. When gene therapy was started, it was favored because people thought that gene therapy could eliminate many diseases. However, at a later date, many ethical issues were raised. Gene therapy can be a one-time procedure, for example, genetic addition of the ADA gene in SCID patients. CF may be treated by providing the patient with normal functioning pulmonary cells. Such cells can be introduced in two ways: by carrying out the lung transplant or by genetically altering the patient's lung cells so they can help defective CF genes produce normal protein. The former is an expensive halfway technology that requires ongoing immunosuppression to prevent rejection and constant alertness to infection that may result from immunosuppression.

Gene therapy can cause a permanent change in CF patients with no risk. However, sometimes gene therapy is disappointing because of a few unexpected complications and side effects. In gene therapy, one must keep a balance between harmful and beneficial effects along with cost-effectiveness.

20.10 Ethical and Social Concerns in Germ-line Gene Therapy

Gene therapy can be either germ-line gene therapy or somatic cell therapy. Germline gene therapy is more difficult as compared to somatic gene therapy because of the insertion of mutations and inadvertent production of chimeras. Germline gene therapy is carried out in embryos; hence only those couples who benefit from undergo *in-vivo* fertilization.

Many factors are involved in gene therapy; some of these are a stable intergenerational transfer of genetic information and accurate gene expression, therefore, will require substitution or repair of a defective gene *in situ*. The only method of therapy recently available for humans is gene addition. In **Table 20.3**, some of the germline therapies are illustrated.

The advantages of prenatal gene therapy are that it can provide early phenotypic corrections, reducing or avoiding otherwise devastating effects of genetic disease.

20.11 Other Modes of Gene Therapy

RNA interference (RNAi) reveals a new approach to gene therapy. It targets the specific genes that downregulate gene expression. Small interfering RNA or siRNA is the most potent form. Small fragments of double-stranded RNA, specific for a particular gene target, are introduced into the cell-specific hybridization between the naturally occurring transcript and the induced siRNA (antisense portion) destroying the message.

This form of RNAi acts at the transcriptional level of gene expression directly. To know the efficacy, it is recommended to create knock-down genes. The specificity and efficiency of in *vivo* models is a big issue. The main aim of this model is to create RNAi from an effective functional genomics tool into a therapeutic delivery system. However, there remain many issues that still need to be addressed.

Table 20.3: Proof of Principle for Therapeutic in Utero Gene Transfer (DG = days of gestation)

Animal Model	Vector	Route of Application	Outcome
Gunn rat, model of Crigler Najjar disease Type I	HIV lentivirus expressing human UDP-glucuronyl-bilirubin transferase	Topical liver injection 17-19 DG	45 per cent reduction of nonconjugated bilirubin for 1 year
RPE65-/- mouse model of Leber's congenital amaurosis	AAV 2 expressing human RPE65	subretinal injection 14DG	54 per cent of treated eyes showed improvement in ERC and 70 per cent in photo-response-sensitivity at 1-2.5 months after birth' RPE65 expression detectable up to 6 months
KO mouse model of GSDII deficiency (Pompe's disease)	AAV2 expressing acid α-glucosidase	intraperitoneal injection on 15 DG	Restoration of normal GAA levels in the diaphragm prevented glycogen accumulation and restored diaphragm contractility up to 6 months postpartum
Factor IX KO mice model for hemophilia B	HIV lentivirus expressing human Factor IX	Yolk sac vessel injection on 15 DG	Permanent therapeutic levels of human factor IX at 18-32 per cent of normal, lifetime phenotype correction of bleeding disorder and tolerance of human factor IX

20.12 CRISPR/Cas, a System Used for Gene Therapy

In the last two years, scientists have understood that many microbes can be used as a system of proteins and RNAs to defend themselves against invading viruses. Researchers have adapted this bacterial immune system, referred to as CRISPR/Cas, to edit single base pairs of the human genome as well as larger stretches of DNA. The CRISPR/Cas system uses a "guide" RNA to bring the DNA-cutting Cas protein to a specific DNA sequence that contains a disease-causing mutation. Once the Cas protein cuts the DNA at a target site, the system replaces the faulty DNA sequence with a healthy version. It is suggested that these technologies can change single base pairs and allow new ways of handling gene therapy. Many inherited diseases, including cystic fibrosis and sickle-cell anemia, are caused by single base pair changes to the DNA sequence of genes; the precise CRISPR/Cas technology could correct these mutations in patients.

Generally, the treatments involve adding a functional copy of a gene to a patient's cells. The healthy copy can either incorporate itself into the patient's genome, or it can remain as a distinct entity, depending on the design of the treatment. However, if the dysfunctional gene produces a dangerous protein version or if other DNA mutations cause a gene to be overproduced to the point of toxicity, adding a healthy copy will not treat the resulting disease. CRISPR/Cas could address diseases like Huntington's disease, an inherited neurodegenerative disorder caused by a toxic protein made by a dysfunctional gene. Traditional methods of gene therapy cannot address this disease, but CRISPR/Cas has the potential to correct the faulty DNA sequence. One of the powerful aspects of this technique is that it allows modifications where other approaches have failed. Another advantage of the gene-editing potential of CRISPR/Cas technology is that the corrected gene remains in its normal chromosome location, which preserves the way the cell normally turns a gene on or off. CRISPR/Cas system can make unwanted cuts in the genome if RNA nearly matches other DNA regions outside the gene that researchers want to treat. Another area to explore is how to deliver the molecular tools into a patient's cells.

21

Bioinformatics: Genomics, Functional Genomics, and Proteomics

Most of the literature on human genetics started pouring in the late 1980s. The gene bank was recognized in 1988. Gene bank contains data about the known sequences. This was associated with NSFnet (National Science Foundation Network). The need for more databases arose as soon as the human genome project came into existence. Databases are required to facilitate data analysis for mutations (gene-specific), enhancer elements, promoters/ repressors, mitochondrial DNA, Y chromosome DNA genetic diseases, protein-protein interactions, protein structures, gene expression, *etc.* Different types of molecular databases have been generated against each aspect, as shown in **Table 21.1**. More than 17 billion nucleotide bases in the GenBank contain about 14 million sequences derived from more than 100,000 different organisms. At present, approximately 400 molecular databases exist; most of these are DNA and protein sequence repositories (*e.g.*, GenBank, Ensembl, UCSC Gene Browser, Genome Database, Protein Information Resource, Swiss- Prot) to bibliographic and information resources (*e.g.*, OMIM, PubMed, RefSeq, Gene Clinics) (**Table 21.2**). These details are available at http://nar.oupjournals.org.

Due to the creation of these databases, it is easy to compare one's data with the available biological and medical data. This information has led to the field of bioinformatics. The databases can be used in different ways; for example, if we want to retrieve the data for CFTR gene, we will first look for the coding sequence cDNA using the Genome Database (GDB). On the GDB home page (http://www.gdb.org.gdb/), type CFTR in the "query" box, and click on the "SUBMIT" box. The default settings have not changed. On the "Genomic Segment Query Results" page, click on GDB: 120584. This page will provide several links related to the CFTR gene, such as chromosomal location, nucleotide acid, protein

Table 21.1 Different Databases

Databases
Bibliographic resources
Cellular regulation chromosome aberration comparative genomics
Gene expression
Gene identification and structure gene mutation
Gene sequences
Genetic and physical maps genetic disorders
Genomic sequences intermolecular interactions metabolic pathways
Protein motifs
Protein sequences
Protein structure
Proteome resources
RNA sequences
Transgenic organisms

Table 21.2 Molecular Databases

Database	URL	Description
ALFRED	http://alfred.med.yale.edu	Allele frequencies and DNA polymorph isms
Alzheimer disease mutations	http://molgen-www.uia.ac. be/ ADmutations/	All gene mutations related to Alzheimer disease
Array express	http://www.ebi.ac.uk/arrayexpress	Microarray gene expression data
Atlas of genetics and cytogenetics in oncology and hematology	http://www.infobiogen.fr/services/ chromcancer/	Genes, cytogenetics, and clinical features of cancer and cancer-prone diseases
Cooperative human linkage center	http://gai.nci.nih.gov/CHLC/	Integrated genetic and marker maps of human chromosomes
Database of interacting proteins (DIP)	http://dip.doe-mbLucla.edu	Protein-protein interactions
dbSNP	http://www.ncbLnlm.nih.gov/SNP/	Single-nucleotide polymorphisms
DNA data bank of Japan (DDBJ)	http://www.ddbj.nig.ac.jp	All known nucleotide and protein sequences
EMBL nucleotide sequence database	http://www.ebLac.uk/embl.html	All known nucleotide and protein sequences
Ensembl	http://www.ensembl.org/	Annotated information on eukaryotic genomes
ExPASy molecular biology server	http://ca.expasy.org	Expert protein analysis system. Links to protein databases
GDB	http://www.gdb.org	Human genes and genomic maps
GenAtlas	http://www.citi2.fr/GENATLAS/	Human genes, markers, and phenotypes
GenBank	http://www.ncbLnlm.nih.gov/	All known nucleotide and protein sequences
GeneCards	http://bioinfo.weizmann.ac.il/cards/	Integrated database of human genes, maps, proteins, and diseases
GeneClinics	http://www.geneclinics.org/	Medical genetics information resource
Gene new	http://www.gene.ucl.ac.uk/cgi-bin/ nomenclature/searchgenes.pl	Nomenclature for human genes
HugeIndex	http:;/hugeindex.org	mRNA expression levels of human genes in normal tissues
HuGeMap	http://www.infobiogen.fr/ servicesjHugemap	Human genome genetic and physical map data
Human gene mutation database	http://www.hgmd.org	Links to locus-specific mutation databases
Kyoto encyclopedia of genes and genomes (KEGG)	http://www.genome.ad.jp/kegg	Metabolic and regulatory pathways
MITOMAP	http://www.gen.emory.edu/mitomap.html	Human mitochondrial genome
Online mendelian inheritance in man	http://www.ncbi.nlm.nih.gov/Omim/	Catalog of human genetic and genomic disorders
PROSITE	http://www.expasy.org/prosite	Biologically significant protein patterns and profiles
Protein information resource (PIR)	http://pir.georgetown.edu	Comprehensive, annotated, nonredundant protein sequences
PubMed	http://www.ncbi.nlm.nih.gov/PubMed/	Abstracts of journal articles
RefSeq	http://www.ncbLnlm.nih.gov/LocusLink/ refseq.html	Non-redundant sequences from genomes, genes, transcripts, and proteins
SNP consortium database	http://snp.cshl.org	Single-nucleotide polymorphisms
Stanford microarray database	http://genome-www.stanford.edu/ microarray	Raw and normalized data from microarray experiments
Swiss-Protein/TrEMBL	http://www.expasy.ch/sprot	Protein sequences
UCSC genome browser	http://genome.ucsc.edu/	Genome assemblies and annotation

sequence data, *etc.* However, for the cDNA sequences, click on "M28668 (cDNA)" to go to the "Nucleic Acid Sequence Database Preference" page. Click on "EMBL sample." This link accesses the complete sequence of the CFTR cDNA in the GenBank format. Mter the CFTR sequence information is loaded, scroll down to the cDNA nucleotide sequence that begins after the SQline with the nucleotides "aattggaagc", highlight the first 600 nucleotides including the numbers, and copy this portion of the CFTR cDNA sequence in the computer.

For the rest of the manifestation, go to the home page of the National Center for Biotechnology Information (http://www.ncbc.nlm.nih.govl) and click on "BLAST" in the menu bar. Then, under the section entitled "Nucleotide", select "nucleotide-nucleotide BLAST (blastn)", which will open the BLAST input page. Paste the CFTR cDNA sequence (query sequence) from the clipboard into the "Search" box. Now the database setting can be left at non-redundant (nr) and click on the "BLAST" button. The "formatting BLAST" page will estimate the time it will take to get the comprehensive analysis. The total time is approximately 60 seconds, then click the "Format!" button. Save the BLASTN results in the computer for further analysis.

BLAST is a sequence tool that has a lot of scope. It is used for correct and fast search of sequence databases considering either a nucleotide or a protein sequence. This tool is also used for nucleotide alignment. The sequence in question matches all the sequences in the database. A raw score is generated on assigned values depending upon mismatches, gaps, and the number of characters in a gap with each alignment between the query sequence and sequences of the database. Normalizing parameters are used and scored as S. This is called the bit score. Depending on the bit score, an expected value (E-value) is generated for each alignment. A high bit score (S) shows a good alignment between the two sequences, but this value is not enough to show the status of alignment significance. The E-value, the length of the query sequence's aligned portion, and the database's total sequence length are considered. Values for S and E with a particular query sequence will change as more and more sequences are added to the database. If an E-value is significantly less than 1.0, then it is unlikely that alignment will occur by chance. Here the value of biological significance is added, revealing that alignment is absolute (http: www.ncbi.nlm.nih.gov/EducationlBLASTinfo/guide. html for BLAST tutorial).

Excellent alignments with CFTR sequences from humans, rhesus monkeys (*Macaca mulatta*), and other primates; cattle (*Bos taurus*); sheep (*Ovis aries*), mouse (*Mus musculus*); and many other animals have been reported. Various laboratory methods have been used to generate the data. These methods are summarized below.

21.1 DNA Microarray

Microarray uses a small platform. Approximately 10,000 different probes are placed in a 1 to 2cm area. It is a high throughput technique and can be used for GWAS data and expression analysis. The basic principle of microarray is demonstrated in (**Figure 21.1**).

In microarray, a vast amount of data is generated through genomic and expression analysis.

21.2 Serial Analysis of Gene Expression

Polyadenylated mRNA is used for serial analysis of gene expression; oligo (dT) sequences are labeled with biotin and attached to a streptavidin-coated magnetic bead cDNA is synthesized (**Figure 21.2**). The magnetic beads and cDNAs are situated in a single reaction tube. The cDNAs are cut with the help of restriction endonuclease NlaIII; this cuts the cDNA to find out the sequence CATG/GTAC and cuts outside to the G: C base pairs, exiting through

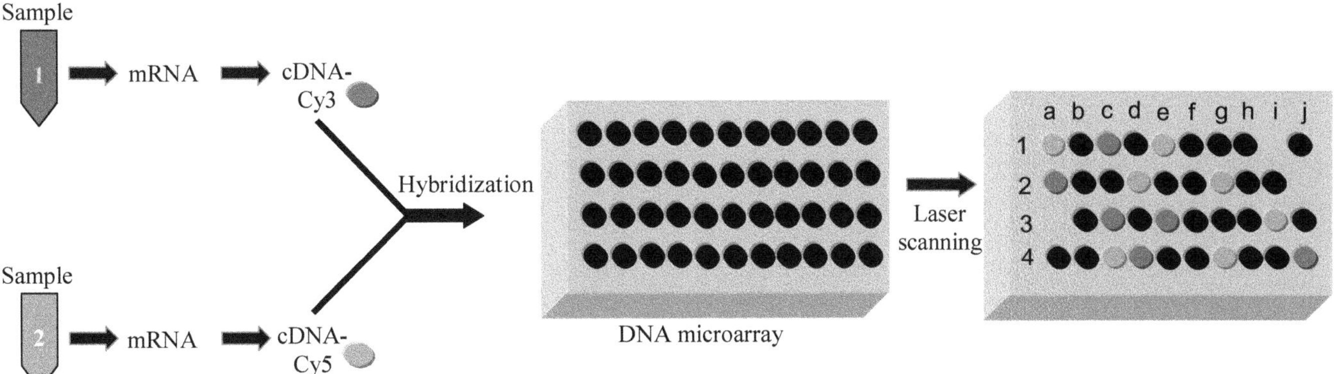

Figure 21.1: Use of DNA Microarrays for Gene Expression Studies using Cy3 and Cy5 Dyes. The signals can be read using a confocal microscope.

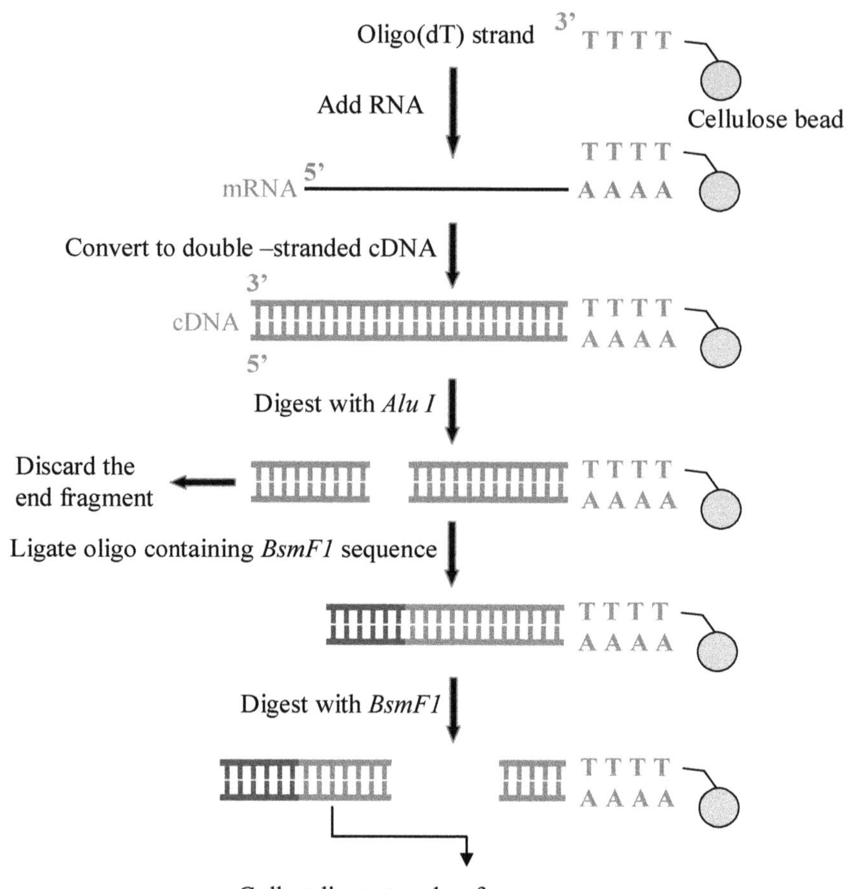

Figure 21.2: Serial Analysis of Gene Expression (SAGE).

a 3′ GTAC extension. The enzyme is an anchoring enzyme. Because NlaIII cuts, on average, 1 in 256 base pairs, there is a high probability that each cDNA will have at least one NlaIII recognition site. The cDNAs are bound to beads at each NlaIII cut site close to the 3′ end of a cDNA. The fragments which remain unbound are washed away. After digestion is complete, the cDNA sample is divided into two parts. One part is ligated with an adapter A and the other with another adapter B. Both adapters have a CATG extension complementary to the extension produced by NlaIII digestion, a 5-base pair recognition site for the restriction endonuclease BsmFI, and a sequence for priming a PCR reaction. The adapters have different primer sequences to prevent base pairing (snap back) during PCR. Mter the adapters are ligated to NlaIII-cut eDNA; the products are treated with BsmFI. Which acts as a tagging enzyme; this type IIS restriction endo nuclease binds to its recognition site, but unlike NlaIII and another type II restriction endonucleases, it cuts ten nucleotides downstream of one DNA strand and 14 nucleotides in the other strand from the recognition sites regardless of the intervening nucleotide sequence. The adapter-tag molecules are released into solution. The 4-nucleotide extension of the BsmFI -cut DNA is filled into and forms a blunt-end molecule. The sample is ligated. Under these conditions, among other possibilities, the blunt ends of two tags are joined to form a two-tag molecule, a ditag that is flanked by primer sequences. Because ditag formation is completely random, tags from different cDNAs are joined during ligation. Of all the ligation products, the ditags are readily amplified during PCR. The amplified ditags are treated with NlaIII to release the adapter sequences and produce ditags with an NlaIII extension at each end. The NlaIII -cut ditags are separated from the adapter sequences and ligated to form randomly joined multiple combinations (concatemers) of ditags. Concatemers that are about 500 base pairs in length are isolated and cloned in an E. coli plasmid. Then, the sequenced "tag to gene" database is evaluated to see the most probable gene. In addition, based on the information from DNA sequencing, the number of times each tag is sequenced is determined, representing its abundance in the initial sample. Up- and down-regulated mRNAs can be recognized by finding out the frequencies of tags in different samples.

Online resources (http://www.ncbc.nlm.nih.gov/SAGE, http://www.cgap.nci.nih.gov/SAGE) are available for matching tags to likely genes, determining the frequency of a tag among various SAGE libraries, and providing other pertinent information.

21.3 Proteomics

The study of proteins is called proteomics. Information about the protein is essential as it can be used for therapeutic purposes. The study of proteins requires high-throughput computers and software for analysis purposes.

It is estimated that nearly 60 per cent to 70 per cent of human genes undergo alternative splicing; in humans, there are more proteins than DNA, and there are approximately 85,000. Analysis of proteins is a little tricky. Various techniques used for this purpose are given below (**Figure 21.3**).

21.4 Separation and Identification of Proteins

Proteins can be separated using two-dimensional gel electrophoresis. It helps in the separation of proteins. A protein sample is subjected to electrophoresis through an immobilized pH gradient in one dimension, the first dimension, and then electrophoresed at right angles to the first dimension, that is, the second dimension, through a polyacrylamide gel. The proteins are separated in the first dimension based on their net charge (isoelectric point) and in the second dimension according to molecular mass. The separated proteins form an array of spots in the gel that is visualized by staining with Comassie Blue or silver. Approximately 2000 proteins are separated. The pattern of stained spots is captured by densitometric scanning. Databases with images of 2D PAGE gels from different cell types have been established. Even the protein with a minimal amount can be resolved by 2D PAGE. Mass spectrometry is commonly used for this purpose when the number of proteins is less.

Mass spectrometers have different configurations according to the nature (wet or dry) of the sample (analyte), the mode of ionization of the analyte, how the electric field(s) is recognized for accelerating the ions, and the method of identification is based on different masses. Peptide masses are usually determined by MALDI-mass spectrometry (MALDI-MS) and amino acid sequences by ESI-tandem mass spectrometry (ESI-MS/MS). Mass spectrometry is an important proteomic tool because an analysis is rapid and accurate and requires small amounts of starting material.

The amino acid sequence of a peptide can be obtained by ESIMS/MS and used to search a protein database to identify an unknown protein.

A non gel system called shotgun proteomics uses liquid chromatography (LC) combined with tandem mass spectrometry (LC-MS/MS) to analyze the proteins of a proteome. In this case, a complete protein mixture is initially treated with a protease. Then, the peptides are separated by liquid chromatography, and MS/MS determines the amino acid sequence of

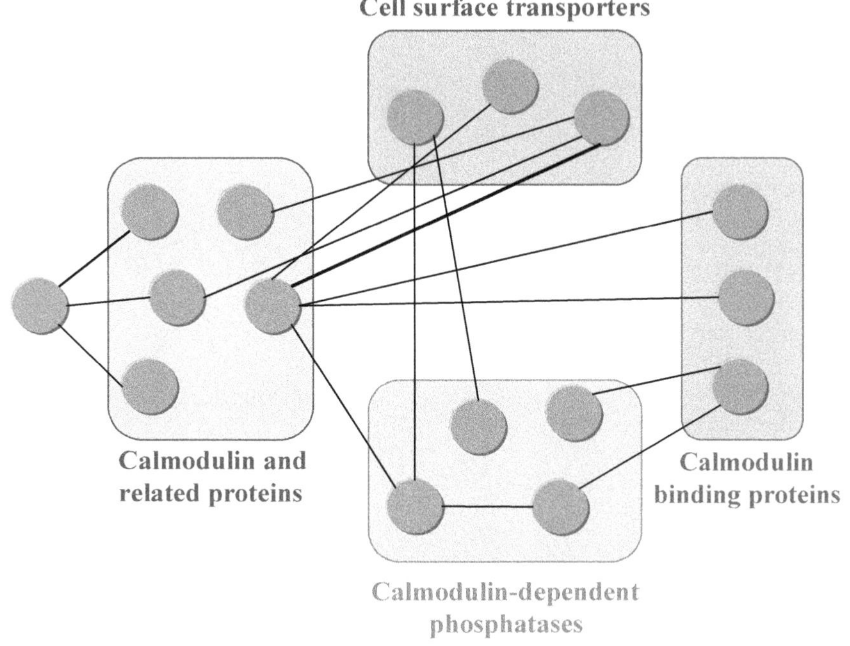

Figure 21.3: Protein Interaction Map of *Drosophila melanogaster* of Calcium Signaling Protein Clusters.

each peptide. In the final setup, data analysis of MS or MS/MS is crucial.

21.5 Protein Expression Profiling

Protein expression profiling is essential in comparing normal and cancerous cells in cancer studies.

21.6 Protein-Protein Interaction Mapping

Many experimental manipulations are required to see protein interaction; however, it is difficult to recognize all interactions.

As a first step for any large-scale protein-protein interaction study, many cDNAs are prepared and cloned adjacent to specified DNA sequences to form hybrid genes. Regardless of the protocol, large numbers of interactions are scored. Specialized computer programs are essential to categorize and map different relationships. As a part of this analysis, rigorous statistical criteria are used to lower the number of possible sham positive interactions in the final data set. With *Drosophila melanogaster*, an overall protein interaction map of 3000 interactions with 3522 proteins has been found. In addition, the nuclear, cytoplasmic, and extracellular locations of 2268 interactions with 2346 proteins were mapped. Finally, smaller interacting sets of proteins within cellular regions were noted (**Figure 21.3**). Also, a network of 5500 protein interactions was constructed for the roundworm (*Caenorhabditis elegans*).

Glossary

Acrocentric: Acrocentric chromosomes are when the centromere is very close to one end. The short arms (p) usually have small dot-like appendages on stalks, known as "chromosomal satillites". In an acrocentric chromosome, the p arm contains repeating sequences, such as the nucleolar organizing regions (NORs). The acrocentric chromosomes in humans are chromosomes 13, 14, 15, 21, and 22.

Achondroplasia: The most common and well known form of short limbed dwarfism characterized by a normal trunk size with disproportionally short arms and legs, and a disproportionally large head. It occurs as an autosomal dominant condition.

Activators: Activator is a transcription regulatory protein found in eukaryotes (transcription factor) that increases gene transcription of a gene or set of genes.

Adaptation: In biology, an adaptation, also called an adaptive trait, is a trait with a current functional role in the life of an organism that is maintained and evolved by means of natural selection

Additive genetic variance: The additive genetic variance, Var(A) is one where the variance is due to the average effects (additive effects) of phenotype, genotype and environment. ($V_P = V_G + V_E + V_{GE,}$ where p stands for phenotype, G for genotype and E for environment. and V_{GE} = variation associated with the genetic and environmental factor interactions

Adenine: A nitrogenous base, one member of the base pair A-T (adenine-thymine).

Adenosine Adenosine is a purine nucleoside composed of a molecule of adenine attached to a ribose sugar molecule (ribofuranose) moiety via a β-N9-glycosidic bond. Advanced maternal age – women over age 34 (age 35 at delivery) at increased risk for non-disjunction trisomy in fetus.

AFLP: PCR or just AFLP is a PCR-based tool used in genetics research, DNA fingerprinting, and in the practice of genetic engineering.

Allele Alternative forms of a genetic locus; a single allele for each locus is inherited separately from each parent (*e.g.*, at a locus for eye color the allele might result in blue or brown eyes).

Allele frequency: The allele frequency represents the incidence of a gene variant in a population. Alleles are variant forms of a gene that are located at the same position.

Alloenzymes: (or also called allozymes) are variant forms of an enzyme that are coded by different alleles at the same locus. These are opposed to isozymes,

Allogamy: is a term used in the field of biological reproduction describing the fertilization of an ovum from one individual with the other individual

Allele specific oligonucleotide hybridization: A procedure using PCR primer, to distinguish alleles that differ by one base pair

Alloploidy: Alloploidy al·lo·ploi·dy (al- ô-ploy′dç) n. A hybrid individual or cell having two or more sets of chromosomes derived from two different species.

Alternative splicing: Different ways of combining a gene's exons to make variants of the complete protein

Alpha-fetoprotein (AFP): A protein excreted by the fetus into the amniotic fluid and from there into the mother's bloodstream through the placenta.

Alu repetitive sequence: the most common dispersed repeated DNA sequence in the human genome accounting for 5 per cent of human DNA. The name is derived from the fact that these sequences are cleaved by the restriction endonuclease Alu.

Amino acid: Any of a class of 20 molecules that are combined to form proteins in living things. The sequence of amino acids in a protein and hence protein function are determined by the genetic code

Amniocentesis: Prenatal diagnosis method using cells in the amniotic fluid to determine the number and kind of chromosomes of the fetus and, when indicated, perform biochemical studies.

Amniocyte: Cells obtained by amniocentesis.

Amplification: Any process by which specific DNA sequences are replicated disproportionately greater than their representation in the parent molecules.

Aneuploidy: State of having variant chromosome number (too many or too few). (*i.e.* Down syndrome, Turner syndrome).

Angelman syndrome: A condition characterized by severe mental deficiency, developmental delay and growth deficiency, puppet-like gait and frequent laughter unconnected to emotions of happiness.

Annotation: Adding pertinent information such as gene coded for, amino acid sequence, or other commentary to the database entry of raw sequence of DNA bases

Apert syndrome: A condition caused by the premature closure of the sutures of the skull bones, resulting in an altered head shape, with webbed fingers and toes.

Artificial insemination: The placement of sperm into a female reproductive tract or the mixing of male and female gametes by other than natural means.

Autosome: a nuclear chromosome other than the X- and Y-chromosomes.

Autoradiograph: a photographic picture showing the position of radioactive substances in tissues.

Alpha helix: The alpha helix (α-helix) is a common secondary structure of proteins and is a right hand-coiled or spiral conformation (helix) in which every backbone N-H group

Amplification: An increase in gene copy number per cell; cytogenetic manifestations include double minutes and homogeneously staining regions.

Anticodon: a sequence of three nucleotides in a region of transfer RNA that recognizes a complementary coding triplet of nucleotides in messenger RNA during translation.

Anaphase: is the fourth phase of mitosis, which is a process that separates the duplicated genetic material carried in the nucleus of a parent cell into two

Antisense strand: It is the DNA antisense strand which serves as the source for the protein code, because, with bases complementary to the DNA sense strand,

Apoptosis: A form of cell death in which a programmed sequence of events leads to the elimination of cells without releasing harmful substances into the surrounding area. Too little or too much apoptosis plays a role in a great many diseases. When programmed cell death does not work properly, cells that should be eliminated may remain and become immortal as in cancers such as leukemia.

Array CGH: This method is advancement to CGH on metaphase chromosomes, as it provides significantly higher resolution. Defined DNA fragments are fixed in a matrix system (array CGH). There are bacterial artificial chromosome (BAC) arrays or even shorter single-stranded oligonucleotide arrays, or the more recent single-nucleotide polymorphism (SNP arrays). SNP arrays have multiple applications for assessing genomic imbalance, not reflected as copy number variations (CNVs).

Autoradiography: A technique that uses X-ray film to visualize radioactively labeled molecules or fragments of molecules; used in analyzing length and number of DNA fragments after they are separated by gel electrophoresis.

Autosome: All chromosomes, excluding sex chromosomes are classified as autosomes. Humans have a total of 22 pairs of autosomes.

Autosomal dominant: A gene on one of the non-sex chromosomes that is always expressed, even if only one copy is present. The chance of passing the gene to offspring is 50 per cent for each pregnancy.

Autosomal recessive: A genetic condition that appears only in individuals who have received two copies of an autosomal gene, one copy from each parent. The gene is on an autosome, a non sex chromosome. The parents are carriers who have only one copy of the gene and do not exhibit the trait because the gene is recessive to its normal counterpart gene.

Bacterial artificial chromosome (BAC)): A vector used to clone DNA fragments (100- to 300-kb insert size; average, 150 kb) in *Escherichia coli* cells. Based on naturally occurring F-factor plasmid found in the bacterium *E. coli*.

Bacteriophage: Also called phage or bacterial virus, any of a group of viruses that infect bacteria.

Base: One of the molecules that form DNA and RNA molecules.

Bar bodies: A highly condensed and transcriptionally inactive X chromosome found in the nuclei of somatic cells of female mammals.

Base sequence: The order of nucleotide bases in a DNA molecule; determines structure of proteins encoded by that DNA.

Base sequence analysis: A method, sometimes automated, for determining the base sequence.

Base excision repair: It corrects DNA damage from oxidation, deamination and alkylation. Such base lesions cause little distortion to the DNA helix structure

Base modifying agent: Base modifying agent is one which directly modifies the structure or chemical properties of bases. An example of this is nitrous oxide which is a deaminating in nature

Base pair substitution is a type of mutation where one base pair is replaced by a different base pair. The term also refers to the replacement of one amino acid. For example an AT is replaced by a GG pair.

Bidirectional replication: a type of DNA replication where replication is moving in both the directions from the starting point. This creates two replication forks, moving in the opposite direction away from the origin of replication

Back-cross: Backcrossing is a crossing of a hybrid with one of its parents or an individual genetically similar to its parent, in order to achieve offspring with a genetic identity which is closer to that of the parent. It is used in horticulture, animal breeding and in production of gene knockout organisms.

Balanced translocation: Occurs when two chromosomes break and exchange places leaving the same amount of genetic material. An individual with a balanced translocation will be unaffected, but children may be affected in a variety of ways.

Base pair (bp): Two nitrogenous bases (adenine and thymine or guanine and cytosine) held together by weak bonds. Two strands of DNA are held together in the shape of a double helix by the bonds between base pairs.

Base sequence: The order of nucleotide bases in a DNA molecule.

Base sequence analysis: A method, sometimes automated, for determining the base sequence

Bivalent during: The prophase of meiosis I, homologous chromosomes pair and form synapses. The paired chromosomes are called bivalents. The bivalent has two chromosomes and four chromatids, with one chromosome coming from each parent

Biotechnology: A set of biological techniques developed through basic research and now applied to research and product development. In particular, biotechnology refers to the use by industry of recombinant DNA, cell fusion, and new bio processing techniques.

BRCA: Most inherited cases of breast cancer are associated with two abnormal genes: BRCA1 (BReast CAncer gene one) and BRCA2 (BReast CAncer gene two). Everyone has BRCA1 and BRCA2 genes. The function of the BRCA genes is to repair cell damage and keep breast cells growing normally. But when these genes contain abnormalities or mutations that are passed from generation to generation, the genes don't function normally and breast cancer risk increases. Abnormal BRCA1 and BRCA2 genes may account for up to 10 per cent of all breast cancers, or 1 out of every 10 cases.

BLAST: A computer program that identifies homologous (similar) genes in different organisms, such as human, fruit fly, or nematode.

Blending theory: An incorrect 19th century theory about the inheritance of characteristics. It was proposed that inherited traits blend from generation to generation. Through his plant cross-breeding experiments, Gregor Mendel proved that this was wrong

Blotting: A technique for transferring a specific protein or nucleic acid from an electrophoresis gel matrix to a microporous membrane.

Breakpoint: Refers to sites of breakage when chromosomes break (and recombine).

Bootstrap: A method for determining confidence levels attached to the branching patterns of a phylogenetic tree chosen by the parsimony approach.

Capillary array: Gel-filled silica capillaries used to separate fragments for DNA sequencing. The small diameter of the capillaries permit the application of higher electric fields, providing high speed, high throughput separations that are significantly faster than traditional slab gels.

Carrier: An individual who possesses an unexpressed, recessive trait.

cDNA library: A collection of DNA sequences that code for genes. The sequences are generated in the laboratory from mRNA sequences.

Cell: The basic unit of any living organism that carries on the biochemical processes of life.

Centimorgan (cM): A unit of measure of recombination frequency. One centimorgan is equal to a 1 per cent chance that a marker at one genetic locus will be separated from a marker at a second locus due to crossing over in a single generation. In human beings, one centimorgan is equivalent, on average, to one million base pairs.

Chromatid: The two identical halves of a chromosome produced for cell division and meiosis.

Chromosome: Structures found in the nucleus of cells composed of DNA and proteins. Normally humans have 46 chromosomes in each cell, 23 from each parent. Of these, 22 are autosomes and 1 is a sex chromosome.

Clone: An exact copy made of biological material such as a DNA segment (*e.g.*, a gene or other region), a whole cell, or a complete organism.

Cloning: Using specialized DNA technology to produce multiple, exact copies of a single gene or other segment of DNA to obtain enough material for further study. This process, used by researchers in the Human Genome Project, is referred to as cloning DNA. The resulting cloned (copied) collections of DNA molecules are called clone libraries. A second type of cloning exploits the natural process of cell division to make many copies of an entire cell. The genetic makeup of these cloned cells, called a cell line, is identical to the original cell. A third type of cloning produces complete, genetically identical animals such as the famous Scottish sheep, Dolly.

Cloning vector: DNA molecule originating from a virus, a plasmid, or the cell of a higher organism into which another DNA fragment of appropriate size can be integrated without loss of the vector's capacity for self-replication; vectors introduce foreign DNA into host cells, where the DNA can be reproduced in large quantities. Examples are plasmids, cosmids, and yeast artificial chromosomes; vectors are often recombinant molecules containing DNA sequences from several sources.

Complementary sequence: Nucleic acid base sequence that can form a double-stranded structure with another DNA fragment by following base-pairing rules (A pairs with T and C with G). The complementary sequence to GTAC for example, is CATG.

Cosmid: Artificially constructed cloning vector containing the cos gene of phage lambda. Cosmids can be packaged in lambda phage particles for infection into *E. coli*; this permits cloning of larger DNA fragments (up to 45kb) than can be introduced into bacterial hosts in plasmid vectors.

Cytosine (C): A nitrogenous base, one member of the base pair GC (guanine and cytosine) in DNA

Clones: A group of cells derived from a single ancestor.

Cloning: The process of asexually producing a group of cells (clones), all genetically identical, from a single ancestor. In recombinant DNA technology, the use of DNA manipulation procedures to produce multiple copies of a single gene or segment of DNA is referred to as cloning DNA.

Cloning vector: DNA molecule originating from a virus, a plasmid, or the cell of a higher organism into which another DNA fragment of appropriate size can be integrated without loss of the vector's capacity for self-replication; vectors introduce foreign DNA into host cells, where it can be reproduced in large quantities. Examples are plasmids, cosmids, and yeast artificial chromosomes; vectors are often recombinant molecules containing DNA sequences from several sources.

Complementary DNA (cDNA): DNA that is synthesized from a messenger RNA template; the single-stranded form is often used as a probe in physical mapping.

Complementary sequences: Nucleic acid base sequences that can form a double-stranded structure by matching base pairs; the complementary sequence to G-T-A-C is C-A-T-G.

Conserved sequence: A base sequence in a DNA molecule (or an amino acid sequence in a protein) that has remained essentially unchanged throughout evolution.

Contig map: A map depicting the relative order of a linked library of small overlapping clones representing a complete chromosomal segment.

Contigs: Groups of clones representing overlapping regions of a genome.

Cosmid: Artificially constructed cloning vector containing the cos gene of phage lambda. Cosmids can be packaged in lambda phage particles for infection into E. coli; this permits cloning of larger DNA fragments (up to 45 kb) than can be introduced into bacterial hosts in plasmid vectors.

Crossing over: The breaking during meiosis of one maternal and one paternal chromosome, the exchange of corresponding sections of DNA, and the rejoining of the chromosomes. This process can result in an exchange of alleles between chromosomes. Compare recombination.

Cytosine (C): A nitrogenous base, one member of the base pair G-C (guanine and cytosine).

CAPS: Cap analysis gene expression it is also known as the CAGE which is the technique used in molecular biology to produce a snapshot of the 52 end of the messenger RNA population in a biological sample. The small fragments (usually 27 nucleotides long) from the very beginnings of mRNAs (5′ ends of capped transcripts) are extracted, reverse-transcribed to DNA, PCR amplified and sequenced.

Cell division: The mechanism by which cells multiply during the growth of tissues or organs. The type of cell division involved in the growth of the body is known as mitosis. The cell division which produces sperm or ova (an egg) in the testis or ovary is known as meiosis.

Centimorgan (cM): A unit of measure of recombination frequency. One centimorgan is equal to a 1 per cent chance that a marker at one genetic locus will be separated from a marker at a second locus due to crossing over in a single generation. In human beings, 1 centimorgan is equivalent, on average, to 1 million base pairs.

Central dogma: The central dogma of molecular biology explains the flow of genetic information, from DNA? to RNA?, to make a functional product, a protein.

Centromere: A region of a chromosome to which spindle traction fibers attach during mitosis and meiosis; the position of the centromere determines whether the chromosome is considered an acrocentric, metacentric or telomeric chromosome.

Centromere interference: Phenomenon in which the presence of one crossover interferes with the formation of another crossover nearby. Mathematically defined as 1 minus the coefficient of coincidence. Also called chiasma interference.

Character: A characteristic that results from gene action and is transmitted from one generation to another.

Chiasma (Plural, Chiasmata): A cross-shaped structure formed during crossing –over and visible during the diplonema stage of meiosis.

Chimera: In medicine, a person composed of two genetically distinct types of cells. Human chimeras were first discovered with the advent of blood typing.

Chloroplast: Organelles found in green plants in which photosynthesis occurs.

Chromatid: One of the two visibly distinct replicated copies of each chromosome that becomes visible between early prophase and metaphase of mitosis and is joined to its sister chromatid at their centromeres.

Chromatin: The DNA-protein complex that constitutes eukaryotic chromosomes and can exist in various degrees of folding or compaction.

Chromomere: One of the serially aligned beadlike granules of concentrated chromatin that constitutes a chromosome during the early phases of cell division.

Chromosome: In eukaryotic cells, a linear structure composed of a single DNA molecule complexed with protein. Each eukaryotic species has a characteristic number of chromosomes in the nucleus of its cells. Most prokaryotic cells contain a single, usually circular chromosome.

Chromosome jumping: A technique of isolating clones from a genomic library that are not contiguous by skipping a region between known points on the chromosome.

Chromosome walking: A method for analyzing long segments of DNA by sequentially identifying adjacent, overlapping clones in a genomic library.

Cis-dominant: Referring to a gene or DNA sequence that can control genes on the same DNA molecule but not on other DNA molecules.

Cistron: The term cistron is used to emphasize that genes exhibit a specific behavior in a cis-trans test;

distinct positions (or loci) within a genome are cistronic.

Cline: A systematic change in allelic frequencies within a continuous population distributed over a geographic region.

Clade: A clade is a group of organisms that consists of a common ancestor and all its lineal descendants, and represents a single "branch" on the "tree of life".

Cladogenesis: is an evolutionary splitting event where a parent species splits into two distinct species, forming aclade.

Clone: An exact copy made of biological material such as a DNA segment (*e.g.*, a gene or other region), a whole cell, or a complete organism.

Co-dominance: The condition in which an individual heterozygous for a gene exhibits the phenotypes of both homozygotes.

Codon: A group of three adjacent nucleotides in an mRNA molecule that specifies either one amino acid in a polypeptide chain or the termination of polypeptide synthesis.

Coefficient of coancestry: The probability fAB that two homologous genes, one from individual A and the other from individual B, are identical by decent, *i.e.*, are descended from the same ancestral gene. The complementary probability, 1-fAB, is the probability that these

Coefficient of inbreeding: The inbreeding coefficient of an individual is approximately half the relationship (R) between the two parents.

Coefficient of parentage: Measure for genetic distance of two individuals or varieties. Values range between 0 (= no common ancestor) to 1 (= same individual or variety).

Coefficient of relationship: The coefficient of relationship is a measure of the degree of consanguinity (or biological relationship) between two individuals.

Comparative mapping: Comparative genetic mapping studies reveal similarities and differences in gene content and gene order between genera belonging to different taxa

Complementary DNA: DNA copies made from RNA templates in a reaction catalyzed by the enzyme reverse transcriptase.

Complementary RNA: Single-stranded RNA whose base sequence is complementary to specific DNA sequences (*e.g.*, genes) or, more rarely, another single- stranded RNA

Complementary sequence: Nucleic acid base sequence that can form a double-stranded structure with another DNA fragment by following base-pairing rules (A pairs with T and C with G). The complementary sequence to GTAC for example, is CATG.

Complementation test: A test used to determine whether two independently isolated mutations that confer the same phenotype are located within the same gene or in two different genes also known as the cis-trans test.

Complete linkage: In genetics Complete Linkage is defined as the state in which two loci are so close together that alleles of these loci are virtually never separated by crossing over. The closer the physical location of two genes on the DNA, the less likely they are to be separated. In case of Drosophila and female Silkworms, where there is complete absence of recombinant types due to absence of crossing over

Compound heterozygote: The presence of two different mutant alleles at a particular gene locus, one on each chromosome of a pair.

Conjugation: In bacteria, process of unidirectional transfer of genetic material through direct cellular contact between a donor ("male") cell and a recipient ("female") cell.

Consanguinity: Genetic relationship. Consanguineous individuals have at least one common ancestor in the preceding few generations.

Consensus sequence: The series of nucleotides found most frequently at each position in a particular DNA sequence among different species.

Conserved sequence: Conserved sequences are similar or identical sequences which occur in DNA, and cause sequences in RNA, proteins and carbohydrates.

Contigs: Groups of clones representing overlapping regions of a genome.

Contig map: A map depicting the relative order of a linked library of small overlapping clones representing a complete chromosomal segment.

Contiguous gene syndrome: A disorder due to deletion of multiple gene loci that is adjacent to one another. Contiguous gene syndromes are characterized by multiple, apparently unrelated, clinical features caused by deletion of the multiple adjacent genes. Each of the individual genes within a contiguous region, when mutated, gives rise to a distinct feature.

Cosmid: Artificially constructed cloning vector containing the cos gene of phage lambda.

Cosmids can be packaged in lambda phage particles for infection into *E. coli;* this permits cloning of larger DNA fragments (up to 45kb) than can be introduced into bacterial hosts in plasmid vectors.

Crossing: In Crossing over the exchange of genetic material between two paired chromosome during meiosis.

Cytogenetics : the study of chromosomes

Cytosine: A nitrogenous base, one member of the base pair GC (guanine and cytosine) in DNA

Darwinian fitness: The relative reproductive ability of individuals with a particular genotype.

Deletion: The loss of a segment of the genetic material from a chromosome.

Diakinesis: The final stage in prophase I of meiosis during which the replicated chromosomes (bivalents) are most condensed, the nuclear envelope breaks down, and the spindle begins to form.

Dicentric: Dicentric chromosomes result from the abnormal fusion of two chromosome pieces, each of which includes a centromere.

Dihybrid cross: A cross between two individuals of the same genotype that is heterozygous for two pairs of alleles at two different loci (*e.g.*, Ss Yy × Ss Yy).

Dioecious: Referring to plant species in which individual plants possess either male or female sex organs. See also monoecious.

Diploid: A full set of genetic material, consisting of paired chromosomes one chromosome from each parental set. Most animal cells except the gametes have a diploid set of chromosomes. The diploid human genome has 46 chromosomes. Compare haploid.

Diplotene: The stage of the first meiotic prophase, following the pachytene, in which the two chromosomes in each bivalent begin to repel one another and a split, occurs between the chromosomes.

Disassortative mating: the reproductive pairing of individuals that have traits more dissimilar than would likely be the case if mating is at random

DNA: The molecule that encodes genetic information. DNA is a double-stranded molecule held together by weak bonds between base pairs of nucleotides. The four nucleotides in DNA contain the bases adenine (A), guanine (G), cytosine (C), and thymine (T). In nature, base pairs form only between A and T and between G and C; thus the base sequence of each single strand can be deduced from that of its partner.

DNA fingerprinting: A method employed to determine differences in amino acid sequences between related proteins; relies upon the presence of simple tandem-repetitive sequences that are scattered throughout the human genome.

DNA polymerase: Any enzyme that catalyzes the polymerization of deoxyribonucleotides into a DNA chain. All DNA polymerases synthesize DNA in the 5? to 3? direction.

DNA replication: The use of existing DNA as a template for the synthesis of new DNA strands. In humans and other eukaryotes, replication occurs in the cell nucleus.

DNA sequence: The relative order of base pairs, whether in a DNA fragment, gene, chromosome, or an entire genome.

DNA typing: Molecular analysis of DNA polymorphisms to identify individuals based on the unique characteristics of their DNA.

Dominant: A discrete portion of a protein with its own function. The combination of domains in a single protein determines its overall function.

Double helix: The twisted-ladder shape that two linear strands of DNA assume when complementary nucleotides on opposing strands bond together.

Duplication: Duplication of a sequence of DNA or section of chromosome.

Electrophoresis: A method of separating large molecules (such as DNA fragments or proteins) from a mixture of similar molecules. An electric current is passed through a medium containing the mixture, and each kind of molecule travels through the medium at a different rate, depending on its electrical charge and size. Agarose and acrylamide gels are the media commonly used for electrophoresis of proteins and nucleic acids.

Endo-nuclease: An enzyme that cleaves nucleotides sequentially from free ends of a linear nucleic acid substrate.

Enhancer: A set of gene regulatory elements in eukaryotic genomes that can act over distances up to thousands of base pairs upstream or downstream from a gene. Most enhancers bind activators and act to stimulate transcription.

Environmental variance: Component of the phenotypic variance for a trait that is due to any non genetic source of variation among individuals in a population. VE includes variation arising from general environmental effects, which permanently influence phenotype; special environmental effects, which temporarily influence phenotype; and family environmental effects, which are shared by family members.

Enzyme: A protein catalyst which is essential to the correct functioning of biochemical reactions.

Epigenetic: Making heritable changes in the way that a gene works or functions, without altering the sequence of the genetic code, or DNA. Ways in which this may be changed is via structural changes

Epistasis: This is a phenomenon that consists of the effect of one gene being dependent on the presence of one or more 'modifier genes' (genetic background).

Eugenics: The practice of trying to influence human heredity by encouraging the transmission of 'desirable' characteristics and discouraging the transmission of 'undesirable' ones.

Eukaryote: Cell or organism with membrane-bound, structurally discrete nucleus and other well-developed subcellular compartments. Eukaryotes include all organisms except viruses, bacteria, and blue-green algae. Compare prokaryote. See chromosomes.

Euploidy: It is the state of a cell or organism having the same number of each homologous chromosome, possibly excluding the sex-determining chromosomes.

Evolution: Genetic change in a population of organisms that occurs over time. The term is also frequently used to refer to the appearance of a new species.

Exogenous DNA: DNA originating outside an organism.

Exon: The part of the DNA message that is translated into a protein.

Exo-nuclease: An enzyme that cleaves nucleotides sequentially from free ends of a linear nucleic acid substrate.

Expressed gene: When the coded information contained in the gene is understood by the cells to produce a product such as a protein.

Expressed sequence tag: An expressed sequence tag or EST is a short sub-sequence of a cDNA sequence. ESTs may be used to identify gene transcripts, and are instrumental in gene discovery and in gene-sequence determination. The identification of ESTs has proceeded rapidly, with approximately 74.2 million ESTs now available in public database

Expressivity: The degree to which an inherited characteristic is expressed in a person. 'Variable expressivity' refers to the variation in expression and severity of particular characteristics or severity of a condition.

F_1 generation: the first offspring (or filial) generation. The next and subsequent generations are referred to as f2, f3, *etc.*

F_2 generation: The offspring that result from crossing F_1 individuals; the second filial generation.

Fecundity: In demography and biology, fecundity is the actual reproductive rate of an organism or population, measured by the number of gametes (eggs).

Fecundity selection: is the process by which differential reproductive success among individuals in a population is the result of phenotypic traits that contribute to the production of a higher number of offspring per reproductive episode.

Filial generation: A generation in a breeding experiment that is successive to a mating between parents of two distinctively different but usually relatively pure genotypes.

FISH (fluorescent in situ hybridization): A technique for detecting chromosomes carrying a particular DNA sequence by hybridizing them with a cloned DNA probe that is tagged with a fluorescent chemical. The chromosomes are treated to separate the double-stranded DNA into single strands, which base-pair with a probe whose sequence is complementary to a region of the chromosomal DNA.

Fitness: The relative reproductive ability of individuals with a particular genotype.

Fixed allele: A fixed allele is an allele that is the only variant that exists for that gene in all the population. A fixed allele is homozygous for all members of the population.

Flanking region: The DNA sequences extending on either side of a specific locus or gene. It could be at 5′ or 3′ end

Flowcytometry: Analysis of biological material by detection of the light-absorbing or fluorescing properties of cells or sub cellular fractions (*i.e.*, chromosomes) passing in a narrow stream through a laser beam. An absorbance or fluorescence profile of the sample is produced. Automated sorting devices, used to fractionate samples, sort successive droplets of the analyzed stream into different fractions depending on the fluorescence emitted by each droplet.

Flow karyotyping: Use of flow cytometry to analyze and/or separate chromosomes on the basis of their DNA content.

Fluorescence in situ hybridization: Fluorescence in situ hybridization (FISH) is a cytogenetic technique that uses fluorescent probes that bind to only those parts of the chromosome with a high degree of sequence complementarity.

Founder effect: The founder effect is the loss of genetic variation that occurs when a new population is established by a very small number of individuals from a larger population

Frame shift: Mutation is a genetic mutation caused by indels (insertions or deletions) of a number of nucleotides in a DNA sequence that is not divisible by three.

Frame shift mutation: A frame shift mutation is a genetic mutation caused by indels (insertions or deletions) of a number of nucleotides in a DNA sequence that is not divisible by three.

Gamete: Mature male or female reproductive cell (sperm or ovum) with a haploid set of chromosomes (23 for humans).

Gene: The basic unit of heredity; a segment of DNA which contains the information for a specific characteristic or function.

Gene expression: The process by which genes coded information is converted into the structures present and operating in the cell. Expressed genes include those that are transcribed into mRNA and then translated into protein and those that are transcribed into RNA but not translated into protein (*e.g.*, transfer and ribosomal RNAs).

Gene family: Groups of closely related genes that make similar products.

Gene flow: The transference of genes from one population to another, usually as a result of migration. The loss or addition of individuals can easily change the gene pool frequencies of both the recipient and donor populations–that is, they can evolve.

Gene mapping: Determining the relative locations of different genes on chromosomes.

Gene pool: All of the genes in all of the individuals in a breeding population. More precisely, it is the collective genotype of a population.

Gene product: The biochemical material, either RNA or protein, resulting from expression of a gene. The amount of gene product is used to measure how active a gene is; abnormal amounts can be correlated with disease-causing alleles.

Gene therapy: Insertion of normal DNA directly into cells to correct a genetic defect.

Genetic code: The sequence of nucleotides, coded in triplets (codons) along the mRNA, that determines the sequence of amino acids in protein synthesis. The DNA sequence of a gene can be used to predict the mRNA sequence, and the genetic code can in turn be used to predict the amino acid sequence.

Genetic counseling: Information and support provided by a specialist doctor, usually a geneticist, to parents who have known conditions in their families or who are concerned about the future possibility of genetically transmitted conditions.

Genetic drift: Evolution, or change in gene pool frequencies, resulting from random chance. Genetic drift occurs most rapidly in small populations. In large populations, random deviations in allele frequencies in one direction are more likely to be cancelled out by random changes in the opposite direction.

Genetic engineering: Laboratory techniques used to alter or manipulate the genetic make-up of cells or an organism by deliberately removing, changing or inserting individual genes.

Genetic engineering technologies: See recombinant DNA technologies.

Genetic mapping: Determination of the relative positions of genes on a chromosome and a measure of the distance between them.

Genetic variance: a phenotypic variance of a trait in a population attributed to genetic heterogeneity.

Genetics: The study of genes and inheritance patterns.

Genome: The complete set of genes carried by an individual or a cell.

Genome project: Research and technology development efforts aimed at mapping and sequencing some or all of the genome of human beings and other organisms.

Genomic library: A collection of clones made from a set of randomly generated overlapping DNA fragments representing the entire genome of an organism. Compare library, arrayed library.

Genotype: The genetic constitution of an individual

Germ line: The family of cells that divide to produce new germ cells

Germ line mosaicism: When the germ cells (sperm or egg cells) have a different genetic make-up to the cells in the rest of the body.

Guanine: One of the four basic building blocks (nucleotide bases) that makes up the genetic code (DNA). It is represented by the letter G.

Haploid: A single set of chromosomes (half the full set of genetic material), present in the egg and sperm cells of animals and in the egg and pollen cells of plants. Human beings have 23 chromosomes in their reproductive cells. Compare diploid.

Haploinsufficient: Describing a gene that can support the normal wild-type phenotype when present in only one copy (heterozygous condition) in a diploid cell. A haplosufficient gene exhibits complete dominance in genetic crosses.

Haplotype: A set of closely linked alleles on a chromosome that is normally inherited as a block.

Hardy-Weinberg equilibrium: The concept that both gene frequencies and genotype frequencies will remain constant from generation to generation in an infinitely large, interbreeding population in which mating is at random and there is no selection, migration or mutation.

Hereditary: The transfer of a gene from parent to child. In mothers, the gene is transferred via the DNA in the egg and in fathers the gene is transferred via the DNA in the sperm.

Heredity: The passing of genetic characteristics form parent to child.

Heritability: The degree to which a characteristic is determined by genetics or genes.

Heterochromatin: Chromatin that remains condensed throughout the cell cycle and is usually not transcribed. See also euchromatin.

Heteroduplex DNA: A region of double-stranded DNA with different sequence information on the two strands.

Heterokaryon: A cell containing two nuclei with different genotypes produced by fusing cells from different sources.

Homeobox: A 180-bp consensus sequence found in many genes that regulate development.

Homologous chromosome: Chromosomes that pair during meiosis; each homologue is a duplicate of one chromosome from each parent.

Homozygote: An individual with both identical alleles (versions of a single gene) at one locus (position).

Human Genome Initiative: Collective name for several projects begun in 1986 by DOE to (1) create an ordered set of DNA segments from known chromosomal locations, (2) develop new computational methods for analyzing genetic map and DNA sequence data, and (3) develop new techniques and instruments for detecting and analyzing DNA. This DOE initiative is now known as the Human Genome Program. The national effort, led by DOE and NIH, is known as the Human Genome Project.

Hybrid: Offspring that are the result of mating between two genetically different kinds of parents-the opposite of purebred.

Hybridization: The pairing of a single-stranded, labeled probe (usually DNA) to its complementary sequence.

Identity: Patterns of inheritance affected by whether the inheritance was from the mother or father.

Imprinting: The 'memory' held by a chromosome as to whether it was inherited from the mother or the father. The memory is chemically 'stamped' into the DNA and can result in chromosomes or the genes located on the chromosomes behaving differently, depending on the parent of origin.

In situ hybridization: Use of a DNA or RNA probe to detect the presence of the complementary DNA sequence in cloned bacterial or cultured eukaryotic cells.

In vitro: Outside a living organism.

Inbreeding: Preferential mating between close relatives.

Incomplete dominance: The condition in which neither of two alleles is completely dominant to the other, so that the heterozygote has a phenotype between the phenotypes of individuals homozygous for either allele involved. Also called partial dominance.

Informatics: The study of the application of computer and statistical techniques to the management of information. In genome projects, informatics includes the development of methods to search databases quickly, to analyse DNA sequence information, and to predict protein sequence and structure from DNA sequence data.

Inheritance: The passing of familial elements from one generation to the next.

Insertion: The addition of a piece of chromosomal material into a chromosome in a place where it is not normally found. This may result in a condition, because the genetic code may then be read or translated incorrectly.

Inter phase: The period in the cell cycle when DNA is replicated in the nucleus; followed by mitosis.

Intron: The DNA base sequences interrupting the protein-coding sequences of a gene; these sequences are transcribed into RNA but are cut out of the message before it is translated into protein. Compare exons.

Inversion: Occurs where a chromosome breaks in two and becomes reattached in reverse orientation. This may or may not affect gene function.

Karyotype: The term used to describe an individual's chromosomes that have been photographed through the microscope and then arranged according to a standard classification based on their group and size.

kb: A segment of DNA which is 1,000 base pairs in length.

Library: An unordered collection of clones (*i.e.*, cloned DNA from a particular organism), whose relationship to each other can be established by physical mapping. Compare genomic library, arrayed library.

Linkage: The proximity of two or more markers (*e.g.*, genes, RFLP markers) on a chromosome; the closer together the markers are, the lower the probability that they will be separated during DNA repair or replication processes (binary fission in prokaryotes, mitosis or meiosis in eukaryotes), and hence the greater the probability that they will be inherited together.

Linkage map: A map of the relative positions of genetic loci on a chromosome, determined on the basis of how often the loci are inherited together. Distance is measured in centimorgans (cM).

Localize: Determination of the original position (locus) of a gene or other marker on a chromosome.

Loci: The position on a chromosome of a gene or other chromosome marker; also, the DNA at that position. The use of locus is sometimes restricted to mean regions of DNA that are expressed.

Locus: The position on a chromosome of a gene or other chromosome marker; also, the DNA at that position. The use of locus is sometimes restricted to mean regions of DNA that are expressed. See gene

LOD score: logarithm of the odd score; a measure of the likelihood of two loci being within a measurable distance of each other.

Macro restriction map: Map depicting the order of and distance between sites at which restriction enzymes cleave chromosomes.

Marker: An identifiable physical location on a chromosome (*e.g.*, restriction enzyme cutting site, gene) whose inheritance can be monitored. Markers can be expressed regions of DNA (genes) or some segment of DNA with no known coding function but whose pattern of inheritance can be determined. See RFLP, restriction fragment length polymorphism.

Megabase: Unit of length for DNA fragments equal to 1 million nucleotides and roughly equal to 1 cM.

Meiosis: Two successive nuclear divisions of a diploid nucleus, following one DNA replication, that result in the formation of haploid gametes or of spores having one-half the genetic material of the original cell.

Messenger RNA: Class of RNA molecules that contain coded information specifying the amino acid sequences of proteins.

Metaphase: A stage in mitosis or meiosis during which the chromosomes are aligned along the equatorial plane of the cell.

Methylation: addition of a methyl group (-CH3) to DNA or RNA.

Mitochondria: Mitochondria are the cell's power sources, compartments within the cell that provide its energy. They contain their own DNA, called mitochondrial DNA, a very small amount of DNA inherited only from your mother.

Mitochondrial DNA: the mitochondrial genome consists of a circular DNA duplex, with 5 to 10 copies per organelle.

Mitosis: Cell division producing two genetically identical cells.

Monoclonal: A group of cells that are all identical copies of an original cell.

Monoecious: Referring to plant species in which individual plants possess either male or female sex organs and thus produce male and female gametes. Monoecious plants are capable of self –fertilization.

Monogenic: A characteristic which is due to the information contained in a single gene.

Monohybrid cross: A cross between two individuals that are both heterozygous for the same pair of alleles (*e.g.*, Aa x Aa). By extension, the term also refers to crosses involving the pure-breeding parents that differ with respect to the alleles of one locus (*e.g.*, AA x aa).

Monosomy: The total loss of one of a pair of chromosomes. This occurs, for example, in Turner Syndrome where one X chromosome is lost leaving a total of 45 chromosomes.

Mosaicism: Where a genetic or chromosomal abnormality does not occur in all body cells. Often related to X chromosome inactivation.

mRNA: messenger RNA; an RNA molecular that functions during translation to specify the sequence of amino acids in a nascent polypeptide.

Multiplexing: A sequencing approach that uses several pooled samples simultaneously, greatly increasing sequencing speed.

Mutation: Any heritable change in DNA sequence. Compare polymorphism.

Natural selection: Differential reproduction of individuals in a population resulting from differences in their genotypes.

Nitrogenous base: A nitrogen-containing purine or pyrimidine that, along with a pentose sugar and a phosphate, is one of the three parts of a nucleotide

Non-disjunction: Where the chromosome pairs fail to separate correctly in meiosis, resulting in sperm or egg cells which have missing or extra chromosomes *e.g.* if chromosome number 21 fails to separate in the formation of an egg (or sperm), one egg (or sperm) will contain an extra copy of chromosome 21 (24 chromosomes) while the other egg (or sperm) will contain only 22 chromosomes.

Nonsense mutation: a mutation in which a codon is changed to a stop codon, resulting in a truncated protein product.

Nuclease: An enzyme that catalyzes the degradation of a nucleic acid by breaking phosphodiester bonds

Nucleic acid: High-molecular-weight polynucleotide. The main nucleic acids in cells are DNA and RNA

Nucleolus: An organelle in the nucleus of eukaryotic cells that is the site of transcription of ribosomal RNA genes and assembly of the ribosomal subunits

Nucleoside: A purine or pyrimidine covalently linked to a sugar.

Nucleosome: The basic structural unit of eukaryotic chromatin, consisting of two molecules each of the four core histones (H2A, H2B, H3, and H4, the histone octamer), a single molecule of the linker histone H1, and about 180 bp of DNA.

Nucleotide: one of the monomeric units from which DNA or RNA polymers are constructed; consists of a purine or pyrimidine base, a pentose sugar and a phosphoric acid group.

Oncogene: A gene, one or more forms of which is associated with cancer. Many oncogenes are involved, directly or indirectly, in controlling the rate of cell growth.

Open reading frame: In a segment of DNA, a potential protein-coding sequence identified by a start codon in frame with a stop codon.

Operator: A short DNA region, adjacent to the promoter of a bacterial operon that binds repressor proteins responsible for controlling the rate of transcription of the operon.

Operon: In bacteria, a cluster of adjacent genes that share a common operator and promoter and are transcribed into a single mRNA. All the genes in an operon are regulated coordinately; that is, all are transcribed or none are transcribed.

Origin: A specific site on a DNA molecule at which the double helix denatures into single strands and replication is initiated.

Ovum: The female reproductive cell or egg which contains 23 chromosomes.

P1: Derived artificial chromosome

Parthenogenesis: The development of an individual from an egg without fertilization.

PCR: Polymerase chain reaction; a technique for copying the complementary strands of a target DNA molecule simultaneously for a series of cycles until the desired amount is obtained.

Pedigree: A diagram of the heredity of a particular trait through many generations of a family.

Penetrance: An "ll or none" reference to clinical expression of a mutant gene.

Phage: A virus for which the natural host is a bacterial cell.

Phenotype: The physical and/or biochemical characteristics of a person, an animal or other organism which are determined by their genetic make-up and/or environment.

Phenotypic variance: A specific site on a DNA molecule at which the double helix denatures into single strands and replication is initiated

Physical map : A map of the locations of identifiable landmarks on DNA (*e.g.*, restriction enzyme cutting sites, genes), regardless of inheritance. Distance is measured in base pairs. For the human genome, the lowest-resolution physical map is the banding patterns on the 24 different chromosomes; the highest-resolution map would be the complete nucleotide sequence of the chromosomes.

Plasmid: Autonomously replicating, extra chromosomal circular DNA molecules, distinct from the normal bacterial genome and nonessential for cell survival under nonselective conditions. Some plasmids are capable of integrating into the host genome. A number of artificially constructed plasmids are used as cloning vectors.

Pleiotropy: The phenomenon of variable phenotypes for a number of distinct and seemingly unrelated phenotypic effects.

Point mutation: A heritable alteration of the genetic material in which one base pair is changed to another.

Polygenic: A condition or characteristic that is caused by many different genes acting together.

Polygenic disorders: Genetic disorders resulting from the combined action of alleles of more than one gene (*e.g.*, heart disease, diabetes, and some cancers). Although such disorders are inherited, they depend on the simultaneous presence of several alleles; thus the hereditary patterns are usually more complex than those of single-gene disorders. Compare single-gene disorders.

Polymerase chain reaction: A method for amplifying a DNA base sequence using a heat-stable polymerase and two 20-base primers, one complementary to the (+)-strand at one end of the sequence to be amplified and the other complementary to the (-)-strand at the other end. Because the newly synthesized DNA strands can subsequently serve as additional templates for the same primer sequences, successive rounds of primer annealing, strand elongation and dissociation produce rapid and highly specific amplification of the desired sequence. PCR also can be used to detect the existence of the defined sequence in a DNA sample.

Polymorphism: Enzymes that catalyse the synthesis of nucleic acids on preexisting nucleic acid templates, assembling RNA from ribonucleotides or DNA from deoxyribonucleotides.

Polypeptide: A linear polymeric molecule consisting of amino acids joined by peptide bonds.

Polyploidy: Condition in which a cell or organism has more than two sets of chromosomes

Population: A specific group of individuals of the same species.

Positional cloning: The isolation of a gene associated with a genetic disease on the basis of its approximate chromosomal position.

Pribnow box: A part of the promoter sequence in prokaryotic genomes that is located at about 10 base pairs upstream from the transcription start site. Also called the 10 box.

Primer: Nucleotides used in the polymerase chain reaction to initiate DNA synthesis at a particular location.

Probability: The long term frequency of an event relative to all alternative events, and usually expressed as decimal fraction.

Probe: Single-stranded DNA or RNA molecules of specific base sequence, labeled either radioactively or immunologically, that are used to detect the complementary base sequence by hybridisation.

Prokaryote: Cell or organism lacking a membrane-bound, structurally discrete nucleus and other subcellular compartments. Bacteria are prokaryotes. Compare eukaryote. See chromosomes.

Promoter: A site on DNA to which RNA polymerase will bind and initiate transcription.

Protein: A large molecule composed of one or more chains of amino acids in a specific order; the order is determined by the base sequence of nucleotides in the gene coding for the protein. Proteins are required for the structure, function, and regulation of the bodys cells, tissues, and organs, and each protein has unique functions. Examples are hormones, enzymes, and antibodies.

Proteomics: The first stage in mitosis or meiosis during which the replicated chromosomes condense and become visible under the microscope.

Purine: A nitrogen-containing, single-ring, basic compound that occurs in nucleic acids. The purines in DNA and RNA are adenine and guanine.

Pyrimidine: A nitrogen-containing, double-ring, basic compound that occurs in nucleic acids. The pyrimidines in DNA are cytosine and thymine; in RNA, cytosine and uracil.

Quantitative trait: A heritable characteristic that shows a continuous variation in phenotype over a range. Also called continuous trait

Random mating: Matings between individuals of the same or different genotypes that occur in proportion to the frequencies of the genotypes in the population

Reading frame: Linear sequence of codons (groups of three nucleotides) in mRNA that specify amino acids during translation beginning at a particular start codon.

Recessive: Every cell contains two copies of each gene. Each gene contains the information for a particular gene product, such as a protein. If a gene is mutated, the gene no longer codes for the gene product. Where an individual has one gene copy or allele mutated and the other copy 'correct', the cell will only be producing half the amount of gene product. If this does not result in any condition for the individual, the mutation is described as being hidden or 'recessive' to the correct copy of the gene. An individual with this genetic constitution is said to be a 'carrier' of a recessive gene mutation. For a recessive gene mutation to result in a particular characteristic or a condition, both copies of the genes must be mutated.

Recombinant DNA molecule: A combination of DNA molecules of different origin that are joined using recombinant DNA technologies.

Recombinant DNA techniques: Procedures used to join together DNA segments in a cell-free system (an environment outside a cell or organism). Under appropriate conditions, a recombinant DNA molecule can enter a cell and replicate there, either autonomously or after it has become integrated into a cellular chromosome.

Recombination: The process by which progeny derive a combination of genes different from that of either parent. In higher organisms, this can occur by crossing over.

Regulatory region: A DNA base sequence that controls gene expression.

Regulatory sequence: A DNA base sequence that controls gene expression.

Repulsion: In individuals heterozygous for two genetic loci, the arrangement in which each homologous chromosome carries the wild-type allele of one gene and the mutant allele of the other gene; also called trans configuration.

Resolution: Degree of molecular detail on a physical map of DNA, ranging from low to high.

Restriction enzyme: Enzymes that can cut DNA into strands at specific places along its length.

Restriction enzyme cutting site: A specific nucleotide sequence of DNA at which a particular restriction enzyme cuts the DNA. Some sites occur frequently in DNA (*e.g.*, every several hundred base pairs), others much less frequently (rare-cutter; *e.g.*, every 10,000 base pairs).

Restriction fragment length polymorphism: Variation between individuals in DNA fragment sizes cut by specific restriction enzymes; polymorphic sequences that result in RFLPs are used as markers on both physical maps and genetic linkage maps. RFLPs are usually caused by mutation at a cutting site. See marker.

Restriction Mapping: Cedure for locating the relative positions of restriction enzyme cleavage sites in a cloned DNA fragment, yielding a restriction map of the fragment.

Restriction site linker: double-stranded oligodeoxyribonucleotide about 8 to 12 base pairs long that contains the cleavage site for a specific restriction enzyme and is used in cloning cDNAs.

Retrotransposon: A type of mobile genetic element, found only in eukaryotes, that encodes reverse transcriptase and moves in the genome via an RNA intermediate

Retrovirus: A virus with a single-stranded RNA genome that replicates via a double-stranded DNA intermediate produced by reverse transcriptase, an enzyme encoded in the viral genome. The DNA integrates into the host's chromosome where it can be transcribed.

Reverse transcriptase: An enzyme (an RNA-dependent DNA polymerase) that makes a double-stranded DNA copy of an RNA strand.

Ribonucleic acid: A chemical found in the nucleus and cytoplasm of cells; it plays an important role in protein synthesis and other chemical activities of the cell. The structure of RNA is similar to that of DNA. There are several classes of RNA molecules, including messenger RNA, transfer RNA, ribosomal RNA, and other small RNAs, each serving a different purpose.

Ribosomal RNA: A class of RNA found in the ribosomes of cells.

Ribosome: Small cellular components composed of specialised ribosomal RNA and protein; site of protein synthesis. See ribonucleic acid (RNA).

RNA: A chemical found in the nucleus and cytoplasm of cells; it plays an important role in protein synthesis and other chemical activities of the cell. The structure of RNA is similar to that of DNA. There are several classes of RNA molecules, including messenger RNA, transfer RNA, ribosomal RNA, and other small RNAs, each serving a different purpose.

RNA polymerase: Any enzyme that catalyzes the synthesis of RNA molecules from a DNA template in a process called transcription

rRNA: Class of RNA molecules of several different sizes that along with ribosomal proteins make up ribosomes of prokaryotes and eukaryotes

Selection: The favoring of particular combinations of genes in a given environment

Selfing: The union of male and female gametes from the same individual

Sequence tagged site: Short (200 to 500 base pairs) DNA sequence that has a single occurrence in the human genome and whose location and base sequence are known. Detectable by polymerase chain reaction, STSs are useful for localising and orienting the mapping and sequence data reported from many different laboratories and serve as landmarks on the developing physical map of the human genome. Expressed sequence tags (ESTs) are STSs derived from cDNAs.

Sequencing: Determination of the order of nucleotides (base sequences) in a DNA or RNA molecule or the order of amino acids in a protein.

Sex chromosome: The X and Y chromosomes in human beings that determine the sex of an individual. Females have two X chromosomes in diploid cells; males have an X and a Y chromosome. The sex chromosomes comprise the 23rd chromosome pair in a karyotype. Compare autosome.

Single nucleotide polymorphism: A DNA sequence variation that involves a change in a single nucleotide. Variations in the genetic code at the level of one nucleotide may be useful in certain applications such as assessing the patterns of inheritance via genetic linkage studies, or forensic DNA testing or DNA fingerprinting.

Single-gene disorder: Hereditary disorder caused by a mutant allele of a single gene (*e.g.*, Duchene muscular dystrophy, retinoblastoma, sickle cell disease). Compare polygenic disorders.

Somatic cell: Any cell in the body except gametes and their precursors.

Somatic hybrid: Hybrid cell line derived from two different species; contains a complete chromosomal complement of one species and a partial chromosomal complement of the other; human/hamster hybrids grow and divide, losing human chromosomes with each generation until they finally stabilize, the hybrid cell line established is then utilized to detect the presence of genes on the remaining human chromosome.

Southern blotting: Transfer by absorption of DNA fragments separated in electrophoretic gels to membrane filters for detection of specific base sequences by radiolabeled complementary probes.

Sperm: (abbreviation of spermatozoon) the male reproductive cell carrying 23 chromosomes.

Sperm cell: mature male gamete, produced by the testes in male animals. Also called spermatozoon (plural, spermatozoa).

Stop codon: One of three codons in mRNA for which no normal tRNA molecule exists and which signals the termination of polypeptide synthesis.

Tandem repeat sequence: Multiple copies of the same base sequence on a chromosome; used as a marker in physical mapping.

Telomerase: An enzyme that adds short, tandemly repeated DNA sequences (simple telomeric sequences) to the ends of eukaryotic chromosomes. It contains an RNA component complementary to the telomeric sequence and has reverse transcriptase activity

Telomere: The terminal or end segment of each chromosome arm.

Telophase: The stage in mitosis or meiosis during which the migration of the daughter chromosomes to the two poles is completed.

Terminator: A DNA sequence located at the distal (downstream) end of a gene that signals the termination of transcription.

Test-cross: A cross of an individual of unknown genotype, usually expressing the dominant phenotype, with a homozygous recessive individual to determine the unknown genotype.

Thymine: A pyrimidine found in DNA but not in RNA. In double-stranded DNA, thymine pairs with adenine, a purine, by hydrogen bonding.

Trait: Any detectable phenotypic property of an organism.

Transcription: the formation of an RNA molecule upon a DNA template by complementary base pairing.

Transduction: A process by which bacteriophages mediate the transfer of pieces of bacterial DNA from one bacterium (the donor) to another (the recipient).

Transfer RNA: A class of RNA having structures with triplet nucleotide sequences that are complementary to the triplet nucleotide coding sequences of mRNA. The role of tRNAs in protein synthesis is to bond with amino acids and transfer them to the ribosomes, where proteins are assembled according to the genetic code carried by mRNA.

Transformation: A process by which the genetic material carried by an individual cell is altered by incorporation of exogenous DNA into its genome.

Transgenic: A gene artificially introduced into the cell or into the genome of an individual.

Translation: The process in which the genetic code carried by mRNA directs the synthesis of proteins from amino acids. Compare transcription.

Translocation: Chromosomal rearrangement in which a piece of one chromosome is transferred to another one.

Transposable element: A DNA segment that can move from one position in the genome to another (non-homologous) position; also called mobile genetic element. Transposable elements are found in both prokaryotes and eukaryotes

Transposase: An enzyme encoded by many types of mobile genetic elements that catalyzes the movement (transposition) of these elements in the genome.

Transposition: The movement of a transposable element within the genome.

Transposon: A mobile genetic element that contains a gene for transposase, which catalyzes transposition, and genes with other functions such as antibiotic resistance

Transversion: A type of base pair substitution mutation that involves a change of a purine-pyrimidine base pair to a pyrimidine-purine base pair (*e.g.*, A-T to T-A or G-C to T-A) at a particular site in the DNA.

Trisomy: 3 copies of a particular chromosome (normally we have only 2.)

Uracil: A nitrogenous base normally found in RNA but not DNA uracil is capable of forming a base pair with adenine.

Variable-number tandem-repeat: A type of DNA polymorphism involving variation in the number of identical sequences (7bp to a few tens of base pairs in length) that are tandemly repeated at a particular locus in the genome. Also called a minisatellite.

Virus: A non-cellular biological entity that can reproduce only within a host cell. Viruses consist of nucleic acid covered by protein; some animal viruses are also surrounded by membrane. Inside the infected cell, the virus uses the synthetic capability of the host to produce progeny virus.

VNTR: Variable number tandem repeats; any gene whose alleles contain different numbers of tandemly repeated oligonucleotide sequences.

Western blotting: a technique used to identify a specific protein; the probe is a radioactively labeled antibody raised against the protein in question.

Wild-type: Term describing an allele or phenotype that is designated as the standard ("normal") for an organism and is usually, but not always, the most prevalent in a "wild" population of the organism; also used in reference to a strain or individual

Yeast artificial chromosome: A vector used to clone DNA fragments (up to 400 kb); it is constructed from the telomeric, centromeric, and replication origin sequences needed for replication in yeast cells. Compare cloning vector, cosmid.

Zygote: The single cell with 46 chromosomes resulting from the fertilisation of an egg (containing 23) by a sperm (containing 23). Through cell division (mitosis), the zygote develops into a multi cellular embryo and then into a fetus.

Multiple-Choice Questions

Q1 If four chromosomes synapse into a cross-shaped configuration during meiotic prophase, the organism is heterozygous for a

(a) Pericentric inversion

(b) Deletion

(c) Translocation

(d) Paracentric inversion

(e) None of the above

Q2 A segment of chromosome may be protected from recombination by

(a) An inversion

(b) A translocation

(c) Balanced lethals

(d) More than one of the above

(e) All of the above

Q3 Although sickle cell disease has negative effects on those who suffer from it, the allele is widespread in many parts of the world. This is because in areas where malaria is a significant danger, due to sickle cell allele?

(a) Ceases to cause symptoms.

(b) Attacks the parasite that causes malaria.

(c) Spreads rapidly in people weakened by malaria.

(d) Conveys a health advantage to those who carry the allele.

(e) None of the above

Q4 Even though sickle-cell anemia is usually fatal if found in homozygous individuals, the disease persists because?

(a) Gene therapy has alleviated the condition

(b) The disease is carried on a dominant allele

(c) Individuals with one allele for sickle-cell anemia are resistant to malaria

(d) A combination of all of the above

(e) None of the above

Q5 What name is given to the collection of traits exhibited by an organism?

(a) Holotype (b) Morphology

(c) Phenotype (d) Genotype

(e) Typology

Q6 A person with sickle cell trait, having one S allele and one normal, will be resistant to malaria and eventually develop sickle cell anemia.

(a) True (b) False

Q7 The term gene was coined by

(a) Darwin (T/F)

(b) Fransis Galton (T/F)

(c) William Bateson (T/F)

(d) Wihelm Johansson (T/F)

(e) Gregor John Mendel (T/F)

Q8 Which one of the following statements are true

(a) Medical Genetics by Peter Harper in 2008 dealt with the history of Medical Geentics (T/F)

(b) Principles and practices of Medical Genetics by Emery and Rimoin 5th Edition 2007 deal with the history of Medical Genetics. (T/F)

(c) Eugenics was not proposed by early greek physicians (T/F)

(d) Theory of Pangenesis was proposed by Athenian philosopher (500-428 BC) (T/F)

(e) Comprehensive theory of inheritance was proposed by Aristotle (T/F)

Q9 Which one of the following statements are true

(a) Mendal used quantitative traits for the three laws of inheritance (T/F)

(b) Sutton suggested the pairing of homologous chromosomes (T/F)

(c) Thomas Hunt Morgan used fruit fly (***Drosophila Melangaster*** to explain X linked inheritance) (T/F)

(d) Mutations are always inherited (T/F)

(e) Frequency of recombination is same for all gene combinations (T/F)

Q10 What is the blending theory of inheritance?

(a) Mendel's theory of how the traits of parents are passed to offspring through the gametes

(b) Darwin's theory of how traits are passed from all parts of the parent's body into the gamete to be transmitted to the offspring

(c) The modern theory of how genetic information is passed from parents to offspring

(d) An old theory that said that offspring show traits intermediate between those of the parents

(e) The idea that some traits prevent others from being expressed

Q11 The most easily recognized characteristic of an inversion heterozygote in plants is

(a) Gigantism (b) Semisterility

(c) A cross-shaped chromosome configuration during meiosis

(d) Pseudodominance

(e) None of the above

Q12 A mechanism that can cause a gene to move from one linkage group to another is

(a) Translocation (b) Inversion

(c) Crossing over (d) Duplication

(e) Dosage compensation

Q13 A person with Klinefelter syndrome is considered to be

(a) Monosomic

(b) Triploid

(c) Trisomic

(d) Deletion heterozygote

(e) None of the above

Q14 If the garden pea has 14 chromosomes in its diploid complement, how many double trisomics could theoretically exist?

(a) 6 (b) 9

(c) 16 (d) 21

(e) None of the above.

Q15 Which characteristic of pea plants was NOT important in their selection as Mendel's research organism?

(a) Most other scientists of the time were also using peas, so a lot was known about them.

(b) Peas are easy to cultivate.

(c) Pea plants have a short generation time.

(d) Pea plants are self-pollinating but can be cross-fertilized easily.

(e) Many true-breeding varieties were available.

Q16 In a Mendelian monohybrid cross, which generation is always completely homozygous?

(a) F_1 generation

(b) F_2 generation

(c) F_3 generation

(d) P generation

(e) None of these

Q17 The symbol "F" in the results of a testcross stands for

(a) Dominant

(b) Recessive

(c) First trait to show up

(d) "Faulty" or unexpected results.

(e) Filial

Q18 The F1 offspring of a monohybrid cross would show the genotype(s)

(a) AA and Aa (b) Aa and aa

(c) AA, Aa, and aa (d) AA only

(e) Aa only

Q19 The offspring of a monohybrid testcross would show the genotype(s)

(a) AA and Aa (b) Aa and aa.

(c) AA, Aa, and aa (d) AA only

(e) aa only.

Q20 Z-DNA have a

(a) Double helical nature

(b) Zig-Zag apperarance

(c) Uracil base

(d) Single stranded nature

(e) All of the above

Q21 The type of sugar in DNA are

(a) Triose (b) Tetrose

(c) Pentose (d) Hexose

(e) None of the above

Q22 The length of one turn of DNA is

(a) 3.4 Ao (b) 34 Ao

(c) 20 Ao (d) 3.04 Ao

(e) None of the above

Q23 A short length of DNA molecule has 80 thymine and 80 guanine bases. The total number of nucleotide in the DNA fragment is

(a) 160 (b) 40

(c) 320 (d) 640

(e) None of the above

Q24 In the experiments of Griffith, the conversion of non-lethal R-strain bacteria to lethal S-strain bacteria:

(a) Was the result of genetic mutation.

(b) Was an example of the genetic exchange known as transformation.

(c) Supported the case for proteins as the genetic material.

(d) Could not be reproduced by other researchers.

(e) Was an example of conjugation.

Q25 Where is translation accomplished?

(a) sER (b) Lysosomes

(c) Peroxisomes (d) Ribosomes

(e) Nucleoli

Q26 Which of the following enzymes is responsible for RNA synthesis?

(a) RNA ligase (b) RNA gyrase

(c) RNA polymerase (d) RNase

(e) RNA helicase

Q27 Peptide bonds form between ________

(a) An mRNA codon and a tRNA anticodon

(b) Amino acids

(c) An mRNA transcript and the small ribosomal subunit

(d) A tRNA and the amino acid it is carrying

(e) The small ribosomal subunit and the large ribosomal subunit

Q28 A mutation within a gene that will insert a premature stop codon in mRNA would ________

(a) Result in a polypeptide that is one amino acid shorter than the one produced prior to the mutation

(b) Result in an amino acid substitution

(c) Result in a shortened polypeptide chain

(d) Alter the reading frame

(e) Alter the location at which transcription of the next gene begins

Q29 The process by which messenger RNA is synthesized by complementary base pairing of ribo-nucleotides with deoxyribo-nucleotides to match a section of DNA (a gene) is called:

(a) Translation

(b) Replication

(c) Transcription

(d) All of the above

(e) None of above

Q30 Small nuclear ribonucleoprotien particles (snRNPs), such as U1, U2 and U6, are involve with which of the following processes?

(a) Initiation of transcription

(b) Initiation of DNA replication

(c) Recruitment of RNA polymerase to the mRNA template

(d) Splicing

(e) Translation

Q31 The essential transcriptional binding factor that binds directly to the TATA box initiating RNA polymerase II transcription in mammalian cells is

(a) TFIIE (b) TFIIS

(c) TFIIB (d) TFIID

(e) All of the above

Q32 Which type of processing is involved causing the apolipoprotein B-100 expression in the liver but apolipoprotein B-48 expression in the intestine? This change in tissue expression results from a C-to-U conversion by deamination?

(a) RNA editing

(b) Polyadenylation

(c) Protein splicing

(d) RNA splicing

(e) 5′ methyl cap

Q33 The Triplet code of CAT in DNA is represented as ________ in mRNA and ________ in tRNA.

(a) GAA, CAT (b) CAT, CAT

(c) GUA, CAU (d) GTA, CAU

(e) None of the above

Q34 In the genetic code, the codon UUU and AUC both code for the amino acid phenylalanine. What does this indicate about genetic code.

(a) The genetic code is ambiguous.

(b) The genetic code is redundant.

(c) All of the above

(d) None of above

Q35 The codons which do not specify an amino acid are called

(a) Initiation code

(b) Termination code

(c) Propagation code

(d) None of the above.

(e) All of the above

Q36 In prokaryotes, AUG encode S

(a) Methoinine

(b) N-formyl methionine

(c) A stop codon

(d) Alanine

(e) None of the above

Q37 A codon bias

(a) Is used for genome mapping

(b) Is found in intergenie regions

(c) Is found in functional RNA

(d) Is used to identify genes.

(e) None of the above

Q31 Genetic code is

(a) Universal

(b) Universal except for rare exception in mitochondria.

(c) Species specific

(d) Kingdom specific

(e) None of the above

Q39 Which of the following is not an effect of a mutation?

(a) Prevents a protein from forming

(b) Lowers the amount of a protein

(c) Adds a function to a protein

(d) Any of the above can occur

(e) None of the above

Q40 Two healthy people have a child expressing a genetic condition caused by a dominant allele. What can you conclude about this situation?

(a) Infidelity- one parent had to contribute the disease allele

(b) The parents are not free from the disorder-one must be affected

(c) The mutation arose spontaneously in the child

(d) The child is not the biological child of the couple described

(e) None of the above

Q41 DNA is repaired by ________ in which a pyrimidine dimer is split.

(a) Excision repair

(b) Photoreactivation

(c) Mismatch repair

(d) Proofreading repair

(e) None of the above

Q42 Which of the following genetic diseases results from a mutation that prevents a protein from forming?

(a) Duchenne muscular dystrophy

(b) Hemophilia A

(c) Epidermolysisbullosa

(d) Huntington disease

(e) All of the above

Q43 A point mutation that changes a codon specifying an amino acid into a stop codon is called a

(a) Missense mutation

(b) Nonsense mutation

(c) Frameshift mutation

(d) Deletion mutation

(e) None of the above

Q44 A plant of genotype C/C; d/d is crossed to c/c; D/D and an F1 testcrossed to c/c ; d/d. If the genes are linked, and 20 map units apart, the percentage of c/c ; d/d recombinants will be

(a) 10 (b) 20

(c) 25 (d) 50

(e) 75

Q45 A ________ mutation originates during meiosis while a ________ mutation originates during mitosis.

(a) Germinal, somatic

(b) Germinal, spontaneous

(c) Somatic, germinal

(d) Spontaneous, point

(e) All of the above

Q46 What is the chromosome aberration of the type ABCD · EFFGH called? (Note: the "dot" denotes the centromere.)

(a) Inversion (b) Translocation

(c) Deletion (d) Duplication

(e) Dicentric

Q47 What is the mode of inheritance for Oculodentodigital Dysplasia?

(a) Autosomal dominant

(b) Autosomal recessive

(c) Arises from novel mutation

(d) Both a and c are correct

(e) Both b and c are correct

Q48 What is the physical basis of mutational hotspots?

(a) Transposons (b) Tautomers

(c) Palindromes (d) Transitions

(e) None of the above

Q49 Northern blots probe ________

(a) DNA and RNA (b) DNA

(c) RNA and proteins (d) Proteins

(e) RNA

Q50 In the purple penguin, a series of alleles occurs at the p locus on an autosome. All alleles affect the color of feathers: pd = dark-purple, pm = medium-purple, pl = light-purple, and pvl = very pale purple (almost white). The order of dominance is pd> pm >pl>pvl. If a light-purple female, heterozygous for very pale purple, is crossed to a dark-purple male, heterozygous for medium-purple, the ratio of phenotypes expected among the baby penguins would be

(a) 2 dark:1 medium:1 light.

(b) 1 dark:1 medium.

(c) 1 dark:1 medium:1 light:1 very pale.

(d) 1 medium:1 light.

(e) 2 dark:1 light:1 very pale.

Q51 The polymerase chain reaction is a technique that ________

(a) Selectively replicates RNA

(b) Probes and selectively replicates DNA

(c) Selectively replicates DNA

(d) Probes DNA

(e) None of the above

Q52 Restriction enzymes are involved in all of the following genetic engineering techniques except ________

(a) Cloning DNA into vectors

(b) Mapping studies

(c) Identification of genetic markers

(d) PCR

(e) None of the above

Q53 A "probe" in molecular biology is ________

(a) An instrument used to manipulate cells in culture

(b) A type of vector system

(c) A DNA or an RNA molecule used in hybridization reactions

(d) Probes are not used in molecular biology.

(e) None of the above

Q54 Shotgun cloning differs from the clone-by-clone method in which of the following ways?

(a) The location of the clone being sequenced is known relative to other clones within the genomic library in shotgun cloning.

(b) Genetic markers are used to identify clones in shotgun cloning.

(c) Computer software assembles the clones in the clone-by-clone method.

(d) No genetic or physical maps of the genome are needed to begin shotgun cloning.

(e) None of the above

Q55 A "YAC" is a useful ________

(a) Vector (b) Screen

(c) Host (d) Library

(e) All of the above

Q56 Why didn't Mendel find linkage?

(a) Some genes were linked, but they were too close together to cross over.

(b) All seven genes were on separate chromosomes.

(c) Mendel did detect linkage. He discovered this genetic phenomenon.

(d) He didn't execute his crosses correctly, so linkage went undetected.

(e) Some genes were linked, but they were too far apart for crossing over to be distinguished from independent assortment, or linked genes were never tested for at the same time in the same cross.

Q57 Chromosome walking is a technique used to ________

(a) Move chromosomes around the nucleus

(b) Move a fragment of chromosomal DNA from one area of a chromosome to another. This is a casual term for the work that genetic engineers do.

(c) Recombination between chromosomal DNA of two different species

(d) A method used to locate a gene using a set of clones from a DNA library

(e) None of the above

Q58 Ancient DNA can sometimes be reconstructed through what technique?

(a) Ancient DNA cannot be recovered

(b) Monosomy

(c) Polymerase chain reaction (PCR)

(d) Meiosis

(e) None of the above

Q59 The mutations blistered (bs) and clot (cl) are located on opposite ends of chromosome II in Drosophila. A blistered female is crossed with a clot male. The F1 female is then backcrossed to a blistered clot male. What ratio will be expected in the offspring of the backcross?

(a) 9/16 wild: 3/16 blistered: 3/16 clot: 1/16 blistered, clot

(b) 1/2 blistered: 1/2 clot

(c) 2/24 wild: 1/4 blistered: 1/4 clot

(d) 1/4 wild: 1/4 clot: 1/4 blistered: 1/4 blistered, clot

(e) 4/10 blistered: 4/10 clot: 1/10 wild: 1/10 blistered, clot

Q60 If you have three genes (A, B, and C) linked on the same chromosome, how can you determine the gene order?

(a) Look for double crossover phenotypes involving the wild type and mutant alleles of genes A, B, and C.

(b) Look for single crossover phenotypes involving the wild type and mutant alleles of genes A, B, and C.

(c) Look for parental phenotypes.

(d) You cannot determine gene order by looking at the results of a cross.

(e) None of the above

Q61 Which is NOT correct regarding nondisjunction?

(a) Occurs more frequently in human females over age 35

(b) May fail to separate maternal from paternal chromatids

(c) May fail to separate maternal chromatids from one another or paternal chromatids from one another

(d) May lead to Down syndrome

(e) Separates maternal from paternal chromatids

Q62 Genes A and B are farther apart than are genes A and C, and all three are linked. What CANNOT be concluded?

(a) B might be between A and C.

(b) C might be between A and B.

(c) A might be between B and C.

(d) More crossovers will occur between A and B than between A and C.

(e) All four statements may be true.

Q63 Nonshared environment

(a) Can not be measured in MZ twins

(b) Is essentially the same thing as the family environment

(c) Is not limited to measures of the family environment

(d) Is primarily important in the behavioral development of infants

(e) None of the above

Q64 Although schizophrenia runs in families, the particular subtype (catatonic, paranoid, disorganized) does not. This means that

(a) If you have the genes, you will be schizophrenic

(b) If you have the genes, you will be schizophrenic, but environmental factors will determine your subtype of schizophrenia

(c) Schizophrenia is not actually a single disorder

(d) As a quantitative disorder, milder forms of schizophrenia will be more heritable because the abnormal is normal

(e) None of the above

Q65 Which of the following is NOT directly related to linkage groups?

(a) A group of genes found on the same chromosome

(b) Members of the same family expressing the same trait

(c) LOD scores

(d) The number of haploid chromosomes in an organism's genome

(e) None of the above

Q66 During meiosis, chromosome pairs can exchange pieces, resulting in new combinations of genes and DNA sequences. This result is known as

(a) Linkage

(b) Crossing over

(c) Recombination

(d) Mutation

(e) None of the above

Q67 Some have argued that environmental influences that affect behavioral development operate on a family-by-family basis. However, it is probably better to view environmental influences as operating on an individual-by-individual basis. This means

(a) Environmental effects are going to be relatively specific to each child, rather than general for all children in a family

(b) The concept of shared environment is antiquated and should be eliminated from behavioral genetics

(c) That family experiences are unimportant in child development

(d) That two children growing up in the same family will be more alike than children growing up in different families

(e) None of the above

Q68 Behavioural genetics work on alcoholism is interesting because it is one of the

(a) Few behaviors that people falling within the autism spectrum disorder continuum do not seem to be susceptible to

(b) Few behaviors that is almost entirely under heritable control

(c) Few types of addiction for which earlier onset is actually less heritable than late onset

(d) Few behaviors showing evidence for shared environment that is shared by siblings but not by parents and offspring

(e) None of the above

Q69 Single-locus polymorphisms can account for

(a) Much of the heritable component of obesity

(b) The regulatory mechanisms of fat storage in the body

(c) Relatively few of the cases of obesity in the population

(d) Ghrelin's role in appetite control

(e) None of the above

Q70 Positional cloning efforts benefit from the human genome project because it has already identified:

(a) Numerous useful DNA markers

(b) The function of every human gene

(c) The function of every human protein

(d) The sequences of all normal and mutant alleles

(e) None of the above

Q71 Mycoplasma genitalium is the smallest microorganism known to be able to reproduce:

(a) True (b) False

Q72 Passive genotype-environment correlations involve interactions between ________ and are typically most significant in ________ because they have ________ to modify their environment.

(a) Genetic relatives, adolescents, increasing ability

(b) Genetic relatives, children, limited ability

(c) Anyone or anything, adults, high ability

(d) Anyone or anything, children, no ability

(e) Non relatives, children, low capacity

Q73 The first step in the positional cloning of a disease gene is to:

(a) Identify a cDNA encoding the gene product

(b) Map the location of a human disease gene

(c) Clone (or copy) the gene

(d) Identify the protein's amino acid sequence to isolate the gene

(e) Cross recessive plants to yield pure breeding strains

Q74 Which genetic map is derived from restriction length polymorphisms (RFLPs) and is used to distinguish genes tens of kb apart?

(a) Cytogenetic map (b) Linkage map

(c) Physical map (d) Sequence map

(e) None of the above

Q75 According to the genomic analysis of Mycoplasma genitalium, what are the minimal number of genes required for life?

(a) 48,000 (b) 480

(c) 265 - 350 (d) 300 Mb

(e) 3

Q76 Which of these is the scientific name for the fruit fly?

(a) Saccharomyces cerevisiae

(b) Caenorhabditiselegans

(c) Drosophila melanogaster

(d) Physcomitrella patens

(e) None of these

Q77 Promoter regions are nucleotide sequences that:

(a) Are involved in the initiation of transcription

(b) Are involved in transcription termination

(c) Contain the code for 1mRNA molecule

(d) Are important to the translation process

(e) None of the above

Q78 Which of these would be considered the simplest organism with a nucleus?

(a) Saccharomyces cerevisiae

(b) Caenorhabditiselegans

(c) Drosophila melanogaster

(d) Physcomitrella patens

(e) None of these

Q79 The size of the E.coli genome is

(a) 4640 bp

(b) 4.64 Kbp

(c) 4.64 Mbp

(d) Not known with certainty

(e) None of the above

Q80 To identify genes that are actively expressed, researchers isolate:

(a) DNA (b) Protein

(c) MRNA (d) CDNA

(e) Amino acids

Q81 Which of the following involve negative control of transcription through a repressor protein:

(a) Enzyme repression

(b) Transcriptional activator

(c) Alternative sigma factor

(d) a and b

(e) None of the above

Q82 The anticodon is in:

(a) DNA (b) mRNA

(c) tRNA (d) rRNA

(e) None of the above

Q83 On of the great mysteries of embryo development has been how to differentiate gene expression in the first cell. Modern biology has shown that.

(a) Gravity affects which part of the fist cell will become the head or the tail.

(b) A concentration gradient of cytoplasmic material is determined by the mother and this affects gene regulation in the developing egg.

(c) Development of the first set of cells is random, then position effects the organogenesis and following development

(d) This statement is false, we have no idea how embriogenesis occurs.

(e) None of the above

Q84 Control of expression of heat shock genes involves :

(a) An alternative sigma factor

(b) A helix-turn-helix DNA binding protein

(c) The response regulator of a two component regulator system

(d) None of the above

(e) All of the above

Q85 The proper sequence after fertilization is

(a) Fertilized egg, gastrula, blastula, mordula

(b) Fertilized egg,, blastula, mordula, gastrula

(c) Blastula, mordula, gastrula fertilized egg.

(d) None of the above

(e) All of the above

Q86 Mechanism of differentiation depends up on ________ organization of unfertilized egg.

(a) Homogenous (b) Heterogeneous

(c) Exogenous (d) All A, B and C

(e) None of the above

Q87 The process that triggers development is

(a) Morphogenesis

(b) Fertilization

(c) Acrosomal reaction

(d) Cortical reaction

(e) None of the above

Q88 The control development of cap in Acetabularia is through production of developmentally active substance in

(a) Nucleus (b) Cytoplasm

(c) Gray crescent (d) Both A and B

(e) All of these

Q89 A tube of tissue formed by a thickening and rolling up of the neural plate during embryonic neurulation. It will later form the brain and spinal cord of the animal. This is called

(a) Neurocoel (b) Neural groove

(c) Neurospore (d) Neural tube

(e) None of the above

Q90 The study of degenerative changes in aging is called

(a) Developmental biology (b) Paedology

(c) Gerontology (d) Choronology

(e) All of these

Q91 The science of studying and treating malformations and monstrosities of organisms is called

(a) Gerontology (b) Teratology

(c) Dermatology (d) Etiology

(e) All of these

Q92 Which of the following supports the hypothesis that says "gene duplication is essential to evolution"?

(a) Gene families

(b) Intercalary deletions

(c) Inversions

(d) Fragile sites

(e) Genetic anticipation

Q93 Mutations in termination codons can cause gene elongation.

(a) True (b) False

Q94 According to the endosymbiotic theory, mitochondrial genes are derived through the compartmentalization of nuclear genes.

(a) True (b) False

Q95 Character data can always be converted into distance data.

(a) True (b) False

Q96 Additive distances are always ultrametric.

(a) True (b) False

Q97 In a rooted ultrametric tree with 4 OTUs (A, B, C, D), the distance between the root and A is equal to the distance between the root and C.

(a) True (b) False

Q98 In a rooted additive tree with 4 OTUs (A, B, C, D), the distance between the root and A is equal to the distance between the root and C.

(a) True (b) False

Q99 The Maximum Parsimony method of phylogenetic reconstruction uses distance matrices.

(a) True (b) False

Q100 UPGMA produces ultrametric trees.

(a) True (b) False

Q101 Which of the following is not a source of variation in a population?

(a) Inherited genetic differences.

(b) Differences due to health.

(c) Differences due to age.

(d) Differences due to accident.

(e) None of the above.

Q102 Why is genetic variation important from an evolutionary standpoint?

(a) If all organisms were the same, the entire population would be vulnerable to particular pathogens, like viruses.

(b) All evolutionary adaptations (*e.g.* the origin of forelimbs) are the result of the gradual build up of genetic differences between organisms over geologic time.

(c) Evolution (at the population level) refers to changes in the frequencies of genes in the population over time.

(d) All of the above.

(e) None of the above.

Q103 Which of the following is an example of environmental variation?

(a) Apu is a tongue roller, but his brother Sanjay is not.

(b) Marge dies her hair blue.

(c) Homer inherited baldness from his father's side of the family.

(d) Patti and Selma have hanging ear lobes.

(e) Bart is a boy because he has both an X and a Y chromosome.

Q104 What's the difference between genetic drift and change due to natural selection?

(a) Genetic drift does not require the presence of variation.

(b) Genetic drift does not involve competition between members of a species.

(c) Genetic drift never occurs in nature, natural selection does.

(d) There is no difference.

(e) None of the above.

Q105 If the theory of natural selection is the survival of the fitness, and the fittest are identified as those who survive, why isn't it regarded as a tautology (a statement that is true only because of the meaning of the terms)?

(a) The effect of traits on the fitness of an organism can be assessed independently of whether the organism indeed survives.

(b) It is regarded as a tautology - the question is based on a false assumption.

(c) Natural selection is true.

(d) There may be some statements in science that are useful even if they are not falsifiable or refutable in principle.

(e) A and D.

Q106 How was Mendel's work ultimately reconciled with Darwin's theory of natural selection during the evolutionary synthesis in the 1930s and 1940s?

(a) Scientists recognized that once one thinks about species as populations, rather than individuals, there is no incompatibility between them.

(b) Mendel's theory was replaced by the mutation theory.

(c) It was recognized much of the variation we observe in nature is due to recombination, rather than mutation.

(d) A and C.

(e) None of the above.

Q107 What is the relationship between the wing of a bird and the wing of a bat?

(a) They are homologous because they represent modified forms of a trait present in a common ancestor (forelimbs).

(b) They are analogous because while each carries out the same function (flight), this trait has arisen independently as a result of convergence (*i.e.* the common ancestor of both did not have a forelimb that allowed it to fly).

(c) A and B.

(d) They represent derived homologies.

(e) None of the above.

Q108 Which of the following is an example of an ancestral homology?

(a) Almost all modern reptiles, birds and mammals have forelimbs, a trait they also share with contemporary amphibians.

(b) The first birds and all their descendant species have feathers, a trait that is unknown in any other group.

(c) Humans and many insect species have eyes.

(d) All of the above.

(e) None of the above.

Q109 Which of the following are difficult to explain in terms of natural selection?

(a) Male peacocks evolve tail feathers that would appear to make them more rather than less vulnerable to predators.

(b) Male deer evolve antlers that are not used to defend themselves against predators.

(c) A bird issues a warning cry that puts it at greater risk of being noticed by a predator.

(d) Some traits appear to have no adaptive value.

(e) All of the above.

Q110 Which of the following are kingdoms?

(a) Monera (b) Protista

(c) Animalae (d) Plantae

(e) All of the above.

Q111 Why is similarity per se misleading when it comes to inferring evolutionary relationships?

(a) Organisms that look alike may be very distantly related to one another.

(b) Similarities between two species may be due to common descent, without indicating how closely the two are related to one another.

(c) A and B only.

(d) The presence of a shared derived character state is often misleading when it comes to inferring relationships between species.

(e) A, B, and D.

Q112 How might an evolutionary biologist explain why a species of birds has evolved a larger beak size?

(a) Large beak size occurred as a result of mutation in each member of the population.

(b) The ancestors of this bird species encountered a tree with larger than average sized seeds. They needed to develop larger beaks in order to eat the larger seeds, and over time, they adapted to meet this need.

(c) Some members of the ancestral population had larger beaks than others. If larger beak size was advantageous, they would be more likely to survive and reproduce. As such, large beaked birds increased in frequency relative to small beaked birds.

(d) The ancestors of this bird species encountered a tree with larger than average sized seeds. They discovered that by stretching their beaks, the beaks would get longer, and this increase was passed on to their offspring. Over time, the bird beaks became larger.

(e) None of the above.

Q113 Which of the following is the most fit in an evolutionary sense?

(a) A lion who is successful at capturing prey but has no cubs.

(b) A lion who has many cubs, eight of which live to adulthood.

(c) A lion who overcomes a disease and lives to have three cubs.

(d) A lion who cares for his cubs, two of who live to adulthood.

(e) A lion who has a harem of many lionesses and one cub.

Q114 How is extinction represented in a tree diagram?

(a) A branch splits.

(b) A branch ends.

(c) A branch shifts along the X axis.

(d) A branch shifts along the Y axis.

(e) E. None of the above.

Q115 Snapdragon flowers may be red (CrCr), pink (CrCW), or white (CWCW). A sample from a population of these plants contained 80 white, 100 pink, and 20 red-flowered plants. A chi-square test of the sample data against the Hardy-Weinberg expectations produces a chi-square value of

(a) 1.96 (b) 2.43

(c) 2.87 (d) 3.02

(e) 3.11

Q116 Black pelage is an autosomal dominant trait in guinea pigs; white is the alternative recessive trait. A Hardy-Weinberg population was sampled and found to contain 336 black and 64white individuals. The frequency of the dominant black gene is estimated to be

(a) 0.60 (b) 0.81

(c) 0.50 (d) 0.89

(e) None of the above

Q117 Black pelage is an autosomal dominant trait in guinea pigs; white is the alternative recessive trait. A Hardy-Weinberg population was sampled and found to contain 336 black and 64white individuals. The probability that a black male crossed to a white female would produce a white offspring is approximately

(a) 0.12 (b) 0.14
(c) 0.16 (d) 0.18
(e) None of the above (0.286)

Q118 Yellow body color in Drosophila is governed by a sex-linked recessive gene; wild-type color is produced by its dominant allele. If the frequency of the yellow allele is 0.01, the percentage of wild-type females expected to carry the yellow allele is

(a) 1.98 (b) 1.67
(c) 2.04 (d) 2.76
(e) None of the above

Q119 A malignant epithelial cell neoplasm derived from any of the three(e) germ layers is referred to as:

(a) Mixed cell tumor (b) Carcinoma
(c) Teratoma (d) Sarcoma
(e) Adenoma

Q120 A benign epithelial cell neoplasm derived from non-glandular surfaces is referred to as:

(a) Papilloma (b) Sarcoma
(c) Adenoma (d) Hamartoma
(e) Squamous cell carcinoma

Q121 Each of the following is an anaplastic change except:

(a) Pleomorphism and hyperchromatism
(b) Increased mitosis and abnormal mitotic figures
(c) Nuclei that vary in shape and size
(d) Presence of undifferentiated cells
(e) Presence of abundant chromatin in cytoplasmic organelles

Q122 Which of the following is least likely to be used as a means of distinguishing a benign from a malignant neoplasm?

(a) Degree of cellular differentiation
(b) Rate of growth
(c) Type and amount of necrosis
(d) Evidence of metastasis
(e) Mode of spread

Q123 Which one of the following features is more characteristic of a benign than a malignant neoplasm?

(a) Grows by expansion and implantation occurs frequently
(b) Metastasizes if the brain is the site of origin
(c) Usually non-encapsulated and necrosis seldom occurs
(d) Tend to recur after surgical removal
(e) Usually occur singly and do not recur after surgical removal

Q124 What causes the effects of albinism?

(a) The lack of pigment production
(b) An extra chromosome 21
(c) The presence of two different codominant alleles
(d) Inability produce normal connective tissue
(e) The environment interacting with the genotype

Q125 Which of the following diseases is most prevalent in the Jewish community?

(a) Neurofibromatosis
(b) Tay-Sachs disease
(c) Phenylketonuria
(d) Albinism
(e) Cystic fibrosis

Q126 Assume your parents have blood types A and B respectively, but you do not know if they are homozygous or heterozygous. At the next family reunion, you get out a notepad and draw a pedigree chart. Whose blood types would provide possible clues to your parents' genotypes?

(a) Your grandparents
(b) Those aunts and uncles that are your parents siblings, especially if your grandparents are not alive
(c) Your brothers and sisters
(d) You
(e) All of the choices are correct.

Q127 Which disease results in deformed red blood cells, poor circulation, and anemia?

(a) Achondroplasia
(b) Sickle-cell disease
(c) Huntington disease
(d) Hemophilia
(e) Phenylketonuria

Q128 In a pedigree chart, which is correct?

(a) Circles = males; squares = females.

(b) A line between a circle and a square represents a mating.

(c) A carrier with a normal phenotype is represented by a black circle or square.

(d) Offspring are represented by triangles.

(e) All of these are true.

Q129 Gene therapy is

(a) Method aim to cure genetic disorders

(b) Method to provide correct version of the defective gene

(c) Method to replace a defective gene with a healthy gene

(d) All of the above

Q130 The introduction of remedial gene to bone marrow cells comes under

(a) Germ line therapy

(b) Somatic cell therapy

(c) Both a and b

(d) Corrective gene therapy

(e) All of the above

Q131 Introduction of healthy gene into cells, tissues or organs cultured in vitro and reimplanting back to the patient is referred as

(a) Germ line therapy

(b) Somatic cell therapy

(c) Ex vivo therapy

(d) In vivo therapy

(e) All of the above

Q132 Embryo therapy was devised by Handyside *et al.*, to cure

(a) Cystic fibrosis

(b) Hemophilia

(c) Thalassemia

(d) Severe combined immunodeficiency disease

(e) All of the above

Q133 The common gene delivery system for In vivo gene therapy is

(a) Micro injection

(b) Lipofecction

(c) Adeno viral vectors

(d) Electroporation

(e) All of the above

Q134 CpG islands and codon bias are tools used in eukaryotic genomics to ________

(a) Identify open reading frames

(b) Differentiate between eukaryotic and prokaryotic DNA sequences

(c) Find regulatory sequences

(d) Look for DNA-binding domains

(e) Identify a gene's function

Q135 Gene duplication has been found to be one of the major reasons for genome expansion in eukaryotes. In general, what would be the selective advantage of gene duplication?

(a) If one gene copy is nonfunctional, a backup is available.

(b) Larger genomes are more resistant to spontaneous mutations.

(c) Duplicated genes will make more of the protein product.

(d) Gene duplication will lead to new species evolution.

(e) All of the above

Q136 Two-dimensional gels are used to ________

(a) Separate DNA fragments

(b) Separate RNA fragments

(c) Separate different proteins

(d) Observe a protein in two dimensions

(e) Separate DNA from RNA

Q137 A protein is poorly expressed in a diseased tissue. To determine whether the defect is at the level of transcription or translation, which of the following blotting techniques would you use?

(a) Southern and Western

(b) Southern and Northern

(c) Northern and Western

(d) Western and South –Western

(e) All of the above

Q138 Genome wise gene expression analysis is performed using.

(a) DNA microarrays

(b) Northern analysis

(c) Real time PCR

(d) RT-PCR

(e) All of the above

Q139 Phenylketonuria is

(a) An eating disorder

(b) A rare inherited disease that is treated through diet

(c) Caused by an accident after you were born

(d) All of the above

(e) None of the above

Q140 Pseudodominance may be observed in heterozygotes for

(a) A deletion

(b) A duplication

(c) A paracentric inversion

(d) A reciprocal translocation

(e) More than one of the above

Q141 Given a normal chromosome with segments labeled C123456 (C = centromere), a homologue containing an inversion including regions 3–5′ and a single two-strand crossover between regions 4 and 5; then the acentric fragment present during first meiotic anaphase is

(a) 63456 (b) 12344321

(c) 65521 (d) 654321

(e) None of the above

Q142 If a pea plant shows a recessive phenotype,

(a) It can be either TT or Tt.

(b) It can be either Tt or tt.

(c) It can be only TT.

(d) It can be only tt.

(e) It can be TT, Tt, or tt.

Q143 The basic repeating units of a DNA molecule is

(a) Nucleoside (b) Nucleotide

(c) Histones (d) Aminoacids

(e) None of the above

Q144 Left handed DNA

(a) A-DNA (b) B-DNA

(c) Z-DNA (d) C-DNA

(e) None of the above

Q145 Transcription is the ________

(a) Modification of a strand of RNA prior to the manufacture of a protein

(b) Manufacture of a strand of RNA complementary to a strand of DNA

(c) Manufacture of a new strand of DNA complementary to an old strand of DNA

(d) Manufacture of two new DNA double helices that are identical to an old DNA double helix

(e) Manufacture of a protein based on information carried by RNA

Q146 The DNA codon AGT codes for an amino acid carried by a tRNA with the anticodon

(a) AGU (b) AGT

(c) TCU (d) TCA

(e) UCA

Q147 The expressed (coding) regions of eukaryotic genes are called ________

(a) Exons (b) Introns

(c) Promoters (d) Caps

(e) Tails

Q148 What type of bond joins the 7-methylguanosine to the mRNA?

(a) 3′-> 3′ linkage (b) 5′-> 2′ linkage

(c) 3′-> 5′ linkage (d) 5′-> 5′ linkage

(e) 5′-> 3′ linkage

Q149 Which of the following is not a eukaryotic regulatory sequeence element found upstream of the mRNA initiation site?

(a) Enhancers (b) GC box

(c) CAAT box (d) TATA box

(e) a-helix

Q150 AUG codes for a methionine act as a

(a) Initiation code

(b) Elongation code

(c) Termination code

(d) Prolongation code

(e) None of the above

Q151 A ________ mutation is intentionally caused.

(a) Spontaneous (b) Somatic

(c) Site-directed (d) Conditional

(e) None of the above

Q152 A deletion on the q arm of what chromosome causes Oculodentodigital Dysplasia.

(a) 1 (b) 9

(c) 6 (d) 20

(e) 17

Q153 In Drosophila, the two genes w and sn are X-linked and 25 map units apart. A female fly of genotype w+sn+/w sn is crossed to a male from a wild-type line. What per cent of male progeny will be w+sn?

(a) 0
(b) 12.5
(c) 25
(d) 37.5
(e) 50

Q154 Recombinant DNA research is dependent on ________

(a) Cloning
(b) Hosts
(c) Restriction endonucleases
(d) All of the above are correct.
(e) None of the above

Q155 None of the genes and signaling molecules in flies and mice found to be involved in learning and memory are specific to learning processes, but are involved in basic cell functions. This means

(a) One probably can not extrapolate these non-human animal studies to human learning and memory
(b) That only modern neuroimaging techniques are going to be capable of elucidating the genetics of learning and memory
(c) That because basic cellular function is very different for different species, it is going to be hard to make cross-species comparisons of the genes and signaling molecules.
(d) That learning and memory likely arises out of complicated interacting networks of neuronal systems.
(e) None of the above

Q156 In terms of understanding the pathways between genes and behavior, it is fairly safe to say that

(a) We know more about the environment than the genes
(b) We know more about the genes than the environment
(c) The new field of molecular genetics is the best way to gain a full understanding of the gene/environment interactions in the pathways
(d) QTL analysis has actually hindered the understanding of these pathways by suggesting so many separate gene contributions to basic behaviour processes.
(e) None of the above

Q157 One surprise from comparative genome projects is that:ÿ

(a) All species examined so far have the same genes
(b) We do not know the function of a great many genes
(c) Small organisms can have large genomes
(d) Large organisms can have small genomes
(e) Genomes seem to be contracting over time

Q158 During transcription:

(a) Nucleotides are polymerized by DNA polymerase
(b) Initiation occurs at a site recognized by the sigma factor
(c) Only single gene-sized mRNA molecules are synthesized
(d) Both DNA strands of a single gene are used as templates simultaneously
(e) None of the above

Q159 There are a number of methods that animals used in the development and Morphogenesis of the embryo. These must function properly or the embryo will die. Which of the following are properly matched?

(a) Movement – improper movement results in the switching of the left and right side
(b) Morphogenesis – improper morphogenesis results in a cleft palate
(c) Programmed cell death – improper morphogenesis results in extra structures liked webbed fingers
(d) Folding - improper folding results in positional switching of arms and legs
(e) All of the above

Q160 The situation where one embryonic tissue influences an other so that the responding tissue differentiates is known as

(a) Instruction
(b) Evocation
(c) Induction
(d) None of these
(e) All of these

Q161 Which of the following examples of variation is not important from an evolutionary standpoint?

(a) Genetic differences between individual organisms comprising the population.
(b) Inherited differences between individual organisms comprising the population.

(c) Differences due to diet, health, age or accident that have no affect on an individual's ability to survive and reproduce.

(d) A and B.

(e) None of the above.

Q162 According to our reading, how did Georges Cuvier account for extinctions in nature?

(a) Extinctions never occur—there are unexplored parts of the globe where organisms that appear to have gone extinct may still live.

(b) Extinctions occur when the slow adaptation of organisms over time to their environment is not quick enough to help them respond to changing conditions.

(c) Extinctions occur at random, they do not reflect God's will.

(d) Extinctions are due to catastrophic events.

(e) All of the above.

Q163 Which of the following is not an example of a macro evolutionary process?

(a) One lion species splits to form two lion species over geological time.

(b) The same trait evolves independently in two different taxa (*e.g.* wings in birds and in insects).

(c) As a result of their activities, humans drive Dodos (a bird species) extinct.

(d) Over a short period of time, the frequency of a single gene declines from 10 to 8 per cent.

(e) All of the above.

Q164 Which of the following is not an example of micro evolutionary change?

(a) The dark form of many moth species has increased in areas darkened by pollution.

(b) Penicillin resistant forms of bacteria have arisen since the introduction of antibiotics.

(c) The proportion of left and right bending moths in cichlid fish remains roughly 50:50.

(d) The last American eagle dies off, leading to the extinction of the species.

(e) All of the above.

Q165 Which of the following must increase over geological time according to evolutionary biologists?

(a) Size

(b) Complexity

(c) Speed of evolutionary processes such as mutation.

(d) Social interactions.

(e) None of the above.

Q166 A biologist is trying to infer how five closely related species of snakes are related to one another. She notices that some of the snakes have forked tongues and others do not. Which of the following would help her distinguish the ancestral state?

(a) She looks among snake fossils for evidence that being forked is a characteristic of the ancestor of this group, but determines no such fossils exist.

(b) She locates a specimen of a more distantly related snake to see if it has a forked tongue.

(c) She looks at a representative mammal species to see if it has a forked tongue.

(d) She flips a coin.

(e) None of the above.

Q167 Snapdragon flowers may be red (CrCr), pink (CrCW), or white (CWCW). A sample from a population of these plants contained 80 white, 100 pink, and 20 red-flowered plants. The percentage of pink-flowered plants expected on the basis of the Hardy-Weinberg equation is approximately

(a) 35 (b) 45

(c) 50 (d) 55

(e) None of the above

Q168 Yellow body color in Drosophila is governed by a sex-linked recessive gene; wild-type color is produced by its dominant allele. A sample from a Hardy-Weinberg population contained 1021 wild-type males, 997 wild-type females, and 3 yellow males. The percentage of the gene pool represented by the yellow allele is estimated to be

(a) 0.04 (b) 0.16

(c) 0.21 (d) 0.42

(e) None of the above (0.29)

Q169 Which one of the following neoplasms is highly invasive but seldom spread by metastisis?

(a) Papillomas of the skin

(b) Squamous cell carcinomas of the skin

(c) Adenocarcinomas of the lungs

(d) Basal cell carcinomas of the skin

(e) Osteogenic sarcomas of the limbs

Q170 The polymerase chain reaction (PCR)

(a) Is used to transcribe specific genes

(b) Amplifies specific DNA sequences

(c) Uses a DNA polymerase that denatures at 550c

(d) Is a method for sequencing DNA

(e) None of the above

Q171 If the remedial gene does the function of defective gene. The approach is called as

(a) Gene replacement therapy

(b) Gene augmentation therapy

(c) Both a and b

(d) Corrective gene therapy

(e) All of the above

Q172 Introduction of healthy gene at specific sites to displace the defective gene is referred as

(a) Germ line therapy

(b) Somatic cell therapy

(c) Both a and b

(d) Corrective gene therapy

(e) None of the above

Q173 How are so many different antibodies produced from fewer than 300 major genes?

(a) Gene duplication

(b) Alternative splicing mechanisms

(c) The formation of polyproteins

(d) The formation of nonspecific b cells

(e) Recombination, deletions, and random assortment of DNA segments

Q174 Dietary treatment of PKU consists of:

(a) Simply eliminating animal products from the diet

(b) A special formula and measured amounts of low protein foods

(c) Taking vitamin/mineral supplements

(d) All of the above

(e) None of the above

Q175 A treatment often used to induce polyploidy experimentally in plants is

(a) X-rays

(b) Gibberellic acid

(c) Colchicine

(d) Acridine dyes

(e) Azothioprene

Q176 Which of the following crosses would always result in offspring that only display the dominant phenotype?

(a) TT x tt

(b) Tt x Tt

(c) TT x TT

(d) Tt x Tt

(e) Both TT x tt and TT x TT

Q177 The F2 offspring of a monohybrid cross would show the genotype(s)

(a) AA and Aa

(b) Aa and aa.

(c) AA, Aa, and aa

(d) AA only.

(e) Aa only.

Q178 Adjacent nucleotides are joined by

(a) Covalent bond

(b) Phosphodiester bond

(c) Ionic bond

(d) Peptide bond

(e) None of the above

Q179 What aspect of Mendel's background gave him the necessary tools to discover the laws of inheritance?

(a) He was a monk.

(b) He was a teacher.

(c) He lived in Austria.

(d) He had studied mathematics and probability.

(e) He corresponded with Charles Darwin.

Q180 How many nucleotides make up a codon?

(a) Two

(b) Four

(c) One

(d) Three

(e) Five

Q181 Failure of an endonuclease to recognize the sequence AAUAAA in the 3′ end of heterogeneous nuclear RNA will cause a defect in which of the following processes involving mRNA?

(a) Initiation of transcription

(b) Splicing

(c) Polyadenylation

(d) Hybridization

(e) Capping

Q182 Introns within primary eukaryotic transcripts ________

(a) Are highly conserved in nucleotide sequence

(b) Vary considerably in size and number among different genes

(c) Often function as exons in other genes
(d) Are joined to form mature mRNA
(e) None of the above

Q183 Translation begins
(a) At the replication fork
(b) On the lagging strand
(c) At the start codon
(d) In nucleus
(e) None of the above

Q184 Which of the following is not a property of genetic code?
(a) Non overlapping
(b) Almost universal
(c) Four stop codons
(d) Redundant
(e) None of the above

Q185 Which of the following genetic diseases results from a mutation that lowers the amount of protein produced?
(a) Duchenne muscular dystrophy
(b) Hemophilia A
(c) Epidermolysisbullosa
(d) Huntington disease
(e) All of the above

Q186 In Drosophila, the genes R and S are linked. Flies of genotype Rs/Rs and rS/rS are crossed and an F1 obtained. The F1 allele arrangement is called
(a) Recombinant.
(b) Complementary.
(c) Coupling (cis).
(d) Repulsion (trans).
(e) Crossover.

Q187 The chromosome constitution of an allotetraploid can be represented by
(a) n1 + n2 (b) 2n1 + 2n2
(c) 2n1 (d) 2n2
(e) n1 + 2n2

Q188 Given what is known about single, double, and noncrossovers in a three-point map, triple crossovers in a four-point map can be predicted to be ________
(a) More frequent than single crossovers.
(b) More frequent than noncrossovers.
(c) More frequent than double crossovers.
(d) Less frequent than single crossovers, but more frequent than double crossovers.
(e) Less frequent than all of the above.

Q189 LOD scores are used to predict ________
(a) Crossover frequency
(b) Gene sequence
(c) Gene linkage
(d) The number of chromosomes in a genome
(e) The number of genes involved in the expression of a given trait

Q190 The proofreading of newly synthesized DNA, to excise incorrect nucleotides which have been inserted, is done by:
(a) A restriction endonucleases
(b) DNA gyrase
(c) DNA ligase
(d) DNA polymerase III
(e) None of the above

Q191 The codon is found in :
(a) DNA (b) rRNA
(c) tRNA (d) mRNA
(e) None of the above

Q192 In which medium would the level of an enzyme of arginine biosynthesis be the lowest
(a) Glucose +salts
(b) Lactose +salts
(c) Glucose +salts + tryptophan
(d) Arginine+salts
(e) All of the above

Q193 The technique of producing a genetically identical copy of an organism by replacing the nucleus of an unfertilized ovum with the nucleus of a body cell from the organism is
(a) Test tube baby (b) Cloning
(c) In vitro fertilization (d) All A, B and C
(e) All of these

Q194 Gene duplication has been found to be one of the major reasons for genome expansion in eukaryotes. In general, what would be the selective advantage of gene duplication?
(a) If one gene copy is nonfunctional, a backup is available.
(b) Larger genomes are more resistant to spontaneous mutations.

(c) Duplicated genes will make more of the protein product.

(d) Gene duplication will lead to new species evolution.

(e) None of the above

Q195 Which of the following is an example of genetic variation?

(a) Two children have different eye colors.

(b) One person is older than another.

(c) One person has a scar, but her friend does not.

(d) Todd eats meat, but his brother Rod is a vegetarian.

(e) None of the above.

Q196 The variation natural selection operates on is due to random mutations. What does this imply about natural selection?

(a) Natural selection is also a random process.

(b) Natural selection is nevertheless a directed process- the likelihood one variant will be favored in a given environment over another is predictable, even if the origin is not.

(c) There is no possibility God could be involved in this process.

(d) A, B and C.

(e) None of the above.

Q197 Which of the following is not an example of a monophyletic taxon?

(a) The first fish species and every living organism that looks like a fish.

(b) The first mammal species and all its descendants.

(c) The first bird species and all its descendants.

(d) All of the above.

(e) None of the above

Q198 Which of the following are the most distantly related to one another?

(a) Sunfish and dolphins.

(b) Tree frogs and snakes.

(c) Vampire bats and birds.

(d) Bears and whales.

(e) Robins and turtles.

Q199 Snapdragon flowers may be red (CrCr), pink (CrCW), or white (CWCW). A sample from a population of these plants contained 80 white, 100 pink, and 20 red-flowered plants. Refer to Table 2-3 to answer this question. Assuming the sample is representative of its population in problem 3 above, it may be said that

(a) The chi-square test is significant and the sampled population is not in genetic equilibrium,

(b) The chi-square test is non-significant and the sampled population is not in genetic equilibrium

(c) The chi-square test is significant and the sampled population is in genetic equilibrium

(d) The chi-square test is non significant and the sampled population is in genetic equilibrium the chi-square value is significant, thereby invalidating the test.

(e) None of the above

Q200 Yellow body color in Drosophila is governed by a sex-linked recessive gene; wild-type color is produced by its dominant allele. If the frequency of the yellow allele is 1.0 in females and 0 in males, the frequency of that allele in males of the next generation is expected to be

(a) 1.0

(b) 0.5

(c) 0.33

(d) 0.67

(e) None of the above

Q201 Which one of the following is not considered to be a distinctive pattern of non-neoplastic growth?

(a) Regeneration
(b) Hypertrophy
(c) Hyperplasia'
(d) Anaplasia
(e) Metaplasia

Q202 Which of the following is least likely to be the primary cause of neoplastic tranformation in animals?

(a) RNA oncogenic viruses

(b) Radiation

(c) Chemical carcinogens

(d) Immunologic reactions mediated by IgE

(e) DNA oncogenic viruses

Q203 The most common lethal genetic disease among Caucasians is ________

(a) Neurofibromatosis

(b) Tay-Sachs disease

(c) Phenylketonuria

(d) Albinism

(e) Cystic fibrosis.

Q204 Which provides protection against malaria in the heterozygote?

(a) Pattern baldness

(b) Sickle-cell disease

(c) Huntington disease

(d) Hemophilia

(e) Duchenne muscular dystrophy

Q205 The possibility of introducing correct version of the defective gene in all cells of the individual is achieved by ________

(a) Germ line therapy

(b) Somatic cell therapy

(c) Gene augmentation therapy

(d) Corrective gene therapy

Q206 If you were using a proteomics approach to find the cause of a muscle disorder, which of the following techniques might you be using?

(a) Creating a genomic library

(b) Sequencing the gene responsible for the disorder

(c) Developing physical maps from genomic clones

(d) Determining which environmental factors influence the expression of your gene of interest

(e) Annotating the gene sequence

Q207 What would be a likely explanation for the existence of pseudogenes?

(a) Gene duplication

(b) Gene duplication and mutation events

(c) Mutation events

(d) Unequal crossing over

(e) Evolutionary pressure

Q208 If during synapsis a certain kind of abnormal chromosome is always forced to bulge away from its normal homologue, the abnormality is classified as ________

(a) An inversion

(b) A duplication

(c) An isochromosome

(d) A deficiency

(e) None of the above

Q209 In a Mendelian monohybrid cross, which generation is always completely heterozygous?

(a) F1 generation (b) F2 generation

(c) F3 generation (d) P generation

(e) All of these are possible

Q210 The width of DNA molecule is ________

(a) 15 Ao (b) 3.4 Ao

(c) 20 Ao (d) 25 Ao

(e) None of the above

Q211 The length of DNA having 23 base pairs is ________

(a) 78 Ao (b) 78.4 Ao

(c) 78.2 Ao (d) 74.8 Ao

(e) All of the above

Q212 How many amino acids are common to all living systems?

(a) 30 (b) 10

(c) 15 (d) 20

(e) 25

Q213 If a strand of DNA has the sequence AAGCTC, transcription will result in a (n) ________

(a) Single RNA strand with the sequence UUCGAG

(b) Single RNA strand with the sequence TTCGAG

(c) RNA double helix with the sequence UUCGAG for one strand and AAGCUC for the complimentary strand

(d) DNA double helix with the sequence AAGCTC for one strand and TTCGAG for the complimentary strand

(e) Single DNA strand with the sequence TTCGAG

Q214 During the initiation of protein synthesis in eukaryotic cells eIF4F, eIF4A and eIF4B associate with which of the following?

(a) The 5′ cap on the mRNA

(b) The Shine-Delgarno sequence

(c) The 10S rRNA molecule in the small ribosomal subunit

(d) The f-met-initiator tRNA molecule

(e) The poly-A tail on the mRNA

Q215 Which of the following is true regarding the post transcriptional processing of tRNAs?

(a) Chemical modification of certain bases, such as those that create unusual bases such as inosine

(b) Replacement of the 3′ nucleotides with GGA

(c) Addition of a 3′-methyl cap

(d) Addition of a 5′ leader sequence

(e) Addition of a 3′ poly-A tail

Q216 How many different codons are possible?

(a) 3

(b) 20

(c) 64

(d) An indefinite number

(e) None of the above

Q217 Fragile-X syndrome and Huntington disease are caused by a (an)

(a) Tandem duplication

(b) Fusion gene

(c) Expanding triplet repeat

(d) Deletion

(e) None of the above

Q218 A plant of genotype C/C; d/d is crossed to c/c; D/D and an F1 test crossed to c/c; d/d. If the genes are unlinked, the percentage of c/c ; d/d recombinants will be ________

(a) 10 (b) 20

(c) 25 (d) 50

(e) 75

Q219 In a certain breed of dog, the alleles B and b determine black and brown coats respectively. However, the allele Q of a gene on a separate chromosome is epistatic to the B and b color alleles resulting in a gray coat (q has no effect on color). If animals of genotype B/b ; Q/q are intercrossed, what phenotypic ratio is expected in the progeny?

(a) 9 gray, 3 brown, 4 black

(b) 1 black, 2 gray, 1 brown

(c) 9 black, 6 brown, 1 gray

(d) 9 black, 4 gray, 3 brown

(e) 12 gray, 3 black, 1 brown

Q220 Which is the most specific recombinant DNA library?

(a) Genomic

(b) Protein

(c) cDNA

(d) Chromosomal

(e) cRNA

Q221 Interference in genetics is ________

(a) The masking of one gene's phenotype, due to the expression of another gene

(b) The phenomenon observed when one recombination event on a chromosome inhibits other closely linked recombination events

(c) Used in "mapping the centromere"

(d) Used in synteny testing

(e) None of the above

Q222 Your textbook argues that broadly speaking with respect to personality, the decreasing order of effect/control on a trait is:

(a) Genes, unique environment, shared environment

(b) Unique environment, genes, shared environment

(c) Genes, shared environment, unique environment

(d) Unique environment, shared environment, genes

(e) None of the above

Q223 Findings from behavioral genetics suggest that the traditional Behaviorist approach that assumes that offspring resemble their parents because parents provide the family environment for their offspring, and that siblings resemble each other because they share that family environment is

(a) Mostly correct

(b) Correct for parent-offspring similarities, but not for similarities/differences between siblings

(c) Correct for MZ twin behavioral traits, but wrong for dz twin behavioral traits

(d) Wrong for many behavioral traits

(e) None of the above

Q224 DNA sequencing can reveal:

(a) Mutations that do not alter phenotype

(b) Mutations that do not alter genotype

(c) Which tissues will express a gene

(d) How many genes a person has

(e) Which protein is misfolded

Q225 What is the anticodon that recognizes CGA:

(a) UGC (b) CGA

(c) GCU (d) GCT

(e) None of the above

Q226 ________ is the process of selection of activation of some genes by a cell, which are not activated by other cells of the embryo.

(a) Cell induction

(b) Cell transformation

(c) Cell differentiation

(d) Cell mediation

(e) None of the above

Q227 The normal process of development is disturbed by abnormalities. These abnormalities are due to

(a) Abnormal functioning of glands

(b) Abnormal Chromosomal number

(c) UV radiations

(d) All of these

(e) None of these

Q228 What's the difference between natural selection and sexual selection?

(a) Sexual selection occurs during sex.

(b) Natural selection is a type of sexual selection.

(c) Sexual selection is a type of natural selection.

(d) Sexual selection occurs within demes, natural selection does not.

(e) None of the above.

Q229 Which of the following is evidence for Darwin's theory of common descent?

(a) There are patterns in the fossil record that suggest other species have diverged from a single ancestor species.

(b) There are biogeographic patterns in the distribution of species, for instance distinct bird species on an island tend to resemble one another, suggesting a common ancestor.

(c) There are common stages in the early embryological development of organisms representing several distinct vertebrate groups.

(d) Anatomical structures, such as forelimbs, in different groups appear to be modified versions of structures that might have been present in a common ancestor.

(e) All of the above.

Q230 How might an evolutionary biologist explain why a species of species of salamander becomes blind after colonizing a cave?

(a) It is possible that in the cave there is a source of pollution that increases the mutation rate for a gene that makes salamanders blind. Over time, due to exposure to this chemical, the members of the population lose their sight.

(b) Members of the ancestral population that colonized the cave differed in their ability to see. If maintaining the ability to see in the cave was a waste of energy, blind salamanders might actually have more offspring than those who could see.

(c) There is no way to explain this in terms of natural selection

(d) The members of this salamander species no longer needed to use their eyes. Over time, due to lack of use, they lost the ability to see.

(e) None of the above.

Q231 Black pelage is an autosomal dominant trait in guinea pigs; white is the alternative recessive trait. A Hardy-Weinberg population was sampled and found to contain 336 black and 64white individuals. The percentage of black individuals that is expected to be heterozygous is approximately

(a) 46 (b) 57

(c) 49 (d) 53

(e) None of the above

Q232 The study of neoplastic growths is referred to as:

(a) Tetralogy (b) Anaplasia

(c) Oncology (d) Neoplasia

(e) Dysplasia

Q233 Liver nodules that consist of congenital localized overgrowth of mature hepatocytes are referred to as:

(a) Hepatomas

(b) Hamartomas

(c) Benign sarcomas

(d) Nodular hyperplasia

(e) Bile duct anaplasia

Q234 Phenylketonuria is

(a) A recessive trait.

(b) Easily detected by high levels of phenylalanine in the blood.

(c) The most common inherited disorder that affects the nervous system.

(d) Due to lack of an enzyme needed to metabolize phenylalanine, and damage can therefore be controlled by a diet low in this amino acid.

(e) All of the choices are correct.

Q235 Gene therapy in humans was first practised by Blease and Andresco to cure.

(a) Cystic fibrosis

(b) Hemophilia

(c) Thalassemia

(d) Severe combined immunodeficiency disease

(e) All of these

Q236 Somatic cell therapy include

(a) Ex vivo therapy

(b) In vivo therapy

(c) Antisense therapy

(d) All of these

(e) None of the above

Answers from 1-100

Q1	(c)	Q9(a)	(T)	Q22	(b)	Q38	(b)	Q54	(d)	Q70	(a)	Q86	(a)
Q2	(e)	Q9(b)	(T)	Q23	(c)	Q39	(d)	Q55	(a)	Q71	(a)	Q87	(b)
Q3	(d)	Q9(c)	(F)	Q24	(b)	Q40	(c)	Q56	(e)	Q72	(b)	Q88	(a)
Q4	(c)	Q9(d)	(F)	Q25	(d)	Q41	(b)	Q57	(d)	Q73	(b)	Q89	(d)
Q5	(c)	Q10	(d)	Q26	(c)	Q42	(a)	Q58	(c)	Q74	(c)	Q90	(c)
Q6	(b)	Q11	(b)	Q27	(b)	Q43	(b)	Q59	(d)	Q75	(c)	Q91	(b)
Q7(a)	(F)	Q12	(a)	Q28	(c)	Q44	(a)	Q60	(a)	Q76	(c)	Q92	(a)
Q7(b)	(F)	Q13	(c)	Q29	(a)	Q45	(a)	Q61	(e)	Q77	(a)	Q93	(a)
Q7(c)	(F)	Q14	(d)	Q30	(d)	Q46	(d)	Q62	(a)	Q78	(a)	Q94	(b)
Q7(d)	(T)	Q15	(a)	Q31	(d)	Q47	(d)	Q63	(c)	Q79	(c)	Q95	(a)
Q7(e)	(F)	Q16	(d)	Q32	(a)	Q48	(c)	Q64	(c)	Q80	(c)	Q96	(b)
Q8(a)	(T)	Q17	(e)	Q33	(c)	Q49	(e)	Q65	(b)	Q81	(d)	Q97	(a)
Q8(b)	(T)	Q18	(e)	Q34	(b)	Q50	(b)	Q66	(c)	Q82	(c)	Q98	(b)
Q8(c)	(F)	Q19	(b)	Q35	(b)	Q51	(c)	Q67	(b)	Q83	(b)	Q99	(b)
Q8(d)	(T)	Q20	(b)	Q36	(b)	Q52	(d)	Q68	(d)	Q84	(a)	Q100	(a)
Q8(e)	(F)	Q21	(c)	Q37	(d)	Q53	(c)	Q69	(c)	Q85	(b)		

Answers from 101-200

Q101	(e)	Q116	(d)	Q131	(c)	Q146	(b)	Q161	(c)	Q176	(e)	Q191	(d)
Q102	(d)	Q117	(b)	Q132	(a)	Q147	(a)	Q162	(d)	Q177	(c)	Q192	(d)
Q103	(b)	Q118	(e)	Q133	(c)	Q148	(d)	Q163	(d)	Q178	(b)	Q193	(b)
Q104	(b)	Q119	(c)	Q134	(a)	Q149	(e)	Q164	(d)	Q179	(d)	Q194	(a)
Q105	(e)	Q120	(a)	Q135	(a)	Q150	(a)	Q165	(e)	Q180	(d)	Q195	(a)
Q106	(d)	Q121	(e)	Q136	(c)	Q151	(c)	Q166	(b)	Q181	(c)	Q196	(d)
Q107	(c)	Q122	(c)	Q137	(d)	Q152	(c)	Q167	(e)	Q182	(b)	Q197	(a)
Q108	(a)	Q123	(e)	Q138	(a)	Q153	(b)	Q168	(e)	Q183	(c)	Q198	(a)
Q109	(e)	Q124	(a)	Q139	(b)	Q154	(d)	Q169	(d)	Q184	(c)	Q199	(a)
Q110	(e)	Q125	(b)	Q140	(a)	Q155	(d)	Q170	(b)	Q185	(b)	Q200	(a)
Q111	(c)	Q126	(e)	Q141	(a)	Q156	(b)	Q171	(c)	Q186	(d)		
Q112	(c)	Q127	(b)	Q142	(d)	Q157	(b)	Q172	(d)	Q187	(b)		
Q113	(b)	Q128	(b)	Q143	(b)	Q158	(b)	Q173	(e)	Q188	(e)		
Q114	(b)	Q129	(d)	Q144	(c)	Q159	(b, c)	Q174	(b)	Q189	(c)		
Q115	(b)	Q130	(b)	Q145	(b)	Q160	(c)	Q175	(c)	Q190	(d)		

Answers from 201-236

Q201	(d)	Q207	(b)	Q213	(a)	Q219	(e)	Q225	(c)	Q231	(a)
Q202	(d)	Q208	(b)	Q214	(a)	Q220	(c)	Q226	(c)	Q232	(c)
Q203	(e)	Q209	(a)	Q215	(a)	Q221	(b)	Q227	(d)	Q233	(b)
Q204	(b)	Q210	(c)	Q216	(c)	Q222	(b)	Q228	(c)	Q234	(e)
Q205	(a)	Q211	(d)	Q217	(c)	Q223	(d)	Q229	(e)	Q235	(d)
Q206	(d)	Q212	(d)	Q218	(c)	Q224	(a)	Q230	(b)	Q236	(d)

Index

I

K

L

M

N

O

P

www.ingramcontent.com/pod-product-compliance
Ingram Content Group UK Ltd.
Pitfield, Milton Keynes, MK11 3LW, UK
UKHW052230270726
14060UKWH00004B/697